Thermodynamics a

Henning Struchtrup

Thermodynamics and Energy Conversion

Henning Struchtrup
Dept. Mechanical Engineering
University of Victoria
British Columbia
Canada

ISBN 978-3-662-43714-8 ISBN 978-3-662-43715-5 (eBook)
DOI 10.1007/978-3-662-43715-5
Springer Heidelberg New York Dordrecht London

Library of Congress Control Number: 2014944090

Printed on acid-free paper

Springer is part of Springer Science+Business Media (www.springer.com)

Preface

This textbook grew out of lecture notes for the thermodynamics courses offered in the Department of Mechanical Engineering at the University of Victoria. Writing my own notes forced me to thoroughly consider how, in my subjective view, engineering thermodynamics should be taught. At the same time I aimed for a concise presentation, with the material of three courses delivered on about 600 pages.[1] My hope in publishing this book is that students of thermodynamics might find the chosen approach accessible, and maybe illuminating, and discover thermodynamics and its interesting applications for themselves.

Probably the biggest difference to standard texts is when and how the second law of thermodynamics and its central quantity, the entropy, are introduced. The second law describes irreversible processes like friction and heat transfer, which are related to a loss in work. For instance, work that is needed to overcome friction in a generator cannot be converted into electricity, hence there is a loss. Accordingly, it should be one of the main goals of a thermal engineer to reduce irreversibility as much as possible. Indeed, the desire to understand and quantify irreversible losses is one of the central themes of the present treatment, it is touched upon in almost all chapters.

The emphasis on irreversibilities requires the introduction of the second law as early as possible. The classical treatment, which is still used in most texts on engineering thermodynamics, is to derive the second law from discussion on thermal engines with and without losses. Obviously, this requires an extensive discussion of thermodynamic processes and thermal engines by means of the first law of thermodynamics—the law of conservation of energy—before the second law can even be mentioned. In the present treatment, entropy and

---

[1] The courses (13 weeks à 3 hours), and the relevant book chapters, as currently taught at the University of Victoria, are:
Thermodynamics (UVic Mech 240): Chapters 1-10
Energy Conversion (UVic Mech 390): Chapters 11-14, 18.1-18.9, 19, 23.1-23.5, 24
Advanced Thermodynamics (UVic Mech 443): Chapters 16-18, 20-26

the second law are introduced directly after the first law, based on observations of rather simple processes, in particular the trend of unmanipulated systems to approach a unique equilibrium state. With this, the complete set of thermodynamic laws is available almost immediately, and the discussion of all thermodynamic processes and engines relies on both laws from the start. All considerations on engines which are typically used to derive the second law, are now a result of the analysis of the engines by means of the first and second law.

As soon as the thermodynamic laws are stated we are in calmer waters. The discussion of property relations, processes in closed and open systems, thermodynamic cycles, mixtures and so on follows established practice, only, perhaps, with the additional emphasis on irreversibility and loss. Some elements that might not be found in other books on engineering thermodynamics concern the microscopic definition of entropy, the afore mentioned emphasis on thermodynamic losses, and the detailed discussion of a number of advanced energy conversion systems such as Atkinson engine, solar tower (updraft power plant), turbo-fan air engine, ramjet and scramjet, compressed air energy storage, osmotic power plants, carbon sequestration, phase and chemical equilibrium, or fuel cells. The principles of non-equilibrium thermodynamics are used to derive transport laws such as Newton's law of cooling, Darcy's law for flow through porous media, and activation losses in fuel cells.

There are about 300 end-of-chapter problems for homework assignments and exams. The problems were chosen in order to emphasize all important concepts and processes. They are accompanied by detailed solved examples in all chapters, and it is recommended to first study the examples and then tackle the problems. Many problems require the use of thermodynamic property tables, which are widely available in print and online.

Any presentation of a large topic such as thermodynamics can never be complete. The choice of topics in this book is a personal one, but I am confident that after studying this book the reader will find easy access to most other thermodynamics texts, be they written for mechanical engineers, chemical engineers, or scientists. *Thermodynamics and Energy Conversion* processes will remain an important part of modern civilization. High energy efficiency can only be obtained from a deep understanding of the Laws of Thermodynamics, which describe the interplay of *Energy, Entropy, and Efficiency.* It is my sincere hope that this book will contribute to this end.

Victoria, BC
Spring 2014

Henning Struchtrup
(struchtr@uvic.ca)

Acknowledgments

My view on thermodynamics has evolved over the years, and I benefitted from discussions with many colleagues and friends, in particular: my teacher Prof. Ingo Müller (Technical University of Berlin, Germany), Prof. Manuel Torrilhon (RWTH Aachen University, Germany), Prof. Hans Christian Öttinger (ETH Zürich, Switzerland), Profs. Signe Kjelstrup and Dick Bedeaux (NTNU Trondheim, Norway).

All chapters of this book went through several runs of the respective course, and each re-run led to additions and deletions, changes and adjustments, more examples and new problems. For feedback, corrections, and, sometimes, critical praise I would like to thank the countless students that went through these courses, as well as the graduate students that served as teaching assistants.

The Department of Mechanical Engineering at the University of Victoria provides a wonderfully collegial atmosphere for which I express my heartfelt thanks to my colleagues.

Finally, I thank my wife, Martina, and our daughter, Nora, for their continuous support, understanding, and love.

Acknowledgements

[illegible]

Contents

Chapter 1
Introduction: Why Thermodynamics?

1.1 Energy and Work in Our World

Mechanical and electrical work is what drives our daily lives: cars, trucks, planes, trains, ships—in short, all transport—require motors which are either based on combustion of fuel or on electric energy. Our home and work environments are unthinkable without the many devices that are powered by electricity: light, microwave and stove, freezer and refrigerator, television and radio, DVD and bluray, CD and MP3 player, smartphone and telephone landline, computer and printer, washer and dryer, air conditioning and (sometimes) heat, power drill and lawn mower; the list goes on. Hospitals and factories are filled to the rim with mechanical and electronic devices and robots that are driven by electrical energy.

Electricity, however, is mainly obtained by converting mechanical work into electrical work in a generator: our lifestyle requires an endless supply of mechanical work.

For most of its history, humankind was only able to harvest mechanical work as nature provided it. Wind and water wheels were used not only to mill grain, but also for other purposes, most importantly for pumping irrigation water.[1] But else, there was little, and an abundance of tasks had to be done by human labor: farming, e.g., harvesting with a scythe, and weaving with a loom come to mind immediately.

Heat, as obtained from combustion of wood, fat, oil or coal, however, was used for cooking, lighting and heating, and other tasks unrelated to mechanical work, most importantly probably the smelting of metals.

The industrial revolution was triggered in the 18th century by the invention of *heat engines*, that is engines that convert heat into work. In particular the development of the *steam engine* by engineers like James Watt led to the lifestyle we enjoy. Now a wide array of heat engines is available. The original

[1] Today, wind turbines and large hydropower dams harvest the same natural powers to directly produce electricity.

H. Struchtrup, *Thermodynamics and Energy Conversion*,
DOI: 10.1007/978-3-662-43715-5_1,

piston steam engines are replaced by steam turbines, while piston engines such as the Diesel and Otto engines are omnipresent on our streets and in ships, gas turbines drive aircraft and run in power plants.

For all engines, the heat is typically created by the burning of a fuel, such as coal, natural gas, oil, etc., or from nuclear power. The fuel is costly, and scarce, and therefore one will aim to make heat engines as efficient as possible. Moreover, combustion of fuels releases carbon dioxide into the atmosphere, which impacts global climate. More efficient use of fuels can at least slow down the rate at which carbon dioxide is added to the atmosphere, and hence high efficiencies have more than pecuniary value.

Thermodynamics was developed out of the desire to understand the limits of heat engine efficiency. Modern power plants run on intricate improvements on the original steam engine process that result from the deeper understanding of thermodynamic processes. Early steam engines had efficiencies of heat to work conversion of only a few percent, while modern combined cycle gas turbine/steam power plants exhibit efficiencies of up to 60%. Moreover, thermodynamic consideration can establish absolute upper bounds on efficiencies for processes, answering questions such as: how much work can be obtained at best from a heat source at a given temperature? or: what is the maximum work that could be obtained from a given amount of fuel? Only comparing actual performance against these theoretical limits can give adequate measures of efficiency. Of course, these questions and the answers will be discussed throughout this book.

Since its beginnings with the industrial revolution, thermodynamics has developed into a science that explains a wide array of natural and technical phenomena. Thermodynamic laws govern a host of processes: heat to work conversion in heat engines, and the inverse, i.e., the work to heat conversion in freezers, refrigerators and air conditioning systems; mixing and separation; transport through membranes (osmosis); chemical reactions and combustion, and so on. All of these will be discussed in this text.

In short, a good understanding of thermodynamics is indispensable in a wide range of fields, in particular mechanical and chemical engineering, chemistry, physics, and life sciences.

1.2 Mechanical and Thermodynamical Forces

Newton's laws of motion describe how a system reacts to an applied force: it moves. For instance, a weight on a coiled-up thread can be used to bring a shaft to rotation. When the shaft is connected to a generator, electricity is produced: the potential energy of the weight is transformed into electrical work. We see that a force, here gravity acting on the weight, can be used to generate mechanical work, here the rotation of the shaft, which then can be transformed again, here into electrical work. As long as the mechanical and electrical systems used are frictionless and resistance free, there is no loss, that is the electrical work produced is equal to the mechanical work done by the weight.

But, of course, there is friction and ohmic resistance, and some of the input work is required to overcome these. Hence, in a system with friction, the amount of electrical work provided is less than the work done by the weight. So where has that work gone? Thermodynamics gives the answer: due to friction and resistance, the system becomes warmer than its environment, and then heat flows into the environment: some of the work is converted to heat. While mechanics can describe friction losses, and electrodynamics can describe ohmic resistance, a full account of the system requires a thermodynamic description, entailing quantities like *temperature* and *heat*, which do not appear in mechanics and electrodynamics.

We all have some idea of what temperature is, since we have a sense for hot and cold. Also, we have the experience that when we put a cold and a hot body in contact, the cold body will become warmer and the warm body will become colder, until they have the same temperature. Just think of a soft drink originally at room temperature and ice: the ice will warm and melt, and the drink will become cooler ... and a bit watery. Or think of a hot cup of coffee left on a table: After sufficiently long time the coffee assumes the temperature of the room around—it has cooled down—while the air in the room has become just a tiny bit warmer. Our iced soft drink, when forgotten on the table, will eventually warm up to the room temperature. In both cases, thermal energy is redistributed between the subsystems we looked at—soft drink, ice, coffee, air in the room. The associated temperature change is also linked to the size of the system: soft drink and ice both experience sensible changes in temperature and state; also the coffee's temperature changes noticeably, while it would require a rather sensitive thermometer to measure the temperature change of the air in the room.

Hot and cold drinks are just an example for a fundamental observation: heat goes from hot to cold in the desire to equilibrate temperature. In analogy to mechanics, where a force causes movement of its point of application, we can say that the temperature difference is a *thermodynamic force* that causes heat to flow. And just as a mechanical force can drive a generator to produce electricity, the thermodynamic force can be used to generate mechanical work, and electricity. This, in fact, is what a heat engine does.

The tendency to equilibrate into a homogeneous state is not observed only for temperature but also for other quantities. For instance, a droplet of ink added to a glass of water will distribute until, after some time, the ink concentration is homogeneous. This desire to mix evenly is driven by another thermodynamic force, which is related to the difference in concentration. Careful analysis will show that the driving force is the difference in a quantity known as the *chemical potential.* Also this force can be harvested for work, e.g. using osmosis, where freshwater is drawn into saltwater through semi-permeable membranes that allow only water, but not salt, to pass.

There are other examples for nature's desire to equilibrate, for instance in chemical reactions. The amounts of reactants and reaction products will assume an equilibrium state that depends on the actual conditions in the reactor, such as pressure and temperature. Any equilibration process can be described by a thermodynamic force, and can be used—at least in principle—to provide work.

Processes opposite to equilibration move against the thermodynamic forces, and hence work must be provided to force these processes to happen. A refrigerator cools only a small part of the kitchen, by forcing heat from the inside to the outside: (electrical) work is required to drive the compressor in the refrigerator. Separation processes require work, or other forms of energy, as well. Using the same semi-permeable membrane as above, one can produce freshwater from saltwater by pressing the latter against the membrane. This requires high pressures, and consumes work.

Chemically fabricated materials are everywhere in our lives from clothing—fleece has replaced wool—to medication. About one percent of the world's energy consumption is used to produce ammonia (NH_3) which is the base product for nitrogen fertilizers and explosives. Widespread availability of fertilizer, together with modern machines—driven by heat engines—for the year-round farm work have increased yield from fields largely, while at the same time the relative number of people working in farming has—in the first world—declined dramatically. For all chemical processes the goal is to run the reactors such that the yield is high. This requires perfect understanding of the thermodynamic forces and equilibria, so that one can set the conditions, e.g., pressure and temperature, accordingly.

1.3 Systems, Balance Laws, Property Relations

In order to describe thermal processes accurately, we require a number of equations and relations to describe the behavior of the *thermal system* under consideration, and the details of the materials contained in the system.

The previous paragraph in fact points to the first requirement of any thermodynamic analysis, which is to chose a well defined system to be described, e.g., the system could be an entire power plant, or just the steam turbine within. In any case, the system boundary must be well defined so that all transport of material, energy etc. across the system boundary is well understood.

The processes within a system are described by *balance laws*, equations that account for all changes within the system as well as the transport across the system boundary. Balance laws are often written as rate equations, where the change of the amount of the balanced quantity over time is equated to causes for change, such as flow over the boundary, or creation/destruction inside the system.

The simplest balance law is the *conservation law for mass*, which states that mass cannot be created or destroyed. Hence, the mass inside the system boundaries can change only due to transfer of mass in or out over the boundaries. A *closed system* is defined as having no mass transfer over the boundaries, accordingly the mass in a closed system is constant. In an *open system* mass can enter or leave; this can lead to changing amount of mass within the system, for instance when a container is filled, or, when the inflow is balanced by the outflow, to an exchange of the material in the system, while the mass in the system is constant.

The *conservation laws for energy* states that energy cannot be created or destroyed. To emphasize its central importance, it is known as the *First Law of Thermodynamics*. Energy exists in different forms; familiar from mechanics are *kinetic* and *potential energy*, and thermodynamics adds *internal* (also called *thermal*) *energy*. The first law describes the conversion between the different forms of energy, and the transport of energy in form of heat and work. While the first law describes conversion from work to heat and vice versa, it cannot distinguish between possible and impossible processes.

Indeed, thermodynamic processes are restricted in many ways, e.g., heat will by itself go from hot to cold and not vice versa, or a mixture will not spontaneously separate. These restrictions are formulated in the *Second Law of Thermodynamics*. The second law introduces a new quantity, the *entropy*, which can only be created, but not destroyed. Accordingly, the second law is a balance law for entropy that describes the change of entropy in the system due to transport across the boundary, and creation inside the system. Since we have no sense for entropy, this quantity is somewhat non-intuitive, however, the second law is seen at work quite easily, for instance in all equilibration processes such as those discussed above. Processes in which entropy is created are called *irreversible*, and any creation of entropy can be related to a loss in work. The second law has far ranging consequences, including the restriction of the *efficiency* of heat engines to values below unity, that is, *heat cannot be fully converted into work.*

As just stated, we have no sense for entropy. But then, we have no sense for energy as well, and our sense for temperature is rather inexact. To fill the thermodynamic laws with life, the quantities appearing in them, most importantly energy and entropy, must be related to measurable quantities.

With temperature playing a prominent role in thermodynamics, temperature and its measurement must be clearly defined, which is done by means of the *Zeroth Law of Thermodynamics*, which states that two bodies in equilibrium have the same temperature. The assigned number (*zero*) indicates that this law now is introduced before the first and second laws, but historically its importance for a sound development of thermodynamic theory was recognized only after these were named.

Measurable quantities are length (and thus area and volume), time (and thus velocity and acceleration), mass, pressure (or force), temperature, and concentration. For specific systems the thermodynamic laws must be furnished with *property relations* that describe the physical behavior of the materials contained in the system, by relating measurable quantities to each other as well as to those that cannot be measured directly (energy, entropy ...). Property relations are laid down in equations or in tables, they result from careful measurements and evaluation of these by means of the thermodynamic laws.

1.4 Thermodynamics as Engineering Science

Typically, scientists and engineers ask different questions. With some simplification, one might say that a scientist asks *Why does this happen?*, while an engineer asks *How can I use it?* But then, the boundaries between science and engineering are rather fluent, and there is significant overlap. For instance, both disciplines will be interested in the basic laws—in form of equations—that describe the observed phenomena. For the scientist this is part of understanding and describing nature, while for the engineer the basic laws are tools to model and improve engineering devices. It is probably fair to say that the deeper an engineer understands the laws of nature, the more use she/he can make of them. Deeper understanding will lead to new ideas that might not be obvious at the first look.

As stated earlier, thermodynamics was developed out of an engineering desire, namely to improve the efficiencies of heat engines. This need resulted in the thermodynamic laws, which were found on purely phenomenological grounds, that is by observation and conclusion. As will be seen, thermal energy and entropy arise as necessary, and rather helpful, quantities, which appear in their respective laws (1st and 2nd). The laws describe work and heat exchange, and the trend to equilibrium, but they do not answer the questions *What is energy? What is entropy?* Indeed, for engineering applications the answer to these questions is not relevant, as long as property relations for energy and entropy can be found from measurements, as is the case.

Nevertheless, a deeper understanding of these quantities can be obtained by looking at the microscopic description of matter, that is on the atomic or molecular level. Thermal energy can be related to microscopic kinetic and potential energies, so that concepts from mechanics can be transferred to some degree.

Entropy can be related to the number of microscopic realizations of the same macroscopic state, as will be discussed for rather simple model examples. The trend to equilibrium as expressed in the second law is then simply a motion of the system towards macroscopic states that have a larger number of microscopic realizations. The final equilibrium state has the largest number of realizations, and thus is—by far—most likely. Often one finds explanations of entropy as a measure for "disorder", but this might be misleading wording, unless a careful definition of "disorder" is provided.

Even somewhat superficial arguments on microscopic behavior can yield deeper insight into entropy, and thermodynamic processes. Hence, the introductory chapter on entropy and the chapters on reacting and non-reacting mixtures contain some descriptions on the microscopic level. Naturally, we can just scratch the surface. The reader might study these sections for some insight—for deeper understanding, we must refer to the relevant scientific literature.

As might become clear from the above, the microscopic description of entropy relies on ideas of statistics. The proper understanding of matter on the microscopic level is subject of *Statistical Mechanics*, a branch of physics which for instance can be used to find property relations.

1.5 Thermodynamic Analysis

After the introduction and general discussion of the basic laws of thermodynamics and the property relations, the study of thermodynamics turns to the thermodynamic analysis of a wide variety of systems.

The system considered, and the goal of the analysis, depends on the field of study. This text focuses on engineering applications of thermodynamics, where the aim is to understand the working principles, and to evaluate the performance of thermodynamic systems such as power plants, refrigerators, chemical reactors and so on. Deep understanding of system behavior from thermodynamic analysis will lead to performance enhancement by proper setting of available parameters, redesign for improved efficiency, replacement by more efficient alternatives, and, possibly, to development of completely new system configurations.

Thermodynamic analysis of a system entails some or all of the following:

- Introductory discussion of the system under consideration. What is the purpose of the system, how is it achieved?
- General discussion of the working principles of the system.
- Clear identification of the system. Decomposition into subsystems for easier evaluation.
- Material considerations. Are there limiting values for system parameters, e.g. maximum temperatures and pressures, that cannot be exceeded?
- Determination of all relevant physical data (pressure, temperature, energy, entropy and so on) at all relevant locations in the system, and its subsystems.
- Computation of all heat and work exchanges for the system, and its subsystems.
- Evaluation of system performance, as expressed through meaningfully defined efficiency measures, both for subsystems and the overall system.
- Analysis of system configuration and performance. Which controllable parameters must be changed, and how, to improve or optimize the system?
- Second law analysis: Identification of irreversible processes in the system. Determination of entropy generation and associated work loss, both within the system and in the exchange between system and its surroundings.

Which processes have the largest losses? Can the system be modified to reduce the loss?
- General analysis. Is the system as evaluated suitable for the chosen purpose? What are the alternative systems, or system configurations? Which system/configuration should be preferred?

1.6 Applications

Before we enter the technical part of our studies of thermodynamics, for a quick overview, we present a list of the engineering applications that will be discussed in the following chapters:

- Hydrostatic and barometric pressure laws.
- Efficiency limits for heat engines, refrigerators and heat pumps.
- Perpetual motion engines.
- Internal combustion engines: Otto, Diesel, Atkinson.
- Simple open systems: compressor, pump, turbine, throttle, nozzle, diffuser.
- Heat exchangers: co- and counter-flow closed heat exchangers, open heat exchangers
- Steam power plants: standard and reheat cycles, advanced cycles with open and closed feedwater heaters.
- Vapor refrigeration systems and heat pumps: standard cycles, advanced multi-stage cycles.
- Linde gas liquefaction process.
- Stirling and Ericsson engines.
- Multi-stage compressors.
- Gas turbine systems for power generation: standard cycle and multi-stage cycles.
- Combined cycle: gas turbine and steam cycle for high efficiency.
- Solar tower: updraft power plant.
- Air engines: standard air turbine and turbo-fan engines.
- Supersonic flows: rockets, ramjet and scramjet.
- Filling and discharge.
- Compressed air energy storage (CEAS): storing energy by compressing air into large caverns.
- Temperature change in throttling (Joule-Thomson coefficient).
- Thermodynamic equilibrium and phase equilibrium.
- Ice skating.
- Mixtures, heat and entropy of mixing.
- Psychrometrics: humidifying and de-humidifying for air conditioning, cooling towers.
- Mixing and separation.
- Osmosis.
- Desalination by reverse osmosis.
- Pressure retarded osmosis: power from mixing freshwater and saltwater.

- Gas separation and CO_2 removal: work requirement for carbon capture and storage.
- Two-phase mixtures: ideal and non-ideal mixtures, activity and fugacity, Raoult's law, phase diagrams.
- Distillation columns.
- Gas solubility: Henry's law, carbonized and nitrogenated drinks.
- Reacting mixtures: law of mass action and Le Chatelier principle.
- Haber-Bosch process for ammonia (NH_3) production.
- Combustion.
- Work potential of a fuel.
- Work losses in a steam power plant.
- Fuel cells: potential and power, losses caused by mass transfer, resistance, activation and crossover.
- Electrolyzers.

Chapter 2
Systems, States, and Processes

2.1 The Closed System

The first step in any thermodynamic consideration is to identify the system that one wishes to describe. Any complex system, e.g., a power plant, can be seen as a compound of some—or many—smaller and simpler systems that interact with each other. For the basic understanding of the thermodynamic laws it is best to begin with the simplest system, and study more complex systems later as assemblies of these simple systems.

The simplest system of interest is the *closed system* where a substance is enclosed by walls, and no mass flows over the system boundaries. The prototype of the closed system is a piston-cylinder device, as depicted in Fig. 2.1. We shall assume that the device contains a fixed amount of a simple substance, that is a substance that does not undergo chemical changes.

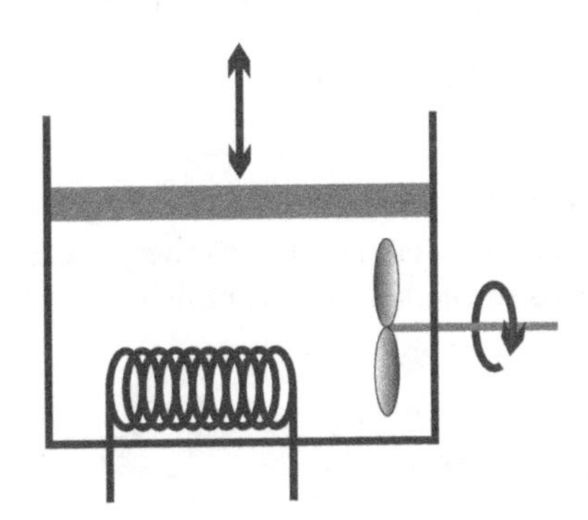

Fig. 2.1 The piston-cylinder device with heat and work exchange is the standard example for closed systems

There is only a small number of manipulations possible to change the state of a closed system, which are indicated in the figure: the volume of

H. Struchtrup, *Thermodynamics and Energy Conversion*,
DOI: 10.1007/978-3-662-43715-5_2, © Springer-Verlag Berlin Heidelberg 2014

the system can be changed by moving the piston, the system can be stirred with a propeller, and the system can be heated or cooled by changing the temperature of the system boundary, as indicated by the heating coil.[1] These actions lead to exchange of energy between the system and its surroundings, either by work in case of piston movement and stirring, or by the exchange of heat. The transfer of energy by work and heat will be formulated in the *First Law of Thermodynamics.*

The change of energy and volume of the system will lead to changes in other properties of the enclosed substance, in particular pressure and temperature. *Thermodynamic laws* and *property relations* are required to predict the changes of the different properties, and the exchange of heat and work.

Most processes have a direction in time. For instance, we can do work to move the propeller and stir a liquid, which increases the liquid temperature due to friction, but we will never observe that a liquid at rest suddenly begins to move a propeller and does work (e.g. the lifting of a weight), see Fig. 2.2. The direction of processes is formulated in the *Second Law of Thermodynamics*, which has, as will be seen, far ranging consequences for technical applications.

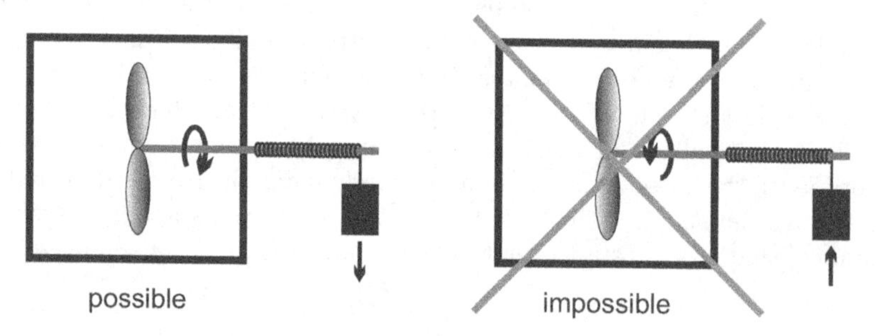

Fig. 2.2 A possible and an impossible process

We shall first consider the complete set of thermodynamic equations for closed systems. In open systems mass crosses the system boundaries, and this leads to additional terms in the thermodynamic laws. These will be discussed in Chapter 9.

2.2 Micro and Macro

A macroscopic amount of matter filling the volume V, say a steel rod or a gas in a box, consists of an extremely large number—to the order of 10^{23}—of atoms or molecules. These are in constant interaction which each other

[1] Another possibility to heat or cool the system is through absorption and emission of radiation, and transfer of radiation across the system boundary.

and exchange energy and momentum, e.g., a gas particle in air at standard conditions undergoes about 10^9 collisions per second.

From the viewpoint of mechanics, one would have to describe each particle by its own (quantum mechanical) equation of motion, in which the interactions with all other particles would have to be taken into account. Obviously, due to the huge number of particles, this is not feasible. Fortunately, the constant interaction between particles leads to a collective behavior of the matter already in very small volume elements dV, in which the state of the matter can be described by few macroscopic properties like pressure, mass density, temperature and others. This allows us to describe the matter not as an assembly of atoms, but as a continuum where the state in each volume element dV is described by these few macroscopic properties.

Note that the underlying assumption is that the volume element contains a sufficiently large number of particles. Indeed, the continuum hypothesis breaks down under certain circumstances, in particular for highly rarefied gases. In all what follows, however, we shall only consider systems in which the assumption is well justified.

2.3 Mechanical State Properties

Of the many state properties that we shall meet, we first introduce those properties that can be easily measured, and are familiar from mechanics.

We consider a system of volume V which is filled by a mass m of substance. To describe variation of properties in space, it is useful to divide the system into infinitesimal elements of size dV and mass dm, as sketched in Fig. 2.3.

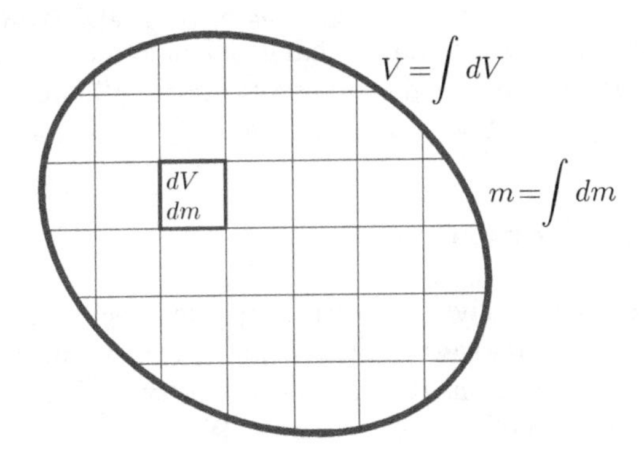

Fig. 2.3 A system of volume V and mass m is divided into infinitesimal elements of size dV and mass dm

The volume $V = \int dV$ filled by the substance can, in principle, be measured by means of a ruler. The SI unit for volume is the cubic meter $[\mathrm{m}^3]$, for smaller volumes one might use the litre, 1 litre $= 10^{-3}\,\mathrm{m}^3$, or the cubic centimetre, $1\,\mathrm{cm}^3 = 10^{-3}$ litre $= 10^{-6}\,\mathrm{m}^3$.

The mass $m = \int dm$ of the substance can be measured using a scale. The SI unit of mass is kilogram [kg]. For small masses it is convenient to use the gram, $1\,\mathrm{g} = 10^{-3}\,\mathrm{kg}$, and for large masses it is convenient to use the metric ton, $1\,\mathrm{t} = 1000\,\mathrm{kg}$.

The pressure p of the substance can be measured as the force required to keep a piston in place, divided by the surface area of the piston. The SI unit for pressure is the Pascal: $1\,\mathrm{Pa} = 1\frac{\mathrm{N}}{\mathrm{m}^2} = 1\frac{\mathrm{kg}}{\mathrm{m\,s}^2}$; one often uses the kilo-Pascal, $1\,\mathrm{kPa} = 1000\,\mathrm{Pa}$, or the mega-Pascal, $1\,\mathrm{MPa} = 10^6\,\mathrm{Pa}$. Two common non-SI units for pressure are the bar, $1\,\mathrm{bar} = 10^5\,\mathrm{Pa} = 0.1\,\mathrm{MPa}$, and the atmosphere (the standard air pressure at sea level), $1\,\mathrm{atm} = 1.01325\,\mathrm{bar}$.

Many pressure measuring devices (manometers) do not measure absolute pressure, but the difference to the local atmospheric pressure p_{atm} (which is normally not 1 atm!), the so-called gauge pressure $p_{\mathrm{gauge}} = p - p_{\mathrm{atm}}$. For pressures below the local atmospheric pressure the gauge pressure would be negative, and it is common to use the vacuum pressure $p_{\mathrm{vac}} = p_{\mathrm{atm}} - p$.

The velocity vector of a mass element is defined as its directed displacement per unit time. Mostly we shall be interested only in the absolute velocity $\mathcal{V}$, the SI unit is meters per second $\left[\frac{\mathrm{m}}{\mathrm{s}}\right]$.

2.4 Extensive and Intensive Properties

It is useful to distinguish between extensive properties, which are related to the size of the system, and intensive properties, which are independent of the size of the system. Mass m and volume V are extensive quantities, e.g., they double when the system is doubled; pressure p and temperature T are intensive properties, they remain unchanged when the system is doubled. As an example Fig. 2.4 shows the combination (or splitting) of a system at pressure p and temperature T, with total mass $m_1 + m_2$ and volume $V_1 + V_2$.

2.5 Specific Properties

A particular class of intensive properties are the specific properties, which are defined as the ratio between an extensive property and the corresponding mass. In general notation, the specific property ϕ corresponding to the extensive property Φ is defined as

$$\phi = \frac{\Phi}{m} \,. \tag{2.1}$$

For instance, the specific volume is

$$v = \frac{1}{\rho} = \frac{V}{m} \,. \tag{2.2}$$

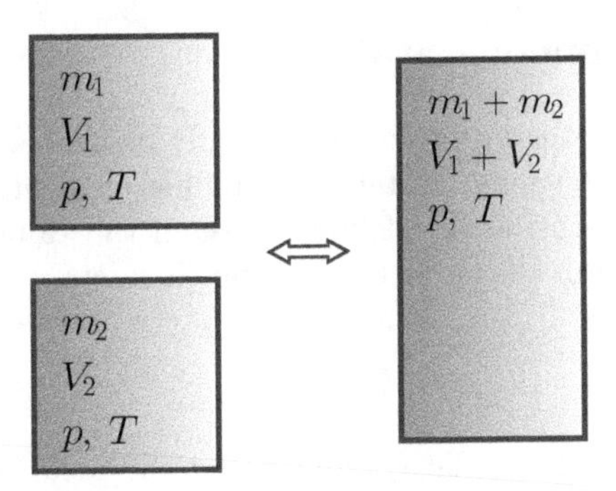

Fig. 2.4 When two systems at same pressure p and temperature T are combined, the intensive properties pressure and temperature remain unchanged, while the extensive properties mass m and volume V add

Here $\rho = \frac{m}{V}$ is the mass density, i.e., the mass per volume.

2.6 Molar Properties

For the thermodynamic discussion of mixtures, reacting or not, it is advantageous to consider the number of particles involved rather than mass. The number of atoms or molecules is rather large and thus it is customary to count the number of particles in moles, with the unit [mol] or [kmol]. One mole is the number of atoms in 12 g of the carbon isotope ^{12}C, which is given by the Avogadro constant (Amedeo Avogadro, 1776-1856)

$$N_A = 6.022 \times 10^{23} \frac{1}{\text{mol}} . \tag{2.3}$$

The mass of one mole of particles is the molar mass M with the unit $1\frac{\text{g}}{\text{mol}} = 1\frac{\text{kg}}{\text{kmol}}$. By definition, the molar mass of ^{12}C is $M_\text{C} = 12\frac{\text{kg}}{\text{kmol}}$, the molar mass for other substances can be found in tables.

The number of moles of substance is related to mass by

$$n = \frac{m}{M} .$$

Mole specific properties will be labeled with an overbar, they are related to extensive and mass specific properties as, e.g.,

$$\bar{\phi} = \frac{\Phi}{n} = \frac{m}{n}\frac{\Phi}{m} = M\phi . \tag{2.4}$$

2.7 Inhomogeneous States

In inhomogeneous states intensive and specific properties vary locally, that is they have different values in different volume elements dV. In this case, the *local* specific properties are defined through the values of the extensive property $d\Phi$ and the mass dm in the volume element,

$$\phi = \frac{d\Phi}{dm} . \tag{2.5}$$

For example, the local specific volume v and the local mass density ρ are defined as

$$v = \frac{1}{\rho} = \frac{dV}{dm} . \tag{2.6}$$

The values of the extensive properties for the full system are determined by integration of the specific properties over the mass elements,

$$\Phi = \int \phi dm , \tag{2.7}$$

or, by means of the relation $dm = \rho dV$, by integration over the volume elements,

$$\Phi = \int \rho\phi dV . \tag{2.8}$$

As an example, Fig. 2.5 shows the inhomogeneous distribution of mass density ρ in a system. Note that due to inhomogeneity, the density is a function of location $\vec{r} = \{x, y, z\}$ of the element dV, hence $\rho = \rho(\vec{r})$.

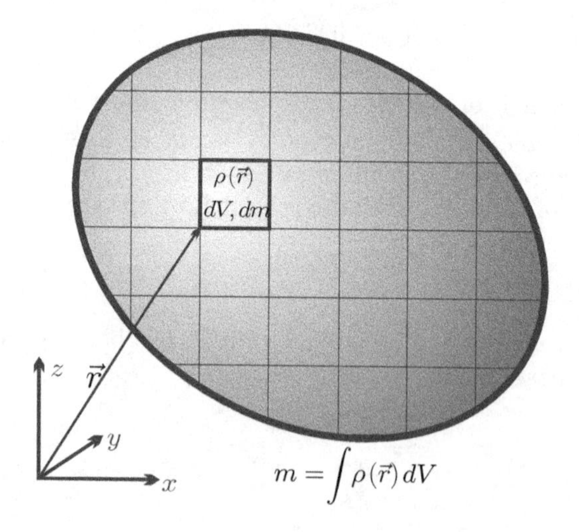

Fig. 2.5 Inhomogeneous distribution of mass density $\rho(\vec{r})$ in a system

2.8 Processes and Equilibrium States

A *process* is any change in one or more properties occurring within a system. The system depicted in Fig. 2.1 can be manipulated by moving the piston or propeller, and by exchanging heat. Any manipulation changes the state of the system locally and globally: a process occurs.

After all manipulation stops, the states in the volume elements will keep changing for a while—that is the process continues—until a stable final state is assumed. This stable final state is called the *equilibrium state.* The system will remain in the equilibrium state until a new manipulation commences.

Simple examples from daily life are: (a) A cup of coffee is stirred with a spoon. After the spoon is removed, the coffee will keep moving for a while until it comes to rest. It will stay at rest indefinitely, unless stirring is recommenced or the cup is moved. (b) Milk is poured into coffee. Initially, there are light-brown regions of large milk content and dark-brown regions of low milk content. After a while, however, coffee and milk are well-mixed, at mid-brown color, and remain in that state. Stirring speeds the process up, but the mixing occurs also when no stirring takes place. (c) A spoon used to stir hot coffee becomes hot at the end immersed in the coffee. A while after it is removed from the cup, it will have assumed a homogeneous temperature. (d) Oil mixed with vinegar by stirring will separate after a while, with oil on top of the vinegar. The last example shows that not all equilibrium states are homogeneous; however, temperature will always be homogeneous in equilibrium.

In short, observation of daily processes, and experiments in the laboratory, show that a system that is left to itself for a sufficiently long time will approach a stable equilibrium state, and will remain in this state as long as the system is not subjected to further action.

The details of the equilibrium state depend on the constraints on the system, in particular material, size and energy. The time required for reaching the equilibrium state depends on the initial deviation from the equilibrium state, the material, and the geometry.[2] A change of pressure at the system boundary propagates with the speed of sound (sound is a pressure wave) into the system, which will reach a new equilibrium pressure relatively fast. On the other hand, a change of temperature at the system boundary diffuses relatively slowly into the system: the spoon that is used to stir hot coffee needs quite a while to feel hot at the side that is not immersed in the cup.

2.9 Quasi-static and Fast Processes

When one starts to manipulate a system that is initially in equilibrium, the equilibrium state is disturbed, and a new process occurs. When the manipulation

[2] Some systems remain in metastable states for very long time, until a bigger disturbance causes them to go into their stable equilibrium state.

happens sufficiently slow, the system can adapt so that it is in an equilibrium state at *any* time. Slow processes that lead the system through a series of equilibrium states are called quasi-static, or quasi-equilibrium, processes.

If the manipulation that causes a quasi-static process stops, the system is already in an equilibrium state, and no further change will be observed.

Equilibrium states are simple, quite often they are homogenous states, or can be approximated as homogeneous states. The state of the system is fully described by few extensive properties, such as mass, volume, energy, and the corresponding pressure and temperature.

When the manipulation is fast, so that the system has no time to reach a new equilibrium state, it will be in non-equilibrium states. If the manipulation that causes a non-equilibrium process stops, the system will undergo changes until it has reached its equilibrium state. The equilibration process takes place while no manipulation occurs, i.e., the system is left to itself. Thus, the equilibration is an uncontrolled process.

Non-equilibrium processes typically are inhomogeneous. Their proper description requires values of the properties at all locations $\vec{r}$ (i.e., in all volume elements dV) of the system. The detailed description of non-equilibrium processes is more complex than the description of quasi-static processes.

All real-life applications of thermodynamics involve some degree of non-equilibrium. Quasi-static processes are an idealization that serves to approximate real-life—i.e., non-equilibrium—processes.

2.10 Reversible and Irreversible Processes

The approach to equilibrium introduces a timeline for processes: As time progresses, an isolated system will always go towards its unique equilibrium state. The opposite will not be observed, that is a system will never be seen spontaneously leaving its equilibrium state when no manipulation occurs.

Indeed, we immediately detect whether a movie of a non-equilibrium process is played forward or backwards: well mixed milk coffee will not separate suddenly into milk and coffee; a spoon of constant temperature will not suddenly become hot at one end, and cold at the other; a propeller immersed in a fluid at rest will not suddenly start to move and lift a weight (Fig. 2.2); oil on top of water will not suddenly mix with the water; etc. We shall call processes with a time-line *irreversible*.

Only for quasi-static processes, where the system is always in equilibrium states, we cannot distinguish whether a movie is played forwards or backwards. We shall call these processes *reversible*. Since equilibration requires time, quasi-static, or reversible, processes typically are slow processes, so that the system always has sufficient time to adapt to an imposed change.

Equilibration processes can have quite different time scales. For instance, pressure changes are transported with the speed of sound ($\sim 350 \frac{\mathrm{m}}{\mathrm{s}}$), and piston cylinder systems can be approximated as quasi-static if the piston velocity is significantly below the speed of sound. The mean piston speed in a car engine, which depends on stroke and speed, is typically below $20\frac{\text{m}}{\text{s}}$, hence compression and expansion processes in a car engine can be considered as quasi-static. Heat transfer, on the other hand, is a very slow process, with a time scale determined by the heat conductivity. Accordingly, quasi-static processes involving heating must be rather slow. For fast processes such as the compression and expansion process in a car engine, there is no time at all for significant heat transfer between the cylinder walls and the gas, and the process can be approximated as quasi-static processes with no heating.

The second law of thermodynamics will be introduced as formalization of the observation that an isolated system is moving towards a unique equilibrium state, and will allow for a more formal definition of *reversible* and *irreversible* processes.

2.11 Temperature and the Zeroth Law

So far we have discussed only properties known from mechanics, namely mass m, volume V, pressure p, and velocity $\mathcal{V}$. Temperature, as a measure of how hot or cold a body is, is the first thermodynamic quantity that we introduce.

Indeed, through touching objects we can distinguish between hot and cold. However, our sense for temperature is relatively inexact, just feel the metal and the wood of your chair, which have the same temperature, but feel different. Objective measurement of temperature requires (a) a proper definition, and (b) a proper device for measurement—a thermometer.

Observation of nature and of processes towards equilibrium have established the following definition of temperature:

Two bodies in thermal equilibrium have the same temperature.

This statement is so important that it is known as the *Zeroth Law of Thermodynamics*. As example, consider two bodies, e.g., a cup of hot coffee and a spoon, or two stones, at different temperatures $\bar{T}_A > \bar{T}_B$ which are brought into thermal contact, see Fig. 2.6 for a schematic representation. An equilibration process occurs, and after a while the system comprised of the two bodies reaches its equilibrium state, with a common temperature T. While we shall need the first law—the conservation of energy—to compute its actual value, we know from experience that the final temperature will lie between the initial temperatures, $\bar{T}_A > T > \bar{T}_B$.

The zeroth law as stated above implies that if body A is in thermal equilibrium with bodies B and C, than also bodies B and C will be in equilibrium. All three will have the same temperature.

Thus, to measure the temperature of a body, all we have to do is to bring a calibrated thermometer into contact with the body and wait until the equilibrium state of the system (body and thermometer) is reached. When the size of the thermometer is small compared to the size of the body, the final

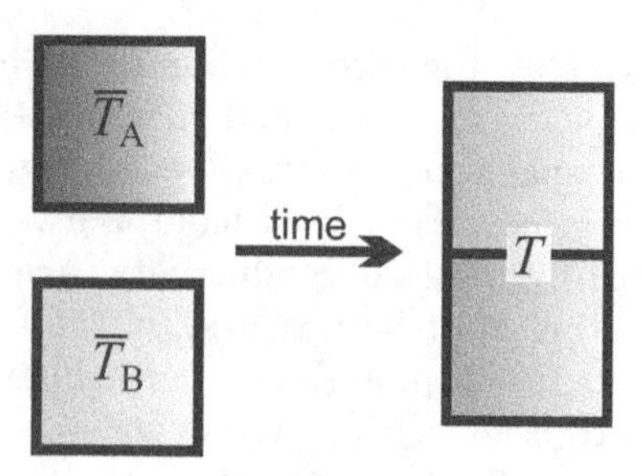

Fig. 2.6 Two bodies of different temperatures $\bar{T}_A$, $\bar{T}_B$ assume a common temperature T a while after they are brought into thermal contact

temperature of body *and* thermometer will be almost equal to the initial temperature of the body, see Sec. 3.12.

2.12 Thermometers and Temperature Scale

So what is a thermometer? Thermometers rely on the change of physical properties with temperature. The volume of most liquids grows with temperature, the volume change is employed in mercury or alcohol thermometers: liquid thermometers rely on the measurement of length. Resistance thermometers rely on the change of ohmic resistance of electric conductors with temperature. Thermocouples use thermoelectric effects—voltage caused by temperature difference—to measure temperatures.

Thermometers must be carefully calibrated, so that different thermometers, and different types of thermometers, will agree in their measurements. The calibration requires reference points that can be reproduced accurately, and a proper definition of the scale between the reference points.

The temperature scale used in daily life is the Celsius scale which measures temperature in degrees Celsius [°C]. The Celsius scale was originally defined based on the boiling and freezing points of water at $p = 1\,\text{atm}$ to define the temperatures of 100 °C and 0 °C. The Fahrenheit scale, which is employed in the USA, assigns these points the temperatures 212 °F and 32 °F. National and international bureaus of standards now use a larger number of well-defined fix points for the calibration of thermometers.

Just having reference points is not enough, there must be a well-defined scale for the temperatures *between* the reference points. As an example we consider two liquid thermometers filled with different liquids A and B, which are build such that their liquid columns have the same heights for the reference points at 0 °C and 100 °C, see Fig. 2.7. However, the change of volume with temperature might follow different non-linear functions $V(T)$ for the two liquids, so that both thermometers show different heights for temperatures between the reference points, as example the figure shows different readings for 50 °C.

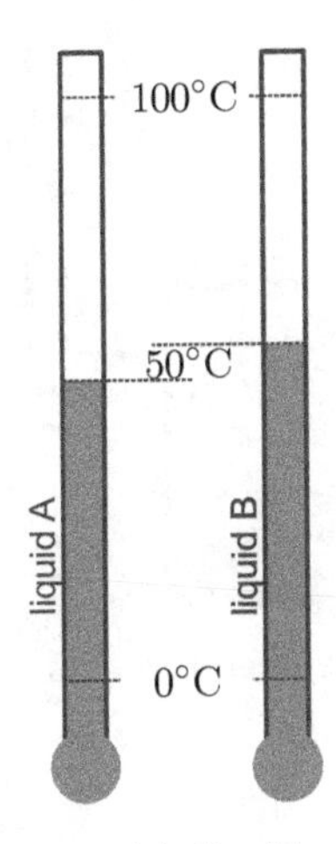

Fig. 2.7 Two liquid thermometers with liquids of different temperature-volume characterisitics

2.13 Gas Temperature Scale

To define a proper temperature scale between the reference points, one has to agree on a particular reference substance, and define the scale for that substance. Thermometers involving other substances, or other physical effects, can then be calibrated based on the reference.

The reference substance used is the ideal gas. Any gas at sufficiently low pressures and large enough temperatures (see Sec. 6.10), behaves as an ideal gas. From experiments one observes that for an ideal gas confined to a fixed volume the pressure increases with temperature. The temperature scale is *defined* such that the relation between pressure and temperature is linear, that is

$$T\,(^\circ\mathrm{C}) = a + bp \tag{2.9}$$

where the two constants a and b can be found from two well-defined reference points. With this, temperature is determined through measurement of pressure, see Fig. 2.8. For the Celsius scale one finds $a = -273.15\,^\circ\mathrm{C}$ independent of the ideal gas used. The constant b depends on the volume, mass and type of the gas in the thermometer.

By shifting the temperature scale by a, one can define an alternative scale, the ideal gas temperature scale, as

$$T\,(\mathrm{K}) = bp\,. \tag{2.10}$$

The ideal gas scale has the unit Kelvin [K, not $^\circ$K] and is related to the Celsius scale as

$$T\,(\mathrm{K}) = T\,(^\circ\mathrm{C})\,\frac{\mathrm{K}}{^\circ\mathrm{C}} + 273.15\,\mathrm{K}\,. \tag{2.11}$$

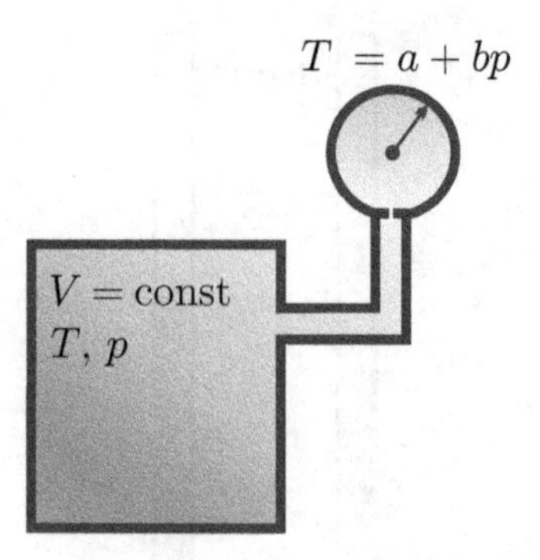

Fig. 2.8 In a gas thermometer, temperature T is determined through measurement of pressure p

For engineering problems one often uses $T\,(\mathrm{K}) = T\,(^{\circ}\mathrm{C})\,\frac{\mathrm{K}}{^{\circ}\mathrm{C}} + 273\,\mathrm{K}$, that is one ignores the difference of $0.15\,\mathrm{K}$ to the exact value. Temperature differences have the same numerical value for both scales, $\Delta T\,(\mathrm{K}) = \Delta T\,(^{\circ}\mathrm{C})\,\frac{\mathrm{K}}{^{\circ}\mathrm{C}}$.

Since pressure cannot be negative, the ideal gas temperature cannot assume negative values, $T\,(\mathrm{K}) \geq 0$. The ideal gas temperature scale fulfills all requirements on the thermodynamic temperature scale that will follow from the second law, and it coincides with the thermodynamic Kelvin scale. Much later, in Sec. 23.6, we will learn about the 3rd law of thermodynamics which, simply put, states that absolute zero $\equiv 0\,\mathrm{K}$ cannot be reached.

Some care must be taken in notation. For convenience temperatures are quite often given on the Celsius scale, but many thermodynamic equations require the thermodynamic temperature in Kelvin. Most often the same symbol, T, is used for temperatures on either scale, one has to be careful to not get confused.

2.14 Thermal Equation of State

Careful measurements on simple substances show that specific volume v (or density $\rho = 1/v$), pressure p and temperature T cannot be controlled independently. Indeed, they are linked through a relation of the form $p = p\,(v, T)$, or $p = p\,(\rho, T)$, known as the *thermal equation of state.* For most substances, this relation cannot be easily expressed as an actual equation, but is laid down in property tables, see Chapter 6.

The thermal equation of state relates measurable properties. It suffices to know the values of two properties to determine the values of others. This will still be the case when we add energy and entropy to the list of thermodynamic properties, which can be determined through measurement of any two of the measurable properties, i.e., (p, T) or (v, T) or (p, v).

For inhomogeneous states, where the properties are space dependent, we assume the validity of the thermal equation of state in the local volume element dV. This assumption reflects our understanding that the atoms and molecules of the considered substance are interacting frequently, and thus behave collectively, see Sec. 2.2.

To summarize: The complete knowledge of the macroscopic state of a system requires the values of two intensive properties in each location (i.e., in each infinitesimal volume element), and the local velocity. The state of a system in equilibrium, where properties are homogeneous, is described by just two intensive properties (plus the size of the system, that is either total volume, or total mass). In comparison, the knowledge of the microscopic state would require the knowledge of location and velocity of each particle.

2.15 Ideal Gas Law

The ideal gas is one of the simplest substances to study, since it has simple property relations. Ideal gases are employed in many engineering applications. Arguably, the most important ideal gas is air, which is the working substance in a large number of systems, including internal combustion engines.

Careful measurements have shown that for an ideal gas pressure p, total volume V, thermodynamic temperature T, and mass m are related by an explicit thermal equation of state, the *ideal gas law*

$$pV = mRT \,. \tag{2.12}$$

Here, R is the *gas constant* that depends on the type of the gas. Alternative forms of the equation result from introducing the specific volume $v = V/m$ or the mass density $\rho = 1/v$ so that

$$pv = RT \quad , \quad p = \rho RT. \tag{2.13}$$

The ideal gas law is our first property relation. According to this equation, the properties appearing in the equation cannot be changed independently: the change of one property must necessarily lead to a change of at least one other property. When the temperature is kept constant, an increase in pressure leads to a reduction of volume; when pressure is kept constant, the increase of temperature leads to increase of volume; when volume is kept constant, reduction of temperature leads to lower pressure. Of course, there can be processes where all three, pressure, volume, and temperature, change.

The gas constant R has the unit $\left[\frac{\mathrm{kJ}}{\mathrm{kg\,K}}\right]$, where $1\,\mathrm{kJ} = 10^3\,\mathrm{N\,m}$ is the energy unit kilo-Joule. Further examination has shown that the gas constant is related to the molar mass of the gas. One finds

$$R = \frac{\bar{R}}{M} \,, \tag{2.14}$$

where

$$\bar{R} = 8.314 \frac{\text{kJ}}{\text{kmol K}} \tag{2.15}$$

is the *universal gas constant.* The following list shows the molar masses of some important gases:[3] air, hydrogen, helium, nitrogen, oxygen, carbon dioxide,

$$\begin{aligned} M_{\text{air}} &= 29 \frac{\text{kg}}{\text{kmol}} , \; M_{\text{H}_2} = 2 \frac{\text{kg}}{\text{kmol}} , \; M_{\text{He}} = 4 \frac{\text{kg}}{\text{kmol}} , \\ M_{\text{N}_2} &= 28 \frac{\text{kg}}{\text{kmol}} , \; M_{\text{O}_2} = 32 \frac{\text{kg}}{\text{kmol}} , \; M_{\text{CO}_2} = 44 \frac{\text{kg}}{\text{kmol}} . \end{aligned} \tag{2.16}$$

Note that air is a mixture of nitrogen (~78% by particle number), oxygen (~21%), and argon (~1%) with traces of other substances including carbon dioxide (~0.04%).

The corresponding values of the gas constant are

$$\begin{aligned} R_{\text{air}} &= 0.287 \frac{\text{kJ}}{\text{kg K}} , \; R_{\text{H}_2} = 4.157 \frac{\text{kJ}}{\text{kg K}} , \; R_{\text{He}} = 2.077 \frac{\text{kJ}}{\text{kg K}} , \\ R_{\text{N}_2} &= 0.297 \frac{\text{kJ}}{\text{kg K}} , \; R_{\text{O}_2} = 0.260 \frac{\text{kJ}}{\text{kg K}} , \; R_{\text{CO}_2} = 0.189 \frac{\text{kJ}}{\text{kg K}} . \end{aligned} \tag{2.17}$$

More values can be found in property tables.

Mass m and mole number n are related as $n = m/M$, so that the ideal gas equation can be written in yet another form, with the universal gas constant,

$$pV = n\bar{R}T . \tag{2.18}$$

This equation does not contain any quantities that depend on the type of gas, accordingly the behavior of ideal gases is universal.

2.16 A Note on Problem Solving

Before we start solving our first problems, it might be worthwhile to briefly list good practices for problem solving. Typically, any engineering problem should be tackled by the following steps:

1. Understand the problem, i.e., read the question carefully. Nothing good can come from a solution that is based on a misunderstanding.
2. Make a sketch of the relevant system, and proper diagrams. A good sketch can summarize a complicatedly worded problem in a far more accessible form.

[3] The given values are rounded for easier memorization. Exact values are $M_{\text{air}} = 28.97 \frac{\text{kg}}{\text{kmol}}$, $M_{\text{H}_2} = 2.01588 \frac{\text{kg}}{\text{kmol}}$, $M_{\text{He}} = 4.002602 \frac{\text{kg}}{\text{kmol}}$, $M_{\text{N}_2} \doteq 28.0134 \frac{\text{kg}}{\text{kmol}}$, $M_{\text{O}_2} = 31.9988 \frac{\text{kg}}{\text{kmol}}$, $M_{\text{CO}_2} = 44.0095 \frac{\text{kg}}{\text{kmol}}$.

3. Indicate all quantities that are known on the sketch, or in a list, so that they are easy to find when needed. List all processes that occur in the system
4. List the quantities that need to be determined.
5. List the relevant thermodynamic equations for their determination.
6. Simplify the equations based on what is known about the processes in the system.
7. Solve the equations for the quantities of interest. Do not insert values for quantities before all manipulation of equations is complete; in other words, solve symbolically, and insert values as late as possible. This simplifies double checking of computations, and makes it far easier to correct errors that occur due to mistyping of values or wrong unit conversions.
8. Carefully consider and simplify all units. Before you have sufficient practice, never assume the outcome of a unit conversion, or the final unit for a value in a computation. Wrong unit conversions are a rather frequent source of major problems: always double-check. Note that each property value must be accompanied by a unit, i.e., never just write a number but [number value×unit].
9. Add comments throughout your treatment of the problem, so that you have text and equations/values on your answer. Written out sentences make the solution accessible, and you can explain assumptions, simplifications etc. This makes it far easier to follow through the line of argument for any reader—including yourself at a later point in time; just equations make for an unreadable submission.
10. Finally, use experience and common sense to scrutinize the final results. Do they make sense, e.g., are the values for temperatures, pressures, energies etc. realistic?

We shall as much as possible adhere to these steps in the examples throughout this book. However, due to space restrictions, we will, e.g., not always have a sketch, and will skip over algebraic reformulations of equations. Moreover, explicit unit conversions will be shown only in few early examples. It is strongly recommended that the reader goes through the examples carefully, including making a sketch, and double checking of all calculations, including the units.

2.17 Example: Air in a Room

A room of dimension $5\,\text{m} \times 10\,\text{m} \times 3\,\text{m}$ is filled with air at $20\,^\circ\text{C}$, $1\,\text{atm}$. Compute the mass of air in the room, the number of moles and the number of particles. If the temperature in the room increases to $25\,^\circ\text{C}$ for the same pressure, what amount of air has left through doors and windows?

We use the ideal gas law (2.12) with the values $V = 150\,\text{m}^3$, $p = 101.325\,\text{kPa}$, $T = 293\,\text{K}$. Note that the ideal gas law requires the Kelvin temperature! We find, with $R = 0.287\frac{\text{kJ}}{\text{kg}\,\text{K}}$ as the gas constant for air,

$$m = \frac{pV}{RT} = \frac{101.325 \times 150}{0.287 \times 293} \frac{\text{kPa m}^3}{\frac{\text{kJ}}{\text{kg K}}\text{K}} = 180.74 \frac{\text{kPa m}^3}{\text{kJ}} \text{kg} = 180.74\,\text{kg} .$$

For the unit conversion we have used that $1\,\text{kPa} = 10^3 \frac{\text{N}}{\text{m}^2}$ and $1\,\text{kJ} = 10^3\,\text{N m}$, hence $1\,\text{kPa} = 1\frac{\text{kJ}}{\text{m}^3}$. The corresponding mole number n and particle number N are, with $M = 29\frac{\text{kg}}{\text{kmol}}$ and the Avogadro constant (2.3),

$$n = \frac{m}{M} = 6.23\,\text{kmol} \ , \ N = nN_A = 3.75 \times 10^{27} .$$

For a temperature of $25\,°\text{C} \equiv 298\,\text{K}$ we find the mass $m = 177.71\,\text{kg}$, that is a mass of $3.03\,\text{kg}$ has left the room.

2.18 Example: Air in a Refrigerator

A refrigerator of volume $V_R = 330$ litre which maintains food at $T_R = 5\,°\text{C}$ is located in a kitchen at $T_0 = 22\,°\text{C}$, $p_0 = 1.02\,\text{atm}$. When the refrigerator door is opened, warm air enters the cooling space, and when the door is closed again, this warm air is cooled to T_R. We ask for the amount of air inside, and for the net force on the door after cooling is complete.

To simplify the problem, we assume that the air in the refrigerator is completely exchanged, so that in the moment of closing all air in the interior is at T_0, p_0. Then, the mass in the interior is, with $p_0 = 103.35\,\text{kPa}$, $V_R = 0.33\,\text{m}^3$, $T_0 = 295\,\text{K}$ and $R = 0.287\frac{\text{kJ}}{\text{kg K}}$,

$$m = \frac{p_0 V_R}{RT_0} = 0.403\,\text{kg} .$$

As long as no air enters during the cooling process, the pressure in the interior after cooling is complete is, with $T_R = 278\,\text{K}$,

$$p_R = \frac{mRT_R}{V_R} = p_0\frac{T_R}{T_0} = 0.961\,\text{atm} = 97.4\,\text{kPa} .$$

When the door has an area of $A = 0.6\,\text{m}^2$, the pressure difference between inside and outside gives the net force

$$F = A\,(p_0 - p_R) = 3.57\,\text{kN} .$$

This force must be overcome to open the door. Note that the calculation assumes perfect sealing, and complete replacement of the cold air with warm air. Actual kitchen refrigerators have imperfect seals, so that some air creeps through during cooling, hence the observed forces are weaker. Nevertheless, in particular not too long after closing, one can observe this effect. Try your refrigerator at home!

2.19 More on Pressure

Pressure p is defined as the force (F) exerted by a fluid per unit area (A), $p = F/A$, in the limit of infinitesimal area. Pressure is isotropic, that is the force on a surface is independent of the orientation of that surface.

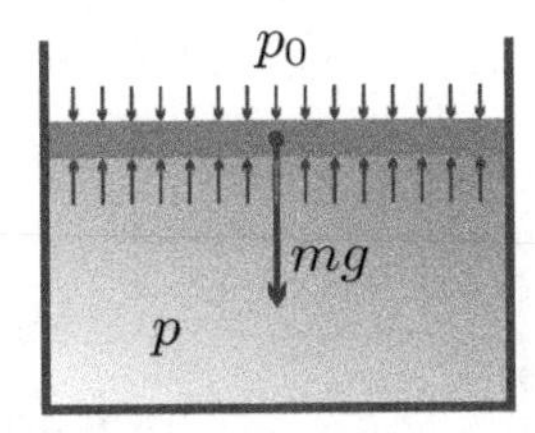

Fig. 2.9 A piston resting on a liquid

A piston of mass m and cross section A rests on a liquid in a cylinder, as depicted in Fig. 2.9; the atmospheric pressure is p_0. We determine the pressure p of the liquid at the piston.

The piston is at rest, in mechanical equilibrium, which implies that all forces F_i on the piston add up to zero, $\sum F_i = 0$. The acting forces are the weight mg of the piston, where $g = 9.81\frac{\text{m}}{\text{s}^2}$ is the gravitational acceleration, and the pressure forces due to atmospheric and liquid pressure, $p_0 A$ and pA, respectively. With the proper signs for the forces we have

$$mg + p_0 A - pA = 0 \quad \Longrightarrow \quad p = p_0 + \frac{mg}{A} \,. \tag{2.19}$$

The system pressure, p, balances the external pressure, p_0, and the weight of the piston.

Gravitation leads to variation of water pressure with depth, and of air pressure with height. We compute both following Fig. 2.10. The water-air interface is at the location $z = 0$ where the pressure is p_0. The insert shows a small layer of substance, air or water, of thickness dz and cross section A. The mass of the layer, dm, follows from the mass density ρ and the layer volume $dV = Adz$ as $dm = \rho Adz$.

The forces acting on the layer are the contributions of the pressures below, $p(z)\,A$, and above, $p(z+dz)\,A$, and the weight $g\,dm = \rho g A\,dz$. We assume the fluid (air or water) is at rest, so that the forces balance,

$$p(z+dz)\,A + \rho g A dz - p(z)\,A = 0 \,. \tag{2.20}$$

For infinitesimal dz we can use Taylor's formula $p(z+dz) = p(z) + \frac{dp(z)}{dz}dz$ and find a differential equation for pressure,

$$\frac{dp(z)}{dz} = -\rho g \,. \tag{2.21}$$

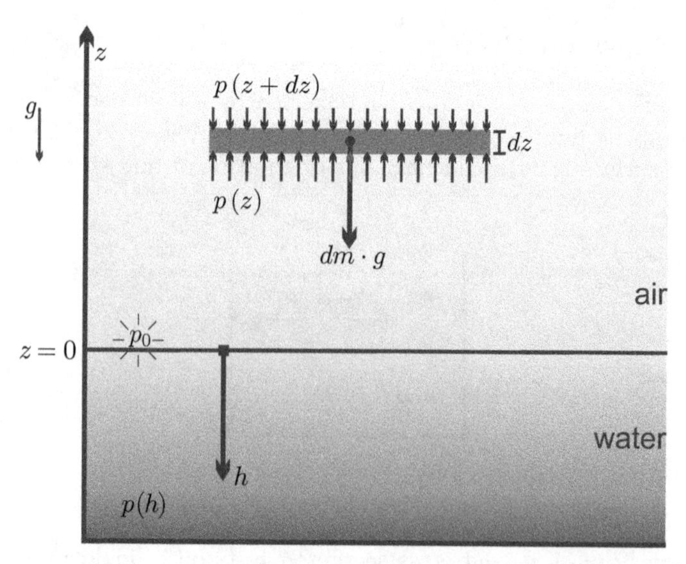

Fig. 2.10 On the computation of pressure variation in the gravitational field

To proceed, we have to differentiate between water and air. First we consider water: Water can be assumed in good approximation to be an incompressible substance, that is the water density is constant, with the well-known value of $\rho_{H_2O} \simeq 1000 \frac{\text{kg}}{\text{m}^3}$. Integration of (2.21) is straightforward, and gives, together with the condition $p(z=0) = p_0$,

$$p(z) = p_0 - \rho g z \,. \tag{2.22}$$

This is the hydrostatic pressure law, which is often written in terms of depth $h = -z$ as

$$p(h) = p_0 + \rho g h \,. \tag{2.23}$$

This relation is valid for all incompressible liquids, where the appropriate mass density ρ must be used. For water, depth increase by $\Delta h = 10.33\,\text{m}$ increases the pressure by about 1 atm. Hydrostatic pressure depends only on depth, not on the actual weight of liquid above. This implies that hydrostatic pressure is independent of the geometry of the container, see Fig. 2.11 for an illustration.

Air, on the other hand, is compressible, it obeys the ideal gas law $p = \rho RT$. Using this to eliminate density from the differential equation for pressure (2.21), we find

$$\frac{dp(z)}{p} = -\frac{g}{RT(z)} dz \,. \tag{2.24}$$

Integration is only possible when we have additional information on the temperature $T(z)$ as a function of height z.

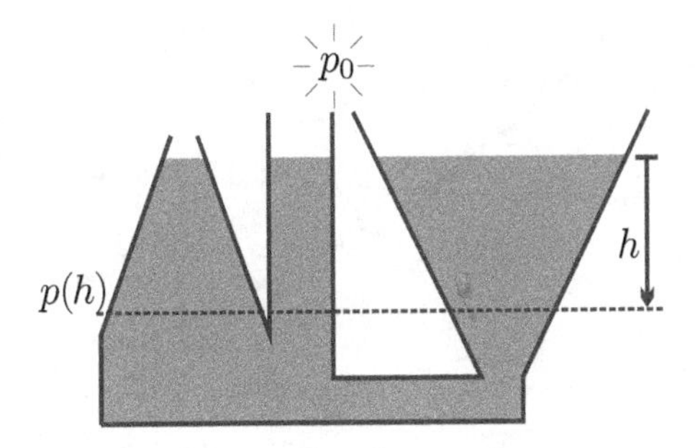

Fig. 2.11 Hydrostatic pressure depends only on depth

When the atmospheric temperature is constant, $T(z) = T_0$, integration of (2.24) gives, together with the boundary condition $p(z=0) = p_0$, the barometric formula

$$p = p_0 e^{-\frac{gz}{RT_0}} . \tag{2.25}$$

This formula describes the exponential decrease of pressure with height in an isothermal atmosphere.

In the actual atmosphere, however, the temperature is not constant, but decreases with height approximately as $T(z) = T_0 - \alpha z$ with $\alpha = 10 \frac{\mathrm{K}}{\mathrm{km}}$. Then, integration of (2.24) gives

$$p = p_0 \left(1 - \frac{\alpha z}{T_0}\right)^{\frac{g}{\alpha R}} . \tag{2.26}$$

Figure 2.12 compares both pressure functions for air and $T_0 = 293\,\mathrm{K}$. For the non-isothermal atmosphere the pressure decreases slightly faster, but at moderate heights the difference is almost not noticeable. The Canadian town of Banff is located at an altitude of 1463 m above sea level. When the sea level pressure is $p_0 = 1\,\mathrm{atm}$ we compute from (2.25) and (2.26) local pressures of 0.843 atm and 0.839 atm, respectively. Note that most weather forecasts do not present the actual local pressure p, but the normalized pressure, that is the corresponding sea level pressure p_0. For instance, when the forecast gives the pressure for Banff as 990 kPa, based on (2.26) the actual pressure in town will be 831 kPa.

The example shows a marked influence of gravitation on pressure for heights on the kilometer scale. Most engineering devices are relatively small, at most on the scale of several meters, and the variation of gas pressure can be safely ignored. Therefore it is sufficient to assign just one value of pressure to a gaseous system in equilibrium.

Gas pressure results from the momentum change of those gas particles that hit the wall and bounce back, and thus exert a force. When a gas filled container is put on a scale, the scale will show the total weight of container *and* gas, although most of the gas particles are not in contact with the container

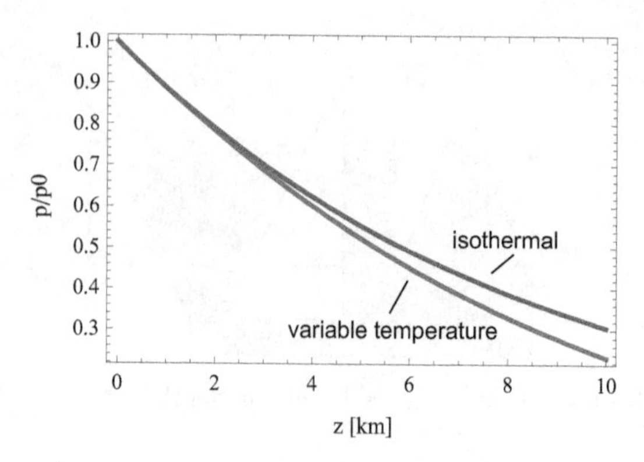

Fig. 2.12 Atmospheric pressure over height for constant and variable temperature

walls. Indeed, it is the small variation with pressure between top and bottom of a container which puts the weight of the gas on the scale.

Problems

2.1. Ideal Gas I

A 5 litre container holds helium at a temperature of 25 °C and a pressure of 2 atm. Determine the mass of gas in the container, the number of moles, and the number of particles.

2.2. Ideal Gas II

A cylinder with radius 5 cm contains 5 g of pure oxygen at a temperature of 200 °C. The cylinder is closed with a freely moving piston, which in equilibrium rests at a height of 50 cm. Determine the mass of the piston.

2.3. Ideal Gas Thermometer

An ideal gas thermometer holds a fixed gas volume of $1000\,\text{cm}^3$. For calibration, the thermometer is brought into contact with melting ice and boiling water, both at 1 atm, where the pressures measured are $p_1 = 51.6\,\text{kPa}$ and $p_2 = 70.5\,\text{kPa}$.

1. For Celsius temperature, assume a linear relation of the form $T(°\text{C}) = a + bp$ and determine the constants a and b.
2. Determine the mole number of particles enclosed.
3. Careful measurement shows that the mass of gas enclosed is 1 g. Find the molar mass—what gas is it most likely?

2.4. Ideal Gas and Spring

The following process is done in a room at a temperature of 20 °C and a pressure of 100 kPa: A container with quadratic base of 10 cm side length is closed by a piston of mass $m_p = 100\,\text{g}$. Initially, the piston is fixed at a height of $H_0 = 10\,\text{cm}$, and the cylinder is filled with 2.5 g of carbondioxide. A spring is attached to the piston from above, so that at the initial state the spring is at its rest length. When the fixing of the piston is removed, the piston moves up, and the spring is compressed. The system comes to an equilibrium state such that the piston has moved up by $\Delta H = 3\,\text{cm}$.

1. Determine the spring constant k.
2. The gas in the container is now heated to $120\,°\text{C}$. Determine the final displacement of the piston.

2.5. Climbing a Mountain
In an atmosphere where the temperature depends on height z as $T = T_0 - \alpha z$, with $T_0 = T(z = 0) = 15\,°\text{C}$ and $\alpha = 8.5 \frac{\text{K}}{\text{km}}$, a climber located at height $z_1 = 500\,\text{m}$ fills a piston-cylinder device with air, so that the device contains $2000\,\text{cm}^3$ of air. At this height, the climber measures an atmospheric pressure of $p_1 = 0.95\,\text{bar}$. The device is closed by a freely moving piston with mass $m_p = 300\,\text{g}$ and surface area of $A_p = 40\,\text{cm}^2$. The climber carries the system to the top of the mountain, at $z_2 = 4810\,\text{m}$.

1. Determine temperature and pressure of the atmosphere at z_1, z_2.
2. Determine the pressure inside the system at z_1, and the mass of air in the system.
3. The climber reaches the top of the mountain at z_2. After the system has established equilibrium with the surrounding air, determine the pressure in the system, and the system volume.

2.6. Ascent of a Balloon
The volume of a closed balloon shell is $V_f = 800\,\text{m}^3$, if completely filled. The mass of the balloon, including basket, but without the gas filling, is $m_B = 500\,\text{kg}$. The temperature of the environment and of the gas filling is $5\,°\text{C}$, and remains constant during the ascent. Initially, the balloon is filled with $V_0 = 500\,\text{m}^3$ of helium at the ground level pressure of $p_0 = 0.98\,\text{bar}$. As the balloon rises, its volume increases until it reaches V_f.

For the solution of the following questions, assume that the air pressure depends on height z according to the barometric formula $p(z) = p_0 \exp\left[-\frac{gz}{R_{air} T_0}\right]$ where R_{air} is the specific gas constant for air, and z is the height above ground.

Helium can be considered as an ideal gas with $M_{He} = 4 \frac{\text{kg}}{\text{kmol}}$.

1. Compute the mass of the helium filling.
2. As the balloon ascents, the volume of the filling increases. Compute the volume of the balloon as function of height. Above which height is the balloon completely filled? (Hint: the pressures of helium filling and the surrounding air are equal as long the balloon is not filled completely).
3. Compute the buoyancy for $V < V_f$, and show that it is independent of height. The buoyancy force is given as $F_B = \rho_{air}(z) g V$ where V is the actual balloon volume, and $\rho_{air}(z)$ the density of the surrounding air at height z.

4. Set up an equation for the buoyancy for the case that the balloon is completely filled. How high will the balloon rise?

2.7. An Experiment
As you accelerate in a car, you are pressed in the seat—what happens to a helium filled balloon? Think about it, or try it, then explain!

Chapter 3
The First Law of Thermodynamics

3.1 Conservation of Energy

It is our daily experience that heat can be converted to work, and that work can be converted to heat. A propeller mounted over a burning candle will spin when the heated air rises due to buoyancy: heat is converted to work. Rubbing your hands makes them warmer: work is converted to heat. Humankind has a long and rich history of making use of both conversions. Friction between a fast spun stick and a resting piece of wood is used since millennia to create a fire. Technical applications of heat to work conversions are abundant through history, and our modern life is unthinkable without heat engines such as steam and gas turbines for generation of electricity, or car and aircraft engines for transport. In cooling engines work is used to withdraw heat, such as in refrigerators or in air conditioning devices.

The evaporation of water to steam by heating provides a large change in volume and/or pressure. Devices using this effect were known already more than 2000 years ago, but they became prevalent with the development of the steam engine. Thermodynamics was initially developed to better understand the processes in steam engines and other conversion devices, so that the understanding can be used to improve the engines.

While the heat-to-work and work-to-heat conversions are readily observable in simple and more complex processes, the governing law is not at all obvious from simple observation. It required groundbreaking thinking and careful experiments to unveil the *law of conservation of energy.* Due to its importance in thermodynamics, it is also known as the *First Law of Thermodynamics.*

Expressed in words, the *First Law of Thermodynamics* reads:

> *Energy cannot be produced nor destroyed, it can only be transferred, or converted from one form to another. In short, energy is conserved.*

It took quite some time to formulate the first law in this simple form, the credit for finding and formulating it goes to Robert Meyer (1814-1878), James

H. Struchtrup, *Thermodynamics and Energy Conversion*,
DOI: 10.1007/978-3-662-43715-5_3,

Prescott Joule (1818-1889), and Hermann Helmholtz (1821-1894). Through careful measurements and analysis, they recognized that thermal energy, mechanical energy, and electrical energy can be transformed into each other, which implies that energy can be transferred by doing work, as in mechanics, and by heat transfer.

The first law is generally valid, no violation was ever observed. As knowledge of physics has developed, other forms of energy had to be included, such as radiative energy, nuclear energy, or the mass-energy equivalence of the theory of relativity, but there is no doubt today that energy is conserved under all circumstances.

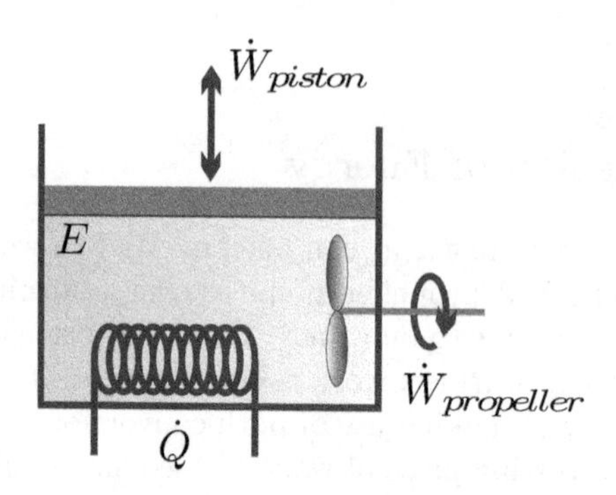

Fig. 3.1 Closed system with energy E exchanging work $\dot{W}$ and heat $\dot{Q}$ with its surroundings

We formulate the first law for the simple closed system, depicted again in Fig. 3.1, where all three possibilities to manipulate the system from the outside are indicated. For this system, the conservation law for energy reads

$$\frac{dE}{dt} = \dot{Q} - \dot{W} , \tag{3.1}$$

where E is the total energy of the system, $\dot{Q}$ is the heat transfer rate in or out of the system, and $\dot{W} = \dot{W}_{piston} + \dot{W}_{propeller}$ is the total power—the work per unit time—exchanged with the surroundings.

This equation states that the change of the system's energy in time (dE/dt) is equal to the energy transferred by heat and work per unit time $(\dot{Q} - \dot{W})$. The sign convention used is such that heat transferred *into* the system is positive, and work done *by* the system is positive.

The SI unit of energy, work, and heat is the Joule, $1\,\mathrm{J} = 1\,\mathrm{N\,m} = 1\,\frac{\mathrm{kg\,m^2}}{\mathrm{s^2}}$; the SI unit of power and heat transfer rate is the Watt, $1\,\mathrm{W} = 1\,\frac{\mathrm{J}}{\mathrm{s}}$.

All contributions to the first law (3.1), i.e., energy, work and heat, will be discussed in detail in the following sections.

3.2 Total Energy

The total energy E of the system is the sum of its kinetic energy E_{kin}, potential energy E_{pot}, and internal—or thermal—energy U,

$$E = U + E_{kin} + E_{pot} . \tag{3.2}$$

Presently, these are the only forms of energy that we need for the description of thermal processes; other forms of energy that can be relevant are chemical energy, nuclear energy, radiative energy and electrical energy, which will be introduced when required.

We address the different contributions to energy in the next sections.

3.3 Kinetic Energy

The kinetic energy is well-known from mechanics. For a homogeneous system of mass m and velocity $\mathcal{V}$, kinetic energy is given by

$$E_{kin} = \frac{m}{2}\mathcal{V}^2 . \tag{3.3}$$

For inhomogeneous states the total kinetic energy of the system is obtained by integration of the specific kinetic energy e_{kin} over all mass elements $dm = \rho dV$; we have

$$e_{kin} = \frac{1}{2}\mathcal{V}^2 \quad \text{and} \quad E_{kin} = \int \rho e_{kin} dV = \int \frac{\rho}{2}\mathcal{V}^2 dV . \tag{3.4}$$

3.4 Potential Energy

Also the potential energy in the gravitational field is well-known from mechanics. For a homogeneous system of mass m , potential energy is given by

$$E_{pot} = mgz , \tag{3.5}$$

where z is the elevation of the system's center of mass over a reference height, and $g = 9.81 \frac{\mathrm{m}}{\mathrm{s}^2}$ is the gravitational acceleration on Earth.

For inhomogeneous states the total potential energy of the system is obtained by integration of the specific potential energy e_{pot} over all mass elements $dm = \rho dV$; we have

$$e_{pot} = gz \quad \text{and} \quad E_{pot} = \int \rho e_{pot} dV = \int \rho gz dV . \tag{3.6}$$

3.5 Internal Energy and the Caloric Equation of State

Even if a macroscopic element of matter is at rest, its atoms move (in a gas or liquid) or vibrate (in a solid) fast, so that each atom has microscopic kinetic energy. Moreover, the atoms are subject to interatomic forces, which contribute microscopic potential energies. These microscopic energies depend on temperature, and the higher the temperature, the higher the average microscopic energy. Since the microscopic kinetic and potential energies cannot be observed macroscopically, one speaks of the *internal energy*, or *thermal energy*, of the material, denoted as U.

For inhomogeneous states the total internal energy of the system is obtained by integration of the specific internal energy u over all mass elements $dm = \rho dV$. For homogeneous and inhomogeneous systems we have

$$U = mu \quad \text{and} \quad U = \int \rho u dV \,. \tag{3.7}$$

Internal energy cannot be measured directly. The *caloric equation of state* relates the specific internal energy u to measurable quantities, it is of the form $u = u(T, v)$, or $u = u(T, p)$. Recall that pressure, volume and temperature are related by the thermal equation of state, $p(v, T)$; therefore it suffices to know two properties in order to determine the others.

The caloric equation of state must be determined by careful measurements, where the response of the system to heat or work supply is evaluated by means of the first law. For most materials the results cannot be easily expressed as equations, and are tabulated in property tables, see Chapter 6. Some simple caloric equations of state will be presented already in Sec. 3.10.

For inhomogeneous states, where the properties are space dependent, we assume the validity of the caloric equation of state in the local volume element dV. This assumption reflects our understanding that the atoms and molecules of the considered substance are interacting frequently, and thus behave as a collective, see Sec. 2.2.

3.6 Work and Power

Work, denoted by W, is the product of a force and the displacement of its point of application. Power, denoted by $\dot{W}$, is work done per unit time, that is the force times the velocity of its point of application. The total work for a process is the time integral of power over the duration $\Delta t = t_2 - t_1$ of the process,

$$W = \int_{t_1}^{t_2} \dot{W} dt \,. \tag{3.8}$$

For the closed system depicted in Fig. 2.1 there are two contributions to work: *moving boundary work*, due to the motion of the piston, and *rotating shaft work,* which moves the propeller. Other forms of work, e.g., spring work or electrical work will be discussed as required.

Work and power can be positive or negative. We follow the sign convention that work done *by* the system is positive and work done *to* the system is negative.

Moving boundary work is best computed from a piston-cylinder system, as depicted in Fig. 3.2; however, the subsequent expressions are valid for arbitrary system geometries. The force on the piston of cross section A is pA and thus the work for an infinitesimal displacement ds is given by $\delta W = pAds = pdV$, where $dV = Ads$ is the volume change associated with the displacement. As the piston is moved, the pressure within the system might change. Thus, the work W_{12} for a finite displacement $V_2 - V_1$ must be computed by summing over the infinitesimal contributions δW, that is by integration, $W_{12} = \int \delta W = \int_1^2 pdV$.

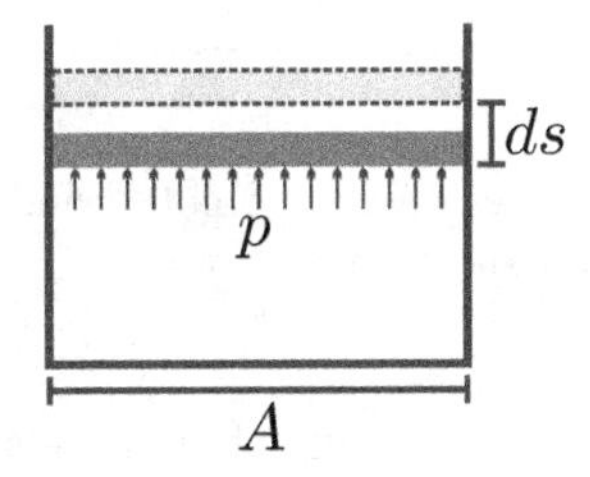

Fig. 3.2 Moving boundary work in a piston-cylinder system at pressure p, piston area A, displacement ds. Work is $\delta W = pdV = pAds$.

The power $\dot{W}$ is obtained from multiplying the force pA with the velocity $\frac{ds}{dt}$ of the piston. Since the cross section does not change, we have $A\frac{ds}{dt} = \frac{dV}{dt}$, and $\dot{W} = p\frac{dV}{dt}$.

Altogether we have the following expressions for moving boundary work with finite and infinitesimal displacement, and for power,

$$W_{12} = \int_1^2 pdV \quad , \quad \delta W = pdV \quad , \quad \dot{W} = p\frac{dV}{dt} \; . \tag{3.9}$$

Here, p is the pressure at the piston; for simplicity we have ignored variations of pressure along the piston surface.

Closed equilibrium systems are characterized by a single homogeneous pressure[1] p, a single homogeneous temperature T, and the volume V. In quasi-static (or reversible) processes, the system passes through a series of equilibrium states which can be indicated in suitable diagrams. Figure 3.3 shows a pressure-volume diagram (p-V-diagram) of two different reversible processes connecting the points $\{p_1, V_1\}$ and $\{p_2, V_2\}$. Due to the relation

[1] Hydrostatic variation ignored, see Sec. 2.19.

$W_{12} = \int_1^2 pdV$, the work is the area below the respective process curves as indicated by hatching. Obviously, the amount of work depends on the process: work is a path dependent function.

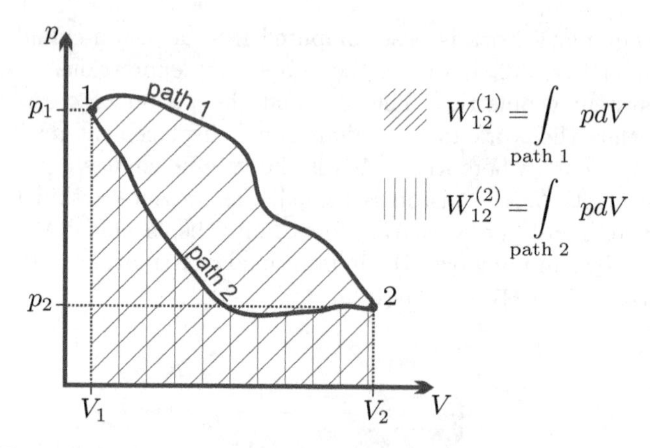

Fig. 3.3 Two reversible processes between points 1 and 2 in the p-V-diagram, and the corresponding moving boundary work

The power transmitted by a rotating shaft is related to the torque $\mathbf{T}$ and the revolutionary speed $\dot{n}$ (revolutions per unit time) as $\dot{W} = 2\pi\dot{n}\mathbf{T}$, the total work transmitted during a finite time interval is, again, $W_{12} = \int_1^2 \dot{W}dt$. The transmission between the shaft and the working fluid is performed by a propeller (turbines, compressors etc.).

In a closed system the propeller stirs the working fluid and creates inhomogeneous states. Fluid friction transmits fluid motion (i.e., momentum and kinetic energy) from the fluid close to the propeller to the fluid further away. Due to the inherent inhomogeneity, stirring of a fluid in a closed system cannot be a quasi-static process.

This is different in open systems, where fluid is entering and leaving the system. The motion of the fluid can be used to drive the propeller, which decelerates the fluid and transmits work out of the system, or the propeller can provide work to accelerate the fluid. These flow processes can be reversible.

In general, there might be several work interactions $\dot{W}_j$ of the system, then the total work for the system is the sum over all contributions; e.g., for power

$$\dot{W} = \sum_j \dot{W}_j \,. \tag{3.10}$$

Finally we note that mechanical work can be transferred without restrictions between systems in mechanical contact:

By using gears and levers, one can transfer work from slow moving to fast moving systems and vice versa, and one can transmit work from high pressure to low pressure systems and vice versa.

3.7 Exact and Inexact Differentials

Above we have seen that work depends on the process path. In the language of mathematics this implies that the work for an infinitesimal step is not an exact differential, and that is why a Greek delta (δ) is used to denote the work for an infinitesimal change as δW. As will be seen in the next section, heat is path dependent as well.

State properties like pressure, temperature, volume and energy describe the momentary state of the system, or, for inhomogeneous states, the momentary state in the local volume element. State properties have exact differentials for which we write, e.g., dE and dV. The energy change $E_2 - E_1 = \int_1^2 dE$ and the volume change $V_2 - V_1 = \int_1^2 dV$ are independent of the path connecting the states.

It is important to remember that work and heat, as path functions, only describe property *changes*, not states. A state is characterized by state properties (pressure, temperature, etc.), it does not possess work or heat.

Quasi-static (reversible) processes go through well defined equilibrium states, so that the whole process path can be indicated in diagrams, e.g., the p-V-diagram.

Non-equilibrium (irreversible) processes, for which typically the states are different in all volume elements, cannot be drawn into diagrams. Often irreversible processes connect homogeneous equilibrium states which can be indicated in the diagram. We shall use dashed lines to indicate non-equilibrium processes that connect equilibrium states. As an example, Fig. 3.4 shows a p-V-diagram of two processes, one reversible, one irreversible, between the same equilibrium states 1 and 2. We emphasize that the dashed line does not refer to actual states of the system. The corresponding work for the non-equilibrium process cannot be indicated as the area below the curve, since its computation requires the knowledge of the—inhomogeneous!—pressures at the piston surface at all times during the process.

3.8 Heat Transfer

Heat is the transfer of energy due to differences in temperature. Experience shows that for systems in thermal contact the direction of heat transfer is restricted:

Heat will always go from hot to cold by itself, but not vice versa.

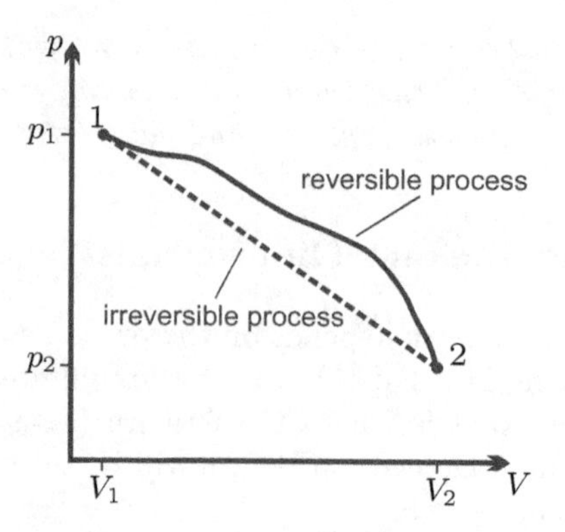

Fig. 3.4 A reversible (quasi-static) and an irreversible (non-equilibrium) process between the equilibrium states 1 and 2

This restriction of direction is an important difference to energy transfer by work between systems in mechanical contact, which is not restricted.

Since heat flows only in response to a temperature difference, a quasi-static (reversible) heat transfer process can only be realized in the limit of infinitesimal temperature differences between the system and the system boundary, and for infinitesimal temperature gradients within the system.

The main heat transfer mechanisms are: (a) Heat conduction, where thermal energy is transmitted by microscopic energy exchange between neighboring particles. (b) Convection, where fluid elements move to hotter or colder parts of the system and then exchange energy with the new neighborhood. (c) Radiative transfer, where electromagnetic radiation that crosses the system boundaries is absorbed or emitted by the matter inside the system. In the present context we do not need to discuss the details of these heat transfer mechanisms, which ultimately describe the same thing: energy transfer driven by temperature difference.

We use the following notation: $\dot{Q}$ denotes the heat transfer rate, that is the amount of energy transferred as heat per unit time. Heat depends on the process path, so that the heat exchanged for an infinitesimal process step, $\delta Q = \dot{Q}dt$, is not an exact differential. The total heat transfer for a process between states 1 and 2 is

$$Q_{12} = \int_1^2 \delta Q = \int_{t_1}^{t_2} \dot{Q} dt \,. \tag{3.11}$$

By convention, heat transferred into the system is positive, heat transferred out of the system is negative.

A process in which no heat transfer takes place, $\dot{Q} = 0$, is called *adiabatic process*.

In general, there might be several heat interactions $\dot{Q}_k$ of the system, then the total heat for the system is the sum over all contributions; e.g., for the heating rate

$$\dot{Q} = \sum_k \dot{Q}_k \,. \tag{3.12}$$

Confusion might result between the use of the word *heat* in everyday language, and its use in thermodynamics. In thermodynamics, *heat* solely describes a means to transfer energy in response to temperature differences. In particular we emphasize that heat is *not* a form of energy, and does not relate to how hot it might be outside. To say "oh, what a heat" is common language, a thermodynamicist will say "oh, it's pretty hot outside", or, even better, "oh, the temperature is pretty high today." A state is characterized by its energy or temperature, a change of state is characterized by heat (transfer).

3.9 The First Law for Reversible Processes

The form (3.1) of the first law is valid for all closed systems. When only reversible processes occur within the system, so that the system is in equilibrium states at any time, the equation can be simplified as follows: From our discussion of equilibrium states we know that for reversible processes the system will be homogeneous, and that all changes must be very slow, which implies very small velocities. Therefore, kinetic energy can be ignored, $E_{kin} = 0$. Stirring, which transfers energy by moving the fluid and friction, is irreversible, hence in a reversible process only moving boundary work can be transferred. As long as the system location does not change, the potential energy does not change, and we can set $E_{pot} = 0$.

With all this, for reversible (quasi-static) processes the first law of thermodynamics reduces to

$$\frac{dU}{dt} = \dot{Q} - p\frac{dV}{dt} \quad \text{or} \quad U_2 - U_1 = Q_{12} - \int_1^2 p dV \,, \tag{3.13}$$

where the second form results from integration over the process duration. We shall later, in particular in Chapter 7, use this equation extensively to study reversible processes in closed systems.

3.10 The Specific Heat at Constant Volume

We consider a closed system heated at constant volume (*isochoric* process), where the first law (3.13) reduces to (recall that $U = mu\,(T, v)$ and $m =$ *const.*)

$$m\left(\frac{\partial u}{\partial T}\right)_v \frac{dT}{dt} = \dot{Q} \,. \tag{3.14}$$

Here, $\left(\frac{\partial u}{\partial T}\right)_v = \frac{\partial u(T,v)}{\partial T}$ denotes the partial derivative of internal energy with temperature at constant specific volume[2] $v = V/m$. This derivative is known as the *specific heat* (or *specific heat capacity*) *at constant volume,*

$$c_v = \left(\frac{\partial u}{\partial T}\right)_v . \tag{3.15}$$

To understand this name for c_v, we rewrite (3.14) as

$$c_v dT = \frac{\dot{Q}dt}{m} = \frac{\delta Q}{m} . \tag{3.16}$$

From this equation we see that c_v is the amount of heat required to increase the temperature of 1 kg of substance by 1 K at constant volume. The specific heat can be measured by controlled heating of a fixed amount of substance in a fixed volume system, and measurement of the ensuing temperature difference; its SI unit is $\left[\frac{\text{kJ}}{\text{kg K}}\right]$.

In general, $c_v(T, v) = \left(\frac{\partial u}{\partial T}\right)_v$ is a function of temperature and specific volume. For incompressible liquids and solids the specific volume is constant, $v = const$, and the specific heat is a function of temperature alone. Interestingly, also for ideal gases the specific heat turns out to be a function of temperature alone, both experimentally and from theoretical considerations. For these materials the internal energy depends only on temperature, and integration gives the caloric equation of state as

$$u(T) = \int_{T_0}^{T} c_v(T')\, dT' + u_0 . \tag{3.17}$$

Only energy differences can be measured, where the first law is used to evaluate careful experiments. The choice of the energy constant $u_0 = u(T_0)$ fixes the energy scale. The actual value of this constant will only become relevant for the discussion of chemical reactions. Note that proper mathematical notation requires to distinguish between the actual temperature T of the system, and the integration variable T'.

For materials in which the specific heat varies only slightly with temperature in the interval of interest, the specific heat can be approximated by a suitable constant average c_v^{avg}, so that the caloric equation of state assumes the particularly simple linear form

[2] Due to the abundance of thermodynamic properties, and the freedom to choose any two of them as variables, one needs to be careful with the notation. In the present context, internal energy depends on two variables, and when a partial derivative is taken with respect to one variable, it is customary to indicate the second variable by a subscript, to condense notation. This notation, where, e.g., $\left(\frac{\partial u}{\partial T}\right)_v = \frac{\partial u(T,v)}{\partial T}$ will be used throughout this text for partial derivatives of properties.

$$u(T) = c_v^{avg}(T - T_0) + u_0 \,. \tag{3.18}$$

This relation for the caloric equation of state will serve us well in our first examples.

For temperatures around the standard environmental temperature $T_0 = 298\,\mathrm{K}$ ($\equiv 25\,°\mathrm{C}$), the specific heat of air is $c_v^{\mathrm{air}} = 0.717\frac{\mathrm{kJ}}{\mathrm{kg\,K}}$, for liquid water one finds $c_w = 4.18\frac{\mathrm{kJ}}{\mathrm{kg\,K}}$. The old unit for heat and thermal energy, the calorie [cal], is defined such that one calorie is the heat required to raise the temperature of one gram of water by one degree Celsius (from 14.5 °C to 15.5 °C at $p_0 = 1\,\mathrm{atm}$), thus $1\,\mathrm{cal} = 4.18\,\mathrm{J}$, and $1\,\mathrm{kcal} = 4.18\,\mathrm{kJ}$.

3.11 Enthalpy

In many thermodynamic calculations one encounters the combination $U+pV$, or the mass divided equivalent $u+pv$, and it is convenient to introduce a name and a symbol for these. We define the total and the specific *enthalpy* as

$$H = U + pV \quad , \quad h = u + pv \,, \tag{3.19}$$

where $H = mh$.

Using enthalpy to replace internal energy, the first law for quasi-static processes assumes the forms[3]

$$\frac{dH}{dt} = \dot{Q} + V\frac{dp}{dt} \quad \text{and} \quad H_2 - H_1 = Q_{12} + \int_1^2 V dp \,. \tag{3.20}$$

As an application we consider a closed system heated at constant pressure (*isobaric* process), so that $\frac{dp}{dt} = 0$. In this case, the first law reduces to $\frac{dH}{dt} = \dot{Q}$, or, since $H = mh(T,p)$,

$$m\left(\frac{\partial h}{\partial T}\right)_p \frac{dT}{dt} = \dot{Q} \,. \tag{3.21}$$

Here $\left(\frac{\partial h}{\partial T}\right)_p = \frac{\partial h(T,p)}{\partial T}$ denotes the partial derivative of specific enthalpy with respect to temperature at constant pressure p. This derivative is known as the *specific heat at constant pressure*

$$c_p = \left(\frac{\partial h}{\partial T}\right)_p \,. \tag{3.22}$$

To understand this name for c_p, we rewrite the above equation as

[3] From the defintion of enthalpy we have $U = H - pV$, hence $\frac{dU}{dt} = \frac{d(H-pV)}{dt} = \frac{dH}{dt} - p\frac{dV}{dt} - V\frac{dp}{dt}$; inserting this into the first law (3.13) gives the shown result.

$$c_p dT = \frac{\delta Q}{m} \, . \tag{3.23}$$

We see that c_p is the amount of heat required to increase the temperature of 1 kg of substance by 1 K at constant pressure; its SI unit is $\left[\frac{\text{kJ}}{\text{kg K}}\right]$.

For an ideal gas the thermal equation of state gives $pv = RT$, and the internal energy $u(T)$ is a function of temperature alone. It follows that for the ideal gas also the enthalpy $h(T) = u + pv = u(T) + RT$ is a function of temperature alone. From the definitions of the specific heats (3.15, 3.22) follows $c_p = c_v + R$; for air at T_0 one finds $c_p^{\text{air}} = 1.004 \frac{\text{kJ}}{\text{kg K}}$. For ideal gases enthalpy and specific heat are related as

$$h(T) = \int_{T_0}^{T} c_p(T') \, dT' + h_0 \quad \text{or} \quad h = c_p^{avg}(T - T_0) + h_0 \, , \tag{3.24}$$

where the latter relation holds in case of constant specific heat.

For incompressible solids and liquids ($v = const.$), the specific heats at constant pressure and constant volume agree, since $\left(\frac{\partial pv}{\partial T}\right)_p = p\left(\frac{\partial v}{\partial T}\right)_p = 0$ (also see Sec. 16.7 for a more detailed argument), and one writes the specific heat without an index, $c = c_v = c_p$. The specific heat for water will be denoted as c_w.

While its internal energy depends only on temperature, the enthalpy of an incompressible substance (constant specific volume v) depends on temperature and pressure. Indeed, by its definition $h = u + pv$, enthalpy depends explicitly on pressure. With h_0 as the enthalpy at a reference point (T_0, p_0), the enthalpy for an incompressible solid or liquid with constant specific heat becomes

$$h(T, p) = c^{avg}(T - T_0) + (p - p_0) v + h_0 \, . \tag{3.25}$$

Note that no substance is truly incompressible, normally the specific volume changes at least a little bit. This leads to small differences between specific heats which can be ignored as long as the compressibility is sufficiently small.

3.12 Example: Equilibration of Temperature

We apply the first law to the situation depicted in Fig. 2.6. Two bodies A and B that are initially at different temperatures $\bar{T}_A$ and $\bar{T}_B$, respectively, are brought into thermal contact. After a sufficiently long time, we find that both bodies have assumed the common temperature T.

For this problem, kinetic energy is zero, and potential energy does not change. When the system $[A + B]$ is adiabatically enclosed ($\dot{Q} = 0$), and no work is done ($\dot{W} = 0$), the first law of thermodynamics simply states that the energy of the system remains constant,

$$\frac{dU}{dt} = 0 \,.$$

Thus, the energy of the end state is equal to the initial energy, $U_{\text{end}} = U_{\text{init}}$.

For simple incompressible solids the internal energy is given by[4] $U = mcT$, where c denotes the average specific heat (assumed to be a constant) and m is the mass. The internal energy of the system consisting of the two bodies is initially

$$U_{\text{init}} = U_A + U_B = m_A c_A \bar{T}_A + m_B c_B \bar{T}_B \,.$$

To emphasize that the first law does not automatically give equal final temperatures for the two bodies, we write $U_{\text{end}} = m_A c_A T_A + m_B c_B T_B$ with different final temperatures T_A and T_B. We solve for T_A,

$$T_A = \bar{T}_A + \frac{m_B c_B}{m_A c_A} \left(\bar{T}_B - T_B \right) \,, \tag{3.26}$$

and see that there are infinitely many solutions for the final temperatures (T_A, T_B) that fulfill the first law: conservation of energy alone is not sufficient to determine the final equilibrium state.

However, our experience, laid down in the zeroth law, tells us that the final temperatures agree: $T_A = T_B = T$. We find the final temperature as the weighted average of the two initial temperatures,

$$T = \frac{m_A c_A \bar{T}_A + m_B c_B \bar{T}_B}{m_A c_A + m_B c_B} \,,$$

with the weights given by the thermal masses $m_A c_A$, $m_B c_B$. We shall later employ the second law to find the same result.

As discussed, a thermometer utilizes the equilibration of temperature. The act of measurement should not affect the result. To study the relevant condition, let body B be the thermometer, used to measure the temperature of body A. The final temperature T of body *and* thermometer can be written in the alternative form

$$T = \bar{T}_A + \frac{m_B c_B \left(\bar{T}_B - \bar{T}_A \right)}{m_A c_A + m_B c_B} \,.$$

The measured temperature T is close to the initial temperature $\bar{T}_A$ of body A when $m_B c_B \ll m_A c_A$. It follows that a thermometer should have considerably smaller thermal mass mc than the body whose temperature is to be measured.

---

[4] With the energy constant $u_0 = cT_0$.

3.13 Example: Uncontrolled Expansion of a Gas

Our next example concerns the uncontrolled expansion of an ideal gas. For this, we consider an ideal gas in a container which is divided by a membrane, see Fig. 3.5. Initially the gas is contained in one part of the container at $\{T_1, p_1, V_1\}$, while the other part is evacuated. The membrane is destroyed, and the gas expands into the container. The fast motion of the gas is slowed down by internal friction, and in the final homogeneous equilibrium state $\{T_2, p_2, V_2\}$ the gas is at rest and distributed over the total volume of the container. Note that we have no control over the flow after the membrane is destroyed: this is an irreversible process.

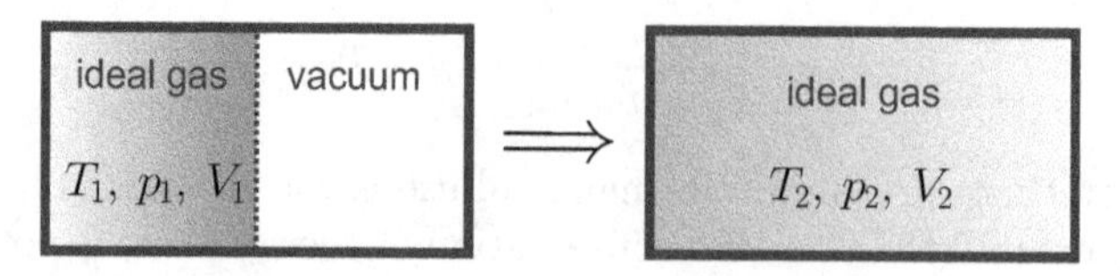

Fig. 3.5 Irreversible adiabatic expansion of an ideal gas

The container is adiabatically enclosed to the exterior, and, since its walls are rigid, no work is transmitted to the exterior. Thus, the first law for closed systems (3.1) reduces to

$$\frac{d\left(U + E_{kin} + E_{pot}\right)}{dt} = 0 \,,$$

or, after integration,

$$U_2 + E_{kin,2} + E_{pot,2} = U_1 + E_{kin,1} + E_{pot,1} \,.$$

Since the gas it at rest initially and in the end, $E_{kin,1} = E_{kin,2} = 0$, and since potential energy has not changed $E_{pot,1} = E_{pot,2}$, the above reduces to $U_2 = U_1$. With $U = mu$, and $m = const.$, the specific internal energy remains unchanged,

$$u\left(T_1, v_1\right) = u\left(T_2, v_2\right) \,.$$

Measurements for ideal gases show that $T_1 = T_2$, that is the initial and final temperatures of the gas are the same. With this, the previous condition becomes

$$u\left(T_1, v_1\right) = u\left(T_1, v_2\right) \,,$$

which can only hold if the internal energy of the ideal gas does not depend on volume. This experiment verifies that the internal energy of the ideal gas is independent of volume, and depends only on temperature, $u = u\left(T\right)$.

3.14 Example: Friction Loss

One litre of water in an adiabatic container is stirred such that the initial average velocity of the water is $\mathcal{V}_1 = 5\,\frac{\text{m}}{\text{s}}$. Stirring stops, and due to internal friction and friction between water and container walls the water will come to rest after a while. The water still moves after the stirrer is removed, but we have no control over the water motion: this is an irreversible process. We compute the change of temperature in the equilibration process.

After stirring stops, the system is isolated, no heat and work are exchanged, $\dot{Q} = \dot{W} = 0$, potential energy remains constant, $\frac{dE_{pot}}{dt} = 0$. The energy balance (3.1) reduces to $\frac{d}{dt}\left(U + E_{kin}\right) = 0$, so that the total energy $U + E_{kin}$ stays constant,

$$U_2 + E_{kin,2} = U_1 + E_{kin,1} \; .$$

From experience we know that in the final homogeneous equilibrium state the water is at rest, $E_{kin,2} = 0$, and we find

$$U_2 - U_1 = mc_w \Delta T = E_{kin,1} = \frac{m}{2}\mathcal{V}_1^2 \; .$$

Mass cancels and we find the temperature difference as

$$\Delta T = \frac{1}{2}\frac{\mathcal{V}_1^2}{c_w} = \frac{1}{2}\frac{25\frac{\text{m}^2}{\text{s}^2}}{4.18\frac{\text{kJ}}{\text{kg}}} = 0.003\,\text{K} \; .$$

For the unit conversion, we have used that $1\frac{\text{kJ}}{\text{kg}} = 10^3\frac{\text{J}}{\text{kg}} = 10^3\frac{\text{m}^2}{\text{s}^2}$. Note that this very small temperature change is due only to the destruction of the initial kinetic energy. Constant stirring of a viscous liquid can increase its temperature considerably; the relevant form for the first law is $\frac{dU}{dt} = -\dot{W}$, or $\frac{dT}{dt} = -\frac{\dot{W}}{mc}$, where $\dot{W}$ is the work required for stirring. A good example from daily life is the kneading of pizza dough, which can become quite warm.

3.15 Example: Heating Problems

3.15.1 Heating of Water

2 litre of water at $T_1 = 20\,^\circ\text{C}$ are heated in a well isolated 2 kW electric kettle. We compute the time required to heat the water to $T_2 = 90\,^\circ\text{C}$.

In this temperature range, liquid water can be well described as an incompressible liquid with mass density $\rho = 1000\frac{\text{kg}}{\text{m}^3}$ and constant specific heat $c_w = 4.18\frac{\text{kJ}}{\text{kg K}}$. The mass of water in the heater is $m = \rho V = 2\,\text{kg}$. Since the volume of the water remains unchanged, there is no work done, and with the internal energy $U = mc_w\left(T - T_0\right) + u_0$, the first law reduces to

$$mc_w\frac{dT}{dt} = \dot{Q} \; ,$$

with $\dot{Q} = 2\,\mathrm{kW}$ as given. Separation of variables and integration over the duration of the process gives $mc_w\,(T_2 - T_1) = \dot{Q}\,(t_2 - t_1)$, or

$$\Delta t = t_2 - t_1 = \frac{mc_w}{\dot{Q}}\,(T_2 - T_1) = 292.6 \frac{\mathrm{kg}\frac{\mathrm{kJ}}{\mathrm{kg\,K}}}{\mathrm{kW}}\,\mathrm{K} = 292.6\,\mathrm{s}\,.$$

For the unit conversion we have used that $1\,\mathrm{kW} = 1\frac{\mathrm{kJ}}{\mathrm{s}}$.

3.15.2 Heating of Water with Heat Loss

We consider the same problem as above, only that now the water in the heater loses heat to the environment at $T_0 = 20\,°\mathrm{C}$ at a rate of $\dot{Q}_{loss} = \alpha\,(T - T_0)$ with a transfer coefficient $\alpha = 25\frac{\mathrm{W}}{\mathrm{K}}$. The heat loss must be added to the heat supplied by the kettle, so that the first law reads (careful with the sign, the heat loss must be subtracted)

$$mc_w \frac{dT}{dt} = \dot{Q} - \alpha\,(T - T_0)\ .$$

Since T_0, $\dot{Q}$ and α are constant in time, this differential equation can be written in the equivalent form

$$mc_w \frac{d\left(T - T_0 - \frac{\dot{Q}}{\alpha}\right)}{dt} = -\alpha\left(T - T_0 - \frac{\dot{Q}}{\alpha}\right)\ .$$

Integration between $\{t_1, T_1\}$ and $\{t_2, T_2\}$ gives the solution

$$\ln\left(T_2 - T_0 - \frac{\dot{Q}}{\alpha}\right) - \ln\left(T_1 - T_0 - \frac{\dot{Q}}{\alpha}\right) = -\frac{\alpha}{mc_w}\Delta t\ ,$$

or, solved for Δt

$$\Delta t = \frac{mc_w}{\alpha}\ln\frac{T_1 - T_0 - \frac{\dot{Q}}{\alpha}}{T_2 - T_0 - \frac{\dot{Q}}{\alpha}} = 695.4\,\mathrm{s}\,.$$

The higher the water temperature becomes, the more heat is lost. With the values given above, for $\Delta t \to \infty$, we find a maximum temperature of $T_2^\infty = T_0 + \frac{\dot{Q}}{\alpha} = 100\,°\mathrm{C}$ (note that $\frac{\dot{Q}}{\alpha} = \frac{2\,\mathrm{kW}}{25\frac{\mathrm{W}}{\mathrm{K}}} = \frac{2000\,\mathrm{W}}{25\frac{\mathrm{W}}{\mathrm{K}}} = 80\,\mathrm{K}$). The chosen value for α is a bit high for a water heater, which normally can bring water to boil and evaporate in finite time.

The heat transfer coefficient α depends on the material, and the system configuration. Our sense of cold or hot is not a sense of temperature, but rather a sense of heat transfer. When we touch an object with large heat transfer coefficient, a large amount of heat is exchanged between our hand

and the object, which feels hotter or colder as an object with smaller heat transfer coefficient at the same temperature. The amount of energy available plays a role as well. A larger amount of heat can be transferred to our hand from an object with large thermal mass mc. Wood feels not as cold as metal of the same temperature.

3.15.3 Isochoric Heating of an Ideal Gas

We consider the air-filled room from a previous example which contains 180.94 kg of air, initially at 20 °C, 1 atm. We now assume the room is perfectly sealed, so that the air volume remains constant, and ask for the total amount of heat that must be supplied to heat the room to 25 °C.

We describe air as an ideal gas with constant specific heat, so its internal energy is given by $U = mc_v (T - T_0) + u_0$, with $c_v = 0.717 \frac{\text{kJ}}{\text{kg K}}$. Since the volume remains constant, no work is done, $W_{12} = \int_1^2 pdV = 0$, and the first law reduces to

$$mc_v \frac{dT}{dt} = \dot{Q} \,.$$

Integration gives

$$Q_{12} = \int_1^2 \dot{Q} dt = mc_v \Delta T = 648.7\,\text{kJ} \,.$$

A 2 kW heater would need $\Delta t = Q_{12}/\dot{Q} = 324\,\text{s}$ to heat the air in the room by 5 °C. The heating of a real room takes longer, since a substantial amount of heat is required to heat the walls, which have a large thermal mass mc, moreover one will expect heat loss through the walls to the colder outside environment.

The pressure after heating is completed, p_2, follows from the ideal gas law $pV = mRT$. Since mass and volume remain constant we have $p/T = mR/V = const$, so that $p_2/T_2 = p_1/T_1$ or $p_2 = p_1 T_2/T_1 = 1.017\,\text{atm} = 1.0305\,\text{bar}$ (temperatures in Kelvin!).

3.15.4 Isobaric Heating of an Ideal Gas

Next we ask for the amount of heat required to heat the same amount of air under constant pressure.

In this case, the heat is best computed from the alternative form (3.20) which for constant pressure reduces to

$$\frac{dH}{dt} = \dot{Q} \,,$$

where $H = mh$ is the enthalpy of the enclosed air, with $h = c_p (T - T_0) + h_0$ and $c_p = 1.004 \frac{\text{kJ}}{\text{kg K}}$. Integration gives

$$Q_{12} = \int_1^2 \dot{Q} dt = mc_p \Delta T = 908.3\,\text{kJ}\,.$$

A 2 kW heater would need $\Delta t = Q_{12}/\dot{Q} = 454\,\text{s}$ to heat the air in the room, provided that no heat loss occurs to and through the walls.

The initial volume is $V_1 = 150\,\text{m}^3$ and the volume after heating follows from the ideal gas law $pV = mRT$. Since pressure and mass remain constant, we have $V_2/T_2 = V_1/T_1$ and find $V_2 = V_1T_2/T_1 = 152.6\,\text{m}^3$. The expansion of the air requires moving boundary work. Since pressure is constant we find

$$W_{12} = \int_1^2 pdV = p\int_1^2 dV = p\,(V_2 - V_1) = 259.4\,\text{kJ}\,.$$

In the isochoric case all heat supplied goes to increase the internal energy. The heat required for isobaric heating is bigger since, while the increase of internal energy is the same, additional energy is required to provide the expansion work.

Problems

3.1. Tank and Contents

A well-insulated copper tank of mass 12 kg at 27 °C is filled with 4 litres of water at 50 °C. The tank is heated with a 1 kW resistance heater for $2\frac{1}{2}$ minutes, and then left alone. Determine the temperature of the system after equilibrium is established. Is the process reversible or irreversible? For copper: $\rho = 8.9\frac{\text{kg}}{\text{litre}}$, $c_p = 0.386\frac{\text{kJ}}{\text{kg K}}$.

3.2. Cooling Process

A $0.5\,\text{m}^3$ block of steel ($\rho = 7.83\frac{\text{kg}}{\text{litre}}$, $c_p = 0.5\frac{\text{kJ}}{\text{kg K}}$) initially at 250 °C is heated with a constant rate of $\dot{Q} = 50\,\text{kW}$. How long does it take until the block's temperature is 600 °C?

3.3. Equilibration of Temperature

To warm the water in your bathtub, you decide to heat it by throwing a block of hot iron into the water. When your bathtub holds 150 litres of water initially at 20 °C, and you can heat the iron to 400 °C, what mass should the iron block have so that you can have a bath at 33 °C? Is the process reversible or irreversible? Assume no heat loss to anywhere, and no boiling, evaporation etc.

Specific heats: $c_w = 4.18\frac{\text{kJ}}{\text{kg K}}$, $c_{iron} = 0.450\frac{\text{kJ}}{\text{kg K}}$.

3.4. Irreversible Expansion

An ideal gas is confined to one side of a rigid, insulated (= no heat transfer, adiabatic) container, divided by a partition. The other side is initially evacuated. The initial state of the gas is $p_1 = 2\,\text{bar}$, $T_1 = 400\,\text{K}$, $V_1 = 0.02\,\text{m}^3$. When the partition is removed, the gas expands to fill the entire container and achieves a final equilibrium pressure of 1.5 bar. Determine the volume of the container.

3.5. Stirring of a Liquid
A thermally insulated 2 litre tank is filled with mercury, which is stirred. When the stirring power is 200 W, how long does it take to raise the temperature of the mercury by 10 °C? Is the process reversible or irreversible?

3.6. Kneading of a Pizza Dough
A high quality kitchen mixer has a 575 W electric motor. Good pizza dough should be kneaded for about 10 minutes. When 2 kg of dough is kneaded in an adiabatically insulated container, and its initial temperature was 20 °C, what temperature will the dough have after kneading?

Assume specific heat of dough as $c = 2.73\frac{\text{kJ}}{\text{kg K}}$.

3.7. Measurement of Specific Heat
To measure the specific heat of light oil (incompressible liquid, mass density $0.91\frac{\text{kg}}{\text{litre}}$) two litres of oil are stirred in a well-insulated container for 12.5 minutes. The stirrer consumes a power of 75 W, and it is observed that the temperature rises from 23 °C to 40 °C. Ignore kinetic and potential energies. and determine the specific heat of the oil.

3.8. Ice Cream Maker
An ice maker stirs 5 kg of a fruit-cream-air mixture ($\rho = 570\frac{\text{kg}}{\text{m}^3}$, $c_p = 1.7\frac{\text{kJ}}{\text{kg K}}$). The electric motor of the stirrer consumes 575 W of power. It is observed that after 10 minutes the temperature of the ice cream has dropped from $T_1 = -2\,°\text{C}$ to $T_2 = -18\,°\text{C}$. Determine the cooling rate of the ice cream maker.

3.9. Heating of a Room
A room of 300 sq.ft. area and 8 ft height is to be maintained at a constant temperature of 68 °F while the outside temperature is 32 °F. The heat transfer rate to the outside is given by Newton's law of cooling, $\dot{Q} = \alpha (T - T_0)$ with $\alpha = 25\frac{\text{W}}{\text{K}}$.

1. Compute the heating power required to maintain the temperature constant.
2. When the heating power is doubled, how long does it take to heat the room from 68 °F to 77 °F?

Convert all results to SI units.

3.10. Isobaric Heating of an Ideal Gas
0.5 kg hydrogen gas (H_2) are enclosed in a piston-cylinder system at 22 °C, 3 atm. In a reversible isobaric process (constant pressure), the hydrogen doubles its volume.

Determine:

1. The initial volume of the system, and the work done in the expansion.
2. The temperature at the end of the expansion, and the heat exchange with the surroundings.

3.11. Isothermal Compression of an Ideal Gas
10 kg helium are enclosed in a piston-cylinder system at 20 °C, 10 bar. In a reversible isothermal process (constant temperature), the helium is compressed to half the original volume. Compute:

1. The initial volume of the system.
2. The work required for compression.
3. The heat exchange with the surroundings.

3.12. Ideal Gas with Non-constant Specific Heat
In a series of experiments you have found that for temperatures in (300 K, 900 K), the specific heat at constant volume of air is $c_v(T) = \left(0.695 + \frac{0.0598T}{1000\,\mathrm{K}}\right)\frac{\mathrm{kJ}}{\mathrm{kg\,K}}$.

1. Make a table with the specific heats $c_v(T)$ and $c_p(T)$, specific internal energy $u(T)$, and specific enthalpy $h(T)$ for temperatures in the range of validity. As reference value chose $u(300\,\mathrm{K}) = 215\frac{\mathrm{kJ}}{\mathrm{kg}}$.
2. 2 kg of air are isobarically heated from 340 K to 860 K. By means of your table, determine the heat supply Q_{12} and the work W_{12}.
3. Redo the calculation under the assumption that the specific heat can be approximated by its value at 300 K (so that it is constant). Determine the relative error for heat and work.

3.13. Work and Heat
A fixed mass m of carbon monoxide (CO) gas at $T_0 = 30\,°\mathrm{C}$ is confined in a piston-cylinder system. The gas undergoes a reversible isothermal process (constant temperature), that is the pressure changes according to the relation $p = mRT_0/V$. The initial and final volumes are $V_1 = 0.1\,\mathrm{m}^3$ and $V_2 = 0.15\,\mathrm{m}^3$ and the initial pressure is $p_1 = 500\,\mathrm{kPa}$.

Consider CO as ideal gas with constant specific heat and molar mass $M = 28\frac{\mathrm{kg}}{\mathrm{kmol}}$. Determine:

1. The mass of CO in the system.
2. The pressure p_2 at the end of the process.
3. The total work required for the process. Show the process in a p-V-diagram.
4. The total heat exchange.

3.14. Work and Heat
Nitrogen (ideal gas with constant specific heats) undergoes a reversible process in a closed system, where the pressure changes according to the relation $p = aV^2 + b$. The initial and final volumes are $V_1 = 0.3\,\mathrm{m}^3$ and $V_2 = 0.1\,\mathrm{m}^3$, and the corresponding pressures are $p_1 = 100\,\mathrm{kPa}$ and $p_2 = 200\,\mathrm{kPa}$; the initial temperature is $T_1 = 30\,°\mathrm{C}$. Determine:

1. The mass of nitrogen in the system.
2. The temperature at the end of the process.
3. The total work required for the process. Show the process in a p-V-diagram.
4. The total heat exchange.

3.15. Work and Heat

Helium, initially at temperature $T_1 = 0\,°\mathrm{C}$ undergoes a reversible process in a closed system, where the pressure changes according to the relation $p = aV^3 + b$. The initial and final volumes are $V_1 = 0.1\,\mathrm{m}^3$ and $V_2 = 0.2\,\mathrm{m}^3$, and the corresponding pressures are $p_1 = 100\,\mathrm{kPa}$ and $p_2 = 40\,\mathrm{kPa}$. For the relevant temperature range helium behaves as an ideal gas. As for all noble gases, its specific heat is constant, $c_v = \frac{3}{2}R$. Determine:

1. The mass of helium in the system.
2. The temperature at the end of the process.
3. The total work required for the process. Show the process in a p-V-diagram.
4. The total heat exchange.

Chapter 4
The Second Law of Thermodynamics

4.1 The Second Law

In our qualitative description of processes we have already emphasized the trend of any isolated system towards an unique and stable equilibrium state. The *Second Law of Thermodynamics* is the quantitative formulation of this observation. Its importance goes well beyond the computation of the unique equilibrium states for isolated systems. In particular, as will be seen, it gives strong restrictions for the efficiency of energy conversion systems, and thus is of enormous importance for engineering applications.

The original derivation of the second law through Rudolf Clausius (1822–1888) was based on the argument that the direction of heat transfer is restricted and then relied heavily on statements on thermodynamic cycles. The following derivation postulates an inequality to describe the trend to equilibrium, and uses arguments on process direction for simple equilibration processes to identify terms in the postulated equation. This approach allows us to introduce the second law quite early, before any thermodynamic processes and cycles are discussed. With this, entropy and the second law will be available for the evaluation of processes and cycles from the start. All equations and conclusions agree to the classical approach, as presented in most textbooks on engineering thermodynamics, just the order of arguments is different.

4.2 Entropy and the Trend to Equilibrium

To set the stage, we briefly summarize our earlier statements on processes in closed systems: a closed system can be manipulated by exchange of work and heat with its surroundings only. In non-equilibrium—i.e., irreversible—processes, when all manipulation stops, the system will undergo further changes until it reaches a final equilibrium state. This equilibrium state is stable, that is the system will not leave the equilibrium state spontaneously.

H. Struchtrup, *Thermodynamics and Energy Conversion*,
DOI: 10.1007/978-3-662-43715-5_4, © Springer-Verlag Berlin Heidelberg 2014

It requires new action—exchange of work or heat with the surroundings—to change the state of the system.

The following non-equilibrium processes are well-known from experience, and will be used in the considerations below: (a) Heat goes from hot to cold. When two bodies at different temperatures are brought into thermal contact, heat will flow from the hotter to the colder body until both reach their common equilibrium temperature. (b) Work can be transferred without restriction, by means of gears and levers. However, in transfer some work might be lost to friction.

The process from an initial non-equilibrium state to the final equilibrium state requires some time. However, if the actions on the system (only work and heat!) are sufficiently slow, the system has enough time to adapt and will be in equilibrium states at all times. We speak of quasi-static—or, reversible—processes. When the slow manipulation is stopped at any time, no further changes occur.

The behavior of isolated systems described above—a change occurs until a stable state is reached—can be described mathematically by an inequality. The final stable state must be a maximum (alternatively, a minimum) of a suitable extensive property describing the system. We call that extensive property *entropy*, denoted S, and write an inequality for the isolated system,

$$\frac{dS}{dt} = \dot{S}_{gen} \geq 0 \,. \tag{4.1}$$

$\dot{S}_{gen}$ is called the *entropy generation rate.* The entropy generation rate is positive in non-equilibrium ($\dot{S}_{gen} > 0$), and vanishes in equilibrium ($\dot{S}_{gen} = 0$). The new equation (4.1) states that in an isolated system the entropy will grow in time ($\frac{dS}{dt} > 0$) until the stable equilibrium state is reached ($\frac{dS}{dt} = 0$). Non-zero entropy generation describes the irreversible process towards equilibrium, e.g., through internal heat transfer and friction. There is no entropy generation in equilibrium. Since entropy only grows before the equilibrium state is reached, the latter is a maximum of entropy.

The above postulation of an inequality is based on phenomenological arguments. The discussion of irreversible processes has shown that all isolated systems will in time evolve to a unique equilibrium state. The first law alone does not suffice to describe this behavior. We have seen this in the description of temperature equilibration in Sec. 3.12, where the first law has infinitely many solutions for the final temperatures T_A, T_B, and additional input is needed to state that $T_A = T_B$ in equilibrium. Above, we relied on experience as additional input, the second law is a formalization of that experience. Non-equilibrium processes aim to reach equilibrium, and the inequality is required to describe the clear direction in time.

In the next sections we will extend the second law to non-isolated system, and identify entropy as a measurable property.

4.3 Entropy Flux

In non-isolated systems, which exchange heat and work with the surroundings, we expect an exchange of entropy with the surroundings which must be added to the entropy inequality. We write

$$\frac{dS}{dt} = \dot{\Psi} + \dot{S}_{gen}, \quad \text{with } \dot{S}_{gen} \geq 0 \,, \tag{4.2}$$

where $\dot{\Psi}$ is the *entropy flux*. This equation states that the change of entropy in time (dS/dt) is due to transport of entropy over the system boundary $(\dot{\Psi})$ and generation of entropy within the system boundaries $(\dot{S}_{gen})$. This form of the second law is valid for all processes in closed systems. The entropy generation rate is positive, $\dot{S}_{gen} > 0$, for irreversible processes, and it vanishes, $\dot{S}_{gen} = 0$, in equilibrium, and for reversible processes, where the system is in equilibrium states at all times.

All real technical processes are somewhat irreversible, since friction and heat transfer cannot be avoided. Reversible processes are idealizations that can be used to study the principle behavior of processes, and best performance limits.

Since a closed system can only be manipulated through the exchange of heat and work with the surroundings, the transfer of any other property, including the transfer of entropy, must be related to heat and work, and must vanish when heat and work vanish. Therefore the entropy flux $\dot{\Psi}$ can only be of the form

$$\dot{\Psi} = \beta\dot{Q} - \gamma\dot{W} \,, \tag{4.3}$$

with coefficients β, γ that must be related to system and process properties.

Equation (4.2) gives the mathematical formulation of the trend to equilibrium for a non-isolated closed system (exchange of heat and work, but not of mass). The next step is to identify entropy S and the coefficients β, γ in the entropy flux $\dot{\Psi}$ in terms of quantities we can measure or control.

4.4 Entropy in Equilibrium

For quasi-static processes, which are in equilibrium states at all times, the entropy generation vanishes, $\dot{S}_{gen} = 0$, and the equation (4.2) for entropy becomes

$$\frac{dS}{dt} = \dot{\Psi} \,; \tag{4.4}$$

in quasi-static processes the entropy of a closed system changes by entropy transfer only.

With this and (4.3), we have for reversible processes, where $\dot{W} = p\frac{dV}{dt}$,

$$\frac{dS}{dt} = \beta\dot{Q} - \gamma p\frac{dV}{dt} \,. \tag{4.5}$$

Eliminating heat $\dot{Q}$ between this, and the first law for quasi-static processes (3.13), $\frac{dU}{dt} = \dot{Q} - p\frac{dV}{dt}$, yields

$$\frac{dS}{dt} = \beta \frac{dU}{dt} + (\beta - \gamma)\, p \frac{dV}{dt} \,. \tag{4.6}$$

This equation relates entropy S to the state properties U and V, and implies that $S(U,V)$ is a state property as well. Since pressure p, volume V, temperature T, and internal energy U are related through the thermal and caloric equations of state, $p = p(T,V)$, $U = U(T,V)$, the knowledge of any two of these determines the others. Thus, entropy, our new property, can be written as a function of any two of the above properties, e.g., $S(T,V)$ or $S(p,T)$ or $S(U,p)$ or $S(U,V)$. From the last form, we compute the time derivative of entropy with the chain rule,

$$\frac{dS}{dt} = \left(\frac{\partial S}{\partial U}\right)_V \frac{dU}{dt} + \left(\frac{\partial S}{\partial V}\right)_U \frac{dV}{dt} \,. \tag{4.7}$$

Comparison of the last two equations relates β and $(\beta - \gamma)$ to the partial derivatives of entropy,

$$\left(\frac{\partial S}{\partial U}\right)_V = \beta \quad , \quad \left(\frac{\partial S}{\partial V}\right)_U = (\beta - \gamma)\, p \,. \tag{4.8}$$

So far, entropy and the coefficients β and γ in the entropy flux are not yet fixed. Since entropy is a state property, also its derivatives $\left(\frac{\partial S}{\partial U}\right)_V$ and $\left(\frac{\partial S}{\partial V}\right)_U$ are state properties, and it follows that β and $(\beta - \gamma)$ are state properties as well.[1] Since S, U and V are extensive, their quotients and derivatives must be intensive quantities; therefore β and γ are intensive quantities. Obviously, we are interested in non-trivial entropy functions, and therefore we must have $\beta \neq 0$, $(\beta - \gamma) \neq 0$.

In anticipation of later discussion we introduce the *thermodynamic temperature* as $T = 1/\beta$. At this point, this is a just a definition, however, it will be shown soon that T has all the characteristics required for the definition of a thermodynamic temperature. In particular, it will be seen that the entropy flux term $\beta\dot{Q} = \dot{Q}/T$ is related to the restriction of the direction of heat transfer: heat flows from warm to cold, not vice versa. No such restriction applies for work, which, by means of gears and levers, can be transmitted from slow to fast and vice versa, or from low force to high force and vice versa. Because of this, γ must be a constant, which can be set to $\gamma = 0$—the interested reader will find the full argument in Sec. 4.17.

With $\beta = 1/T$ and $\gamma = 0$, we have the partial derivatives of entropy expressed through measurable quantities,

[1] Note that, when the entropy flux (4.3) was introduced, this was not clear: β and $(\beta - \gamma)$ could in principle depend on work and heat. The argument presented here shows that this is not so, at least in equilibrium.

$$\left(\frac{\partial S}{\partial U}\right)_V = \frac{1}{T} \,, \quad \left(\frac{\partial S}{\partial V}\right)_U = \frac{p}{T} \,. \tag{4.9}$$

The entropy flux (4.3) is

$$\dot{\Psi} = \frac{\dot{Q}}{T} \,. \tag{4.10}$$

4.5 Entropy as Property: The Gibbs Equation

With the partial derivatives of entropy as above, the differential $dS = \left(\frac{\partial S}{\partial U}\right)_V dU + \left(\frac{\partial S}{\partial V}\right)_U dV$ becomes

$$TdS = dU + pdV \,. \tag{4.11}$$

This relation is known as the Gibbs equation, named after Josiah Willard Gibbs (1839 - 1903). The Gibbs equation is a differential relation between properties and valid for *all* simple substances.

Since T and p are intensive, and U and V are extensive properties, entropy is extensive. The specific entropy $s = S/m$ can be computed from the Gibbs equation for specific properties, which is obtained by division of (4.11) with the constant mass m,

$$Tds = du + pdv \,. \tag{4.12}$$

Replacing internal energy by enthalpy, $u = h - pv$, gives an alternative form of the Gibbs equation,

$$Tds = dh - vdp \,. \tag{4.13}$$

The Gibbs equation gives a large number of relations and restrictions between properties, in particular it allows to determine property relations for entropy.

Entropy, just as internal energy, cannot be measured directly. Property relations for entropy are computed from the Gibbs equation, and the thermal and caloric equations of state, $p\,(T,v)$ and $u\,(T,v)$. Here, we consider this for incompressible substances and for ideal gases.

For incompressible liquids and solids, the specific volume is constant, hence $dv = 0$. The caloric equation of state (3.18) implies $du = cdT$ and the Gibbs equation reduces to $Tds = cdT$. For constant specific heat, $c = const.$, integration gives entropy as explicit function of temperature,

$$s\,(T) = c \ln \frac{T}{T_0} + s_0 \,, \tag{4.14}$$

where s_0 is the entropy at the reference temperature T_0. As long as no chemical reactions are involved, the definition of the entropy scale, i.e., the value of s_0, can be freely chosen; the third law of thermodynamics will fix the scale properly.

For an ideal gas we have $du = c_v dT$ and $v = RT/p$ so that the Gibbs equation (4.13) becomes

$$ds = c_p \frac{dT}{T} - R \frac{dp}{p} \, . \tag{4.15}$$

For a gas with temperature dependent specific heat, integration yields

$$s(T, p) = s^0(T) - R \ln \frac{p}{p_0} \, , \tag{4.16}$$

where

$$s^0(T) = \int_{T_0}^{T} \frac{c_p(T')}{T'} dT' + s_0 \tag{4.17}$$

is the—temperature dependent—entropy at reference pressure p_0, and s_0 is the reference entropy at $\{T_0, p_0\}$.

For a gas with constant specific heat, the integration can be performed to give

$$s(T, p) = c_p \ln \frac{T}{T_0} - R \ln \frac{p}{p_0} + s_0 \, . \tag{4.18}$$

The entropy $s(T, v)$ follows from this either by replacing the pressure with the ideal gas equation ($p = RT/v$) or from integrating (4.12) as (for constant specific heat)

$$s(T, v) = c_v \ln \frac{T}{T_0} + R \ln \frac{v}{v_0} + s_0 \tag{4.19}$$

Property relations for other substances will be presented in Chapter 6.

4.6 T-S-Diagram

Solving the first law for reversible processes (3.13) for heat and comparing the result with the Gibbs equation we find, with $\dot{Q} dt = \delta Q$,

$$dS = \frac{1}{T}(dU + pdV) = \frac{1}{T} \delta Q \, . \tag{4.20}$$

We recall that heat is a path function, i.e., δQ is an inexact differential, but entropy is a state property, i.e., dS is an exact differential. In the language of mathematics, the inverse thermodynamic temperature $\frac{1}{T}$ serves as an integrating factor for δQ, such that $dS = \frac{1}{T} \delta Q$ becomes an exact differential.

From the above, we see that for reversible processes $\delta Q = TdS$. Accordingly, the total heat exchanged in a reversible process can be computed from temperature and entropy as the area below the process curve in the temperature-entropy diagram (T-S-diagram),

$$Q_{12} = \int_1^2 TdS \, . \tag{4.21}$$

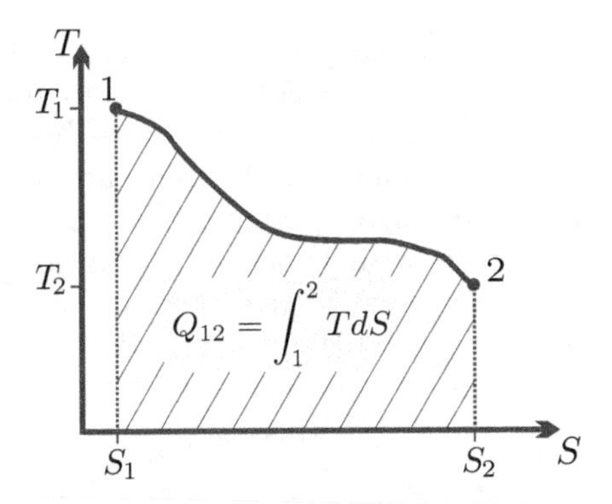

Fig. 4.1 Heat as the area below the reversible process curve in the T-S-diagram

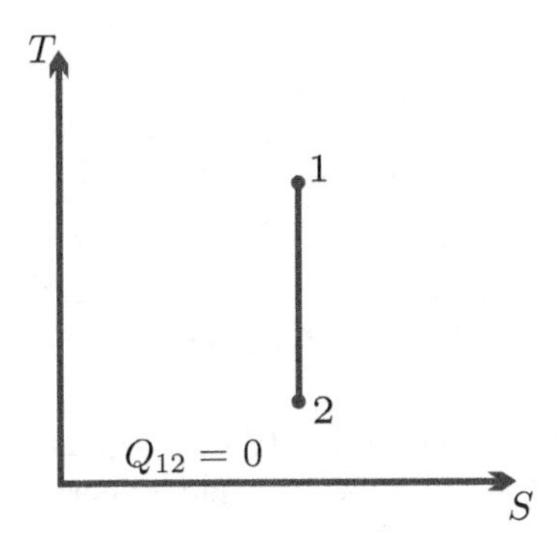

Fig. 4.2 Isentropic (adiabatic reversible) process in the T-S-diagram

This is analog to the computation of the work as $W_{12} = \int_1^2 pdV$, Fig. 4.1 gives an illustration.

For a reversible adiabatic process $\delta Q = TdS = 0$, that is the entropy is constant in the process. We say a reversible adiabatic process is *isentropic.* The process curve in the T-S-diagram is a vertical line, see Fig. 4.2.

4.7 The Entropy Balance

In the previous sections, we considered homogeneous systems that undergo equilibrium processes. To generalize for processes in inhomogeneous systems, we consider the system as a compound of sufficiently small subsystems. The key assumption is that each of the subsystems is in local equilibrium, so that it can be characterized by the same state properties as a macroscopic equilibrium system. To simplify the proceedings somewhat, we consider numbered subsystems of finite size, and summation. A more refined argument would consider infinitesimal cells dV, and integration.

Figure 4.3 indicates the splitting into subsystems, and highlights a subsystem i inside the system and a subsystem k at the system boundary. Temperature and pressure in the subsystems are given by T_i, p_i and T_k, p_k, respectively. Generally, temperature and pressure are inhomogeneous, that is adjacent subsystems have different temperatures and pressures. Accordingly, each subsystem interacts with its neighborhood through heat and work transfer as indicated by the arrows. Heat and work exchanged with the surroundings of the system are indicated as $\dot{Q}_k$ and $\dot{W}_k$.

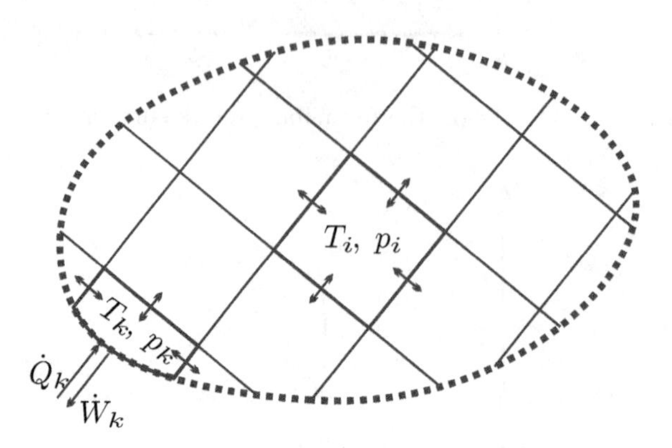

Fig. 4.3 Non-equilibrium system split into small equilibrium subsystems. Arrows indicate work and heat exchange between neighboring elements, and the surroundings.

Internal energy and entropy in a subsystem i are denoted as E_i and S_i, and, since both are extensive, the corresponding quantities for the complete system are obtained by summation over all subsystems, $E = \sum_i E_i$, $S = \sum_i S_i$. Note that in the limit of infinitesimal subsystems the sums become integrals, as in Sec. 2.7. The balances of energy and entropy for a subsystem i read

$$\frac{dE_i}{dt} = \dot{Q}_i - \dot{W}_i \quad , \quad \frac{dS_i}{dt} = \frac{\dot{Q}_i}{T_i} + \dot{S}_{gen,i} \, , \tag{4.22}$$

where $\dot{Q}_i = \sum_j \dot{Q}_{i,j}$ is the net heat exchange, and $\dot{W}_i = \sum_j \dot{W}_{i,j}$ is the net work exchange for the subsystem. Here, the summation over j indicates the exchange of heat and work with the neighboring cells, such that, e.g., $\dot{Q}_{i,j}$ is the heat that i receives from the neighboring cell j.

To obtain first and second law for the compound system, we have to sum the corresponding laws for the subsystems, which gives

$$\frac{dE}{dt} = \dot{Q} - \dot{W} \quad \text{with} \quad \dot{Q} = \sum_k \dot{Q}_k \, , \; \dot{W} = \sum_k W_k \tag{4.23}$$

and

$$\frac{dS}{dt} = \sum_k \frac{\dot{Q}_k}{T_k} + \dot{S}_{gen} \quad \text{with} \quad \dot{S}_{gen} \geq 0 \,. \tag{4.24}$$

In the above $\dot{Q}_k$ is the heat transferred over a system boundary which has temperature T_k. As will be explained next, the summation over k concerns only heat and work exchange with the surroundings.

Since energy is conserved, the internal exchange of heat and work between subsystems cancels in the conservation law for energy (4.23). For instance, in the exchange between neighboring subsystems i and j, $Q_{i,j}$ is the heat that i receives from j and $W_{i,j}$ is the work that i does on j. Moreover, $Q_{j,i}$ is the heat that j receives from i and $W_{j,i}$ is the work that j does on i. Since energy is conserved, no energy is added or lost in transfer between i and j, that is $Q_{i,j} = -Q_{j,i}$ and $W_{i,j} = -W_{j,i}$. Accordingly, the sums vanish, $Q_{i,j} + Q_{j,i} = 0$ and $W_{i,j} + W_{j,i} = 0$. Extension of the argument shows that the internal exchange of heat and work between subsystems adds up to zero, so that only exchange with the surroundings, indicated by subscript k, appears in (4.23).

Entropy, however, is not conserved, but may be produced. Exchange of heat and work between subsystems, if irreversible, will contribute to the entropy generation rate $\dot{S}_{gen}$. Thus, the total entropy generation rate $\dot{S}_{gen}$ of the compound system is the sum of the entropy generation rates in the subsystems $\dot{S}_{gen,i}$ plus additional terms related to the energy transfer between subsystems, $\dot{S}_{gen} = \sum_i \dot{S}_{gen,i} + \sum_{i,j} \frac{\dot{Q}_{i,j}}{T_i}$. In simple substances, entropy generation occurs due to internal heat flow and internal friction. We repeat that entropy generation is strictly positive, $\dot{S}_{gen} > 0$, in irreversible processes, and is zero, $\dot{S}_{gen} = 0$, in reversible processes.

To fully quantify entropy generation, that is to compute its actual value, requires the detailed local computation of all processes inside the system from the conservation laws and the second law as partial differential equations. The derivation and analysis of the local laws is a topic of *Non-equilibrium Thermodynamics*.

The above derivation of the second law equation (4.24) relies on the assumption that the equilibrium property relations for entropy are valid locally also for non-equilibrium systems. This local equilibrium hypothesis—equilibrium in a subsystem, but not in the compound system—works well for most systems in technical thermodynamics. It should be noted that the assumption breaks down for extremely strong non-equilibria; this lies outside the scope of our endeavours.

4.8 The Direction of Heat Transfer

A temperature reservoir is defined as a large body whose temperature does not change when heat is removed or added. Figure 4.4 shows heat transfer between two reservoirs of temperatures T_H and T_L, where T_H is the

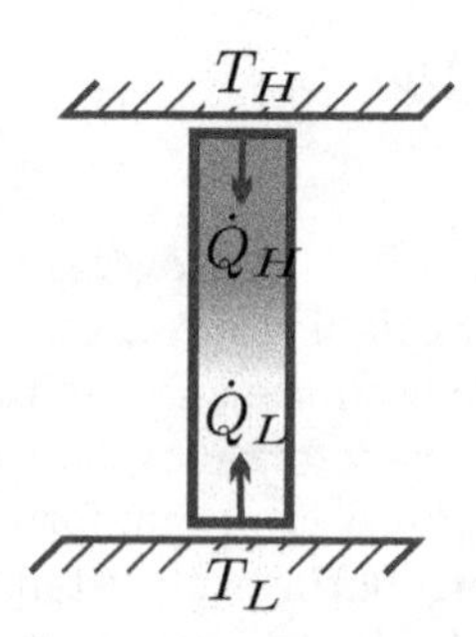

Fig. 4.4 Heat transfer between two reservoirs at T_1 and T_2. In steady state the heat conductor does not accumulate energy, therefore $\dot{Q}_L = -\dot{Q}_H$.

temperature of the hotter system. The heat is transferred through a heat conductor, which is the thermodynamic system to be evaluated. A pure heat transfer problem is studied, where the conductor receives the heat flows $\dot{Q}_H$ and $\dot{Q}_L$, and exchanges no work with the surroundings, $\dot{W} = 0$. The first and second law (4.23, 4.24) applied to the heat conductor read

$$\frac{dU}{dt} = \dot{Q}_L + \dot{Q}_H \quad , \quad \frac{dS}{dt} - \frac{\dot{Q}_L}{T_L} - \frac{\dot{Q}_H}{T_H} = \dot{S}_{gen} \geq 0 \, . \tag{4.25}$$

For steady state conditions no changes over time are observed in the conductor, so that $\frac{dU}{dt} = \frac{dS}{dt} = 0$. The first law shows that the heat fluxes must be equal in absolute value, but opposite in sign,

$$\dot{Q}_H = -\dot{Q}_L = \dot{Q} \, . \tag{4.26}$$

With this, the second law reduces to the inequality

$$\dot{Q}\left(\frac{1}{T_L} - \frac{1}{T_H}\right) = \dot{S}_{gen} \geq 0 \, . \tag{4.27}$$

Clausius' original derivation of the second law is based on the statement that *heat will go from hot to cold by itself, but not vice versa.* We shall use this statement to learn more on thermodynamic temperature T. Since we declared T_H as the temperature of the hotter reservoir, heat should go from the reservoir at T_H to the reservoir at T_L. According to Fig. 4.4 the proper direction of heat transfer in accordance to Clausius' statement is for positive $\dot{Q}_H$, that is for $\dot{Q} > 0$, which implies $\dot{Q}_L < 0$. With $\dot{Q} > 0$ the inequality (4.27) holds for $\left(\frac{1}{T_L} - \frac{1}{T_H}\right) > 0$, which is fulfilled if (a) $T_H > T_L$, and (b) T_H and T_L have the same sign, i.e., T is either always positive or always negative. The discussion of friction in the next section will show that T must be positive.

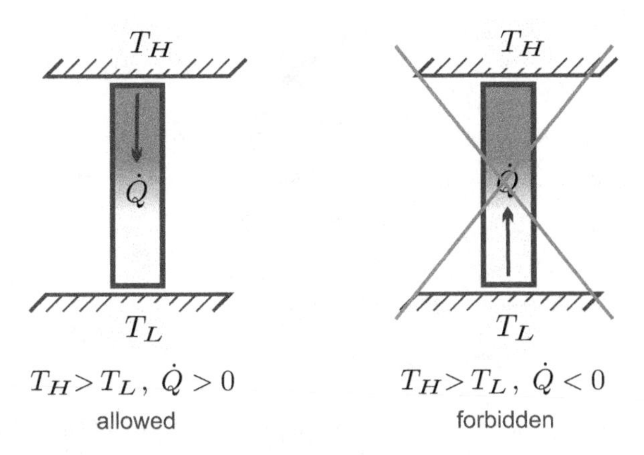

Fig. 4.5 Heat transfer between two reservoirs with $T_H > T_L$. Heat must go from warm to cold.

Figure 4.5 gives an illustration of the allowed process, where heat goes from hot to cold, and the forbidden process, where heat would go from cold to hot by itself.

4.9 Internal Friction

The sign of temperature follows from the observation that a stirred substance will come to rest due to friction with the container walls, and within the fluid. When coffee, or any other liquid, is stirred, it will spin a while after the spoon is removed. The motion will slow down because of internal friction, and finally the coffee will be at rest in the cup. The second law should describe this well-known behavior, which is observed in all viscous fluids.

With the fluid in motion, we have to account for the kinetic energy of the swirling, which must be computed by summation (i.e., integration), of the local kinetic energies $\frac{\rho(\vec{r})}{2}\mathcal{V}\left(\vec{r}\right)^2$ in all volume elements; see Fig. 4.6. The first and second law read

$$\frac{d\left(U+E_{kin}\right)}{dt}=\dot{Q}-\dot{W} \quad , \quad \frac{dS}{dt}-\sum\frac{\dot{Q}_k}{T_k}\geq 0\,. \tag{4.28}$$

We assume adiabatic systems ($\dot{Q}=0$) without any work exchange ($\dot{W}=0$, this implies constant volume), so that

$$\frac{d\left(U+E_{kin}\right)}{dt}=0 \quad , \quad \frac{dS}{dt}\geq 0\,. \tag{4.29}$$

If we ignore local temperature differences within the stirred substance, we have with (4.9)

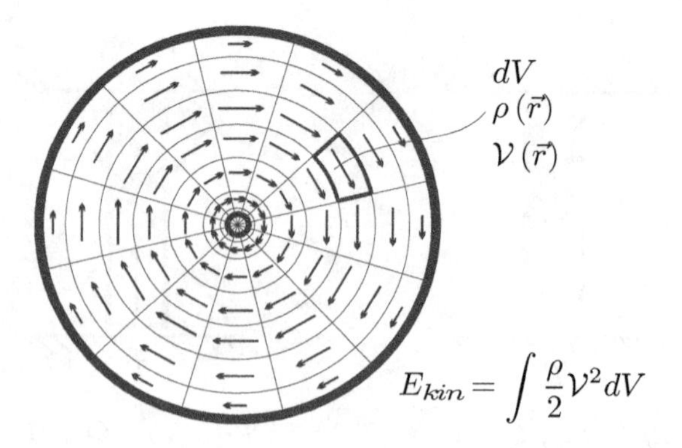

Fig. 4.6 The kinetic energy E_{kin} of a stirred fluid is the sum of the kinetic energies in all volume elements. Friction with the container wall, and within the fluid, will slow down the fluid until it comes to rest in the final equilibrium state.

$$\frac{dS}{dt} = \left(\frac{\partial S}{\partial U}\right)_V \frac{dU}{dt} = \frac{1}{T}\frac{dU}{dt} = -\frac{1}{T}\frac{dE_{kin}}{dt} \geq 0 . \tag{4.30}$$

Experience shows that over time the fluid slows down, hence the kinetic energy $E_{kin} = \int \frac{\rho}{2}\mathcal{V}^2 dV$ decreases over time, and will be zero in equilibrium, where the stirred substance comes to rest, $\mathcal{V} = 0$; this implies

$$\frac{dE_{kin}}{dt} \leqslant 0 . \tag{4.31}$$

The latter inequality is compatible with the 2nd law in the form (4.30) only if the thermodynamic temperature is non-negative, $T \geq 0$.

An equivalent experience is that work in transmission can be lost to friction, but not gained. Figure 4.7 shows the work and heat flows for a gearbox operating at constant temperature T, at steady state. The gearbox receives the work $-\left|\dot{W}_{in}\right|$, and delivers the work $\dot{W}_{out} > 0$. Moreover, the gearbox is in thermal contact with the environment from which it receives the heat $\dot{Q}$. The figure shows absolute values for work, the arrows indicate the direction of the work flows. The first law is straightforward to evaluate: Since the gear box operates at steady state, the energy supply must equal the energy loss. The statement of the first law can be read straight from the figure: Energy flow in (arrows pointing towards the gearbox) must be equal to energy flow out (arrows out of the gearbox),

$$\left|\dot{W}_{in}\right| + \dot{Q} = \dot{W}_{out} . \tag{4.32}$$

There is only a single heat flow contribution, therefore the second law becomes

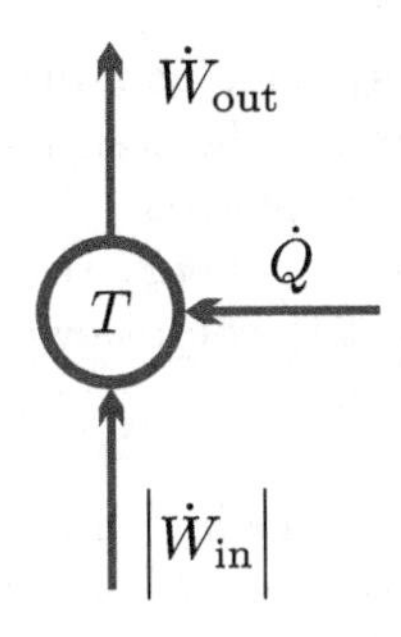

Fig. 4.7 Work and heat flow in a gearbox operating at constant temperature T. The heat is expected to be negative, $\dot{Q} < 0$.

$$-\frac{\dot{Q}}{T} = \frac{\left|\dot{W}_{in}\right| - \dot{W}_{out}}{T} = \dot{S}_{gen} \geq 0\,. \tag{4.33}$$

We consider it a general experience that some work is lost in the transmission through a gearbox, and that the box sheds heat into the environment,[2]

$$\dot{W}_{out} \leq \left|\dot{W}_{in}\right| \quad , \quad \dot{Q} < 0\,. \tag{4.34}$$

Comparison between the two last equations shows, again, that the thermodynamic temperature must not be negative, $T \geq 0$.

The lost work leaves the gearbox in form of heat $\left|\dot{Q}_{out}\right| = -\dot{Q} = \left|\dot{W}_{in}\right| - \dot{W}_{out} > 0$, which is transmitted into the environment. The reason for the loss is friction within the gearbox.

4.10 Newton's Law of Cooling

We return to the discussion of heat transfer. The inequality (4.27) requires that $\dot{Q}$ has the same sign as $\left(\frac{1}{T_L} - \frac{1}{T_H}\right)$, a requirement that is fulfilled for a heat transfer rate

$$\dot{Q} = \alpha A \left(T_H - T_L\right) \tag{4.35}$$

with a positive heat transfer coefficient $\alpha > 0$, and the heat exchange surface area A. This relation, which we already used in an example, is known as Newton's law of cooling, and is often used in heat transfer problems. The values of the positive coefficient α must be found from the detailed configuration and

[2] This is tantamount to the statement that a system exchanging heat with a single reservoir cannot produce work (see Sec. 5.3 further below). Indeed, if the system would *receive* the heat $\dot{Q} > 0$, the first law would require $\dot{W}_{out} > \left|\dot{W}_{in}\right|$: more work would leave the system than enter, which means that the single incoming heat $\dot{Q}$ would be converted to work.

conditions in the heat transfer system. The surface area A appears due to the intuitive expectation that enlarging the transfer area leads to a proportional increase in the amount of heat transferred.

Heat transfer was introduced as energy transfer due to temperature difference with heat going from hot to cold, Newton laws of cooling states that as a result of the temperature difference one will observe a response, namely the heat flux.

The procedure to derive Newton's law of cooling can be described as follows: The entropy generation rate (4.27) is interpreted as the product of a thermodynamic force—here, the temperature difference $(T_H - T_L)$—and a corresponding flux—here, the heat flux $\dot{Q}$. To ensure positivity of the entropy generation rate, the flux must be proportional to the force, with a positive factor αA that must be measured. The same strategy can be used for other force-flux pairs.

With Newton's law of cooling it is easy to see that heat transfer over finite temperature differences is an irreversible process. Indeed, the second law (4.27) gives with (4.35)

$$\dot{S}_{gen} = \dot{Q}\left(\frac{1}{T_L} - \frac{1}{T_H}\right) = \alpha A \frac{(T_H - T_L)^2}{T_L T_H} > 0 \,. \tag{4.36}$$

Only when the temperature difference is infinitesimal, i.e., $T_H = T_L + dT$, entropy generation can be ignored, and heat transfer can be considered as a reversible process. This can be seen as follows: For infinitesimal dT the entropy generation rate becomes $\dot{S}_{gen} = \alpha A \left(\frac{dT}{T_L}\right)^2$. Since quadratic terms in infinitesimal differences can be ignored, this implies $\dot{S}_{gen} = 0 \ (dT \to 0)$. In this case, to have a finite amount of heat transferred, the heat exchange area A must go to infinity.

4.11 Zeroth Law and Second Law

While above we considered heat transfer between reservoirs, the conclusion is valid for heat conduction between arbitrary systems: As long as the systems are in thermal contact through heat conductors, and their temperatures are different, there will be heat transfer between the systems. Only when the temperatures of the systems are equal, heat transfer will cease. This is the case of thermal equilibrium, where no change in time occurs anymore. This includes that the temperature of an isolated body in thermal equilibrium will be homogeneous, where equilibration occurs through heat transfer within the system.

The zeroth law states: In equilibrium systems in thermal contact assume the same temperature. Thus, the zeroth law of thermodynamics might appear as a special case of the second law. It stands in its own right, however, since it *defines* temperature as a measurable quantity.

4.12 Example: Equilibration of Temperature

We return to the problem considered in Sec. 3.12, the equilibration of temperature between two bodies A and B, with initial temperatures $\bar{T}_A$ and $\bar{T}_B$. The first law alone was not sufficient to find the final common temperature, which will now be obtained from the second law. The compound system $A+B$ is adiabatic to the outside, so that the second law becomes

$$\frac{dS}{dt} = \frac{d\left(S_A + S_B\right)}{dt} \geq 0 \,.$$

Thus the total entropy $S = S_A + S_B$ of the system will grow in time until it will assume its maximum in equilibrium, when no further changes occur.

For the simple solids under consideration, by (4.14) the entropy is $S = mc \ln \frac{T}{T_0}$, so that

$$S = S_A + S_B = m_A c_A \ln \frac{T_A}{T_0} + m_B c_B \ln \frac{T_B}{T_0} \,.$$

The first law relates the actual temperatures T_A and T_B of the two bodies and their initial temperatures $\bar{T}_A$, $\bar{T}_B$ through (3.26). With this, the entropy of the system becomes a function of T_B only,

$$S = m_A c_A \ln \left(\frac{\bar{T}_A}{T_0} + \frac{m_B c_B}{m_A c_A} \frac{\bar{T}_B - T_B}{T_0} \right) + m_B c_B \ln \frac{T_B}{T_0} \,.$$

Since the entropy of the compound system $A+B$ can only grow, in equilibrium the entropy assumes the largest possible value, which is obtained from the condition $\frac{dS}{dT_B} = 0$. The evaluation, left as an exercise for the reader, gives the expected result for the common equilibrium temperature,

$$T_B = T_A = T = \frac{m_A c_A \bar{T}_A + m_B c_B \bar{T}_B}{m_A c_A + m_B c_B} \,.$$

4.13 Example: Uncontrolled Expansion of a Gas

We consider the entropy change for the uncontrolled expansion of an ideal gas in Sec. 3.13, for which the first law gave $T_1 = T_2$. The second law for this adiabatic process simply reads

$$\frac{dS}{dt} = \dot{S}_{gen} \geq 0.$$

Integration over the process duration yields

$$S_2 - S_1 = \int_{t_1}^{t_2} \dot{S}_{gen} dt = S_{gen} \geq 0 \,.$$

The total change of entropy follows from the ideal gas entropy (with constant specific heat), Eq. (4.19) as

$$S_2 - S_1 = m(s_2 - s_1) = mR\ln\frac{V_2}{V_1} = mR\ln\frac{p_1}{p_2} \geq 0\,.$$

Since in this process the temperature of the ideal gas remains unchanged, the growth of entropy is only attributed to the growth in volume: by filling the larger volume V_2, the gas assumes a state of larger entropy. Since the container is adiabatic, there is no flux of entropy over the boundary (i.e., $\sum \frac{\dot{Q}_k}{T_k} = 0$), and all entropy generated stays within the system, $S_{gen} = S_2 - S_1$.

4.14 What Is Entropy?

The arguments that gave us the second law and entropy as a property centered around the trend to equilibrium observed in any system left to itself (isolated system). Based on the derivation, the question *What is entropy?* can be answered simply by saying *It's a quantity that arises when one constructs an inequality that describes the trend to equilibrium.* Can there be a deeper understanding of entropy?

Before we try to answer, we look at internal energy: When the first law of thermodynamics was found, the concept of internal energy was new, and it was difficult to understand what it might describe. At that time, the atomic structure of matter was not known, and internal energy could not be interpreted—it appeared because it served well to describe the phenomena. Today we know more, and we understand internal energy as the kinetic and potential energies of atoms and molecules on the microscopic level. Thus, while the concept of internal energy arose from the desire to describe phenomena, today it is relatively easy to understand, because it has a macroscopic analogue in mechanics.

Entropy also came into play to describe the phenomena, but it is a new quantity, without a mechanical analogue. A deeper understanding of entropy can be gained, as for internal energy, from considerations on the atomic scale. Within the framework of his *Kinetic Theory of Gases*, Ludwig Boltzmann (1844-1905) found a microscopic interpretation of entropy, where entropy is related to concepts of probability. A not too precise description of this interpretation follows below.

Macroscopically, a state is described by only a few macroscopic properties, e.g., temperature, pressure, volume. Microscopically, a state is described through the location and momentum of all atoms within the system. The microscopic state is constantly changing due to the microscopic motion of the atoms, and there are many microscopic states that describe the same macroscopic state. If we denote the total number of all microscopic states that describe the same macroscopic state by Ω, then the entropy of the macroscopic state according to Boltzmann is

$$S = k_B \ln \Omega \, . \tag{4.37}$$

The constant $k_B = \bar{R}/A = 1.3804 \times 10^{-23} \frac{\text{J}}{\text{K}}$ is the Boltzmann constant, which can be interpreted as the gas constant per particle.

The growth of entropy in an isolated system, $\frac{dS}{dt} \geq 0$, thus means that the system shifts to macrostates which have larger numbers of microscopic realizations. Equilibrium states have particularly large numbers of realizations, and this is why they are observed.

To make the ideas somewhat clearer, we consider the expansion of a gas when a barrier is removed, see Secs. 3.13, 4.13. This is a particularly simple case, where the internal energy, and thus the distribution of energy over the particles, does not change. Hence, we can ignore the distribution of thermal energy over the particles, and the exchange of energy between them.

We assume a system of N gas particles in a volume V. The volume of a single particle is v_0, and in order to be able to compute the number Ω, we "quantize" the accessible volume V into $n = V/v_0$ boxes that each can accommodate just one particle. Note that in a gas, where the distance between individual particles is relatively large, most boxes are empty. Due to their thermal energy, the atoms move from box to box. The number of microstates is simply given by the number of realizations of a state with N filled boxes and $(n - N)$ empty boxes, which is

$$\Omega(N, V) = \frac{n!}{N!\,(n-N)!} \, . \tag{4.38}$$

By means of Stirling's formula $\ln x! = x \ln x - x$ (for $x \gg 1$) the entropy (4.37) for this state becomes

$$S(N, V) = k_B \left[-N \ln \frac{N}{n} - (n - N) \ln \left(1 - \frac{N}{n} \right) \right] . \tag{4.39}$$

Now we can compute the change of entropy with volume. For this, we consider the same N particles in two different volumes, $V_1 = n_1 v_0$ and $V_2 = n_2 v_0$. The entropy difference $S_2 - S_1 = S(N, V_2) - S(N, V_1)$ between the two states can be written as

$$\begin{aligned} S_2 - S_1 = k_B \Bigg[& N \ln \frac{n_2}{n_1} + n_1 \ln \left(1 - \frac{N}{n_1} \right) \\ & - n_2 \ln \left(1 - \frac{N}{n_2} \right) + N \ln \frac{\left(1 - \frac{N}{n_2} \right)}{\left(1 - \frac{N}{n_1} \right)} \Bigg] . \end{aligned} \tag{4.40}$$

In an ideal gas the number of possible positions n is much bigger than the number of particles N, that is $\frac{N}{n_1} \ll 1, \frac{N}{n_2} \ll 1$. Taylor expansion yields the entropy difference to leading order as

$$S_2 - S_1 = k_B N \ln \frac{n_2}{n_1} = mR \ln \frac{V_2}{V_1} , \tag{4.41}$$

where we reintroduced volume ($V_{1,2} = n_{1,2}v_0$), and introduced the mass as $m = M\,N/A$; $R = \bar{R}/M$ is the gas constant. This is just the change of entropy computed in Sec. 4.13.

It is instructive to compare the number of realizations for the two cases, for which we find

$$\frac{\Omega_2}{\Omega_1} = \exp \frac{S_2 - S_1}{k} = \exp\left(N \ln \frac{V_2}{V_1}\right) = \left(\frac{V_2}{V_1}\right)^N . \tag{4.42}$$

For a macroscopic amount of gas, the particle number N is extremely large (order of magnitude $\sim 10^{23}$), so that already for a small difference in volume the ratio of microscopic realization numbers is enormous. For instance for $V_2 = 2V_1$, we find $\frac{\Omega_2}{\Omega_1} = 2^N$.

Microscopic states change constantly due to travel of, and collisions between, particles. Each of the Ω microstates compatible with the given macrostate is observed with the same probability, $1/\Omega$. The Ω_1 microstates in which the gas is confined in the volume V_1 are included in the Ω_2 microstates in which the gas is confined in the larger volume V_2. Thus, after removal of the barrier, there is a finite, but extremely small probability of $P = \frac{\Omega_1}{\Omega_2} = \left(\frac{V_1}{V_2}\right)^N$ to find all gas particles in the initial volume V_1. This probability is so small that the expected waiting time for observing a return into the original volume exceeds the lifetime of the universe by many orders of magnitude. If we do not want to wait that long for the return to initial state, we have to push the gas back into the initial volume, which requires work.

In generalization of the above, we can conclude that it is quite unlikely that a portion V_ν of the volume is void of particles. The corresponding probability is $P_\nu = \left(\frac{V - V_\nu}{V}\right)^N$. The average volume available for one particle is $\bar{V} = \frac{V}{N}$, and when $V_\nu = \nu \bar{V}$ we find, for the large particle numbers in an macroscopic amount of gas, $P_\nu = \left(1 - \frac{\nu}{N}\right)^N \simeq e^{-\nu}$. Thus, as long as V_ν is bigger than the average volume for a single particle, so that $\nu > 1$, the probability for a void is very small. Moreover, inhomogeneous distributions are rather unlikely, since the number of homogeneous distributions is far larger than the number of strongly inhomogeneous distributions. This is why we observe homogeneous distributions in equilibrium.

Figure 4.8 gives an illustration of microstates for a rather small system. The system of $N = 9$ particles with $n = 81$ boxes allows for $\Omega = \frac{81!}{9!(81-9)!} = 2.61 \times 10^{11}$ microstates, three of which are shown in the figure. Microstate A is one of the $\Omega_L = \frac{27!}{9!(27-9)!} = 4.69 \times 10^6$ microstates in which the gas in confined to the left third of the system. Microstates B and C are more homogeneous distributions.

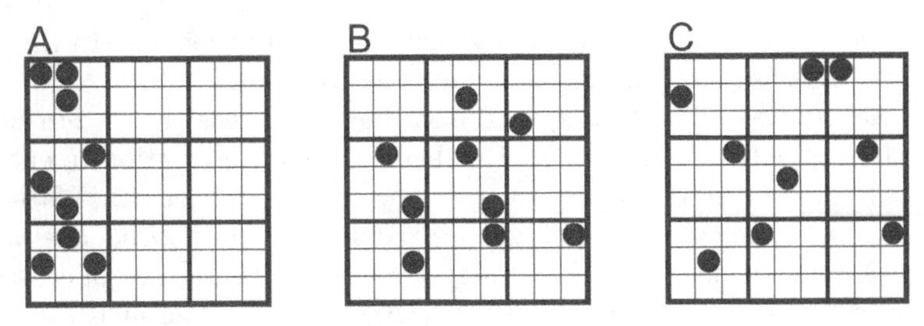

Fig. 4.8 A system of 9 particles with 81 accessible positions in three different microstates A, B, C

4.15 Entropy and Disorder

Often it is said that entropy is a measure for disorder, where disorder has a higher entropy. This can be related to the above discussion by means of the following analogy: The ordered state of an office is the state where all papers, folders, books and pens are in their designated shelf space. Thus, they are confined to a relatively small initial volume of the shelf, V_1. When work is done in the office, all these papers, folders, books and pens are removed from their initial location, and, after they are used, are dropped somewhere in the office—now they are only confined to the large volume of the office, V_2. The actions of the person working in the office constantly change the microstate of the office (the precise location of that pen ... where is it now?), in analogy to thermal motion.

At the end of the day, the office looks like a mess and needs work to clean up. Note, however, that the final state of the office—which appears to be so disorderly—is just *one* accessible microstate, and therefore it has the same probability as the fully ordered state, where each book and folder is at its designated place on the shelf. A single microstate, e.g., a particular distribution of office material over the office in the evening, has no entropy. Entropy is a macroscopic property that counts the number of all possible microstates, e.g., all possible distributions of office material.

A macroscopic state which puts strong restrictions on the elements has a low entropy, e.g., when all office material is in shelves behind locked doors. When the restrictions are removed—the doors are unlocked—the number of possible distributions grows, and so does entropy. Thermal motion leads to a constant change of the distribution within the inherent restrictions.

To our eye more restricted macroscopic states—all gas particles only in a small part of the container, or all office material behind closed doors—appear more orderly, while less restricted states generally appear more disorderly. In this sense one can say that entropy is a measure for disorder.

In the office, every evening the disordered state differs from that of the previous day. Over time, one faces a multitude of disordered states, that is the disordered office has many realizations, and a large entropy. In the end, this makes cleaning up cumbersome, and time consuming.

Our discussion focussed on spatial distributions where the notion of order is well-aligned with our experience. The thermal contribution to entropy is related to the distribution of microscopic energy e_m over the particles, where e_m is the microscopic energy per particle. In *Statistical Thermodynamics* one finds that in equilibrium states the distribution of microscopic energies between particles is exponential, $A \exp\left[-\frac{e_m}{kT}\right]$. The factor A must be chosen such that the sum over all particles gives the internal energy, $U = \sum_m A e_m \exp\left[-\frac{e_m}{k_B T}\right]$. One might say that the exponential itself is an orderly function, so that the equilibrium states are less disordered than non-equilibrium states. Moreover, for lower temperatures the exponential is more narrow, the microscopic particle energies are confined to lower values, and one might say that low temperature equilibrium states are more orderly than high temperature equilibria. And indeed, we find that entropy grows with temperature, that is colder systems have lower entropies.

4.16 Entropy and Life

The second law states that systems left to themselves tend to disorder, in the non-trivial sense discussed above. To leave a system to itself, it must be isolated from its surroundings, so that no transport of mass and energy over the system boundaries occurs. For such a system the second law reads $\frac{dS}{dt} \geq 0$, entropy—disorder—must increase. A system which is not isolated can have decreasing entropy. Indeed, for a closed system the second law reads $\frac{dS}{dt} - \sum \frac{\dot{Q}_k}{T_k} \geq 0$; thus, by suitable manipulation of the system, in particular cooling ($\dot{Q}_k < 0$), its entropy can decrease, more ordered states are possible.

Earth itself is not isolated, since it receives an abundance of high temperature energy from the sun in form of radiation (sun surface temperature $T_S \simeq 5700\,\text{K}$). At the same time Earth emits low temperature energy, also in form of radiation (Earth surface temperature $T_E \simeq 300\,\text{K}$). This exchange of energy with Earth's surrounding allows decreasing entropy locally on the planet. When we assume that the amount of heat received and emitted by radiation is the same ($\left|\dot{Q}\right|$), the second law for Earth reads $\frac{dS}{dt} \geq \left|\dot{Q}\right| \left(\frac{1}{T_S} - \frac{1}{T_E}\right)$. Since $T_S > T_E$, the left hand side is negative, Earth's entropy may, but must not, decrease.

If entropy is decreasing within a system (which cannot be isolated!), entropy must be growing somewhere else. When a sufficient portion of the surroundings are included in the system, entropy must grow. The entropy in the universe, which is a rather large isolated system, is increasing. The processes in the sun create entropy locally, in the sun.

Life, most importantly, is fed by the sun. Just think of the human body: we grow, we learn, and thus keep disorder within the confines of our body rather small. As humans are open systems, we maintain a low entropy level by exchange of mass and energy with our surroundings. Within the larger system around us, entropy grows, but within the smaller boundaries of our bodies (and minds!), entropy decreases, or is at least maintained at the same level.

The sun is the source of life, since it provides the energy we need to lower entropy in our open system Earth, and in our open system human body. Evolution, as an increase of order, does not contradict the second law.

4.17 The Entropy Flux Revisited

When we discussed the possible form of the entropy flux $\dot{\Psi}$ in Sec. 4.3, we introduced two coefficients β, γ but we soon set them to $\beta = 1/T$ and $\gamma = 0$ in order to simplify the proceedings. In this section, we run briefly through the proper line of arguments that show that γ must be a constant, which can be set to zero. The argument also shows that β must depend only on temperature, must grow inversely to temperature, and must be positive. Thus β behaves like inverse thermodynamic temperature, which agrees with our statements above.

For the argument we split the inhomogeneous system under consideration into a large number of small subsystems, each with their individual properties, see Fig. 4.3. With the entropy flux $\dot{\Psi} = \beta\dot{Q} - \gamma\dot{W}$ we find the second law for non-equilibrium systems from summing over subsystems as

$$\frac{dS^{(\gamma)}}{dt} + \sum \gamma_k \dot{W}_k - \sum \beta_k \dot{Q}_k = \dot{S}_{gen} \geq 0 \,, \tag{4.43}$$

where $S^{(\gamma)}$ is the entropy for this choice of flux. As before, $\dot{Q}_k$, $\dot{W}_k$ denote the exchange of heat and work with the surroundings of the system, and β_k, γ_k are the corresponding values of the unknown coefficients in the sub-systems at the system boundaries. All internal exchange of heat and work between the subsystems must be such that entropy is generated. The corresponding terms are absorbed in the entropy generation rate $\dot{S}_{gen}$. The first law for the system is given in (4.23).

We consider the above form (4.43) of the second law for a heat conductor. For steady state heat transfer without any work exchange between a hot reservoir (H) and a cold reservoir (L) through the heat conductor, the above reduces to

$$-\beta_H \dot{Q}_H - \beta_L \dot{Q}_L = \dot{S}_{gen} \geq 0 \,. \tag{4.44}$$

Here, β_H and β_L are the values of β at the hot and cold sides of the conductor, respectively. The first law gives $\dot{Q}_H = -\dot{Q}_L = \dot{Q}$ so that

$$(\beta_L - \beta_H)\,\dot{Q} = \dot{S}_{gen} \geq 0 \,. \tag{4.45}$$

Since heat must go from hot to cold, the heat must be positive, $\dot{Q} > 0$, which requires $(\beta_L - \beta_H) > 0$. Thus, the coefficient β must be smaller for the part of the system which is in contact with the hotter reservoir. This must be so irrespective of the values of any other properties at the system boundaries (L, H). Therefore β must depend on temperature *only*.

It follows that β must be a decreasing function of temperature alone, if temperature of hotter states is defined to be higher. The left part of Fig. 4.9 shows a schematic of the heat transfer process.

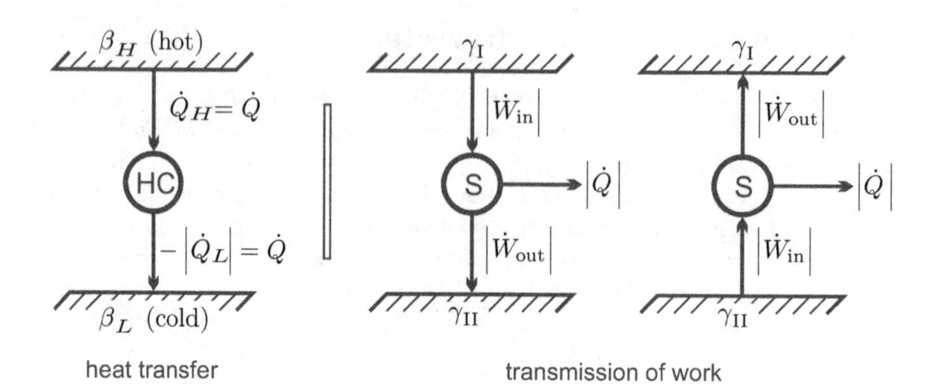

Fig. 4.9 Heat transfer through a heat conductor HC (left) and transmission of work through a steady state system S (right)

We proceed with the discussion of the coefficient γ. For this, we turn our attention to the transmission of work. The right part of Fig. 4.9 shows two "reservoirs" characterized by different values $\gamma_{\rm I}$, $\gamma_{\rm II}$ between which work is transmitted by a steady state system S. The direction of work transfer is not restricted: by means of gears and levers work can be transmitted from low to high force and vice versa, and from low to high velocity and vice versa. Therefore, transmission might occur from I to II, and as well from II to I. Accordingly, there is no obvious interpretation of the coefficient γ.

Friction might occur in the transmission. Thus, in the transmission process we expect some work lost to frictional heating, therefore $\left|\dot{W}_{out}\right| \leq \left|\dot{W}_{in}\right|$. In order to keep the transmission system at constant temperature, some heat must be removed. Work and heat for both cases are indicated in the figure, the arrows indicate the direction of transfer.

The first law for both transmission processes reads (steady state, $\frac{dU}{dt} = 0$)

$$0 = -\left|\dot{Q}\right| - \left|\dot{W}_{out}\right| + \left|\dot{W}_{in}\right| , \tag{4.46}$$

where the signs account for the direction of the fluxes. Since work loss in transmission means $\left|\dot{W}_{out}\right| \leq \left|\dot{W}_{in}\right|$, this implies that heat must leave the system, $\dot{Q} = -\left|\dot{Q}\right| \leq 0$.

Due to the different direction of work in the two processes considered, the second law (4.43) gives different conditions for both situations,

$$-\gamma_{\mathrm{I}}\left|\dot{W}_{in}\right|+\gamma_{\mathrm{II}}\left|\dot{W}_{out}\right|+\beta\left|\dot{Q}\right| \geq 0\,, \quad \gamma_{\mathrm{I}}\left|\dot{W}_{out}\right|-\gamma_{\mathrm{II}}\left|\dot{W}_{in}\right|+\beta\left|\dot{Q}\right| \geq 0\,, \tag{4.47}$$

where, as we have just seen, β is a measure for the temperature of the transmission system. Elimination of the heat between first and second laws gives two inequalities,

$$(\gamma_{\mathrm{II}}-\beta)\left|\dot{W}_{out}\right|-(\gamma_{\mathrm{I}}-\beta)\left|\dot{W}_{in}\right| \geq 0\,,\ (\gamma_{\mathrm{I}}-\beta)\left|\dot{W}_{out}\right|-(\gamma_{\mathrm{II}}-\beta)\left|\dot{W}_{in}\right| \geq 0, \tag{4.48}$$

or, after some reshuffling,

$$(\beta-\gamma_{\mathrm{II}})\frac{\left|\dot{W}_{out}\right|}{\left|\dot{W}_{in}\right|} \leq (\beta-\gamma_{\mathrm{I}}) \quad , \quad (\beta-\gamma_{\mathrm{I}})\frac{\left|\dot{W}_{out}\right|}{\left|\dot{W}_{in}\right|} \leq (\beta-\gamma_{\mathrm{II}})\,. \tag{4.49}$$

Combining the two equations gives the two inequalities

$$(\beta-\gamma_{\mathrm{I}})\left(\frac{\left|\dot{W}_{out}\right|}{\left|\dot{W}_{in}\right|}\right)^2 \leq (\beta-\gamma_{\mathrm{I}}) \quad , \quad (\beta-\gamma_{\mathrm{II}})\left(\frac{\left|\dot{W}_{out}\right|}{\left|\dot{W}_{in}\right|}\right)^2 \leq (\beta-\gamma_{\mathrm{II}})\,. \tag{4.50}$$

From the latter follows, since $0 \leq \frac{\left|\dot{W}_{out}\right|}{\left|\dot{W}_{in}\right|} \leq 1$, that $(\beta-\gamma)$ must be non-negative.

Both inequalities (4.49) must hold for arbitrary transmission systems, that is for all $0 \leq \frac{\left|\dot{W}_{out}\right|}{\left|\dot{W}_{in}\right|} \leq 1$, and all β. For a reversible transmission, where $\frac{\left|\dot{W}_{out}\right|}{\left|\dot{W}_{in}\right|} = 1$, both inequalities (4.49) can only hold for $\gamma_{\mathrm{I}} = \gamma_{\mathrm{II}}$. Accordingly, $\gamma_{\mathrm{I}} = \gamma_{\mathrm{II}} = \gamma$ must be a constant, and $(\beta-\gamma) \geq 0$ for all β.

With γ as a constant, the entropy balance (4.43) becomes

$$\frac{dS^{(\gamma)}}{dt} + \gamma\dot{W} - \sum \beta_k \dot{Q}_k = \dot{S}_{gen} \geq 0\,. \tag{4.51}$$

The energy balance $\dot{W} = \sum \dot{Q}_k - \frac{dE}{dt}$ allows to eliminate work,

$$\frac{d\left(S^{(\gamma)}-\gamma E\right)}{dt} - \sum\left(\beta_k-\gamma\right)\dot{Q}_k = \dot{S}_{gen} \geq 0\,. \tag{4.52}$$

With $S = S^{(\gamma)} - \gamma E$ as the standard entropy, and $T = 1/(\beta - \gamma)$ as the positive thermodynamic temperature, we find the second law in the form (4.24). This is equivalent to setting $\gamma = 0$, and $\beta = 1/T$ as was done in Sec. 4.4.

Problems

4.1. Isothermal Stirring of Mercury
2 litre of mercury confined in a container in thermal contact to an environment at 15 °C are stirred with a 200 W stirrer. How much entropy is created in 20 minutes of stirring? Does the entropy of the mercury change? What happens to the entropy created?

4.2. Adiabatic Stirring of Mercury
2 litre of mercury confined in an isolated container are stirred with a 200 W stirrer. When the mercury was at 15 °C initially, what is its temperature after 20 minutes of stirring? How much entropy is created in the process? What happens to the entropy created?

4.3. Kneading of Pizza Dough
2 kg of dough ($c = 2.73 \frac{\mathrm{kJ}}{\mathrm{kg\,K}}$) confined in a container are kneaded with a 350 W kitchen mixer.

1. The container is in thermal contact to an environment at 25 °C so that the temperature of the dough is 25 °C at all times. How much entropy is created in 10 minutes of kneading?
2. The container is thermally isolated. When the dough was at 25 °C initially, how long does it take until the temperature is 40 °C? How much entropy is created in the process?
3. Both processes are irreversible, hence entropy is created. Explain where the produced entropy goes.

4.4. Stirring of Petroleum
4 litre of petroleum ($\rho = 640 \frac{\mathrm{kg}}{\mathrm{m}^3}$, $c_p = 2.0 \frac{\mathrm{kJ}}{\mathrm{kg\,K}}$) confined in an isolated rigid container are stirred by an electric motor which consumes 50 W of electrical power.

1. How long does it take until the temperature of the petroleum is raised by 5 °C?
2. What is the relation between entropy generation and power? Is the process reversible or irreversible?

4.5. Industrial Stirrer
During manufacture, 2 tons of polyethylene (incompressible liquid, specific heat $c = 2.9308 \frac{\mathrm{kJ}}{\mathrm{kg\,K}}$) are stirred in a well-insulated container for 20 minutes. It is observed that the temperature rises from 42 °C to 49 °C. Ignoring kinetic and potential energies, determine the power demand of the stirrer, and the entropy generated during the process.

4.6. A Brick Falls

A 2 t brick cube falls to the ground on a planet without atmosphere. The gravitational acceleration is $1\frac{\mathrm{m}}{\mathrm{s}^2}$. The cube crashes on the ground and comes to rest. From what height must the cube fall to increase its temperature by 10 K? When the brick's initial temperature was 200 K, how much entropy is created in the process? How much work could be obtained in a reversible process? Brick: $\rho = 1922\frac{\mathrm{kg}}{\mathrm{m}^3}$, $c = 0.79\frac{\mathrm{kJ}}{\mathrm{kg\,K}}$

4.7. A Bad Accident

A 2t truck running at a speed of 120 km/ h crashes against a concrete wall and comes to rest. Assume that the truck is made of steel ($\rho = 7830\frac{\mathrm{kg}}{\mathrm{m}^3}$, $c = 0.5\frac{\mathrm{kJ}}{\mathrm{kg\,K}}$), and that all energy stays in the truck.

1. By what amount will the average temperature of the truck change?
2. How much entropy is created in the process? Assume initial temperature is $T_0 = 20\,°\mathrm{C}$.
3. How much work could have been obtained in a reversible process, e.g., by electromagnetic brakes that charge a battery? Compare the possible work to $T_0 S_{gen}$.

4.8. Dissipation of Kinetic Energy

In Sec. 4.9 it was shown that in an isolated system kinetic energy will vanish in equilibrium. Repeat the proof for a non-adiabatic system.

4.9. Tank and Contents

A well-insulated steel tank of mass 10 kg contains 5 litre of liquid water. Initially, the temperature of the tank is 7 °C, and the temperature of the water is 90 °C. Specific heats: $c_{steel} = 0.5\frac{\mathrm{kJ}}{\mathrm{kg\,K}}$, $c_{water} = 4.18\frac{\mathrm{kJ}}{\mathrm{kg\,K}}$

1. What is the temperature of the system after equilibrium is established?
2. Compute the change of entropy of the tank.
3. Compute the change of entropy of the water.
4. How much entropy is created in the process? Is the process reversible or irreversible?

4.10. Property Change in Argon

The state of argon (ideal gas with constant specific heats) is changed by heating and compression from initial state $p_1 = 1\,\mathrm{bar}$, $T_1 = 230\,\mathrm{K}$ to the final state $p_2 = 20\,\mathrm{bar}$, $T_2 = 400\,\mathrm{K}$. Compute the change of internal energy and the change of entropy of the gas. Do you have enough information to compute the heat and work exchanged? Why not?

4.11. Work and Heat

Krypton (Kr) gas at $T_1 = 230\,°\mathrm{C}$ is confined in a piston-cylinder system. The gas undergoes a reversible process where the pressure changes according to the relation $p = p_1 (V_1/V)^2$. The initial and final volumes are $V_1 = 0.2\,\mathrm{m}^3$ and $V_2 = 0.1\,\mathrm{m}^3$ and the initial pressure is $p_1 = 4\,\mathrm{bar}$.

As all monatomic gases, krypton behaves as an ideal gas with constant specific heat; its molar mass is $M = 83.8\frac{\mathrm{kg}}{\mathrm{kmol}}$. Determine:

1. The mass of Kr in the system.
2. Pressure p_2 and temperature T_2 at the end of the process.
3. The total work for the process.
4. The total heat exchange.
5. The change of entropy of the gas.

4.12. Irreversible Expansion of Xenon
Xenon (ideal gas with constant specific heats) is confined in one half of a 2.5 litre container. The other half of the container is evacuated, and the container is well-insulated. When the partition is removed, the gas expands irreversibly to fill the whole container. Initially, the xenon is at $p_1 = 20\,\text{bar}$, $T_1 = 400\,\text{K}$. Compute the final state p_2, T_2 and the entropy generated.

4.13. Irreversible Expansion of Neon
Neon (ideal gas with constant specific heats) is confined in a 1 litre gas container at $p_1 = 13\,\text{bar}$, $T_1 = 500\,\text{K}$. This container is enclosed in an evacuated rigid container of unknown volume, which is well-insulated. The inner container becomes defect, and the neon expands irreversibly to fill the accessible volume. The final pressure is measured as 4 bar. From the first and second law determine the final temperature T_2, the volume of the bigger container, and the entropy generated.

4.14. Ideal Gas with Non-constant Specific Heat
We go back to problem 3.12, where you made a table of values for $c_v(T)$, $c_p(T)$, $u(T)$ and $h(T)$ for air, when the specific heat at constant volume is $c_v(T) = \left(0.695 + \frac{0.0598T}{1000\,\text{K}}\right)\frac{\text{kJ}}{\text{kg K}}$.

1. To your table, add a column for the entropy at standard pressure p_0, defined as $s^0(T) = \int_{T_0}^{T} \frac{c_p(T')}{T'} dT + s_0$ for temperatures in the range $(300\,\text{K}, 900\,\text{K})$. As reference value chose $s^0(300\,\text{K}) = 7.14\frac{\text{kJ}}{\text{kg K}}$.
2. 3 kg of air are heated in a reversible isochoric process (constant volume) from 320 K, 2 bar to 800 K. By means of your table, determine the work W_{12}, the heat supply Q_{12}, and the change in entropy, $S_2 - S_1$.
3. Redo the calculation of 2. under the assumption that the specific heat can be approximated by its value at 300 K (so that it is constant). Determine the relative errors for heat, work and entropy difference.

4.15. Equilibrium State I
N blocks of different metals with masses m_i, specific heats c_i and temperatures T_i are enclosed in an adiabatic rigid chamber. All blocks are brought into thermal contact. Use the first and second law to show that in equilibrium all blocks must have the same temperature.

Hint: In equilibrium entropy must be a maximum. Since energy is conserved, entropy must be maximized under the constraint of given energy. The most elegant way to solve the problem is using the method of Lagrange multipliers to take care of the constraint.

4.16. Equilibrium State II

An insulated container holds the mass $m_0 = \int \rho dV$ of an ideal gas, and the overall energy is fixed at $E_0 = \int \rho \left(u(T) + \frac{1}{2}\mathcal{V}^2 + gz\right) dV$. Note that in general density ρ, temperature T and velocity $\mathcal{V}$ depend on location $\vec{r}$. Show that in equilibrium temperature is homogeneous, density follows the barometric law, and velocity vanishes.

Hint: Here you have to maximize total entropy $S = \int \rho s(T, \rho)\, dV$ under constraints of given mass and energy. Use Lagrange multipliers and Euler's equation of variational calculus.

4.17. Equilibrium State III

An insulated room contains a rigid shelf on which rests a metal ball (mass m, specific heat c, initial temperature T). The shelf is at height H above the floor. By using first and second law, answer the following questions: Is the system in a thermodynamic equilibrium state, and if so, why? If not, what is the system's thermodynamic equilibrium state, and why? Does your answer depend on whether the room is evacuated, or filled with air? If you find the system is not in thermodynamic equilibrium, why do we find it in the unstable configuration?

Chapter 5
Energy Conversion and the Second Law

5.1 Energy Conversion

In the preceding sections we have evaluated the second law with respect to its ability for the description of basic equilibration processes, e.g., the equilibration of temperature, the direction of heat transfer, the dissipation of kinetic energy, and friction losses in gears. Now we shall apply thermodynamic analysis to conversion processes between work and heat.

The science of thermodynamics emerged from an engineering question: How much work can be obtained from a given amount of heat? This question arose when the first steam engines were built, which had efficiencies of only a few percent. The question is still of outmost importance, as a sustainable way of living requires the optimal use of resources. Having a good understanding of the possibilities and limitations in energy conversion processes is the first step in building better—more efficient—engines.

Before we discuss more complex energy conversion processes, we consider a relatively simple problem: Energy conversion processes between two thermal reservoirs at different temperatures T_H and T_L, with $T_H > T_L$.

The natural environment, usually assumed to be at $T_0 = 25\,^\circ\mathrm{C}$, is the prototype of a thermal reservoir. Due to its size, the environment has almost infinite thermal mass mc_v, and hence it can provide or accept a large amount of heat without changing its temperature. Ultimately, all systems are in thermal contact with the natural environment, and it serves as heat sink or source for most energy conversion processes.

Many of today's heat engines rely on the combustion of a fuel (coal, oil, gas). Combustion processes do not create a reservoir of constant high temperature, but rather a flow of hot combustion gases that provides heat at varying temperature. Therefore, the following considerations are not always directly applicable to combustion systems. Nevertheless, the subsequent sections give important and relevant insights, to which we shall come back again and again.

H. Struchtrup, *Thermodynamics and Energy Conversion*,
DOI: 10.1007/978-3-662-43715-5_5, © Springer-Verlag Berlin Heidelberg 2014

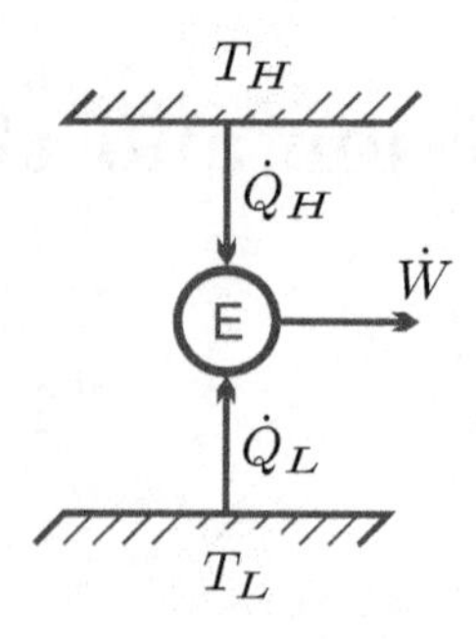

Fig. 5.1 Two heat reservoirs at T_H and T_L connected by a thermal engine E

Pure heat transfer between the two reservoirs was discussed already in Sec. 4.8, with the statement that by itself heat goes from warm to cold, but cannot go from cold to warm. We now consider processes that involve heat and work. The systems considered are engines that operate at steady state, that is they do not accumulate or loose energy or entropy over time, $\frac{dE}{dt} = \frac{dS}{dt} = 0$. The detailed processes inside the engines will be discussed extensively later. For the present overall evaluation, however, they are of no concern, and thus the set-up considered is as simple as shown in Fig. 5.1: The thermal engine E exchanges heat with both reservoirs, and produces or consumes power. The direction of the arrows in Fig. 5.1 simply indicates the convention for heat and work: heat in and work out are positive. In the following figures, however, we will use absolute values of heat and work, and the direction of the flows will be indicated by the directions of the arrows.

For steady state processes, the first and second law for this set-up read

$$0 = \dot{Q}_H + \dot{Q}_L - \dot{W} \quad , \quad -\frac{\dot{Q}_H}{T_H} - \frac{\dot{Q}_L}{T_L} = \dot{S}_{gen} \geq 0 \,. \tag{5.1}$$

5.2 Heat Engines

First we consider power generation, that is the conversion of heat into work in a *heat engine*, so that $\dot{W} = \left|\dot{W}\right| > 0$. Elimination of $\dot{Q}_L$ between the first and second law (5.1) gives the work

$$\dot{W} = \left(1 - \frac{T_L}{T_H}\right) \dot{Q}_H - T_L \dot{S}_{gen} \,. \tag{5.2}$$

Since we require $\dot{W} > 0$, the right hand side of this equation must be positive as well. The last term, $-T_L \dot{S}_{gen}$, is zero or negative, since thermodynamic temperature and entropy generation rate are both non-negative; therefore, the first term, $\left(1 - \frac{T_L}{T_H}\right) \dot{Q}_H$, must be positive. Since the bracket is always positive, this implies positive heat input from the hot reservoir, $\dot{Q}_H = \left|\dot{Q}_H\right| > 0$. The heat rejected to the colder reservoir is $\dot{Q}_L = -\frac{T_L}{T_H}\dot{Q}_H - T_L\dot{S}_{gen} < 0$. Figure 5.2 shows the direction of heat and work flow for a heat engine between the reservoirs.

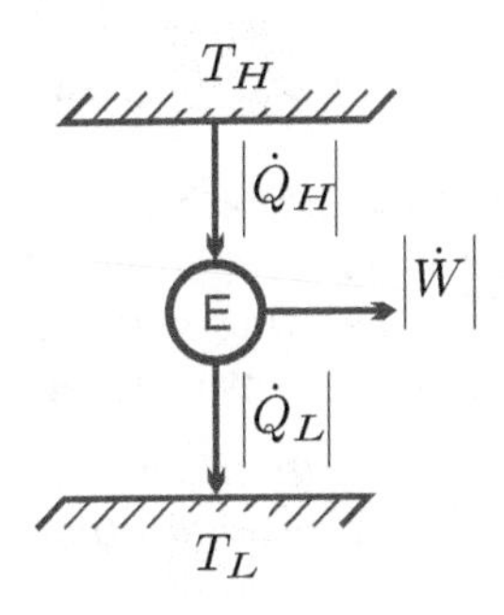

Fig. 5.2 Heat and work directions in a heat engine

According to (5.2), for given T_H, T_L, the work output is larger for smaller entropy generation rate $\dot{S}_{gen} \geq 0$. Entropy generation is due to heat transfer and friction processes within the engine, and between engine and reservoirs, and cannot be totally avoided in real engines. Instead, the engineering task is to minimize entropy generation within the system as much as possible, in order to achieve the best possible performance of the engine. The work loss to irreversibilities is proportional to the entropy generation,

$$\dot{W}_{\text{loss}} = T_L\dot{S}_{gen} \geq 0\,. \tag{5.3}$$

The theoretical limit for the power generated from two reservoirs with constant temperatures is obtained for $\dot{S}_{gen} = 0$, that is for a fully reversible engine, as

$$\dot{W}_C = \left(1 - \frac{T_L}{T_H}\right)\dot{Q}_H\,. \tag{5.4}$$

This is the work output of a Carnot engine, named after Sadi Carnot (1796-1832), who established this theoretical limit. Any entropy generation $\dot{S}_{gen}$ in the engine reduces the work output by $T_L\dot{S}_{gen}$.

To quantify the performance of engines, it is useful to define dimensionless efficiency measures that compare the output ("what you get") to the input ("what you pay for"). For heat engines, accordingly, one defines the thermal efficiency η_{th} as the ratio between work output and heat input. For heat engines operating between two reservoirs, we obtain

$$\eta_{th} = \frac{\dot{W}}{\dot{Q}_H} = 1 - \frac{T_L}{T_H} - \frac{T_L\dot{S}_{gen}}{\dot{Q}_H} < 1 \tag{5.5}$$

for irreversible engines, and

$$\eta_C = \frac{\dot{W}_C}{\dot{Q}_H} = 1 - \frac{T_L}{T_H} < 1 \tag{5.6}$$

for the Carnot engine.

The Carnot efficiency η_C is the efficiency of a fully reversible engine operating between two reservoirs at constant temperatures. Since it was computed from general considerations, its value is completely independent of the details of the engine, i.e., it does not depend on the working fluid used, nor on the realization of the engine. The Carnot efficiency is a universal limit for the thermal efficiency *any* engine operating between two reservoirs at T_H, T_L can have. We summarize the above in two statements:

(a) *The thermal efficiency of a fully reversible engine operating between two reservoirs is independent of the realization of the engine; it is given by the Carnot efficiency* η_C.

(b) *Any engine operating between two reservoirs in which irreversible processes occur has a thermal efficiency below that of a fully reversible engine.*

The amount of work produced grows with the temperature ratio T_H/T_L between the reservoirs. In technical energy conversion processes one will aim for high upper temperature T_H to ensure high energy conversion efficiency. At high temperatures material strength is limited, so that the upper temperatures are limited through the materials used for building the engines. Typically, the lower temperature T_L is the temperature of the environment, T_0.

For temperature ratios T_H/T_L close to unity, i.e., small temperature differences, the thermal efficiency is small, and only little power can be produced. Hence, low temperature waste heat (low T_H) is relatively useless for power production, and, if possible, should rather be used for space heating. High temperature waste heat (high T_H), however, has considerable work potential that should be used. In other words:

Energy at high temperature is more valuable than energy at low temperature, since more work can be extracted from it.

5.3 The Kelvin-Planck Statement

Even the—fully reversible—Carnot engine has a thermal efficiency η_C below unity: Not all heat received from the hot reservoir can be converted into work, some heat must be rejected to a colder reservoir. The Kelvin-Planck formulation of the second law states this as follows:

No steady state thermodynamic process is possible in which heat is completely converted into work.

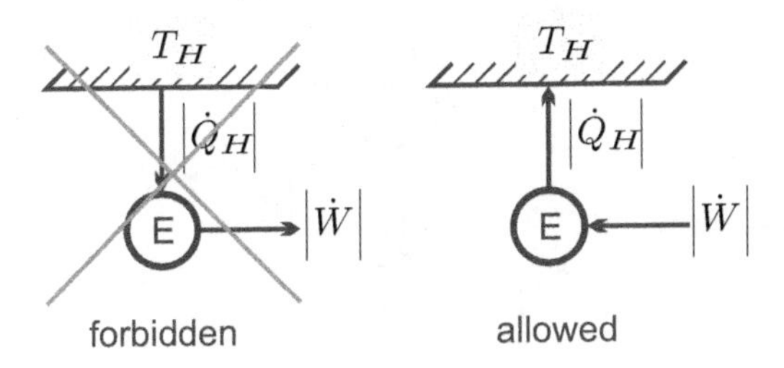

Fig. 5.3 Heat cannot be completely converted into work, but work can be completely converted to heat

This statement is a direct consequence of the first and second law. For a steady state process with just one heat exchange the laws require

$$-\frac{\dot{Q}_H}{T_H} = -\frac{\dot{W}}{T_H} \geq 0 \,, \tag{5.7}$$

hence heat and work must both be negative. Figure 5.3 shows the forbidden process, and also the—allowed—inverse process, the complete conversion of work into heat through friction. A typical example for the latter are resistance heaters in which electrical work is converted to heat through electric resistance.

5.4 Refrigerators and Heat Pumps

While heat cannot go from cold to warm by itself, one can use work consuming devices to perform this task, a refrigerator or heat pump as depicted in Fig. 5.4.

A refrigerator removes heat from a cold reservoir, e.g., the interior of a freezer, at T_L, and rejects heat to the environment at T_H—the goal is to cool the cold reservoir. A heat pump is used for space heating, it takes heat from the environment at T_L, and rejects heat into the room that is being heated at T_H. While the values of the temperatures T_L, T_H differ for refrigerator and heat pump, both operate according to the same principles.

With heat being removed from the colder reservoir, and heat rejected into the warm reservoir, we have $\dot{Q}_H = -\left|\dot{Q}_H\right| < 0$, and $\dot{Q}_L = \left|\dot{Q}_L\right| > 0$. From combining first and second law by eliminating $\dot{Q}_H$, we find the condition

$$-\dot{W} - \left(\frac{T_H}{T_L} - 1\right)\left|\dot{Q}_L\right| = T_H \dot{S}_{gen} \geq 0 \,. \tag{5.8}$$

Since $\left(\frac{T_H}{T_L} - 1\right) > 0$, the sign requirement can only be fulfilled if work is done on the system, $\dot{W} = -\left|\dot{W}\right| < 0$, where

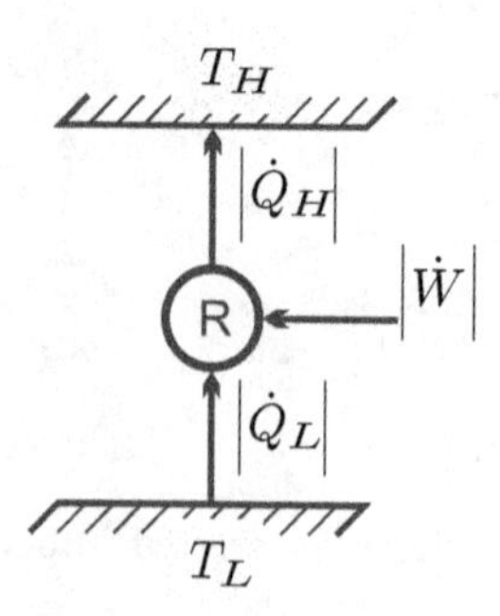

Fig. 5.4 Heat and work directions in a refrigerator/heat pump

$$\left|\dot{W}\right| = \left(\frac{T_H}{T_L} - 1\right)\left|\dot{Q}_L\right| + T_H \dot{S}_{gen} \,. \tag{5.9}$$

This equation relates the work requirement, $\dot{W}$, to the heat removed from the colder reservoir, $\dot{Q}_L$; it is well suited for evaluating refrigerators.

For heat pump systems one is interested in the work required in relation to the heat supply $\dot{Q}_H$ to the warmer reservoir. Eliminating $\left|\dot{Q}_L\right|$ one finds

$$\left|\dot{W}\right| = \left(1 - \frac{T_L}{T_H}\right)\left|\dot{Q}_H\right| + T_L \dot{S}_{gen} \,. \tag{5.10}$$

Since $T\dot{S}_{gen} \geq 0$, any generation of entropy within a refrigeration or heat pump system increases the work requirement $\left|\dot{W}\right|$, and thus the operating cost. The extra work to overcome irreversibilities is $T_H \dot{S}_{gen}$ for a refrigerator and $T_L \dot{S}_{gen}$ for a heat pump. One will aim at reducing all causes for entropy generation, i.e., friction, heat transfer over finite temperature difference, etc., as much as possible.

The theoretical limit for the work of the refrigeration and heat pump systems are obtained for fully reversible engines, for which $\dot{S}_{gen} = 0$. This results in the expressions for a Carnot refrigerator and a Carnot heat pump, respectively, which read

$$\left|\dot{W}\right|_{R,C} = \left(\frac{T_H}{T_L} - 1\right)\left|\dot{Q}_L\right| \quad , \quad \left|\dot{W}\right|_{HP,C} = \left(1 - \frac{T_L}{T_H}\right)\left|\dot{Q}_H\right| \,. \tag{5.11}$$

Also the performance of refrigerators and heat pumps is measured by dimensionless efficiency measures that compare the output ("what you get") to the input ("what you pay for"), which here are the ratios of heat removed/supplied to the work required to run the device, known as the *coefficients of performance* (COP). We obtain, for refrigerator and heat pump operating between two reservoirs,

$$\mathrm{COP_R} = \frac{\left|\dot{Q}_L\right|}{\left|\dot{W}\right|} = \frac{1}{\frac{T_H}{T_L} - 1 + \frac{T_H \dot{S}_{gen}}{\left|\dot{Q}_L\right|}} \lesseqgtr 1\,, \tag{5.12}$$

$$\mathrm{COP_{HP}} = \frac{\left|\dot{Q}_H\right|}{\left|\dot{W}\right|} = \frac{1}{1 - \frac{T_L}{T_H} + \frac{T_L \dot{S}_{gen}}{\left|\dot{Q}_H\right|}} \geq 1\,. \tag{5.13}$$

The COP of a refrigerator can be above or below unity, but the COP of a heat pump is never below unity. A resistance heater (RH), which converts electrical power $\dot{W}_{RH}$ fully into heat $\dot{Q}_{RH} = \dot{W}_{RH}$ has a COP of unity, $\mathrm{COP_{RH}} = 1$, which is the lower bound for heat pumps. A typical heat pump has a COP above unity and gives more efficient heating.

Irreversible processes in engines lead to entropy generation and reduce the COP. For fully reversible engines we find the COP of Carnot engines,

$$\mathrm{COP}_{\mathrm{R},C} = \frac{1}{\frac{T_H}{T_L} - 1} \gtrless 1 \quad , \quad \mathrm{COP}_{\mathrm{HP},C} = \frac{1}{1 - \frac{T_L}{T_H}} > 1\,. \tag{5.14}$$

The COPs for the Carnot refrigerator and Carnot heat pump are the maximum possible COP for refrigeration or heat pump processes between two heat reservoirs at T_H, T_L.

5.5 Kelvin-Planck and Clausius Statements

Clausius' statement of the second law says that *heat will not go from cold to warm by itself.* Note that the two words "by itself" are important here: a heat pump system can transfer heat from cold to warm, but work must be supplied, so the heat transfer is not "by itself."

The Kelvin-Planck statement of the second law says that it is *impossible to construct a device operating at steady state that receives heat from a single reservoir and produces work.* In other words, no heat engine can be build that has a thermal efficiency of $\eta_{th} = 1$. In our treatment, this statement followed from the evaluation of the second law, while the Clausius statement was used explicitly in its development.

The Clausius statement is a daily experience—when we touch a hot plate, we do not expect to get colder hands—but the Kelvin-Planck statement might be more difficult to grasp. It is instructive to show that both statements are equivalent. To this end, we consider the setting shown in Fig. 5.5, consisting of an engine I that completely converts the heat $\left|\dot{Q}_H\right|$ to power $\left|\dot{W}\right| = \left|\dot{Q}_H\right|$, and an engine II, a heat pump that consumes the work produced by engine I. Engine I is forbidden by the Kelvin-Planck statement while engine II is allowed by the Clausius statement. As the figure shows, the net effect of the

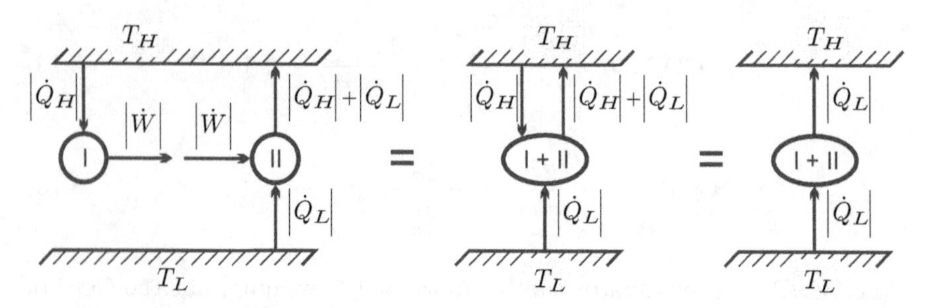

Fig. 5.5 The equivalency of the Kelvin-Planck (K/P) and Clausius (C) statements of the second law

combined system [I + II] is heat transfer from cold to warm "by itself", which is forbidden by the Clausius statement. Both statements are equivalent.

5.6 Thermodynamic Temperature

In the derivation of the second law we have introduced thermodynamic temperature T as the factor of proportionality between the heat transfer rate $\dot{Q}$ and the entropy flux $\dot{\Psi}$.

In previous sections we have seen that this definition of thermodynamic temperature stands in agreement with the direction of heat transfer: heat flows from hot (high T) to cold (low T) by itself. The heat flow aims at equilibrating the temperature within any isolated system that is left to itself, so that two systems in thermal equilibrium have the same thermodynamic temperature. Moreover, the discussion of internal friction showed that thermodynamic temperature must be positive.

The discussion of energy conversion processes between two reservoirs adds another requirement for thermodynamic temperature: For any reversible engine operating between two reservoirs, it must fulfill the relation

$$\frac{T_H}{T_L} = -\frac{\dot{Q}_H}{\dot{Q}_L} \,. \tag{5.15}$$

This relation follows from $(5.1)_2$ for the case of a fully reversible engine, $\dot{S}_{gen} = 0$, independent of the realization of the reversible engine, or the working substance employed.

It is therefore possible, at least in principle, to measure temperature ratios through measurement of the heat exchange in fully reversible engines. Accordingly, to define the thermodynamic temperature scale, only a single reference temperature is required.

The Kelvin temperature scale, named after William Thomson, Lord Kelvin (1824 - 1907), uses the triple point of water (611 kPa, 0.01 °C) as reference. The triple point is the state at which a substance can coexist in all three phases, solid, liquid and vapor, see Sec. 6.3. The Kelvin scales assigns the value of $T_{Tr} = 273.16\,\mathrm{K}$ to this unique point, which can be reproduced easily in laboratories.

Since thermodynamic temperature cannot be negative, the smallest possible thermodynamic temperature is $0\,\mathrm{K}$, known as *absolute zero*.

The ideal gas temperature scale, introduced in Sec. 2.13, coincides with the Kelvin scale. This will be seen later, in Sec. 8.2, when we explicitly compute the thermal efficiency of a Carnot cycle operating with an ideal gas.

5.7 Perpetual Motion Engines

Perpetual motion engines are engines that violate the first or the second law of thermodynamics, or both. Naturally, one will never meet these engines since they are impossible to build—the thermodynamic laws are not to be violated! One might meet inventors, however, who claim to have invented engines that do miraculous things. Inevitably, the inventors will never be able to show their engines in working condition, and their claims remain eternally unproven.

A *perpetual motion engine of the first kind* is an engine that violates the first law of thermodynamics, e.g., an engine that produces more work than the net heat exchange, $\left|\dot{W}\right| > \left|\dot{Q}_H\right| - \left|\dot{Q}_L\right|$.

A *perpetual motion engine of the second kind* is an engine that violates the second law of thermodynamics, e.g., a heat engine operating between two reservoirs at T_L, T_H with an efficiency above the Carnot efficiency, $\eta > 1 - \frac{T_L}{T_H}$. Violations of the second law are sometimes difficult to understand, and thus perpetual motion engines of the second kind are more difficult to identify for not so clever inventors, and their gullible investors.

5.8 Reversible and Irreversible Processes

Irreversible processes are associated with entropy generation which reduces the performance of engines. So far the terms *reversible* and *irreversible* were rather loosely defined in Sec. 2.10. A more exact definition of these terms will make it easier to identify irreversible processes, and the related losses. We define:

> *A thermodynamic process from state 1 to state 2 is* reversible, *if the process can be inverted so that the system returns to its initial state (state 1), and no changes remain in its surroundings.*
>
> *A thermodynamic process from state 1 to state 2 is* irreversible, *if, when the system is brought back into its initial state (state 1), changes remain in its surroundings.*

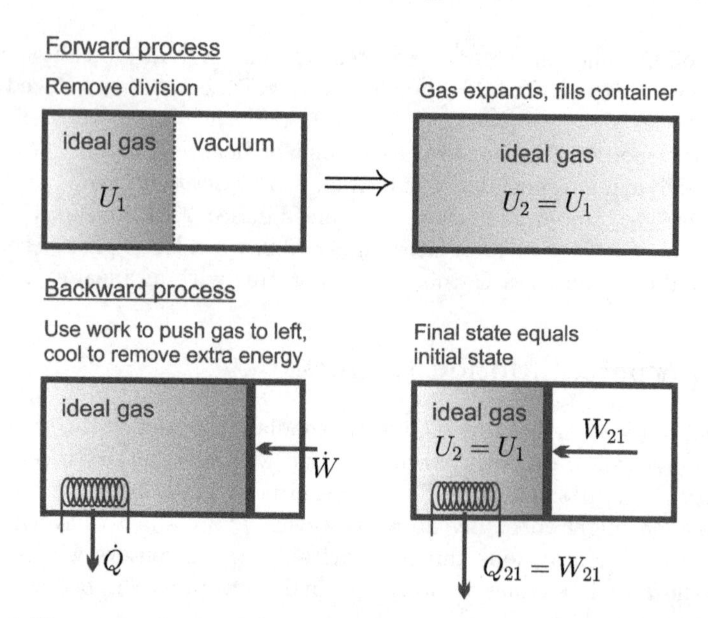

Fig. 5.6 Upper row: Irreversible expansion of a gas into vacuum. Lower row: The intial state is recovered by pushing the piston and removing heat. Since the heat added to the surroundings, Q_{21}, cannot be fully converted into the work needed to push, W_{21}, changes remain in the surroundings.

For an example, we return to the uncontrolled expansion of an ideal gas, which is shown again in Fig. 5.6. We found, in Sec. 3.13, that the internal energies of initial and final states are the same, $U_2 = U_1$, while the gas fills a bigger volume in the final state, $V_2 > V_1$. To return the gas to the initial state, its volume must be reduced by compression, which requires the (reversible) work $W_{21} = \int_{V_2}^{V_1} pdV < 0$. The first law for any process from state 2 back to the initial state 1 reads

$$U_1 - U_2 = Q_{21} - W_{21} = 0 \, ,$$

so that the heat $Q_{21} = W_{21} < 0$ must be removed from the system, as shown in the figure. Thus, the process 2-1 draws the work W_{21} from the surroundings and transfers the heat Q_{21} to the surroundings. The process would be reversible, if the heat Q_{21} could be completely converted to the work W_{21} by an engine residing in the surroundings. This, however, is forbidden by the Kelvin-Planck statement of the second law, which states that only some of the heat can be converted to work, but not all. Thus, some extra work has to be provided to return the system to its original state, the system's surroundings have changed: the original process is irreversible.

Heat transfer serves as another example: The heat $|Q_{AB}|$ has flown from a hot body A to a cold body B by itself. To return both bodies to their original state one can use a heat pump, which consumes the work W_{HP}, removes the heat Q_{AB} from the colder body B, and delivers the heat $|Q'_{AB}| = |Q_{AB}|+|W_{HP}|$ to the warmer body A. After this, body B is in its initial state. Body A received too much heat, however. To return A into its initial state, the heat $|Q''_{AB}| = |W_{HP}|$ must be moved from A to the surroundings. Due to the Kelvin-Planck statement, the heat added to the surroundings $|Q''_{AB}|$ can only provide part of the work $|W_{HP}|$ required to drive the heat pump: the process is irreversible.

5.9 Internally and Externally Reversible Processes

For a sound thermodynamic evaluation of processes it is important to identify and understand *all* causes for work loss to irreversible processes. Even if a process is reversible within the boundaries of the system considered, there might be associated irreversible processes outside the system boundaries. For the thorough evaluation of the performance of a system, in particular for accounting for the associated work losses inside and outside the system, the following definitions are useful:

Internally reversible process: No irreversible processes occur inside the system boundaries.

Externally reversible process: No irreversibilities occur outside the system boundaries.

Fully reversible process: A process which is both, externally and internally reversible.

5.10 Irreversibility and Work Loss

The thermodynamic laws for closed systems that exchange heat with an arbitrary number of reservoirs read

$$\frac{d\left(U+E_{kin}\right)}{dt}=\dot{Q}_0+\sum\dot{Q}_k-\dot{W} \quad , \quad \frac{dS}{dt}-\frac{\dot{Q}_0}{T_0}-\sum\frac{\dot{Q}_k}{T_k}=\dot{S}_{gen}\geq 0 \; , \tag{5.16}$$

where the heat exchange $\dot{Q}_0$ with a reservoir at T_0 is highlighted. Most thermodynamic engines utilize the environment as heat source or sink, and in this case $\dot{Q}_0$ should be considered as the heat exchanged with the environment. Note that the environment is freely available, and no cost is associated with removing heat from, or rejecting heat into, the environment. For the heat engines of Sec. 5.2 and the heat pumps of Sec. 5.4 the environmental temperature is $T_0 = T_L$, while for the refrigerators of Sec. 5.4 we have $T_0 = T_H$.

Elimination of $\dot{Q}_0$ between the two laws and solving for work gives

$$\dot{W} = \sum \left(1 - \frac{T_0}{T_k}\right) \dot{Q}_k - \frac{d\left(U + E_{kin} - T_0 S\right)}{dt} - T_0 \dot{S}_{gen} \,. \tag{5.17}$$

This equation generalizes the findings of the previous sections to arbitrary processes in closed systems: The generation of entropy in irreversible processes reduces the work output of work producing devices (where $\dot{W} > 0$, e.g., heat engines) and increases the work requirement of work consuming devices (where $\dot{W} < 0$, e.g., heat pumps and refrigerators). We note the appearance of the Carnot factor $\left(1 - \frac{T_0}{T_k}\right)$ multiplying the heating rates $\dot{Q}_k$.

The amount of work lost to irreversible processes is

$$\dot{W}_{\text{loss}} = T_0 \dot{S}_{gen} \geq 0 \,, \tag{5.18}$$

sometimes it is denoted as the *irreversibility*. It is an important engineering task to identify and quantify the irreversible work losses, and to reduce them by redesigning the system, or use of alternative processes.

5.11 Examples

5.11.1 *Entropy Generation in Cooling*

A 2 kg block of copper at $T_1 = 250\,°\mathrm{C}$ equilibrates with the environment at $T_0 = 20\,°\mathrm{C}$ through heat transfer. The left part of Fig. 5.7 shows a sketch of the process, where the system boundary is chosen such that heat is transferred at the environmental temperature T_0. Copper can be considered as an incompressible solid with constant specific heat $c = 0.4 \frac{\mathrm{kJ}}{\mathrm{kg\,K}}$, specific internal energy $u = c\left(T - T_0\right)$, and specific entropy $s = c \ln \frac{T}{T_0}$. We compute the amount of heat transferred into the environment, and the total entropy generated.

The first and second law for this process read

$$\frac{dU}{dt} = \dot{Q} \quad , \quad \frac{dS}{dt} = \frac{\dot{Q}}{T_0} + \dot{S}_{gen} \,.$$

Integrating over time between initial state (T_1) and final state ($T_2 = T_0$) gives

$$U_2 - U_1 = Q_{12} \quad , \quad S_2 - S_1 = \frac{Q_{12}}{T_0} + S_{gen} \,,$$

where $S_{gen} = \int_1^2 \dot{S}_{gen} dt$ is the total entropy generation. With the given property relations we find

$$mc\left(T_0 - T_1\right) = Q_{12} \quad , \quad mc \ln \frac{T_0}{T_1} = \frac{Q_{12}}{T_0} + S_{gen}$$

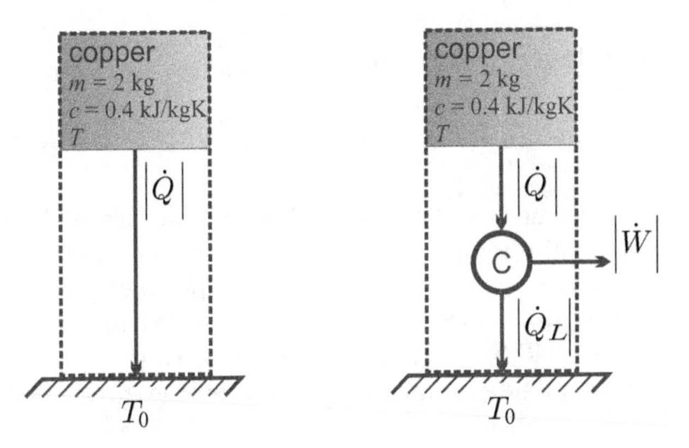

Fig. 5.7 A block of copper initially at T_1 cools to environmental temperature T_0 by heat transfer (left), or by driving a reversible engine (Carnot engine, right); T is the actual temperature at time t.

and thus, with $T_0 = 293\,\mathrm{K}$, $T_1 = 523\,\mathrm{K}$ (thermodynamic temperature must be used for entropy and the second law!),

$$Q_{12} = -184\,\mathrm{kJ} \quad , \quad S_{gen} = \frac{mc}{T_0}\left[(T_1 - T_0) - T_0 \ln \frac{T_1}{T_0}\right] = 0.163 \frac{\mathrm{kJ}}{\mathrm{K}} \, .$$

Since entropy is generated, an irreversible loss is associated with the process. The entropy generating process is heat transfer over the finite temperature difference between copper block and environment.

5.11.2 Work Generation in Cooling

In the this example we determine the amount of work that could have been obtained if the heat was not just transferred, but used to drive a heat engine. We consider the same block of copper as before, but now the heat is used to drive a Carnot engine in contact with the environment, as shown in the right part of Fig. 5.7. In this case, there is no entropy generation, since the Carnot engine is fully reversible. Thus, the first and second laws read (system boundaries include the Carnot engine)

$$\frac{dU}{dt} = \dot{Q}_L - \dot{W} \quad , \quad \frac{dS}{dt} = \frac{\dot{Q}_L}{T_0} \, .$$

Integration gives

$$U_2 - U_1 = Q_L - W \quad , \quad S_2 - S_1 = \frac{Q_L}{T_0} \, ,$$

so that

$$Q_L = mcT_0 \ln \frac{T_0}{T_1} = -135.8\,\text{kJ}\,,$$

$$W = Q_L - mc\,(T_0 - T_1) = mc\left[T_1 - T_0 - T_0 \ln \frac{T_1}{T_0}\right] = 48.2\,\text{kJ}\,.$$

A temperature difference can be used to drive a heat engine. If heat is just transferred over a finite temperature difference, entropy is created, and the opportunity to provide work is lost. In this example about 26% of the heat leaving the copper ($Q_H = 184\,\text{kJ}$) could be converted to work in the best case. Note that $W = T_0 S_{gen}$, where S_{gen} is the entropy generation in case that no work is produced as computed in the previous section.

5.11.3 Perpetual Motion Engines

We consider some perpetual motion engines.

(a) A company claims to produce a power generation device that produces 7 kW of power, takes in 11 kW of heat at a temperature of 840 K and rejects 8 kW of heat at 280 K.

We evaluate this claim: The work and heat flows are as in Fig. 5.2. Evaluation of the first law shows that $\left|\dot{Q}_H\right| = \left|\dot{Q}_L\right| + \left|\dot{W}\right|$ should hold. With $\left|\dot{Q}_H\right| = 11\,\text{kW}$, $\left|\dot{W}\right| = 7\,\text{kW}$ and $\left|\dot{Q}_L\right| = 8\,\text{kW}$ the first law is *not* fulfilled—the device is a perpetual motion engine of the first kind.

(b) Another company claims to produce a power generation device that produces 7 kW of power, takes in 10 kW of heat at a temperature of 840 K and rejects 3 kW of heat at 280 K.

We evaluate this claim: The first law is balanced now, we need to check the second law. The thermal efficiency of the device would be $\eta = \dot{W}/\dot{Q}_H = 0.7$. The work of a Carnot engine operating between the same temperatures is $\eta_C = 1 - T_L/T_H = 2/3$. Thus the efficiency claimed is bigger than the Carnot efficiency, which violates the second law—this engine is a perpetual motion engine of the second kind.

(c) Yet another company markets a refrigeration device that removes 1 kW of heat from a cold space that is kept at $-10\,°\text{C}$, and rejects heat into an environment at $22\,°\text{C}$. The company claims a power consumption of 122 W.

We evaluate this claim: The coefficient of performance for a Carnot refrigeration device operating between the same temperatures is $\text{COP}_{\text{R},C} = \frac{|\dot{Q}|}{|\dot{W}|} = \frac{1}{T_H/T_L - 1} = 8.219$ which would give a power consumption of $\dot{W}_C = 122\,\text{W}$. Thus, the company claims to have a Carnot refrigeration device. While this claim does not violate the first or the second law, it stands in contrast to the fact that *any real process is irreversible.* Thus, for an actual device one must expect efficiencies and COP's below the Carnot values, which are the maxima obtained for fully reversible processes. The company's claim must be wrong.

5.11.4 A Heat Engine

An engine that operates at steady state between two reservoirs at $T_H = 750\,°\mathrm{C}$ and $T_L = 15\,°\mathrm{C}$, has a heat intake of 0.1 MW, and rejects 50 kW of heat to the low temperature environment. We compute the power produced, the thermal efficiency, the entropy generation rate, and the work loss to irreversibilities.

We identify $\dot{Q}_H = 100\,\mathrm{kW}$ and $\dot{Q}_L = -50\,\mathrm{kW}$. The first law applied to the engine gives $\dot{W} = \dot{Q}_H + \dot{Q}_L = 50\,\mathrm{kW}$. Accordingly, the engine's thermal efficiency is $\eta = \frac{\dot{W}}{\dot{Q}_H} = 0.5$.

The second law gives $\frac{\dot{Q}_H}{T_H} + \frac{\dot{Q}_L}{T_L} + \dot{S}_{gen} = 0$ and thus the entropy generation rate is $\dot{S}_{gen} = \frac{|\dot{Q}_L|}{T_L} - \frac{\dot{Q}_H}{T_H} = 0.076\frac{\mathrm{kW}}{\mathrm{K}}$. Since $\dot{S}_{gen} > 0$, the second law is fulfilled.

The thermal efficiency for a Carnot engine operating between the same temperatures is $\eta_C = 1 - \frac{T_L}{T_H} = 0.718$ (Kelvin temperatures!) which is above the efficiency of the engine, as it must be. A Carnot engine would produce the power $\dot{W}_C = \eta_C \dot{Q}_H = 71.8\,\mathrm{kW}$. The work loss to irreversibilities is $\dot{W}_{loss} = \dot{W}_C - \dot{W} = 21.8\,\mathrm{kW}$.

A more instructive way to compute the work loss is as follows: Eliminating the heat exchange with the environment, $\dot{Q}_L$, between first and second law gives

$$\dot{W} = \left(1 - \frac{T_L}{T_H}\right)\dot{Q}_H - T_L \dot{S}_{gen} \,.$$

The work loss to irreversibilities is $\dot{W}_{\mathrm{loss}} = T_L \dot{S}_{gen} = 21.8\,\mathrm{kW}$.

5.11.5 Refrigerator

A restaurant refrigerator located in a kitchen at 21 °C maintains its interior at 4 °C. The refrigerator consumes 300 W of power with a coefficient of performance $\mathrm{COP_R} = 3$. We compute entropy generation and work loss.

The heat withdrawn from the interior is $\dot{Q}_L = \mathrm{COP_R}\left|\dot{W}\right| = 900\,\mathrm{W}$. According to the first law, the heat rejected into the kitchen is $\left|\dot{Q}_H\right| = \left|\dot{W}\right| + \dot{Q}_L = 1200\,\mathrm{W}$. The entropy generation rate follows from the second law as

$$\dot{S}_{gen} = -\frac{\dot{Q}_H}{T_H} - \frac{\dot{Q}_L}{T_L} = \frac{\left|\dot{Q}_H\right|}{T_H} - \frac{\dot{Q}_L}{T_L} = 0.083\frac{\mathrm{W}}{\mathrm{K}}\,,$$

where $T_L = 277\,\mathrm{K}$ and $T_H = 294\,\mathrm{K}$. Eliminating the heat rejected into the kitchen between first and second law yields

$$\left|\dot{W}\right| = -\dot{W} = \left(\frac{T_H}{T_L} - 1\right)\dot{Q}_L + T_H \dot{S}_{gen} \,.$$

The work loss to irreversibilities is $\dot{W}_{\text{loss}} = T_H \dot{S}_{gen} = 247.7\,\text{W}$; this work is required as input to overcome irreversibilities. A fully reversible refrigerator, i.e., a Carnot refrigerator, which removes the same amount of heat $\dot{Q}_L$ has a $\text{COP}_{\text{R},C} = 1/\left(\frac{T_H}{T_L} - 1\right) = 16.3$, and would consume 55 W of electrical power.

Note that efficient operation of a refrigerator is not only achieved by increasing its COP, but also by improving the thermal insulation. Indeed, the heat $\dot{Q}_L$ that is removed from the interior has crept in through the insulated walls of the refrigerator. Better insulation reduces the amount of heat that must be removed, and thus the work consumption of the refrigerator.

5.11.6 Heat Pump with Internal and External Irreversibilities

A heat pump is used to keep a home at 20 °C. The heat pump draws heat from the outside environment at 0 °C; its heating power is $\left|\dot{Q}_H\right| = 2\,\text{kW}$ for a power consumption of $\left|\dot{W}\right| = 0.5\,\text{kW}$. In order to facilitate sufficient heat transfer, a temperature difference of 10 K is required between the working substance of the heat pump and the respective environments. Figure 5.8 gives a sketch of the heat and work flows, and the relevant temperature levels.

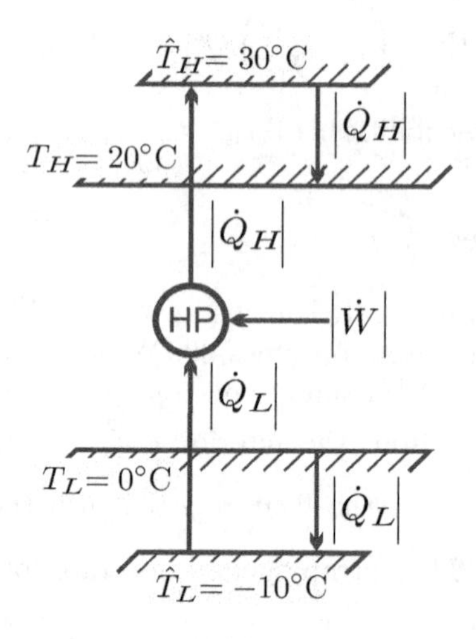

Fig. 5.8 A heat pump that requires a finite temperature difference of 10K for heat exchange

We evaluate the process step by step. Let us first consider a perfectly reversible Carnot heat pump, that is a device that can operate at the actual temperatures of the two environments, $T_H = 293\,\text{K}$ and $T_L = 273\,\text{K}$. The coefficient of performance of such an ideal engine is

$$\text{COP}_{\text{HP},C} = \frac{\left|\dot{Q}_H\right|}{\left|\dot{W}_C\right|} = \frac{1}{1-\frac{T_L}{T_H}} = 14.65\,.$$

For the given heating power the fully reversible heat pump would consume $\left|\dot{W}_C\right| = 0.137\,\text{kW}$ of power.

A internally reversible heat pump with external irreversibilities due to heat transfer over finite temperatures is a Carnot heat pump operating between the temperatures $\hat{T}_H = 303\,\text{K}$ and $\hat{T}_L = 263\,\text{K}$. This engine would have a coefficient of performance

$$\text{COP}_{\text{HP},C-int} = \frac{1}{1-\frac{\hat{T}_L}{\hat{T}_H}} = 7.58$$

and would consume $\dot{W}_{C-int} = 0.264\,\text{kW}$ of power. The internally reversible engine requires more work than the fully reversible engine, since a bigger temperature interval is bridged. Entropy is generated in the heat transfer over finite temperature differences, with the generation rate

$$\dot{S}_{gen} = \left|\dot{Q}_H\right|\left(\frac{1}{T_H}-\frac{1}{\hat{T}_H}\right) + \left|\dot{Q}_L\right|\left(\frac{1}{\hat{T}_L}-\frac{1}{T_L}\right)\,.$$

Since the engine is internally reversible, the relation $\frac{\left|\dot{Q}_L\right|}{\hat{T}_L} = \frac{\left|\dot{Q}_H\right|}{\hat{T}_H}$ holds, so that

$$\dot{S}_{gen} = \frac{\left|\dot{Q}_H\right|}{\hat{T}_H}\left[\frac{\hat{T}_H}{T_H}-\frac{\hat{T}_L}{T_L}\right] = 0.467\frac{\text{W}}{\text{K}}\,.$$

As always, the work loss is more interesting than the entropy generation rate. We find the work loss to *external irreversibilities* as

$$\dot{W}_{\text{loss}-ext} = \dot{W}_{C-int} - \dot{W}_C = T_L \dot{S}_{gen} = 0.127\,\text{kW}\,.$$

The actual engine consumes $\left|\dot{W}\right| = 0.5\,\text{kW}$ of power, that is it loses an additional $0.236\,\text{kW}$ to *internal irreversibilities.* Its coefficient of performance, $\text{COP}_{\text{HP}} = \dot{Q}_H \Big/ \left|\dot{W}\right| = 4$, is typical for a commercial heat pump system.

The realistic heat pump system is 4 times more efficient than a resistance heater ($\text{COP}_{\text{RH}} = 1$), but the perfect—i.e., fully reversible—Carnot heat pump is 3.7 times more efficient than the real engine.

We note that the heat required to keep the home at a comfortable temperature needs to be provided since the home loses the same amount of heat through its walls. Better insulation significantly reduces the heat requirement, and thus the heating costs.

Problems

5.1. Heat Engines
An engine that operates between two reservoirs at $T_H = 500\,°\mathrm{C}$ and $T_L = 25\,°\mathrm{C}$ produces 1 MW of power from a heat intake of 2.5 MW. Compute the heat rejected, the thermal efficiency, the entropy generation rate, and the work loss to irreversibilities.

Another engine operates between two reservoirs at $T_H = 1000\,°\mathrm{C}$ and $T_L = 10\,°\mathrm{C}$, and has a thermal efficiency of 45% and its heat rejection rate is 1.76 MW. Compute the power produced, the heat intake rate, the entropy generation rate, and the work loss to irreversibilities.

5.2. Investment Advice
A friend asks whether he should invest in a new start-up company. The company claims to sell a power generation device that produces 12.5 kW of power, takes in 21 kW of heat at a temperature of 800 K and rejects 13 kW of heat at 300 K. What advice do you give? Why?

5.3. More Investment Advice
Another friend asks whether she should invest in a company which claims to make a power generation device that produces 12 kW of power, takes in heat at a temperature of 800 K and rejects 8 kW of heat at 350 K. What advice do you give? Why?

5.4. Your Friends Keep Asking You for Advice
A neighbor would like to have an air conditioning system. He finds a product with the following specifications: For keeping a room at 20 °C when the outside temperature is 30 °C the product consumes 0.5 kW to remove 9 kW of heat. What's your advice here, and why?

5.5. The Perfect Heater?
A relative needs a new heating system. She shows you a flyer from a company marketing baseboard heaters. The flyer claims a 100% heating efficiency. Is that a valid claim? Can your relative find a more efficient alternative? If so, what would it be?

5.6. A Refrigerator
Yet another friend asks whether he should invest in a company which claims to produce a refrigeration device with a COP of 7, that consumes 0.9 kW of power to keep the inside at 4 °C, and rejects of heat to the warm environment at 32 °C. What advice do you give? Why?

5.7. A Heat Pump
An off-the shelf heat pump system has a COP of 3.5 for operation between 30 °C and −10 °C. Determine the entropy generation per kW of heating, and the percentage of consumed power required to overcome irreversibilities.

5.8. Another Heat Pump
A heat pump providing 2 kW of heat operates between the temperatures of 23 °C and −2 °C ; its entropy generation rate is $\dot{S}_{gen} = 1.3\frac{\text{W}}{\text{K}}$. Determine the power needed to drive the heat pump, and its COP.

5.9. Heat Engine with External Irreversibilities
A internally reversible heat engine operates between two reservoirs at 300 K and 400 K; the engine produces 40 kW of power. The heat exchangers between the engine and the reservoirs require a temperature difference of 20 K. Determine the heat exchanged with the two environments, the entropy generated in heat transfer, and the work loss.

5.10. Refrigerator with Internal and External Irreversibilities
In a frozen pizza factory, the freezing compartment is kept at a temperature of −30 °C, while the outside temperature is 25 °C. The cooling system removes 2.25 MW of heat, and consumes 1.5 MW of power. Measurements show that both heat exchangers operate at a temperature difference of 12 °C to their respective environments.

1. Determine the COP of the refrigeration system, and the COP and power requirement of a fully reversible system used for the same cooling purpose.
2. Determine the work losses to internal and external irreversibilities.

5.11. Heat for Cooling
A chemical plant rejects 1 MW of waste heat at 400 °C. Elsewhere in the plant, 5 MW of heat have to be removed from a warehouse at −10 °C. Can the waste heat be used to cool the warehouse when the environment is at 17 °C? If so, how? Give arguments based on 1st and 2nd law, discuss your assumptions.

5.12. Entropy Generation
In an industrial process, a device conducts heat between two hot reservoirs, which are at 200 °C and 400 °C, and the environment at 23 °C. Specifically, the conductor exchanges 4 kW of heat with the hottest reservoir, and 6 kW of heat with the environment. Determine the entropy generation, and the respective work loss.

5.13. Heat in the T-S-Diagram
In a reversible process in a closed system the heat is given as the area below the process curve in the T-S-diagram, $Q_{12} = \int_1^2 TdS$, or, when we divide by mass, $q_{12} = \frac{Q_{12}}{m} = \int_1^2 Tds$. To make use of this formula, one therefore needs temperature as a function of entropy, $T(s)$, for the process.

Consider a reversible process in air, as ideal gas with constant specific heats, for which pressure and temperature are related as $p = p_1 \left(\frac{T}{T_1}\right)^n$ with a constant n. A process of this kind is called a polytropic process.

1. Find the function $T(s, p)$ by inverting the property relation for entropy, $s(T, p)$.
2. Simplify for the polytropic process to obtain $T(s)$.
3. Make a sketch of the curve for polytropic processes with various values of n. How does it change when n gets bigger?
4. Find heat by integration: $q_{12} = \int_1^2 T ds$. Also compute the work per unit mass, $w_{12} = \frac{W_{12}}{m}$.
5. Specify for a polytropic process with $n = 2$ that starts at $20\,°\mathrm{C}$, $7\,\mathrm{bar}$ (state 1) and proceeds until pressure has doubled.

Chapter 6
Properties and Property Relations

6.1 State Properties and Their Relations

The thermodynamic laws contain many state properties, e.g. [SI units in brackets]

T	temperature $[\mathrm{K}]$
p	pressure $[\mathrm{kPa}]$
m	mass $[\mathrm{kg}]$
V	volume $[\mathrm{m}^3]$
$v = V/m$	specific volume $[\frac{\mathrm{m}^3}{\mathrm{kg}}]$
$\rho = \frac{1}{v}$	mass density $[\frac{\mathrm{kg}}{\mathrm{m}^3}]$
$\mathcal{V}$	velocity $[\frac{\mathrm{m}}{\mathrm{s}}]$
u	specific internal energy $[\frac{\mathrm{kJ}}{\mathrm{kg}}]$
$h = u + pv$	specific enthalpy $[\frac{\mathrm{kJ}}{\mathrm{kg}}]$
s	specific entropy $[\frac{\mathrm{kJ}}{\mathrm{kg\,K}}]$

However, only few properties $(T, p, m, V, \mathcal{V})$ can be measured directly, while many of the quantities that appear in the thermodynamic laws $(u, h, s, \ldots)$ cannot be measured directly.

Experience shows that state properties are not independent, but are related through property relations, which depend on the substance. By means of property relations, thermodynamic quantities $(u, h, s, \ldots)$ can be determined indirectly, through measurement of $(T, p, m, V, \mathcal{V})$.

Measurements show that for simple substances it is sufficient to know two properties to find all others. This implies property relations of the form

$p = p(T, v)$	thermal equation of state
$v = v(T, p)$	thermal equation of state
$u = u(T, v)$	caloric equation of state
$h = h(T, p)$	caloric equation of state
$s = s(T, p)$	entropy

H. Struchtrup, *Thermodynamics and Energy Conversion*,
DOI: 10.1007/978-3-662-43715-5_6,

and so on. The thermal and caloric equations of state, $p(T,v)$ and $u(T,v)$, must be determined in careful measurements, where the measurement of the latter relies on the first law. In most cases, the equations of state are not given as explicit equations, but in form of tables. The best known exception is the ideal gas law, $p = RT/v$.

Entropy must be determined from the thermal and caloric equations of state through integration of the Gibbs equation, which gives a differential relation between properties, and holds for *all* simple substances in the form

$$Tds = du + pdv \,, \tag{6.1}$$

or, with $h = u + pv$ and thus $dh = du + pdv + vdp$, in the alternative form

$$Tds = dh - vdp \,. \tag{6.2}$$

Property relations can be formulated between any set of three properties. For instance: Considering the entropy as function of temperature and pressure, $s(T,p)$, together with the thermal equation of state, $p(T,v)$, both can be combined to $s(T,p(T,v)) = s(T,v)$, that is entropy as function of temperature and volume. Inversion of the caloric equation of state $u(T,v)$ for temperature yields temperature as a function of energy and volume, $T(u,v)$. Considering the latter in the entropy expression $s(T,v)$ yields entropy as function of energy and volume, $s(u,v)$. Solving this relation for energy, yields energy as a function of entropy and volume, $u(s,v)$. And so on. These are just some examples of variable changes in property relations. A detailed analysis of property relations, where variable changes are used to identify deeper relations between properties can be found in Chapter 16, where it will be seen that the Gibbs equation substantially reduces the measurements necessary to produce thermodynamic tables.

6.2 Phases

Depending on the conditions, e.g., the values of pressure and temperature, a substance assumes different phases—solid, liquid, vapor—which can also coexist. We shall need property relations for all individual phases as well as for the coexisting states.

Atoms and molecules interact through interatomic potentials $\phi(r)$ of the form depicted in Fig. 6.1. For intermediate particle distances around d, the particles attract each other, while they repel each other when they are pushed very close together ($r < d$). For large distances ($r \gg d$), the particles do not notice each others presence ($\phi(r) \to 0$ for $r \to \infty$).

In a solid, the particles sit at fixed locations in the atomic compound, e.g., a crystal lattice, and oscillate around the minimum of the potential. The interatomic forces are strong, and keep the solid together.

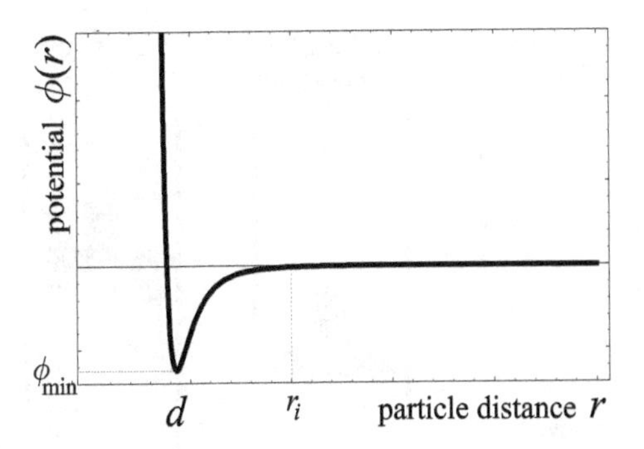

Fig. 6.1 Interparticle potential ϕ as function of interparticle distance r

When the temperature is increased, the oscillations become stronger, and the particles have enough energy to split the molecular bonds with their neighbors, while the attractive forces are still significant. The particles can move freely, but are densely packed with distances close to d. This is the liquid state.

At even higher temperatures the particle energies exceed the attractive potentials which cannot hold the particles together anymore. The particles move fast at greater average distances. This is the gaseous, or vapor, state.

In solid and liquid states, the particles are in permanent contact and interaction. While gas particles have a large average distance, they nevertheless interact through frequent collisions. The interaction between particles leads to microscopic exchange of energy and momentum which facilitates the macroscopic transfer of energy and momentum. The constant redistribution of momentum and energy between particles drives the system towards the equilibrium state.

6.3 Phase Changes

It is a daily experience that matter changes between phases: ice will melt, water will boil and evaporate, dew will condense out of moist air, and so on.

We study the evaporation of liquid water at constant pressure $p = 1\text{atm}$, as depicted in Fig. 6.2. Water is confined in a piston-cylinder system with a moving piston, the mass of the piston fixes the pressure in the system.

We go through the figure from left to right: At temperatures below 100 °C (and above 0 °C) only the liquid phase is found, we speak of *compressed liquid.* Isobaric heat supply increases the temperature of the compressed liquid. When the temperature reaches 100 °C, the water starts to evaporate. Further heat supply does not increase the temperature, which still is 100 °C, but leads to more evaporation. As evaporation occurs, liquid and vapor are in an equilibrium state where both phases coexist, the *saturated state.* The corresponding liquid and vapor states are denoted as *saturated liquid* and

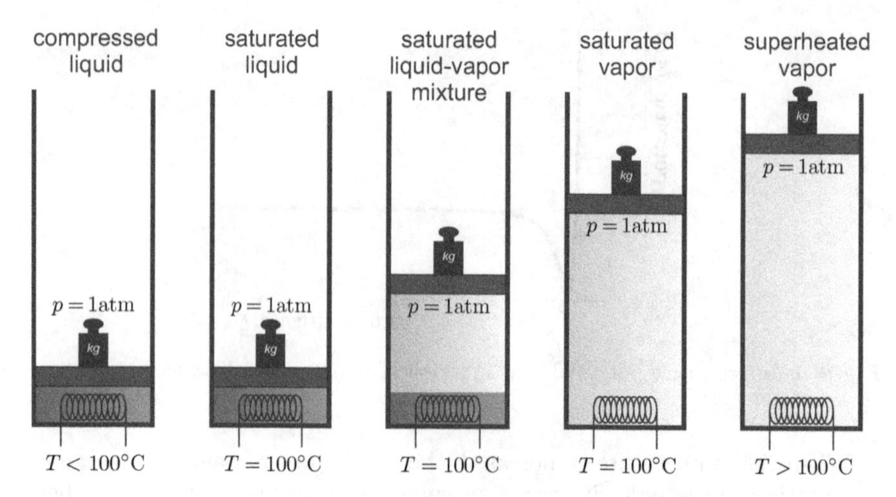

Fig. 6.2 Constant pressure evaporation of water at $p = 1\,\mathrm{atm}$

saturated vapor, respectively. Finally, when all liquid is evaporated, further heat supply increases the temperature of the vapor above 100 °C, we speak of *superheated vapor.*

When heat is withdrawn, the opposite process happens: the superheated vapor will cool down until it reaches 100 °C, then vapor will start to condense. After all vapor is condensed, the compressed liquid cools to lower temperatures.

The *saturation temperature* depends on pressure, we write $T_{\mathrm{sat}}(p)$. The inversion gives the *saturation pressure*, denoted as $p_{\mathrm{sat}}(T)$. In the example we have $T_{\mathrm{sat}}(1\,\mathrm{atm}) = 100\,°\mathrm{C}$ and $p_{\mathrm{sat}}(100\,°\mathrm{C}) = 1\,\mathrm{atm}$. Figure 6.3 shows a sketch of the saturation curve of water in the p-T-diagram. The curve begins in the *triple point* $(611\,\mathrm{Pa}, 0.01\,°\mathrm{C})$ and ends in the *critical point* $(22.09\,\mathrm{MPa},\ 374.14\,°\mathrm{C})$.

For temperatures above the critical temperature, and for pressures above the critical pressure, a saturated liquid-vapor equilibrium is not possible. In the critical point all properties agree between vapor and liquid, and above the critical point only one phase exists, one speaks of *supercritical fluid.*

The triple point gives the lowest temperature/pressure at which a saturated liquid-vapor equilibrium is possible; only at this point all three phases, solid, liquid and vapor, can coexist.

Apart from the liquid-vapor phase change, i.e., evaporation and condensation, one observes the phase changes between solid and liquid, i.e., melting and freezing (solidification), and between solid and vapor, i.e., sublimation and deposition. For each, phase equilibrium is only possible for values of pressure and temperature T and pressure p on the corresponding *saturation curve*, $p_{\mathrm{sat}}(T)$.

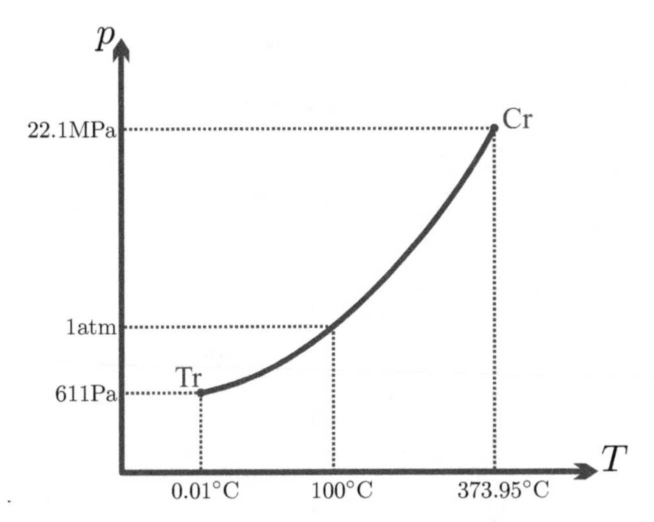

Fig. 6.3 Liquid-vapor saturation curve for water in the p-T-diagram with data for triple point (Tr), critical point (Cr), and the boiling point of water at standard pressure

Figure 6.4 shows the saturation curves for water as ice, liquid, and vapor in a p-T-diagram. Note the large number of different ice phases, which reflect different lattice configurations.[1] Phase equilibria (coexistence of two phases) are only possible on those curves which are given by the saturation pressure $p_{\text{sat}}(T)$ for the respective phase equilibrium, or, alternatively, by the saturation temperature $T_{\text{sat}}(p)$ which is the inverse function. All three phases can coexist in only one point, the triple point. Away from the saturation lines the substance will be in just one of the phases as indicated in the figure. An interesting information that can be drawn from the diagram is that no liquid water exists at temperatures below $-23\,^{\circ}\text{C}$.

A particular feature of water is the negative slope of its melting curve which implies that ice will melt under pressure. This behavior is related to the volume change: A given amount of ice has a larger volume than the same amount of liquid water, as can be seen by ice swimming on water. Melting reduces the volume and thus counteracts the pressure increase. Melting under pressure might play a role in the flow of glaciers, but does not explain the slipperiness of ice, see Sec. 17.12.

Sublimation can be observed in winter, where snow evaporates, in particular on dry sunny days, without melting. An industrial application of

[1] Read about Kurt Vonnegut's fictitious *ice-nine* in his book *Cat's Cradle*. Fortunately, the real ice IX (not included in the diagram) has properties that differ from those fabled by Vonnegut. Everything you want to know about water (including full phase diagrams up to ice XV) can be found on Martin Chaplin's water site at http://www1.lsbu.ac.uk/water.

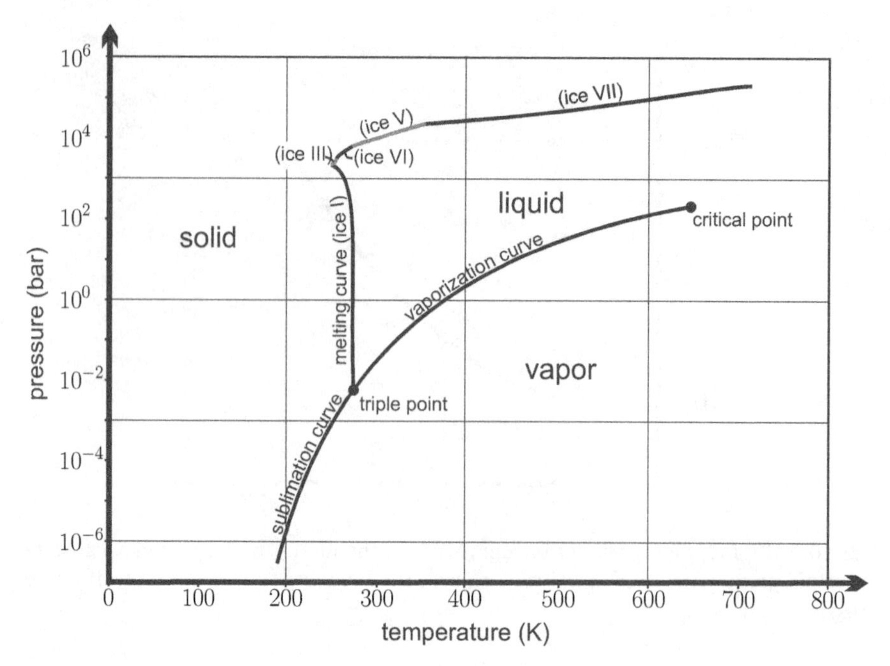

Fig. 6.4 Phase diagram of water (after chart from http://www.chemicalogic.com). Note that the pressure axis is logarithmic.

sublimation is the process of freeze-drying which is used to produce instant coffee: coffee is frozen at a temperature T_C, and then subjected to a pressure p_C below the sublimation pressure, $p_C < p_{\text{sub}}(T_C)$; this forces direct evaporation of ice.

Saturation curves for other substances show the same principal characteristics as those for water, in particular the existence of critical and triple points. However, for almost all other substances the solid has a smaller volume than the liquid, and the solid-liquid line has a positive slope. Figure 6.5 shows p-T-diagrams with the saturation lines for sublimation, melting and vaporization, and indication of the solid, liquid, and vapor regions. For supercritical fluid there is no distinction between liquid and vapor.

Phase changes are related to volume changes. For most substances the volume of the liquid is larger than that of the solid (see the left Fig. 6.5), with water being an exception (see the right Fig. 6.5). Other substances that exhibit expansion on freezing are silicon, gallium and bismuth. Vapor volume is always larger than liquid volume at the same pressure. The volume differences do not become apparent in the p-T-diagram, where the saturated states appear as lines, but in the pressure-volume diagram (p-v-diagram). For a substance that contracts on freezing, such a diagram is sketched in Fig. 6.6. Saturated state lines in the diagram are indicated. There are two lines for saturated liquid, one describes phase equilibrium with saturated vapor, the

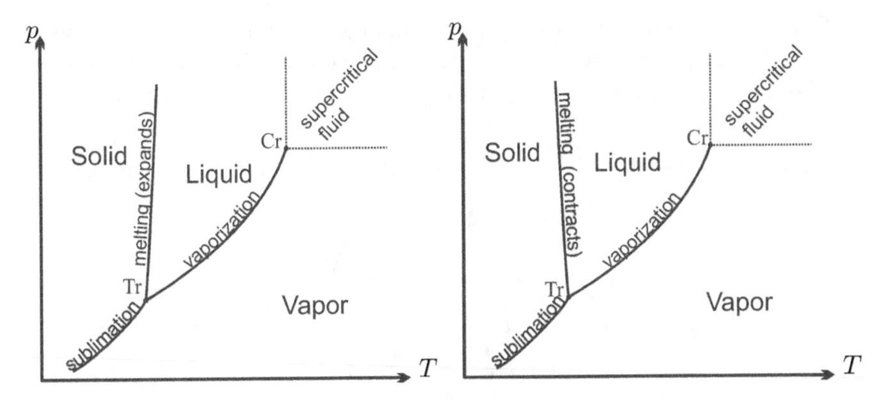

Fig. 6.5 Saturation lines and phases in the p-T-diagram. Left: Ordinary substance, which expands on melting. Right: Water, which contracts on melting.

other phase equilibrium with saturated solid. In the two-phase regions (solid + liquid, liquid + vapor, solid + vapor) one observes mixtures of saturated states, as discussed in the next section. On the triple line, one observes mixtures of all three phases, solid (volume v_s^{tr}), liquid (v_l^{tr}) and vapor (v_v^{tr}) where all three phases are at triple point pressure and temperature, p_{tr}, T_{tr}.

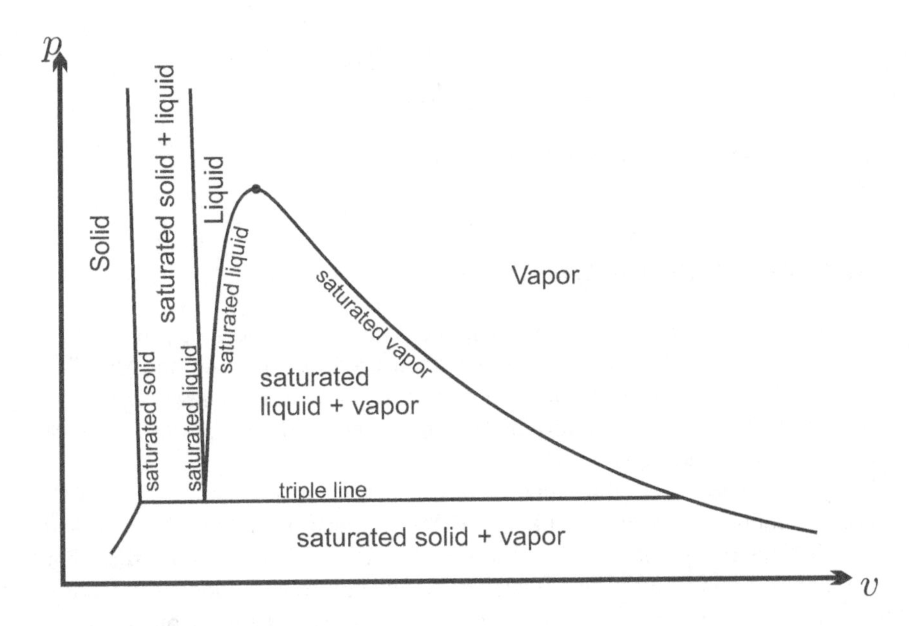

Fig. 6.6 p-v-diagram for an ordinary substance

We could also plot a T-v-diagram, but instead we show, in Fig. 6.7, the p-v-T-surface of an ordinary substance (contracts on freezing). The p-T-, p-v-, and T-v-diagrams are just the appropriate projections of the surface.

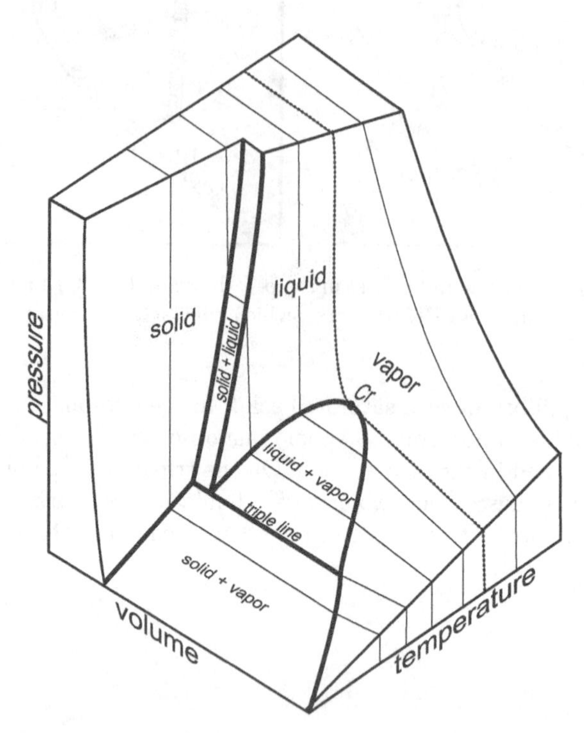

Fig. 6.7 p-v-T-surface of an ordinary substance

6.4 p-v- and T-s-Diagrams

An indispensable tool for thermodynamic analysis are plots of processes in suitable diagrams. The diagrams most often used are the p-v- and the T-s-diagram. For most processes only liquid and vapor or gas phases are encountered, and thus one uses diagrams that only show liquid and vapor states, and the corresponding two-phase region.

Figure 6.8 shows both diagrams including saturation lines and critical point. Isothermal lines (constant temperature) are sketched in the p-v-diagram, and isobaric lines (constant pressure) are sketched in the T-s-diagram. Note that both are horizontal in the two-phase region, where pressure and temperature are related through the saturation equation $p = p_{\text{sat}}(T)$. Obviously, in the p-v-diagram constant pressure lines are horizontal, and constant volume lines are vertical; in the T-s-diagram constant temperature lines are horizontal, and constant entropy lines are vertical.

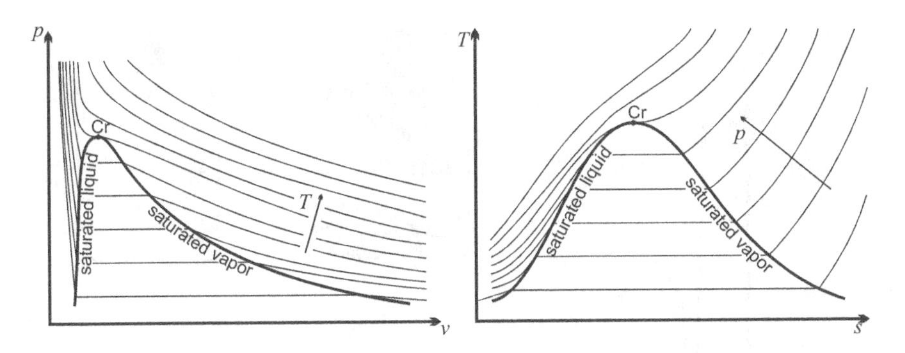

Fig. 6.8 p-v-diagram with two-phase region and isothermal lines (left), and T-s-diagram with two-phase region and isobaric lines (right)

6.5 Saturated Liquid-Vapor Mixtures

For technical applications the most important phase change is that between liquid and vapor; it is, e.g., employed in steam power plants and vapor refrigeration systems. We describe the properties of liquid-vapor mix in detail. Other phase equilibria, e.g., liquid-solid equilibrium, can be treated along the same lines.

We consider a mass m of a substance at temperature T and saturation pressure $p_{\text{sat}}(T)$ in liquid-vapor equilibrium. In phase equilibrium, saturated liquid and vapor can either be separated, with the liquid on the bottom of the container, or they can be mixed, with the liquid dispersed as droplets in the vapor, see Fig. 6.9. The mass of substance in the liquid phase is m_f, and the mass of substance in the vapor phase is m_g, where $m_f + m_g = m$. The use of the indices f (for *fluid*) and and g (for *gaseous*) stems from a time when the word *fluid* was synonymous with *liquid*, while the word today includes gaseous states as well.

The specific volumes of the saturated liquid and vapor are $v_f(T)$ and $v_g(T)$, respectively,[2] and thus the total volume of the saturated mixture is

$$V = m_f v_f + m_g v_g \,. \tag{6.3}$$

The specific volume of the mixture is obtained by division with the total mass,

$$v = \frac{V}{m} = \frac{m_f}{m} v_f + \frac{m_g}{m} v_g = (1 - x)\, v_f + x v_g \,. \tag{6.4}$$

Here, we have introduced

[2] Normally, specific volume is a function of temperature and pressure, $v(T,p)$. For saturated states, however, the pressure is the saturation pressure $p_{\text{sat}}(T)$ which is a function of temperature. Therefore the specific volume of a saturated state is a function only of temperature. The same holds for other specific quantities (energy, enthalpy, entropy).

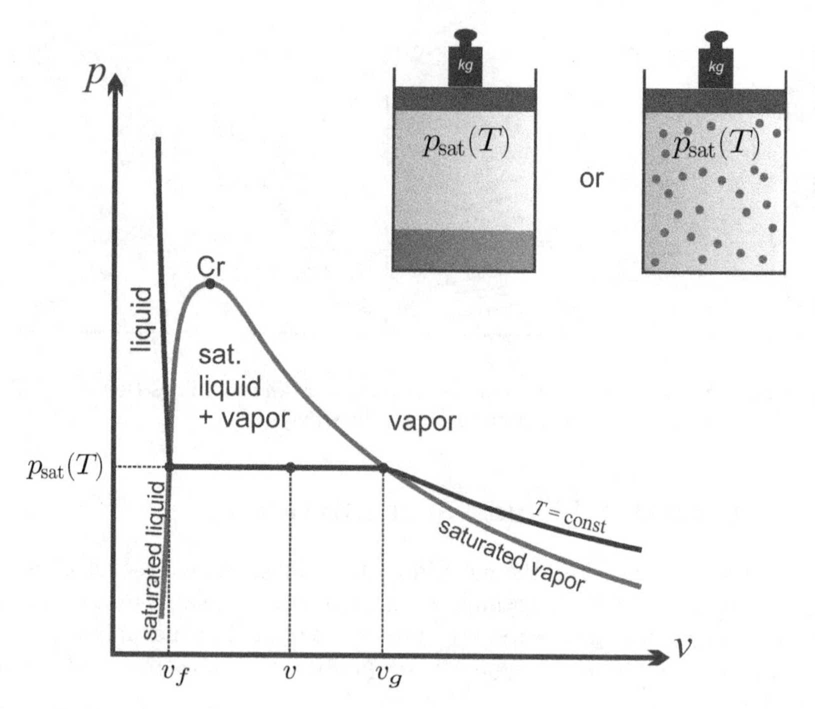

Fig. 6.9 Saturated state in p-v-diagram. The liquid might collect on the container bottom, or might be dispersed as droplets.

$$x = \frac{m_g}{m} = \frac{m_g}{m_f + m_g} \tag{6.5}$$

as the *quality* of the saturated liquid-vapor mixture, defined as the relative mass of saturated vapor. Note that $\frac{m_f}{m} = \frac{m-m_g}{m} = 1 - x$.

Other extensive quantities, e.g., internal energy U, enthalpy H, or entropy S, are computed from the specific properties of the saturated liquid and vapor states just like volume. The specific energy, enthalpy, entropy of the saturated liquid are denoted as $u_f(T)$, $h_f(T)$, $s_f(T)$, and those of the saturated vapor as $u_g(T)$, $h_g(T)$, $s_g(T)$. Total energy, enthalpy, entropy of the mixture are

$$\begin{aligned} U &= m_f u_f + m_g u_g \,, \\ H &= m_f h_f + m_g h_g \,, \\ S &= m_f s_f + m_g s_g \,. \end{aligned} \tag{6.6}$$

The corresponding specific properties, $u = U/m$ etc., are weighted averages,

$$\begin{aligned} v &= (1-x)\, v_f + x v_g \,, \\ u &= (1-x)\, u_f + x u_g = u_f + x u_{fg} \,, \\ h &= (1-x)\, h_f + x h_g = h_f + x h_{fg} \,, \\ s &= (1-x)\, s_f + x s_g = s_f + x s_{fg} \,. \end{aligned} \tag{6.7}$$

Here,

$$u_{fg} = u_g - u_f , \quad h_{fg} = h_g - h_f , \quad s_{fg} = s_g - s_f \tag{6.8}$$

are the energy of vaporization, the enthalpy of vaporization, and the entropy of vaporization. For the quality the above implies the identities

$$x = \frac{m_g}{m_f + m_g} = \frac{v - v_f}{v_g - v_f} = \frac{u - u_f}{u_{fg}} = \frac{h - h_f}{h_{fg}} = \frac{s - s_f}{s_{fg}} \,. \tag{6.9}$$

Property data for saturated states are listed in tables, either ordered by temperature ("temperature table", with $p = p_{sat}(T)$) or by pressure ("pressure table", with $T = T_{sat}(p)$). Figure 6.10 shows an excerpt of a temperature table and Fig. 6.11 shows an excerpt of a pressure table, both for water. Saturation tables for other substances are widely available.

Property data for internal energy and enthalpy is determined from experiments by evaluating the first law, which only allows to determine energy or enthalpy *differences*. Therefore, in designing a property table, one has the freedom to choose the value of a reference energy. For the tables shown, the internal energy of the saturated liquid at the triple point was chosen as $u_f(T_{Tr}) = 0$. All other energy and enthalpy values refer to this choice. Entropy is determined from integration of the Gibbs equation, and one has a choice of an integrating constant, which was chosen here such that, $s_f(T_{Tr}) = 0$. Often, the reference value used in tables is determined from the third law (Sec. 23.6).

Care has to be taken when one uses data from different tables, since these might rely on different choices for the energy and entropy references, which will lead to errors, if not properly corrected.

6.6 Identifying States

Quality can only have values between 0 and 1. If one finds values outside this range, one either has compressed liquid, or superheated vapor.

A state of given temperature T for which another property (v or u or h or s) is known, is compressed liquid for

$$v < v_f(T) \quad \text{or} \quad u < u_f(T) \quad \text{or} \quad h < h_f(T) \quad \text{or} \quad s < s_f(T) \,,$$

and it is superheated vapor if

$$v > v_g(T) \quad \text{or} \quad u > u_g(T) \quad \text{or} \quad h > h_g(T) \quad \text{or} \quad s > s_g(T) \,.$$

Liquid-vapor saturation states of water, temperature table

T	psat	vf	vg	uf	ug	hf	hfg	hg	sf	sfg	sg
deg-C	kPa	m3/kg	m3/kg	kJ/kg	kJ/kg	kJ/kg	kJ/kg	kJ/kg	kJ/kgK	kJ/kgK	kJ/kgK
0.01	0.6113	0.001000	206.14	0.00	2375.3	0.00	2501.4	2501.4	0.0000	9.1562	9.1562
10	1.2276	0.001000	106.38	42.00	2389.2	42.01	2477.8	2519.8	0.1510	8.7498	8.9008
20	2.339	0.001002	57.79	83.95	2402.9	83.96	2454.1	2538.1	0.2966	8.3706	8.6672
30	4.246	0.001004	32.89	125.78	2416.6	125.79	2430.5	2556.3	0.4369	8.0164	8.4533
40	7.384	0.001008	19.52	167.56	2430.1	167.57	2406.7	2574.3	0.5725	7.6845	8.2570
50	12.35	0.001012	12.03	209.32	2443.5	209.33	2382.8	2592.1	0.7038	7.3725	8.0763
60	19.94	0.001017	7.671	251.11	2456.6	251.13	2358.5	2609.6	0.8312	7.0784	7.9096
70	31.19	0.001023	5.042	292.95	2469.6	292.98	2333.8	2626.8	0.9549	6.8004	7.7553
80	47.39	0.001029	3.407	334.86	2482.2	334.91	2300.4	2635.3	1.0753	6.5369	7.6122
90	70.14	0.001036	2.361	376.85	2494.5	376.92	2283.2	2660.1	1.1925	6.2866	7.4791
	MPa										
100	0.10135	0.001044	1.6729	418.94	2506.5	419.04	2257.1	2676.1	1.3069	6.0480	7.3549
110	0.14327	0.001052	1.2102	461.14	2518.1	461.30	2230.2	2691.5	1.4185	5.8202	7.2387
120	0.19853	0.001060	0.89190	503.50	2529.3	503.71	2202.6	2706.3	1.5276	5.6020	7.1296
130	0.2701	0.001070	0.66850	546.02	2539.9	546.31	2174.2	2720.5	1.6344	5.3925	7.0269
140	0.3613	0.001080	0.50890	588.74	2550.0	589.13	2144.8	2733.9	1.7391	5.1908	6.9299
150	0.4758	0.001091	0.39280	631.68	2559.5	632.20	2114.3	2746.5	1.8418	4.9961	6.8379
160	0.6178	0.001102	0.30710	674.87	2568.4	675.55	2082.6	2758.1	1.9427	4.8075	6.7502
170	0.7917	0.001114	0.24280	718.33	2576.5	719.21	2049.5	2768.7	2.0419	4.6244	6.6663
180	1.0021	0.001127	0.19405	762.09	2583.7	763.22	2015.0	2778.2	2.1396	4.4461	6.5857
190	1.2544	0.001141	0.15654	806.19	2590.0	807.62	1978.8	2786.4	2.2359	4.2720	6.5079
200	1.5538	0.001157	0.12736	850.65	2595.3	852.45	1940.8	2793.2	2.3309	4.1014	6.4323
210	1.9062	0.001173	0.10441	895.53	2599.5	897.76	1900.7	2798.5	2.4248	3.9337	6.3585
220	2.318	0.001190	0.08619	940.87	2602.4	943.62	1858.5	2802.1	2.5178	3.7683	6.2861
230	2.795	0.001209	0.07158	986.74	2603.9	990.12	1813.9	2804.0	2.6099	3.6047	6.2146
240	3.344	0.001229	0.05976	1033.21	2604.0	1037.32	1766.5	2803.8	2.7015	3.4422	6.1437
250	3.973	0.001251	0.05013	1080.39	2602.4	1085.36	1716.1	2801.5	2.7927	3.2803	6.0730
260	4.688	0.001276	0.04221	1128.39	2599.0	1134.37	1662.5	2796.9	2.8838	3.1181	6.0019
270	5.499	0.001302	0.03564	1177.36	2593.7	1184.51	1605.2	2789.7	2.9751	2.9550	5.9301
280	6.412	0.001332	0.03017	1227.46	2586.1	1235.99	1543.6	2779.6	3.0668	2.7903	5.8571
290	7.436	0.001366	0.02557	1278.92	2576.0	1289.07	1477.1	2766.2	3.1594	2.6227	5.7821
300	8.581	0.001404	0.02167	1332.0	2563.0	1344.0	1405.0	2749.0	3.2534	2.4511	5.7045
310	9.856	0.001447	0.018350	1387.1	2546.4	1401.3	1326.0	2727.3	3.3493	2.2737	5.6230
320	11.27	0.001499	0.015488	1444.6	2525.5	1461.5	1238.6	2700.1	3.4480	2.0882	5.5362
330	12.85	0.001561	0.012996	1505.3	2498.9	1525.3	1140.6	2665.9	3.5507	1.8910	5.4417
340	14.59	0.001638	0.010797	1570.3	2464.6	1594.2	1027.8	2622.0	3.6594	1.6763	5.3357
350	16.51	0.001740	0.008813	1641.9	2418.4	1670.6	893.3	2563.9	3.7777	1.4335	5.2112
360	18.65	0.001893	0.006945	1725.2	2351.5	1760.5	720.5	2481.0	3.9147	1.1379	5.0526
370	21.03	0.002213	0.004925	1844.0	2228.5	1890.5	441.6	2332.1	4.1106	0.6865	4.7971
374.14	22.09	0.003155	0.003155	2029.6	2029.6	2099.3	0.0	2099.3	4.4298	0.0000	4.4298

source: http://www.thermofluids.net/

Fig. 6.10 Saturation table for water (temperature table)

A state of given pressure p for which another property (v or u or h or s) is known, is compressed liquid for

$$v < v_f(p) \quad \text{or} \quad u < u_f(p) \quad \text{or} \quad h < h_f(p) \quad \text{or} \quad s < s_f(p) \ ,$$

and it is superheated vapor if

$$v > v_g(p) \quad \text{or} \quad u > u_g(p) \quad \text{or} \quad h > h_g(p) \quad \text{or} \quad s > s_g(p) \ .$$

A state of given pressure p and temperature T is compressed liquid for

$$T < T_{\text{sat}}(p) \quad \text{or} \quad p > p_{\text{sat}}(T) \ ,$$

Liquid-vapor saturation states of water, pressure table

p kPa	Tsat deg-C	vf m3/kg	vg m3/kg	uf kJ/kg	ug kJ/kg	hf kJ/kg	hfg kJ/kg	hg kJ/kg	sf kJ/kgK	sfg kJ/kgK	sg kJ/kgK
0.6113	0.01	0.001000	206.14	0.00	2375.3	0.00	2501.4	2501.4	0.0000	9.1562	9.1562
1	6.98	0.001000	129.21	29.30	2385.0	29.30	2484.9	2514.2	0.1059	8.8697	8.9756
2	17.50	0.001001	67.00	73.48	2399.5	73.48	2460.0	2533.5	0.2607	8.4630	8.7237
3	24.08	0.001003	45.67	101.04	2408.5	101.05	2444.5	2545.5	0.3545	8.2231	8.5776
5	32.88	0.001005	28.19	137.81	2420.5	137.82	2423.7	2561.5	0.4764	7.9187	8.3951
7.5	40.29	0.001008	19.24	168.78	2430.5	168.79	2406.0	2574.8	0.5764	7.6751	8.2515
10	45.81	0.001010	14.67	191.82	2437.9	191.83	2392.9	2584.7	0.6493	7.5009	8.1502
20	60.06	0.001017	7.649	251.38	2456.7	251.40	2358.3	2609.7	0.8320	7.0765	7.9085
30	69.10	0.001022	5.229	289.20	2468.4	289.23	2336.1	2625.3	0.9439	6.8247	7.7686
50	81.33	0.001030	3.240	340.44	2483.9	340.49	2305.4	2645.9	1.0910	6.5029	7.5939
75	91.78	0.001037	2.217	384.31	2496.7	384.39	2278.6	2663.0	1.2130	6.2434	7.4564
MPa											
0.100	99.63	0.001043	1.694	417.36	2506.1	417.46	2258.0	2675.5	1.3026	6.0568	7.3594
0.150	111.37	0.001053	1.1593	466.94	2519.7	467.11	2226.5	2693.6	1.4336	5.7897	7.2233
0.200	120.23	0.001061	0.8857	504.49	2529.5	504.70	2202.0	2706.7	1.5301	5.5970	7.1271
0.250	127.44	0.001067	0.7187	535.10	2537.2	535.37	2181.5	2716.9	1.6072	5.4455	7.0527
0.300	133.55	0.001073	0.6058	561.15	2543.6	561.47	2163.8	2725.3	1.6718	5.3201	6.9919
0.350	138.88	0.001079	0.5243	583.95	2548.9	584.33	2148.1	2732.4	1.7275	5.2130	6.9405
0.400	143.63	0.001084	0.4625	604.31	2553.6	604.74	2133.9	2738.6	1.7766	5.1193	6.8959
0.500	151.86	0.001093	0.3749	639.68	2561.2	640.23	2108.5	2748.7	1.8607	4.9606	6.8213
0.600	158.85	0.001101	0.3157	669.90	2567.4	670.56	2086.2	2756.8	1.9312	4.8288	6.7600
0.700	164.97	0.001108	0.2729	696.44	2572.5	697.22	2066.3	2763.5	1.9922	4.7158	6.7080
0.800	170.43	0.001115	0.2404	720.22	2576.8	721.11	2048.0	2769.1	2.0462	4.6166	6.6628
0.900	175.38	0.001121	0.2150	741.83	2580.5	742.83	2031.1	2773.9	2.0946	4.5280	6.6226
1.0	179.91	0.001127	0.19444	761.68	2583.6	762.81	2015.3	2778.1	2.1387	4.4478	6.5865
1.5	198.32	0.001154	0.13177	843.16	2594.5	844.89	1947.3	2792.2	2.3150	4.1298	6.4448
2.0	212.42	0.001177	0.09963	906.44	2600.3	908.79	1890.7	2799.5	2.4474	3.8935	6.3409
3.0	233.90	0.001217	0.06668	1004.78	2604.1	1008.42	1795.8	2804.2	2.6457	3.5412	6.1869
3.5	242.60	0.001235	0.05707	1045.43	2603.7	1049.75	1753.7	2803.4	2.7253	3.4000	6.1253
4.0	250.40	0.001252	0.04978	1082.31	2602.3	1087.31	1714.1	2801.4	2.7964	3.2737	6.0701
6.0	275.64	0.001319	0.03244	1205.44	2589.7	1213.35	1571.0	2784.3	3.0267	2.8625	5.8892
8.0	295.06	0.001384	0.02352	1305.57	2569.8	1316.64	1441.4	2758.0	3.2068	2.5364	5.7432
10	311.06	0.001452	0.018026	1393.04	2544.4	1407.56	1317.1	2724.7	3.3596	2.2545	5.6141
12	324.75	0.001527	0.014263	1473.0	2513.7	1491.3	1193.6	2684.9	3.4962	1.9962	5.4924
14	336.75	0.001611	0.011485	1548.6	2476.8	1571.1	1066.5	2637.6	3.6232	1.7485	5.3717
16	347.44	0.001711	0.009306	1622.7	2431.7	1650.1	930.5	2580.6	3.7461	1.4994	5.2455
18	357.06	0.001840	0.007489	1698.9	2374.3	1732.0	777.1	2509.1	3.8715	1.2329	5.1044
20	365.81	0.002036	0.005834	1785.6	2293.0	1826.3	583.4	2409.7	4.0139	0.9130	4.9269
22.09	374.14	0.003155	0.003155	2029.6	2029.6	2099.3	0.0	2099.3	4.4298	0.0000	4.4298

source: http://www.thermofluids.net/

Fig. 6.11 Saturation table for water (pressure table)

and it is superheated vapor if

$$T > T_{\text{sat}}(p) \quad \text{or} \quad p < p_{\text{sat}}(T) \ .$$

It is a useful exercise to verify the above conditions by means of p-v-, T-s-, and p-T-diagrams!

6.7 Example: Condensation of Saturated Steam

As an example we consider the isochoric (constant volume) condensation of saturated steam from an initial temperature of $T_1 = 280\,°\text{C}$ to $T_2 = 200\,°\text{C}$. In the initial state, the properties are just at the saturation values, which can be read from Fig. 6.10 as

$$p_1 = p_{\text{sat}}(T_1) = 64.12\,\text{bar}\,, \quad v_1 = v_g(T_1) = 0.03017\frac{\text{m}^3}{\text{kg}}\,,$$
$$u_1 = u_g(T_1) = 2586.1\frac{\text{kJ}}{\text{kg}}\,, \quad h_1 = h_g(T_1) = 2779.6\frac{\text{kJ}}{\text{kg}}\,,$$
$$s_1 = s_g(T_1) = 5.8571\frac{\text{kJ}}{\text{kg}\,\text{K}}\,.$$

The values of two properties—two bits of information—are required to fix a state. In state 1 these are the temperature and the knowledge that the steam is saturated. For state 2, we know its temperature T_2, and its volume, which is unchanged, $v_2 = v_1$. To learn more about the final state, it is best to draw the process into a p-v-diagram. As shown in Fig. 6.12, the isochoric process to lower temperature is a vertical line downwards from the saturated vapor curve, and the final state 2 lies in the two-phase region between the saturation lines. Hence, this state is a mixture of saturated liquid at volume $v_f(T_2)$, and saturated vapor at volume $v_g(T_2)$, which we find from the table as $v_f(T_2) = 0.001157\frac{\text{m}^3}{\text{kg}}$ and $v_g(T_2) = 0.12736\frac{\text{m}^3}{\text{kg}}$.

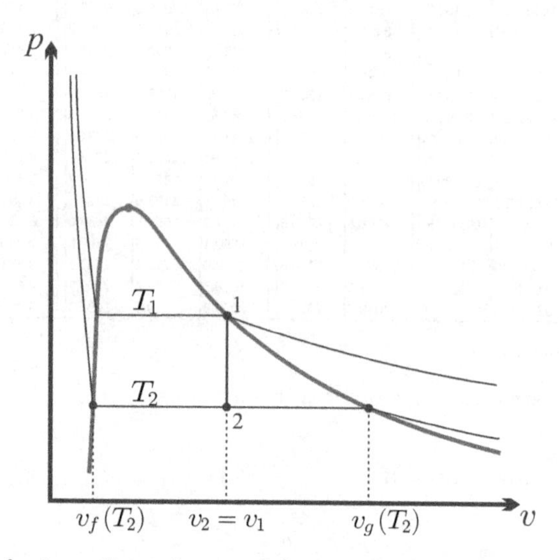

Fig. 6.12 Isochoric cooling of saturated vapor between T_1 and T_2 in the p-v-diagram. The final state 2 is in the two-phase region (mixture of saturated liquid and saturated vapor).

Since $v_2 = v_1 = v_g(T_1)$, the quality of the final state is

$$x_2 = \frac{v_2 - v_f(T_2)}{v_g(T_2) - v_f(T_2)} = \frac{0.03017 - 0.001157}{0.12736 - 0.001157} = 0.23\,.$$

With this value for quality we find the properties at the end point as

$$
\begin{aligned}
p_2 &= p_{\text{sat}}(T_2) = 15.54\,\text{bar}\,, \\
v_2 &= v_1 = 0.03017\frac{\text{m}^3}{\text{kg}}\,, \\
u_2 &= u_f(T_2) + x_2 u_{fg}(T_2) = 1251.7\frac{\text{kJ}}{\text{kg}}\,, \\
h_2 &= h_f(T_2) + x_2 h_{fg}(T_2) = 1298.6\frac{\text{kJ}}{\text{kg}}\,, \\
s_2 &= s_f(T_2) + x_2 s_{fg}(T_2) = 3.274\frac{\text{kJ}}{\text{kg K}}\,.
\end{aligned}
$$

The values for $u_f(T_2), u_{fg}(T_2)$ etc. are taken from the table. The verification of the above results is left to the reader.

We recall that quality *must* have values between 0 and 1. If one computes a quality outside this range, the corresponding state is *not* a saturated state, but either compressed liquid or superheated vapor, for which the property data must be found in the appropriate tables.

6.8 Superheated Vapor

For superheated vapors the equations of state depend on two properties, and are normally laid down in extensive tables, or in computer software. Figure 6.13 shows an excerpt of a table with data for water vapor at some pressures between 10 kPa and 20 MPa.

As an example we consider the adiabatic reversible compression of saturated vapor at $T_1 = 100\,°\text{C}$ to a pressure $p_2 = 3\,\text{MPa}$. From the second law for reversible processes, $\delta q = Tds$ follows that such a process is isentropic (constant entropy), and thus it is a natural choice to draw the process curve in a T-s-diagram as depicted in Fig. 6.14. Clearly, the final state 2 is outside the two phase region, to the right, which means the final state is superheated vapor. The properties of state 1 can be read from the saturation table in Fig. 6.10 as

$$
\begin{aligned}
p_1 &= p_{\text{sat}}(T_1) = 1.014\,\text{bar}\,, \\
v_1 &= v_g(T_1) = 1.673\frac{\text{m}^3}{\text{kg}}\,, \\
u_1 &= u_g(T_1) = 2506.5\frac{\text{kJ}}{\text{kg}}\,, \\
h_1 &= h_g(T_1) = 2676.1\frac{\text{kJ}}{\text{kg}}\,, \\
s_1 &= s_g(T_1) = 7.3549\frac{\text{kJ}}{\text{kg K}}\,.
\end{aligned}
$$

superheated water vapor

deg-C	m3/kg	kJ/kg	kJ/kg	kJ/kg K	m3/kg	kJ/kg	kJ/kg	kJ/kg K	m3/kg	kJ/kg	kJ/kg	kJ/kg K
	p = 0.01 MPa (45.81 °C)				p = 0.10 MPa (99.63 °C)				p = 1.00 MPa (179.91 °C)			
T	v	u	h	s	v	u	h	s	v	u	h	s
Sat.	14.674	2437.9	2584.7	8.1502	1.694	2506.1	2675.5	7.3594	0.19444	2583.6	2778.1	6.5865
50	14.869	2443.9	2592.6	8.1749								
100	17.196	2515.5	2687.5	8.4479	1.696	2506.7	2676.2	7.3614				
150	19.512	2587.9	2783.0	8.6882	1.936	2582.8	2776.4	7.6143				
200	21.825	2661.3	2879.5	8.9038	2.172	2658.1	2875.3	7.8343	0.2060	2621.9	2827.9	6.6940
250	24.136	2736.0	2977.3	9.1002	2.406	2733.7	2974.3	8.0333	0.2327	2709.9	2942.6	6.9247
300	26.445	2812.1	3076.5	9.2813	2.639	2810.4	3074.3	8.2158	0.2579	2793.2	3051.2	7.1229
400	31.063	2968.9	3279.6	9.6077	3.103	2967.9	3278.2	8.5435	0.3066	2957.3	3263.9	7.4651
500	35.679	3132.3	3489.1	9.8978	3.565	3131.6	3488.1	8.8342	0.3541	3124.4	3478.5	7.7622
600	40.295	3302.5	3705.4	10.1608	4.028	3301.9	3704.4	9.0976	0.4011	3296.8	3697.9	8.0290
700	44.911	3479.6	3928.7	10.4028	4.490	3479.2	3928.2	9.3398	0.4478	3475.3	3923.1	8.2731
800	49.526	3663.8	4159.0	10.6281	4.952	3663.5	4158.6	9.5652	0.4943	3660.4	4154.7	8.4996
900	54.141	3855.0	4396.4	10.8396	5.414	3854.8	4396.1	9.7767	0.5407	3852.2	4392.9	8.7118
1000	58.757	4053.0	4640.6	11.0393	5.875	4052.8	4640.3	9.9764	0.5871	4050.5	4637.6	8.9119
1100	63.372	4257.5	4891.2	11.2287	6.337	4257.3	4891.0	10.1659	0.6335	4255.1	4888.6	9.1017
1200	67.987	4467.9	5147.8	11.4091	6.799	4467.7	5147.6	10.3463	0.6798	4465.6	5145.4	9.2822
1300	72.602	4683.7	5409.7	11.5811	7.260	4683.5	5409.5	10.5183	0.7261	4681.3	5407.4	9.4543

	p = 2.00 MPa (212.42 °C)				p = 3.00 MPa (233.90 °C)				p = 5.0 MPa (263.99 °C)			
T	v	u	h	s	v	u	h	s	v	u	h	s
Sat.	0.09963	2600.3	2799.5	6.3409	0.06668	2604.1	2804.2	6.1869	0.03944	2597.1	2794.3	5.9734
225	0.10377	2628.3	2835.8	6.4147								
250	0.11144	2679.6	2902.5	6.5453	0.07058	2644.0	2855.8	6.2872				
300	0.12547	2772.6	3023.5	6.7664	0.08114	2750.1	2993.5	6.5390	0.04532	2698.0	2924.5	6.2084
350	0.13857	2859.8	3137.0	6.9563	0.09053	2843.7	3115.3	6.7428	0.05194	2808.7	3068.4	6.4493
400	0.15120	2945.2	3247.6	7.1271	0.09936	2932.8	3230.9	6.9212	0.05781	2906.6	3195.7	6.6459
500	0.17568	3116.2	3467.6	7.4317	0.11619	3108.0	3456.5	7.2338	0.06857	3091.0	3433.8	6.9759
600	0.19960	3290.9	3690.1	7.7024	0.13243	3285.0	3682.3	7.5085	0.07869	3273.0	3666.5	7.2589
700	0.2232	3470.9	3917.4	7.9487	0.14838	3466.5	3911.7	7.7571	0.08849	3457.6	3900.1	7.5122
800	0.2467	3657.0	4150.3	8.1765	0.16414	3653.5	4145.9	7.9862	0.09811	3646.6	4137.1	7.7440
900	0.2700	3849.3	4389.4	8.3895	0.17980	3846.5	4385.9	8.1999	0.10762	3840.7	4378.8	7.9593
1000	0.2933	4048.0	4634.6	8.5901	0.19541	4045.4	4631.6	8.4009	0.11707	4040.4	4625.7	8.1612
1100	0.3166	4252.7	4885.9	8.7800	0.21098	4250.3	4883.3	8.5912	0.12648	4245.6	4878.0	8.3520
1200	0.3398	4463.3	5142.9	8.9607	0.22652	4460.9	5140.5	8.7720	0.13587	4456.3	5135.7	8.5331
1300	0.3631	4679.0	5405.1	9.1329	0.24206	4676.6	5402.8	8.9442	0.14526	4672.0	5398.2	8.7055

	p = 8.0 MPa (295.06 °C)				p = 12.5 MPa (327.89 °C)				p = 20.0 MPa (365.81 °C)			
T	v	u	h	s	v	u	h	s	v	u	h	s
Sat.	0.02352	2569.8	2758.0	5.7432	0.013495	2505.1	2673.8	5.4624	0.005834	2293.0	2409.7	4.9269
300	0.02426	2590.9	2785.0	5.7906								
350	0.02995	2747.7	2987.3	6.1301	0.016126	2624.6	2826.2	5.7118				
400	0.03432	2863.8	3138.3	6.3634	0.02000	2789.3	3039.3	6.0417	0.009942	2619.3	2818.1	5.5540
450	0.03817	2966.7	3272.0	6.5551	0.02299	2912.5	3199.8	6.2719	0.012695	2806.2	3060.1	5.9017
500	0.04175	3064.3	3398.3	6.7240	0.02560	3021.7	3341.8	6.4618	0.014768	2942.9	3238.2	6.1401
550	0.04516	3159.8	3521.0	6.8778	0.02801	3125.0	3475.2	6.6290	0.016555	3062.4	3393.5	6.3348
600	0.04845	3254.4	3642.0	7.0206	0.03029	3225.4	3604.0	6.7810	0.018178	3174.0	3537.6	6.5048
700	0.05481	3443.9	3882.4	7.2812	0.03460	3422.9	3855.3	7.0536	0.02113	3386.4	3809.0	6.7993
800	0.06097	3636.0	4123.8	7.5173	0.03869	3620.0	4103.6	7.2965	0.02385	3592.7	4069.7	7.0544
900	0.06702	3832.1	4368.3	7.7351	0.04267	3819.1	4352.5	7.5182	0.02645	3797.5	4326.4	7.2830
1000	0.07301	4032.8	4616.9	7.9384	0.04658	4021.6	4603.8	7.7237	0.02897	4003.1	4582.5	7.4925
1100	0.07896	4238.6	4870.3	8.1300	0.05045	4228.2	4858.8	7.9165	0.03145	4211.3	4840.2	7.6874
1200	0.08489	4449.5	5128.5	8.3115	0.05430	4439.3	5118.0	8.0937	0.03391	4422.8	5101.0	7.8707
1300	0.09080	4665.0	5391.5	8.4842	0.05813	4654.8	5381.4	8.2717	0.03636	4638.0	5365.1	8.0442

source: http://www.thermofluids.net/

Fig. 6.13 Excerpt from a property table for superheated water vapor for a variety of pressures. The temperature in brackets is the saturation temperature.

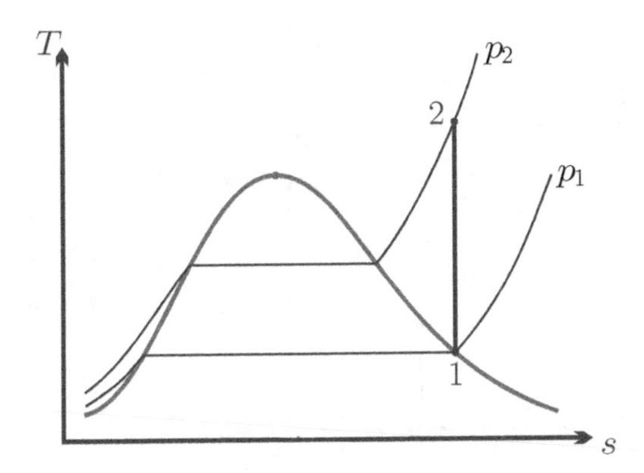

Fig. 6.14 Isentropic compression of saturated vapor from p_1 to p_2 in the T-s - diagram

Two bits of information are required to identify state 2, and here these are its pressure, p_2, and its entropy, since the process is isentropic, $s_2 = s_1 = 7.3549\frac{\text{kJ}}{\text{kg K}}$. In the table for superheated water vapor, Fig. 6.13, we have to consider the center box which refers to the pressure 3 MPa. The required value for entropy cannot be found in the table, but lies between values given. The values closest above and below the required value of $s_2 = 7.3549\frac{\text{kJ}}{\text{kg K}}$ in the table are

$$s_a = s\left(p_2 = 3\,\text{MPa}, T_a = 500\,^\circ\text{C}\right) = 7.2338\frac{\text{kJ}}{\text{kg K}}\,,$$

$$s_b = s\left(p_2 = 3\,\text{MPa}, T_b = 600\,^\circ\text{C}\right) = 7.5085\frac{\text{kJ}}{\text{kg K}}\,.$$

Figure 6.15 shows a sketch of the function $s\left(p_2, T\right)$ in a diagram, with the tabled data points s_a, s_b and the target point s_2 indicated. Assuming that the line $a - 2 - b$ can be well approximated by a straight line, we find the target temperature T_2 by linear interpolation as

$$T_2 = T_a + \frac{s_2 - s_a}{s_b - s_a}\left(T_b - T_a\right) = 543.1\,^\circ\text{C}\,.$$

Correspondingly, the values for volume, internal energy, and enthalpy are computed by interpolation as

$$v_2 = v_a + \frac{s_2 - s_a}{s_b - s_a}\left(v_b - v_a\right) = 0.12337\frac{\text{m}^3}{\text{kg}}\,,$$

$$u_2 = u_a + \frac{s_2 - s_a}{s_b - s_a}\left(u_b - u_a\right) = 3186.0\frac{\text{kJ}}{\text{kg}}\,,$$

$$h_2 = h_a + \frac{s_2 - s_a}{s_b - s_a}\left(h_b - h_a\right) = 3556.0\frac{\text{kJ}}{\text{kg}}\,.$$

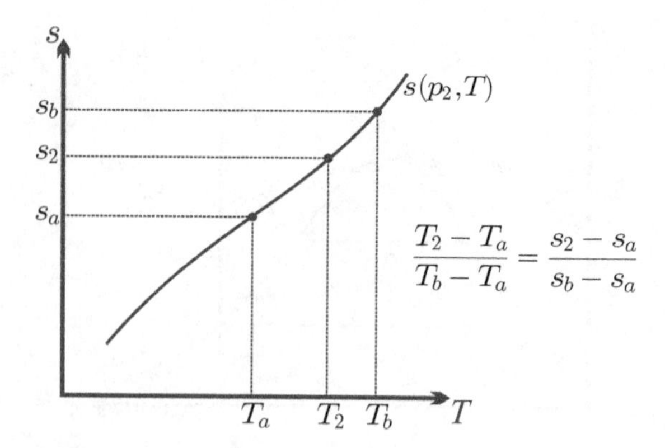

Fig. 6.15 Linear interpolation

Here, $v_a = v(p_2, T_a)$, $u_a = u(p_2, T_a)$ etc. are the appropriate data values from the table.

Since the process is adiabatic, we have $q_{12} = 0$ and the work per unit mass can be computed from the first law as $w_{12} = u_1 - u_2 + q_{12} = -679.5 \frac{\text{kJ}}{\text{kg}}$.

Thermodynamic properties are often listed in tables as discrete values, and interpolation must be frequently used. Typically, tabulated values are spaced such that the assumption of linearity is valid in good approximation.

6.9 Compressed Liquid

For compressed liquid, i.e., the pure liquid state, only few tables are available, Fig. 6.16 shows a table for compressed liquid water.

Most liquids, including water, are almost incompressible for a wider range of pressures, and this allows us to develop useful approximations that relate compressed liquid properties to those of saturated liquid.

For incompressible fluids a change of pressure does not lead to a change of volume, so that the volume can be approximated by the volume of the saturated liquid,

$$v(T,p) \simeq v(T) \simeq v_f(T) \,, \tag{6.10}$$

that is the volume is independent of pressure, but not of temperature. Incompressibility refers to changes at constant temperature, while thermal expansion or contraction are allowed. With this approximation, isothermal lines for the compressed liquid in the p-v-diagram are vertical lines upwards from the saturated liquid line.

Internal energy seen as a function of temperature and volume can then be reduced to its saturated liquid value as well:

$$u(T,v) \simeq u(T, v_f(T)) = u_f(T) \,. \tag{6.11}$$

Compressed Liquid Water (H2O) Table

deg-C	m3/kg	kJ/kg	kJ/kg	kJ/kg K	m3/kg	kJ/kg	kJ/kg	kJ/kg K	m3/kg	kJ/kg	kJ/kg	kJ/kg K
	p = 5 MPa (263.99 C)				p =10 MPa (311.06 C)				p = 15 MPa (342.24 C)			
T	v	u	h	s	v	u	h	s	v	u	h	s
sat.	0.0012859	1147.8	1154.2	2.9202	0.0014524	1393.0	1407.6	3.3596	0.0016581	1585.6	1610.5	3.6848
0	0.0009977	0.0	5.0	0.0001	0.0009952	0.1	10.0	0.0002	0.0009928	0.2	15.1	0.0004
20	0.0009995	83.7	88.7	0.2956	0.0009972	83.4	93.3	0.2945	0.0009950	83.1	98.0	0.2934
40	0.0010056	167.0	172.0	0.5705	0.0010034	166.4	176.4	0.5686	0.0010013	165.8	180.8	0.5666
60	0.0010149	250.2	255.3	0.8285	0.0010127	249.4	259.5	0.8258	0.0010105	248.5	263.7	0.8232
80	0.0010268	333.7	338.9	1.0720	0.0010245	332.6	342.8	1.0688	0.0010222	331.5	346.8	1.0656
100	0.0010410	417.5	422.7	1.3030	0.0010385	416.1	426.5	1.2992	0.0010361	414.7	430.3	1.2955
120	0.0010576	501.8	507.1	1.5233	0.0010549	500.1	510.6	1.5189	0.0010522	498.4	514.2	1.5145
140	0.0010768	586.8	592.2	1.7343	0.0010737	584.7	595.4	1.7292	0.0010707	582.7	598.7	1.7242
160	0.0010988	672.6	678.1	1.9375	0.0010953	670.1	681.1	1.9317	0.0010918	667.7	684.1	1.9260
180	0.0011240	759.6	765.3	2.1341	0.0011199	756.7	767.8	2.1275	0.0011159	753.8	770.5	2.1210
200	0.0011530	848.1	853.9	2.3255	0.0011480	844.5	856.0	2.3178	0.0011433	841.0	858.2	2.3104
220	0.0011866	938.4	944.4	2.5128	0.0011805	934.1	945.9	2.5039	0.0011748	929.9	947.5	2.4953
240	0.0012264	1031.4	1037.5	2.6979	0.0012187	1026.0	1038.1	2.6872	0.0012114	1020.8	1039.0	2.6771
260	0.0012749	1127.9	1134.3	2.8830	0.0012645	1121.1	1133.7	2.8699	0.0012550	1114.6	1133.4	2.8576
280					0.0013216	1220.9	1234.1	3.0548	0.0013084	1212.5	1232.1	3.0393
300					0.0013972	1328.4	1342.3	3.2469	0.0013770	1316.6	1337.3	3.2260
320									0.0014724	1431.1	1453.2	3.4247
340									0.0016311	1567.5	1591.9	3.6546

	p = 20 MPa (365.81 C)				p = 30 Mpa				p = 50 MPa			
T	v	u	h	s	v	u	h	s	v	u	h	s
sat.	0.0020360	1785.6	1826.3	4.0139								
0	0.0009904	0.2	20.0	0.0004	0.0009856	0.3	29.8	0.0001	0.0009766	0.2	49.0	0.0014
20	0.0009928	82.8	102.6	0.2923	0.0009886	82.2	111.8	0.2899	0.0009804	81.0	130.0	0.2848
40	0.0009992	165.2	185.2	0.5646	0.0009951	164.0	193.9	0.5607	0.0009872	161.9	211.2	0.5527
60	0.0010084	247.7	267.9	0.8206	0.0010042	246.1	276.2	0.8154	0.0009962	243.0	292.8	0.8052
80	0.0010199	330.4	350.8	1.0624	0.0010156	328.3	358.8	1.0561	0.0010073	324.3	374.7	1.0440
100	0.0010337	413.4	434.1	1.2917	0.0010290	410.8	441.7	1.2844	0.0010201	405.9	456.9	1.2703
120	0.0010496	496.8	517.8	1.5102	0.0010445	493.6	524.9	1.5018	0.0010348	487.7	539.4	1.4857
140	0.0010678	580.7	602.0	1.7193	0.0010621	576.9	608.8	1.7098	0.0010515	569.8	622.4	1.6915
160	0.0010885	665.4	687.1	1.9204	0.0010821	660.8	693.3	1.9096	0.0010703	652.4	705.9	1.8891
180	0.0011120	751.0	773.2	2.1147	0.0011047	745.6	778.7	2.1024	0.0010912	735.7	790.3	2.0794
200	0.0011388	837.7	860.5	2.3031	0.0011302	831.4	865.3	2.2893	0.0011146	819.7	875.5	2.2634
220	0.0011695	925.9	949.3	2.4870	0.0011590	918.3	953.1	2.4711	0.0011408	904.7	961.7	2.4419
240	0.0012046	1016.0	1040.0	2.6674	0.0011920	1006.9	1042.6	2.649	0.0011702	990.7	1049.2	2.6158
260	0.0012462	1108.6	1133.5	2.8459	0.0012303	1097.4	1134.3	2.8243	0.0012034	1078.1	1138.2	2.7860
280	0.0012965	1204.7	1230.6	3.0248	0.0012755	1190.7	1229.0	2.9986	0.0012415	1167.2	1229.3	2.9537
300	0.0013596	1306.1	1333.3	3.2071	0.0013307	1287.9	1327.8	3.1741	0.0012860	1258.7	1323.0	3.1200
320	0.0014437	1415.7	1444.6	3.3979	0.0013997	1390.7	1432.7	3.3539	0.0013388	1353.3	1420.2	3.2868
340	0.0015684	1539.7	1571.0	3.6075	0.0014920	1501.7	1546.5	3.5426	0.0014032	1452.0	1522.1	3.4557
360	0.0018226	1702.8	1739.3	3.8772	0.0016265	1626.6	1675.4	3.7494	0.0014838	1556.0	1630.2	3.6291
380					0.0018691	1781.4	1837.5	4.0012	0.0015884	1667.2	1746.6	3.8101

source: http://www.thermofluids.net/

Fig. 6.16 Excerpt from a property table for compressed liquid water for a variety of pressures. The temperature in brackets is the saturation temperature.

For consistency, enthalpy needs to be treated differently. Due to the definition $h = u + pv$, the above approximations give in a first step $h(T,p) = u_f(T) + pv_f(T)$. For the saturated liquid at the same temperature we have $h_f(T) = u_f(T) + p_{\rm sat}(T)\, v_f(T)$. Combining both by eliminating $u_f(T)$, we find the approximation for the enthalpy of compressed liquid as

$$h(T,p) \simeq h_f(T) + (p - p_{sat}(T))\, v_f(T) \ . \tag{6.12}$$

For small enough pressures, the correction term for enthalpy can be ignored, so that $h(T,p) \simeq h_f(T)$.

Finally, entropy can be treated similar to internal energy,

$$s(T,v) \simeq s(T, v_f(T)) = s_f(T) \ . \tag{6.13}$$

With this approximation, isobaric lines for the compressed liquid in the T-s-diagram lie on the saturated liquid line.

As an example, we consider compressed liquid water at $p = 10\,\text{MPa}$ and $T = 200\,°\text{C}$, for which the table in Fig. 6.16 gives

$$v(T,p) = v(200\,°\text{C}, 10\,\text{MPa}) = 0.001148\frac{\text{m}^3}{\text{kg}} ,$$

$$u(T,p) = u(200\,°\text{C}, 10\,\text{MPa}) = 844.5\frac{\text{kJ}}{\text{kg}} ,$$

$$h(T,p) = h(200\,°\text{C}, 10\,\text{MPa}) = 856.0\frac{\text{kJ}}{\text{kg}} ,$$

$$s(T,p) = s(200\,°\text{C}, 10\,\text{MPa}) = 2.3178\frac{\text{kJ}}{\text{kg}\,\text{K}} .$$

With the above approximations, we find the corresponding values from the saturation table in Fig. 6.10 as

$$v(T,p) \simeq v_f(T) = v_f(200\,°\text{C}) = 0.001157\frac{\text{m}^3}{\text{kg}} ,$$

$$u(T,v) \simeq u_f(T) = u_f(200\,°\text{C}) = 850.65\frac{\text{kJ}}{\text{kg}} ,$$

$$h(T,p) \simeq h_f(T) = h_f(200\,°\text{C}) = 852.45\frac{\text{kJ}}{\text{kg}} ,$$

$$h(T,p) \simeq h_f(T) + (p - p_{sat}(T))\, v_f(T) = 862.2\frac{\text{kJ}}{\text{kg}} ,$$

$$s(T,v) \simeq s_f(T) = s_f(200\,°\text{C}) = 2.3309\frac{\text{kJ}}{\text{kg}\,\text{K}} .$$

For this particular example, the approximations yield relative errors below 1%, and even smaller at lower pressures. For higher pressures, however, the relative errors are larger, since compressibility affects all property values, hence these approximations should be used with care. Whenever a full table for compressed liquid states is available, that table should be used. If a table for the liquid states is not available, as is often the case for relatively low pressures, the approximations are quite useful.

6.10 The Ideal Gas

When the temperature of a vapor is sufficiently above the critical temperature or when the pressure is sufficiently below the critical pressure, it will obey the ideal gas law

$$pv = RT \,, \tag{6.14}$$

where $R = \bar{R}/M$ is the gas constant. We have discussed ideal gases already in Sec. 2.15, and used the ideal gas law and the caloric equation of state in examples. We repeat some of the property relations and add new ones.

Experiments and theoretical considerations (see Sec. 16.3) show that for ideal gases internal energy u and enthalpy $h = u + pv = u + RT$ depend on temperature *only*. Therefore, also their derivatives, the specific heats at constant volume, c_v, and at constant pressure, c_p, defined in (3.15, 3.22), depend only on temperature,

$$\begin{aligned} c_v &= \left(\frac{\partial u}{\partial T}\right)_v = \frac{du}{dT} = c_v\left(T\right) \,, \\ c_p &= \left(\frac{\partial h}{\partial T}\right)_p = \frac{dh}{dT} = c_p\left(T\right) \,. \end{aligned} \tag{6.15}$$

Since $h = u + RT$, it follows

$$c_p = c_v + R \,. \tag{6.16}$$

Integration of the specific heats gives energy and enthalpy,

$$\begin{aligned} u\left(T\right) &= \int_{T_0}^{T} c_v\left(T'\right) dT' + u_0 \,, \\ h\left(T\right) &= \int_{T_0}^{T} c_p\left(T'\right) dT' + h_0 \,, \end{aligned} \tag{6.17}$$

with reference energy u_0 and, for consistency, reference enthalpy $h_0 = u_0 + RT_0$.

The entropy of an ideal gas is determined from integration of the Gibbs equation (6.2). With $dh = c_p dT$ and the ideal gas law, the Gibbs equation assumes the form

$$ds = \frac{c_p}{T} dT - \frac{v}{T} dp = \frac{c_p}{T} dT - \frac{R}{p} dp \,. \tag{6.18}$$

The entropy for the state (T, p) follows by integration between (T, p) and (T_0, p_0) as

$$s\left(T, p\right) = s^0\left(T\right) - R \ln \frac{p}{p_0} \,, \tag{6.19}$$

where we introduced the abbreviation

$$s^0(T) = \int_{T_0}^{T} \frac{c_p(T')}{T'} dT' + s(T_0, p_0) \ . \tag{6.20}$$

The constant of integration is chosen such that $s^0(T)$ is the ideal gas entropy at reference pressure $p_0 = 1\,\text{bar}$. The value of the reference entropy $s(T_0, p_0)$ can be obtained from the third law, which will be discussed later (Sec. 23.6). As long as non-reacting mixtures are considered, its value is unimportant, since it cancels in all calculations. Indeed, only entropy differences are relevant, for which we find

$$s(T_2, p_2) - s(T_1, p_1) = s^0(T_2) - s^0(T_1) - R \ln \frac{p_2}{p_1} \ . \tag{6.21}$$

When one is not interested in entropy as a function of T and p, but as a function of T and v, the ideal gas law can be used to eliminate pressure,

$$s(T_2, v_2) - s(T_1, v_1) = s^0(T_2) - s^0(T_1) - R \ln \frac{T_2 v_1}{T_1 v_2} \ . \tag{6.22}$$

In summary, energy and enthalpy of the ideal gas depend only on temperature, and its entropy depends explicitly on pressure (6.21) or volume (6.22), and on temperature through the function $s^0(T)$. The temperature dependent quantities $u(T), h(T), s^0(T)$ are tabulated.[3]

As an example we consider a property table for air. The molar specific heat of air can be approximated by the Shomate equation

$$\bar{c}_p = a_0 + a_1 T + a_2 T^2 + a_3 T^3 + \frac{a_4}{T^2} \ , \tag{6.23}$$

with (for $T \leq 1000\,\text{K}$)

$$\begin{aligned} a_0 &= 30.0051 \frac{\text{kJ}}{\text{kmol K}} \ , \quad a_1 = -8.86766 \times 10^{-3} \frac{\text{kJ}}{\text{kmol K}^2} \ , \\ a_2 &= 2.21273 \times 10^{-5} \frac{\text{kJ}}{\text{kmol K}^3} \ , \quad a_3 = -1.02450 \times 10^{-8} \frac{\text{kJ}}{\text{kmol K}^4} \ , \\ a_4 &= 838.737 \frac{\text{kJ K}}{\text{kmol}} \ . \end{aligned} \tag{6.24}$$

The mass based specific heats are $c_p = \bar{c}_p / M$ and $c_v = c_p - R$. Internal energy u, enthalpy h and entropy function $s^0(T)$ follow from integration using the formulas above. Figure 6.17 shows the resulting table.

Tables for other gases are widely available, or can be easily produced from the Shomate equation with the appropriate data for the coefficients, which can be found, e.g., from NIST (http://webbook.nist.gov/).

[3] Some tables list molar quantities $\bar{u} = uM$, $\bar{h} = hM$, $\bar{s}^0 = s^0 M$.

Property table for AIR as ideal gas

T	cv [kJ/kg]	cp [kJ/kg]	u [kJ/kg]	h [kJ/kg]	s0 [kJ/kgK]	Pr	Vr
220	0.715	1.002	157.68	220.81	6.826	0.346	636.0
230	0.715	1.002	164.83	230.83	6.870	0.404	569.3
240	0.715	1.002	171.98	240.85	6.913	0.469	512.1
250	0.715	1.002	179.13	250.87	6.954	0.540	462.5
260	0.715	1.002	186.28	260.89	6.993	0.620	419.5
270	0.715	1.002	193.43	270.91	7.031	0.707	381.8
273	0.715	1.002	195.57	273.92	7.042	0.735	371.5
280	0.716	1.003	200.58	280.94	7.067	0.803	348.7
290	0.716	1.003	207.74	290.96	7.103	0.908	319.5
298.15	**0.716**	**1.003**	**213.57**	**299.14**	**7.130**	**1.000**	**298.2**
300	0.716	1.003	214.90	300.99	7.137	1.022	293.6
310	0.717	1.004	222.07	311.03	7.169	1.146	270.5
320	0.718	1.005	229.24	321.08	7.201	1.281	249.9
330	0.718	1.005	236.42	331.13	7.232	1.426	231.4
340	0.719	1.006	243.61	341.18	7.262	1.584	214.7
350	0.720	1.007	250.81	351.25	7.291	1.753	199.6
360	0.721	1.008	258.01	361.33	7.320	1.935	186.0
370	0.722	1.009	265.23	371.42	7.347	2.131	173.6
380	0.724	1.011	272.46	381.52	7.374	2.341	162.3
390	0.725	1.012	279.70	391.63	7.401	2.565	152.0
400	0.726	1.013	286.96	401.75	7.426	2.805	142.6
410	0.727	1.014	294.22	411.89	7.451	3.060	134.0
420	0.729	1.016	301.51	422.04	7.476	3.333	126.0
430	0.730	1.017	308.80	432.21	7.500	3.622	118.7
440	0.732	1.019	316.11	442.39	7.523	3.930	112.0
450	0.734	1.021	323.44	452.59	7.546	4.257	105.71
460	0.735	1.022	330.79	462.80	7.569	4.603	99.93
470	0.737	1.024	338.15	473.03	7.591	4.970	94.56
480	0.739	1.026	345.53	483.28	7.612	5.358	89.58
490	0.741	1.028	352.92	493.55	7.633	5.769	84.94
500	0.743	1.030	360.34	503.83	7.654	6.202	80.62
510	0.744	1.031	367.77	514.14	7.674	6.659	76.59
520	0.746	1.033	375.23	524.46	7.694	7.141	72.82
530	0.749	1.036	382.70	534.81	7.714	7.648	69.30
540	0.751	1.038	390.20	545.17	7.734	8.182	66.00
550	0.753	1.040	397.72	555.56	7.753	8.744	62.90
560	0.755	1.042	405.25	565.97	7.771	9.33	59.99
570	0.757	1.044	412.81	576.40	7.790	9.95	57.26

T	cv [kJ/kg]	cp [kJ/kg]	u [kJ/kg]	h [kJ/kg]	s0 [kJ/kgK]	Pr	Vr
580	0.759	1.046	420.39	586.85	7.808	10.61	54.69
590	0.761	1.048	428.00	597.32	7.826	11.29	52.27
600	0.764	1.051	435.62	607.82	7.844	12.00	49.98
610	0.766	1.053	443.27	618.34	7.861	12.75	47.83
620	0.768	1.055	450.95	628.88	7.878	13.54	45.80
630	0.771	1.058	458.64	639.44	7.895	14.36	43.87
640	0.773	1.060	466.36	650.03	7.912	15.22	42.05
650	0.775	1.062	474.10	660.64	7.928	16.12	40.33
660	0.778	1.065	481.87	671.28	7.944	17.06	38.70
670	0.780	1.067	489.66	681.94	7.960	18.04	37.15
680	0.783	1.070	497.47	692.62	7.976	19.06	35.68
690	0.785	1.072	505.31	703.33	7.992	20.12	34.29
700	0.787	1.074	513.17	714.06	8.007	21.24	32.96
710	0.790	1.077	521.06	724.82	8.023	22.40	31.70
720	0.792	1.079	528.97	735.60	8.038	23.61	30.50
730	0.795	1.082	536.91	746.41	8.053	24.86	29.36
740	0.797	1.084	544.87	757.24	8.067	26.17	28.27
750	0.800	1.087	552.85	768.09	8.082	27.54	27.24
760	0.802	1.089	560.86	778.97	8.096	28.95	26.25
770	0.805	1.091	568.90	789.87	8.111	30.43	25.31
780	0.807	1.094	576.95	800.80	8.125	31.96	24.40
790	0.809	1.096	585.03	811.75	8.139	33.55	23.54
800	0.812	1.099	593.14	822.73	8.152	35.21	22.72
810	0.814	1.101	601.27	833.73	8.166	36.92	21.94
820	0.816	1.103	609.42	844.75	8.180	38.70	21.19
830	0.819	1.106	617.59	855.79	8.193	40.55	20.47
840	0.821	1.108	625.79	866.86	8.206	42.47	19.78
850	0.823	1.110	634.01	877.95	8.219	44.46	19.12
860	0.826	1.112	642.26	889.07	8.232	46.52	18.49
870	0.828	1.115	650.52	900.20	8.245	48.65	17.88
880	0.830	1.117	658.81	911.36	8.258	50.86	17.30
890	0.832	1.119	667.12	922.54	8.271	53.15	16.74
900	0.834	1.121	675.45	933.74	8.283	55.52	16.21
910	0.836	1.123	683.81	944.96	8.295	57.97	15.70
920	0.838	1.125	692.18	956.21	8.308	60.51	15.20
930	0.840	1.127	700.57	967.47	8.320	63.13	14.73
940	0.842	1.129	708.98	978.75	8.332	65.84	14.28
950	0.844	1.131	717.41	990.05	8.344	68.64	13.84

Fig. 6.17 Property data for air: specific heats $c_v(T)$ and $c_p(T)$, internal energy $u(T)$, enthaply $h(T)$ and entropy function $s^0(T)$ as functions of temperature

6.11 Monatomic Gases (Noble Gases)

For monatomic gases, i.e., the noble gases helium (He), neon (Ne), argon (Ar), krypton (Kr), xenon (Xe), and radon (Rn), the specific heats are true constants with the values

$$c_v = \frac{3}{2}R \quad , \quad c_p = c_v + R = \frac{5}{2}R \tag{6.25}$$

and the caloric equation of state follows from straightforward integration as

$$\begin{aligned} u(T) &= c_v(T - T_0) + u_0 \, , \\ h(T) &= c_p(T - T_0) + h_0 \, . \end{aligned} \tag{6.26}$$

With $c_p = const$, the integration in (6.20) can be performed easily, and the entropy becomes

$$s(T,p) = c_p \ln \frac{T}{T_0} - R \ln \frac{p}{p_0} + s_0 \, . \tag{6.27}$$

Since the resulting expressions for the thermodynamic quantities of monatomic gases are rather simple, these are typically not tabulated.

6.12 Specific Heats and Cold Gas Approximation

The value of the specific heat is related to the degrees of freedom of a molecule. Specifically, each degree of freedom contributes $\frac{1}{2}R$ to the specific heat at constant volume (equipartition of energy). The atoms of monatomic gases are essentially spheres that can translate in three directions (up/down, right/left, forward/backward); accordingly, the specific heat of monatomic gases is $c_v = 3 \times \frac{1}{2}R$.

For diatomic gases like oxygen (O_2), nitrogen (N_2), hydrogen (H_2), the molecules are shaped like dumb-bells. At low temperatures these have, in addition to their three translational degrees of freedom, two rotational degrees of freedom for the rotation about two principal axes—there is no rotation around the longitudinal axis. More complex molecules like carbondioxid (CO_2) and water (H_2O) have three translational and three rotational degrees of freedom. Moreover, the molecules can oscillate, the more complicated a molecule is, the more oscillating modes are observed.

At sufficiently low temperatures only translational and rotational modes are excited. With each mode contributing $\frac{1}{2}R$ to the specific heat, we have at low T for a diatomic gas $c_v = \frac{5}{2}R$, $c_p = \frac{7}{2}R$, and for a polyatomic gas $c_v = 3R$, $c_p = 4R$. Oscillatory modes obey quantum mechanical laws; they are not excited at low temperatures and contribute in a temperature dependent manner for higher temperatures. Figure 6.18 shows the molar specific heat $\bar{c}_p = Mc_p$ for a variety of ideal gases. Note the temperature independent value $\bar{c}_p = \frac{5}{2}\bar{R} = 20.8\frac{\text{kJ}}{\text{kg K}}$ for monatomic gases, and the common low temperature value of $\bar{c}_p = \frac{7}{2}\bar{R} = 29.1\frac{\text{kJ}}{\text{kg K}}$ for diatomic gases.

Air, as a mixture of roughly 78% N_2, 21% O_2 and 1% Ar, behaves essentially like a diatomic gas, with the low temperature specific heats $c_v^{air} = \frac{5}{2}R_{air}$, $c_p^{air} = \frac{7}{2}R_{air}$. As air temperatures rises, so do the specific heats.

To simplify computations, one frequently assumes constant specific heats. To not deviate too much from the actual states, one should use suitable average values c_v^{avg}, c_p^{avg} for the temperature interval under consideration, or, alternatively, the values at room temperature. In the latter case one speaks of the *cold-gas-approximation*, or, for air, *cold-air-approximation*. Internal energy, enthalpy and entropy are

$$
\begin{aligned}
u(T) &= c_v^{avg}(T - T_0) + u_0 \,, \\
h(T) &= c_p^{avg}(T - T_0) + h_0 \,, \\
s(T,p) &= c_p^{avg}\ln\frac{T_2}{T_0} - R\ln\frac{p_2}{p_0} + s_0 \,.
\end{aligned}
\tag{6.28}
$$

The cold-gas-approximation, where one uses $c_v^{avg} = c_v(T_0)$, works best for relatively low temperatures (e.g., $T < 600\,\text{K}$ for air), but is highly useful to understand the basic behavior of thermodynamic systems. Constant specific heats allow analytical calculations that give, e.g., explicit expressions for

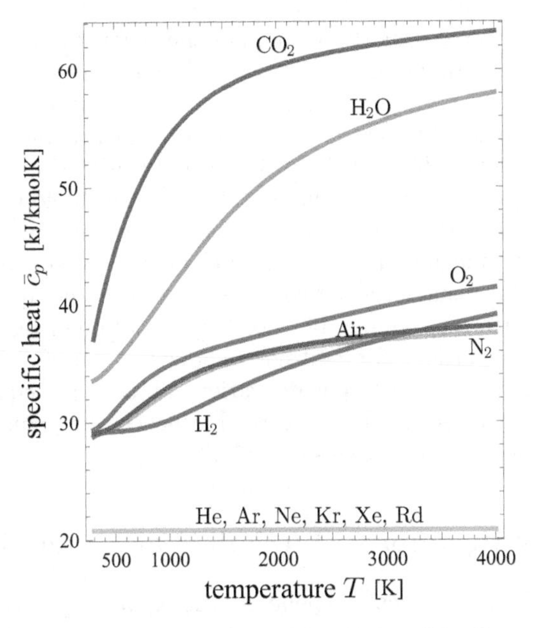

Fig. 6.18 Molar specific heat at constant pressure $\bar{c}_p = Mc_p$ for various ideal gases as function of temperature. Note that specific heat of monatomic gases (noble gases) is constant. Based on specific heat data from NIST.

efficiencies that help to further the understanding. Exact engineering calculations must use variable specific heats, of course, and tabulated data *must* be used, unless the gas is monatomic and the specific heat independent of temperature!

6.13 Real Gases

Gases (or vapors) at relatively high pressures or relatively low temperatures do not obey the ideal gas law. To understand why that is the case, it is helpful to know a little bit about the derivation of the ideal gas law with the tools of *Statistical Thermodynamics*, which relies on two assumptions: (a) Gas particles are mass points, that is their volume can be ignored. (b) There are no long-distance forces between the particles, they only interact in short collisions, and travel most distance between collisions in free flight.

J. D. van der Waals (1837-1923) derived an equation that modifies the ideal gas equation to address both points. The van der Waals equation reads

$$p = \frac{RT}{v-b} - \frac{a}{v^2} \, . \tag{6.29}$$

The constant b accounts for the volume of the particles, where $v - b$ is the volume accessible to an individual particle. The constant a accounts for long-range attractive forces between the particles, which reduce the pressure. The constants a, b can be obtained from fitting to critical point data. For large values of the specific volume v the equation reduces to the ideal gas law. A deeper discussion of the van der Waals equation can be found in Sec. 16.8, where it will be seen that the equation gives a good qualitative description of real gas effects and liquid-vapor phase change. However, its quantitative agreement with gas behavior is not so good. Therefore, the equation is mainly used as an educational example, but not for simulation of real processes.

Since explicit equations for real gas behavior are useful for simulations and calculations, there exist a wide variety of real gas equations, which can be found in the technical literature (Redlich-Kwong equation, Beattie-Bridgeman equation, virial expansions, etc.).

6.14 Fully Incompressible Solids and Liquids

Also for solids the specific heats depend in general on temperature and volume (or any other pair of properties), and must be collected in tables. Quite often it is possible to treat the solids to be fully incompressible (no change of volume, $v = const$), and to assume constant specific heat. Then, internal energy, enthalpy and entropy are

$$\begin{aligned} u(T) &= c(T - T_0) + u_0 , \\ h(T,p) &= c(T - T_0) + v(p - p_0) + h_0 , \\ s(T) &= c \ln \frac{T}{T_0} + s_0 . \end{aligned} \tag{6.30}$$

As always, u_0, h_0 and s_0 are suitable reference values. Due to incompressibility, the specific heats at constant volume and constant pressure agree, as the following line of equations shows:

$$c_p = \left(\frac{\partial h}{\partial T}\right)_p = \left(\frac{\partial (u + pv)}{\partial T}\right)_p = \left(\frac{\partial u}{\partial T}\right) + \left(v\frac{\partial p}{\partial T}\right)_p = \left(\frac{\partial u}{\partial T}\right) = c_v . \tag{6.31}$$

The same approximations can be used for fully incompressible liquids.

Problems

6.1. Property Diagrams and Data (Water)

Draw schematic p-T, p-v, T-v and T-s-diagrams for water, and mark the following points in the diagrams.

CR) critical point TR) triple point

1) $p = 1\,\text{bar},\ v = 0.85\frac{\text{m}^3}{\text{kg}}$ 2) $p = 1\,\text{bar},\ h = 3400\frac{\text{kJ}}{\text{kg}}$

3) $p = 20\,\text{MPa},\ v = 0.0012\frac{\text{m}^3}{\text{kg}}$ 4) $h = 2700\frac{\text{kJ}}{\text{kg}},\ x = 1$

5) $p = 20\,\text{MPa}, u = 3100\frac{\text{kJ}}{\text{kg}}$ 6) $s = 3\frac{\text{kJ}}{\text{kg K}},\ T = 255\,°\text{C}$

Also, determine temperature, quality, specific internal energy, specific enthalpy, and specific volume for each point, and say whether you have compressed liquid, saturated state, or superheated vapor.

6.2. Property Diagrams and Data (R134a)
Consider cooling fluid R134a. Based on the posted tables, determine temperature, pressure, quality, specific internal energy, specific enthalpy, and specific volume for each point, and say whether you have compressed liquid, saturated state, or superheated vapor. Put all values in a table.

1. $T = -4\,°\text{C}, h = 178.2\frac{\text{kJ}}{\text{kg}}$, 2. $T = -24\,°\text{C}, p = 0.2\,\text{MPa}$,
3. $T = 20\,°\text{C}, s = 0.9883\frac{\text{kJ}}{\text{kg K}}$

6.3. Boiling Temperature
Water in a 5 cm deep pan is observed to boil at 98 °C. At what temperature will the water in a 50 cm deep pan boil? Assume both pans are filled to the rim.

6.4. Food Preservation
To preserve fruit or vegetables (canning), the food is cooked in a jar which is covered by a lid, resting on a rubber seal. As water is evaporated during cooking, vapor escapes and carries air out. After a while, only food, liquid water and vapor are left in the jar. Then cooking stops, and as the jar cools, the pressure in the jar drops, tightly sealing the jar. Consider a jar of 20 cm diameter at 15 °C, and determine the force necessary to pull of the lid.

6.5. Cooling of Steam
2 kg of superheated steam at 2 bar, 300 °C (state 1) are isobarically cooled to the saturated vapor state (state 2). Then, the volume of the container is fixed and the steam is cooled further until the temperature is 20 °C (state 3).

1. Draw the process into a p-v-diagram with respect to saturations lines. Mark the critical point.
2. Compute heat and work exchanged for the processes 1-2 and 2-3.

6.6. Isentropic Expansion of R-134a Vapor
Refrigerant R-134a at 1.2 MPa, 50 °C (state 1) enclosed in a piston-cylinder device expands in an adiabatic reversible (i.e., isentropic) process to 100 kPa (state 2). Determine specific volume, internal energy, enthalpy, entropy at both points. Compute heat and work exchanged between refrigerant and surroundings.

6.7. Evaporation of Water
3 kg of saturated liquid water at 70 °C (state 1) are isobarically heated until the volume reaches $0.921\,\frac{\mathrm{m}^3}{\mathrm{kg}}$ (state 2). Then, the volume of the container is fixed and the heating continues until all liquid is just evaporated (state 3).

1. Draw the process into a p-v-diagram with respect to saturation lines. Mark the critical point.
2. Compute heat and work exchanged for the processes 1-2, and 2-3.

6.8. Condensation of R134a
500 g of cooling fluid R134a are enclosed in a piston cylinder system at 3.2 bar, 25 °C. The system is isobarically cooled until the cooling fluid assumes a temperature of −8 °C.

1. Draw the process into p-v- and T-s-diagram with respect to saturation lines.
2. Determine heat and work exchanged.
3. Determine the change of entropy.

Chapter 7
Reversible Processes in Closed Systems

7.1 Standard Processes

In Chapter 8 we shall study thermodynamic cycles in closed systems which model thermal engines, including internal combustion engines. The focus will lie on the understanding of the working principles of the cycles, and on the main parameters that determine their efficiency. For this it is customary to base the analysis on reversible processes, which allow a full analysis.

There are a number of processes that are often realized (at least approximately) in thermodynamic systems: processes at constant volume, constant pressure, constant temperature, or adiabatic processes. Typical thermodynamic cycles consist of closed chains of several of these processes. In this chapter we compute work and heat for these standard processes as a reference for the discussion of cycles.

7.2 Basic Equations

Figure 7.1 shows, again, a piston-cylinder device as the prototypical closed system. In reversible (quasi-static) processes, the system exchanges energy

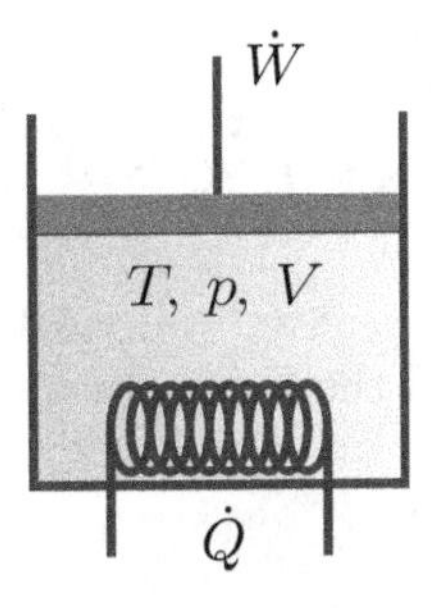

Fig. 7.1 Closed system with piston work and heat exchange. In this chapter we are interested in reversible processes only, so there is not stirring.

H. Struchtrup, *Thermodynamics and Energy Conversion*,
DOI: 10.1007/978-3-662-43715-5_7, © Springer-Verlag Berlin Heidelberg 2014

through heating and piston work only; stirring (propeller work) as an irreversible process is excluded. All movement of the material in the system is so slow that velocity and kinetic energy can be ignored. For a stationary system, potential energy is constant and can be ignored as well. Thus, at all times the system is in homogeneous equilibrium states which are characterized by the temperature T, the pressure p, and the volume V.

We list the relevant equations from previous chapters. Under the above simplifications, the first law for closed systems reduces to

$$\frac{dU}{dt} = \dot{Q} - \dot{W} , \tag{7.1}$$

where $\dot{Q}$ is the heat transfer rate, and $\dot{W}$ denotes power. Integration over the duration of the process gives the time-integrated energy balance

$$U_2 - U_1 = Q_{12} - W_{12} , \tag{7.2}$$

where

$$Q_{12} = \int_{t_1}^{t_2} \dot{Q} dt \quad \text{and} \quad W_{12} = \int_{t_1}^{t_2} \dot{W} dt \tag{7.3}$$

are the total amounts of heat and work exchanged between the states 1 (at time t_1) and 2 (at time t_2).

For an infinitesimal step of the process (duration dt) we have the differential form of the first law

$$dU = \delta Q - \delta W , \tag{7.4}$$

where $\delta Q = \dot{Q} dt$ and $\delta \dot{W} = \dot{W} dt$ are heat and work exchanged during dt. The notation implies that work and heat have inexact differentials, since they are process dependent quantities.

For a reversible process in a closed system, the work is just the piston work,

$$\dot{W} = p\frac{dV}{dt} \quad \text{or} \quad W_{12} = \int_{t_1}^{t_2} \dot{W} dt = \int_1^2 \delta W = \int_1^2 pdV , \tag{7.5}$$

and the heat can be computed from the second law, which for reversible processes ($\dot{S}_{gen} = 0$) reduces to

$$\dot{Q} = T\frac{dS}{dt} \quad \text{or} \quad Q_{12} = \int_{t_1}^{t_2} \dot{Q} dt = \int_1^2 \delta Q = \int_1^2 TdS . \tag{7.6}$$

Thus, for reversible processes, heat and work are the areas below the process curves in the p-V-diagram and the T-S-diagram, respectively, as depicted in Fig. 7.2.

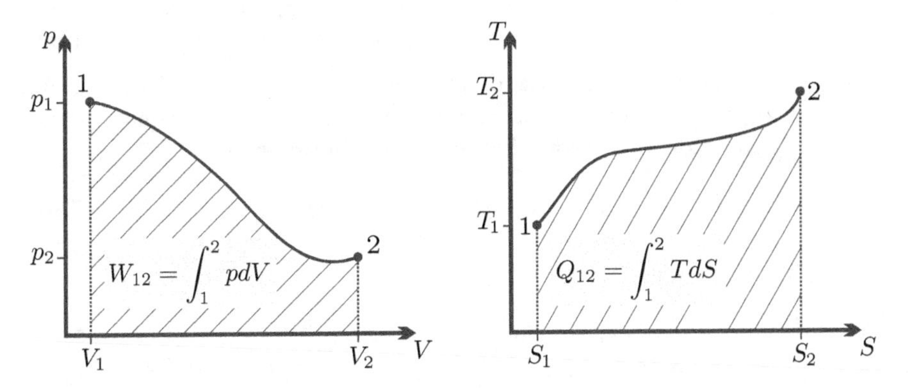

Fig. 7.2 Heat and work in reversible processes as areas below the process curves in the p-V- and the T-S-diagram

In the following sections we shall compute work and heat per unit mass, which for reversible processes are given by

$$w_{12} = \frac{W_{12}}{m} = u_1 - u_2 + q_{12} = \int_1^2 pdv \ , \tag{7.7}$$

$$q_{12} = \frac{Q_{12}}{m} = u_2 - u_1 + w_{12} = \int_1^2 Tds \ . \tag{7.8}$$

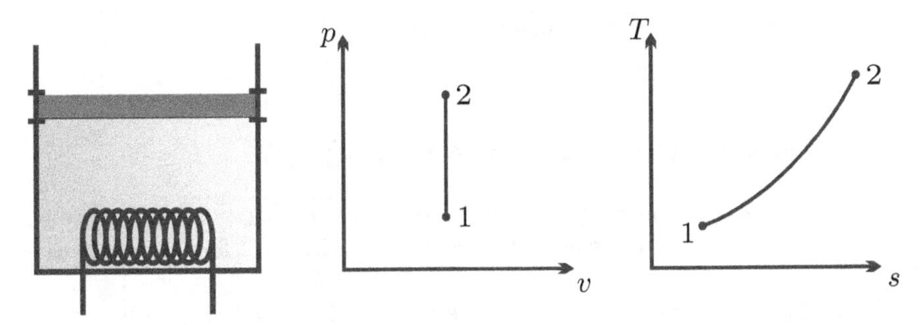

Fig. 7.3 Isochoric process: Realization, p-v- and T-s-diagrams

7.3 Isochoric Process: $v = const., dv = 0$

Isochoric processes (constant volume) can be easily realized by fixing the volume, e.g., by clamping the piston, see Figure 7.3 for process sketch and diagrams.

With $dv = 0$ in the constant volume process, heat and work follow from (7.7, 7.8) as

$$w_{12} = 0 \quad , \quad q_{12} = u_2 - u_1 \; . \tag{7.9}$$

We compute the process curve of an isochoric process in the T-s-diagram for an ideal gas with constant specific heats. From the Gibbs equation and the caloric equation of state we find for the isochoric process

$$Tds = du + pdv = du = c_v dT \; , \tag{7.10}$$

so that upon integration

$$s - s_1 = c_v \ln \frac{T}{T_1} \quad \text{or} \quad T = T_1 e^{\frac{s-s_1}{c_v}} \; . \tag{7.11}$$

Thus, for an ideal gas, the isochoric process in the T-s-diagram follows an exponential, as indicated in the T-s-diagram.

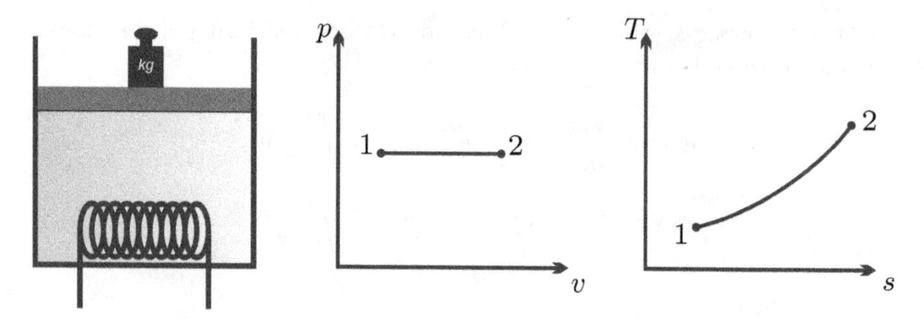

Fig. 7.4 Isobaric process: Realization, p-v- and T-s-diagrams

7.4 Isobaric Process: $p = const.$, $dp = 0$

Isobaric processes (constant pressure) are easily realized by free pistons, where the piston weight controls the pressure; see Fig. 7.4 for process sketch and diagrams.

With $dp = 0$ in the constant pressure process, heat and work follow from (7.7, 7.8) as

$$\begin{aligned} w_{12} &= \int_1^2 pdv = p \int_1^2 dv = p\,(v_2 - v_1) \; , \\ q_{12} &= \int_1^2 Tds = \int_1^2 (dh - vdp) = \int_1^2 dh = h_2 - h_1 \; . \end{aligned} \tag{7.12}$$

Here we have used the Gibbs equation in the form (4.13), $Tds = dh - vdp$.

Again we compute the process curve in the T-s-diagram for an ideal gas with constant specific heats. From the Gibbs equation and the caloric equation of state we find for the isobaric process

$$Tds = dh - vdp = dh = c_p dT \ , \tag{7.13}$$

so that upon integration

$$s - s_1 = c_p \ln \frac{T}{T_1} \quad \text{or} \quad T = T_1 e^{\frac{s-s_1}{c_p}} \ . \tag{7.14}$$

This was used for drawing the curve in the diagram. For an ideal gas, the isobaric process in the T-s-diagram follows an exponential. Since $c_p = c_v + R > c_v$, isobaric lines in the T-s-diagram are not as steep as isochoric lines starting at the same point.

7.5 Isentropic Process: $q_{12} = \delta q = ds = 0$

A system that is insulated against heat transfer is adiabatic. However, adiabatic processes are also realized if the process is sufficiently fast, so that there is no time to exchange heat. A pressure disturbance at the boundary, i.e., induced by the moving piston, travels with the speed of sound, and accordingly mechanical equilibrium in the working fluid is assumed rather fast. A temperature disturbance at the boundary, however, diffuses slowly into the working fluid. In other words, pressure equilibration and heat transfer occur on quite distinct time scales. Accordingly, a compression (i.e., pressure increase) or expansion (i.e., pressure decrease) process may be slow enough to allow for pressure equilibration in the system, but at the same time may be so fast that there is no time to exchange heat between the working fluid and the system walls, even if they have different temperatures. Such a process can be modelled to be (approximately) adiabatic.

From the relation between entropy and heat for reversible processes (7.6) follows for an adiabatic process

$$\delta q = 0 = Tds \Longrightarrow ds = 0 \ , \ s = const. \tag{7.15}$$

The reversible adiabatic process is isentropic, see Fig. 7.5 for process sketch and diagrams.

The work is best computed from (7.8) which gives

$$w_{12} = u_1 - u_2 \ . \tag{7.16}$$

We study the isentropic process in the ideal gas in more detail: From (6.21) follows

$$s_2 - s_1 = s^0 (T_2) - s^0 (T_1) - R \ln \frac{p_2}{p_1} = 0 \tag{7.17}$$

or

$$\frac{p_2}{p_1} = \frac{a \exp\left[\frac{s^0(T_2)}{R}\right]}{a \exp\left[\frac{s^0(T_1)}{R}\right]} = \frac{p_r (T_2)}{p_r (T_1)} \ . \tag{7.18}$$

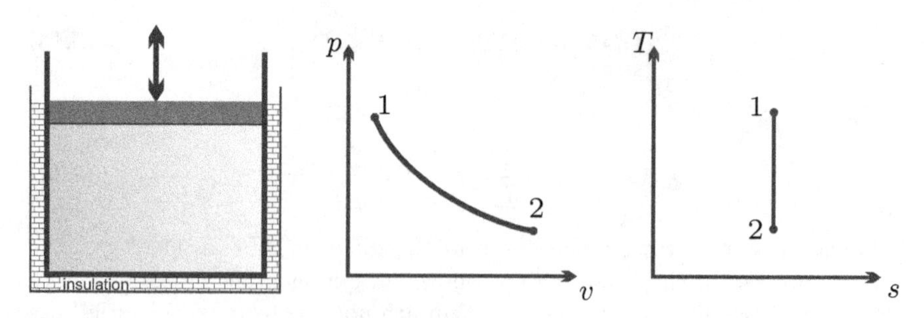

Fig. 7.5 Isentropic process: Realization, p-v- and T-s-diagrams

$p_r(T) = a \exp\left[\frac{s^0(T)}{R}\right]$ is called the relative pressure and often is tabulated (e.g., for air); a is a constant used for scaling of $p_r(T)$, its value does not affect the relation (7.18). With the ideal gas law $p = \frac{RT}{v}$ we can rewrite (7.18) as

$$\frac{v_2}{v_1} = \frac{b\frac{T_2}{p_r(T_2)}}{b\frac{T_1}{p_r(T_1)}} = \frac{v_r(T_2)}{v_r(T_1)}, \tag{7.19}$$

where $v_r(T) = b\frac{T}{p_r(T)}$ is called the relative volume, and might be tabulated as well; b is another scaling constant. The ideal gas table for air in Fig. 6.17 includes columns for $p_r(T)$ and $v_r(T)$, where a was chosen such that $p_r(298.15\,\text{K}) = 1$, i.e., $a = \exp\left[-\frac{s^0(298.15\,\text{K})}{R}\right]$, and $b = 1$; other tables might use other values of both constants.

In case of constant specific heats, the entropy is given by (6.27), which for an isentropic process gives

$$s_2 - s_1 = c_p \ln\frac{T_2}{T_1} - R\ln\frac{p_2}{p_1} = 0\,. \tag{7.20}$$

Solving for the temperature ratio we find, with $R = c_p - c_v$,

$$\frac{T_2}{T_1} = \left(\frac{p_2}{p_1}\right)^{\frac{k-1}{k}}, \tag{7.21}$$

where k denotes the ratio of specific heats $k = \frac{c_p}{c_v}$. By means of the ideal gas law we find the alternative relations

$$\frac{p_2}{p_1} = \left(\frac{v_2}{v_1}\right)^{-k}, \quad \frac{T_2}{T_1} = \left(\frac{v_2}{v_1}\right)^{1-k}. \tag{7.22}$$

The above relations can be expressed in compact from as

$$Tp^{\frac{1-k}{k}} = const.\,, \quad pv^k = const.\,, \quad Tv^{k-1} = const. \tag{7.23}$$

The value of the ratio of specific heats is $k = 1.667$ for monatomic gases, and, under the cold-gas approximation, $k = 1.4$ for diatomic gases, and $k = 1.333$ for polyatomic gases. Equations (7.23) are the adiabatic relations for ideal gases with constant specific heats.

7.6 Isothermal Process: $T = const$, $dT = 0$

Isothermal processes require exchange of heat with a large reservoir at constant temperature. Since heat exchange is slow, isothermal processes must be rather slow and therefore they are not found in the most common thermodynamic cycles. Figure 7.6 shows process sketch and diagrams.

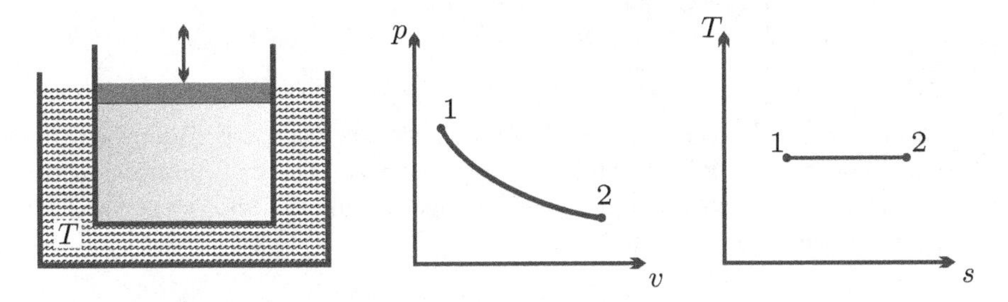

Fig. 7.6 Isothermal process: Realization, p-v- and T-s-diagrams

Since temperature is constant, heat and work can be computed as

$$
\begin{aligned}
q_{12} &= \int_1^2 T ds = T \int_1^2 ds = T\left(s_2 - s_1\right) , \\
w_{12} &= u_1 - u_2 + q_{12} = u_1 - u_2 + T\left(s_2 - s_1\right) .
\end{aligned}
\tag{7.24}
$$

We consider the special case of the ideal gas where $p = \frac{RT}{v}$, which was used to draw the curve in the p-v-diagram. With the thermal equation of state explicitly known, the work can also be determined by integration (recall T is constant!),

$$
w_{12} = \int_1^2 p dv = RT \int_1^2 \frac{dv}{v} = RT \ln \frac{v_2}{v_1} = -RT \ln \frac{p_2}{p_1} . \tag{7.25}
$$

For the ideal gas the internal energy depends only on temperature, that is $du = 0$ when $dT = 0$, and thus the heat exchange is equal to the work,

$$
q_{12} = w_{12} = RT \ln \frac{v_2}{v_1} = -RT \ln \frac{p_2}{p_1} . \tag{7.26}
$$

It is left to the reader to confirm that (7.24) evaluated with the property relations for an ideal gas yields the same result.

7.7 Polytropic Process (Ideal Gas): $pv^n = const$

Processes in actual applications might differ from those discussed above. A useful approximate description of a wide variety of processes in ideal gases is offered by the polytropic process, which is a generalization of the adiabatic relations (7.23) to arbitrary exponents n,

$$Tp^{\frac{1-n}{n}} = const. \quad , \quad pv^n = const. \quad , \quad Tv^{n-1} = const. \tag{7.27}$$

Special choices for the polytropic exponent n refer to the previously discussed processes as follows

$$\begin{array}{lll} n = 0 & \Rightarrow p = const & \text{isobaric} \\ n = 1 & \Rightarrow pv = RT = const. & \text{isothermal} \\ n = k & \Rightarrow pv^k = const & \text{isentropic (const. } c_p) \\ n = \infty & \Rightarrow v = const. & \text{isochoric} \end{array}$$

Often one uses values of n in the interval $1 \leq n \leq k$ to describe processes that are not fully adiabatic and not fully isothermal, e.g., compression or expansion processes with small heat exchange. Figure 7.7 shows the various processes in the two diagrams.

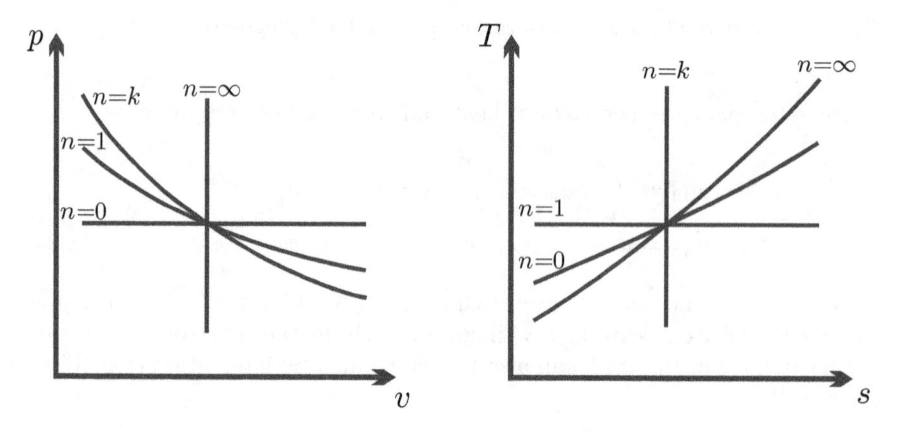

Fig. 7.7 Polytropic processes with $n = 0, 1, k, \infty$ in p-v- and T-s-diagram

Since $pv^n = p_1 v_1^n$, the work follows by integration as

$$w_{12} = \int_1^2 p dv = p_1 v_1^n \int_1^2 \frac{dv}{v^n} = \frac{p_1 v_1}{1-n} \left[\left(\frac{v_2}{v_1} \right)^{1-n} - 1 \right] = \frac{R}{1-n} (T_2 - T_1) \ , \tag{7.28}$$

which holds for all $n \neq 1$. The work for the case $n = 1$ (isothermal) can be found from the above by using l'Hôpital's rule:

$$\lim_{n\to 1} w_{12} = \lim_{n\to 1} \frac{RT_1}{1-n}\left[\left(\frac{v_2}{v_1}\right)^{1-n} - 1\right] = RT_1 \lim_{n\to 1} \frac{\frac{d}{dn}\left(\frac{v_2}{v_1}\right)^{1-n}}{\frac{d}{dn}(1-n)} = RT_1 \ln\frac{v_2}{v_1}\,. \tag{7.29}$$

The heat exchanged follows from the first law, $q_{12} = u_2 - u_1 + w_{12}$.

7.8 Summary

For easy reference, we collect the results of this section in a table,

isochoric	$dv = 0$	$w_{12} = 0$	$q_{12} = u_2 - u_1$
isobaric	$dp = 0$	$w_{12} = p\,(v_2 - v_1)$	$q_{12} = h_2 - h_1$
isentropic	$ds = 0$	$w_{12} = u_1 - u_2$	$q_{12} = 0$
isothermal	$dT = 0$	$w_{12} = u_1 - u_2 + q_{12}$	$q_{12} = T\,(s_2 - s_1)$
isothermal (id. gas)	$dT = 0$	$w_{12} = RT\ln\frac{v_2}{v_1}$	$q_{12} = w_{12}$
polytropic (id. gas)	$pv^n = const.$	$w_{12} = \frac{R}{1-n}\,(T_2 - T_1)$	$q_{12} = u_2 - u_1 + w_{12}$

7.9 Examples

7.9.1 *Isochoric Process for Ideal Gas*

Carbon dioxide is confined in a 10 litre tank at a pressure of $p_1 = 10\,\text{bar}$. In an isochoric heat transfer process the temperature drops from $T_1 = 670\,\text{K}$ to $T_2 = 25\,°\text{C}$. We compute the heat transferred from the system and the entropy change.

The mass of carbon dioxide in the system is ($M_{CO_2} = 44\frac{\text{kg}}{\text{kmol}}$, $R_{CO_2} = 0.189\frac{\text{kJ}}{\text{kg K}}$)

$$m = \frac{pV}{RT} = 79.0\,\text{g}\,.$$

Since volume is constant, the ideal gas law gives the final pressure as

$$p_2 = p_1\frac{T_2}{T_1} = 4.45\,\text{bar}\,.$$

Specific internal energy and specific entropy in initial and final state can be read from an appropriate table as

$$u_1 = \frac{\bar{u}(T_1)}{M} = 467.9\frac{\text{kJ}}{\text{kg}}\,,\quad s_1 = \frac{\bar{s}^0(T_1)}{M} - R\ln\frac{p_1}{p_0} = 5.231\frac{\text{kJ}}{\text{kg K}}\,,$$

$$u_2 = \frac{\bar{u}(T_2)}{M} = 156.4\frac{\text{kJ}}{\text{kg}}\,,\quad s_2 = \frac{\bar{s}^0(T_2)}{M} - R\ln\frac{p_2}{p_0} = 4.579\frac{\text{kJ}}{\text{kg K}}\,.$$

Since the process is isochoric, the work is zero, $w_{12} = 0$. The heat withdrawn is

$$Q_{12} = mq_{12} = m(u_2 - u_1) = -23.8\,\mathrm{kJ}\,,$$

and the total entropy change is

$$S_2 - S_1 = m(s_2 - s_1) = -51.48\frac{\mathrm{J}}{\mathrm{kg}}\,.$$

7.9.2 Isochoric Heating of Water

Saturated liquid-vapor mix at 100 °C with quality $x = 0.1$ is isochorically heated until the pressure is 2.5 MPa. We compute the final state, and the heat supplied per unit mass.

From a steam table we find initial volume, specific energy and specific entropy as

$$\begin{aligned} v_1 &= [(1 - x_1)\, v_f + x_1 v_g]_{T=100\,^\circ\mathrm{C}} \\ &= 0.9 \times 0.001044\frac{\mathrm{m}^3}{\mathrm{kg}} + 0.1 \times 1.673\frac{\mathrm{m}^3}{\mathrm{kg}} = 0.168\frac{\mathrm{m}^3}{\mathrm{kg}}\,, \end{aligned}$$

$$u_1 = [u_f + x_1 u_{fg}]_{T=100\,^\circ\mathrm{C}} = 418.94\frac{\mathrm{kJ}}{\mathrm{kg}} + 0.1 \times 2087.5\frac{\mathrm{kJ}}{\mathrm{kg}} = 627.7\frac{\mathrm{kJ}}{\mathrm{kg}}\,,$$

$$s_1 = [s_f + x_1 s_{fg}]_{T=100\,^\circ\mathrm{C}} = 1.3069\frac{\mathrm{kJ}}{\mathrm{kg\,K}} + 0.1 \times 6.048\frac{\mathrm{kJ}}{\mathrm{kg\,K}} = 1.912\frac{\mathrm{kJ}}{\mathrm{kg\,K}}\,.$$

The final state is superheated vapor of the same volume at 2.5 MPa. Using steam tables (with interpolation) we find

$$T_2 = T\left(2.5\,\mathrm{MPa}, 0.168\frac{\mathrm{m}^3}{\mathrm{kg}}\right) = 645.8\,\mathrm{K}\,,$$

$$u_2 = u\left(2.5\,\mathrm{MPa}, 0.168\frac{\mathrm{m}^3}{\mathrm{kg}}\right) = 3370.7\frac{\mathrm{kJ}}{\mathrm{kg}}\,,$$

$$s_2 = s\left(2.5\,\mathrm{MPa}, 0.168\frac{\mathrm{m}^3}{\mathrm{kg}}\right) = 7.709\frac{\mathrm{kJ}}{\mathrm{kg\,K}}\,.$$

Since the process is isochoric, the work is zero, $w_{12} = 0$. The heat supplied per unit mass for this evaporation process is

$$q_{12} = u_2 - u_1 = 2743.0\frac{\mathrm{kJ}}{\mathrm{kg}}\,.$$

7.9.3 Isobaric Heating of Ideal Gas

200 kg of air are isobarically heated from the initial state $p_1 = 15\,\text{bar}$, $T_1 = 440\,\text{K}$ until the volume has doubled. We compute the final state, the heat supplied, and the work done by the gas.

The initial volume is

$$V_1 = \frac{mRT_1}{p_1} = 16.84\,\text{m}^3 \;;$$

specific energy and enthalpy are (from table)

$$u_1 = u\,(440\,\text{K}) = 316.11\frac{\text{kJ}}{\text{kg}} \;\;,\;\; h_1 = 442.39\frac{\text{kJ}}{\text{kg}} \;.$$

The initial entropy is

$$s_1 = s^0\,(T_1) - R\ln\frac{p_1}{p_0} = 7.523\frac{\text{kJ}}{\text{kg}\,\text{K}} - 0.287\frac{\text{kJ}}{\text{kg}\,\text{K}}\ln\frac{15}{1.01325} = 6.750\frac{\text{kJ}}{\text{kg}\,\text{K}} \;.$$

With the volume doubled, and the pressure constant, the final temperature is

$$T_2 = \frac{pV_2}{mR} = 2\frac{pV_1}{mR} = 2T_1 = 880\,\text{K} \;.$$

With the temperature known, the other properties are found in the table as

$$u_2 = u\,(T_2) = 658.81\frac{\text{kJ}}{\text{kg}} \;\;,\;\; h_2 = h\,(T_2) = 911.36\frac{\text{kJ}}{\text{kg}} \;,$$

$$s_2^0 = s_2^0\,(T_2) = 8.258\frac{\text{kJ}}{\text{kg}\,\text{K}} \;\;,\;\; s_2 = s_2^0 - R\ln\frac{p_2}{p_0} = 7.485\frac{\text{kJ}}{\text{kg}\,\text{K}} \;.$$

The work done by the system is

$$W_{12} = p\,(V_2 - V_1) = 25.26\,\text{MJ} \;,$$

while the heat exchanged is

$$Q_{12} = m\,(h_2 - h_1) = 93.79\,\text{MJ} \;.$$

7.9.4 Isobaric Cooling of R134a

Superheated cooling fluid R134a at initial state $p_1 = 0.18\,\text{MPa}$ and $T_1 = 40\,°\text{C}$ is isobarically cooled until the temperature is $T_2 = -24\,°\text{C}$. We determine initial and end state properties, the heat transfer per unit mass, and the work per unit mass.

From a vapor table we find the initial data as

$$v_1 = v\,(0.18\,\text{MPa}, 40\,°\text{C}) = 0.1373\frac{\text{m}^3}{\text{kg}}\,,$$
$$u_1 = u\,(0.18\,\text{MPa}, 40\,°\text{C}) = 261.53\frac{\text{kJ}}{\text{kg}}\,,$$
$$h_1 = h\,(0.18\,\text{MPa}, 40\,°\text{C}) = 286.24\frac{\text{kJ}}{\text{kg}}\,,$$
$$s_1 = s\,(0.18\,\text{MPa}, 40\,°\text{C}) = 1.0898\frac{\text{kJ}}{\text{kg}\,\text{K}}\,.$$

The saturation temperature for 0.18 MPa is $T_{\text{sat}} = -12.7\,°\text{C}$ and since the final temperature lies below this value, the final state is compressed liquid. We use the approximations for compressed liquid to determine[1]

$$v_2 \simeq v_f\,(T_2) = 0.00073\frac{\text{m}^3}{\text{kg}}\,, \quad u_2 \simeq u_f\,(T_2) = 19.21\frac{\text{kJ}}{\text{kg}}\,,$$
$$h_2 \simeq h_f\,(T_2) = 19.29\frac{\text{kJ}}{\text{kg}}\,, \quad s_2 \simeq s_f\,(T_2) = 0.0798\frac{\text{kJ}}{\text{kg}}\,.$$

Work and heat per unit mass are obtained as

$$w_{12} = p\,(v_2 - v_1) = -24.58\frac{\text{kJ}}{\text{kg}}\,,$$
$$q_{12} = h_2 - h_1 = -266.95\frac{\text{kJ}}{\text{kg}}\,.$$

7.9.5 *Isentropic Compression of Ideal Gas*

We consider the isentropic (adiabatic reversible) compression of air with a compression ratio $V_1/V_2 = 8$. The initial state of the air is $T_1 = 290\,\text{K}$ and $p_1 = 95\,\text{kPa}$, so that $s_1^0 = s^0\,(T_1) = 7.103\frac{\text{kJ}}{\text{kg}\,\text{K}}$ (from the Table in Fig. 6.17). The final temperature must be obtained from (7.19), which reads

$$8 = \frac{v_1}{v_2} = \frac{v_r\,(T_1)}{v_r\,(T_2)} = \frac{T_1 \exp\left[\frac{s^0(T_2)}{R}\right]}{T_2 \exp\left[\frac{s^0(T_1)}{R}\right]}\,. \tag{7.30}$$

The table includes values for $v_r\,(T)$ which we use now. From the table we find $v_r\,(T_1) = 319.5$, so that $v_r\,(T_2) = v_r\,(T_1)\,/8 = 39.93$. To find T_2 from the table, we have to interpolate between 650 K and 660 K, which gives

$$T_2 = 650\,\text{K} + \frac{v_r\,(T_2) - v_r\,(650\,\text{K})}{v_r\,(660\,\text{K}) - v_r\,(650\,\text{K})}10\,\text{K} = 652.4\,\text{K}\,.$$

[1] The saturation pressure at T_2 is $p_{\text{sat}}\,(-30°\text{C}) = 84.4\text{kPa}$. The derived approximation for enthalpy adds the term $v_f\,(T_2)\,(p - p_{\text{sat}}\,(T_2)) = 0.069\frac{\text{kJ}}{\text{kg}}$ to enthalpy—here this term contributes little, and can be safely ignored.

To find the proper temperature value when v_r is not provided in a table, one has to use trial and error. A first guess can be obtained from assuming constant specific heats, which yields $\hat{T}_2 = T_1 \left(\frac{v_1}{v_2}\right)^{k-1} = 666\,\mathrm{K}$ (with $k = 1.4$). With data from the table we find for $T_2^{(1)} = 650\,\mathrm{K}$ and $T_2^{(2)} = 660\,\mathrm{K}$

$$\frac{T_1 \exp\left[\frac{s^0\left(T_2^{(1)}\right)}{R}\right]}{T_2^{(1)} \exp\left[\frac{s^0(T_1)}{R}\right]} = 7.905 \quad , \quad \frac{T_1 \exp\left[\frac{s^0\left(T_2^{(2)}\right)}{R}\right]}{T_2^{(2)} \exp\left[\frac{s^0(T_1)}{R}\right]} = 8.231 \, .$$

Linear interpolation gives a value below the estimate, $T_2 = 642.9\,\mathrm{K}$.

The internal energies for the two states are read from the table as

$$u_1 = u\left(T_1\right) = 207.74 \frac{\mathrm{kJ}}{\mathrm{kg}} \quad , \quad u_2 = u\left(T_2\right) = 475.97 \frac{\mathrm{kJ}}{\mathrm{kg}} .$$

The work required for compression is

$$w_{12} = u_1 - u_2 = -268.2 \frac{\mathrm{kJ}}{\mathrm{kg}} \, .$$

Since the process is adiabatic, $q_{12} = 0$.

7.9.6 Reversible and Irreversible Adiabatic Expansion

Reversible processes require control over the process at all times. The slow expansion of a gas in a piston-cylinder system is the prototypical example for reversible processes.

To further our understanding, we consider the adiabatic expansion—reversible and irreversible—of air as ideal gas at initial state $p_1 = 10\,\mathrm{bar}$, $T_1 = 500\,\mathrm{K}$ to an end state of half the pressure, so that $p_2 = \frac{1}{2} p_1$.

For the adiabatic reversible case we can refer to the above table which tells us that

$$w_{12} = u\left(T_1\right) - u\left(T_2\right) \, ,$$

where T_2 follows from isentropicity of the process,

$$0 = s\left(T_2, p_2\right) - s\left(T_1, p_1\right) = s^0\left(T_2\right) - s^0\left(T_1\right) - R \ln \frac{p_2}{p_1} \, .$$

With the relative pressure $p_r\left(T\right) = a \exp\left[\frac{s^0(T)}{R}\right]$, this relation assumes the form

$$\frac{p_r\left(T_2\right)}{p_r\left(T_1\right)} = \frac{p_2}{p_1} = \frac{1}{2} \, .$$

The table in Fig. 6.17 gives $p_r\,(T_1 = 500\,\mathrm{K}) = 6.202$, hence $p_r\,(T_2) = 3.101$, and interpolation in the table yields $T_2 = 412\,\mathrm{K}$.

The tabulated relative pressure simplifies the determination of the final state. We now show how one has to proceed when p_r is not in the table: The property table gives $s^0\,(T_1 = 500\,\mathrm{K}) = 7.654\frac{\mathrm{kJ}}{\mathrm{kg\,K}}$ and with $R_{air} = 0.287\frac{\mathrm{kJ}}{\mathrm{kg\,K}}$ and $\frac{p_2}{p_1} = \frac{1}{2}$ the above equation gives

$$s^0\,(T_2) = s^0\,(T_1) + R\ln\frac{p_2}{p_1} = 7.455\frac{\mathrm{kJ}}{\mathrm{kg\,K}} .$$

Interpolation in the table gives $T_2 = 412\,\mathrm{K}$. The corresponding work per unit mass is

$$w_{12} = u\,(500\,\mathrm{K}) - u\,(412\,\mathrm{K}) = 360.34\frac{\mathrm{kJ}}{\mathrm{kg}} - 295.38\frac{\mathrm{kJ}}{\mathrm{kg}} = 64.96\frac{\mathrm{kJ}}{\mathrm{kg}} .$$

We compare the reversible adiabatic process to the fully irreversible process, which was discussed in Secs. 3.13, 4.13. There we saw that the temperature of the ideal gas remained constant. If the initial and final pressures are the same as for the reversible process, we compute the change of entropy as

$$s_2 - s_1 = s\,(T_1, p_2) - s\,(T_1, p_1) = s^0\,(T_1) - s^0\,(T_1) - R\ln\frac{p_2}{p_1} = 0.199\frac{\mathrm{kJ}}{\mathrm{kg\,K}} .$$

In the irreversible process no useful work is produced and entropy is generated.

7.9.7 Isentropic Expansion of Compressed Water

Water at $T_1 = 300\,°\mathrm{C}$, $p_1 = 20\,\mathrm{MPa}$ expands in an isentropic (adiabatic and reversible) process to $p_2 = 1\,\mathrm{atm}$. We determine the final temperature, the volume change, and the work.

The saturation temperature for the initial pressure is $T_{\mathrm{sat}}\,(p_1) = 365.8\,°\mathrm{C} > T_1$. Accordingly the initial state is compressed liquid. As will be seen, the process ends in the two phase region. Figure 7.8 shows the p-v- and T-s-diagrams.

From the property table in Fig. 6.16 we find the initial properties

$$v_1 = v\,(20\,\mathrm{MPa}, 300\,°\mathrm{C}) = 0.0013596\frac{\mathrm{m}^3}{\mathrm{kg}} ,$$

$$u_1 = u\,(20\,\mathrm{MPa}, 300\,°\mathrm{C}) = 1306.1\frac{\mathrm{kJ}}{\mathrm{kg}} ,$$

$$h_1 = h\,(20\,\mathrm{MPa}, 300\,°\mathrm{C}) = 1333.3\frac{\mathrm{kJ}}{\mathrm{kg}} ,$$

$$s_1 = s\,(20\,\mathrm{MPa}, 300\,°\mathrm{C}) = 3.2071\frac{\mathrm{kJ}}{\mathrm{kg\,K}} ,$$

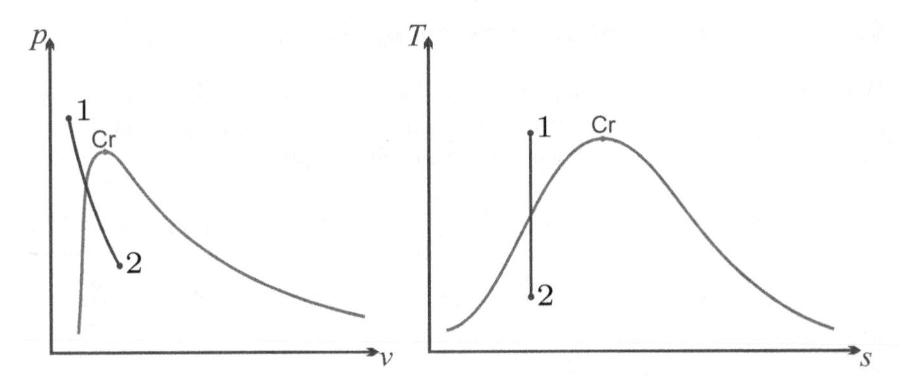

Fig. 7.8 Isentropic expansion of compressed liquid into the two-phase region in p-v- and T-s-diagram

while the approximations of Sec. 6.9 give

$$v_1 \simeq v_f(T_1) = 0.001404\frac{\text{m}^3}{\text{kg}},$$
$$u_1 \simeq u_f(T_1) = 1332.7\frac{\text{kJ}}{\text{kg}},$$
$$h_1 \simeq h_f(T_1) + v_f(T_1)(p_1 - p_{sat}(T_1)) = 1360.0\frac{\text{kJ}}{\text{kg}},$$
$$s_1 \simeq s_f(T_1) = 3.2534\frac{\text{kJ}}{\text{kg K}}.$$

Comparison shows that the approximations introduce errors of {3.2%, 2.0%, 2%, 1.4%}. For the computation of the final state, we use the table values.

The final state is given by its entropy, $s_2 = s_1 = 3.2071\frac{\text{kJ}}{\text{kg K}}$ and its pressure, $p_2 = 1\,\text{atm}$. Since $s_f(p_2) < s_2 < s_g(p_2)$ this state is saturated liquid-vapor mixture with the quality

$$x_2 = \frac{s_2 - s_f(p_2)}{s_{fg}(p_2)} = 0.314,$$

and the properties

$$T = 100\,^\circ\text{C}, \quad v_2 = 0.526\frac{\text{m}^3}{\text{kg}}, \quad u_2 = 1074.8\frac{\text{kJ}}{\text{kg}},$$
$$h_2 = 1128.2\frac{\text{kJ}}{\text{kg}}, \quad s_2 = 3.2071\frac{\text{kJ}}{\text{kg K}}.$$

The volume changes quite a bit due to evaporation, $v_2 - v_1 = 0.525\frac{\text{m}^3}{\text{kg}}$, and the expansion gives the work $w_{12} = u_1 - u_2 = 231.3\frac{\text{kJ}}{\text{kg}}$.

7.9.8 Isothermal Expansion of Steam

Water vapor at $p_1 = 200\,\text{bar}$ and $T = 400\,°\text{C}$ is isothermally expanded to $p_2 = 1\,\text{bar}$. To determine work and heat we require internal energy and entropy at the two states. From a steam table we find

$$u_1 = u\,(20\,\text{MPa}, 400\,°\text{C}) = 2619.3\frac{\text{kJ}}{\text{kg}}\,, \quad s_1 = s\,(20\,\text{MPa}, 400\,°\text{C}) = 5.554\frac{\text{kJ}}{\text{kg K}}\,,$$

$$u_2 = u\,(1\,\text{bar}, 400\,°\text{C}) = 2967.9\frac{\text{kJ}}{\text{kg}}\,, \quad s_2 = s\,(1\,\text{bar}, 400\,°\text{C}) = 8.5435\frac{\text{kJ}}{\text{kg K}}\,,$$

so that heat and work are

$$q_{12} = T\,(s_2 - s_1) = 2012.4\frac{\text{kJ}}{\text{kg}}\,,$$

$$w_{12} = u_1 - u_2 + q_{12} = 1663.8\frac{\text{kJ}}{\text{kg}}\,;$$

note that for the computation of heat the thermodynamic temperature of 673.15 K must be taken.

It is interesting to compare this result with that obtained under the assumption that water vapor can be described as an ideal gas ($R = 0.462\frac{\text{kJ}}{\text{kg K}}$), which yields

$$q_{12} = w_{12} = -RT\ln\frac{p_2}{p_1} = 1647.8\frac{\text{kJ}}{\text{kg}}\,.$$

We see clear differences, in particular for the heat q_{12}, which are due to real gas effects. Clearly, the assumption of ideal gas behavior of steam at these conditions is not justified.

7.9.9 Polytropic Process

A mass of $m = 3\,\text{kg}$ of neon (monatomic ideal gas with $M = 20.18\frac{\text{kg}}{\text{kmol}}$, $R = 0.412\frac{\text{kJ}}{\text{kg K}}$, $c_v = \frac{3}{2}R = 0.618\frac{\text{kJ}}{\text{kg K}}$) is compressed from $V_1 = 2\,\text{m}^3$, $T_1 = 450\,°\text{C}$ to $V_2 = 0.5\,\text{m}^3$ in a polytropic process with polytropic exponent $n = 1.3$.

The polytropic exponent lies between unity and the ratio of specific heats, $k = 1.67$, and thus the process curve must lie between isothermal and isentropic lines as indicated in Fig. 7.9.

The final temperature follows from the polytropic relation as

$$T_2 = T_1\left(\frac{V_1}{V_2}\right)^{n-1} = 1095.9\,\text{K}\,.$$

The work can be obtained from (7.28) as

$$W_{12} = \frac{mR}{1-n}\,(T_2 - T_1) = -1536.2\,\text{kJ}\,.$$

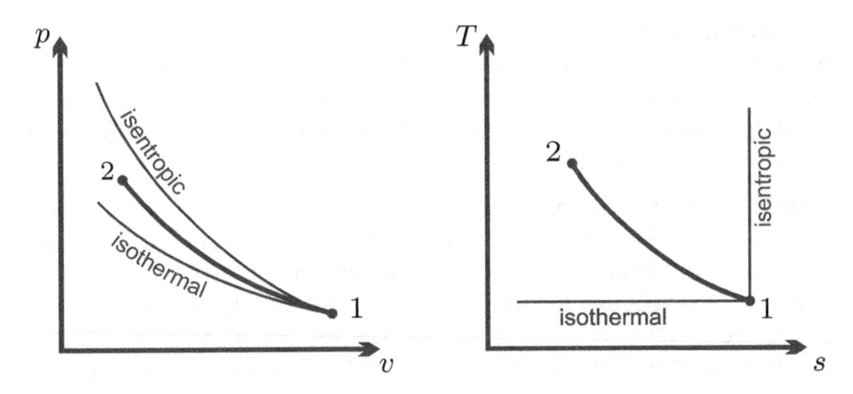

Fig. 7.9 Polytropic process with $1 < n < k$

The heat removed from the system follows, with $U = mc_v (T - T_0)$, as

$$Q_{12} = U_2 - U_1 + W_{12} = m \left[c_v + \frac{R}{1-n} \right] (T_2 - T_1) = -845.0\,\mathrm{kJ}\,.$$

To draw the proper curve in the T-s-diagram, it is best to compute the entropy change,

$$\begin{aligned} S_2 - S_1 &= m \left[c_v \ln \frac{T_2}{T_1} + R \ln \frac{V_2}{V_1} \right] \\ &= mR \left[\frac{1}{k-1} - \frac{1}{n-1} \right] \ln \frac{T_2}{T_1} = -0.942 \frac{\mathrm{kJ}}{\mathrm{K}}\,. \end{aligned}$$

Thus, in this compression process, entropy decreases, and temperature grows, as indicated in the diagram.

Problems

7.1. Water in Tank

A closed rigid tank contains 3 kg of saturated water vapor, initially at 140 °C. Heat transfer occurs, and the pressure drops to 200 kPa. Kinetic and potential energy effects are negligible. Determine heat and work exchanged during the process.

7.2. Isochoric Heating of Air

2 kg of air are heated in a reversible process at constant volume. The initial temperature and pressure are $T_1 = 20\,°\mathrm{C}$ and $p_1 = 2\,\mathrm{bar}$, and the final temperature is $T_2 = 500\,\mathrm{K}$. Compute heat and work exchanged, and the change in entropy. Draw the process in p-v- and T-s-diagrams.

7.3. Water in Tank

A closed rigid tank contains 2 kg of saturated water (liquid and vapor), initially at 0.2 MPa with a quality of 4.65%. How much heat must be added so that the final state is saturated vapor? What is the final temperature, and how much work is required?

7.4. Heating and Melting of Ice

2 kg of ice are initially at −20 °C and 1 bar. The ice is isobarically heated, then melted and further heated until a temperature of 20 °C is reached. Determine the heat required for this process, the volume change, and the work. Determine also the heat required to heat the ice to 0 °C and for melting at 0 °C. The heat of melting at 1 bar is $h_{sf} = 333.1 \frac{\text{kJ}}{\text{kg}}$, and the specific heat of ice is $c_{ice} = 2.1 \frac{\text{kJ}}{\text{kg K}}$.

7.5. Freezing of Water

1.6 kg of liquid water are initially at 15 °C and 1 bar. The water is isobarically cooled, then frozen and further cooled until a temperature of −15 °C is reached. Determine the heat required for this process, the volume change, and the work. The heat of melting at 1 bar is $h_{sf} = 333.1 \frac{\text{kJ}}{\text{kg}}$, and the specific heat of ice is $c_{ice} = 2.1 \frac{\text{kJ}}{\text{kg K}}$.

7.6. Condensation of Steam

Steam (water vapor) initially at 30 bar, 450 °C is isobarically cooled until the volume is one half of the initial volume.

1. Draw the process in a p-v- and in a T-s-diagram with respect to saturation lines.
2. Determine heat and work for the process when the initial volume was 2 m^3.
3. Now the volume is fixed and heat is supplied. At what temperature is the saturated vapor state reached?

7.7. Lowering of a Piston

A freely moving piston with cross section $A = 0.1\,\text{m}^2$ and mass $m = 2\,\text{t}$ closes a cylinder filled with air; the external pressure is 1 atm. The initial state in the cylinder is $V_1 = 0.3\,\text{m}^3$, $T_1 = 500\,\text{K}$. Heat is withdrawn, and the piston moves down as the volume of the gas decreases. The piston movement stops when the volume reaches 2/3 of the original volume, but there is further cooling until the temperature is 270 K. Compute the mass of air in the cylinder, and the total amounts of work and heat exchanged. Draw the process in p-v and T-s-diagrams.

7.8. Cooling of Air

10 grams of air at 1400 K, 150 bar are cooled in a closed system. The total heat withdrawn is 7936 J and the final temperature is 600 K. The cooling occurs first at constant pressure (from state 1 to state 2), and then at constant volume (from state 2 to the final state 3). Compute first the temperature at state 2, and then the pressure at state 3. Also determine the work done by the process.

7.9. Isentropic Compression of Saturated Liquid-Vapor Mixture
Saturated liquid-vapor mixture of water at 25 °C with a quality of $x = 0.9$ is compressed in an adiabatic reversible process to 175 bar. Determine the temperature of the final state, and work and heat per unit mass.

7.10. Isentropic Expansion of Air
Air is isentropically expanded in a closed system from $T_1 = 25\,°\mathrm{C}$ and $p_1 = 1\,\mathrm{MPa}$ to $p_2 = 2.5\,\mathrm{bar}$. Determine heat and work exchanged per unit mass. Draw the process in p-v and T-s-diagrams.

7.11. Isentropic Expansion
Neon and air are expanded isentropically from 1000 kPa and 500 °C to 100 kPa in a piston-cylinder device. Which gas has the lower temperature after expansion? Why? Compute the work per unit mass for both.

7.12. Isentropic Compression
Which of the two gases—neon or air—has the higher final temperature as it is compressed isentropically from 100 kPa and 450 K to 1000 kPa in a piston-cylinder device? Compute the work per unit mass for both cases.

7.13. Isentropic Expansion of Superheated R134a Vapor
Cooling fluid R134a in a closed system is initially at 1.2 MPa, 50 °C. Then the cooling fluid is expanded in an adiabatic reversible process to 0.12 MPa. Determine the temperature of the final state, and work and heat per unit mass.

7.14. Isentropic Expansion of R134a Vapor
Cooling fluid R134a in a closed system is initially at 1.6 MPa, 60 °C. Then the cooling fluid is expanded in an adiabatic reversible process to 0.32 MPa. Determine the temperature of the final state, and work and heat per unit mass.

7.15. Expansion of Air
Air (ideal gas with variable specific heats) at 1400 K, 50 bar is expanded in a piston-cylinder system until its volume is 12 times the initial volume. Determine work and heat per unit mass (a) when the expansion is isentropic, (b) when the expansion is isothermal.

7.16. Isothermal Compression of Water Vapor
In a piston-cylinder system, a mass of 20 kg of water vapor initially at 3 bar, 1200 °C is isothermally compressed to 50 bar.

1. Determine heat and work for this process based on the property tables of water.
2. Assume water vapor at these conditions can be described as an ideal gas and compute work and heat based on this assumption. Compare with the result of the exact calculation and discuss the differences.

7.17. Evaporation and Expansion
As part of the processes in a low temperature Carnot engine, R134a undergoes the following process in a piston-cylinder system:

1-2: Isothermal evaporation and heating from saturated liquid state at $T_1 = 60\,°\mathrm{C}$ until the volume is 13 times the initial volume.

2-3: Isentropic expansion to $p_3 = 0.28\,\mathrm{MPa}$.

1. Draw the process in a p-v- and in a T-s-diagram with respect to saturation lines.
2. Determine heat and work for the process when the initial volume was $V_1 = 20\,\mathrm{litres}$.
3. What would be the thermal efficiency of the corresponding Carnot engine?

7.18. Polytropic Compression of Oxygen
Pure oxygen is compressed in a polytropic process with polytropic exponent $n = 1.25$ so that the final volume is half the original volume. The initial temperature is 300 K, the final pressure is 10 bar, and the work done is 40 kJ. Determine the final temperature, the initial pressure, the mass of oxygen, the heat exchanged in the process, and the change in entropy. Draw the process in p-v and T-s-diagrams.

7.19. Polytropic Compression
Argon gas, initially at 1 bar, 100 K, undergoes a polytropic process with $n = 1.5$ to a final pressure of 17 bar. Determine the specific work and heat transfer for the process. Argon can be treated as an ideal gas; recall that it is a monatomic gas, so the specific heats are constant.

7.20. Polytropic Expansion
Helium gas, initially at 20 bar, 200 K, undergoes a polytropic process with $n = 1.2$ to a final pressure of 2 bar. Determine the specific work and heat transfer for the process. Helium can be treated as an ideal gas, recall that it is a monatomic gas, so the specific heats are constant.

7.21. Polytropic Compression
Radon gas (Rn, $M_{\mathrm{Rn}} = 222\frac{\mathrm{g}}{\mathrm{mol}}$) initially at 4 bar, 400 K, is compressed in a piston cylinder system. After compression the measured pressure and temperature are 12 bar and 600 K, respectively. Assume that the process can be described as being polytropic, and determine the polytropic exponent n. Then determine the specific work and heat transfer for the process. Radon can be treated as an ideal gas; it is monatomic, hence the specific heats are constant.

7.22. Compression of Air
Air at $T_1 = 227\,°\mathrm{C}$, $p_1 = 1\,\mathrm{atm}$ is compressed in a piston-cylinder device to 1/3 of its original volume. Compute the work and the heat transfer per kg of air when the compression process is (a) isothermal, (b) isentropic, (c) isentropic with constant specific heats (cold air approximation), (d) polytropic

with $n = 1.4$, (e) polytropic with $n = 1.1$. Draw the process curves in p-v and T-s-diagrams.

7.23. Irreversible Expansion of Helium
An adiabatic and rigid container is divided by a membrane so that one third of the container holds 1 kg of helium at 300 K and 100 Pa while the other part is evacuated. The membrane is destroyed, and the gas undergoes a rather fast and irreversible process until it assumes its stable equilibrium state.

1. Compute temperature and pressure in the equilibrium state, and the change of entropy for the process.
2. Design a reversible compression process that will bring the gas back to its original state (i.e. filling 1/3 of the container, 300 K, 100 Pa) and compute the work and heat exchange required.

7.24. Ice and Saturated Liquid-Vapor Mixture
An insulated piston–cylinder device initially contains 0.01 m^3 of saturated liquid–vapor mixture with a quality of 0.2 at 120 °C. How much ice at 0 °C must be added isobarically to the cylinder so that after equilibrium is reached the cylinder contains saturated liquid at 120 °C? Hint: The process is isobaric, work is done.

Chapter 8
Closed System Cycles

8.1 Thermodynamic Cycles

In previous chapters we discussed thermal efficiency of heat engines and coefficient of performance of refrigerators and heat pumps from a general viewpoint, without asking for the processes occurring within the engines. Now we will discuss the working principles of several closed system cycles.

All engines considered are steady state devices that do not accumulate energy or mass over time. To realize a steady state thermodynamic engine, a working fluid is subjected to a series of processes such that the process curve in a property diagram forms a closed loop. As the engine operates, the working fluid runs through the same cycle of processes again and again.

The closed loop integral of the energy vanishes, $\oint dE = 0$, since energy is a state property;[1] the engine does not accumulate energy. Thus, integration of the differential energy balance $dE = \delta Q - \delta W$ over the full cycle yields

$$W_{\odot} = \oint \delta W = \oint \delta Q = Q_{\odot} = Q_{in} - |Q_{out}| \ , \tag{8.1}$$

where $W_{\odot}$ and $Q_{\odot}$ are the total net work and net heat exchanged for the cycle. Moreover, $Q_{in} > 0$ is the total heat transferred into the cycle, and $Q_{out} < 0$ is the total heat transferred out.

Net work and net heat are positive for a clockwise process—a heat engine—and are negative for a counter-clockwise process—a refrigerator or a heat pump.

For a closed system, the engine contains the constant mass m of the working substance, and the net work and heat per unit mass are

$$w_{\odot} = \frac{W_{\odot}}{m} = \frac{Q_{\odot}}{m} = q_{\odot} = q_{in} - |q_{out}| \ . \tag{8.2}$$

[1] We have $\int_1^2 dE = E_2 - E_1$. For a closed loop inital and endpoint are the same, that is $E_2 = E_1$.

H. Struchtrup, *Thermodynamics and Energy Conversion*,
DOI: 10.1007/978-3-662-43715-5_8,

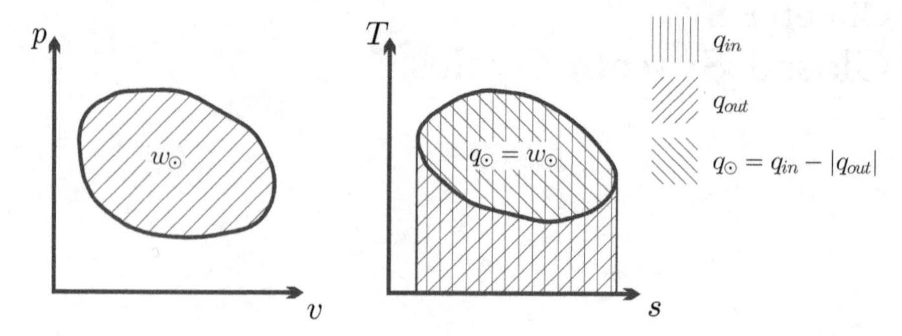

Fig. 8.1 Thermodynamic cycle in p-v- and T-s-diagram. Heat in, q_{in}, and heat out, q_{out}, are indicated for a heat engine, which runs clockwise.

For an engine that runs through the cycle with the frequency $\dot{n}$ (measured for instance in rounds per minute, rpm) the net power and the heat transfer rates are

$$\dot{W}_\odot = \dot{n} m w_\odot = \dot{n} m q_\odot = \dot{Q}_\odot \quad , \quad \dot{Q}_{in} = \dot{n} m q_{in} \quad , \quad \dot{Q}_{out} = \dot{n} m q_{out} \; . \tag{8.3}$$

Figure 8.1 shows a reversible thermodynamic cycle in the p-v- and T-s-diagrams. For a reversible process in a closed system, the net work of the cycle per unit mass of working fluid is

$$w_\odot = \oint p dv \; . \tag{8.4}$$

In the p-v-diagram, the net work is just the area enclosed by the process curve as indicated in the figure. Note that the integral is positive for a clockwise cycle, and negative for a counter-clockwise cycle.

Similarly, the net heat exchanged for a reversible cycle is the area enclosed by the cycle in the T-s-diagram,

$$q_\odot = \oint T ds \; , \tag{8.5}$$

where the heat is positive for a clockwise cycle. Heat in and heat out,

$$q_{in} = \int_{ds>0} T ds \quad , \quad q_{out} = \int_{ds<0} T ds \; , \tag{8.6}$$

can be read from the T-s-diagram as the areas below the respective process curves. This is indicated in the figure for a heat engine (clockwise cycle). With this, the net work ($w_\odot = q_\odot$), the heat in, and the heat out can all be read from the T-s-diagram. This implies that the thermal efficiency or the coefficient of performance for a reversible cycle can be completely determined from the T-s-diagram, by means of the relations

$$\begin{aligned}
\eta_{th} &= \frac{w_\odot}{q_{in}} = \frac{\oint T ds}{\int_{ds>0} T ds} = 1 - \frac{\left|\int_{ds<0} T ds\right|}{\int_{ds>0} T ds} , \\
\mathrm{COP_R} &= \frac{q_{in}}{|w_\odot|} = \frac{\int_{ds>0} T ds}{\left|\oint T ds\right|} = \frac{1}{\frac{\left|\int_{ds<0} T ds\right|}{\int_{ds>0} T ds} - 1} , \qquad (8.7) \\
\mathrm{COP_{HP}} &= \frac{q_{out}}{|w_\odot|} = \frac{\int_{ds<0} T ds}{\left|\oint T ds\right|} = \frac{1}{1 - \frac{\int_{ds>0} T ds}{\left|\int_{ds<0} T ds\right|}} .
\end{aligned}$$

While in this chapter we discuss only reversible cycles, we note that the processes in real engines always suffer from irreversibilities, so that their thermal efficiencies or COP's will be smaller than those that will be computed below.

8.2 Carnot Cycle

As a first example for the evaluation of a thermodynamic cycle we consider the Carnot cycle. We introduced the Carnot engine as a fully reversible engine—no internal or external irreversibilities—operating between two reservoirs of temperatures T_H, T_L. The Carnot cycle is *one* possible realization of such an engine. For a heat engine, it consists of the following four processes

1-2 rev. isothermal compression at T_L
2-3 rev. adiabatic (isentropic) compression from T_L to T_H
3-4 rev. isothermal expansion at T_H
4-1 rev. adiabatic (isentropic) expansion from T_H to T_L

Heat is only exchanged with the reservoirs during the isothermal processes, at which the working substance is at the temperature of the reservoirs, and therefore there are no external irreversibilities associated with the cycle. Since there are no internal or external irreversibilities, the above cycle is a fully reversible cycle that exchanges heat only with the two reservoirs.

The T-s-diagram allows us to compute the thermal efficiency of the cycle: The area enclosed by the cycle is the net work,

$$w_\odot = q_\odot = (T_H - T_L)\Delta s . \qquad (8.8)$$

The area below the curve 3-4 is the heat in,

$$q_{in} = q_{34} = T_H \Delta s . \qquad (8.9)$$

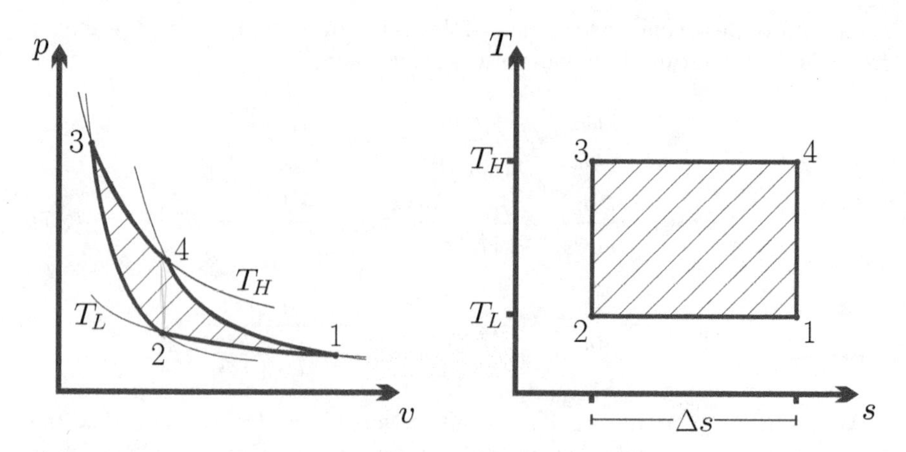

Fig. 8.2 Carnot cycle in the p-v- and T-s-diagrams

Accordingly, the thermal efficiency is, as expected, the Carnot efficiency

$$\eta_C = \frac{w_\odot}{q_{in}} = 1 - \frac{T_L}{T_H} \,. \tag{8.10}$$

We now compute the efficiency by considering the cycle for an ideal gas. With the work and heat of reversible processes computed in the last chapter, we find the values for heat and work of the individual processes as

$$\begin{array}{lll} \text{1-2 isothermal:} & w_{12} = -RT_L \ln \frac{p_2}{p_1} & , \; q_{12} = -RT_L \ln \frac{p_2}{p_1} \,, \\ \text{2-3 isentropic:} & w_{23} = u\left(T_L\right) - u\left(T_H\right) & , \; q_{23} = 0 \,, \\ \text{3-4 isothermal:} & w_{34} = -RT_H \ln \frac{p_4}{p3} & , \; q_{34} = -RT_H \ln \frac{p_4}{p3} \,, \\ \text{4-1 isentropic:} & w_{41} = u\left(T_H\right) - u\left(T_L\right) & , \; q_{41} = 0 \,, \end{array} \tag{8.11}$$

Since the processes (2-3) and (4-1) are isentropic, we have

$$\begin{aligned} 0 &= s_3 - s_2 = s^0\left(T_H\right) - s^0\left(T_L\right) - R \ln \frac{p_3}{p_2} \,, \\ 0 &= s_4 - s_1 = s^0\left(T_H\right) - s^0\left(T_L\right) - R \ln \frac{p_4}{p_1} \,. \end{aligned} \tag{8.12}$$

From comparison of the two equations we find that the pressures at the corner points of the process are related as $\frac{p_3}{p_2} = \frac{p_4}{p_1}$, or, alternatively, $\frac{p_3}{p_4} = \frac{p_2}{p_1}$.

The thermal efficiency of the cycle is

$$\eta_C = \frac{w_\odot}{q_{in}} = \frac{w_{12} + w_{23} + w_{34} + w_{41}}{q_{34}} \,. \tag{8.13}$$

With the above results for work and heat, and the relation between the pressures, we find, once more, the Carnot efficiency,

$$\eta_C = \frac{-T_L \ln \frac{p_2}{p_1} - T_H \ln \frac{p_4}{p_3}}{-T_H \ln \frac{p_4}{p_3}} = 1 - \frac{T_L}{T_H} . \tag{8.14}$$

That we found the well-known result again from detailed calculations for an ideal gas proves that the ideal gas temperature scale (Sec. 2.13) is identical to the thermodynamic temperature scale (Sec. 5.6).

The net work of the ideal gas Carnot cycle,

$$w_\odot = R\,(T_H - T_L) \ln \frac{p_2}{p_1} \tag{8.15}$$

grows with the temperature difference $(T_H - T_L)$ and the pressure ratio $\frac{p_2}{p_1}$. Thus, for large efficiency and large work output the Carnot cycle should operate at large temperature difference $T_H - T_L$ and at large pressure ratios $\frac{p_2}{p_1}$. Then, the volume ratio between smallest and largest volume,[2] $\frac{v_1}{v_3} = \frac{p_2}{p_1}\left(\frac{T_H}{T_L}\right)^{\frac{1}{k-1}}$, becomes large, which is quite unpractical for designing a compact engine. Moreover, the overall pressure ratio $\frac{p_3}{p_1} = \frac{v_1}{v_3}\frac{T_H}{T_L} = \frac{p_2}{p_1}\left(\frac{T_H}{T_L}\right)^{\frac{k}{k-1}}$ becomes large, which makes effective sealing difficult. For example, an engine with air as working gas operating at $T_L = 300\,\mathrm{K}$, $T_H = 750\,\mathrm{K}$, and a pressure ratio of $\frac{p_2}{p_1} = 5$, has the thermal efficiency $\eta_C = 1 - \frac{T_L}{T_H} = 0.6$, the overall volume ratio $\frac{v_1}{v_3} = 49.5$, and the overall pressure ratio $\frac{p_3}{p_1} = 124$; it produces the net work $w_\odot = 208\frac{\mathrm{kJ}}{\mathrm{kg}}$. Internal combustion engines with comparable net work operate with significantly smaller volume and pressure ratios, and thus are more compact and lighter, and suffer less from sealing problems.

Another problem for the Carnot cycle is that the isothermal heat exchange processes (1-2, 3-4) require slow processes, so that the cycle frequency $\dot{n}$ must be low. To produce significant amounts of power $\dot{W} = \dot{n} m w_\odot$ the engine would have to contain a large mass m, that is it must be large. Engines that operate on higher frequencies $\dot{n}$ can be more compact.

The Carnot engine operates between two reservoirs of constant temperature. If the engine is to be heated by burning of a fuel, one does not have a constant high temperature reservoir, but a flow of hot combustion product at flame temperature T_F which is gradually cooled to T_H in the heat exchange. This implies that there will be a temperature difference between engine and combustion gas, and thus an external irreversibility. If one uses such a hot flow to heat the engine, one will have warm exhaust which has still work potential. If the exhaust is expelled into the environment, the equilibration of temperature between the warm exhaust (at T_H or higher) and the environment (at T_L or lower) is an external irreversibility. To eliminate, or at least

[2] For the isothermal process 1-2 we have $p_1 v_1 = p_2 v_2$ which gives $\frac{v_1}{v_3} = \frac{p_2}{p_1}\frac{v_2}{v_3}$; the adiabatic relation for the process 2-3 gives $\frac{v_2}{v_3} = \left(\frac{T_H}{T_L}\right)^{\frac{1}{k-1}}$.

reduce this loss, the exhaust must be used for preheating of the combustion air. We shall come back to this point in Secs. 11.7 and 12.1.

In principle, one could make an effort to design an engine that follows the Carnot cycle. Obviously, due to irreversibilities, the real engine would have an efficiency below the Carnot efficiency. Moreover, for the reasons listed above, such an engine would be relatively large and heavy in relation to the amount of power it could deliver. Thus, in fact, one does not try to build engines that follow the Carnot cycle for practical applications.

The Stirling cycle with an ideal gas, which will be discussed in Sec. 13.1, is an alternative realization of a Carnot engine, but compared to the Carnot cycle it has a significantly smaller volume ratio. Stirling engines, i.e., engines designed to follow the Stirling cycle, are commercially available. Naturally, these have efficiencies below the Carnot efficiency due to unavoidable internal and external irreversibilities.

8.3 Carnot Refrigeration Cycle

Inversion of the Carnot cycle gives the Carnot refrigeration (or heat pump) cycle with the following processes:

1-2 rev. adiabatic (isentropic) compression from T_L to T_H
2-3 rev. isothermal compression at T_H
3-4 rev. adiabatic (isentropic) expansion from T_H to T_L
4-1 rev. isothermal expansion at T_L

The process curves in the p-v- and T-s-diagrams are depicted in Fig. 8.3. As before, the curve in the T-s-diagram is independent of the working fluid, while the p-v-diagram is sketched for an ideal gas as working fluid.

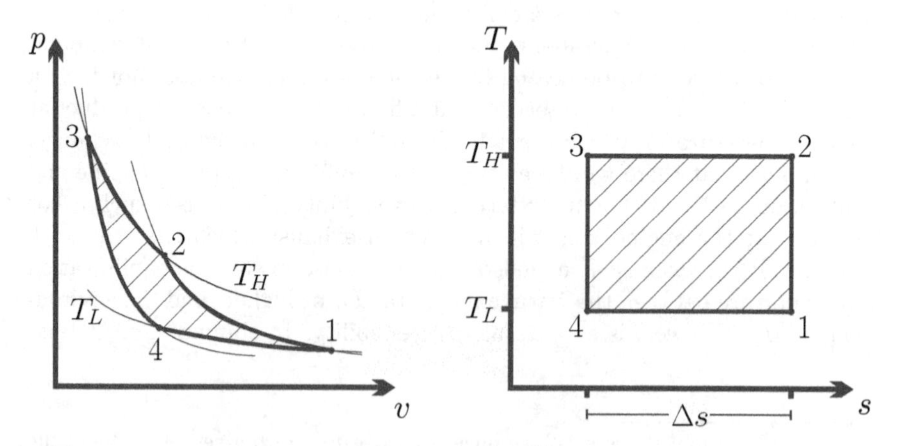

Fig. 8.3 Carnot refrigeration cycle (inverse Carnot cycle)

The analysis of the inverse cycle is just analog to the analysis in the preceding section, and we do not go through the details. The well-known coefficients of performance for a Carnot refrigerator and a Carnot heat pump can be easily read from the T-s-diagram as

$$\mathrm{COP}_{\mathrm{R},C} = \frac{|q_{in}|}{|w_{\odot}|} = \frac{T_L \Delta s}{(T_H - T_L)\,\Delta s} = \frac{1}{\frac{T_H}{T_L} - 1}\,, \tag{8.16}$$

$$\mathrm{COP}_{\mathrm{HP},C} = \frac{|q_{out}|}{|w_{\odot}|} = \frac{T_H \Delta s}{(T_H - T_L)\,\Delta s} = \frac{1}{1 - \frac{T_L}{T_H}}\,. \tag{8.17}$$

8.4 Internal Combustion Engines

In an internal combustion engine, heat is provided to the system by burning an air-fuel mixture *inside* the system. Engine operation requires the exchange of the working fluid after one cycle is completed, to bring in new fuel and oxygen. Accordingly, internal combustion engines exchange mass with their surroundings. Nevertheless, during the working cycle the system is closed, and thus internal combustion engines can be analyzed as closed systems.

Internal combustion engines are the dominant power source for cars, trucks, ships and non-electric trains. In cars one usually finds Otto engines, while in trucks, ships and trains Diesel engines are used.

Figure 8.4 shows a sketch of a single piston-cylinder assembly of an internal combustion engine. The piston is connected through rods to the crankshaft. The crankshaft is driven through the expansion process, which pushes the piston down, and it provides the work for the compression processes. Most engines have several cylinders that run through the cycle with a phase shift to ensure even load on the crankshaft; in single cylinder engines a fly wheel might be mounted to the crankshaft.

As the crank shaft turns, the piston moves between *bottom dead center* and *top dead center*. The volume the piston moves through is known as the *swept volume* V_s, the remaining volume at top dead center is the *clearance volume* V_c. Valves allow the exchange of working fluid with the surroundings.

The main difference between Otto and Diesel engine is that an Otto engine draws in air-fuel-mix while in a Diesel engine the fuel is injected into the cylinder later in the process. The combustion process in the Otto engine is triggered by a spark plug, while in the Diesel engine the fuel begins to burn as soon as it is injected. Accordingly, the figure shows spark plug (for Otto engine) and injector (for Diesel engine).

We now follow through the processes in a four-stroke engine, starting at top dead center:

Stroke I: The first stroke is the intake stroke. The valves are open, and as the piston moves towards bottom dead center air-fuel-mix (Otto) or air (Diesel)

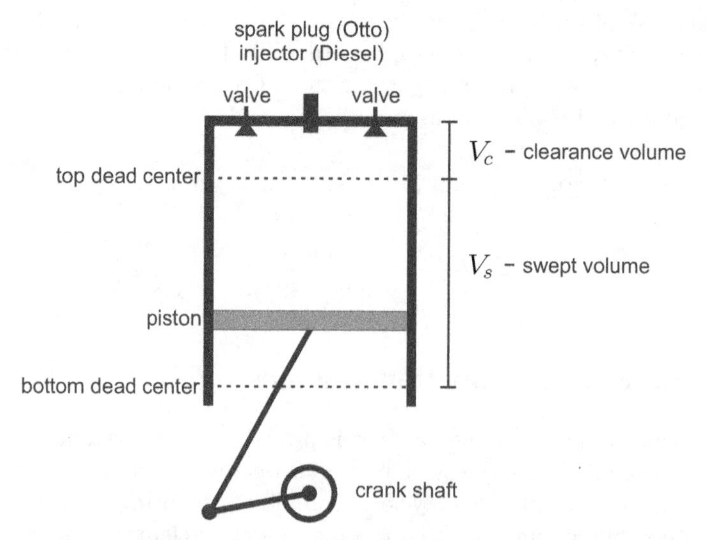

Fig. 8.4 Cylinder and piston of an internal combustion engine

enters the cylinder. Since the valves are open, the pressure in the cylinder is nearly constant. Whence bottom dead center is reached, the valves close.

Stroke II: In the second stroke the piston returns to top dead center. Since the valves are closed, the working fluid (air-fuel-mix or air) is compressed. The process is fast, and nearly adiabatic, pressure and temperature increase. The compression work is provided by the crankshaft. Shortly before the piston reaches top dead center, the combustion is triggered, either by firing the spark plug (Otto), or by injecting fuel which begins to burn in the hot air (Diesel).

Stroke III: As the fuel burns, the temperature of the working fluid is further increased. The hot combustion gas expands in the third stroke, as the piston returns to bottom dead center. The expansion is fast, nearly adiabatic, pressure and temperature decrease. The expansion work is transferred to the crankshaft.

Stroke IV: For the last of the four strokes the valves open, which leads to a sudden pressure drop. The piston returns once more to top dead center and pushes the expanded combustion products out. This is the exhaust stroke.

We summarize the processes in the four strokes in the following list:

Stroke I :	intake	valve open
Stroke II :	compression, combustion starts	valve closed
Stroke III :	combustion continues, expansion	valve closed
Stroke IV :	exhaust	valve open

A rough schematic p-V-diagram for all four strokes is shown in Fig. 8.5.

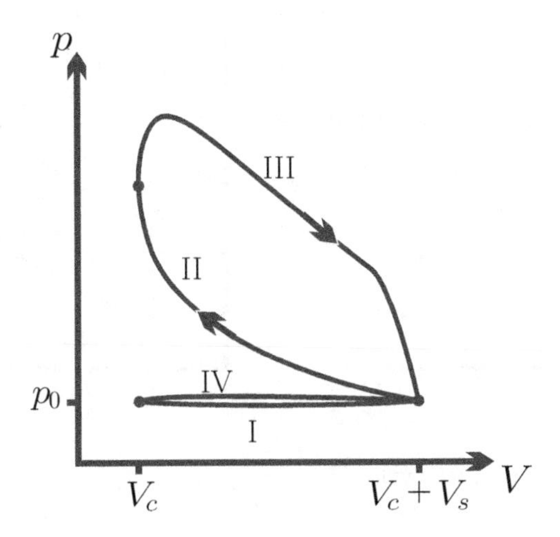

Fig. 8.5 Schematic p-V-diagram for an internal combustion engine

During strokes I and IV the valves are open, and the mass in the cylinder changes. As indicated in the p-V-diagram, in real engines there is a small depression during intake, and a small compression during exhaust, which we ignore. Thus, assuming that during intake and exhaust the pressure equals the exterior pressure p_0, the piston work for the two processes is $W_I = p_0 V_s$ and $W_{IV} = -p_0 V_s$. The work for these two strokes just cancels, $W_I + W_{IV} = 0$, and does not need to be considered further.

During strokes II and III the valves are closed, and the working fluid goes through a closed system cycle. Due to the chemical processes occurring in the combustion processes, the full analysis of this closed cycle is difficult. However, the ratio between the amounts of air and fuel is rather large. Therefore, the amount of fuel can be ignored for a basic analysis, and the working fluid can be considered as air for the complete cycle. This leads to the following modelling assumptions:

(a) The working fluid is air.

(b) The energy that is supplied through the combustion of fuel can be described as a heat transfer into the system.

(c) The exchange of the working fluid in exhaust and intake during which hot expanded combustion product is exchanged against cold precombustion working fluid can be considered as a heat exchange with the surroundings.

This *air standard analysis* allows to study internal combustion engines in a simplified yet significant manner. Further simplifications will appear in the following sections which deal with the Otto and Diesel cycles, and some variants.

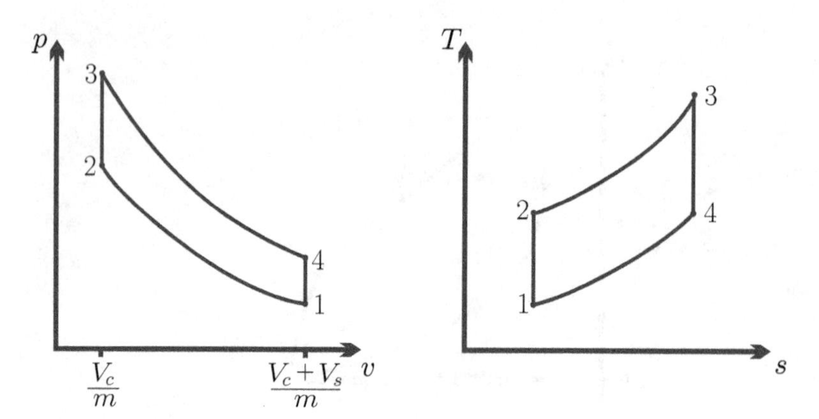

Fig. 8.6 Air-standard Otto cycle

8.5 Otto Cycle

The Otto cycle is named after Nicolaus.A. Otto (1832-1891), who build the first working internal combustion engine. In the Otto cycle, the fuel is mixed into the air outside the cylinder by injection, so that air-fuel mix enters the engine. The compression and expansion processes are assumed to be adiabatic—the processes are fast, and there is hardly time for effective heat exchange. When the spark plug ignites the compressed air-fuel-mix, the mixture explodes. The reaction is so fast, that the piston does not move much, and for modelling we describe this as an isochoric heat transfer into the working fluid. Thus, the compression and expansion strokes of the Otto engine can be modelled as

Stroke II:	1-2	adiabatic compression
	2-3	isochoric heating (air-fuel mix explodes)
Stroke III:	3-4	adiabatic expansion
	4-1	isochoric cooling (exchange of working fluid)

The respective process curves are shown in Fig. 8.6. For further analysis we assume reversible processes for which we find heat and work as

$$\begin{aligned}
&\text{1-2 isentropic: } w_{12} = u_1 - u_2 < 0 \ , \quad q_{12} = 0 \ , \\
&\text{2-3 isochoric: } w_{23} = 0 \ , \quad q_{23} = u_3 - u_2 > 0 \ , \\
&\text{3-4 isentropic: } w_{34} = u_3 - u_4 > 0 \ , \quad q_{34} = 0 \ , \\
&\text{4-1 isochoric: } w_{41} = 0 \ , \quad q_{41} = u_1 - u_4 < 0 \ .
\end{aligned} \tag{8.18}$$

The thermal efficiency of the Otto cycle is

$$\eta_{Otto} = \frac{w_\odot}{q_{in}} = \frac{w_{12} + w_{23} + w_{34} + w_{41}}{q_{23}} = 1 - \frac{u_4 - u_1}{u_3 - u_2} \ . \tag{8.19}$$

Under cold-air assumptions the internal energy is $u(T) = c_v(T - T_0) + u_0$ so that

$$\eta_{Otto} = 1 - \frac{T_4 - T_1}{T_3 - T_2} = 1 - \frac{T_1}{T_2}\frac{\frac{T_4}{T_1} - 1}{\frac{T_3}{T_2} - 1} . \tag{8.20}$$

For the two isentropic processes we have, still using the cold-air assumption,

$$\frac{T_2}{T_1} = \left(\frac{v_1}{v_2}\right)^{k-1} = \left(\frac{V_c + V_s}{V_c}\right)^{k-1} = \left(\frac{v_4}{v_3}\right)^{k-1} = \frac{T_3}{T_4} . \tag{8.21}$$

We conclude that $\frac{T_4}{T_1} = \frac{T_3}{T_2}$, and write the thermal efficiency of the cold-air-standard Otto cycle as

$$\eta_{Otto} = 1 - \frac{1}{r^{k-1}} , \tag{8.22}$$

where we have introduced the compression ratio

$$r = \frac{V_c + V_s}{V_c} . \tag{8.23}$$

The simplified analysis shows that the compression ratio r is the most important parameter for the evaluation of the Otto cycle. Under the cold air assumption the thermal efficiency depends solely on the compression ratio, and it grows with increasing compression ratio. Obviously, one will aim for large compression ratios to have efficient engines.

The compression ratio cannot be arbitrarily large. As the air-fuel-mix is compressed, its temperature increases, $T_2 = T_1 r^{k-1}$. When the temperature of the air-fuel-mix reaches its auto-ignition temperature, it will start to react before the spark plug induces the explosion at the appropriate point in the process. Premature combustion, known as engine knocking, reduces the power output since it leads to increased pressure during compression, and, even more importantly, it damages the engine. To prevent knocking, the compression ratio must be limited to values which guarantee that auto-ignition cannot occur. High grade fuel, i.e., gasoline with larger octane numbers, has a higher auto-ignition temperature, and must be used in engines with larger compression ratios.

Typical values for the compression ratio in Otto engines are between 8 and 12. These values yield efficiencies for the idealized cycle discussed above between 0.565 and 0.63. Real engines have about 30% loss to irreversible processes (friction, heat transfer) within the engine, and another 30% loss to friction in the drive train, so that their actual efficiency is $\eta = 0.7 \times 0.7 \times \eta_{Otto} = 0.28$ (for $r = 8$).

Since the working cycle of the engine happens during only two of the four strokes (recall that inlet and exhaust work cancel), the power delivered by an engine which runs at a frequency $\dot{n}$ is

$$\dot{W} = \frac{\dot{n}}{2} m w_{\odot} \,, \tag{8.24}$$

where $m = p_1 \left(V_c + V_s\right) / RT_1$ is the mass of air in the cylinder.

8.6 Example: Otto Cycle

As an example we consider an engine operating on the ideal Otto cycle with a compression ratio $r = 9$. The total swept volume of all cylinders is $V_s =$ 3 litres, and the engine runs at $\dot{n} = 3000$ rpm. The intake is at $T_1 = 300$ K, $p_1 = 0.98$ bar, and the heat transfer through the combustion process is $q_{23} = 717 \frac{\text{kJ}}{\text{kg}}$. We determine temperatures, pressures and specific volume for the four corner points of the process, and compute heat and work for all subprocesses. For the computation we treat the working fluid as air with constant specific heats (cold-air approximation), $R = 0.287 \frac{\text{kJ}}{\text{kg K}}$, $c_v = 0.717 \frac{\text{kJ}}{\text{kg K}}$, $k = 1.4$.

We begin with the computation of the measurable properties, their numerical values are found in the table below this paragraph. From the ideal gas law we have $v_1 = \frac{RT_1}{p_1}$. Since the process 4-1 is isochoric, we have $v_4 = v_1$, and the other two volumes follow from the compression ratio as $v_2 = v_3 = \frac{v_1}{r}$. The compression process 1-2 is isentropic, so that $p_2 = p_1 r^k$; the ideal gas law gives $T_2 = \frac{p_2 v_2}{R}$. The heat for the isochoric heating process (the explosion) is $q_{23} = u_3 - u_2 = c_v \left(T_3 - T_2\right)$, so that $T_3 = T_2 + \frac{q_{23}}{c_v}$; from the ideal gas law $p_3 = \frac{RT_3}{v_3}$. Since the process 3-4 is isentropic, the pressure at the end of the expansion is $p_4 = p_1 r^{-k}$; the temperature follows again from the ideal gas law, $T_4 = \frac{p_4 v_4}{R}$. The numerical values are

	$v/\frac{\text{m}^3}{\text{kg}}$	T/K	p/bar
1	0.879	300.0	0.98
2	0.098	725.3	21.2
3	0.098	1725.3	50.5
4	0.879	714.0	2.33

The clearance volume follows from the compression ratio $r = \frac{V_c + V_s}{V_c}$ as $V_c = \frac{V_s}{r-1} = 0.375$ litres. The mass in the cylinders is $m = \frac{V_c}{v_2} = 3.83$ g.

Work and heat for the four processes are, with $u_i - u_j = c_v \left(T_i - T_j\right)$,

$$\begin{aligned}
w_{12} &= c_v \left(T_1 - T_2\right) = -304.9 \tfrac{\text{kJ}}{\text{kg}} &, \quad q_{12} &= 0 \,, \\
w_{23} &= 0 &, \quad q_{23} &= c_v \left(T_3 - T_2\right) = 717 \tfrac{\text{kJ}}{\text{kg}} \,, \\
w_{34} &= c_v \left(T_3 - T_4\right) = 725.1 \tfrac{\text{kJ}}{\text{kg}} &, \quad q_{34} &= 0 \,, \\
w_{41} &= 0 &, \quad q_{41} &= c_v \left(T_1 - T_4\right) = -296.8 \tfrac{\text{kJ}}{\text{kg}} \,.
\end{aligned}$$

The net work for the cycle is

$$w_\odot = w_{12} + w_{23} + w_{34} + w_{41} = 420.2\frac{\text{kJ}}{\text{kg}} ,$$

the power produced is

$$\dot{W} = \frac{\dot{n}}{2} m w_\odot = 40.3\,\text{kW} ,$$

and the thermal efficiency of the cycle is

$$\eta = \frac{w_\odot}{q_{23}} = 1 - \frac{1}{r^{k-1}} = 0.585 .$$

8.7 Diesel Cycle

In the Diesel cycle, named after its inventor Rudolf Diesel (1858-1913), only air is drawn in and compressed, and the fuel is injected after the compression. Here, one utilizes the temperature increase of the air in compression: as the fuel is injected it starts to burn in the hot compressed air. The compression ratios of Diesel engines can be substantially higher than those of Otto engines, with values of up to $r = 30$.

However, injection of the fuel is slow, and the injected fuel disperses into droplets which are not as well mixed with the air as the gasified fuel in the Otto engine. Therefore combustion is slower than in the Otto engine, the piston moves as the fuel burns. This process can be modelled as an isobaric heat transfer process, so that the Diesel cycle is modelled as follows:

Stroke II: 1-2 adiabatic compression
Stroke III: 2-3 isobaric heating (slow combustion during fuel injection)
3-4 adiabatic expansion
4-1 isochoric cooling (exchange of working fluid)

The respective process curves are shown in Fig. 8.7, which also indicates clearance volume, swept volume, and the injection volume V_i, which is the volume at the end of the isobaric heating process. For further analysis we assume reversible processes for which we find heat and work as

$$\begin{aligned}
&\text{1-2 isentropic: } w_{12} = u_1 - u_2 < 0 &&, \; q_{12} = 0 , \\
&\text{2-3 isobaric: } w_{23} = p_3 \left(v_3 - v_2\right) > 0 &&, \; q_{23} = h_3 - h_2 > 0 , \\
&\text{3-4 isentropic: } w_{34} = u_3 - u_4 > 0 &&, \; q_{34} = 0 , \\
&\text{4-1 isochoric: } w_{41} = 0 &&, \; q_{41} = u_1 - u_4 < 0 .
\end{aligned} \tag{8.25}$$

The thermal efficiency of the Diesel cycle is

$$\eta_{Diesel} = \frac{w_\odot}{q_{in}} = \frac{w_{12} + w_{23} + w_{34} + w_{41}}{q_{23}} = 1 - \frac{u_4 - u_1}{h_3 - h_2} . \tag{8.26}$$

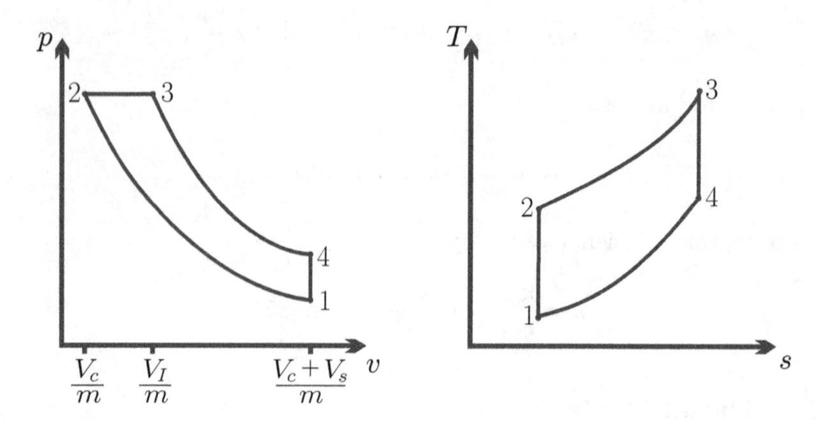

Fig. 8.7 Air-standard Diesel cycle. Note that in the T-s-diagram the isochoric line has a larger slope than the isobaric line.

Under cold-air assumptions energy and enthalpy are $u(T) = c_v(T - T_0) + u_0$, $h(T) = c_p(T - T_0) + h_0$, so that

$$\eta_{Diesel} = 1 - \frac{c_v}{c_p}\frac{T_1}{T_2}\frac{\frac{T_4}{T_1} - 1}{\frac{T_3}{T_2} - 1} . \tag{8.27}$$

For the two isentropic processes we have, still using the cold-air assumption,

$$\frac{T_2}{T_1} = \left(\frac{V_1}{V_2}\right)^{k-1} = r^{k-1} , \quad \frac{T_3}{T_4} = \left(\frac{V_4}{V_3}\right)^{k-1} = \left(\frac{V_4}{V_2}\frac{V_2}{V_3}\right)^{k-1} = \frac{r^{k-1}}{r_c^{k-1}} , \tag{8.28}$$

where $r_c = V_3/V_2 = V_i/V_c$ is the cut-off ratio. For the isobaric process we have (ideal gas law!) $\frac{T_3}{T_2} = \frac{V_3}{V_2} = r_c$, so that the thermal efficiency of the Diesel process becomes[3]

$$\eta_{Diesel} = 1 - \frac{1}{r^{k-1}}\frac{1}{k}\frac{r_c^k - 1}{r_c - 1} . \tag{8.29}$$

For the *same* compression ratio r the Diesel efficiency is below the Otto efficiency. However, the Diesel cycle allows significantly *higher* compression ratios than the Otto cycle, and that is the reason why it has a larger efficiency. For a cycle with cut-off ratio $r_c = 2$ and compression ratio $r = 20$, we find the thermal efficiency as 0.65. Irreversible losses to friction in the engine and drive train reduce the thermal efficiency, actual engines have efficiencies around 37%.

From the p-v-diagram we can see that in a Diesel cycle the maximum pressure is sustained for a longer period, in contrast to the Otto engine, where

[3] With $\frac{T_4}{T_1} = \frac{T_4}{T_3}\frac{T_3}{T_2}\frac{T_2}{T_1} = \frac{r_c^{k-1}}{r^{k-1}} r_c r^{k-1} = r_c^k$.

the maximum pressure is just a short spike. Due to this, Diesel engines must be build more sturdily, which is the main reason why they are more expensive. On the other hand, they are more efficient, and thus require less fuel. The cost of gasoline and Diesel fuel, respectively, also depends on taxation, and thus it might be more cost effective to drive a Diesel powered car in some countries, while an Otto powered car might be more cost effective in other countries.

8.8 Example: Diesel Cycle

An air standard four-stroke Diesel cycle has a clearance volume $V_c = 0.5$ litre, and a compression ratio $r = 16.16$. Intake temperature and pressure are 17 °C and 1 bar, respectively, and the temperature at the end of the expansion is 707 °C. The engine runs at 4400 rpm. We determine temperatures, pressures and specific volume for the four corner points of the process, and compute heat and work for all subprocesses. For the computation we treat the working fluid as air with variable specific heats.

We outline how the various properties are determined, and collect their numerical values in the table below this paragraph. The values for T_1, p_1 and T_4 are given above. From the ideal gas law we have $v_1 = v_4 = \frac{RT_1}{p_1}$; from the compression ratio $v_2 = \frac{v_1}{r}$. The compression process 1-2 is isentropic, so that for the relative volume $v_r(T_2) = \frac{v_r(T_1)}{r}$. With tabulated data for $v_r(T)$ follows T_2, while the ideal gas law gives the pressures $p_2 = p_3 = \frac{RT_2}{v_2}$. With T_4 given, the pressure at 4 follows from the ideal gas law, $p_4 = \frac{RT_4}{v_4}$. The expansion 3-4 is isentropic, so that $p_r(T_3) = p_r(T_4)\frac{p_3}{p_4}$. With tabulated data follows T_3, and from the ideal gas law $v_3 = \frac{RT_3}{p_3}$. With this, all points are identified, and energies and enthalpies can be found from the property table. Altogether we have

	$v/\frac{\text{m}^3}{\text{kg}}$	T/K	p/bar	v_r	p_r	$u/\frac{\text{kJ}}{\text{kg}}$	$h/\frac{\text{kJ}}{\text{kg}}$
1	0.832	290	1.0	319.5		208	
2	0.0515	840	46.81	19.77		626	867
3	0.113	1845	46.81		1075	1531	2060
4	0.832	980	3.38		77.61	743	

The swept volume follows from the compression ratio $r = \frac{V_c+V_s}{V_c}$ as $V_s = V_c(r-1) = 7.58$ litre, the injection volume is $V_i = \frac{v_3}{v_2}V_c = 1.1$ litre, and the cut-off ratio is $r_c = \frac{V_i}{V_c} = 2.2$. The mass in the cylinders during the working cycle is $m = \frac{V_c}{v_2} = 9.71$ g.

Work and heat for the four processes are

$$
\begin{aligned}
w_{12} &= u_1 - u_2 = -418\tfrac{\text{kJ}}{\text{kg}} &, \; q_{12} &= 0\,, \\
w_{23} &= p\,(v_3 - v_2) = 288\tfrac{\text{kJ}}{\text{kg}} &, \; q_{23} &= h_3 - h_2 = 1193\tfrac{\text{kJ}}{\text{kg}}\,, \\
w_{34} &= u_3 - u_4 = 788\tfrac{\text{kJ}}{\text{kg}} &, \; q_{34} &= 0\,, \\
w_{41} &= 0 &, \; q_{41} &= u_1 - u_4 = -535\tfrac{\text{kJ}}{\text{kg}}\,.
\end{aligned}
$$

The net work for the cycle is

$$w_{\odot} = w_{12} + w_{23} + w_{34} + w_{41} = 658\frac{\text{kJ}}{\text{kg}}\,,$$

the power produced is

$$\dot{W} = \frac{\dot{n}}{2} m w_{\odot} = 234.3\,\text{kW}\,,$$

and the thermal efficiency of the cycle is

$$\eta = \frac{w_{\odot}}{q_{23}} = 0.552\,.$$

This is somewhat lower as the efficiency under the cold-air approximation, which is $\eta = 1 - \frac{1}{r^{k-1}} \frac{1}{k} \frac{r_c^k - 1}{r_c - 1} = 0.606$. The difference is due to the effect of temperature dependent specific heats.

8.9 Dual Cycle

In real engines the combustion process is neither just an instantaneous, i.e., isochoric, explosion as assumed above for the Otto cycle, nor is it a slow isobaric combustion as assumed above for the Diesel cycle. To have a somewhat better model, one can modify the description of the fuel combustion process as a combination of isochoric and isobaric heating. The resulting cycle is the *dual cycle*:

Stroke II:	1-2	adiabatic compression
	2-3'	isochoric heating
Stroke III:	3'-3	isobaric heating
	3-4	adiabatic expansion
	4-1	isochoric cooling

Figure 8.8 shows the corresponding diagrams. Depending on the choice of parameters, one can model Otto and Diesel engines as a dual cycle. The dual step would have a more pronounced isochoric and smaller isobaric step for Otto engines, and a smaller isochoric but more pronounced isobaric step for Diesel engines. Moreover, Otto engines have compression ratios below 12, while Diesel engines have higher compression ratios.

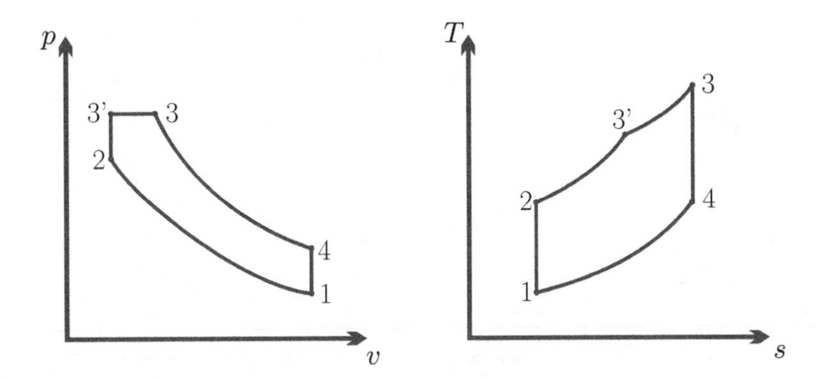

Fig. 8.8 Dual cycle

8.10 Atkinson Cycle

At the end of the expansion process in the Otto and Diesel cycles, the working fluid is at elevated pressure and temperature, and more work could be extracted. One method to use this work is to drive a turbo charger, which essentially is a compressor that is used to fill the cylinders with air-fuel mix (Otto) or air (Diesel) at elevated pressures. With this, the mass of air in the engine is higher than when it only draws air at environmental pressure p_0. Since the power output of the engine is proportional to mass, $\dot{W} = \frac{\dot{n}}{2} m w_{\odot}$, turbo charging increases the power output, or gives the same power from a smaller engine.

An alternative method to make use of the work potential is used in Atkinson engines. The Atkinson cycle is a modification of the Otto cycle where the compression and expansion strokes have different lengths. Specifically, the expansion stroke is longer than the compression stroke, so that—in the best case—the pressure at the end of expansion is equal to the intake pressure. The early engines build by James Atkinson (1846–1914) relied on mechanical valve control. Since valves can be controlled electronically, Atkinson's ideas become more prominent, and engines based on the Atkinson cycle are routinely used in modern hybrid cars.

To achieve the Atkinson cycle in an actual engine one uses clever control of the valves. In the Otto engine the valves close at the end of the intake stroke (Stroke I), and as the piston reverses its direction, compression starts immediately (Stroke II). In an Atkinson engine the valve remains open during the beginning of the second stroke, so that some of the intake is pushed back into the intake manifold. The valve closes a bit later (at volume V_1) and compression commences. Thus, compression occurs only on part of Stroke II. The valves remain closed for the full expansion (Stroke III), so that the expansion stroke is longer than the compression stroke. At the end of expansion, the valves open and the exhaust stroke (Stroke IV) begins. Computerized control

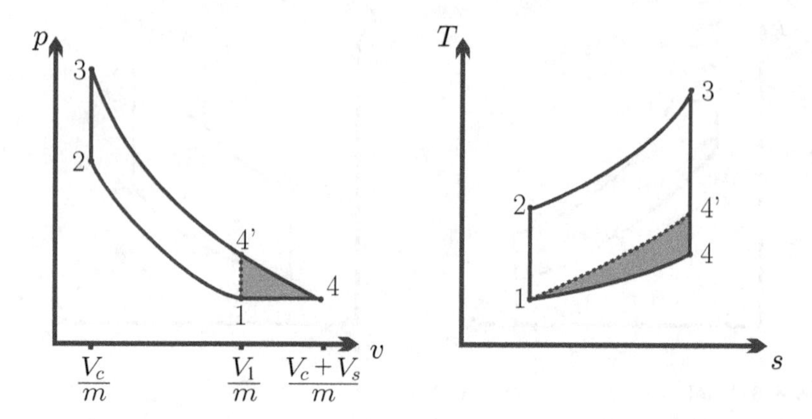

Fig. 8.9 Atkinson cycle 1-2-3-4-1 compared to Otto cycle 1-2-3-4'-1. The shaded area is the extra work delivered by the Atkinson cycle for the same heat input.

of the valve timing allows to vary the length of the compression stroke to optimize engine performance for the current driving conditions.

The processes in the ideal air-standard Atkinson cycle are:

1-2 adiabatic compression
2-3 isochoric heating (air-fuel mix explodes)
3-4 adiabatic expansion
4-1 isobaric cooling (exchange of working fluid)

The corresponding process diagrams are depicted in Fig. 8.9. The figure also indicates the point $4'$ at which the expansion would finish in the Otto cycle. The shaded areas in the two diagrams indicate the additional work generated in the Atkinson cycle as compared to the Otto cycle. The compression ratio is $r = \frac{V_1}{V_c}$ and the expansion ratio is $r_e = \frac{V_c+V_s}{V_c}$.

Work and heat for the processes are

$$\begin{aligned}
&\text{1-2 isentropic: } w_{12} = u_1 - u_2 < 0 &&, \; q_{12} = 0 \,, \\
&\text{2-3 isochoric: } w_{23} = 0 &&, \; q_{23} = u_3 - u_2 > 0 \,, \\
&\text{3-4 isentropic: } w_{34} = u_3 - u_4 > 0 &&, \; q_{34} = 0 \,, \\
&\text{4-1 isobaric: } w_{41} = p_1 (v_1 - v_4) < 0 &&, \; q_{41} = h_1 - h_4 < 0 \,.
\end{aligned} \tag{8.30}$$

Accordingly, the net work and the thermal efficiency of the engine are

$$w_\odot = w_{12} + w_{23} + w_{34} + w_{41} = h_1 - u_2 + u_3 - h_4 \,, \tag{8.31}$$

and

$$\eta_{Atkinson} = \frac{w_\odot}{q_{23}} = 1 - \frac{h_4 - h_1}{u_3 - u_2} \,. \tag{8.32}$$

As before, we evaluate the process under the cold-air approximation, where

$$h_4 - h_1 = c_p (T_4 - T_1) \quad , \quad u_3 - u_2 = c_v (T_3 - T_2) \ , \tag{8.33}$$

so that

$$\eta_{Atkinson} = 1 - k \frac{T_4 - T_1}{T_3 - T_2} \ . \tag{8.34}$$

For the temperatures at the corner points of the process we find

$$T_2 = T_1 r^{k-1} \quad , \quad T_4 = T_1 \frac{v_4}{v_1} = T_1 \frac{r_e}{r} \quad , \quad T_3 = T_4 r_e^{k-1} = T_1 \frac{r_e^k}{r} \ , \tag{8.35}$$

so that net work and thermal efficiency become

$$w_\odot = c_v T_1 \left[k \left(1 - \frac{r_e}{r} \right) + \frac{r_e^k - r^k}{r} \right] \ , \tag{8.36}$$

$$\eta_{Atkinson} = 1 - \frac{1}{r^{k-1}} k \frac{\frac{r_e}{r} - 1}{\left(\frac{r_e}{r}\right)^k - 1} \ . \tag{8.37}$$

It is easy to see that for $r_e > r$ the factor $\left[k \frac{\frac{r_e}{r} - 1}{\left(\frac{r_e}{r}\right)^k - 1} \right]$ is less than unity. Therefore, the thermal efficiency of the Atkinson cycle is larger than that of the Otto cycle with the same compression ratio.

In order to have the ideal Atkinson cycle performed, the heat addition q_{23} must be such that $T_3 = T_1 \frac{r_e^k}{r}$. If more heat is added, so that the temperature T_3 lies above this value, the pressure p_4 at the end of expansion will be above the inlet pressure p_1. In this case, the cooling process (which models exhaust and intake) would include an isochoric process:

1-2 adiabatic compression
2-3 isochoric heating
3-4' adiabatic expansion
4'-4 isochoric cooling
4-1 isobaric cooling

The corresponding process diagrams are shown in Fig. 8.10. Work and heat for the processes are

$$\begin{array}{lll}
\text{1-2 isentropic:} & w_{12} = u_1 - u_2 < 0 & , q_{12} = 0 \\
\text{2-3 isochoric:} & w_{23} = 0 & , q_{23} = u_3 - u_2 > 0 \\
\text{3-4' isentropic:} & w_{34'} = u_3 - u_{4'} > 0 & , q_{34'} = 0 \\
\text{4'-4 isochoric} & w_{4'4} = 0 & q_{4'4} = u_4 - u_{4'} < 0 \\
\text{4-1 isobaric:} & w_{41} = p_1 (v_1 - v_4) < 0 & , q_{41} = h_1 - h_4 < 0
\end{array} \tag{8.38}$$

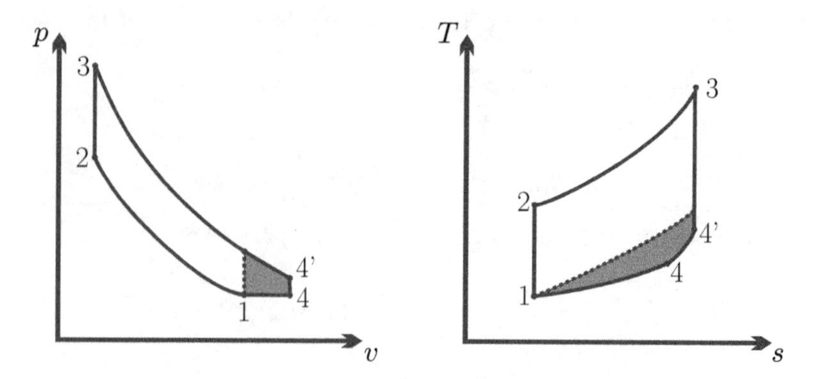

Fig. 8.10 Overheated Atkinson cycle with excess pressure at the end of expansion

The thermal efficiency of the overheated Atkinson cycle is below that of the ideal Atkinson cycle. In the cold-air approximation one finds

$$\eta = 1 - \frac{1}{r^{k-1}} \left[k \frac{T_3 r_e^{1-k} - T_1}{T_3 r^{1-k} - T_1} + (k-1) \frac{T_3 r_e^{1-k} - T_1 \frac{r_e}{r}}{T_3 r^{1-k} - T_1} \right] , \tag{8.39}$$

which reduces to (8.37) for $T_3 = T_1 \frac{r_e^k}{r}$.

Problems

8.1. Power Cycle in the T-s-Diagram
Compute the efficiency of the cycle in the sketch, and compare with the efficiency of the Carnot cycle.

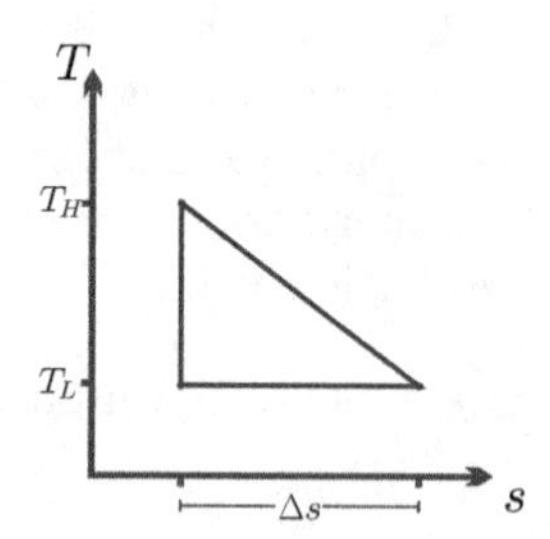

8.2. Refrigeration Cycle in the T-s-Diagram
A closed system runs on the process indicated in the sketch as a refrigeration cycle. Is the cycle running clockwise or counterclockwise? Use the sketch to determine the COP and compare with the Carnot cycle. Draw the p-v-diagram of the cycle for the case that the working fluid is an ideal gas. Assume $T_M = (T_H + T_L)/2$.

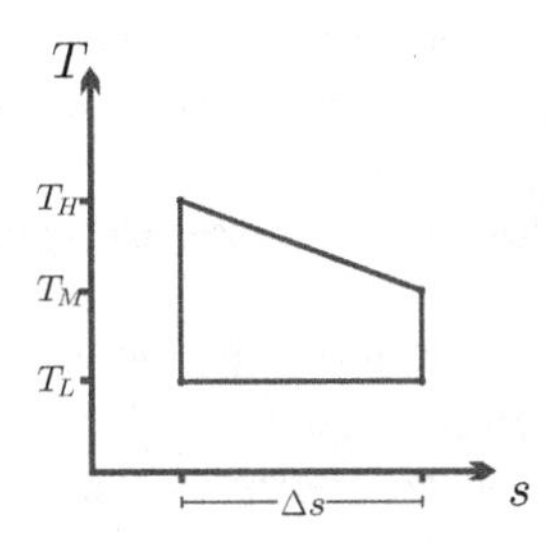

8.3. Inventors

Two inventors have developed heat engines that operate between the temperatures of 900 K and 300 K. One inventor claims an efficiency of 50% for his engine, the other claims an efficiency of 66%. If you had some money to invest, which inventors start-up would you invest in? Explain!

8.4. Carnot Heat Engine

A heat engine operates on a fully reversible Carnot cycle between two reservoirs at 25 °C and 527 °C. The engine contains 4 g of air, and the largest volume is 2 litres. The pressure ratio for isothermal compression is equal to the pressure ratio for isentropic compression.

1. Draw the process diagrams, and make a table with temperature, pressure, specific volume, and specific internal energy at the corner points of the process.
2. Determine specific heat and work for the four processes in the cycle, the net work for the cycle, and the thermal efficiency.
3. When the engine runs at 350 rpm, determine the power produced.

8.5. Carnot Cycle

A Carnot power cycle using hydrogen as working fluid operates between the temperatures 320 K and 1000 K with maximum and minimum pressures given by 40 bar and 0.2 bar, respectively. The maximum volume of the hydrogen in the engine is 6.65 litres and the engine runs at 700 rpm. Draw the process in a T-s-diagram and in a p-v-diagram, then make a table with pressures, specific volumes, internal energies, and entropies at all corner points of the process. Compute the thermal efficiency, the mass of hydrogen in the engine, the maximum volume ratio, and the power produced. Based on the data obtained, discuss the feasibility of the process (apart from the fact that it would be impossible to build a fully reversible engine).

8.6. Carnot Heat Pump

A food drying unit requires a heating power of 200 kW which must be supplied at $T_H = 75\,°\text{C}$, while the exterior temperature is $T_L = 15\,°\text{C}$. For this purpose consider a closed system Carnot heat pump using xenon (ideal gas, monatomic, $M = 131.3 \frac{\text{kg}}{\text{kmol}}$) as working fluid. The ratio between maximum and minimum volume of the unit is 40, and the largest pressure that can occur is 80 bar.

1. Draw the process in a T-s-diagram and in a p-v-diagram.
2. Make a table with pressures, temperatures and specific volumes, at all corner points.
3. Compute heat and work for all processes. Determine the net work and the COP.
4. When the engine operates at 500 rpm, determine the mass of xenon in the engine and the maximum volume (which could be distributed over several cylinders). Based on the data obtained, discuss the feasibility of the process.

8.7. Otto Cycle

The air entering a 3 litre Otto engine (ideal air-standard cycle, air as ideal gas with variable specific heats) with compression ratio $r = 9.2$ is at environmental conditions (98 kPa, 300 K). After the heat supply, the pressure has doubled. Determine heat added, net work output, thermal efficiency, and power output when the engine runs at 2000 rpm. Draw diagrams.

8.8. Otto Engine

An air standard four-stroke Otto engine has a swept volume of 2.5 litres, and a clearance volume of 0.4 litres. Temperature and pressure at intake are given by $T_1 = 290\,\mathrm{K}$, $p_1 = 0.7\,\mathrm{bar}$. The temperature at the end of combustion is $T_3 = 1400\,\mathrm{K}$. Consider the working fluid to be air, as ideal gas with $R = 0.287 \frac{\mathrm{kJ}}{\mathrm{kg\,K}}$, and with variable specific heats.

1. Draw the p-V- and T-s-diagrams for the process.
2. Determine the values of temperature, pressure, specific volume, and internal energy at the corners of the cycle. Make a table with these values.
3. Determine the net work per unit mass and the thermal efficiency.
4. Determine the mass of air in the cylinders and net power output of the engine when it runs at 4500 rpm.

8.9. Otto Engine

An air standard four-stroke Otto engine has a compression ratio of 9.4 and a clearance volume of 0.3 litre. Temperature and pressure at intake are $T_1 = 280\,\mathrm{K}$, $p_1 = 1.1\,\mathrm{bar}$, and the pressure after expansion is $p_4 = 2.984\,\mathrm{bar}$.

Consider the working fluid to be air, as ideal gas with variable specific heats.

1. Draw the p-V- and T-s-diagrams for the process.
2. Determine the values of temperature, pressure, specific volume, and internal energy at the corners of the cycle. Make a table with these values.
3. Determine the net work per unit mass and the thermal efficiency.
4. Determine the mass of air in the cylinders and net power output of the engine when it runs at 1200 rpm.

8.10. Diesel Engine

An air standard four-stroke Diesel engine has a swept volume of 7 litres, and a clearance volume of 0.5 litre. The volume at the end of the fuel injection (isobaric heat supply) is 1 litre.

Temperature and pressure at intake are given by $T_1 = 300\,\mathrm{K}$, $p_1 = 1\,\mathrm{bar}$. Consider the working fluid to be air, as ideal gas with constant specific heats, $R = 287\frac{\mathrm{J}}{\mathrm{kg\,K}}$, $c_v = \frac{5}{2}R$, $k = \frac{c_p}{c_v} = 1.4$.

1. Draw a p-V-diagram of the process and then determine:
2. The values of all temperatures, pressures and specific volumes at the corners of the cycle. Make a table with these values.
3. The net work per unit mass and the thermal efficiency.
4. The mass of air in the cylinder and net power output of the engine when it runs at 2500 rpm.

8.11. Diesel Cycle

A 16 cylinder 170-litre Diesel engine operating on the ideal air-standard Diesel cycle has a compression ratio of 17, and a cut-off ratio of 2.2. Determine the amount of power delivered when the engine runs at 900 rpm based on the air-standard cycle, under cold-air assumption (that is: constant specific heats). Consider the following two cases for outside temperature and pressure: (a) $T_0 = 280\,\mathrm{K}$, $p = 1\,\mathrm{bar}$. (b) $T = 305\,\mathrm{K}$, $p = 0.9\,\mathrm{bar}$. Draw diagrams.

8.12. Diesel and Otto cycle

Draw a schematic, and the process curves in a p-V-diagram and a T-s-diagram for a Diesel and an Otto cycle. Mark swept volume, clearance volume etc. These are four-stroke engines: what are the four strokes? Indicate them in the diagram. Discuss the difference between Diesel and Otto cycles. Why can the Diesel have a higher compression?

8.13. Dual Cycle

The processes in a 4-stroke Diesel cycle with compression ratio of 14 are modeled as dual cycle with the following data: The engine draws air at 1 bar, 27 °C, the maximum temperature reached in the cycle is 2200 K, and the total amount of heat added is $q_{23} = 1520.4\frac{\mathrm{kJ}}{\mathrm{kg}}$. The working fluid can be considered as air (ideal gas, variable specific heats).

1. Draw the process in a p-V-diagram, and a T-s-diagram. Include intake and exhaust, and indicate the 4 strokes in the p-V-diagram.
2. Determine temperatures, pressures, and specific volumes in the points 1,2,3',3,4.
3. Determine the thermal efficiency of the cycle.

8.14. Atkinson Engine

A four-stroke-engine operating on the Atkinson cycle draws air at 0.9 bar, 17 °C. The working cycle consists of the following reversible processes:

1-2: Adiabatic compression of air with compression ratio 8.6
2-3: Isochoric heating to $T_3 = 1500\,\mathrm{K}$
3-4: Adiabatic expansion
4-1: Isobaric cooling

Consider the working fluid to be air, as ideal gas with $R = 0.287\frac{\mathrm{kJ}}{\mathrm{kg\,K}}$, and variable specific heats.

1. Draw p-v- and T-s-diagram for the cycle.
2. Make a list with temperature, pressure, specific volume at the four corner points of the cycle.
3. Determine the expansion ratio of the engine.
4. Determine net work and thermal efficiency of the cycle.
5. The engine runs at 2000 rpm and the engine delivers 28.75 kW. Determine the gas volume at bottom dead center.

8.15. Atkinson Engine

A four-stroke-engine operating on the ideal Atkinson cycle draws air at 0.9 bar, 17 °C. The working cycle consists of the following reversible processes:

1-2: Adiabatic compression of air with compression ratio 10.08.
2-3: Isochoric heating.
3-4: Adiabatic expansion with expansion ratio 16.
4-1: Isobaric cooling.

Consider the working fluid to be air, as ideal gas with $R = 0.287 \frac{\text{kJ}}{\text{kg K}}$, and variable specific heats.

1. Draw p-V- and T-s-diagram for the cycle.
2. Make a list with specific volume, temperature, pressure, at the four corner points of the cycle.
3. Determine net work and thermal efficiency of the cycle.
4. The engine runs at 2000 rpm and the engine delivers 18 kW. Determine the mass in the cylinders, and the swept volume.

8.16. Overheated Atkinson Engine

Show that the thermal efficiency of the overheated Atkinson cycle under the cold-air approximation is given by Eq. (8.39).

8.17. Overheated Atkinson Engine

An overheated 4-stroke Atkinson cycle has a compression ratio of 10, and an expansion ratio of 14.09; its swept volume is 1.2 litres. The cycle draws air at $p = 0.9$ bar, $T_0 = 0\,°\text{C}$, and the total heat rejected into the environment is $397.38 \frac{\text{kJ}}{\text{kg}}$. Assume air-standard conditions, with air as ideal gas with variable specific heat, and reversible processes.

1. Draw the process into p-V- and T-s-diagrams.
2. Determine pressure, temperature, specific volume, specific internal energy and specific enthalpy at all relevant process points.
3. Determine the net work, the heat addition, and the thermal efficiency of the cycle.
4. The engine runs at 1750 rpm, determine the power output.

Chapter 9
Open Systems

9.1 Flows in Open Systems

So far we have considered only closed systems, which do not exchange mass. We shall now extend the discussion to systems which exchange mass with their surroundings. Figure 9.1 shows a generic *open system* with two inflows and two outflows. In the following, and in the figure, the amount of mass exchanged per unit time, the mass transfer rate or *mass flow*, is denoted by $\dot{m}$. The system also exchanges propeller and piston work, $\dot{W} = \dot{W}_{propeller} + \dot{W}_{piston}$, and heat, $\dot{Q} = \dot{Q}_1 + \dot{Q}_2$, with its surroundings, just as a closed system does. Below, we shall add the appropriate terms to the thermodynamic laws to account for mass transport over the system boundary.

States in open systems are normally inhomogeneous. One might think of a mass element entering the system of Fig. 9.1 on the left. As the element of mass travels through the system, it constantly changes its state: When it passes the heating, its temperature changes, when it passes the propeller its

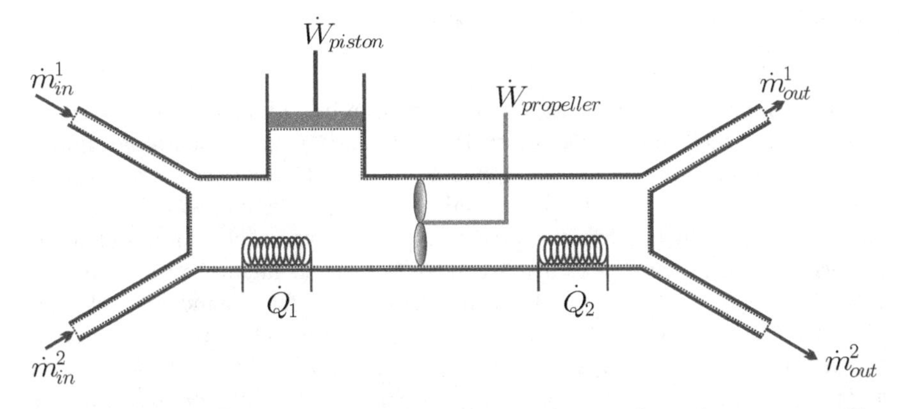

Fig. 9.1 Open system with two inflows, two outflows and two heat sources. The dotted line indicates the system boundary.

H. Struchtrup, *Thermodynamics and Energy Conversion*,
DOI: 10.1007/978-3-662-43715-5_9,

pressure and temperature change and so on. Thus, at each location within the system one finds different properties. As discussed earlier, an inhomogeneous system is in a non-equilibrium state. In an open system the non-equilibrium is maintained through the exchange of mass, heat and work with the surroundings.

9.2 Conservation of Mass

Mass cannot be created or destroyed, that is mass is conserved. Chemical reactions change the composition of the material, but not its mass.[1] In a closed system, the law of mass conservation states that the total mass m in the system does not change in time, i.e., it simply reads $\frac{dm}{dt} = 0$. In an open system, where mass enters or leaves over the system boundaries, the conservation law for mass states that the change of mass in time is due to inflow—which increases mass—and outflow—which decreases system mass. In other words, the rate of change in mass is due to the net difference of mass flows entering and leaving the system,

$$\frac{dm}{dt} = \sum_{\text{in}} \dot{m}_i - \sum_{\text{out}} \dot{m}_e \,. \tag{9.1}$$

Here, $\dot{m}_i$ denotes incoming mass flows and $\dot{m}_e$ denotes exiting mass flows, as indicated in Fig. 9.1. By definition, the mass flow is positive, the direction of flow is made explicit by the signs in the equation.

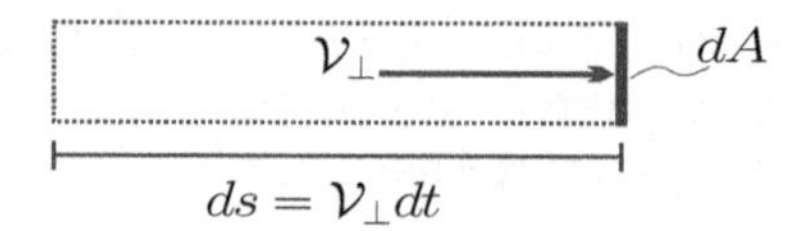

Fig. 9.2 Mass flow through the cross section A

Figure 9.2 shows an element dA of the system boundary that is crossed by mass flowing with the local velocity $\mathcal{V}_\perp$. All velocity components parallel to the surface do not play a role in transport through the surface elements, and it suffices to consider only the normal (i.e., perpendicular to dA) velocity $\mathcal{V}_\perp$ as indicated in the figure. During the time interval dt a mass element travels the distance $ds = \mathcal{V}_\perp dt$. All mass elements that are initially in the volume element $dsdA = \mathcal{V}_\perp\, dA\, dt$ will cross the surface element dA during dt. With the local density ρ, the mass in the volume element is $\rho \mathcal{V}_\perp\, dA\, dt$, which thus is the mass crossing dA during dt. Division by the time interval dt gives the amount of mass crossing the surface element per unit time as

[1] As long as one ignores the very small relativistic mass defect.

$\rho\mathcal{V}_\perp\, dA$. The macroscopic mass flow through a finite cross section A results from integration over all surface elements,

$$\dot{m} = \int_A \rho\mathcal{V}_\perp\, dA\,. \tag{9.2}$$

For a finite cross section one will expect spatial variations of the density ρ and velocity $\mathcal{V}_\perp$ over the cross section which would have to be taken into account in the integral. As an example Fig. 9.3 shows a typical velocity profile in pipe flow, where the fluid sticks to the wall, so that the velocity at the wall is zero, and the flow is fastest in the middle. However, for many engineering applications it is not necessary to resolve the velocity profile, or the profiles of other properties. Instead, it suffices to work with local averages, and this will be done from now on. When the average values are inserted in (9.2), they can be pulled out of the integral, and we find

$$\dot{m} = \rho\,\mathcal{V}\,A\,, \tag{9.3}$$

where ρ and $\mathcal{V}$ are averages over the cross section A.

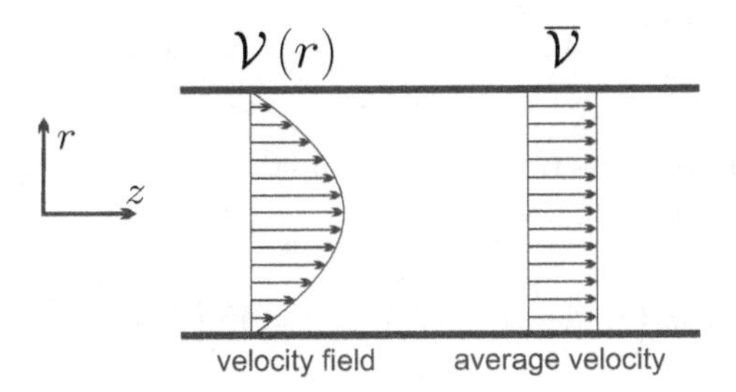

Fig. 9.3 Velocity field and average velocity in pipe flow

Sometimes it is convenient to use the volume flow $\dot{V}$ which is given by

$$\dot{V} = \frac{\dot{m}}{\rho} = \mathcal{V}A\,. \tag{9.4}$$

9.3 Flow Work and Energy Transfer

The total energy E of an open system changes due to exchange of heat and work, and due to *convective energy transport*, that is energy carried in or out by the mass crossing the system boundary. Apart from propeller and volume change work one has also to account for *flow work*, which is the work required to push mass over the system boundaries.

We consider convective transfer first: The specific energy $e = u + \frac{1}{2}\mathcal{V}^2 + gz$ is the local energy per unit mass. Any element of mass crossing the boundary carries this specific energy along, and thus a mass flow $\dot{m}$ carries the convective energy flux

$$\dot{E} = \dot{m}e = \dot{m}\left(u + \frac{1}{2}\mathcal{V}^2 + gz\right) \tag{9.5}$$

over the system boundary.

Now we compute flow work. The power required to push mass over the system boundary is, as always, the force required times the velocity. The force is the local pressure times the cross section, thus

$$\dot{W}_{flow} = -(pA)\,\mathcal{V} = -\frac{p}{\rho}\dot{m}\,. \tag{9.6}$$

Work is done to the system when mass is entering, then $\dot{W}_{flow}$ must be negative. The system does work to push leaving mass out, then $\dot{W}_{flow}$ must be positive. Accordingly, flow work points opposite to mass flow, which is ensured by the minus sign in the equation.

Thus, in comparison to the energy balance for closed systems, the energy balance for the general open system of Fig. 9.1 has additional contributions to account for convective energy transport and flow work. In condensed notation it reads

$$\frac{dE}{dt} = \dot{Q} - \dot{W} + \sum_{\text{in/out}} \dot{E} - \sum_{\text{in/out}} \dot{W}^{flow}\,, \tag{9.7}$$

where the sums have to be taken over all flows crossing the system boundary. From the above expressions we find

$$\dot{E} - \dot{W}^{flow} = \dot{m}\left(u + \frac{p}{\rho} + \frac{1}{2}\mathcal{V}^2 + gz\right) = \dot{m}\left(h + \frac{1}{2}\mathcal{V}^2 + gz\right)\,. \tag{9.8}$$

The enthalpy $h = u + \frac{p}{\rho}$ arises through the combination of convective transport of internal energy and flow work.

Explicitly accounting for mass flows leaving and entering the system, the 1st law—the balance of energy—for the general open system becomes

$$\frac{dE}{dt} = \sum_{\text{in}} \dot{m}_i\left(h + \frac{1}{2}\mathcal{V}^2 + gz\right)_i - \sum_{\text{out}} \dot{m}_e\left(h + \frac{1}{2}\mathcal{V}^2 + gz\right)_e + \dot{Q} - \dot{W}\,. \tag{9.9}$$

The indices (i, e) indicate the values of the properties at the location where the respective flows cross the system boundary, that is their average values at the inlets and outlets, respectively. This equation states that the energy E within the system changes due to convective inflow and outflow, as well as due to heat transfer and work. Note that the flow energy includes the work required to move the mass across the boundaries (flow work). Moreover, there can be several contributions to work and heat transfer, that is $\dot{W} = \sum_j \dot{W}_j$ and $\dot{Q} = \sum_k \dot{Q}_k$.

9.4 Entropy Transfer

All mass that is entering or leaving the system carries entropy. The entropy flow associated with a mass flow is simply $\dot{S} = \dot{m}s$, where s is the specific entropy. Adding the appropriate terms for inflow and outflow to the 2nd law (4.24) for closed systems yields the 2nd law—the balance of entropy—for open systems as

$$\frac{dS}{dt} = \sum_{\text{in}} \dot{m}_i s_i - \sum_{\text{out}} \dot{m}_e s_e + \sum_k \frac{\dot{Q}_k}{T_k} + \dot{S}_{gen} \quad \text{with} \quad \dot{S}_{gen} \geq 0 \,. \tag{9.10}$$

This equation states that the entropy S within the system changes due to convective inflow and outflow, as well as due to entropy transfer caused by heat transfer ($\dot{Q}_k/T_k$) and entropy generation due to irreversible processes inside the system ($\dot{S}_{gen} \geq 0$). If all processes within the system are reversible, the entropy generation vanishes ($\dot{S}_{gen} = 0$). Recall that $\dot{Q}_k$ is the heat that crosses the system boundary where the boundary temperature is T_k.

9.5 Open Systems in Steady State Processes

Many thermodynamic systems and cycles are composed of open system devices. Since the cycles (e.g., power plants or refrigerators) run continuously, they are mainly operating at steady state. In the following sections we shall study the basic equations for steady open systems, and tabulate work and heat for the most important devices.

In steady state processes no changes occur over time at a given location. Then all time derivatives vanish and mass balance (9.1), energy balance (9.9), and entropy balance (9.10) assume the forms

$$\sum_{\text{out}} \dot{m}_e = \sum_{\text{in}} \dot{m}_i \,, \tag{9.11}$$

$$\sum_{\text{out}} \dot{m}_e \left(h + \frac{1}{2}\mathcal{V}^2 + gz\right)_e - \sum_{\text{in}} \dot{m}_i \left(h + \frac{1}{2}\mathcal{V}^2 + gz\right)_i = \dot{Q} - \dot{W} \,, \tag{9.12}$$

$$\sum_{\text{out}} \dot{m}_e s_e - \sum_{\text{in}} \dot{m}_i s_i - \sum_k \frac{\dot{Q}_k}{T_k} = \dot{S}_{gen} \geq 0 \,. \tag{9.13}$$

The interpretation of these equations is straightforward: The mass balance (9.11) states that, in steady state, as much mass leaves the system as enters. The energy balance (9.12) states that a change of the flow energy $\dot{m}\left(h + \frac{1}{2}\mathcal{V}^2 + gz\right)$ between outflow and inflow is effected by exchange of heat and work. The last equation, Eq. (9.13), states that the difference of entropy flowing in and out of the system is due to transfer of entropy caused by heat transfer ($\dot{Q}/T$) and entropy generation inside the system ($\dot{S}_{gen}$). Entropy is

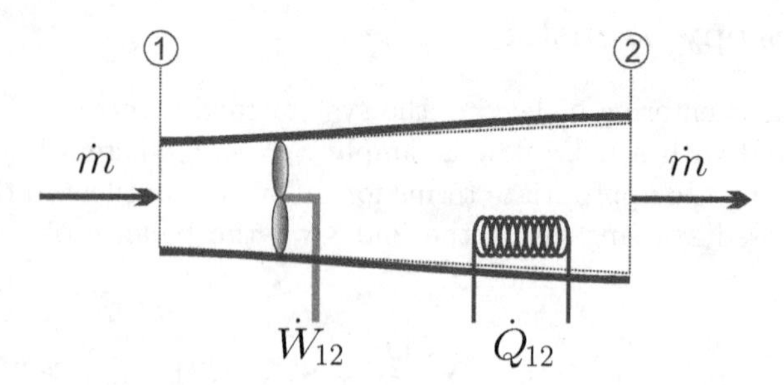

Fig. 9.4 Typical one-inlet-one-exit system

created, if the processes are irreversible ($\dot{S}_{gen} > 0$), and entropy is conserved when all processes are reversible ($\dot{S}_{gen} = 0$).

Note that in a steady state process the volume of the system must remain unchanged, which implies that no piston work occurs: all work exchange is due to propeller work.

9.6 One Inlet, One Exit Systems

A case of particular importance are systems with only one inlet and one exit, as sketched in Fig. 9.4, for which the mass balance reduces to

$$\dot{m}_{in} = \dot{m}_{out} = \dot{m} \, . \tag{9.14}$$

There is just one constant mass flow $\dot{m}$ flowing through each cross section of the system.

The corresponding forms for energy and entropy balance are[2]

$$\dot{m}\left[h_2 - h_1 + \frac{1}{2}\left(\mathcal{V}_2^2 - \mathcal{V}_1^2\right) + g\left(z_2 - z_1\right)\right] = \dot{Q}_{12} - \dot{W}_{12} \, , \tag{9.15}$$

$$\dot{m}\left(s_2 - s_1\right) - \sum_k \frac{\dot{Q}_k}{T_k} = \dot{S}_{gen} \geq 0 \, . \tag{9.16}$$

It is instructive to study the equations for an infinitesimal step within the system, i.e., for infinitesimal system length dx, where the differences reduce to differentials,

[2] The subscripts refer to properties at different locations within the device: "1" denotes the inlet, "2" denotes the outlet. This must be distinguished from the analysis of closed systems, where the subscripts normally refer to states assumed at different times t_1, t_2.

$$\dot{m}\left(dh + \frac{1}{2}d\mathcal{V}^2 + gdz\right) = \delta\dot{Q} - \delta\dot{W}\,, \qquad (9.17)$$

$$\dot{m}ds - \frac{\delta\dot{Q}}{T} = \delta\dot{S}_{gen}\,. \qquad (9.18)$$

Heat and power exchanged, and entropy generated, in an infinitesimal step along the system are process dependent, and as always we write $(\delta\dot{Q}, \delta\dot{W}, \delta\dot{S})$ to indicate that these quantities are not exact differentials. Use of the Gibbs equation in the form $Tds = dh - vdp$ allows to eliminate dh and $\delta\dot{Q}$ between the two equations to give an expression for power,

$$\delta\dot{W} = -\dot{m}\left(vdp + \frac{1}{2}d\mathcal{V}^2 + gdz\right) - T\delta\dot{S}_{gen}\,. \qquad (9.19)$$

The total power for the finite system follows from integration over the length of the system as

$$\dot{W}_{12} = -\dot{m}\int_1^2\left(vdp + \frac{1}{2}d\mathcal{V}^2 + gdz\right) - \int_1^2 T\delta\dot{S}_{gen}\,. \qquad (9.20)$$

The above equation has several implications: First, since $T\delta\dot{S}_{gen} \geq 0$, we see—again—that irreversibilities reduce the power output of a power producing device (where $\dot{W}_{12} > 0$), and increase the power demand of a power consuming device (where $\dot{W}_{12} < 0$). Efficient energy conversion requires to reduce irreversibilities as much as possible.

When we consider (9.19) for a flow without work, we find Bernoulli's equation (Daniel Bernoulli, 1700-1782) for pipe flows as

$$vdp + \frac{1}{2}d\mathcal{V}^2 + gdz = -\frac{1}{\dot{m}}T\delta\dot{S}_{gen}\,. \qquad (9.21)$$

The Bernoulli equation is probably easier to recognize in its integrated form for incompressible fluids (where $v = \frac{1}{\rho} = const.$),

$$H_2 - H_1 = \frac{p_2 - p_1}{\rho} + \frac{1}{2}\left(\mathcal{V}_2^2 - \mathcal{V}_1^2\right) + g\left(z_2 - z_1\right) = -\frac{1}{\dot{m}}\int_1^2 T\delta\dot{S}_{gen}\,. \quad (9.22)$$

Here, $H = \frac{p}{\rho} + \frac{1}{2}\mathcal{V}^2 + gz$ denotes hydraulic head. The right hand side describes loss of hydraulic head due to irreversible processes, in particular friction.

Finally, for reversible processes—where $\delta\dot{S}_{gen} = 0$ in (9.20)—we find the reversible steady-flow *work*

$$\dot{W}_{12}^{rev} = -\dot{m}\int_1^2\left(vdp + \frac{1}{2}d\mathcal{V}^2 + gdz\right)\,. \qquad (9.23)$$

For flows at relatively low velocities and without significant change of level the above relation can be simplified to

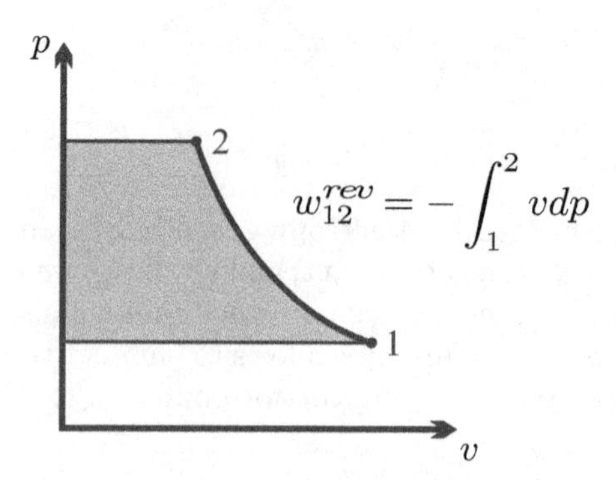

Fig. 9.5 Specific reversible steady state flow work is the area to the left of the process curve

$$\dot{W}_{12}^{rev} = -\dot{m} \int_1^2 vdp \,. \tag{9.24}$$

The *specific reversible flow work*, i.e., the work per unit mass flowing through the device, is

$$w_{12}^{rev} = \frac{\dot{W}_{12}^{rev}}{\dot{m}} = -\int_1^2 vdp \,. \tag{9.25}$$

In a p-v-diagram, w_{12}^{rev} is the area to the left of the process curve, see Fig. 9.5 for an illustration.

The heat exchanged in a reversible process in a steady-state, one inlet, one exit system follows from the integration of the second law (9.18) with $\delta\dot{S}_{gen} = 0$ as

$$\dot{Q}_{12}^{rev} = \int_1^2 \delta\dot{Q}^{rev} = \dot{m} \int_1^2 Tds \,. \tag{9.26}$$

Correspondingly, the specific reversible heat, i.e., the heat per unit mass flowing through the device, is

$$q_{12}^{rev} = \frac{\dot{Q}_{12}^{rev}}{\dot{m}} = \int_1^2 Tds \,. \tag{9.27}$$

In a T-s-diagram, q_{12}^{rev} is the area below the process curve, just as in a closed system.

9.7 Entropy Generation in Mass Transfer

Friction in flows leads to loss of pressure and corresponding entropy generation. When we consider a simple flow with no work added or withdrawn, Eq. (9.21) gives the entropy generated in dx as

$$\delta\dot{S}_{gen} = -\frac{\dot{m}}{T}\left(vdp + \frac{1}{2}d\mathcal{V}^2 + gdz\right) .$$

The total entropy generated in a finite system is

$$\dot{S}_{gen} = -\dot{m}\int_{in}^{out}\frac{1}{T}\left(vdp + \frac{1}{2}d\mathcal{V}^2 + gdz\right) . \tag{9.28}$$

For a system where kinetic and potential energy are unimportant, this reduces to

$$\dot{S}_{gen} = -\dot{m}\int_{in}^{out}\frac{v}{T}dp . \tag{9.29}$$

We interpret the entropy generation rate as the product of a flux, the mass flow $\dot{m}$, and a thermodynamic force, namely the integral over $-\frac{v}{T}dp$. Since specific volume v and thermodynamic temperature T are strictly positive, the force $\int_{in}^{out}\frac{v}{T}dp$ is proportional to the pressure difference, $-\int_{in}^{out}\frac{v}{T}dp \propto (p_{in} - p_{out})$. In order to obtain a positive entropy generation rate, the mass flow must be proportional to the force, which is the case for

$$\dot{m} = \beta A\left(p_{in} - p_{out}\right) = \beta A\Delta p . \tag{9.30}$$

Here, A is the mass transfer area and $\beta > 0$ is a positive transport coefficient that must be measured.

One particular example for this law is the Hagen-Poiseuille relation (Gotthilf Hagen, 1797-1884; Jean Poiseuille, 1797-1869) of fluid dynamics which gives the volume flow $\dot{V} = \dot{m}/\rho$ of a fluid with shear viscosity η through a pipe of radius R and length L as

$$\dot{V} = \frac{\pi R^4}{8\eta L}\Delta p . \tag{9.31}$$

Another example for (9.30) is Darcy's law (Henry Darcy, 1803-1858) that describes flow through porous media. Then A is the cross section of the porous medium considered, and β is a coefficient of permeability.

Real processes are irreversible, and produce entropy. For a simple flow, the work loss to irreversibilities is

$$\dot{W}_{\text{loss}} = \int_{in}^{out} T\delta\dot{S}_{gen} . \tag{9.32}$$

Since $\delta\dot{S}_{gen} = -\frac{\dot{m}}{T}vdp$, for isothermal flow of an incompressible liquid, entropy generation and work loss are

$$\dot{S}_{gen} = \frac{\dot{V}}{T}\left(p_{in} - p_{out}\right) \quad , \quad \dot{W}_{\text{loss}} = \dot{V}\left(p_{in} - p_{out}\right) , \tag{9.33}$$

where $\dot{V} = \dot{m}v$ is the volume flow.

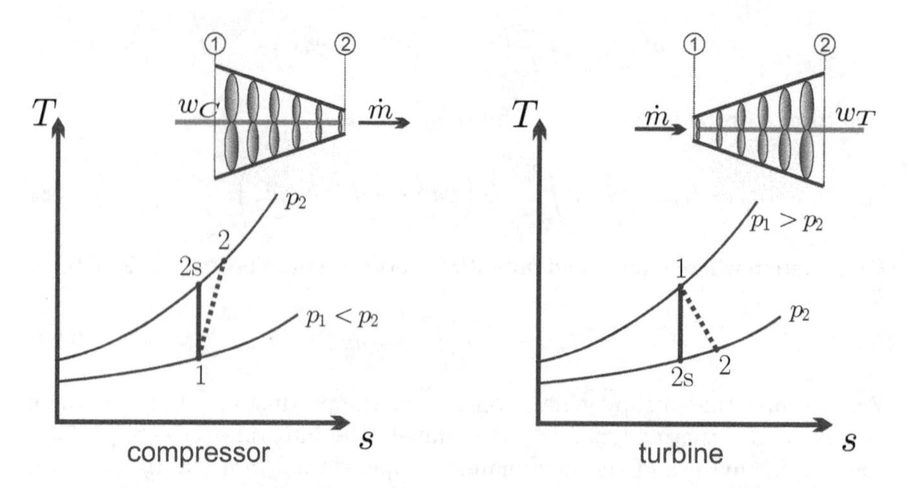

Fig. 9.6 Compressor (left) and turbine (right): Sketch and T-s-diagrams

For an ideal gas flow, we have instead

$$\dot{S}_{gen} = \dot{m} R \ln \frac{p_{in}}{p_{out}} \quad , \quad \dot{W}_{\text{loss}} = -\dot{m} \int_1^2 v dp \,. \tag{9.34}$$

9.8 Adiabatic Compressors, Turbines and Pumps

In a turbine, high pressure flow drives propeller blades that are attached to a rotating shaft. The flow does work on the blades, the rotating shaft delivers the work to the surroundings. As the flow does work, the pressure drops, that is turbines are driven by pressure differences. Compressors perform the opposite task: the shaft is driven, and the rotating blades pressurize the flow: compressors create a pressure difference and consume work.

The flow through most turbines (T) and compressors (C), as sketched in Fig. 9.6, is so fast that there is no time to exchange heat with the surroundings, and thus they are often treated as being adiabatic. Although the flows are fast, the flow velocities are normally ignored, since the kinetic energy does not contribute significantly to work. With these simplifications (9.15, 9.16) reduce to

$$\begin{aligned} &\dot{W}_C = \dot{m}\,(h_1 - h_2) < 0 &&, \quad \dot{W}_T = \dot{m}\,(h_1 - h_2) > 0 \,, \\ &w_C = \frac{\dot{W}_C}{\dot{m}} = h_1 - h_2 < 0 &&, \quad w_T = \frac{\dot{W}_T}{\dot{m}} = h_1 - h_2 > 0 \,, \\ &s_2 - s_1 \geq 0 &&, \quad s_2 - s_1 \geq 0 \,, \end{aligned} \tag{9.35}$$

which hold for reversible and irreversible adiabatic turbines and compressors. In reversible adiabatic compressors and turbines the flow is isentropic (con-

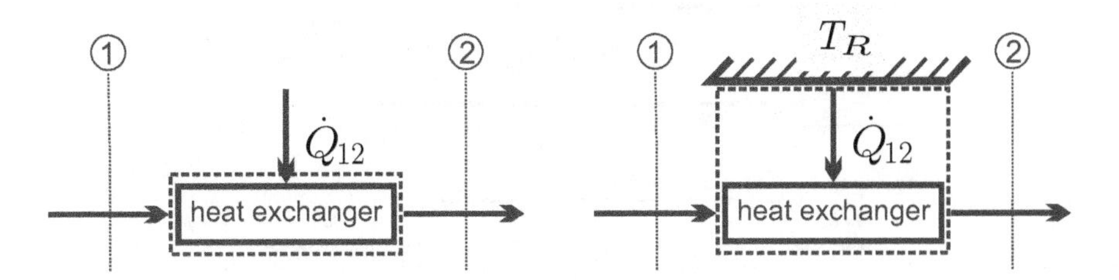

Fig. 9.7 Flow heat exchanger. Left: The system boundary is just around the flow. Right: The system includes the heat transfer to a reservoir R, the system boundary is extended.

tinuous lines 1-2s), while in irreversible adiabatic compressors and turbines the entropy must grow (dashed lines 1-2s), as illustrated in the figure.

Pumps serve to increase the pressure in liquids. Many liquids can be assumed to be incompressible in good approximation, so that $v = const$. The reversible pump work can then easily be computed from the flow work (9.24),

$$w_P^{rev} = -\int_1^2 vdp = -v\int_1^2 dp = v\,(p_1 - p_2) < 0\,. \tag{9.36}$$

9.9 Heating and Cooling of a Pipe Flow

In flow heat exchangers, the working fluid flows through pipes that are exposed to an environment at different temperature. For a simple pipe flow with heat exchange but no work, as depicted in Fig. 9.7, one can usually ignore the kinetic and potential energies, thus the heat exchanged is obtained from the first law as

$$\dot{Q}_{12} = \dot{m}\,(h_2 - h_1) \quad , \quad q_{12} = \frac{\dot{Q}_{12}}{\dot{m}} = h_2 - h_1\,. \tag{9.37}$$

For further insight, we consider the associated entropy generation. First we apply the second law to the flow alone, that is the system boundary is directly at the pipe, see left of Fig. 9.7. The appropriate form of the second law for an element dx of the flow is the Bernoulli equation (9.21). Ignoring kinetic and potential energy, we see that the pressure must drop along the flow, $dp = -\frac{1}{\dot{m}v}T\delta\dot{S}_{gen} \leq 0$. The pressure drop is due to friction in the flow. In our calculations we normally ignore friction effects in heat exchangers, so that $\delta\dot{S}_{gen} = 0$, and thus the pipe flow in heat exchangers is assumed to be isobaric: $dp = 0$, or $p = const$.

While the flow itself is reversible, typically there is irreversibility associated with the heat exchange to the external environment: the isobaric flow heat exchanger is internally reversible, but externally irreversible. The schematic

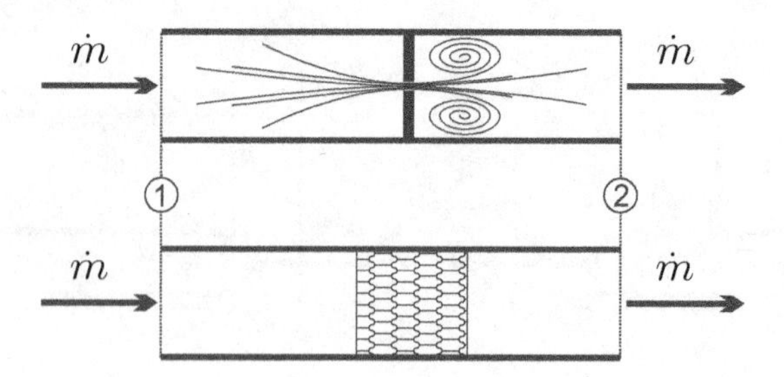

Fig. 9.8 Typical throttling devices: orifice, porous plug

on the right of Fig. 9.7 shows heat exchange with a reservoir at temperature T_R. Application of the second law to the system with boundary at the reservoir, where the temperature is T_R, yields

$$\dot{m}(s_2 - s_1) - \frac{\dot{Q}_{12}}{T_R} = \dot{m}\left(s_2 - s_1 - \frac{h_2 - h_1}{T_R}\right) = \dot{S}_{gen} \, . \tag{9.38}$$

The above equations are valid for heating ($\dot{Q}_{12} > 0$) and cooling ($\dot{Q}_{12} < 0$). Heat exchange between flows in closed and open heat exchangers will be discussed in Secs. 9.15, 9.16.

9.10 Throttling Devices

Throttling devices are used to create significant pressure drops by pressing the fluid through a small orifice or a porous plug, as depicted in Fig. 9.8, or other narrow obstacles which induce friction losses to the flow.

No work is exchanged, and the flow can be considered as adiabatic; flow velocities are small so that kinetic energy can be ignored. Due to high friction losses the process is irreversible, and the pressure drops, $dp = -\frac{1}{\dot{m}v}T\delta\dot{S}_{gen} < 0$, or $p_2 - p_1 < 0$. Then first and second law (9.15, 9.16) reduce to

$$h_2 - h_1 = 0 \quad , \quad s_2 - s_1 > 0 \, . \tag{9.39}$$

Throttling devices are isenthalpic (i.e., constant enthalpy), adiabatic, and irreversible.

9.11 Adiabatic Nozzles and Diffusers

Nozzles and diffusers have no moving mechanical parts, they change flow properties through changing the cross section. Nozzles are used to accelerate a flow, for instance in order to produce thrust for an airplane or a rocket. Diffusers are used to slow down flows and increase the pressure. As can be

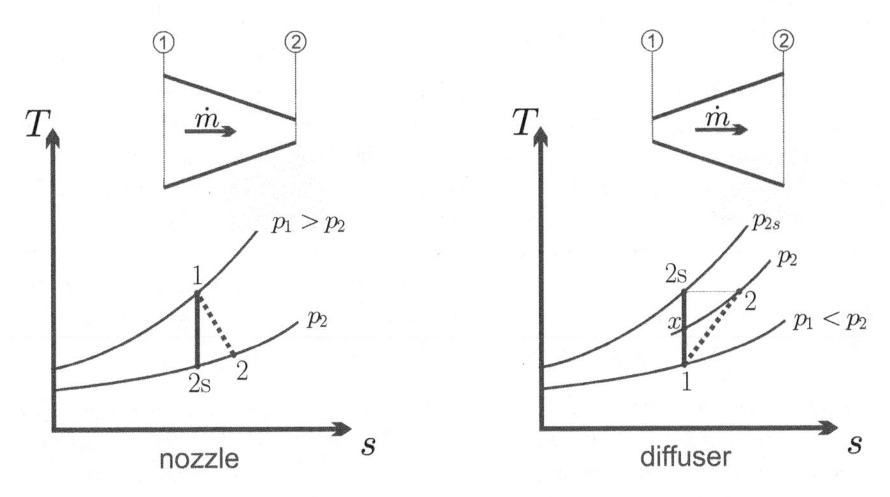

Fig. 9.9 Nozzle and diffuser: sketch and T-s-diagrams

seen from Fig. 9.9, neither work (no moving boundaries!) nor heat (fast flow!) are exchanged, and thus the first law reduces to

$$h_2 + \frac{1}{2}\mathcal{V}_2^2 = h_1 + \frac{1}{2}\mathcal{V}_1^2 . \tag{9.40}$$

For nozzles the inflow velocity is normally much less than the outflow velocity and can be ignored, so that

$$\text{nozzle: } \mathcal{V}_2^2 \gg \mathcal{V}_1^2 \Rightarrow \mathcal{V}_2 = \sqrt{2\left(h_1 - h_2\right)} . \tag{9.41}$$

For diffusers, on the other hand, the inflow velocity normally is high, while the outflow velocity is slow, thus

$$\text{diffuser: } \mathcal{V}_2^2 \ll \mathcal{V}_1^2 \Rightarrow h_2 = h_1 + \frac{1}{2}\mathcal{V}_1^2 . \tag{9.42}$$

To better understand the relation between changes of pressure and velocity we employ the Bernoulli equation (9.21) for reversible processes and constant potential energy, $dz = 0$, which gives $vdp = -\mathcal{V}d\mathcal{V}$. For a nozzle $dp < 0$ and thus $d\mathcal{V} > 0$ (acceleration in pressure gradient), while for a diffuser $d\mathcal{V} < 0$ and thus $dp > 0$ (pressure increase by deceleration).

In reversible adiabatic nozzles and diffusers the flow is isentropic (continuous lines 1-2s), while in irreversible adiabatic nozzles and diffusers the entropy must grow (dashed lines 1-2), as illustrated in Fig. 9.9.

9.12 Isentropic Efficiencies

Many thermodynamic processes are adiabatic, since they happen so fast that no heat can be exchanged during the process. When the process is adiabatic and reversible, the second law tells us that the process is isentropic as well, since $\dot{Q} = 0$ and $\dot{S}_{gen} = 0$. Real processes, however, will always be irreversible, so that real adiabatic processes are accompanied by entropy generation. A useful measure for the quality of the performance of irreversible devices is given by efficiency measures that compare adiabatic irreversible devices to their isentropic counterparts. In the following we discuss isentropic efficiencies for compressors, pumps, turbines, nozzles and diffusers.

Compressor: The left T-s-diagram in Fig. 9.6 shows, for an ideal gas, two isobaric lines. An adiabatic and reversible compression of the gas between pressures p_1 and p_2 follows the isentropic path 1-2s; the required compressor work per unit mass is $w_{C,rev} = h_1 - h_{2s}$. A real compressor is irreversible due to internal friction and internal heat transfer, and the compressed state will have a larger entropy. The endpoint of the compression is point 2 with $s_2 > s_{2s} = s_1$ and the required compressor work is $w_C = h_1 - h_2$. Note that in the figure the line connecting points 1 and 2 is dotted to indicate that during compression the gas is in non-equilibrium states that cannot be marked as a path in the T-s-diagram, only the initial and endpoints are known.

The performance of an adiabatic compressor can be measured by comparing its actual work requirement with the best case by means of the isentropic compressor efficiency

$$\eta_C = \frac{w_{C,rev}}{w_C} = \frac{h_1 - h_{2s}}{h_1 - h_2} \leq 1 \,. \tag{9.43}$$

The real compressor requires more work input, since some work is needed to overcome friction and other irreversibilities. The isentropic efficiency is defined such that it has values between 0 and 1; realistic compressors have efficiencies $\eta_C = 0.8 - 0.9$. This definition can be extended to pumps, which do not compress a gas, but increase the pressure in a liquid.

Turbine: The expansion in an adiabatic and reversible turbine follows the isentropic path 1-2s in the right T-s-diagram of Fig. 9.6; the work delivered per unit mass is $w_{T,rev} = h_1 - h_{2s}$. A real turbine is irreversible, and will expand to state 2 producing the work $w_T = h_1 - h_2$.

The performance of an adiabatic turbine can be measured by comparing its actual work requirement with the best case by means of the isentropic turbine efficiency

$$\eta_T = \frac{w_T}{w_{T,rev}} = \frac{h_1 - h_2}{h_1 - h_{2s}} \leq 1 \,. \tag{9.44}$$

The real turbine delivers less work, since some work is consumed to overcome friction and other irreversibilities. The isentropic efficiency is defined such that it has values between 0 and 1; realistic turbines have efficiencies $\eta_T = 0.85 - 0.95$.

Nozzle: An adiabatic nozzle accelerates the flow, while no work is exchanged. For a nozzle expanding from p_1 to p_2, the outflow velocity is $\mathcal{V} = \sqrt{2\,(h_1 - h_2)}$ (inflow velocity ignored). The nozzle efficiency is usually defined not by the velocity, but the specific kinetic energy of the outflow, as

$$\eta_N = \frac{\frac{1}{2}\mathcal{V}^2}{\frac{1}{2}\mathcal{V}_{rev}^2} = \frac{h_1 - h_2}{h_1 - h_{2s}} \leq 1 \,. \tag{9.45}$$

Typical nozzle efficiencies are above 95%.

Diffuser: Isentropic diffuser efficiency cannot be defined through enthalpies that easily. The first law for the diffuser states $h_2 = h_1 + \frac{1}{2}\mathcal{V}_1^2$, which implies that for an ideal gas, where $h_2 = h\,(T_2)$, the exit temperature T_2 is the same for reversible and irreversible diffusers. However, irreversibilities lead to lower exit pressure p_2 as compared to the reversible exit pressure p_{2s}, see the T-s-diagram in Fig. 9.9. Not all of the available kinetic energy $\frac{1}{2}\mathcal{V}_1^2$ is used for compression, i.e., pressure increase, but some is lost to irreversibility. A reversible diffuser that gives the end pressure p_2 of the irreversible diffuser would convert the kinetic energy $\frac{1}{2}\mathcal{V}_x^2 = h_x - h_1$, where $h_x = h\,(T_x)$, and T_x is the temperature at the end of an isentropic compression between the inlet state $\{p_1, T_1\}$ and $\{p_2, T_x\}$. With the help of this artificial state, isentropic diffuser efficiency can be defined as

$$\eta_D = \frac{h_x - h_1}{h_2 - h_1} \,. \tag{9.46}$$

The definition becomes more transparent for an ideal gas with constant specific heats, for which we find

$$\eta_D = \frac{T_x - T_1}{T_2 - T_1} = \frac{\left(\frac{p_2}{p_1}\right)^{\frac{k-1}{k}} - 1}{\frac{T_2}{T_1} - 1} \,. \tag{9.47}$$

9.13 Summary: Open System Devices

For later reference, we list work, heat and operating conditions for the most common one-inlet-one-exit devices in a table:

adiabatic compressor:	$s_2 - s_1 \geq 0$	$w_C = h_1 - h_2$	$q = 0$
adiabatic turbine:	$s_2 - s_1 \geq 0$	$w_T = h_1 - h_2$	$q = 0$
rev. adiabatic pump:	$s_2 - s_1 = 0$	$w_P = v\,(p_1 - p_2)$	$q = 0$
heating and cooling:	$p_2 - p_1 = 0$	$w = 0$	$q_{12} = h_2 - h_1$
throttling device:	$h_2 - h_1 = 0$	$w = 0$	$q = 0$
adiabatic nozzle:	$\mathcal{V}_2 = \sqrt{2\,(h_1 - h_2)}$	$w = 0$	$q = 0$
adiabatic diffuser:	$h_2 = h_1 + \frac{1}{2}\mathcal{V}_1^2$	$w = 0$	$q = 0$

9.14 Examples: Open System Devices

9.14.1 Reversible Turbine with Kinetic Energy

Steam (water vapor) at 15 MPa, 500 °C is expanded in a steady state adiabatic reversible turbine to 40 kPa. The inlet diameter of the turbine is $d_1 = 0.1\,\mathrm{m}$, the exit diameter is $d_2 = 1.0\,\mathrm{m}$, and the entry velocity is $\mathcal{V}_1 = 40\frac{\mathrm{m}}{\mathrm{s}}$. We determine the properties of the end state and the power produced. As we do this, we consider the question whether the kinetic energy of the flow can be ignored.

From the steam table for water we find specific volume, enthalpy and entropy of the inlet state as

$$v_1 = v\,(15\,\mathrm{MPa}, 500\,^\circ\mathrm{C}) = 0.0208\frac{\mathrm{m}^3}{\mathrm{kg}}\,,$$

$$h_1 = h\,(15\,\mathrm{MPa}, 500\,^\circ\mathrm{C}) = 3309\frac{\mathrm{kJ}}{\mathrm{kg}}\,,$$

$$s_1 = s\,(15\,\mathrm{MPa}, 500\,^\circ\mathrm{C}) = 6.344\frac{\mathrm{kJ}}{\mathrm{kg\,K}}\,.$$

The mass flow through the turbine is ($A_1 = \frac{\pi}{4}d_1^2 = 7.85 \times 10^{-3}\,\mathrm{m}^2$)

$$\dot{m} = \frac{\mathcal{V}_1 A_1}{v_1} = 15.1\frac{\mathrm{kg}}{\mathrm{s}}\,.$$

Since the process is adiabatic and reversible, it is isentropic, so that $s_{2s} = s_1 = 6.344\frac{\mathrm{kJ}}{\mathrm{kg\,K}}$ at $p_2 = 40\,\mathrm{kPa}$. This is a saturated state with the quality

$$x_{2s} = \frac{s_{2s} - s_f}{s_{fg}}_{|p_2 = 40\,\mathrm{kPa}} = \frac{6.344 - 1.026}{6.644} = 0.80\,.$$

The corresponding values for volume and enthalpy are

$$v_{2s} = (1 - x_{2s})\, v_f + x_{2s} v_g = [0.2 \times 0.00103 + 0.8 \times 3.993]\, \frac{\mathrm{m}^3}{\mathrm{kg}} = 3.195 \frac{\mathrm{m}^3}{\mathrm{kg}} \,,$$

$$h_{2s} = h_f + x_{2s} h_{fg} = [317.6 + 0.8 \times 2319]\, \frac{\mathrm{kJ}}{\mathrm{kg}} = 2173 \frac{\mathrm{kJ}}{\mathrm{kg}} \,.$$

Since the mass flux is constant throughout the turbine, the exit velocity is ($A_2 = \frac{\pi}{4} d_2^2 = 0.785\,\mathrm{m}^2$)

$$\mathcal{V}_{2s} = \frac{\dot{m}\, v_{2s}}{A_2} = 61.5 \frac{\mathrm{m}}{\mathrm{s}} \,.$$

From the first law for an adiabatic turbine we find the specific work as

$$w_{T,rev} = h_1 - h_{2s} + \frac{1}{2}\left(\mathcal{V}_1^2 - \mathcal{V}_{2s}^2\right) = 1136 \frac{\mathrm{kJ}}{\mathrm{kg}} - 1091 \frac{\mathrm{m}^2}{\mathrm{s}^2} = 1135 \frac{\mathrm{kJ}}{\mathrm{kg}} \,.$$

Note that $1 \frac{\mathrm{kJ}}{\mathrm{kg}} = 1000 \frac{\mathrm{m}^2}{\mathrm{s}^2}$, thus the contribution of kinetic energy $\frac{1}{2}\left(\mathcal{V}_1^2 - \mathcal{V}_2^2\right)$ is far smaller than the contribution of thermal energy $(h_1 - h_{2s})$. This example supports our statement that in the computation of turbines and compressors kinetic energy can be ignored.

The power produced by the reversible turbine is

$$\dot{W} = \dot{m} w_{T,rev} = 17.2\ \mathrm{MW} \,.$$

The T-s-diagram for the turbine is depicted as process $(1 - 2s)$ in Fig. 9.10.

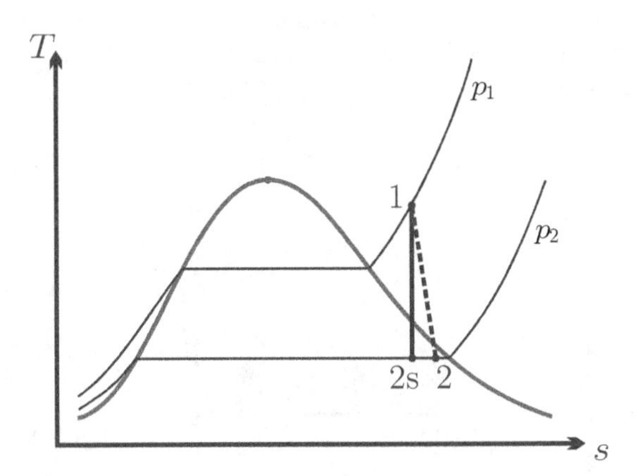

Fig. 9.10 T-s-diagram for a reversible and an irreversible steam turbine between pressures p_1 and p_2

9.14.2 Irreversible Turbine

We consider the turbine of the previous example for the irreversible case, with an isentropic efficiency of $\eta_T = 0.8$. The T-s-diagram for the turbine is depicted as process $(1-2)$ in Fig. 9.10.

Due to the definition of isentropic efficiency, the specific work and the power produced for this turbine is

$$w_T = \eta_T w_{T,rev} = 911.2 \frac{\text{kJ}}{\text{kg}} , \quad \dot{W} = \dot{m} w_T = 13.8\,\text{MW} .$$

As suggested by the previous example, we ignore kinetic energy. Then, the enthalpy of the final state is

$$h_2 = h_1 - w_T = 2398 \frac{\text{kJ}}{\text{kg}} ,$$

which corresponds to the quality at the turbine exit

$$x_2 = \frac{h_2 - h_f}{h_{fg}}_{|p_2} = 0.9 .$$

9.14.3 Irreversible Compressor

In a refrigeration cycle, cooling fluid R134a enters the adiabatic compressor as saturated vapor at $p_1 = 0.14\,\text{MPa}$. The compressor with isentropic efficiency of $\eta_C = 0.85$ compresses the vapor to $p_2 = 1\,\text{MPa}$. We compute the specific work required. The irreversible compression process $(1-2)$, and the reversible ideal process $(1-2s)$ are indicated in the T-s-diagram of Fig. 9.11.

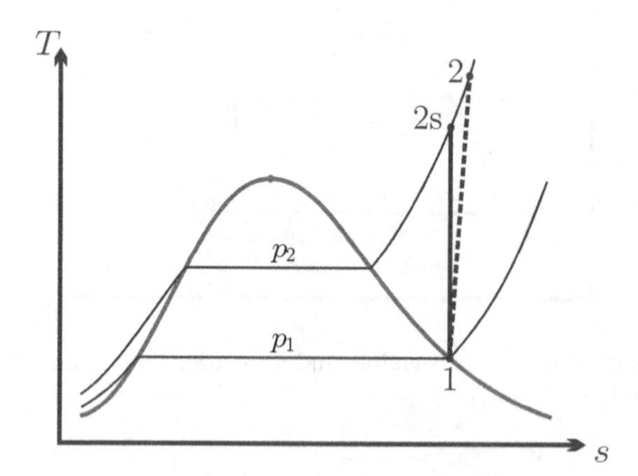

Fig. 9.11 Reversible and irreversible compression of R134a

From the table we find the data for the inlet state as

$$T_1 = T_{\text{sat}}\left(p_1\right) = T_{\text{sat}}\left(0.14\,\text{MPa}\right) = -18.8\,°\text{C}\,,$$
$$h_1 = h_g\left(p_1\right) = h_g\left(0.14\,\text{MPa}\right) = 236.0\frac{\text{kJ}}{\text{kg}}\,,$$
$$s_1 = s_g\left(p_1\right) = s_g\left(0.14\,\text{MPa}\right) = 0.9322\frac{\text{kJ}}{\text{kg}\,\text{K}}\,.$$

The enthalpy at the exit of the reversible compression to p_2 is (interpolation in superheated vapor table)

$$h_{2s} = h\left(p_2, s_2 = s_1\right) = h\left(1\,\text{MPa}, s = 0.9322\frac{\text{kJ}}{\text{kg}\,\text{K}}\right) = 276.8\frac{\text{kJ}}{\text{kg}}\,,$$

the corresponding temperature is $T_{2s} = 47.1\,°\text{C}$. The work for the reversible compressor is

$$w_{C,rev} = h_1 - h_{2s} = -40.8\frac{\text{kJ}}{\text{kg}}\,.$$

The definition of the isentropic compressor efficiency (9.43) yields the work for the irreversible compressor as

$$w_C = \frac{w_{C,rev}}{\eta_C} = -48.0\frac{\text{kJ}}{\text{kg}}\,.$$

Since $w_C = h_1 - h_2$, the enthalpy at the compressor exit is

$$h_2 = h_1 - w_C = 284.0\frac{\text{kJ}}{\text{kg}}\,;$$

interpolation in the steam table yields the exit temperature $T_2 = 53.4\,°\text{C}$.

9.14.4 Irreversible Pump

A water flow of $\dot{m} = 10\frac{\text{kg}}{\text{s}}$ enters a pump at 1 bar, 20 °C, and leaves at 100 bar. The power consumption of the pump is 120 kW. We determine the enthalpy of the compressed liquid at the pump exit, and the isentropic efficiency of the pump.

The water at the pump inlet is in the compressed liquid state, for which we can approximate $h_1 \simeq h_f\left(T_1\right) = 84.0\frac{\text{kJ}}{\text{kg}}$. The power consumption of the pump is

$$\dot{W}_P = \dot{m}w_P = \dot{m}\left(h_1 - h_2\right)\,,$$

which implies that (power consumed is negative, $\dot{W}_P = -120\,\text{kW}$!)

$$h_2 = h_1 - \frac{\dot{W}_P}{\dot{m}} = 96.0\frac{\text{kJ}}{\text{kg}}\,.$$

Considering water as incompressible liquid with $v_1 \simeq v_f(T_1) = 0.001\frac{\mathrm{m}^3}{\mathrm{kg}}$, we have for a reversible pump between the two pressures

$$w_{P,rev} = -v_1 (p_2 - p_1) = -9.9\frac{\mathrm{kJ}}{\mathrm{kg}} .$$

Thus, the isentropic efficiency of the pump is

$$h_P = \frac{w_{P,rev}}{w_P} = \frac{-v_1 (p_2 - p_1)}{h_1 - h_2} = 0.825 .$$

9.14.5 Isobaric Evaporation

A mass flow of $36\frac{\mathrm{t}}{\mathrm{h}}$ compressed water at 25 °C, 80 bar is isobarically heated and evaporated to 550 °C. The heat is provided from combustion of coal which delivers $q_{\mathrm{coal}} = 20\frac{\mathrm{MJ}}{\mathrm{kg}}$ of heat per kg of coal. We determine the mass flow of coal required for this process. Figure 9.12 shows the process diagrams.

With the enthalpies of water

$$h_1 \simeq h_f(25\,^\circ\mathrm{C}) = 104.9\frac{\mathrm{kJ}}{\mathrm{kg}} \quad , \quad h_2 = h(8\,\mathrm{MPa}, 550\,^\circ\mathrm{C}) = 3521\frac{\mathrm{kJ}}{\mathrm{kg}} ,$$

the heat required for heating and evaporation is

$$\dot{Q}_{12} = \dot{m}_{\mathrm{w}} (h_2 - h_1) = 34.2\,\mathrm{MW} .$$

This heat is provided through the combustion of coal, that is

$$\dot{Q}_{12} = \dot{m}_{\mathrm{coal}} q_{\mathrm{coal}} \implies \dot{m}_{\mathrm{coal}} = \frac{\dot{Q}_{12}}{q_{\mathrm{coal}}} = 1.71\frac{\mathrm{kg}}{\mathrm{s}} = 6.15\frac{\mathrm{t}}{\mathrm{h}} .$$

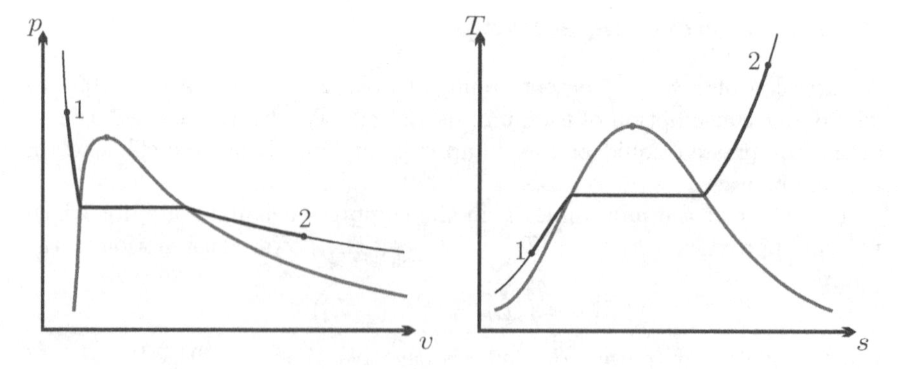

Fig. 9.12 Isobaric heating and evaporation of compressed liquid

We further assume that the heat generated by combustion of coal provides a hot environment (the boiler) at temperature $T_B = 1100\,\mathrm{K}$. The entropies of in- and outflow are

$$s_1 \simeq s_f\,(25\,^\circ\mathrm{C}) = 0.3674\frac{\mathrm{kJ}}{\mathrm{kg\,K}}\ ,\quad s_2 = s\,(8\,\mathrm{MPa}, 550\,^\circ\mathrm{C}) = 6.8778\frac{\mathrm{kJ}}{\mathrm{kg\,K}}\ ,$$

and the associated entropy generation rate is

$$\dot{S}_{gen} = \dot{m}_{\mathrm{w}}\left(s_2 - s_1 - \frac{h_2 - h_1}{T_B}\right) = 34.05\frac{\mathrm{kW}}{\mathrm{K}}\ .$$

9.14.6 Throttling of Compressed Liquid

Cooling fluid R134a at $T_1 = 16\,^\circ\mathrm{C}$, $p_1 = 0.8\,\mathrm{MPa}$, runs through an adiabatic throttle, which it leaves at $p_2 = 140\,\mathrm{kPa}$. We determine the exit state of the flow.

Since the saturation pressure $p_{\mathrm{sat}}\,(T_1) = 0.504\,\mathrm{MPa}$ is less than the actual pressure p_1, the initial state is compressed liquid. The throttle is isenthalpic, that is $h_2 = h_1$. With the approximation for compressed liquid, we have for the latter

$$h_1 \simeq h_f\,(T_1) + v_1\,(p_1 - p_{\mathrm{sat}}\,(T_1)) \simeq h_f\,(T_1) = 71.7\frac{\mathrm{kJ}}{\mathrm{kg}}\ .$$

At 140 kPa, the enthalpy $h_2 = h_1 = 71.7\frac{\mathrm{kJ}}{\mathrm{kg}}$ lies between the enthalpies of the saturated states, $h_f < h_2 < h_g$, so that the exit state is in the two-phase region. The quality is

$$x_2 = \left.\frac{h_2 - h_f}{h_{fg}}\right|_{p_2} = \frac{73.7 - 27.1}{212.1} = 0.22\ .$$

The temperature at the exit is the saturation temperature, $T_2 = T_{\mathrm{sat}}\,(p_2) = -18.8\,^\circ\mathrm{C}$.

From the compressed liquid approximation we find the inlet entropy as $s_1 \simeq s_f\,(T_1) = 0.274\frac{\mathrm{kJ}}{\mathrm{kg\,K}}$. The entropy of the end state is

$$s_2 = s_f + x_2 s_{fg} = 0.1055\frac{\mathrm{kJ}}{\mathrm{kg\,K}} + 0.22 \times 0.827\frac{\mathrm{kJ}}{\mathrm{kg\,K}} = 0.287\frac{\mathrm{kJ}}{\mathrm{kg\,K}}\ .$$

As expected, the entropy grows in the process, $s_2 > s_1$; the T-s-diagram is shown in Fig. 9.13.

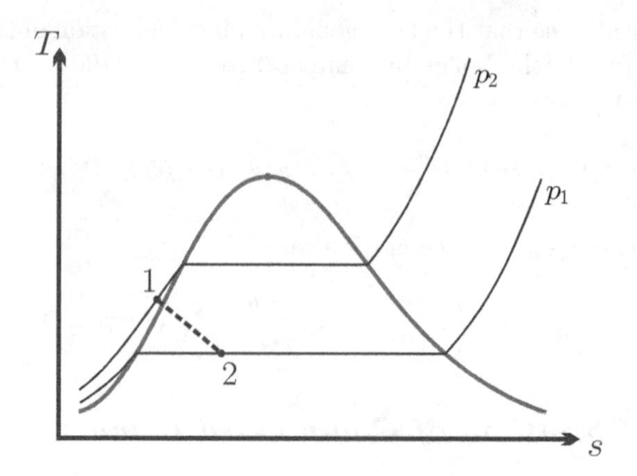

Fig. 9.13 Throttling of compressed R134a liquid into the two phase region

9.14.7 Nozzle

Argon at $T_1 = 600\,\mathrm{K}$, $p_1 = 15\,\mathrm{bar}$ enters an adiabatic nozzle in which it is expanded to $p_2 = 150\,\mathrm{kPa}$. The nozzle efficiency is 92%. We determine the maximum possible exit velocity, the actual exit velocity and the exit temperature.

Argon is a monatomic gas, and thus has constant specific heat, $c_p = \frac{5}{2}R = 0.52\frac{\mathrm{kJ}}{\mathrm{kg\,K}}$, $k = 1.67$. For an isentropic process between the two pressures we find the exit temperature as

$$T_{2s} = T_1 \left(\frac{p_1}{p_2}\right)^{\frac{1-k}{k}} = 238.9\,\mathrm{K}\,.$$

When we ignore the inlet velocity, the exit velocity for the isentropic nozzle is given by

$$\mathcal{V}_{2s} = \sqrt{2\,(h_1 - h_{2s})} = \sqrt{2c_p\,(T_1 - T_{2s})} = 613\frac{\mathrm{m}}{\mathrm{s}}\,.$$

The isentropic nozzle is the best possible case, since it has no losses to friction or other irreversibilities. Thus, the above is the maximum velocity possible.

The definition of the isentropic efficiency of a nozzle (9.45) gives the exit velocity as

$$\mathcal{V}_2 = \sqrt{\eta_N \mathcal{V}_{2s}^2} = 588\frac{\mathrm{m}}{\mathrm{s}}\,.$$

Since $\mathcal{V} = \sqrt{2\,(h_1 - h_2)} = \sqrt{2c_p\,(T_1 - T_2)}$, the exit temperature is

$$T_2 = T_1 - \frac{\mathcal{V}_2^2}{2c_p} = 267.7\,\mathrm{K}\,.$$

9.14.8 *Diffuser*

Air enters the adiabatic irreversible diffuser of an air engine with a velocity of $\mathcal{V}_1 = 300\frac{\text{m}}{\text{s}}$, at $p_1 = 0.1\,\text{bar}$, $T_1 = 230\,\text{K}$. The inlet cross section is $A_1 = 1\,\text{m}^2$. The process can be described as polytropic with $n = 1.5$. We determine the pressure at the end of the diffuser, where the flow velocity can be ignored ($\mathcal{V}_2 \simeq 0$), and the mass flow.

Throughout the process the temperature is so low that constant specific heats can be assumed. We set the enthalpy constants such that $h = c_p T$ where $c_p = 1.004\frac{\text{kJ}}{\text{kg K}}$ and T is the temperature in Kelvin.

The mass flow into the diffuser is

$$\dot{m} = \rho_1 \mathcal{V}_1 A_1 = \frac{p_1}{RT_1}\mathcal{V}_1 A_1 = 45.48\frac{\text{kg}}{\text{s}} \,.$$

The first law for the diffuser gives

$$h_2 = h_1 + \frac{1}{2}\mathcal{V}_1^2 = h\left(T_1\right) + \frac{1}{2}\mathcal{V}_1^2 = 275.9\frac{\text{kJ}}{\text{kg}} \,.$$

With the enthalpy constants chosen here, the corresponding temperature is

$$T_2 = \frac{h_2}{c_p} = 274.8\,\text{K} \,.$$

Since the process is polytropic, the exit pressure is

$$p_2 = p_1 \left(\frac{T_2}{T_1}\right)^{\frac{n}{n-1}} = 0.171\,\text{bar} \,.$$

The entropy generation per unit mass follows from the 2nd law as

$$\frac{\dot{S}_{gen}}{\dot{m}} = s_2 - s_1 = c_p \ln\frac{T_2}{T_1} - R\ln\frac{p_2}{p_1} = \left(c_p - \frac{n}{n-1}R\right)\ln\frac{T_2}{T_1} = 0.025\frac{\text{kJ}}{\text{kg K}} \,.$$

For a reversible adiabatic diffuser, the entropy would stay constant, and the exit pressure would be higher:

$$p_{2s} = p_1 \left(\frac{T_2}{T_1}\right)^{\frac{k}{k-1}} = 0.186\,\text{bar} \,.$$

The isentropic diffuser efficiency (9.46) is

$$\eta_D = \frac{\left(\frac{p_2}{p_1}\right)^{\frac{k-1}{k}} - 1}{\left(\frac{p_{2s}}{p_1}\right)^{\frac{k-1}{k}} - 1} = \frac{\frac{T_2}{T_1}^{\frac{n}{n-1}\frac{k-1}{k}} - 1}{\frac{T_2}{T_1} - 1} = 0.846 \,.$$

9.15 Closed Heat Exchangers

Closed heat exchangers serve to transfer heat between different fluids, for instance from the hot combustion gas to evaporating water in steam power plants, or from the condenser in the power plant to the cooling water. We ignore friction losses in the flow, so that the flows are isobaric. Figure 9.14 shows the exchange of heat between two fluids A and B in co- and counter-flow heat exchangers, where superscripts i and e denote incoming and exiting flows, respectively. Since the fluids flow through different piping systems, they might have different pressures p_A, p_B.

The conservation laws for mass and energy give for both set-ups (adiabatic to the outside, steady state)

$$\dot{m}_A = const. \, , \quad \dot{m}_B = const. \, , \quad \dot{m}_B \left(h_{B,e} - h_{B,i} \right) = -\dot{m}_A \left(h_{A,e} - h_{A,i} \right) \, . \tag{9.48}$$

Due to finite temperature differences, the heat exchange will be irreversible. The second law for externally adiabatic heat exchange reads

$$\dot{m}_A \left(s_{A,e} - s_{A,i} \right) + \dot{m}_B \left(s_{B,e} - s_{B,i} \right) = \dot{S}_{gen} \geq 0 \, . \tag{9.49}$$

For a closer look, we consider the first and second law for an infinitesimal section dx of the heat exchanger (no work, kinetic and potential energies can be ignored), which read

$$\dot{m}_A dh_A + \dot{m}_B dh_B = 0 \quad , \quad \dot{m}_A ds_A + \dot{m}_B ds_B = \delta \dot{S}_{gen} \tag{9.50}$$

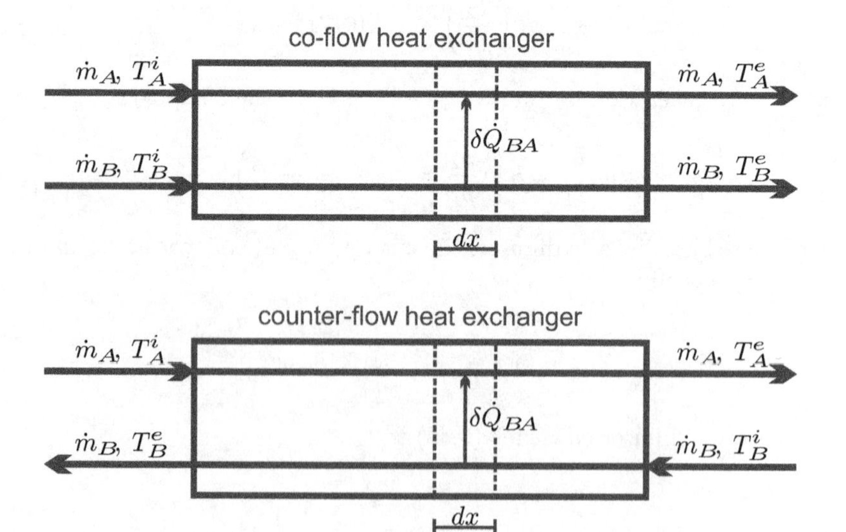

Fig. 9.14 Co-flow and counter flow heat exchangers (schematic)

For isobaric processes the Gibbs equation gives $dh = Tds$, which allows to combine the above to

$$\delta \dot{S}_{gen} = \dot{m}_A \left[\frac{1}{T_A} - \frac{1}{T_B} \right] dh_A \,. \tag{9.51}$$

Accordingly, entropy generation and lost work depend on the local temperature difference between the streams. The total entropy generation is given by the integral over the length of the heat exchanger (HE),

$$\dot{S}_{gen} = \int_{\text{HE}} \delta \dot{S}_{gen} = \dot{m}_A \int_{\text{HE}} \left[\frac{1}{T_A} - \frac{1}{T_B} \right] dh_A \,. \tag{9.52}$$

To see that the entropy generation is positive, we note that, under the given simplifications, the first law for stream A reads $\dot{m}_A dh_A = \delta \dot{Q}_{BA}$ where $\delta \dot{Q}_{BA}$ is the heat going from B to A (heat into the system is positive). Since heat goes from warm to cold, we have $\delta \dot{Q}_{BA} > 0$ if $T_B > T_A$.

A detailed discussion of counter-flow and co-flow heat exchangers shows that the local temperature difference between the two flows is smaller for the counter-flow design, which therefore is thermodynamically preferable (see Sec. 15.2).

9.16 Open Heat Exchangers: Adiabatic Mixing

Another interesting process that occurs quite often is heat exchange between two flows by adiabatic mixing of two (or more) streams, as depicted in Fig. 9.15. To prevent backflow, the mixed streams should have the same pressure, but their temperatures and/or phase states are different. For the computation, pressure losses through the mixing chamber are normally ignored, so that the outflow pressure equals the inflow pressure. Moreover, kinetic and potential energies can be ignored since their changes are much smaller than changes in enthalpy.

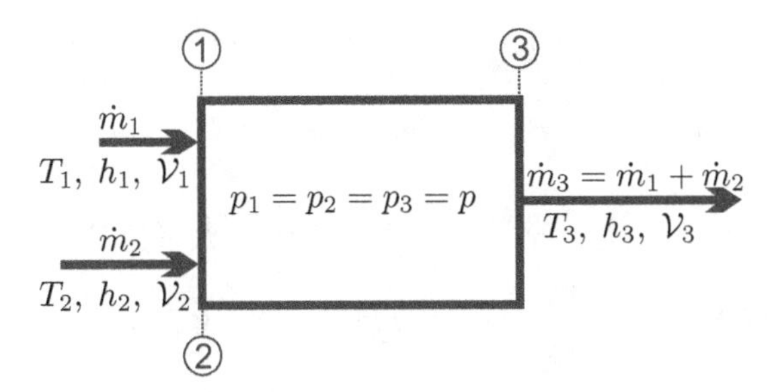

Fig. 9.15 Adiabatic and isobaric mixing of two streams

The steady state mass and energy balances (9.11, 9.12) reduce to

$$\dot{m}_3 = \dot{m}_1 + \dot{m}_2 \quad , \quad \dot{m}_3 h_3 = \dot{m}_1 h_1 + \dot{m}_2 h_2 \, , \tag{9.53}$$

or

$$\dot{m}_2 \left(h_3 - h_2 \right) = \dot{m}_1 \left(h_1 - h_3 \right) \, . \tag{9.54}$$

That is, the mass flow and the total enthalpy flow are conserved in mixing. Since mixing occurs uncontrolled, the process is irreversible, and the entropy will increase; from (9.13) we find

$$\dot{S}_{gen} = \dot{m}_1 \left(s_3 - s_1 \right) + \dot{m}_2 \left(s_3 - s_2 \right) \geq 0 \, . \tag{9.55}$$

There is a corresponding work loss, since the temperature difference between the flows could be used to drive heat engines, but is not.

9.17 Examples: Heat Exchangers

9.17.1 Closed Heat Exchanger

In the condenser of a steam power plant, a mass flow $\dot{m}_c = 10 \frac{\mathrm{t}}{\mathrm{h}}$ saturated liquid-vapor mixture at 35 °C, $x = 0.9$ is isobarically condensed to the saturated liquid state by heat exchange with cooling water. We determine the mass flow $\dot{m}_w$ of cooling water which enters at $T_{w1} = 20\,°\mathrm{C}$ and leaves at $T_{w2} = 30\,°\mathrm{C}$.

The cooling water can be described as incompressible liquid with constant specific heat $c_w = 4.18 \frac{\mathrm{kJ}}{\mathrm{kg\,K}}$. From (9.48) we obtain, with A for the cooling water and B for the condensate,

$$\dot{m}_w = \dot{m}_c \frac{\left(h_{c1} - h_{c2} \right)}{c_w \left(T_{w2} - T_{w1} \right)} \, .$$

The entropy generation is

$$\dot{S}_{gen} = \dot{m}_w c_w \ln \frac{T_{w2}}{T_{w1}} + \dot{m}_c \left(s_{c2} - s_{c1} \right) \, .$$

Enthalpies and entropies of the condensate are

$$h_{c1} = h \left(35\,°\mathrm{C}, x = 0.9 \right) = 2323 \frac{\mathrm{kJ}}{\mathrm{kg}} \, , \quad s_{c1} = s \left(35\,°\mathrm{C}, x = 0.9 \right) = 7.57 \frac{\mathrm{kJ}}{\mathrm{kg\,K}} \, ,$$

$$h_{c2} = h_f \left(35\,°\mathrm{C} \right) = 146.7 \frac{\mathrm{kJ}}{\mathrm{kg}} \, , \quad s_{c2} = s_f \left(35\,°\mathrm{C} \right) = 0.505 \frac{\mathrm{kJ}}{\mathrm{kg\,K}} \, ,$$

so that the required mass flow of cooling water is

$$\dot{m}_w = \dot{m}_c \frac{h_{c1} - h_{c2}}{c_w \left(T_{w2} - T_{w1}\right)} = 521 \frac{\text{t}}{\text{h}} = 145 \frac{\text{kg}}{\text{s}} .$$

The corresponding entropy generation is

$$\dot{S}_{gen} = 20.3 \frac{\text{kW}}{\text{K}} - 19.6 \frac{\text{kW}}{\text{K}} = 0.67 \frac{\text{kW}}{\text{K}} .$$

9.17.2 Adiabatic and Isobaric Mixing

Compressed water at $T_1 = 20\,°\text{C}$ is mixed with superheated steam at $T_2 = 400\,°\text{C}$; both streams are at $p = 2\,\text{bar}$. We determine the mass flow ratio $\dot{m}_2/\dot{m}_1$ so that the outflow is just saturated liquid at $p = 2\,\text{bar}$ (state 3), and the entropy generation rate.

Enthalpy and entropy of the three states are

$$\begin{aligned}
h_1 &= h_f\left(20\,°\text{C}\right) = 84.0 \tfrac{\text{kJ}}{\text{kg}} \quad &, \quad s_1 &= s_f\left(20\,°\text{C}\right) = 0.297 \tfrac{\text{kJ}}{\text{kg K}} ,\\
h_2 &= h\left(2\,\text{bar}, 400\,°\text{C}\right) = 3277 \tfrac{\text{kJ}}{\text{kg}} \quad &, \quad s_2 &= s\left(2\,\text{bar}, 400\,°\text{C}\right) = 8.222 \tfrac{\text{kJ}}{\text{kg}} ,\\
h_3 &= h_f\left(2\,\text{bar}\right) = 504.7 \tfrac{\text{kJ}}{\text{kg}} \quad &, \quad s_3 &= s_f\left(2\,\text{bar}\right) = 1.530 \tfrac{\text{kJ}}{\text{kg K}} .
\end{aligned}$$

From (9.54) we find the mass flow ratio as

$$\frac{\dot{m}_2}{\dot{m}_1} = \frac{h_1 - h_3}{h_3 - h_2} = 0.152 .$$

The entropy generation per unit mass of outflow follows from (9.55) as

$$\frac{\dot{S}_{gen}}{\dot{m}_1 + \dot{m}_2} = \frac{\left(s_3 - s_1\right) + \frac{\dot{m}_2}{\dot{m}_1}\left(s_3 - s_2\right)}{1 + \frac{\dot{m}_2}{\dot{m}_1}} = 0.187 \frac{\text{kJ}}{\text{kg K}} \geq 0 .$$

Problems

9.1. Joule's Honeymoon

When J. P. Joule went into Switzerland for his honeymoon in 1847, he took some rather precise thermometers along. These he used to measure the change of temperatures in waterfalls between top and bottom. For a waterfall of 300 m height, which change in temperature would he have measured?

9.2. Grand Coulee Dam

The Grand Coulee Dam is the largest concrete structure in the US, and the sixth largest production site for electrical energy worldwide. It has 24 turbines and can produce up to 6.8 GW electrical power. The relevant change of height for power production is about 110 m. Compute the mass flow through the turbines when all run at full load. Assume adiabatic flow conditions, neglect kinetic energy (discuss why you can do that!) and assume that the temperature stays constant.

9.3. Seven Mile Dam

Seven Mile Dam in B.C. has a maximum capacity of 848 MW and generates 3200 GWh of energy per year. The height for power generation is 65 m: Determine the mass flow rate at maximum capacity, the total amount of water running through the turbines per year, and the average mass flow rate.

9.4. Taking a Shower

A typical shower head has a flow rate of 8 litres/ min at a pressure of 4 bar. Determine the power required to provide the water for a shower in the top floor of a 400 m high-rise building. Compare with the energy demand for heating the water from 10 °C to 35 °C.

9.5. Air Turbine

The adiabatic turbine in a gas power cycle delivers 25 MW. Air enters the turbine at 1227 °C, 18 bar and the pressure ratio over the turbine is 16.7. Assume reversible operation and determine the mass flow through the turbine. Determine the inlet velocity when the inlet cross section is $900\,\mathrm{cm}^2$.

9.6. Reversible Turbine

Air at 1427 °C, 25 bar enters an adiabatic turbine at a rate of $25\frac{\mathrm{kg}}{\mathrm{s}}$ with a velocity of $60\frac{\mathrm{m}}{\mathrm{s}}$. The exiting air is at the local atmospheric pressure of 0.91 bar and its velocity is $120\frac{\mathrm{m}}{\mathrm{s}}$.

1. Determine the power delivered by the turbine.
2. Determine the cross sections at inlet and exit.
3. Compare the change in enthalpy with the change in kinetic energy and discuss whether the kinetic energy could have been ignored.

9.7. Irreversible Turbine

The irreversible turbine in a gas turbine power plant expands air from $p_1 =$ 12 bar, 1200 K to 1 bar; the isentropic efficiency is 0.9. Compute the work output per kg of air.

9.8. Irreversible Pump

A mass flow of $200\frac{\mathrm{kg}}{\mathrm{s}}$ of liquid water is pumped from 1 bar to 200 bar. Compute the power consumption of the pump for an isentropic pump efficiency of 75%.

9.9. Irreversible Turbine

A mass flow of $44\frac{\mathrm{kg}}{\mathrm{s}}$ steam passes through a well-insulated (i.e., adiabatic) turbine operating at steady state; the turbine develops 34.04 MW of power. The steam enters at 450 °C, and exits as saturated vapor at 0.05 bar. The inlet velocity is $50\frac{\mathrm{m}}{\mathrm{s}}$, and the outlet velocity is $100\frac{\mathrm{m}}{\mathrm{s}}$.

1. Determine the inlet pressure.
2. Compare the change in enthalpy with the change in kinetic energy and discuss whether the kinetic energy could have been ignored.

3. Compute the power that could be obtained in an adiabatic reversible turbine operating between the same inlet condition, and the same exit pressure.
4. Determine the isentropic efficiency of the turbine, η_T.

9.10. Air Compressor
A compressor for air operates on a polytropic process with $n = 1.272$. The state at the inlet is 1 bar, 300 K, and the pressure rises to 6 bar. Heat transfer occurs at a rate of $46.95\frac{\text{kJ}}{\text{kg}}$ of air flowing through the compressor, since the casing is cooled to reduce the work needed for compression. Compute the power required if the mass flow is $4\frac{\text{kg}}{\text{s}}$.

9.11. R134a Compressor
The adiabatic compressor in a refrigeration plant consumes 2 MW. Refrigerant R134a enters the compressor as saturated vapor with a temperature of −18 °C, and is compressed to 1.0 MPa; you can assume reversible operation, and ignore kinetic and potential energies. Determine the mass flow through the turbine, and the inlet cross section when the inlet velocity is $40\frac{\text{m}}{\text{s}}$.

9.12. Irreversible Compressor
The adiabatic compressor in a gas power cycle consumes 12 MW. Air enters the compressor at 320 K, 1.04 bar and the pressure ratio over the compressor is 13. The compressor is irreversible with isentropic efficiency of 0.85. Determine the mass flow through the turbine and the exit cross section when the exit velocity is $100\frac{\text{km}}{\text{h}}$.

9.13. Heat Exchanger in Frozen Pizza Factory
Consider a refrigeration plant with a COP of 3.25. The plant's power consumption is $\dot{W} = 215\,\text{kW}$. The waste heat is rejected into an isobaric stream of water at 1 bar that enters the heat exchanger at 12.5 °C, and leaves at 25 °C. Compute the mass flow of cooling water.

9.14. Cooling of an Air Stream
A mass flow of $20\frac{\text{kg}}{\text{s}}$ of air is isobarically cooled from 200 °C to 50 °C by heat transfer to the outside environment at 18 °C. Determine the cooling rate, entropy generation rate and the work loss for this process.

9.15. Groundwater as Heat Source
Groundwater is used as a heat source for a heat pump. The heat pump has a COP of 4.25, and provides 2.5 kW of heat. The heat pump draws heat from groundwater which comes in at a rate of of $5\frac{\text{kg}}{\text{min}}$ at $T_{in} = 15\,°\text{C}$. Determine the exit temperature of the water flow.

9.16. Adiabatic Throttle
To relieve a duct, superheated water vapor at 10 MPa, 600 °C is throttled into the environment where the pressure is 1 bar. Determine the temperature of the vapor entering the environment, and the entropy generation per unit mass. Estimate work loss.

9.17. Adiabatic Nozzle

Superheated water vapor at 10 MPa, 600 °C is expanded into the environment, where the pressure is 1 bar. The isentropic efficiency of the nozzle is 95%. Determine the velocity and temperature at the nozzle exit.

9.18. Nozzle

To drive a rocket, the combustion product should be accelerated in a nozzle to a velocity of $1.769\frac{\text{km}}{\text{s}}$ with a mass flow of $20\frac{\text{kg}}{\text{s}}$. The exit pressure is 0.2 bar, and the nozzle has an isentropic efficiency of 0.96. Assume negligible inlet velocity and determine the inlet pressure when the inlet temperature is 2300 K, and the exit cross section. Consider the combustion product as air (ideal gas with variable specific heats).

9.19. Irreversible Nozzle Flow of Air

In a jet engine, hot combustion air leaving the turbine at 1200 K, 8 bar is adiabatically expanded in an irreversible nozzle with isentropic efficiency of 0.929. The temperature at the nozzle exit is 600 K. Consider the air as ideal gas with variable specific heats. Compute the exit velocity and the exit pressure.

9.20. Diffuser

Air enters the reversible diffuser of a jet engine that flies with a velocity of $900\frac{\text{km}}{\text{h}}$ at a height of 10 km above sea level, where the temperature is −50 °C and the pressure is 35 kPa. Determine the pressure at the diffuser exit. Consider air as ideal gas with constant specific heats.

9.21. Diffuser

The inlet ducting to a jet engine forms a diffuser that steadily decelerates the entering air to negligible velocity relative to the engine. Consider an airplane flying at $1000\frac{\text{km}}{\text{h}}$ in a height where the pressure is 0.6 bar, and the temperature is −3 °C. Assume ideal gas behavior with variable specific heats, adiabatic reversible flow conditions, and neglect potential energy effects. Compute the temperature and the pressure at the exit of the diffuser.

9.22. Ramjet

A ramjet is a simple engine for aircraft propulsion which works without moving parts. In a simple thermodynamic model it works as follows (seen from an observer resting with the aircraft): Air enters the diffuser where it is decelerated and the pressure increases. Next, the air is heated by combustion of fuel, and then the compressed hot gas is expanded through a nozzle.

Consider this process for air as ideal gas with variable specific heats, and the following sub-processes:

1-2: Outside air at $T_1 = 240\,\text{K}$, $p_1 = 0.2\,\text{bar}$ enters the diffuser with velocity $\mathcal{V}_1 = 3040\frac{\text{km}}{\text{h}}$; this is the velocity of the supersonic aircraft relative to the air.

2-3: The compressed air is isobarically heated to 2300 K.

3-4: The heated air is expanded through an adiabatic and reversible nozzle to the outside pressure $p_4 = p_1 = 0.3\,\text{bar}$.

1. Make a sketch of this series of processes in p-v- and T-s-diagrams.
2. Determine temperature T_2 and pressure p_2 at the diffuser exit. The diffuser can be considered to operate on an adiabatic reversible process, and at the diffuser exit the flow velocity is negligibly small.
3. Determine the heat added per unit mass of air flowing through, q_{23}.
4. Determine the exit velocity $\mathcal{V}_4$.
5. The mass flow of air is $\dot{m} = 1\frac{\text{kg}}{\text{s}}$. Determine the cross sections at diffuser inlet and nozzle exit, A_1 and A_4.
6. Determine the thrust of the engine, $F = \dot{m}\left(\mathcal{V}_4 - \mathcal{V}_1\right)$ and the propulsive power $\dot{W} = F\mathcal{V}_1$.

9.23. Adiabatic Polytropic Process
Consider an irreversible adiabatic compression $p_1 \to p_2$ of an ideal gas with constant specific heats. The process can be described as polytropic with exponent n.

1. Which of the following restrictions apply to the coefficient n? Give your arguments.

$$a)\; 1 \leq n \leq k\,, \quad b)\; 0 \leq n \leq \infty\,, \quad c)\; 0 \leq n \leq 1\,,$$
$$d)\; k \leq n \leq \infty\,, \quad e)\; 0 \leq n \leq k$$

2. What is the equivalent condition for an adiabatic polytropic expansion?

Chapter 10
Basic Open System Cycles

10.1 Steam Turbine: Rankine Cycle

About 70-75% of the World's electrical energy are produced in steam cycles. An external heat source is used to evaporate pressurized water, and then the high pressure vapor is expanded in steam turbines. The fuel for most steam power plants is coal, followed by nuclear power. Since the heat is supplied externally, many other heat sources can be used, including oil, gas and heat from solar radiation.

The Rankine cycle, which we shall discuss now, has been the basic steam cycle for power generation, it is named after William Rankine (1820-1872). Modern steam power plants use more efficient variations of the Rankine cycle that will be discussed in Sec. 12.2.

Figure 10.1 shows a schematic of the Rankine cycle. Saturated liquid water is pressurized in an adiabatic pump (1-2). In the steam generator, the high pressure water is heated, evaporated and superheated (2-3). The superheated steam is expanded in the steam turbine (3-4), which generates work; part of the turbine work is used to run the pump, the net work is delivered to the generator. The turbine discharges into the condenser (4-1), in which the steam is condensed back to the initial state. Pump and turbine may be irreversible. Figure 10.2 shows the Rankine cycle in the diagrams with respect to saturation lines. Work and heat for the four processes are (see Sec. 9.13)

$$\begin{array}{lll} \text{1-2 adiabatic pump:} & w_{12} = h_1 - h_2 & , \; q_{12} = 0 \, , \\ \text{2-3 isobaric heating:} & w_{23} = 0 & , \; q_{23} = h_3 - h_2 \, , \\ \text{3-4 adiabatic turbine:} & w_{34} = h_3 - h_4 & , \; q_{34} = 0 \, , \\ \text{4-1 isobaric cooling:} & w_{41} = 0 & , \; q_{41} = h_1 - h_4 \, . \end{array} \tag{10.1}$$

The net work of the cycle is

$$w_{\odot} = w_{12} + w_{23} + w_{34} + w_{41} = h_1 - h_2 + h_3 - h_4 \, , \tag{10.2}$$

and the heat supply is

H. Struchtrup, *Thermodynamics and Energy Conversion*,
DOI: 10.1007/978-3-662-43715-5_10, © Springer-Verlag Berlin Heidelberg 2014

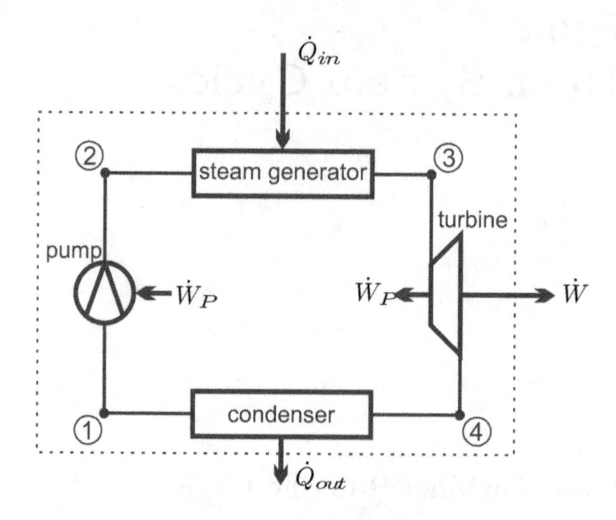

Fig. 10.1 Schematic of Rankine cycle. The dotted line shows the overall system boundary for the cycle. Part of the turbine work is used to drive the pump.

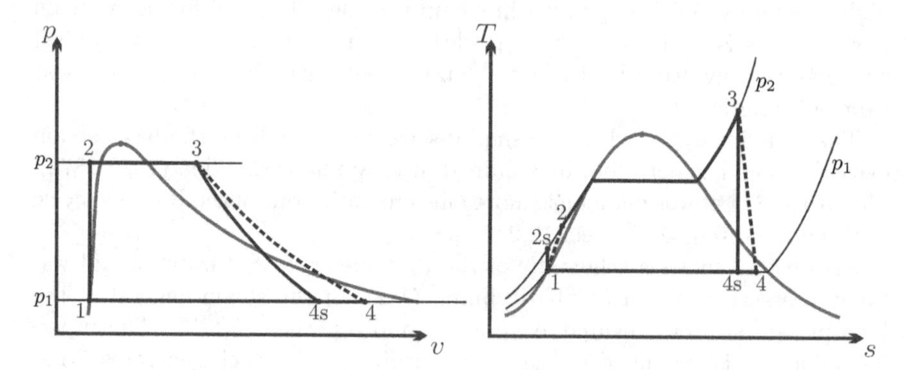

Fig. 10.2 Rankine cycle: p-v- and T-s-diagrams

$$q_{in} = q_{23} = h_3 - h_2 \, . \tag{10.3}$$

Accordingly, the thermal efficiency of the Rankine cycle is

$$\eta_R = \frac{w_\odot}{q_{in}} = \frac{h_1 - h_2 + h_3 - h_4}{h_3 - h_2} = 1 - \frac{h_4 - h_1}{h_3 - h_2} \, . \tag{10.4}$$

The total power produced, the heat consumed, and the heat rejected by the cycle follow after multiplication with the mass flow $\dot{m}$ as

$$\dot{W} = \dot{m} w_\odot \quad , \quad \dot{Q}_{in} = \dot{m} q_{23} \quad , \quad \dot{Q}_{out} = \dot{m} q_{41} \, . \tag{10.5}$$

In early steam engines the steam was expanded in piston-cylinder devices, not in turbines. Reciprocating piston engines have large load changes, and are more bulky, while turbines are running at constant loads, and can deliver the same amount of power with a significantly smaller footprint.

The condenser is James Watt's (1736-1819) most important contribution—of many—to the improvement of steam engines. The temperature T_c in the condenser is prescribed through heat exchange with the environment, so that T_c is not much above the environmental temperature T_0. The condenser pressure is the corresponding saturation pressure $p_{\text{sat}}(T_c)$ which lies substantially below the environmental pressure p_0. Since the work delivered by a turbine grows with the pressure ratio,[1] a steam cycle with a condenser will have a significantly larger work output than a cycle that discharges into the environment. The gain of work through the condenser is illustrated in the T-s-diagram of Fig. 10.3.

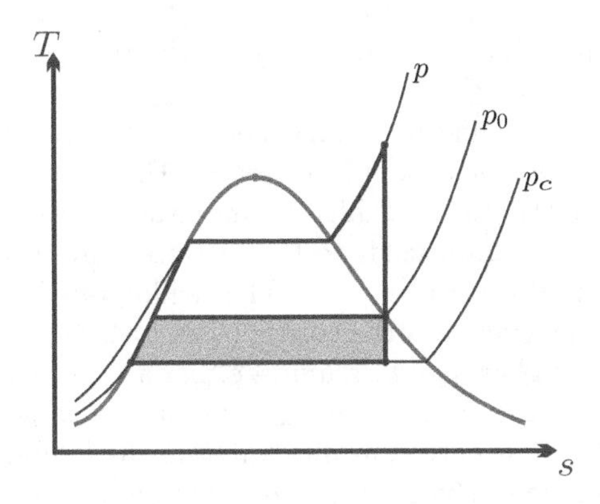

Fig. 10.3 Rankine cycle with and without condenser. The shaded area is the gain of work through the condenser.

A condenser requires a significant amount of cooling which usually is provided by a cooling water cycle that employs cooling towers. Due to the difficulty of providing sufficient cooling, most steam locomotives do not have condensers, and discharge into the environment. Therefore steam locomotives have low thermal efficiencies, and must be supplied with fresh water frequently.[2] In a steam power plant with condenser the working fluid runs continuously through the system in a closed loop, which allows the use of

[1] This can best be seen for an ideal gas with constant specific heats, for which an isentropic turbine delivers the work $w_T = c_p T_1 \left(1 - \left(\frac{p_2}{p_1}\right)^{\frac{k-1}{k}}\right)$.

[2] The water supply for steam locomotives is a driving force in Sergio Leone's wonderful "spaghetti western" *Once Upon a Time in the West* (feat. Henry Fonda, Claudia Cardinale, Jason Robards, Charles Bronson).

purified water to reduce pipe corrosion. Some of the cooling water, however, evaporates in the open cooling towers from which warm moist air rises. As the rising moist air cools down by heat exchange with the surrounding air, some of the water condenses to clouds (Sec. 19.9).

Would the pump be fed with saturated liquid-vapor mix, the sudden collapse of vapor bubbles during the compression process (cavitation) would induce shock waves that lead to material damage and, ultimately, pump failure. Cooling into the compressed liquid region only increases the heat removal, and has no benefit. Thus, the pump should be fed with saturated liquid.

As the steam expands in the turbine, it crosses the saturation line and small liquid droplets form. These droplets hit the fast rotating turbine blades and cause corrosion. On the other hand, a smaller quality x_4 reduces the amount of heat rejected in the condenser, and thus improves thermal efficiency. To obtain a good balance between efficiency and prevention of corrosion, one aims at having the quality at the turbine exit at $x_4 = 0.9$ or higher.

The pipes of the steam generator are normally made from standard steel. At the high pressures that occur in the cycle, the temperature for the pipes should not exceed $\sim 560\,°\text{C}$.

Increase of the pressure in the steam generator improves efficiency since the average temperature for heat supply increases. However, when the pressure becomes large, and the turbine inlet temperature is capped at $T_{\max}$, the expansion into the condenser leads to low qualities, and thus damage of the turbine blades due to droplet formation. This is illustrated in Fig. 10.4, where the standard Rankine cycle has the corner points 1-2-3-4'. To shift the point 4' towards values of higher quality requires either a turbine inlet temperature T_3 above the maximum temperature $T_{\max}$ for the steam generator, or lower pressure p_3. Another alternative, as illustrated in the figure, is to expand the steam in a first turbine to the intermediate pressure p_5, reheat back to $T_{\max}$, and then expand in a low pressure turbine to the condenser pressure p_c. Net work, heat in and thermal efficiency for the reheat cycle are

$$\begin{aligned} w_{\odot} &= h_1 - h_2 + h_3 - h_4 + h_5 - h_6 \,, \\ q_{in} &= h_3 - h_2 + h_5 - h_4 \,, \\ \eta_{reheat} &= 1 - \frac{h_6 - h_1}{h_3 - h_2 + h_5 - h_4} \,. \end{aligned} \tag{10.6}$$

More complex steam cycles involve multiple turbine stages, and internal heat regeneration to improve efficiency (Sec. 12.2).

10.2 Example: Rankine Cycle

As an example we consider a standard Rankine cycle with specifications based on the discussion in the previous section: The condenser temperature is $T_1 = 40\,°\text{C}$, the upper pressure is $p_2 = p_3 = 80\,\text{bar}$, pump and turbine are

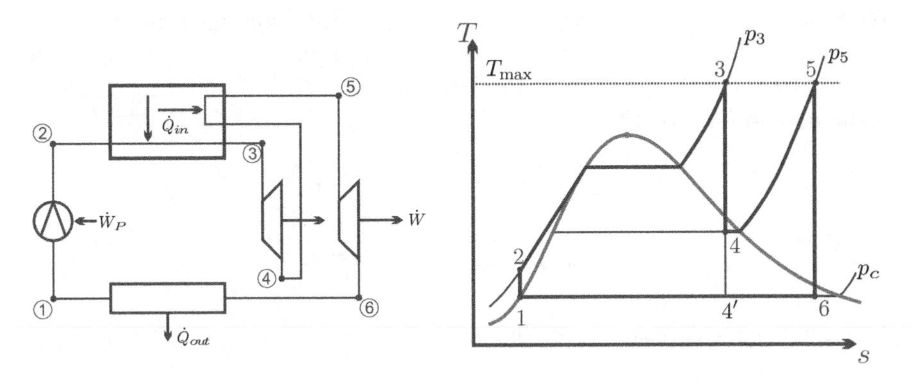

Fig. 10.4 Schematic and T-s-diagram of Rankine cycle with reheat

irreversible with isentropic efficiencies $\eta_P = 0.85$ and $\eta_T = 0.88$, respectively, and the quality at the turbine exit is $x_4 = 0.9$. Schematic and thermodynamic diagrams for this cycle are as in Figs. 10.1 and 10.2.

We begin with the computation for the pump. State 1 is saturated liquid at T_1 with the properties

$$v_1 = v_f(T_1) = 0.001008\frac{\mathrm{m}^3}{\mathrm{kg}} ,$$
$$p_1 = p_{\mathrm{sat}}(T_1) = 7.384\,\mathrm{kPa} ,$$
$$h_1 = h_f(T_1) = 167.57\frac{\mathrm{kJ}}{\mathrm{kg}} .$$

A reversible pump between p_1 and p_2 requires the work

$$w_{p,rev} = h_1 - h_{2s} = -v_1(p_2 - p_1) = -8.057\frac{\mathrm{kJ}}{\mathrm{kg}} .$$

The work for the irreversible pump is

$$w_p = h_1 - h_2 = \frac{w_{p,rev}}{\eta_P} = -9.48\frac{\mathrm{kJ}}{\mathrm{kg}} ,$$

and thus the enthalpy after the pump is

$$h_2 = h_1 - w_P = 177.05\frac{\mathrm{kJ}}{\mathrm{kg}} .$$

Next we study the turbine. The turbine exit state is

$$T_4 = T_1 = 40\,°\mathrm{C} \quad , \quad x_4 = 0.9 ,$$
$$h_4 = (h_f + x_4 h_{fg})_{|T_1} = 2333.6\frac{\mathrm{kJ}}{\mathrm{kg}} .$$

We have to find the corresponding turbine inlet state, for which the pressure $p_3 = 8\,\mathrm{MPa}$ is known, but not the temperature. For the solution we use a trial and error strategy: In the first step, we try $T_3^{(a)} = 550\,°\mathrm{C}$, for which enthalpy and entropy are

$$h_3^{(a)} = h\,(8\,\mathrm{MPa},550\,°\mathrm{C}) = 3521.0\frac{\mathrm{kJ}}{\mathrm{kg}}\,,$$

$$s_3^{(a)} = s\,(8\,\mathrm{MPa},550\,°\mathrm{C}) = 6.8778\frac{\mathrm{kJ}}{\mathrm{kg\,K}}\,.$$

The isentropic expansion from this point to the condenser pressure yields

$$x_{4s}^{(a)} = \frac{s_3^{(a)} - s_f}{s_{fg}}_{|T_1} = 0.821\,,$$

$$h_{4s}^{(a)} = \left(h_f + x_{4s}^{(1)} h_{fg}\right)_{|T_1} = 2142.3\frac{\mathrm{kJ}}{\mathrm{kg}}\,.$$

From the definition of the turbine efficiency (9.44) we finally find the exit enthalpy for the first guess as

$$h_4^{(a)} = h_3^{(a)} - \eta_T\left(h_3^{(a)} - h_{4s}^{(a)}\right) = 2307.7\frac{\mathrm{kJ}}{\mathrm{kg}}\,.$$

Since this value lies below the target value of $h_4 = 2333.6\frac{\mathrm{kJ}}{\mathrm{kg}}$, for the second try we use a higher temperature, $T_3^{(b)} = 600\,°\mathrm{C}$. Following the same line of arguments we find

$$h_3^{(b)} = h\,(8\,\mathrm{MPa},600\,°\mathrm{C}) = 3642.0\frac{\mathrm{kJ}}{\mathrm{kg}}\,,$$

$$s_3^{(b)} = s\,(8\,\mathrm{MPa},600\,°\mathrm{C}) = 7.0206\frac{\mathrm{kJ}}{\mathrm{kg\,K}}\,,$$

$$x_{4s}^{(b)} = \frac{s_3^{(b)} - s_f}{s_{fg}}_{|T_1} = 0.839\,,$$

$$h_{4s}^{(b)} = \left(h_f + x_{4s}^{(b)} h_{fg}\right)_{|T_1} = 2187.0\frac{\mathrm{kJ}}{\mathrm{kg}}\,,$$

$$h_4^{(b)} = h_3^{(b)} - \eta_T\left(h_3^{(b)} - h_{4s}^{(b)}\right) = 2361.6\frac{\mathrm{kJ}}{\mathrm{kg}}\,.$$

The last value lies above the target value of $h_4 = 2333.6\frac{\mathrm{kJ}}{\mathrm{kg}}$, and thus the actual temperature T_3 lies between the two guesses (550 °C,600 °C). Linear interpolation between $\left(T_3^{(a)}, h_4^{(a)}\right)$, $\left(T_3^{(b)}, h_4^{(b)}\right)$, and h_4 gives

$$T_3 = 574\,°\mathrm{C}\,,$$

from which we find

$$h_3 = h\,(8\,\mathrm{MPa},574\,^\circ\mathrm{C}) = 3579.1\frac{\mathrm{kJ}}{\mathrm{kg}}\,,$$

$$s_3 = s\,(8\,\mathrm{MPa},574\,^\circ\mathrm{C}) = 6.946\frac{\mathrm{kJ}}{\mathrm{kg\,K}}\,,$$

$$x_{4s} = \frac{s_3 - s_f}{s_{fg}}_{|T_1} = 0.829\,,$$

$$h_{4s} = (h_f + x_{4s}h_{fg})_{|T_1} = 2163.8\frac{\mathrm{kJ}}{\mathrm{kg}}\,,$$

$$h_4 = h_3 - \eta_T\,(h_3 - h_{4s}) = 2333.6\frac{\mathrm{kJ}}{\mathrm{kg}}\,,$$

$$s_4 = s_f + x_4 s_{fg} = 7.489\frac{\mathrm{kJ}}{\mathrm{kg\,K}}\,.$$

Although some small inaccuracy can be expected due to interpolation, the enthalpy h_4 is at the target value.

With this, all four enthalpy values are determined, and we can compute the net work, the heat in, and the thermal efficiency of the cycle 1-2-3-4 as

$$w_\odot = h_1 - h_2 + h_3 - h_4 = 1236.0\frac{\mathrm{kJ}}{\mathrm{kg}}\,,$$

$$q_{in} = h_3 - h_2 = 3402.1\frac{\mathrm{kJ}}{\mathrm{kg}}\,,$$

$$\eta = \frac{w_\odot}{q_{in}} = 1 - \frac{h_4 - h_1}{h_3 - h_2} = 36.3\%\,.$$

For a power generation of $\dot{W} = 50\,\mathrm{MW}$, the circulating mass flow is

$$\dot{m} = \frac{\dot{W}}{w_\odot} = 40.5\frac{\mathrm{kg}}{\mathrm{s}} = 11.2\frac{\mathrm{t}}{\mathrm{h}}\,.$$

To obtain an idea on the losses to irreversibilities we compute the efficiency for the reversible cycle 1-2s-3-4s,

$$\eta_{rev} = 1 - \frac{h_{4s} - h_1}{h_3 - h_{2s}} = 41.3\%\,.$$

Thus, the irreversibilities in pump and turbine reduce the cycle efficiency by 5%.

Another interesting value is the *back-work-ratio*, defined as the portion of the turbine work that is required to drive the pump, for which we find

$$\mathrm{bwr} = \frac{|w_P|}{w_T} = \frac{h_2 - h_1}{h_3 - h_4} = 0.76\%\,.$$

Less than one percent of the turbine work is required for the pump. To understand the low back-work-ratio we recall that for a reversible adiabatic process the work for both, pump and turbine, is given by $w = -\int v dp$. The overall pressure difference for both devices are the same, but the volumes for both processes differ considerably: The pump is fed with liquid water, while the turbine is fed with vapor which has a substantially larger volume than the liquid.

10.3 Vapor Refrigeration/Heat Pump Cycle

The vapor refrigeration cycle is based on the inversion of the Rankine cycle. However, as we have seen in the last example, the back-work-ratio for the Rankine cycle is very small, that is only a small portion of the turbine work is required for the pump. In the inversion of the cycle the turbine becomes a compressor which consumes power. Due to the small back-work-ratio, only a very small portion of the compressor work could be regained in the inverse pump, and thus one uses a throttling valve instead.

We consider a refrigerator or heat pump, exchanging heat with a cold and a warm external environment at temperatures T_L, T_H, respectively (recall Sec. 5.4). Heat pump and refrigerator follow the same cycle, but they differ in the external temperatures, and in the temperatures and pressures that occur in the process.

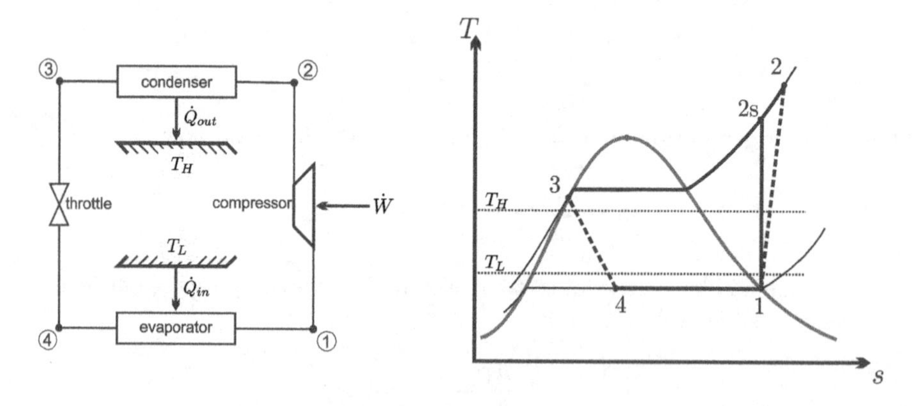

Fig. 10.5 Schematic and T-s-diagram for a standard vapor refrigeration cycle exchaning heat with environments at T_L, T_H

The standard vapor compression cycle operates as shown in the schematic and the T-s-diagram of Fig. 10.5: Saturated or superheated vapor at low temperature (state 1) is adiabatically compressed to higher pressure (state 2). The compressed vapor is cooled and condensed (state 3) in the condenser which exchanges heat with the high temperature environment. The liquid at state 3, either saturated or compressed, is then expanded in the throttling device to the lower pressure (state 4). In the throttling process some of the liquid evaporates, and the temperature drops. The low temperature saturated mixture at state 4 receives heat from the low temperature environment and evaporates to the compressor inlet condition (state 1).

Work and heat for the four processes are

$$\begin{array}{llll} \text{1-2 adiabatic compressor:} & w_{12} = h_1 - h_2 < 0 & , & q_{12} = 0 \, , \\ \text{2-3 isobaric cooling:} & w_{23} = 0 & , & q_{23} = h_3 - h_2 \, , \\ \text{3-4 adiabatic throttle:} & w_{34} = 0 & , & q_{34} = 0 \, , \\ \text{4-1 isobaric heating:} & w_{41} = 0 & , & q_{41} = h_1 - h_4 \, . \end{array} \tag{10.7}$$

The expense for refrigerator and heat pump is the work required to drive the compressor. The gain for the refrigerator is the heat taken in from the cold environment ($q_{in} = q_{41}$), and the gain for the heat pump is the heat rejected into the warm environment ($q_{out} = q_{23}$). Thus, depending on whether we consider a refrigeration device or a heat pump we find the coefficients of performance, as gain/expense,

$$\text{refrigerator:} \qquad \text{COP}_{\text{R}} = \frac{q_{in}}{|w_\odot|} = \frac{h_1 - h_4}{h_2 - h_1} \, , \tag{10.8}$$

$$\text{heat pump:} \qquad \text{COP}_{\text{HP}} = \frac{|q_{out}|}{|w_\odot|} = \frac{h_2 - h_3}{h_2 - h_1} \, . \tag{10.9}$$

The total power consumed, the cooling power, and the heating power of the cycle follow after multiplication with the mass flow $\dot{m}$ as

$$\dot{W} = \dot{m} w_\odot \quad , \quad \dot{Q}_{in} = \dot{m} q_{41} \quad , \quad \dot{Q}_{out} = \dot{m} q_{23} \, . \tag{10.10}$$

Working fluids employed for vapor compression cycles must have good temperature-pressure characteristics. For the low temperatures reached, the pressures should be relatively high, and the specific volumes small, so that the pipes and the compressors must not be too voluminous. The critical point should be high, so that the process runs through the 2-phase region. Moreover, the operating temperatures must lie above the triple point, so that solid formation (freezing) will not occur. The heat of evaporation should be large. Finally, at least for household applications, one will prefer a non-toxic and non-flammable working fluid.

Water, obviously, is not suitable, due to extremely low vapor pressures at low temperatures, and the relatively high triple point temperature $T_{tr} = 0.01\,°\text{C}$. Chlorofluorocarbons (e.g., refrigerant R12, CCl_2F_2) are now phased out since they destroy the ozone layer, and are presently replaced by fluorocarbons (no chlorine) like R134a (CF_3CH_2F). Efficient alternatives are ammonia (NH_3) which is poisonous, and propane (C_3H_8) and methane (CH_4) which are flammable.

10.4 Example: Vapor Compression Refrigerator

A refrigerator operating with R134a maintains the cold environment at 0 °C and rejects heat into an environment at 26 °C, the cooling power is 13 kW.

The coefficient of performance for a Carnot refrigerator operating between these temperatures is

$$\mathrm{COP}_{\mathrm{R},C} = \frac{1}{\frac{T_H}{T_L} - 1} = \frac{1}{\frac{299}{273} - 1} = 10.5 \,,$$

and the Carnot refrigerator would consume a power of $\dot{W}_C = \dot{Q}/\mathrm{COP}_{\mathrm{R},C} = 1.24\,\mathrm{kW}$.

We shall study three increasingly realistic variants of the vapor refrigeration cycle and compare their performance among each other, and with the above COP of the Carnot refrigerator. For all three cycles we assume that the compressor draws saturated vapor, and the throttle draws saturated liquid. The three cycles considered are drawn into the T-s-diagram in Fig. 10.6.

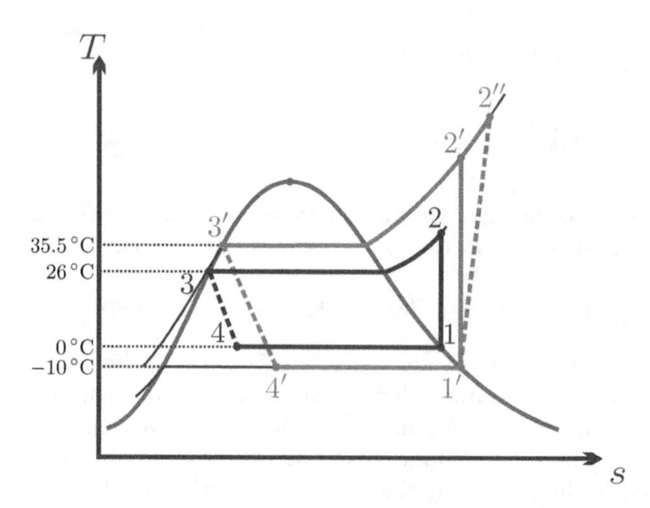

Fig. 10.6 T-s-diagrams for three refrigeration cycles (1-2-3-4, 1′-2′-3′-4′, 1′-2″-3′-4′)

The cycle 1-2-3-4 operates without temperature difference between the cold environment and the evaporator. For the heat transfer between condenser and warm environment (2-3), there is a finite temperature difference for the cooling before the saturation temperature is reached, but the condensation occurs with infinitesimal temperature difference. Moreover, the compressor is reversible.

The cycle 1′-2′-3′-4′ operates with a temperature difference of about 10 °C to facilitate heat transfer between condenser and evaporator and their respective environments. Moreover, its compressor is reversible, while the cycle 1′-2″-3′-4′ has an irreversible compressor with isentropic efficiency $\eta_C = 0.8$.

We first analyze the cycle 1-2-3-4: From the tables we find the following property data for the corner points:[3]

$$
\begin{aligned}
h_1 &= h_g\left(0\,^\circ\mathrm{C}\right) = 247.23\frac{\mathrm{kJ}}{\mathrm{kg}} \quad , \quad s_1 = s_g\left(0\,^\circ\mathrm{C}\right) = 0.9190\frac{\mathrm{kJ}}{\mathrm{kg\,K}} \,, \\
p_2 &= p_3 = p_{\mathrm{sat}}\left(26\,^\circ\mathrm{C}\right) = 685.3\,\mathrm{kPa} \,, \\
h_2 &= h\left(685.3\,\mathrm{kPa}, 0.9190\frac{\mathrm{kJ}}{\mathrm{kg\,K}}\right) = 263.5\frac{\mathrm{kJ}}{\mathrm{kg}} \,, \\
h_3 &= h_4 = h_f\left(26\,^\circ\mathrm{C}\right) = 85.75\frac{\mathrm{kJ}}{\mathrm{kg}} \,.
\end{aligned}
$$

The COP for this cycle is

$$
\mathrm{COP_R} = \frac{h_1 - h_4}{h_2 - h_1} = 9.93 \,.
$$

This value is below the COP of the Carnot refrigerator since irreversible losses occur in the throttle and in heat transfer over finite temperature difference for the first part (before the 2-phase region is reached) of the cooling process 2-3. With the required cooling power $\dot{Q}_{in} = 13\,\mathrm{kW}$ the mass flow and power consumption are

$$
\dot{m} = \frac{\dot{Q}_{in}}{h_1 - h_4} = 0.081\frac{\mathrm{kg}}{\mathrm{s}} \quad , \quad \dot{W} = \frac{\dot{Q}_{in}}{\mathrm{COP_R}} = 1.31\,\mathrm{kW} \,.
$$

Next we consider the cycle 1′-2′-3′-4′ for which we find the data[4]

$$
\begin{aligned}
h_{1'} &= h_g\left(-10\,^\circ\mathrm{C}\right) = 241.3\frac{\mathrm{kJ}}{\mathrm{kg}} \quad , \quad s_{1'} = s_g\left(-10\,^\circ\mathrm{C}\right) = 0.9253\frac{\mathrm{kJ}}{\mathrm{kg\,K}} \,, \\
p_{2'} &= p_{3'} = p_{\mathrm{sat}}\left(35.5\,^\circ\mathrm{C}\right) = 900\,\mathrm{kPa} \,, \\
h_{2'} &= h\left(900\,\mathrm{kPa}, 0.9253\frac{\mathrm{kJ}}{\mathrm{kg\,K}}\right) = 272.4\frac{\mathrm{kJ}}{\mathrm{kg}} \,, \\
h_{3'} &= h_{4'} = h_f\left(900\,\mathrm{kPa}\right) = 99.56\frac{\mathrm{kJ}}{\mathrm{kg}} \,.
\end{aligned}
$$

The COP for this cycle is

$$
\mathrm{COP_{R'}} = \frac{h_{1'} - h_{4'}}{h_{2'} - h_{1'}} = 4.56 \,.
$$

---

[3] To find h_2 one needs to interpolate several times. First find $h\left(600\,\mathrm{kPa}, 0.919\frac{\mathrm{kJ}}{\mathrm{kg\,K}}\right) = 261.97\frac{\mathrm{kJ}}{\mathrm{kg}}$, $h\left(700\,\mathrm{kPa}, 0.919\frac{\mathrm{kJ}}{\mathrm{kg\,K}}\right) = 265.16\frac{\mathrm{kJ}}{\mathrm{kg}}$, then interpolate in pressure.

[4] To reduce the amount of interpolation, we assume a minimum temperature difference of 9.5°C for the condenser.

The finite temperature difference for heat transfer reduces the COP considerably. For this process mass flow and power consumption are

$$\dot{m}' = \frac{\dot{Q}_{in}}{h_{1'} - h_{4'}} = 0.092\frac{\text{kg}}{\text{s}} \quad , \quad \dot{W}' = \frac{\dot{Q}_{in}}{\text{COP}_{\text{R}'}} = 2.85\,\text{kW} \,.$$

For the irreversible compressor 1′-2″ we find the enthalpy at the exit as

$$h_{2''} = h_{1'} + \frac{h_{2'} - h_{1'}}{\eta_C} = 280.2\frac{\text{kJ}}{\text{kg}} \,.$$

The COP for the cycle with irreversible compressor is

$$\text{COP}_{\text{R}''} = \frac{h_{1'} - h_{4'}}{h_{2''} - h_{1'}} = 3.64 \,.$$

The irreversible loss in the compressor reduces the COP further. For this process the mass flow is unchanged, $\dot{m}'' = \dot{m}'$ and the power consumption is

$$\dot{W}'' = \frac{\dot{Q}_{in}}{\text{COP}_{\text{R}''}} = 3.57\,\text{kW} \,.$$

This example shows explicitly how *internal* (throttle, compressor) and *external* (heat transfer over finite temperature differences) irreversibilities lead to a significant reduction of the performance characteristics of a cycle in comparison to the best possible case (here: the Carnot refrigerator).

The realistic cooling cycle 1′-2″-3′-4′ has a COP that is not much more than 1/3 of the Carnot cycle. Advanced cooling cycles use process modifications like cascade refrigeration that increase the COP (Sec. 12.5).

To understand the irreversibilities better, we determine entropy generation and work loss for the four processes in the refrigeration cycle. To compute these, we require the entropies which are

$$\begin{aligned}
s_{1'} &= s_g\left(-10\,^\circ\text{C}\right) = 0.9253\frac{\text{kJ}}{\text{kg}\,\text{K}} \,, \\
s_{2''} &= s\left(900\,\text{kPa}, h = 280.2\frac{\text{kJ}}{\text{kg}}\right) = 0.9499\frac{\text{kJ}}{\text{kg}\,\text{K}} \,, \\
s_{3'} &= s_f\left(900\,\text{kPa}\right) = 0.3656\frac{\text{kJ}}{\text{kg}\,\text{K}} \,, \\
s_{4'} &= s_f\left(-10\,^\circ\text{C}\right) + x_{4'} s_{fg}\left(-10\,^\circ\text{C}\right) = 0.3863\frac{\text{kJ}}{\text{kg}\,\text{K}} \,,
\end{aligned}$$

where we have used that $x_{4'} = \left[\frac{h_{4'} - h_f}{h_{fg}}\right]_{|T=-10\,^\circ\text{C}} = 0.306.$

The 2nd law gives the entropy generation rates in compressor, condenser, throttle and evaporator as

$$\dot{S}_{gen,comp} = \dot{m}\left(s_{2''} - s_{1'}\right) \ ,$$
$$\dot{S}_{gen,cond} = \dot{m}\left(s_{3'} - s_{2''}\right) - \frac{\dot{Q}_{2''3'}}{T_H} = \dot{m}\left[s_{3'} - s_{2''} - \frac{h_{3'} - h_{2''}}{T_H}\right] \ ,$$
$$\dot{S}_{gen,throt} = \dot{m}\left(s_{4'} - s_{3'}\right) \ ,$$
$$\dot{S}_{gen,evap} = \dot{m}\left(s_{1'} - s_{4'}\right) - \frac{\dot{Q}_{1'4'}}{T_L} = \dot{m}\left[s_{1'} - s_{4'} - \frac{h_{1'} - h_{4'}}{T_L}\right] \ .$$

With $T_L = 273\,\mathrm{K}$, $T_H = 299\,\mathrm{K}$ we find the entropy generation rates

$$\dot{S}_{gen,comp} = 2.263\frac{\mathrm{W}}{\mathrm{K}} \ , \quad \dot{S}_{gen,cond} = 1.826\frac{\mathrm{W}}{\mathrm{K}} \ ,$$
$$\dot{S}_{gen,throt} = 1.904\frac{\mathrm{W}}{\mathrm{K}} \ , \quad \dot{S}_{gen,evap} = 1.822\frac{\mathrm{W}}{\mathrm{K}} \ .$$

As was shown in Secs. 5.4,5.10, refrigerator work loss is obtained by multiplying entropy generation with the temperature of the environment,[5] $\dot{W}_{\mathrm{loss},R} = T_H \dot{S}_{gen}$, which gives the individual contributions

$$\dot{W}_{\mathrm{loss},comp} = 0.677\,\mathrm{kW} \ , \quad \dot{W}_{\mathrm{loss},cond} = 0.546\,\mathrm{kW} \ ,$$
$$\dot{W}_{\mathrm{loss},throt} = 0.569\,\mathrm{kW} \ , \quad \dot{W}_{\mathrm{loss},evap} = 0.545\,\mathrm{kW} \ .$$

We see that each of the four sources of irreversibility—two internal, and two external—contributes about one quarter of the total work loss of $\dot{W}_{loss,R} = 2.337\,\mathrm{kW}$. This result clearly shows that for a proper thermodynamic assessment of a system one must consider both, internal *and* external irreversibilities. Note that the work loss is the difference between the actual system work and the work of the Carnot refrigerator (apart from round-off errors).

10.5 Gas Turbine: Brayton Cycle

The Brayton cycle (George Brayton, 1830-1892) is an internal combustion cycle based on a gas turbine. Schematic and T-s-diagram are depicted in Fig. 10.7. Environmental air at $p_1 = p_0$, $T_1 = T_0$ enters the compressor in which it is adiabatically compressed to the pressure p_2. As the air flows through the combustion chamber, fuel is injected and burnt with some of the air's oxygen which leads to heating of the air. The pressurized hot combustion product expands in the turbine and delivers work; some of the work is required to drive the compressor. The expanded combustion product exhausts to the environment.

[5] Also for a heat pump the work loss is entropy generation times the temperature of the environment, only that the latter is at T_L, so that $\dot{W}_{loss,HP} = T_L \dot{S}_{gen}$; re-read Secs. 5.4, 5.10 for details.

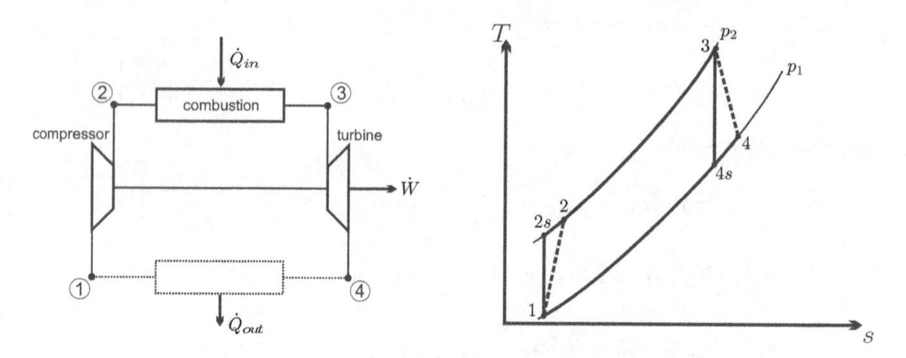

Fig. 10.7 Schematic and T-s-diagram for standard Brayton cycle. Process 4-1 is the equilibration of the exhaust to envrionmental temperature.

As for the reciprocating internal combustion engines, we shall ignore the addition of fuel, and the change of composition through the chemical reaction, and treat the working fluid as air. Then, the combustion is described as an isobaric heating process. The turbine exhaust is warmer than the environment, and exchanges heat with the environment at T_0. The discharge of hot air (state 4) and the intake of cool air (state 1) can be described as an isobaric cooling process at $p_1 = p_4 = p_0$; this process is indicated in the schematic as dotted line.

The Brayton cycle is the equivalent to the Rankine cycle for the ideal gas, and the basic analysis is similar. The main differences are that the Brayton cycle operates without phase change, has no cooler or condenser, and that its back-work-ratio is much larger than that of the Rankine cycle.

Work and heat for the four processes are

$$\begin{array}{llll} \text{1-2 adiabatic compressor:} & w_{12} = h_1 - h_2 & , & q_{12} = 0 \,, \\ \text{2-3 isobaric heating:} & w_{23} = 0 & , & q_{23} = h_3 - h_2 \,, \\ \text{3-4 adiabatic turbine:} & w_{34} = h_3 - h_4 & , & q_{34} = 0 \,, \\ \text{4-1 isobaric cooling:} & w_{41} = 0 & , & q_{41} = h_1 - h_4 \,. \end{array} \tag{10.11}$$

The net work of the cycle is

$$w_{\odot} = w_{12} + w_{23} + w_{34} + w_{41} = h_1 - h_2 + h_3 - h_4 \,, \tag{10.12}$$

and the heat supply is

$$q_{in} = q_{23} = h_3 - h_2 \,. \tag{10.13}$$

Accordingly, the thermal efficiency of the Brayton cycle is

$$\eta_B = \frac{w_{\odot}}{q_{in}} = \frac{h_1 - h_2 + h_3 - h_4}{h_3 - h_2} = 1 - \frac{h_4 - h_1}{h_3 - h_2} \,. \tag{10.14}$$

The total power produced, the heat consumed, and the heat rejected by the cycle follow after multiplication with the mass flow $\dot{m}$ as

$$\dot{W} = \dot{m}w_\odot \quad , \quad \dot{Q}_{in} = \dot{m}q_{23} \quad , \quad \dot{Q}_{out} = \dot{m}q_{41} \; . \tag{10.15}$$

To gain further insight into which parameters are most important, we consider the reversible cycle operating under the cold-air approximation (air as ideal gas with constant specific heats), for which we find the thermal efficiency as

$$\eta_{B,rev} = \frac{h_1 - h_{2s} + h_3 - h_{4s}}{h_3 - h_{2s}} = 1 - \frac{T_{4s} - T_1}{T_3 - T_{2s}} \; . \tag{10.16}$$

The processes 1-2s and 3-4s are isentropic so that

$$\frac{T_{2s}}{T_1} = \left(\frac{p_2}{p_1}\right)^{\frac{k-1}{k}} = P^{\frac{k-1}{k}} = \left(\frac{p_3}{p_4}\right)^{\frac{k-1}{k}} = \frac{T_3}{T_{4s}} \; . \tag{10.17}$$

This yields

$$\eta_{B,rev} = 1 - \frac{T_1}{T_2} = 1 - \frac{1}{P^{\frac{k-1}{k}}} \; . \tag{10.18}$$

The thermal efficiency of the gas turbine increases with the pressure ratio $P = p_2/p_1$.

We proceed by studying operating conditions of gas turbine systems. Obviously, the inlet pressure and temperature, p_1 and T_1, are set by the environmental conditions, p_0 and T_0. The highest temperature in the cycle is the turbine inlet temperature T_3 which is limited by the materials used for the turbine. For fixed values of T_1 and T_3, the work per unit mass of air for the gas turbine is

$$w_\odot = h_1 - h_{2s} + h_3 - h_{4s} = c_p T_1 \left(1 - P^{\frac{k-1}{k}} + \frac{T_3}{T_1}\left(1 - P^{\frac{1-k}{k}}\right)\right) \; . \tag{10.19}$$

Figure 10.8 shows Brayton cycles in the T-s-diagram. All cycles share the intake pressure p_1 and the maximum temperature T_{max}, but their upper pressures $(p_a, p_b, \ldots)$ differ. The specific work is the area enclosed by the cycle in the T-s-diagram, and the figure shows that the cycle with the largest upper pressure (p_h), which has the largest efficiency, has a low work output. Also the cycle with the smallest pressure (p_a), which has the smallest efficiency, has low work output, while a cycle with intermediate pressure (p_d) has the largest work output.

The goal is to produce a certain amount of power $\dot{W} = \dot{m}w_\odot$. A gas turbine with small specific work but large efficiency must have a larger mass flow than a turbine with larger specific work. Larger mass flow can only be achieved by building a bigger system, or several smaller systems, which adds capital and maintenance costs. Normally, one will chose to operate a gas turbine close

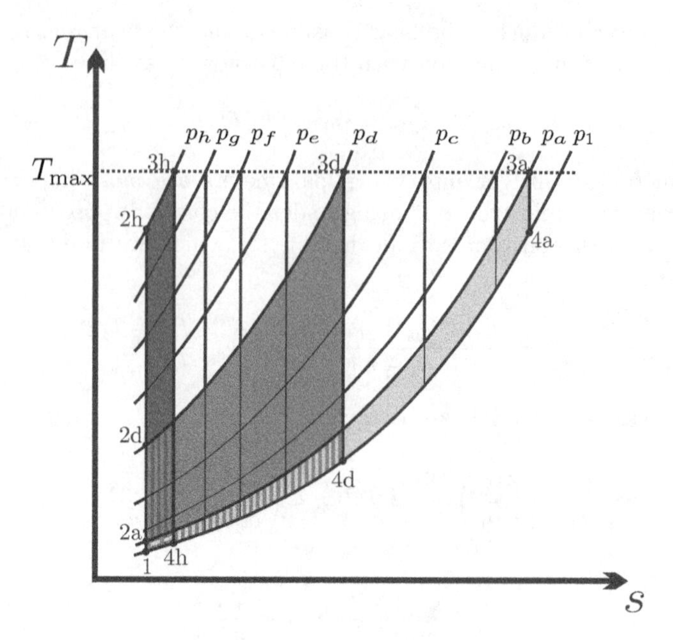

Fig. 10.8 Brayton cycle between pressure p_1 and pressures $p_a \cdots p_h$ with maximum temperature $T_{\max}$. Cycle (1-2d-3d-4d) produces significantly more work than cycles (1-2a-3a-4a) and (1-2h-3h-4h).

to the maximum work output, where the thermal efficiency is smaller, but capital and maintenance costs are smaller as well.

The maximum of the specific work is found, from the condition $\frac{dw_{\odot}}{dP} = 0$, for the pressure ratio

$$P_{\max} = \left(\frac{T_3}{T_1}\right)^{\frac{k}{2(k-1)}} ; \tag{10.20}$$

the corresponding specific work and thermal efficiency are

$$w_{\odot} = c_p T_1 \left(\sqrt{\frac{T_3}{T_1}} - 1\right)^2 , \quad \eta_{B,\max} = 1 - \sqrt{\frac{T_1}{T_3}} . \tag{10.21}$$

From the above it is clear that increase of the turbine inlet temperature T_3 will increase work output *and* thermal efficiency. The use of modern materials and the cooling of turbine blades by pressing cold air through small channels in the blades have led to dramatic increases of turbine inlet temperatures which can be as high as 1700 K for airplanes at take-off.

For simple gas turbine systems, the hot exhaust is just expelled into the environment, where it equilibrates by heat transfer to the environment at T_0. This process is associated with external entropy generation, the corresponding work loss is

$$\dot{W}_{\text{loss},41} = T_0 \dot{S}_{gen,41} = T_0 \dot{m} \left(s_1 - s_4 - \frac{h_1 - h_4}{T_0} \right) . \qquad (10.22)$$

Figure 10.8 shows that the turbine exit temperature T_4 decreases with increasing pressure ratio. This implies that the external loss becomes smaller with increasing pressure ratio, and hence explains why the cycle efficiency increases.

As can be seen from the T-s-diagrams in Figs. 10.7, 10.8, the exhaust temperature T_4 lies above the environmental temperature T_1, and may also lie above the pre-combustion temperature T_2. In advanced gas turbine cycles, the hot exhaust is used to preheat the compressed gas before combustion (regenerative Brayton cycle, Sec. 13.4), or as heat source for a steam power plant (combined cycle, Sec. 13.6). With this, the hot exhaust is used, and the external loss is reduced.

10.6 Example: Brayton Cycle

We assume a gas turbine operating with the compressor inlet temperature $T_1 = 300\,\text{K}$ and the turbine inlet temperature $T_3 = 1400\,\text{K}$. The working medium is air under cold-air approximation, with $R = 0.287 \frac{\text{kJ}}{\text{kg K}}$ and $k = 1.4$.

We first consider a reversible system. From (10.20) we obtain the optimum pressure ratio

$$\frac{p_2}{p_1} = \left(\frac{T_3}{T_1} \right)^{\frac{k}{2(k-1)}} = 14.82$$

and find specific net work and thermal efficiency as

$$w_{\odot} = c_p T_1 \left(\sqrt{\frac{T_3}{T_1}} - 1 \right)^2 = 405.5 \frac{\text{kJ}}{\text{kg}} \quad , \quad \eta_B = 1 - \sqrt{\frac{T_1}{T_3}} = 0.537 \, .$$

The two remaining temperatures are obtained as

$$T_{2s} = T_{4s} = \sqrt{T_1 T_3} = 648.1\,\text{K} \, .$$

The back-work ratio for this cycle is

$$\text{bwr} = \frac{|w_C|}{w_T} = \frac{h_{2s} - h_1}{h_3 - h_{4s}} = \frac{T_{2s} - T_1}{T_3 - T_{4s}} = 46.3\% \, .$$

We compare the above result with that for a gas turbine system with internal irreversibilities operating at the same values for p_1, p_2, T_1, T_3 but with isentropic efficiencies for compressor and turbine given as $\eta_C = \eta_T = 0.9$. From (9.43, 9.44) we obtain the temperatures T_2 and T_4 as

$$T_2 = T_1 - \frac{T_1 - T_{2s}}{\eta_C} = 686.8\,\text{K} \quad , \quad T_4 = T_3 - \eta_T \left(T_3 - T_{4s} \right) = 723.3\,\text{K} \, .$$

The thermal efficiency of the irreversible gas turbine system is

$$\eta_B = \frac{h_1 - h_2 + h_3 - h_4}{h_3 - h_2} = 1 - \frac{T_4 - T_1}{T_3 - T_2} = 40.6\% \, .$$

The internal irreversibilities reduce the thermal efficiency by 25%.

The back-work ratio for the irreversible cycle is

$$\text{bwr} = \frac{|w_C|}{w_T} = \frac{h_2 - h_1}{h_3 - h_4} = \frac{T_2 - T_1}{T_3 - T_4} = 57.2\% \, ;$$

more than 50% of the turbine work is needed to drive the compressor.

10.7 Gas Refrigeration System: Inverse Brayton Cycle

The inversion of the Brayton cycle results in a gas cooling system as depicted in Fig. 10.9. Gas is compressed adiabatically (1-2), and then cooled by heat exchange with the warm environment (2-3). The cooled gas is expanded adiabatically in a turbine, and assumes a low temperature (3-4). Finally, the gas is heated by drawing heat from the cold environment (4-1). As always, we consider a cooling system exchanging heat with reservoirs at T_L, T_H. Then, to have heat transfer in the proper direction, the compressor inlet temperature T_1 must not be above T_L, and the turbine inlet temperature T_3 must not be smaller than T_H. These temperature requirements are shown in the T-s-diagram in Fig. 10.9.

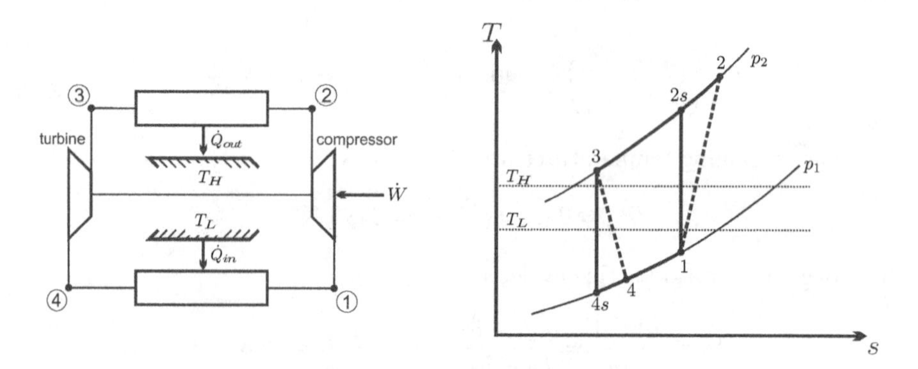

Fig. 10.9 Inverse Brayton cycle: Schematic and T-s-diagram

The processes in the inverse Brayton cycle are

$$\begin{array}{lll} \text{1-2 adiabatic compressor:} & w_{12} = h_1 - h_2\ , & q_{12} = 0\ , \\ \text{2-3 isobaric cooling:} & w_{23} = 0\ , & q_{23} = h_3 - h_2\ , \\ \text{3-4 adiabatic turbine:} & w_{34} = h_3 - h_4\ , & q_{34} = 0\ , \\ \text{4-1 isobaric heating:} & w_{41} = 0\ , & q_{41} = h_1 - h_4\ . \end{array} \tag{10.23}$$

and the coefficient of performance is

$$\text{COP}_\text{R} = \frac{q_{in}}{|w_\odot|} = \frac{h_1 - h_4}{h_2 - h_3 + h_4 - h_1} = \frac{1}{\frac{h_2 - h_3}{h_1 - h_4} - 1} . \tag{10.24}$$

Gas cooling cycles are mainly used at low temperatures, and thus we proceed by assuming constant specific heats for the ideal gas that is used as cooling fluid (cold-gas assumption); then

$$\text{COP}_\text{R} = \frac{q_{in}}{|w_\odot|} = \frac{1}{\frac{T_2 - T_3}{T_1 - T_4} - 1} . \tag{10.25}$$

We consider the special case of reversible compressor and turbine, for which we find

$$\text{COP}_\text{R} = \frac{1}{\frac{T_{2s} - T_3}{T_1 - T_{4s}} - 1} = \frac{1}{\frac{T_{2s}}{T_1} - 1} = \frac{1}{\left(\frac{p_2}{p_1}\right)^{\frac{k-1}{k}} - 1} . \tag{10.26}$$

Here, we used the adiabatic relations for compressor and turbine,

$$\frac{T_{2s}}{T_1} = \left(\frac{p_2}{p_1}\right)^{\frac{k-1}{k}} = \frac{T_3}{T_{4s}} . \tag{10.27}$$

Since $T_1 \leq T_L$ and $T_2 > T_H$, the pressure ratio must be sufficiently large,

$$\frac{p_2}{p_1} > \left(\frac{T_H}{T_L}\right)^{\frac{k}{k-1}} . \tag{10.28}$$

Larger pressure ratios give smaller COP, but increase the cooling power. In particular for large pressure ratios there are large external irreversibilities associated with the heat transfer over finite temperature differences between the cooling fluid and the two environments (see T-s-diagram in Fig. 10.9). Nevertheless, gas refrigeration cycles offer a relatively simple means to achieve low temperatures ($\sim 130\,\text{K}$). Advanced gas cooling systems use internal heat exchange (regeneration) to increase the COP.

Problems

10.1. Rankine Cycle

A small steam power plant produces 50 MW from a simple Rankine cycle operating between an evaporator pressure of 60 bar and a condenser pressure of 10 kPa. The turbine inlet temperature is 550 °C and the quality at its exit is 0.9. Assume that the adiabatic pump is reversible.

1. Draw a sketch of the cycle, and the corresponding p-v- and T-s-diagrams.
2. Make a list with the values of the relevant enthalpies of the cycle.
3. Determine the isentropic efficiency of the turbine.
4. Determine the mass flow of steam in the cycle and the thermal efficiency of the system.

10.2. Steam Cycle

Consider a simple steam power plant that develops a power of 100 MW. The condenser pressure is 10 kPa, the pressure in the steam generator is 8 MPa, and the temperature at the turbine inlet is 500 °C. The isentropic efficiency of the pump is 75%. At the turbine exit the quality is measured as $x = 0.91$.

1. Draw a schematic, a T-s-diagram, and a p-v-diagram (both with respect to saturation lines) of the plant.
2. Determine the enthalpies at the corner points, and the thermal efficiency of the cycle.
3. Determine the isentropic efficiency of the turbine.
4. Determine the mass flow of water through the cycle.
5. In the condenser the heat is transferred to a stream of cooling water which changes its temperature by 10 °C. Compute the mass flow of cooling water (incompressible liquid with specific heat $c_w = 4.2 \frac{\text{kJ}}{\text{kg K}}$).

10.3. Geothermal Steam Power Plant

In the Larderello (Italy) steam power plant, steam at 5 bar, 220 °C is produced by geothermal heating. Assume that the processes in the plant follow the basic Rankine cycle with irreversible turbine ($\eta_T = 0.85$) and reversible pump. The condenser temperature is 40 °C.

1. Determine the enthalpies at the corner points.
2. Compute the thermal efficiency, and discuss its value as compared to standard steam power plants, which operate at higher pressures and temperatures.

10.4. Reheat Rankine Cycle

A steam power plant with a power output of 80 MW operates on a reheat Rankine cycle:

1-2: adiabatic irreversible pump from condenser pressure $p_1 = 10$ kPa to $p_2 = 10$ MPa, state 1 is saturated liquid, isentropic pump efficiency is 95%

2-3: isobaric heating to 500 °C

3-4: adiabatic turbine, expansion to $p_4 = 1\,\text{MPa}$, isentropic turbine efficiency is 80%

4-5: isobaric reheat to 500 °C

5-6: adiabatic turbine, expansion to $p_6 = p_1$, isentropic turbine efficiency is 80%

1. Draw a schematic, and the corresponding p-v- and T-s-diagrams with respect to saturation lines.
2. Determine the thermal efficiency of the system.
3. Determine the mass flow rate.

10.5. Another Reheat Cycle

For an ideal (i.e., reversible) Rankine cycle with reheat, the minimum and maximum pressures reached are 10 kPa and 9 MPa, respectively. Moreover, the turbine inlet temperatures of both turbines are 500 °C, the quality at the condenser inlet is 90% and the mass flow is $25\,\frac{\text{kg}}{\text{s}}$ of steam. Determine:

1. The reheat pressure.
2. The heat input and the power produced.
3. The thermal efficiency.

10.6. Refrigerators, Heat Pumps

Draw schematics for vapor refrigeration cycles and heat pumps, as well as the corresponding T-s- and p-v-diagrams (include irreversible processes for compressors). Explain the difference in operating conditions between heat pumps and refrigerators. Also draw T-s-diagrams for Carnot heat pump and refrigerator, and use the diagrams to compute their COP's.

10.7. Refrigerator

A vapor-compression refrigeration system uses R134a as working substance. The pressure in the evaporator is 1.4 bar, and the condenser pressure is 7 bar. The temperature at the compressor inlet is −10 °C, and the working fluid leaves the condenser at a temperature of 24 °C. Moreover, the mass flow rate is $0.1\,\frac{\text{kg}}{\text{s}}$, and the isentropic efficiency of the compressor is 67%.

1. Draw a sketch, a p-v-diagram, a T-s-diagram (with respect to saturation lines).
2. Determine the COP of the system.
3. Compute the refrigeration capacity, and the power consumption.
4. Compute COP and power consumption if the isentropic efficiency of the compressor is 85%.

10.8. Refrigeration Cycle

A frozen pizza factory requires a refrigeration rate of 200 kW to maintain the storage facility at −15 °C. Cooling is performed by a standard vapor-compression cycle, using R134a with the following data: condenser pressure:

700 kPa, evaporator temperature: −20 °C, isentropic efficiency of compressor: 75%. The condenser is cooled by liquid water. Use the log p-h diagram for R134a for the solution of this problem.

Determine the mass flow rate of the refrigerant, the power input to the compressor, the COP, and the mass flow rate of the cooling water when its temperature changes by 10 °C.

10.9. Vapor Refrigeration Cycle

A refrigerator uses R134a as working fluid which undergoes the following cycle:

1-2: Adiabatic irreversible compression of saturated vapor at $T_1 = -18\,°\mathrm{C}$ to $p_2 = 9\,\mathrm{bar}$; the isentropic efficiency of the compressor is $\eta_C = 0.85$

2-3: Isobaric cooling and condensation to compressed liquid state at $T_3 = 32\,°\mathrm{C}$

3-4: Adiabatic throttling to evaporator pressure $p_4 = p_1$

4-1: Isobaric evaporation to state 1

1. Draw a schematic, and the process in a T-s-diagram, with respect to saturation lines.
2. Make a table with the values of temperature, pressure, and enthalpy at points 1-4.
3. Compute the coefficient of performance (COP).
4. The refrigerator draws a power of 4 kW. Compute the mass flow and the cooling power.

10.10. Heat Pump

A heat pump that operates on the ideal vapor-compression cycle with R134a is used to heat water from 15 to 54 °C at a rate of $0.24\frac{\mathrm{kg}}{\mathrm{s}}$. The condenser and evaporator pressures are 1.4 and 0.32 MPa, respectively. Determine the COP and the power input to the heat pump.

10.11. Heat Pump

A vapor compression heat pump with R134a as cooling fluid is used to keep a house at 20 °C. The heat pump has a compressor with isentropic efficiency of 85%, and it draws heat from groundwater which has a temperature of 12 °C. The condenser and evaporator pressures are 900 kPa and 320 kPa, respectively, the temperature at the inlet of the throttling valve is 30 °C, and the compressor draws saturated vapor.

Compute the COP, the mass flow, and the power consumption if the heating power is 10 kW. Don't forget to draw schematic and diagrams.

10.12. Vapor Heat Pump Cycle

An air conditioning system sucks in a mass flow of $5000\frac{\mathrm{kg}}{\mathrm{h}}$ of outside air at 10 °C, 0.95 bar, and heats it isobarically to 24 °C. The air is heated by means of an standard vapor heat pump cycle (R134a), whose compressor has an isentropic efficiency of 0.85. The condenser pressure is 800 kPa. The evaporator is outside the building, and the minimum temperature difference for heat transfer is 10 °C. Consider the air as ideal gas with constant specific heats ($c_v = 0.717 \frac{\text{kJ}}{\text{kg K}}$, $R = 0.287 \frac{\text{kJ}}{\text{kg K}}$).

1. Make a sketch of the system, and draw the corresponding T-s-diagram.
2. Make a table with the values of pressure, temperature and enthalpy at the corner points.
3. Determine the COP of the cycle.
4. Determine the work required to run the heat pump.

10.13. Gas Turbine (Ideal Brayton Cycle)
An ideal Brayton cycle with air as working fluid (variable specific heats) is to be designed such that the minimum and maximum temperatures in the cycle are 300 K and 1500 K, respectively. The pressure ratio is 16.7. Compressor and turbine are both irreversible, with an isentropic efficiency of 0.9 for the turbine and 0.85 for the compressor.

Compute compressor and turbine work per unit mass of air, and the thermal efficiency of the cycle.

10.14. Brayton Cycle
This problem compares the calculation with constant and non-constant specific heats, to give an idea of the differences. Air enters the compressor of a simple Brayton cycle gas turbine power plant at 95 kPa and 290 K. The heat transfer rate is $50 \frac{\text{MJ}}{\text{s}}$ and the turbine entry temperature is 1400 K. The pressure ratio is $P = \left(\frac{T_{\text{max}}}{T_{\text{min}}}\right)^{\frac{k}{2(k-1)}}$ for maximum power output. Compressor and turbine are reversible. Compute the power delivered and the thermal efficiency for:

1. Cold air approximation, that is constant specific heats with values at room temperature.
2. Variable specific heats.

Make tables for the values of pressure and temperature at the relevant process points, and draw diagrams and schematic.

10.15. Gas Turbine (Brayton Cycle)
A Brayton cycle delivers a power of 150 MW. The working fluid can be considered as air (ideal gas with variable specific heats), and the following data are known: inlet state $p_1 = 0.9$ bar, $T_1 = 280$ K; state after adiabatic compression $p_2 = 17.06$ bar, $T_2 = 690$ K, maximum temperature in the cycle 1600 K, heating rate 354 MW.

Determine the thermal efficiency of the cycle, compressor and turbine work per unit mass of air, mass flow of air, and the isentropic efficiencies of compressor and turbine.

10.16. Brayton Cooling Cycle
A small gas-cooling system operates on the inverse Brayton cycle. The cycle uses argon as cooling fluid. The cycle is used for maintaining a small cold space at $-60\,^\circ\mathrm{C}$, and rejects heat into the environment at $25\,^\circ\mathrm{C}$. Both heat exchangers require a temperature difference of at least $5\,^\circ\mathrm{C}$ for operation. The cycle operates between the pressures 1 bar and 4 bar, and isentropic efficiencies of compressor and turbine are 0.75 and 0.85, respectively. Draw schematic and diagrams and determine:

1. The COP.
2. The mass flow required for a cooling power of 0.5 kW, and the required power to run the system.
3. The work losses to irreversibilities in turbine, compressor and both heat exchangers. Discuss the results.

10.17. Gas Refrigeration Cycle
A refrigerator for cryogenic applications uses helium as working fluid which undergoes the cycle described below.

1-2: Adiabatic irreversible compression of helium at $T_1 = -75\,^\circ\mathrm{C}$ and $p_1 = 1\,\mathrm{bar}$ to $p_2 = 10\,\mathrm{bar}$, the isentropic efficiency of the compressor is 0.85

2-3: Isobaric cooling until the temperature reaches $T_3 = 30\,^\circ\mathrm{C}$

3-4 : Adiabatic irreversible expansion in turbine to pressure $p_4 = p_1$, the isentropic efficiency of the turbine is 0.8

4-1: Isobaric heating to state 1

Helium is an ideal gas with constant specific heats, with $c_p = 5.196 \frac{\mathrm{kJ}}{\mathrm{kg\,K}}$, $R = 2.0785 \frac{\mathrm{kJ}}{\mathrm{kg\,K}}$.

1. Draw a schematic, and the process in a T-s-diagram.
2. Make a table with the values of temperature and pressure at points 1-4.
3. Compute the coefficient of performance (COP).
4. The mass flow is $1.5 \frac{\mathrm{kg}}{\mathrm{s}}$. Compute the cooling power, and the power needed to run the refrigerator.

10.18. Gas Cooling System
A gas refrigeration system operates on the inverse Brayton cycle (1-2: adiabatic compression, 2-3: isobaric heat exchange, 3-4: adiabatic expansion, 4-1: isobaric heat exchange) with air as the working fluid. The compressor pressure ratio is 3. This system is used to maintain a refrigerated space at $-23\,^\circ\mathrm{C}$ and rejects heat to the environment at $27\,^\circ\mathrm{C}$. The isentropic efficiency of the compressor is 0.8, but the turbine exhibits no losses. The temperature difference for heat transfer is $10\,^\circ\mathrm{C}$. Consider air as an ideal gas with variable specific heats.

1. Draw a schematic, and the corresponding T-s-diagram.
2. Make a list with the enthalpies and temperatures at the corner points of the process.
3. Compute the COP.
4. For the computation above, the knowledge of pressure ratio was enough, so nothing was said about the value of p_1. Discuss the choice of this pressure (should it be high or low . . .).

Chapter 11
Efficiencies and Irreversible Losses

11.1 Irreversibility and Work Loss

In the discussion of work losses for closed and open systems we found that irreversibilities reduce the efficiency of energy conversion, see Sec. 5.10. We shall study this now in greater detail. The arguments used in this section are similar to "exergy accounting" (see Sec. 11.8 further below).

We consider a general thermodynamic system as in Fig. 9.1 which is described through the balance laws for mass, energy, and entropy (9.1, 9.9, 9.10). Part of the heat exchange between the system and its surroundings will take place at a temperature T_0, and we write

$$\dot{Q} = \dot{Q}_0 + \sum_{k=1} \dot{Q}_k \quad , \quad \sum_k \frac{\dot{Q}_k}{T_k} = \frac{\dot{Q}_0}{T_0} + \sum_{k=1} \frac{\dot{Q}_k}{T_k} \, , \tag{11.1}$$

where $\dot{Q}_k$ is heat transferred over the system boundary at temperature T_k; in particular, Q_0 is the heat exchanged at T_0. With that, the first and second laws of thermodynamics (9.9, 9.10) can be written as

$$\frac{dE}{dt} + \sum_{\text{out}} \dot{m}_e \left(h_e + \frac{1}{2}\mathcal{V}_e^2 + gz_e \right) - \sum_{\text{in}} \dot{m}_\alpha \left(h_i + \frac{1}{2}\mathcal{V}_i^2 + gz_i \right) = \\ = \dot{Q}_0 + \sum_{k=1} \dot{Q}_k - \dot{W} \, , \tag{11.2}$$

$$\frac{dS}{dt} + \sum_{\text{out}} \dot{m}_e s_e - \sum_{\text{in}} \dot{m}_i s_i = \frac{\dot{Q}_0}{T_0} + \sum_{k=1} \frac{\dot{Q}_k}{T_k} + \dot{S}_{gen} \, . \tag{11.3}$$

The highlighted temperature T_0 can be freely chosen according to the details of the actual system considered. Since most thermodynamic systems interact with the environment, one most often chooses T_0 to be the temperature of the environment, usually $T_0 = 298\,\text{K}$ for the standard reference

H. Struchtrup, *Thermodynamics and Energy Conversion*,
DOI: 10.1007/978-3-662-43715-5_11,

environment. Indeed, the environment acts as an infinite heat reservoir, and no cost is associated with heat drawn from, or dumped into, the environment.

Elimination of the heat exchange with the environment, $\dot{Q}_0$, between the first and second laws (11.2, 11.3) yields an equation for power,

$$\dot{W} = -T_0 \dot{S}_{gen} + \sum_{k=1} \left(1 - \frac{T_0}{T_k}\right) \dot{Q}_k + \sum_{\text{in}} \dot{m}_i \left(h_i - T_0 s_i + \frac{1}{2}\mathcal{V}_i^2 + g z_i\right) - \sum_{\text{out}} \dot{m}_e \left(h_e - T_0 s_e + \frac{1}{2}\mathcal{V}_e^2 + g z_e\right) - \frac{d\left(E - T_0 S\right)}{dt} \,. \tag{11.4}$$

This equation is the generalization of (5.17) in Sec. 5.10 to include open system boundaries. The equation is valid for any system, open or closed, transient or steady state, that exchanges heat at least at T_0, and possibly at other temperatures.

Note, that for $\dot{Q} = \dot{Q}_0 = 0$ there is nothing to be eliminated between the two equations, so that (11.4) is not relevant for fully adiabatic processes.

The factor T_0 relates power loss to entropy generation; the lost power is sometimes denoted as *irreversibility*,

$$\dot{W}_{\text{loss}} = T_0 \dot{S}_{gen} \,. \tag{11.5}$$

Equation (11.4) allows to relate irreversibility as measured by entropy generation $\dot{S}_{gen}$ to power losses for complete thermodynamic systems, such as power and refrigeration cycles, in contact with the environment. Since $\dot{S}_{gen} \geq 0$ and[1] $T_0 > 0$, this equation shows that irreversibilities reduce work output for a power producing system (where $\dot{W} > 0$, e.g., a power plant) or increase work demand for a power consuming system (where $\dot{W} < 0$, e.g., a heat pump or a refrigerator).

The task of a thermal engineer can be described as to improve efficiency of a thermal system as much as possible. This requires to identify and reduce—as much as possible—causes of losses, i.e., irreversibilities. This will lead to a redesign of the system, which in turn leads to a change of the inflow, outflow and boundary conditions of the system.

A proper understanding of the losses associated with a system requires that *all* sources for irreversibility are considered. For a proper accounting of losses, the system boundary should be wide enough to include all causes for loss, so that internal and external irreversibilities are accounted for.

The redesign should start with removing the main causes for losses. If we indicate the different causes for entropy generation by Greek indices, we can write

$$\dot{W}_{\text{loss}} = \sum_{\alpha} T_0 \dot{S}_{gen}^{\alpha} \,. \tag{11.6}$$

[1] T is the thermodynamic temperature, of course!

The importance of the various entropy generating processes can be evaluated by a relative measure, e.g., the ratio between lost power and the total power exchange for the process

$$\chi_\alpha = \frac{T_0 \dot{S}_{gen}^\alpha}{\left|\dot{W}\right|} . \tag{11.7}$$

One will not be able to avoid losses altogether, and thus will have to accept values of a few percent (or more) for χ_α. Nevertheless, *any* reduction in irreversibilities that can be attained at reasonable cost for redesign and construction will lead to increased power production or decreased power consumption, and thus offer substantial savings in the long run of operation.

11.2 Reversible Work and Second Law Efficiency

A sometimes useful measure[2] for the performance of a system is the second law efficiency $\eta_{\rm II}$, which compares the actual performance to the reversible work, i.e., the best case scenario for the same process parameters.

The *reversible work* $\dot{W}_{\rm rev}$ is the power that would be obtained from a reversible process operating under the same boundary conditions as the system considered, and exchanging heat with the environment at T_0. Its value follows from (11.4) simply by setting the irreversibility to zero, $T_0 \dot{S}_{gen} = 0$,

$$\dot{W}_{\rm rev} = \sum_{k=1} \left(1 - \frac{T_0}{T_k}\right) \dot{Q}_k + \sum_{\alpha,in} \dot{m}_\alpha \left(h_\alpha - T_0 s_\alpha + \frac{1}{2}\mathcal{V}_\alpha^2 + g z_\alpha\right) \\ - \sum_{\alpha,out} \dot{m}_\alpha \left(h_\alpha - T_0 s_\alpha + \frac{1}{2}\mathcal{V}_\alpha^2 + g z_\alpha\right) - \frac{dE - T_0 S}{dt} . \tag{11.8}$$

Since all irreversibilities reduce the process performance, for a work generating process the reversible work is the maximum work that could be obtained, and for a work consuming process it is the minimum work required. The actual work can be expressed as the difference between reversible work and work loss, $\dot{W} = \dot{W}_{\rm rev} - \dot{W}_{\rm loss}$.

The second law efficiency is defined as the ratio between the actual work of the process and the reversible work. For a power producing system one defines

$$\eta_{\rm II} = \frac{\dot{W}}{\dot{W}_{\rm rev}} = \frac{\dot{W}_{\rm rev} - \dot{W}_{\rm loss}}{\dot{W}_{\rm rev}} = 1 - \frac{T_0 \dot{S}_{gen}}{\dot{W}_{\rm rev}} < 1 \quad \left(\dot{W} > 0\right) , \tag{11.9}$$

and for a power consuming system, one defines

[2] *All* efficiencies are useful only when they are defined and applied in a meaningful way!

$$\eta_{\rm II} = \frac{\left|\dot{W}_{\rm rev}\right|}{\left|\dot{W}\right|} = \frac{\left|\dot{W}_{\rm rev}\right|}{\left|\dot{W}_{\rm rev}\right| + \dot{W}_{\rm loss}} = \frac{1}{1 + \frac{T_0 \dot{S}_{gen}}{\left|\dot{W}_{\rm rev}\right|}} < 1 \quad \left(\dot{W} < 0\right) . \tag{11.10}$$

The simplest examples are given by steady state engines operating between two reservoirs at constant temperatures, and we consider heat engine, refrigerator, and heat pump, as depicted in Fig. 11.1. Note that all engines exchange heat with the environment at T_0, while the temperatures of the other reservoirs are different. The fully reversible engines operating between two reservoirs are Carnot engines.

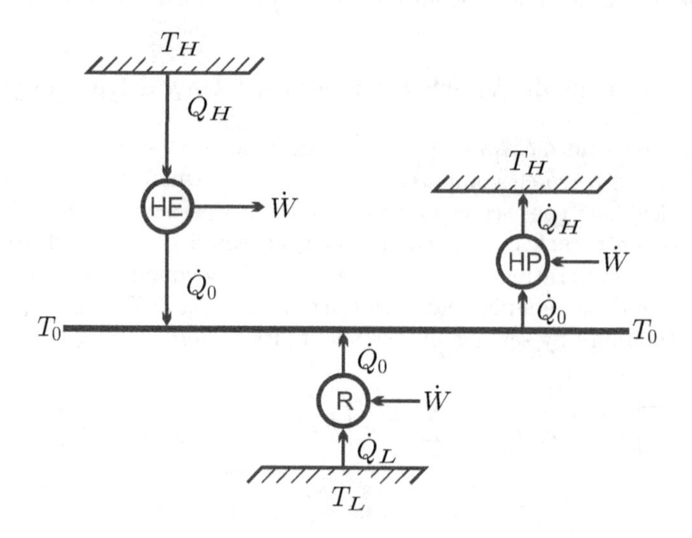

Fig. 11.1 Heat engine (HE), refrigerator (R), and heat pump (HP) in contact with the environment at T_0

For a heat engine that receives the heat $\dot{Q}_H$ from a hot reservoir at $T_H > T_0$, and rejects heat into the environment at T_0, actual and reversible work are $\dot{W} = \left(1 - \frac{T_0}{T_H}\right)\dot{Q}_H - T_0\dot{S}_{gen} > 0$ and $\dot{W}_{\rm rev} = \left(1 - \frac{T_0}{T_H}\right)\dot{Q}_H > 0$. For the second law efficiency we obtain

$$\eta_{\rm II} = \frac{\dot{W}}{\dot{W}_{\rm rev}} = \frac{\dot{W}/\dot{Q}_H}{\dot{W}_{\rm rev}/\dot{Q}_H} = \frac{\dot{W}/\dot{Q}_H}{1 - \frac{T_0}{T_H}} = \frac{\eta}{\eta_C} , \tag{11.11}$$

where $\eta = \dot{W}/\dot{Q}_H$ is the thermal efficiency of the actual engine, and $\eta_C = \dot{W}_{\rm rev}/\dot{Q}_H = 1 - \frac{T_0}{T_H}$ is the thermal efficiency of a fictional Carnot heat engine operating between the same temperatures. It is straightforward to conclude that for a more complex power producing system $\eta_{\rm II} = \eta/\eta_{\rm rev}$ where $\eta_{\rm rev}$ is the thermal efficiency associated with the reversible work.

For a refrigerator that removes the heat $\dot{Q}_L$ from a cold space at $T_L < T_0$ and rejects heat into the environment at T_0, actual and reversible work are $\dot{W} = \left(1 - \frac{T_0}{T_L}\right)\dot{Q}_L - T_0\dot{S}_{gen} < 0$ and $\dot{W}_{\text{rev}} = \left(1 - \frac{T_0}{T_L}\right)\dot{Q}_L < 0$. For the second law efficiency we obtain

$$\eta_{\text{II}} = \frac{\left|\dot{W}_{\text{rev}}\right|}{\left|\dot{W}\right|} = \frac{\dot{Q}_L/\left|\dot{W}\right|}{\dot{Q}_L/\left|\dot{W}_{\text{rev}}\right|} = \left(\frac{T_0}{T_L} - 1\right)\text{COP}_{\text{R}} = \frac{\text{COP}_{\text{R}}}{\text{COP}_{\text{R},C}}, \quad (11.12)$$

where $\text{COP}_{\text{R}} = \dot{Q}_L/\left|\dot{W}\right|$ is the actual coefficient of performance, and $\text{COP}_{\text{R},C} = \dot{Q}_L/\left|\dot{W}_{\text{rev}}\right| = 1\Big/\left(\frac{T_0}{T_L} - 1\right)$ is the COP of the Carnot refrigerator.

For a heat pump that supplies the heat $\dot{Q}_H$ to a warm space at $T_H > T_0$ and draws heat from the environment at T_0, actual and reversible work are $\dot{W} = \left(1 - \frac{T_0}{T_H}\right)\dot{Q}_H - T_0\dot{S}_{gen} < 0$ and $\dot{W}_{\text{rev}} = \left(1 - \frac{T_0}{T_H}\right)\dot{Q}_H < 0$. Also in this case the reversible heat pump is a Carnot engine. For the second law efficiency we obtain

$$\eta_{\text{II}} = \frac{\left|\dot{W}_{\text{rev}}\right|}{\left|\dot{W}\right|} = \frac{\left|\dot{Q}_H\right|/\left|\dot{W}\right|}{\left|\dot{Q}_H\right|/\left|\dot{W}_{rev}\right|} = \left(1 - \frac{T_0}{T_H}\right)\text{COP}_{\text{HP}} = \frac{\text{COP}_{\text{HP}}}{\text{COP}_{\text{HP},C}},$$

where $\text{COP}_{\text{HP}} = \left|\dot{Q}_H\right|/\left|\dot{W}\right|$ is the coefficient of performance, and $\text{COP}_{\text{HP},C} = \left|\dot{Q}_H\right|/\left|\dot{W}_{\text{rev}}\right| = 1\Big/\left(1 - \frac{T_0}{T_H}\right)$ is the COP of the Carnot heat pump.

To summarize, we state that the second law efficiency gives a qualitative measure for the overall quality of a process by comparing actual performance to that of a fictional reversible system with the same parameters..

11.3 Example: Carnot Engine with External Irreversibility

Fully reversible (e.g., Carnot) engines cannot be build since any real process is associated with some irreversibilities. These can be reduced, but not avoided. For instance, heat transfer requires finite temperature differences, which are accompanied by entropy generation, i.e., work loss. As an example we study a reversible Carnot heat engine with external irreversibilities that occur in transferring heat in and out of the engine, see Fig. 11.2.

We compare a fully reversible engine (I), which does not require temperature differences for heat transfer, and an engine (II) that requires finite temperature differences. When both engines take in the same amount of heat, their power outputs are

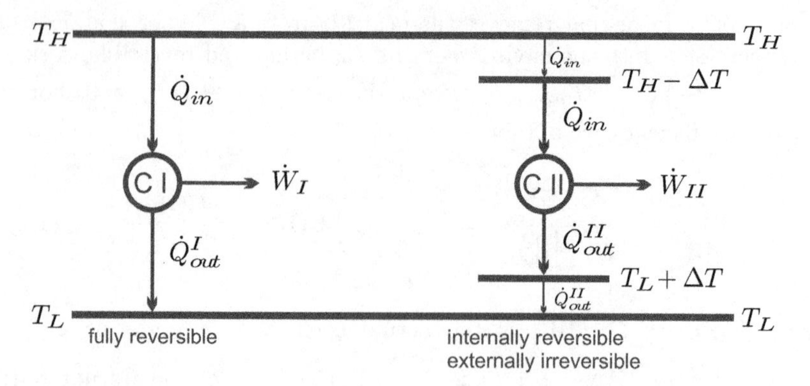

Fig. 11.2 Heat engine between two reservoirs: Fully reversible cycle (I), and internally reversible cycle with external irreversibilities (II)

$$\dot{W}_{\mathrm{I}} = \left(1 - \frac{T_L}{T_H}\right)\dot{Q}_{in} \quad , \quad \dot{W}_{\mathrm{II}} = \left(1 - \frac{T_L + \Delta T}{T_H - \Delta T}\right)\dot{Q}_{in} \; ; \tag{11.13}$$

obviously $\dot{W}_{\mathrm{I}} > \dot{W}_{\mathrm{II}}$. The heat rejected to the environment by the two engines is

$$\left|\dot{Q}^{\mathrm{I}}_{out}\right| = \dot{Q}_{in} - \dot{W}_I = \frac{T_L}{T_H}\dot{Q}_{in} \quad , \quad \left|\dot{Q}^{\mathrm{II}}_{out}\right| = \dot{Q}_{in} - \dot{W}_{\mathrm{II}} = \frac{T_L + \Delta T}{T_H - \Delta T}\dot{Q}_{in} \, . \tag{11.14}$$

For engine II, the entropy generation due to irreversible heat transfer at the higher and lower temperatures are

$$\dot{S}^H_{gen} = \left(\frac{1}{T_H - \Delta T} - \frac{1}{T_H}\right)\dot{Q}_{in} = \frac{\dot{Q}_{in}}{T_H}\frac{\Delta T}{T_H - \Delta T} > 0 \, ,$$

$$\dot{S}^L_{gen} = \left(\frac{1}{T_L} - \frac{1}{T_L + \Delta T}\right)\left|\dot{Q}^{II}_{out}\right| = \frac{\left|\dot{Q}^{II}_{out}\right|}{T_L}\frac{\Delta T}{T_L + \Delta T} = \frac{\dot{Q}_{in}}{T_L}\frac{\Delta T}{T_H - \Delta T} > 0 \, ;$$

the corresponding work loss to heat transfer is

$$\dot{W}_{\mathrm{loss}} = \dot{W}_{\mathrm{I}} - \dot{W}_{\mathrm{II}} = T_L \frac{\dot{Q}_{in}\Delta T}{T_H - \Delta T}\left(\frac{1}{T_L} + \frac{1}{T_H}\right) = T_L\left(\dot{S}^L_{gen} + \dot{S}^H_{gen}\right) \, . \tag{11.15}$$

While the internally reversible engine II has the optimum efficiency with respect to its boundary temperatures $T_L + \Delta T$ and $T_H - \Delta T$, the engine does not have optimum efficiency with respect to the available boundary temperatures T_L and T_H, due to external irreversibilities in heat transfer.

This simple example shows once more the importance of considering external losses. In order to obtain a comprehensive picture of thermodynamic performance, one cannot restrict the attention to the performance of an isolated system, say a heat engine, but one has to account for the interaction of the system with its surroundings as well.

A perfect heat exchanger, which operates at infinitesimal temperature difference, and hence does not generate entropy, requires an infinite exchange surface, and thus can only be thought of, but not be build. Thus, in any existing heat exchanger the temperature difference is finite, and entropy is generated. The art of building heat exchangers, or choosing heat exchangers for a particular system, is to make the temperature difference, and thus the work losses, as small as technology and purchase or construction costs allow.

11.4 Example: Space Heating

Another instructive example is given by the heating of a house with a device powered by electricity, e.g., a resistance heater or a heat pump. The house sits in an external environment at temperature T_0, and its temperature is maintained at $T_{\text{house}} > T_0$. Due to the temperature difference between the warm house and the cooler environment there is a heat loss $\dot{Q}_{\text{loss}}$ to the environment. The heating system must provide the same amount of heat to compensate for the loss, so that the inside temperature remains at T_{house}.

We first examine the case where the heat is provided by a heat pump which draws the power $\dot{W}$ from the electrical grid, draws heat $\dot{Q}_L$ from the environment, and provides the heat $\dot{Q}_H = \dot{Q}_{\text{loss}}$ at a temperature $T_H > T_{\text{house}}$; the temperature difference ensures heat transfer from the heat pump to the house. Figure 11.3 shows the energy flows and the temperature levels for this process.

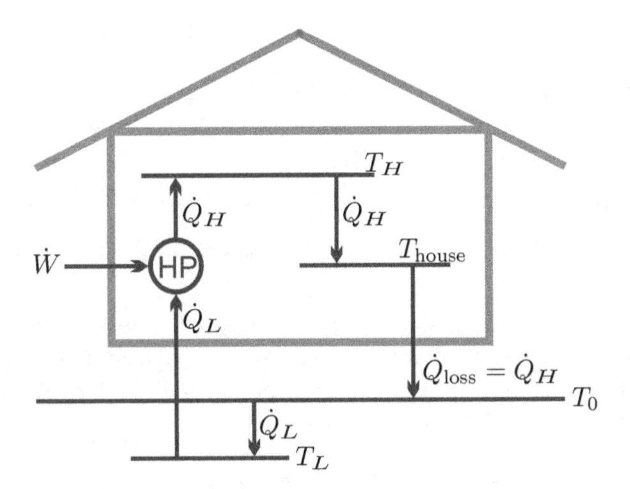

Fig. 11.3 Energy flows for the heating of a house by means of a heat pump

There are two sources of entropy generation, that is, ultimately, work loss: the heat pump itself, and the heat transfer from the house to the environment. The latter is given, from (11.4), as

$$\dot{S}_{\text{loss}} = \left(\frac{1}{T_0} - \frac{1}{T_{\text{house}}}\right) \left|\dot{Q}_{\text{loss}}\right| .$$

This loss is independent of the method used for heating, and can be reduced by reducing the heat loss $\dot{Q}_{\text{loss}}$, or the inside temperature T_{house}. By Newton's law of cooling the heat loss is given by

$$\dot{Q}_{\text{loss}} = \alpha A \left(T_{\text{house}} - T_0\right) ,$$

where A is the outer surface of the house, and α is an overall heat transfer coefficient. The heat loss can be reduced by better insulation of the exterior walls and the roof, which gives smaller value for α, but also by reduction of the inside temperature T_{house}. For a house exposed to an environment at 0 °C, the heating cost can be reduced by 10% when the setting of the thermostat is reduced from 22 °C to 20 °C.

Next we consider losses associated with the heat pump. To ensure heat transfer from the environment and into the house, the heat pump operates between temperatures $T_L < T_0$ and $T_H > T_{\text{house}}$. The first law relates power and heat transfer rates as

$$\left|\dot{W}_{\text{HP}}\right| + \left|\dot{Q}_L\right| = \left|\dot{Q}_H\right| = \left|\dot{Q}_{\text{loss}}\right| .$$

The heat supplied to the house, $\left|\dot{Q}_H\right|$, is the sum of the power to run the heat pump $\left|\dot{W}_{\text{HP}}\right|$, which must be paid for, and the heat intake from the environment $\left|\dot{Q}_L\right|$, which is freely available.

The combined first and second law applied to the heat pump gives the required work as (with $\dot{W}_{\text{HP}} = -\left|\dot{W}_{\text{HP}}\right|$)

$$\left|\dot{W}_{\text{HP}}\right| = \left(1 - \frac{T_0}{T_{\text{house}}}\right) \left|\dot{Q}_{\text{loss}}\right| + T_0 \dot{S}_{gen} ,$$

where $\dot{S}_{gen} = \dot{S}_{gen}^{int} + \dot{S}_{gen}^{ext}$ denotes the associated internal and external irreversibilities. The external irreversibilities due to heat transfer from the environment to the cold side of the heat pump, and heat transfer from the hot side of the heat pump to the house, are

$$\dot{S}_{gen}^{ext} = \left(\frac{1}{T_L} - \frac{1}{T_0}\right) \left|\dot{Q}_L\right| + \left(\frac{1}{T_{\text{house}}} - \frac{1}{T_H}\right) \left|\dot{Q}_H\right| .$$

If these are inserted explicitly into the above equation for power, it assumes the form

$$\left|\dot{W}_{\text{HP}}\right| = \left(1 - \frac{T_L}{T_H}\right)\left|\dot{Q}_{\text{loss}}\right| + T_L \dot{S}_{gen}^{int} \; ;$$

here $T_L \dot{S}_{gen}^{int}$ is the work loss to internal irreversibilities which are unavoidable in a real-life heat pump.

The most efficient heating method is given by a fully reversible heat pump which would require the work $\left|\dot{W}_{\text{rev}}\right| = \left(1 - \frac{T_0}{T_{\text{house}}}\right)\left|\dot{Q}_{\text{loss}}\right|$. Real heat pumps have higher power consumption, and require finite temperature differences for heat exchange, so that irreversible processes are present.

The least efficient heating method are resistance heaters, for which $\dot{Q}_L = 0$, and the heating rate is equal to the power, $\left|\dot{W}_{\text{RH}}\right| = \left|\dot{Q}_H\right|$.

The coefficients of performance (COP) for heat pump and resistance heater are

$$\text{COP}_{\text{HP}} = \frac{\left|\dot{Q}_{house}\right|}{\left|\dot{W}_{\text{HP}}\right|} = \frac{1}{\left(1 - \frac{T_0}{T_{\text{house}}}\right) + \frac{T_0 \dot{S}_{gen}}{\left|\dot{Q}_{\text{loss}}\right|}} > 1 \; , \quad \text{COP}_{\text{RH}} = \frac{\left|\dot{Q}_{house}\right|}{\left|\dot{W}_{\text{RH}}\right|} = 1 \; .$$

Thus, vendors for resistance heaters are right when they claim that their product has an "efficiency" of 100%, but they conceal that a heat pump can have a much higher COP.

A Carnot heat pump operating without temperature differences in heat transfer has the maximum coefficient of performance, $\text{COP}_{\text{HP},C} = \frac{1}{1 - \frac{T_0}{T_{\text{house}}}}$. When the outside temperature is 0 °C and the house is kept at 20 °C, the maximum COP is 14.65.

If the heat pump requires temperature differences of 5 °C for heat transfer, but is internally reversible, its coefficient of performance is $\text{COP}_{\text{HP}} = \frac{1}{1 - \frac{T_L}{T_H}} = \frac{1}{1 - \frac{T_0 - \Delta T}{T_{\text{house}} + \Delta T}} = 9.93$. In this case, the second law efficiency is

$$\eta_R^{\text{II}} = \frac{\text{COP}_{\text{HP}}}{\text{COP}_{\text{HP},C}} = 0.68 \, .$$

A resistance heater has a $\text{COP}_{\text{RH}} = 1$ independent of the temperatures, and, for the same temperatures, the second law efficiency

$$\eta_R^{\text{II}} = \frac{\text{COP}_{\text{RH}}}{\text{COP}_{\text{HP},C}} = 1 - \frac{T_0}{T_{\text{house}}} = 0.068 \, .$$

Obviously, heat pumps have a considerably smaller power demand than resistance heaters, and thus are a better choice for space heating than resistance heaters. Indeed, resistance heaters are common only where electricity is cheap, e.g., in Norway and British Columbia where hydropower is the main source for generation.

11.5 Example: Entropy Generation in Heat Transfer

Heat exchange over finite temperatures is irreversible, the associated entropy generation is related to a work loss. The reason for the work loss is that any temperature difference could be used to drive a heat engine. In heat exchange, there is no engine, hence the loss. As we have seen, energy at higher temperature is more valuable, since more work can be extracted (larger Carnot efficiency). Heat transfer over finite temperature difference conserves the amount of energy transferred (heat), but after transfer the energy is at lower temperature, where the energy is less valuable, since less work can be extracted. We now discuss why the lost work for a system in contact with the environment is $T_0 \dot{S}_{gen}$ (11.5).

We consider heat transfer $\dot{Q}$ between reservoirs at T_H and T_L. The associated entropy generation is

$$\dot{S}_{gen} = \dot{Q}\left(\frac{1}{T_L} - \frac{1}{T_H}\right) .$$

A Carnot heat engine operating between the two reservoirs and receiving the heat $\dot{Q}$ from the hot reservoir could produce the power

$$\dot{W}_C = \left(1 - \frac{T_L}{T_H}\right)\dot{Q} = T_L \dot{S}_{gen} .$$

Note that the Carnot engine involves heat exchange, which must be done by perfect heat exchangers operating at infinitesimal temperature difference, which is impossible in practice.

We recall that the computation of work loss and reversible work relies on the assumption that all boundary parameters remain unchanged. The hypothetical Carnot engine consumes the heat $\dot{Q}_H = \dot{Q}$, but rejects the heat

$$\left|\dot{Q}'_L\right| = \dot{Q} - \dot{W}_C = \frac{T_L}{T_H}\dot{Q}$$

into the cold reservoir, which therefore receives less heat than in the case of pure heat conduction—the difference is just the portion of heat that is converted to power. In order to compensate for this, there must be a second reversible engine employed which rejects the heat

$$\Delta\dot{Q} = \dot{Q} - \left|\dot{Q}'_L\right| = \left(1 - \frac{T_L}{T_H}\right)\dot{Q} = \dot{W}_C$$

at the temperature T_L.

If $T_0 > T_L$, a reversible heat engine operating between T_0 and T_L is employed. An engine that delivers the work $\dot{W}' = \left(\frac{T_0}{T_L} - 1\right)\Delta\dot{Q}$ rejects the required heat $\Delta\dot{Q}$, and the reversible work is

$$\dot{W}_{\text{rev}} = \dot{W}_C + \dot{W}' = \frac{T_0}{T_L}\left(1 - \frac{T_L}{T_H}\right)\dot{Q} = T_0\dot{S}_{gen} \,.$$

If $T_0 < T_L$, a portion of the work $\dot{W}_C$ must be used to drive a Carnot heat pump operating between the environment at T_0 and T_L. To deliver the heat $\Delta\dot{Q}$, the heat pump requires the power $\dot{W}'' = \left(1 - \frac{T_0}{T_L}\right)\Delta\dot{Q}$. Accordingly, the reversible work is, again,

$$\dot{W}_{\text{rev}} = \dot{W}_C - \dot{W}'' = T_0\dot{S}_{gen} \,.$$

In both cases, the second engine is exchanging heat with the environment at T_0. It should be noted that fully reversible engines are not available, so that the discussed systems are only of theoretical interest. Nevertheless, wherever heat transfer over finite temperature differences occurs, there is the potential to do work. Whether it is feasible to do this depends on the individual circumstances of the heat transfer process, in particular on the temperature difference. The larger the temperature difference, the larger is the entropy generation, and the bigger is the work potential.

11.6 Work Potential of a Flow (Exhaust Losses)

Many engines, in particular combustion engines, discard warm or hot exhaust. We ask how much work could be extracted by bringing the exhaust into equilibrium with the environment by reversible processes. For this we consider Fig. 11.4, which shows a system to extract work from an available mass flow $\dot{m}$ at T_1, p_1, $\mathcal{V}_1$. The figure indicates that we can obtain work from propellers inside the flow and by heat transfer through reversible engines which discard

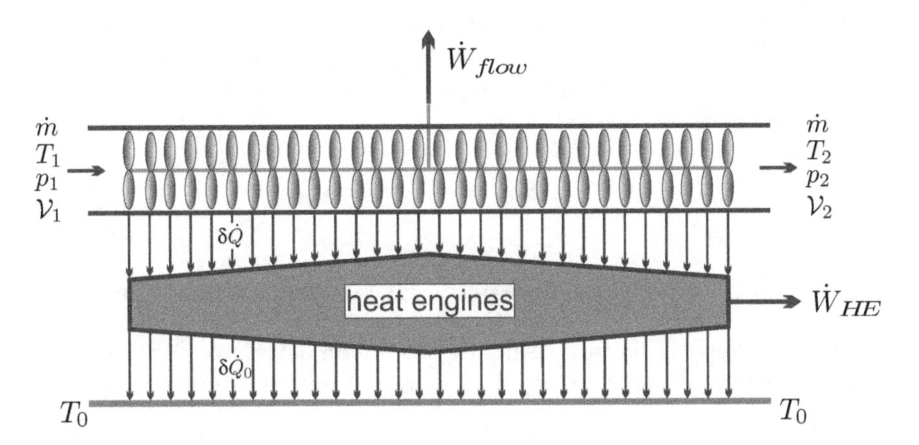

Fig. 11.4 A system to extract work out of a flow by equilibrating it with the environment

heat into the environment at T_0. We consider steady state operation in a one-inlet-one-exit system for which the combined law (11.4) reduces to

$$\dot{W} = -T_0\dot{S}_{gen} - \dot{m}\int_1^2 \left(dh - T_0 ds + d\left(\frac{1}{2}\mathcal{V}^2\right) + gdz\right) . \tag{11.16}$$

With the Gibbs equation $Tds = dh - vdp$ we can write instead

$$\dot{W} = -T_0\dot{S}_{gen} + \dot{m}\int_2^1 \left(vdp + d\left(\frac{1}{2}\mathcal{V}^2\right) + gdz\right) + \dot{m}\int_2^1 \left(1 - \frac{T_0}{T}\right) Tds . \tag{11.17}$$

The first term on the right, $-T_0\dot{S}_{gen}$, describes irreversible losses anywhere in the system, the second term is the reversible flow work extracted from the propellers inside the flow, and the third term is the reversible work available from the external heat engines. Indeed, with $\dot{m}Tds = \delta\dot{Q}$ the last term can be written as $-\int_1^2 \left(1 - \frac{T_0}{T}\right)\delta\dot{Q}$ where $-\delta\dot{Q}$ is the heat supplied to the heat engines on an infinitesimal step of the flow. Thus, if the entropy generation vanishes, the heat engines are a series of infinitesimal Carnot engines.

The maximum work is extracted when all processes are reversible, so that $\dot{S}_{gen} = 0$, and when no external irreversibilities occur, which requires that the exhaust is in equilibrium with the environment, i.e.,[3] $T_2 = T_0$, $p_2 = p_0$, $\mathcal{V}_2 = \mathcal{V}_0 = 0$. Then,

$$\dot{W}_{\text{rev}} = \dot{m}\psi_1 = \dot{m}\left[h_1 - h_0 - T_0(s_1 - s_0) + \frac{1}{2}\mathcal{V}_1^2 + g(z_1 - z_0)\right] . \tag{11.18}$$

Here we have defined the flow exergy (or availability) ψ as the maximum work per unit mass that can be extracted from a flow by equilibrating it with the environment.

The exhaust of a turbine or nozzle has work potential as measured by its exergy. If this work potential is not used, exergy is destroyed and entropy produced. For the case of a single outflow into the environment, the corresponding entropy generation is

$$\dot{S}_{gen} = \frac{\dot{m}\psi}{T_0} . \tag{11.19}$$

11.7 Heat Engine Driven by Hot Combustion Gas

As a relevant application we consider the maximum amount of work that can be obtained from combustion of a fuel in a fully reversible process. We will

[3] The environment is at rest and work could be extracted from any flow faster than the environment. Thus, for no external irreversibilities to occur, one must set $\mathcal{V}_2 = 0$. However, then one could not remove the exhaust—for real applications one will assume $\mathcal{V}_2 \ll \mathcal{V}_1$.

compare the result to that of a case where a single reversible Carnot engine is employed to convert combustion heat into power.

Fuel must be mixed with air and, after its work is done, the combustion product must be removed. Thus, for combustion processes one does not have a single hot reservoir, but the heat is taken from a stream that is gradually cooled as heat is withdrawn.

To simplify the discussion, we ignore the amount of fuel mass added, and treat the combustion product as air; we shall also simplify for constant specific heats to obtain explicit formulae for work and efficiencies. Pressure losses are ignored, and the combustion is assumed to take place at environmental pressure p_0.

First we consider the isobaric combustor: fuel and air enter the combustor at environmental temperature T_0, and the hot combustion gas leaves at the flame temperature T_F. Since we ignore the mass flow of fuel, the heat added to the air is

$$\dot{Q}_{\text{fuel}} = \dot{m}\left[h\left(T_F\right) - h\left(T_0\right)\right] = \dot{m}c_p\left(T_F - T_0\right) , \tag{11.20}$$

where the flame temperature T_F depends on the amount of fuel added. The amount of heat supplied by combustion of the fuel is the product of the mass flow of fuel and the fuel's heating value q_{HV} (measured in kJ/ kg fuel), $\dot{Q}_{\text{fuel}} = \dot{m}_{\text{fuel}} q_{\text{HV}}$. Thus, the cost associated with the process is proportional to $\dot{Q}_{\text{fuel}}$.

$\dot{Q}_{\text{fuel}}$ is also the maximum heat available to convert into power in a heat engine. If $\dot{Q}_{\text{fuel}}$ is completely consumed by the heat engine, the exhaust of the power plant is at T_0. The maximum amount of work that can be obtained from the hot combustion gas[4] follows from (11.18), after ignoring kinetic and potential energies, as

$$\begin{aligned} \dot{W}_{\text{rev}} &= \dot{m}\left[h\left(T_F\right) - h\left(T_0\right) - T_0\left(s\left(T_F, p_0\right) - s\left(T_0, p_0\right)\right)\right] \\ &= \dot{m}c_p\left[T_F - T_0 - T_0 \ln \frac{T_F}{T_0}\right] , \end{aligned} \tag{11.21}$$

with the corresponding thermal efficiency

$$\eta_{\text{max}} = \frac{\dot{W}_{\text{rev}}}{\dot{Q}_{\text{fuel}}} = 1 - \frac{\ln \frac{T_F}{T_0}}{\frac{T_F}{T_0} - 1} . \tag{11.22}$$

This thermal efficiency for power extraction from a hot gas flow, valid only for constant specific heat, is the equivalent to the Carnot efficiency, which describes processes between reservoirs at constant temperatures.

---

[4] This is also the maximum amount of work that can be obtained from the fuel in a combustion process, but, since combustion is irreversible, it is *not* the maximum amount of work available from the fuel—this will be discussed later, when reacting mixtures are discussed.

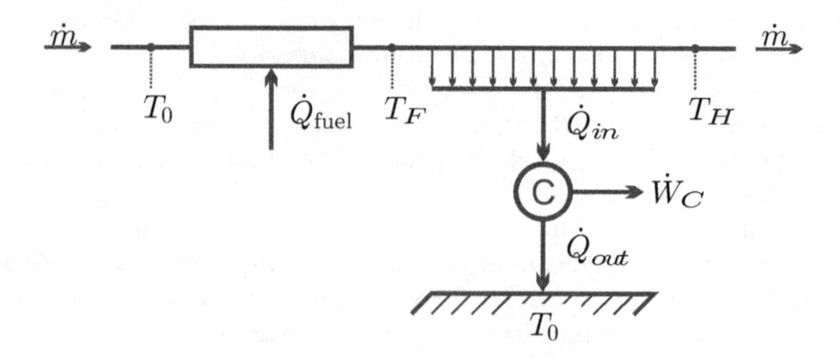

Fig. 11.5 Carnot engine heated by continuous stream of combustion gas

Now we compare this hypothetical best case to the case where the heat extracted from the flow is transferred into a single Carnot engine operating between the temperatures T_0 and T_H. The engine is heated by the hot combustion gas that enters its heat exchanger at flame temperature T_F and leaves at T_H. For now we assume that the exhaust at T_H is not further utilized, but dumped into the environment. The Carnot engine is internally reversible, but obviously there are external irreversibilities associated with the heat transfer into the engine, and with heat transfer between the exhaust gas and the environment. Figure 11.5 shows the corresponding system, including the air temperatures before and after combustion, and before and after heat exchange with the Carnot engine. The heat withdrawn from the combustion gas and added to the Carnot engine is

$$\dot{Q}_{\text{in}} = \dot{m}\left[h\left(T_F\right) - h\left(T_H\right)\right] . \tag{11.23}$$

Accordingly, the work produced by the—reversible—Carnot engine is

$$\begin{aligned} \dot{W}_C &= \left(1 - \frac{T_0}{T_H}\right)\dot{Q}_{\text{in}} = \left(1 - \frac{T_0}{T_H}\right)\dot{m}\left[h\left(T_F\right) - h\left(T_H\right)\right] \\ &= \dot{m}c_p\left(1 - \frac{T_0}{T_H}\right)\left(T_F - T_H\right) . \end{aligned} \tag{11.24}$$

The temperature T_H at the hot side of the engine is a variable of the process. A larger value of T_H increases the thermal efficiency of the Carnot engine, but also leads to a larger exergy of the exiting flow, and thus to a larger external loss. Small values of T_H lead to small thermal efficiency, and to large temperature difference between engine and hot gas flow, which implies large entropy generation and work loss in heat transfer. Closer examination shows that for a given flame temperature T_F and the combustion flow $\dot{m}$, the power produced by the Carnot engine, $\dot{W}_C$, has a maximum at $T_{H,\max} = \sqrt{T_0 T_F}$, where the power produced is

$$\dot{W}_{C,\max} = \dot{m} c_p \left(\sqrt{T_F} - \sqrt{T_0} \right)^2 . \tag{11.25}$$

Next we consider the efficiency of this conversion process. There are several efficiency measures that can be defined. The thermal efficiency of the Carnot engine alone is based on the heat intake of the engine,

$$\eta_{C,\max} = \frac{\dot{W}_{C,\max}}{\dot{Q}_{\text{in}}} = 1 - \frac{T_0}{T_{H,\max}} = 1 - \sqrt{\frac{T_0}{T_F}} . \tag{11.26}$$

Interestingly, this is just the thermal efficiency (10.21) of the ideal Brayton cycle at maximum work. However, this efficiency is not a good measure for the system performance, since it ignores that more heat is available from the exhaust gas at T_H, which is not further used, and, according to the assumptions made, dumped into the environment.

Since the total heat available from cooling the hot flow to environmental temperature is[5] $\dot{Q}_{\text{fuel}} = \dot{m} c_p (T_F - T_0)$, the proper efficiency measure for the conversion of combustion heat into power in this set-up is

$$\eta_{\text{comb}} = \frac{\dot{W}_{C,\max}}{\dot{Q}_{\text{fuel}}} = \frac{\left(\sqrt{T_F} - \sqrt{T_0} \right)^2}{T_F - T_0} = 1 - \frac{2\sqrt{T_0}}{\sqrt{T_F} + \sqrt{T_0}} . \tag{11.27}$$

Figure 11.6 shows the efficiencies $\eta_{\max}$, $\eta_{C,\max}$ and η_{comb} for temperatures T_F between $T_0 = 298\,\text{K}$ and $1500\,\text{K}$. We compute the efficiencies for $T_F = 1500\,\text{K}$ as $\eta_{\max} = 0.599$, $\eta_{C,\max} = 0.554$, and $\eta_{\text{comb}} = 0.383$ ($T_H = 668\,\text{K}$ for the Carnot engine).

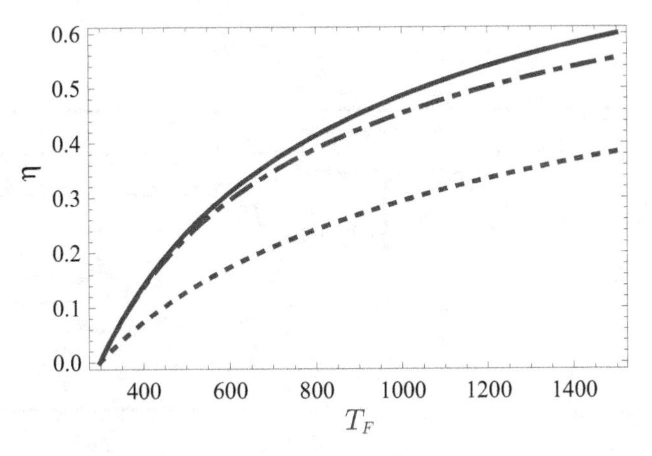

Fig. 11.6 Efficiencies $\eta_{\max}$ (continuous), $\eta_{C,\max}$ (dash-dotted), and η_{comb} (dashed) as defined and discussed in the text

[5] Recall that the fuel cost is proportional to $\dot{Q}_{\text{fuel}}$.

The efficiency $\eta_{\max}$ describes the best possible, i.e., fully reversible, system to produce work from the combustion gas. This efficiency can be used to define the second law efficiency for other processes as $\eta^{\rm II} = \frac{\eta}{\eta_{\max}}$. For the data given above, we find the second law efficiencies $\eta^{\rm II}_{C,\max} = 0.925$, and $\eta^{\rm II}_{\rm comb} = 0.639$.

The fully reversible system has the highest efficiency, $\eta_{\max}$. The single Carnot engine looses work to irreversibilities in heat transfer between hot gas and engine, and in the dumping of hot exhaust into the environment; accordingly it has a significantly lower efficiency $\eta_{\rm comb}$. The efficiency $\eta_{C,\max}$ considers only the actual heat entering the engine, $\dot{Q}_{in}$, not the total heat produced in combustion, $\dot{Q}_F$. This efficiency is higher than $\eta_{\rm comb}$ since it ignores the energy lost with the hot exhaust at T_H.

It is particularly important to note that the efficiency $\eta_{C,\max}$, which ignores the exhaust losses, leads to a far better impression of the quality of the process than the use of the efficiency $\eta_{\rm comb}$ which includes the exhaust losses. The important lesson to be learned here, is that one has to be rather careful about efficiency values presented anywhere, since one can always find an efficiency measure that lets a process appear to be better than it actually is, simply by excluding some, or all, external irreversibilities associated with the process.

The discussed process invites the question what one could do to utilize the exhaust at T_H. For an answer, we recognize that the fuel is needed to heat the air before it exchanges heat with the heat engine. Fuel consumption can be reduced by using the exhaust air (at T_H) to preheat the incoming air before combustion. The heat exchanger used for this is called a regenerator, and Fig. 11.7 shows the system with the regenerator added. When we assume perfect heat exchange between the incoming and the exiting air streams (both

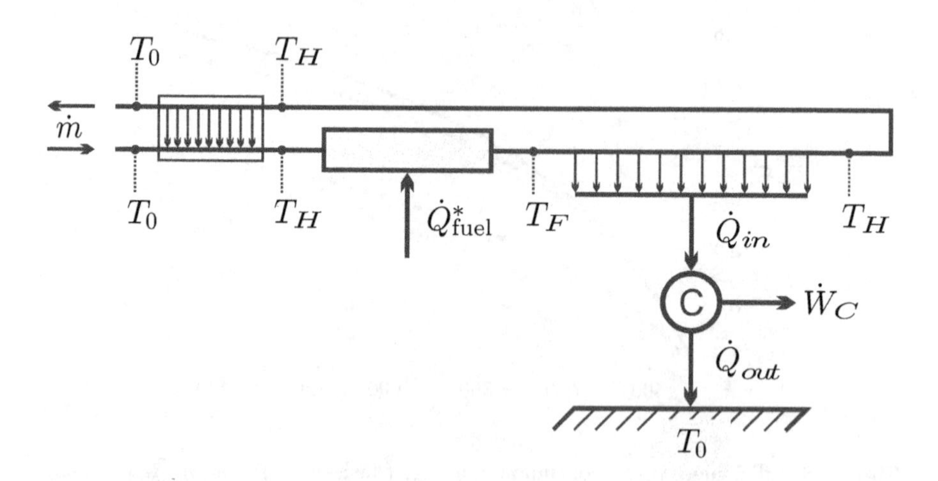

Fig. 11.7 Carnot engine heated by continuous stream of combustion gas with regenerator added

with mass flow $\dot{m}$), the incoming air can be heated to T_H while the exiting air cools down to T_0. In this case, only the heat

$$\dot{Q}^*_{\text{fuel}} = \dot{m}\left[h\left(T_F\right) - h\left(T_H\right)\right] = \dot{m}c_p\left(T_F - T_H\right) = \dot{Q}_{\text{in}} \tag{11.28}$$

must be provided from the combustion of fuel. As indicated, this is just the heat going into the heat engine, $\dot{Q}_{\text{in}}$, so that now all heat from the fuel is used in the engine. The exhaust is in equilibrium with the environment, and hence there is no external entropy generation—and no external loss. The only irreversibility in the considered process is the heat transfer between the hot gas and the engine. The efficiency for the engine with regenerator, and optimal choice for T_H, is

$$\eta_{C,reg} = \frac{\dot{W}_{C,\text{max}}}{\dot{Q}^*_{\text{fuel}}} = \frac{\dot{W}_{C,\text{max}}}{\dot{Q}_{\text{in}}} = \eta_{C,\text{max}} \,. \tag{11.29}$$

Regeneration will be discussed in more detail in subsequent chapters, including the consideration of imperfect heat exchange.

11.8 Exergy

Exergy, also known as "availability", is defined as the maximum amount of work that can be extracted from a flow or an amount of substance by only exchanging heat with the environment until equilibrium with the environment is reached.

Thus, exergy describes work potential, and can be a useful concept to answer questions like whether it is worthwhile to harvest energy from a system. As an example one might think of the exhaust of fuel fired power plants as discussed in Section 11.6.

Exergy analysis is now an accepted method within the field of thermodynamics. We shall introduce the concept in this section, but since we prefer to focus on entropy generation and lost work arguments, we shall use exergy only occasionally.

To compute exergy, the combined first and second law (11.4) is simplified for the case where heat is exchanged only with the environment at T_0 and all processes are reversible, that is by setting $\sum_{k\neq 0}\left(1 - \frac{T_0}{T_k}\right)\dot{Q}_k - T_0\dot{S}_{gen} = 0$.

We shall distinguish between flow exergy, and closed system exergy.

In a closed system, where all mass flows vanish, Eq. (11.4) reduces further to

$$-\frac{d\left(E - T_0 S\right)}{dt} = \dot{W} \,. \tag{11.30}$$

Integrating between the actual state $\{E_a, S_a\}$ and the final equilibrium state $\{E_0, S_0\}$, and subtracting the work done to the environment, which has constant pressure p_0, yields the closed system exergy as

$$\Xi_a = \int \dot{W} dt - \int_a^0 p_0 dV = - \int_a^0 (dE - p_0 dV - T_0 dS)$$
$$\Xi_a = E_a - E_0 + p_0 (V_a - V_0) - T_0 (S_a - S_0) \ . \tag{11.31}$$

The specific closed system exergy is

$$\xi = \frac{\Xi}{m} = e - e_0 + p_0 (v - v_0) - T_0 (s - s_0) \ . \tag{11.32}$$

Flow exergy ψ is defined as the maximum work per unit mass that can be extracted from a single flow in a steady state process as it is equilibrated with the environment, that is heat is exchanged only with the environment and the outflow is in equilibrium with the environment. Thus, from (11.8),

$$\psi = \frac{\dot{W}}{\dot{m}} = h - h_0 - T_0 (s - s_0) + \frac{1}{2} \left(\mathcal{V}^2 - \mathcal{V}_0^2\right) + g (z - z_0) = \xi + (p - p_0) v \ . \tag{11.33}$$

With these definitions, the combined first and second law (11.4) can be written in form of an exergy balance. Use of the mass balance (9.1) to eliminate some terms with the constant factor $(e_0 + p_0 v_0 - T_0 s_0)$, yields

$$\frac{d\Xi}{dt} + \sum_{out} \dot{m}_e \psi_e - \sum_{in} \dot{m}_i \psi_i = \sum_{k \neq 0} \left(1 - \frac{T_0}{T_k}\right) \dot{Q}_k - \left(\dot{W} - p_0 \frac{dV}{dt}\right) - T_0 \dot{S}_{gen} \ . \tag{11.34}$$

This equation describes the change of exergy of a system due to convective transport ($\dot{m}\psi$), heat transfer ($\dot{Q}_k$) at temperatures $T_k \neq T_0$, work ($\dot{W} - p_0 \frac{dV}{dt}$), and exergy destruction due to irreversibilities ($-T_0 \dot{S}_{gen}$).

The combination $\left(\dot{W} - p_0 \frac{dV}{dt}\right)$ is called the useful work. Note that for typical open systems, e.g., turbines, the volume V stays constant, so that $p_0 \frac{dV}{dt}$ is zero. Moreover, most relevant closed system engines are reciprocating, so that for one cycle $\oint p_0 \frac{dV}{dt} dt = p_0 \oint dV = 0$. This is reflected in our discussion of the Otto and Diesel cycles, where we did not consider the work done on the environment.

Problems

11.1. A Heat Engine

A heat engine that operates between two reservoirs at $T_H = 500\,°\mathrm{C}$ and $T_L = 25\,°\mathrm{C}$ produces 1.25 MW of power from a heat intake of 2.5 MW.

Compute the heat rejected, the thermal efficiency, the entropy generation rate, the work loss to irreversibilities, and the 2nd law efficiency of the engine.

11.2. A Heat Pump

An off-the-shelf heat pump system has a COP of 3.4 for operation between 25 °C and −5 °C. Determine the entropy generation per kW of heating, the percentage of power consumed required to overcome irreversibilities, and the 2nd law efficiency of the system.

11.3. A Refrigerator

A refrigeration system has a COP of 2 for operation between 20 °C and −10 °C and consumes 1.5 kW of electrical power. Determine the second law efficiency of the system, the entropy generation rate, and the amount of power required to overcome irreversibilities.

11.4. Heating of House

A small house is exposed to an environment of $T_0 = -5\,°\mathrm{C}$, the temperature inside is to be maintained at $T_h = 22\,°\mathrm{C}$. The heat loss is given by Newton laws of cooling as $\dot{Q}_{loss} = \alpha A(T_h - T_0)$, where $A = 260\,\mathrm{m}^2$ is the outside surface of the house and $\alpha = 1\frac{\mathrm{W}}{\mathrm{m}^2\,\mathrm{K}}$ is an overall heat transfer coefficient. Determine the amount of electrical work for heating the house for the following cases:

1. With a resistance heater.
2. With an internally and externally reversible Carnot heat pump.
3. With an externally irreversible Carnot heat pump with 15 K temperature difference for heat transfer.

11.5. Space Heating

A friend who is going to build a house asks you for advice on heating systems. His contractor has offered the following choices: (a) baseboard resistance heaters, (b) heat pump with hot water radiators (circulating water heated to 60 °C), (c) heat pump with warm water floor heating (circulating water heated to 35 °C).

Based on your knowledge of thermodynamics, which option should your friend chose? Present your arguments. Assume outside temperature −10 °C and inside temperature 20 °C.

11.6. A Cycle

A closed piston-cylinder engine with helium as working medium operates on the following reversible cycle

1-2: Isentropic compression from $p_1 = 10\,\mathrm{bar}$, $T_1 = 300\,\mathrm{K}$
2-3: Isobaric heat addition until $T_3 = 1200\,\mathrm{K}$
3-1: Isochoric cooling to the initial state

1. Draw p-V-diagram and T-s-diagram for the cycle.
2. Determine the thermal efficiency of the cycle and the net work output per unit mass.
3. How much work per unit mass could be obtained from the heat rejected into the environment in the best case? Assume the environment is at $T_0 = 300\,\mathrm{K}$, $p_0 = 1\,\mathrm{bar}$.

11.7. Exhaust of a Car Engine
The engine of a car delivers a net work of $848.4\frac{\text{kJ}}{\text{kg}}$ from an heat intake of $1520.4\frac{\text{kJ}}{\text{kg}}$ (reversible operation). The state at the end of the expansion stroke of the engine is 1146.6 K and 3.822 bar. In the engine, air in this state is exhausted into the environment which is at 300 K, 1 bar.

The exhaust process is modelled as isochoric cooling to the environment. Use the combined first and second law to compute the amount of work that could be obtained from the exhaust in a fully reversible process and compare to the work delivered by the engine.

What would be the thermal efficiencies for the engine alone, and for a system that also provides the work obtainable from the exhaust?

11.8. Exhaust of a Car Engine (Continuous)
The state at the end of the expansion stroke of a car engine is 1140 K and 3.8 bar. In the actual engine, air in this state is exhausted into the environment which is at 300 K, 1 bar.

Assume that the exhaust is leaving the engine as a continuous steady flow, and determine the work that could be obtained from the exhaust in a fully reversible process. Compare to the work delivered by the engine.

11.9. Use of Waste Heat
A chemical plant rejects 2 MW of waste heat at 350 °C and 4 MW at 200 °C. Moreover the plant consumes 5 MW of electrical energy. In the past, the waste heat was just dumped into the environment (20 °C), but now there is a plan to use it for power production to reduce the electricity bill. Estimate what percentage of the electric power consumed in the plant could be produced by a suitable system, when 20% of the maximum possible is lost to irreversibilities.

11.10. Reversible Heat Transfer
Consider a steady state isobaric flow (pressure p, mass flow $\dot{m}$) of a fluid that enters the system at a temperature T_1 and leaves at the environmental temperature T_0. The heat withdrawn is used to drive a series of infinitely many infinitesimal Carnot engines. All processes are fully reversible. Compute the total power output of the system.

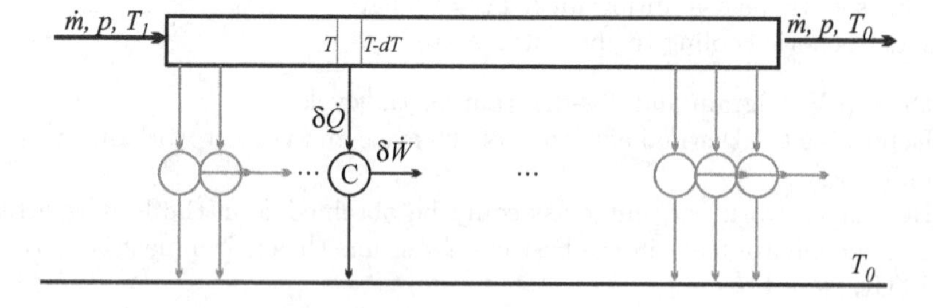

11.11. Entropy Generation in Mixing

In an adiabatic mixing chamber, a mass flow of $200\frac{\text{kg}}{\text{s}}$ compressed liquid water at 10 bar, 40 °C is mixed isobarically with saturated steam so that the exiting flow is in the saturated liquid state. Determine the mass flow of saturated steam that must be added, the entropy generation rate, and the work loss for the process (with respect to standard environment).

11.12. Entropy Generation in Steam Generator

The temperature in the boiler of a big steam power plant is constant at 700 °C. The pipes of the steam generator run trough the boiler, with the inlet state being compressed liquid at 200 bar, 50 °C, and the exit state being at 200 bar, 550 °C. For a mass flow of $1150\frac{\text{t}}{\text{h}}$, determine the heating rate, the entropy generation rate and the associated work loss (with respect to standard environment).

11.13. Entropy Generation in Steam Generator

Consider a 250 MW nuclear power plant with thermal efficiency of 0.32. The steam generator is kept at a pressure of 17.5 MPa. The incoming feedwater is in the compressed liquid state at 40 °C, and the exiting steam is superheated vapor at 400 °C. The heat is provided by a counter-flow of molten sodium (ideal incompressible liquid, mass density $0.927\frac{\text{g}}{\text{cm}^3}$, specific heat $1.26\frac{\text{kJ}}{\text{kg K}}$) which enters at 500 °C and leaves at 350 °C.

1. Determine the mass flows of steam and sodium.
2. Determine the total entropy generation rate in steam generator, and the corresponding work loss.

11.14. Entropy Generation in Condenser

The condenser of a small steam power plant is kept at a temperature of 45 °C. The inlet state is at a quality of 90%, and the exit state is saturated liquid. The condenser rejects heat to the environment which is at 5 °C. For a mass flow of $75\frac{\text{t}}{\text{h}}$, determine the cooling rate, the entropy generation rate and the associated work loss.

11.15. Entropy Generation in Throttling

Cooling fluid R134a in compressed liquid state at 1 MPa, 26 °C is throttled adiabatically to a pressure of 0.14 MPa. For a mass flow of $1\frac{\text{kg}}{\text{s}}$, determine the entropy generation rate, and the associated work loss.

11.16. Work Potential of a Hot Rock

A 2t block of granite (specific heat $c = 0.79\frac{\text{kJ}}{\text{kg K}}$) is initially at a temperature of 500 °C. How much work could be obtained from equilibrating the rock with the environment at 15 °C?

Chapter 12
Vapor Engines

12.1 Boiler Exhaust Regeneration

The discussion of losses in combustion driven systems in the last chapter has shown that regeneration, i.e., use of exhaust energy by means of heat exchange within the system, can yield dramatic improvement of engine efficiency. In direct continuation of the argument, we first discuss regeneration in steam cycles, which rely on external combustion. For this, we need to consider not only the steam cycle, but also its heat source, which is hot combustion air.

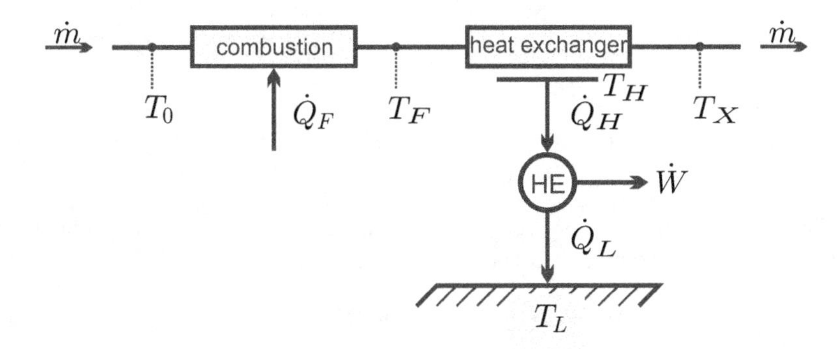

Fig. 12.1 Heat engine driven by external combustion without exhaust regeneration

Figure 12.1 shows a heat engine (HE) which is driven by heat exchange with a hot combustion product. Air at T_0, flowing at rate $\dot{m}$, is mixed with fuel and burned so that the combustion product has the temperature T_F. The heat supplied to the air from the combustion is (air standard approximation, i.e., fuel mass ignored)

$$\dot{Q}_F = \dot{m}\left(h_F - h_0\right) . \tag{12.1}$$

The hot gas runs through the heat exchanger which it leaves at temperature T_X, so the heat supplied to the heat engine is

H. Struchtrup, *Thermodynamics and Energy Conversion*,
DOI: 10.1007/978-3-662-43715-5_12, © Springer-Verlag Berlin Heidelberg 2014

$$\dot{Q}_H = \dot{m}\,(h_F - h_X)\ . \tag{12.2}$$

Since the exhaust leaves at temperature T_X, the heat

$$\dot{Q}_E = \dot{m}\,(h_X - h_0) = \dot{Q}_F - \dot{Q}_H \tag{12.3}$$

remains unused; this is just the heat added to the environment when the exhaust equilibrates.

Earlier, we have discussed this set up when the heat engine is a Carnot engine, and have found the exhaust temperature T_X for optimum work output, see Sec. 11.7. The discussion showed that the simplest way to utilize the exhaust heat $\dot{Q}_E$ is regeneration by preheating the air before combustion.

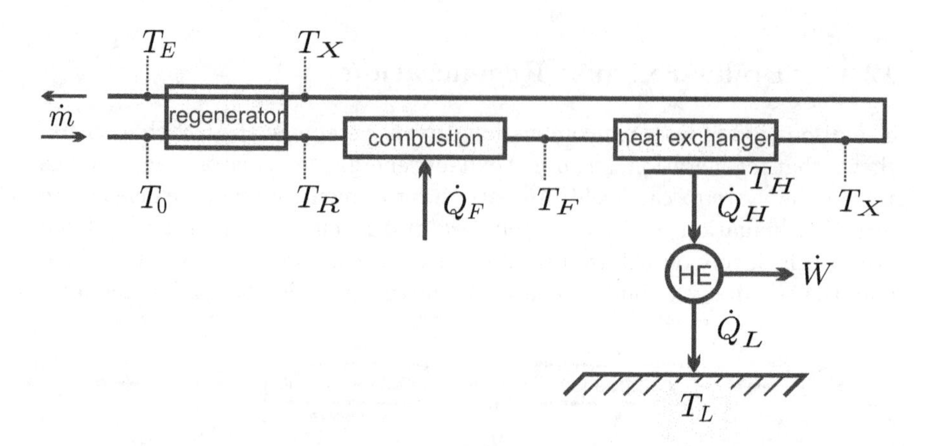

Fig. 12.2 Heat engine driven by external combustion with exhaust regeneration

Figure 12.2 shows the system with an added regenerator for preheating the air. The heat exchange in the regenerator is, from the first law,

$$\dot{Q}_R = \dot{m}\,(h_R - h_0) = \dot{m}\,(h_X - h_E)\ , \tag{12.4}$$

where T_E is the final exhaust temperature, and T_R the preheat temperature. In a perfect regenerator the preheat temperature would be T_X, and the exhaust would leave at T_0. Accordingly, the regenerator effectiveness is defined as the ratio between the heat *used* for preheating, $(h_R - h_0)$, and the heat *available* for preheating, $(h_X - h_0)$, that is

$$\eta_{reg} = \frac{h_R - h_0}{h_X - h_0}\ . \tag{12.5}$$

With regenerator, the heat addition from the fuel becomes

$$\dot{Q}_F = \dot{m}\,(h_F - h_R) = \dot{Q}_H + (1 - \eta_{reg})\,\dot{m}\,(h_X - h_0)\ , \tag{12.6}$$

where $\dot{Q}_H = \dot{m}\,(h_F - h_X)$ is the heat supplied to the heat engine as before. Thus, with a perfect regenerator ($\eta_{reg} = 1$), we have $\dot{Q}_H = \dot{Q}_F$, i.e., all the heat provided from the fuel arrives in the engine.

A realistic regenerator has effectiveness of about 80%, and still leads to a much better fuel usage compared to direct exhaust into the environment. It must be noted that for several reasons a somewhat elevated exhaust temperature T_E is beneficial: The combustion of fossil fuels generates water and sulfur oxides; the exhaust temperature must be high enough to avoid water condensation and subsequent formation of sulfuric acid. Also, the combustion air must be moved through the system, either by means of fans, or by natural draught chimneys, which rely on the buoyancy of warm air (Sec. 13.8). Since effective natural draught requires relatively warm exhaust, there is a marked loss. Therefore, modern power plants use fans.

From our previous discussion of heat engines we know that efficiency is high when heat is added at larger temperatures. Thus, for the heat engine one will aim at having the average temperature for heat addition T_H as high as possible. The temperature T_H is limited by the temperature-pressure characteristics of the working fluid and the materials used for construction. The maximum steam temperature in steam cycles using steel pipes in the steam generator is 560 °C. The regenerative steam cycles discussed below aim at raising the average temperature for heat addition, and thus increasing efficiency.

External (to the heat engine) irreversibilities occur in the combustion chamber, and in heat transfer to the heat engine. Our discussion of combustion processes in Chapter 25 will show that combustion irreversibility decreases with increasing flame temperature T_F. On the other hand, heat transfer irreversibility grows with the temperature difference between combustion product (T_F) and the heat engine (T_H). If T_H is limited, as is the case in steam power plants, reduction of T_F decreases heat transfer irreversibility, but increases combustion irreversibility, with the total irreversibility staying relatively constant. Heat is transferred more easily at larger temperature differences, and one can adjust T_F for efficient heat transfer. More efficient use of the fuel is made when the heat engine temperature T_H is increased, as in the combined cycle of Sec. 13.6.

12.2 Regenerative Rankine Cycle

With a high average temperature of the combustion gas in the boiler possible by exhaust regeneration, we now turn to the question of how to raise the average temperature during heat addition in a steam power plant. The basic Rankine cycle was already discussed in Sec. 10.1, where we introduced reheating between turbine stages as one means to this end, Fig. 12.3 repeats the T-s-diagram and the schematic, with air preheater added to the sketch. A higher pressure in the steam generator implies higher average temperature.

However, direct expansion from high pressure into the condenser (3-4') results in unacceptably low values for steam quality at the turbine exit, and turbine blade damage through droplet formation. Reheat at intermediate pressure is used so that the quality at turbine exit is larger, which implies fewer droplet, and low blade damage.

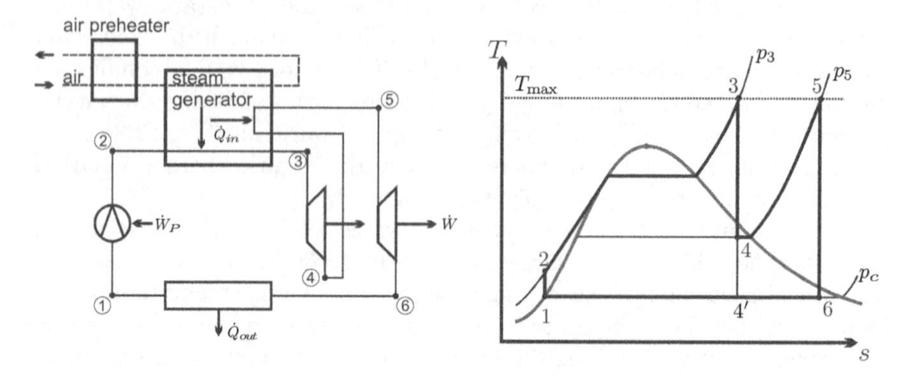

Fig. 12.3 Schematic and T-s-diagram of Rankine cycle with reheat and preheating of combustion air

Heat is added between states 2 and 3, with rather low temperatures in the liquid region right after the pump, the boiler feedwater. In a regenerative Rankine cycle internal heat exchange is used to preheat the feedwater between pump and boiler. For this, some steam is bled-off after the high pressure turbine stage (state 4), and then used to heat the feedwater.

One distinguishes between open feedwater heaters, in which vapor and water are mixed isobarically, and closed feedwater heaters, where vapor and feedwater are running through a heat exchanger at different pressures. In both cases modifications of the cycle are necessary, in order to feed the bled-off flow back into the cycle, and to adjust pressures. Real life power plants employ an array of closed and open feedwater heaters to optimize efficiency and thus increase fuel utilization as much as possible.

12.2.1 Open Feedwater Heater

A reheat steam cycle with a single open feedwater heater is depicted in Fig. 12.4 together with the corresponding T-s-diagram. In the open feedwater heater, compressed liquid water coming from the low pressure pump (P1, state 2) is mixed isobarically with some of the steam leaving the high pressure turbine (T1, state 6). The mixing ratio is adjusted such that the resulting mixture (state 3) is saturated liquid at the mixing pressure $p_3 = p_2 = p_6$. The high pressure pump (P2) compresses this liquid to p_4 before it is fed into the boiler where it is heated, evaporated and superheated to state 5.

The high-pressure, high-temperature steam is expanded in the high pressure turbine and then split into the stream to the feedwater heater, and the main stream. The latter is reheated in the boiler (6-7), expanded in the low pressure turbine (7-8), condensed (8-1), and compressed (1-2), before it enters the feedwater heater.

As the figure indicates, the working fluid undergoes two different cycles. The main stream $\dot{m}_A$ undergoes the full cycle 1-2-3-4-5-6-7-8-1 (continuous line), while the mass flow $\dot{m}_B$ bled-off after the high pressure turbine undergoes the cycle 3-4-5-6-3 (dashed line).

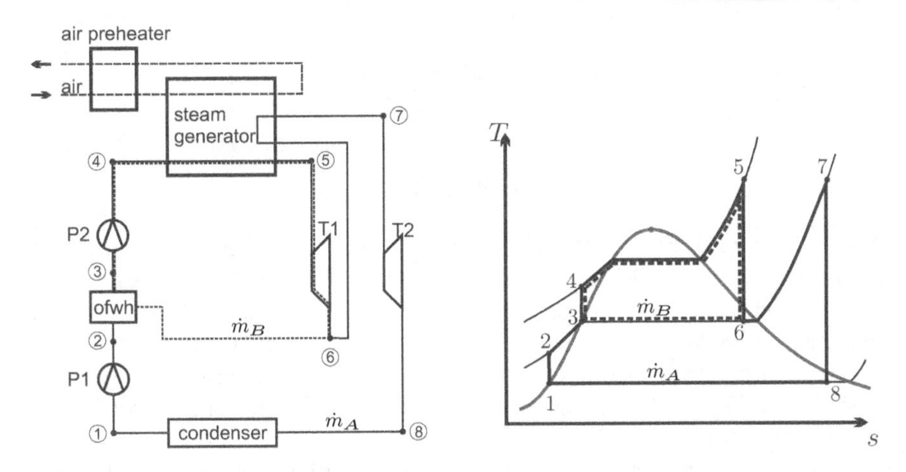

Fig. 12.4 Schematic and T-s-diagram for a steam power plant with reheat and one open feed water heater (o.f.w.h.). Regeneration of combustion air in the air preheater is indicated as well. To not overload the T-s-diagram, only reversible process curves are indicated for pumps and turbines.

The heat added to this process from the fuel is

$$\dot{Q}_{in} = (\dot{m}_A + \dot{m}_B)(h_5 - h_4) + \dot{m}_A (h_7 - h_6) , \tag{12.7}$$

and takes place at relatively high temperature. The low temperature heating 2-3 occurs through internal heat exchange in the open feedwater heater. Increase of the average temperature for external heat addition increases the thermal efficiency, and this is why the open feedwater heater gives better efficiency.

The mass flow ratio is determined from the energy balance for the feedwater heater. We assume that the feedwater heater has no heat loss to the environment, so that

$$\dot{m}_A h_2 + \dot{m}_B h_6 = (\dot{m}_A + \dot{m}_B) h_3 , \tag{12.8}$$

which gives the mass flow ratio as

$$y = \frac{\dot{m}_B}{\dot{m}_A + \dot{m}_B} = \frac{h_3 - h_2}{h_6 - h_2} . \tag{12.9}$$

Together, the two turbines produce the power

$$\dot{W}_T = (\dot{m}_A + \dot{m}_B)(h_5 - h_6) + \dot{m}_A (h_7 - h_8) > 0 , \tag{12.10}$$

while the pumps consume

$$\dot{W}_P = \dot{m}_A (h_1 - h_2) + (\dot{m}_A + \dot{m}_B)(h_3 - h_4) < 0 . \tag{12.11}$$

Accordingly, the thermal efficiency of the cycle is

$$\eta = \frac{\dot{W}}{\dot{Q}_{in}} = \frac{(1-y)(h_1 - h_2 + h_7 - h_8) + h_3 - h_4 + h_5 - h_6}{h_5 - h_4 + (1-y)(h_7 - h_6)} . \tag{12.12}$$

12.2.2 Closed Feedwater Heater

With closed feedwater heaters the streams that exchange heat can be at different pressures, which gives some additional flexibility for process design. The bled-off steam must re-enter the main flow after the feedwater heater, and this can be done either by pumping it into the boiler flow, as depicted in Fig. 12.5, or by throttling it to lower pressure, for instance into the condenser as shown in Fig. 12.6. The latter solution is less efficient, since throttling is highly irreversible.

We consider the case with pump in detail. Steam is bled-off after the high pressure turbine (state 6) and runs through the closed feedwater heater (cfwh)

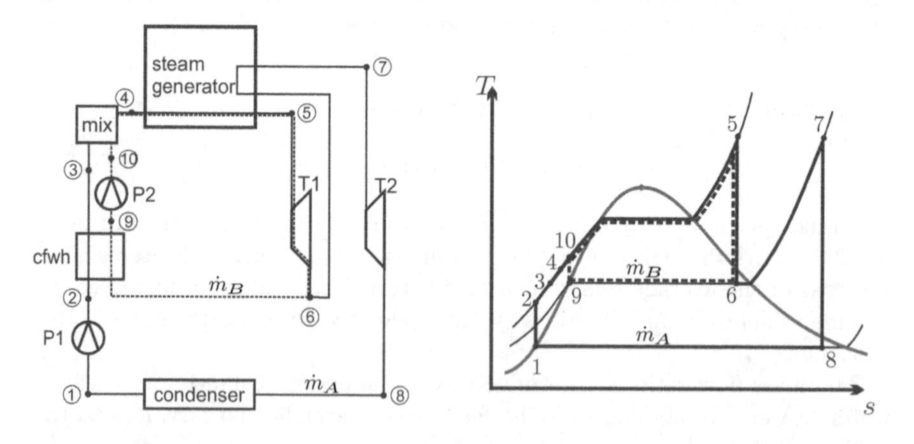

Fig. 12.5 Schematic and T-s-diagram for a steam power plant with reheat and one closed feed water heater (cfwh). Pump 2 only compresses the bled-off flow before it is mixed into the main feedwater flow in the mixing chamber (mix). Combustion air flow is not shown.

where it exchanges heat with the pressurized feedwater, and condenses (state 9). The condensate is brought to the turbine inlet pressure by the second pump (P2, state 10), and then mixed into the main flow to state 4.

Again, the working fluid undergoes two different cycles. The main stream $\dot{m}_A$ runs through the full cycle 1-2-3-4-5-6-7-8-1 (continuous line), while the mass flow $\dot{m}_B$ bled-off after the high pressure turbine runs through the cycle 9-10-4-5-6-9 (dashed line).

The energy balance for the (adiabatic) feedwater heater reads

$$\dot{m}_A (h_3 - h_2) = \dot{m}_B (h_6 - h_9) \ , \tag{12.13}$$

so that the mass flow ratio is

$$y = \frac{\dot{m}_B}{\dot{m}_A + \dot{m}_B} = \frac{h_3 - h_2}{h_6 - h_9 + h_3 - h_2} \ . \tag{12.14}$$

Note that the temperature of the main flow after the feedwater heater, T_3, will be below the temperature of the condensing heating flow, T_9.

The energy balance for the adiabatic mixing chamber reads

$$\dot{m}_A h_3 + \dot{m}_B h_{10} = (\dot{m}_A + \dot{m}_B)\, h_4 \ . \tag{12.15}$$

Heat input, turbine power and pump power are

$$\begin{aligned} \dot{Q}_{in} &= (\dot{m}_A + \dot{m}_B)(h_5 - h_4) + \dot{m}_A (h_7 - h_6) \ , \\ \dot{W}_T &= (\dot{m}_A + \dot{m}_B)(h_5 - h_6) + \dot{m}_A (h_7 - h_8) > 0 \ , \\ \dot{W}_P &= \dot{m}_A (h_1 - h_2) + \dot{m}_B (h_9 - h_{10}) < 0 \ . \end{aligned} \tag{12.16}$$

and thus the thermal efficiency is

$$\eta = \frac{(1-y)(h_1 - h_2 + h_7 - h_8) + y\,(h_9 - h_{10}) + (h_5 - h_6)}{h_5 - h_4 + (1-y)(h_7 - h_6)} \ . \tag{12.17}$$

If the bled-off flow is simply throttled into the condenser, the total mass flow $\dot{m}_A + \dot{m}_B$ must be pumped up and preheated in the feedwater heater. In this case the energy balance for the feedwater heater is

$$(\dot{m}_A + \dot{m}_B)(h_3 - h_2) = \dot{m}_B (h_5 - h_8) \ ,$$

which gives the mass flow ratio as

$$y = \frac{\dot{m}_B}{\dot{m}_A + \dot{m}_B} = \frac{h_3 - h_2}{h_5 - h_8} \ .$$

The thermal efficiency for this cycle can be read from the schematic as

$$\eta_{cwfh\_th} = \frac{h_1 - h_2 + h_4 - h_5 + (1-y)(h_6 - h_7)}{h_4 - h_3 + (1-y)(h_6 - h_5)} \ .$$

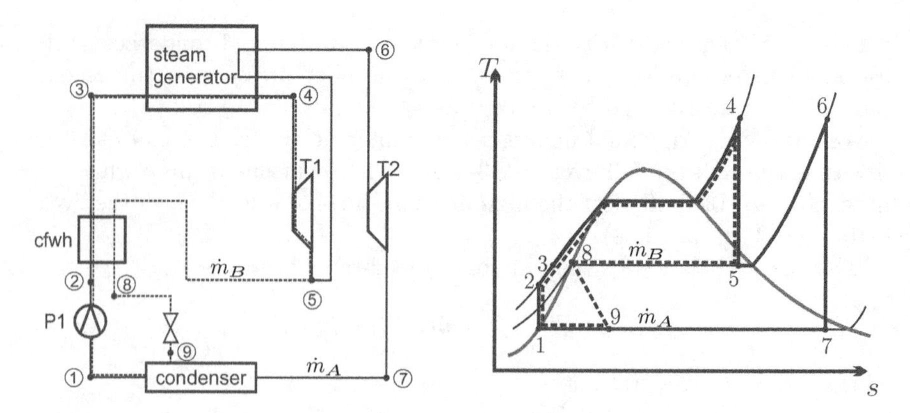

Fig. 12.6 Schematic and T-s-diagram for a steam power plant with reheat and one closed feed water heater (cfwh). The bled-off flow is throttled into the condenser.

12.2.3 Several Feedwater Heaters

The example below will give further evidence that regeneration through feedwater heaters yields improvement of thermal efficiency. Already small improvements of thermal efficiency lead to significant savings (or increased profit), and therefore one aims for optimum process configurations. Modern power plants employ arrays of open and closed feedwater heaters operating at various pressures.

Feedwater heaters reduce the external irreversibility in the heat transfer between the combustion gas and the working fluid in the boiler steam generators. However, feedwater heaters add internal irreversibilities, due to heat transfer over finite temperature differences, mixing, and throttling. The subsequent examples will show that the overall irreversibilities of cycles with feedwater heaters are smaller. With multiple feedwater heaters, the internal irreversibilities become smaller, since the temperature differences for heat exchange become smaller. Since irreversibilities imply work loss, plants with multiple feedwater heaters have higher thermal efficiency, due to smaller irreversibilities. While efficiency grows with the number of feedwater heaters, the increase slows down with the number of heaters. Above a certain number of heaters, the small increase of efficiency cannot offset the cost for construction and maintenance, and thus one will limit their number.

The design of a large power plant with multiple feedwater heaters requires optimization of the process to determine the number of feedwater heaters, and the optimal values for the pressures and mass flow rates of the bled-off flows. This multi-parameter optimization is done by means of computer programs.

Figure 12.7 shows a schematic of a 750MW power plant in Germany. The plant employs three turbine stages at high pressure (HP), intermediate pressure (MP) and low pressure (LP), which drive the generator. The power

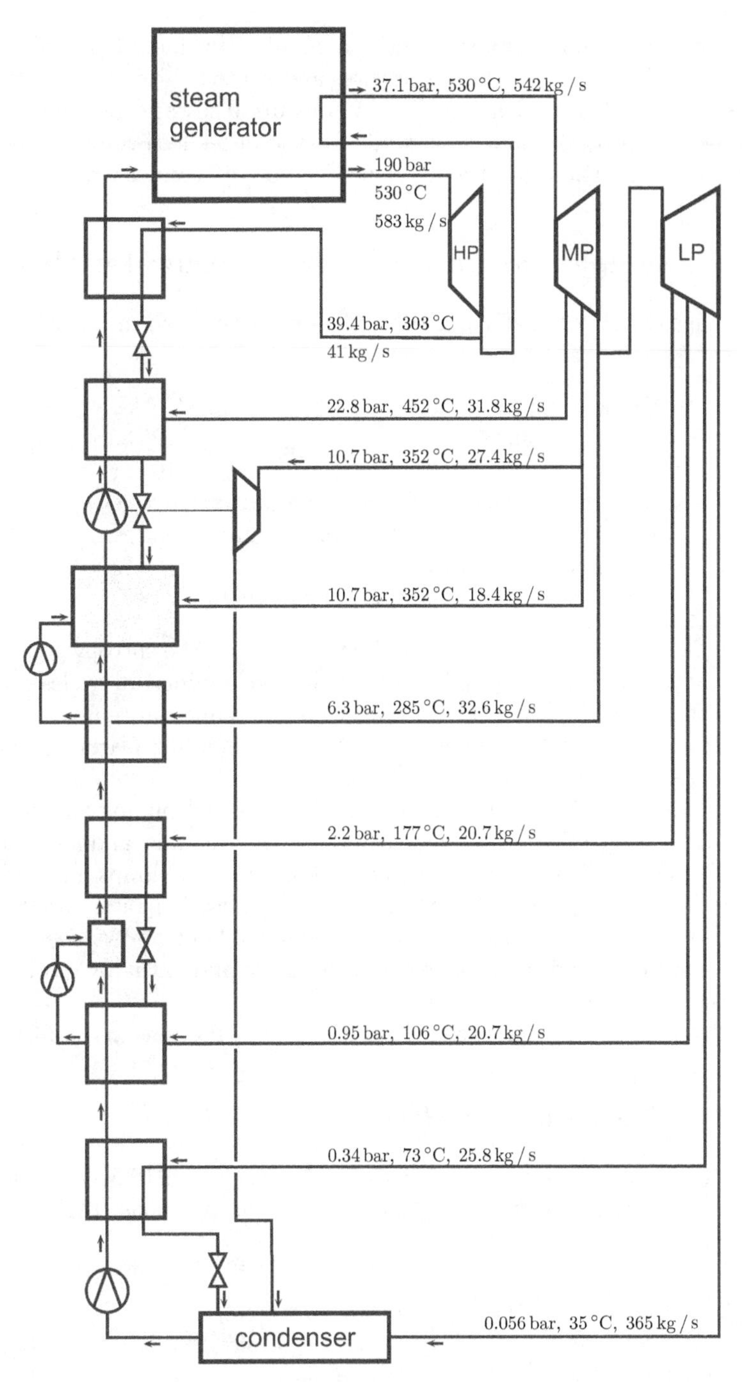

Fig. 12.7 Schematic of 750 MW power plant Bexbach, Germany. The plant has several feed water heaters, and three turbine stages. Note the extra turbine to drive the pump [simplified from H.D. Baehr: Thermodynamik, Springer 1996].

for the feedwater pumps is provided by a smaller turbine. Steam is bled-off from the turbines at a variety of pressures and routed through six feedwater heaters. The condensate leaving the low pressure feedwater heaters 2 and 4 is pumped up into the main flow, while for the other feedwater heaters the condensate is throttled into the lower heaters, or the condenser.

12.3 Example: Steam Cycles with Feedwater Heaters

We consider the thermal efficiencies for the cycles depicted in Figs. 12.3-12.6 with the following base data:

condenser pressure:	7.5 kPa
boiler pressure:	125 bar
reheat pressure:	10 bar
high pressure turbine inlet temperature:	500 °C
low pressure turbine inlet temperature:	500 °C
boiler temperature T_B:	900 K
standard environmental temperature T_0:	298 K

The boiler temperature will be required to estimate the entropy generation in the heat transfer between boiler and steam cycle. Since all cycles operate between the boiler temperature T_B and the environmental temperature T_0, their efficiencies must be compared to the corresponding Carnot efficiency $\eta_C = 1 - \frac{T_0}{T_B} = 0.669$.

To simplify the discussion, we shall assume that all pumps and turbines are reversible. Realistic pumps and turbines are irreversible, and add another source of irreversibility to the discussed cycles. Since all pumps and turbines are considered to be adiabatic, their entropy generation rates are $\dot{S}_{gen} = \dot{m}\,(s_{out} - s_{in})$, where the appropriate mass flow and entropy values must be used. The latter would be obtained from pressure and enthalpy data, based on the isentropic irreversibilities.

It is left to the reader to verify the data and calculations in detail.

12.3.1 No Feedwater Heater

We first examine the basic reheat cycle as shown in Fig. 12.3, for which we find the following property values (point 4 is in the two-phase region)

$$\begin{aligned}
h_1 &= h_f\,(7.5\,\text{kPa}) = 168.8\,\tfrac{\text{kJ}}{\text{kg}} &&, \quad s_1 = s_f\,(7.5\,\text{kPa}) = 0.5764\,\tfrac{\text{kJ}}{\text{kg K}}\,, \\
h_2 &= h\,(125\,\text{bar}, s_1) = 181.4\,\tfrac{\text{kJ}}{\text{kg}} &&, \quad s_2 = s_1\,, \\
h_3 &= h\,(125\,\text{bar}, 500\,^\circ\text{C}) = 3342\,\tfrac{\text{kJ}}{\text{kg}} &&, \quad s_3 = s\,(125\,\text{bar}, 500\,^\circ\text{C}) = 6.462\,\tfrac{\text{kJ}}{\text{kg K}}\,, \\
h_4 &= h\,(10\,\text{bar}, s_3) = 2722\,\tfrac{\text{kJ}}{\text{kg}} &&, \quad s_4 = s_3\,, \\
h_5 &= h\,(10\,\text{bar}, 500\,^\circ\text{C}) = 3479\,\tfrac{\text{kJ}}{\text{kg}} &&, \quad s_5 = s\,(10\,\text{bar}, 500\,^\circ\text{C}) = 7.762\,\tfrac{\text{kJ}}{\text{kg K}}\,, \\
h_6 &= h\,(7.5\,\text{kPa}, s_5) = 2421\,\tfrac{\text{kJ}}{\text{kg}} &&, \quad s_6 = s_5\,.
\end{aligned}$$

The work per unit mass is

$$\frac{\dot{W}}{\dot{m}} = h_1 - h_2 + h_3 - h_4 + h_5 - h_6 = 1665.4 \frac{\mathrm{kJ}}{\mathrm{kg}} ,$$

and the heat intake is

$$\frac{\dot{Q}_{in}}{\dot{m}} = h_3 - h_2 + h_5 - h_4 = 3917.6 \frac{\mathrm{kJ}}{\mathrm{kg}} ,$$

so that the thermal efficiency of this simple reheat cycle is

$$\eta_R = \frac{\dot{W}}{\dot{Q}_{in}} = 0.425 .$$

The entropy generation rates for the two heat transfer steps in the boiler follow from the second law as

$$\dot{S}_{gen,23} = \dot{m}\left(s_3 - s_2\right) - \frac{\dot{Q}_{23}}{T_B} = \dot{m}\left[s_3 - s_2 - \frac{h_3 - h_2}{T_B}\right] ,$$

$$\dot{S}_{gen,45} = \dot{m}\left(s_5 - s_4\right) - \frac{\dot{Q}_{45}}{T_B} = \dot{m}\left[s_5 - s_4 - \frac{h_5 - h_4}{T_B}\right] .$$

The entropy generation in the condenser due to heat transfer to the standard environment at T_0 is

$$\dot{S}_{gen,61} = \dot{m}\left(s_1 - s_6\right) - \frac{\dot{Q}_{61}}{T_0} = \dot{m}\left[s_1 - s_6 - \frac{h_1 - h_6}{T_0}\right] .$$

The total entropy generation for the system is

$$\dot{S}_{gen} = \dot{S}_{gen,23} + \dot{S}_{gen,45} + \dot{S}_{gen,61} ,$$

and from the data we find the entropy generation per unit as

$$\frac{\dot{S}_{gen}}{\dot{m}} = 3.20 \frac{\mathrm{kJ}}{\mathrm{kg\,K}} .$$

As a measure for the associated work loss, we consider the standard work loss $T_0\dot{S}_{gen}$ relative to the power output:

$$\frac{\dot{W}_{\mathrm{loss}}}{\dot{W}} = \frac{T_0\dot{S}_{gen}}{\dot{W}} = 57\% .$$

12.3.2 Open Feedwater Heater

We next examine the reheat cycle with open feedwater heater as shown in Fig. 12.4, for which we find

$$
\begin{aligned}
h_1 &= h_f\,(7.5\,\text{kPa}) = 168.8\tfrac{\text{kJ}}{\text{kg}} &&, \; s_1 = s_f\,(7.5\,\text{kPa}) = 0.5764\tfrac{\text{kJ}}{\text{kg K}}\;, \\
h_2 &= h\,(10\,\text{bar}, s_1) = 169.9\tfrac{\text{kJ}}{\text{kg}} &&, \; s_2 = s_1\;, \\
h_3 &= h_f\,(10\,\text{bar}) = 762.8\tfrac{\text{kJ}}{\text{kg}} &&, \; s_3 = s_f\,(10\,\text{bar}) = 2.139\tfrac{\text{kJ}}{\text{kg K}}\;, \\
h_4 &= h\,(125\,\text{bar}, s_3) = 775.8\tfrac{\text{kJ}}{\text{kg}} &&, \; s_4 = s_3\;, \\
h_5 &= h\,(125\,\text{bar}, 500\,{}^\circ\text{C}) = 3342\tfrac{\text{kJ}}{\text{kg}} &&, \; s_5 = s\,(10\,\text{bar}, 500\,{}^\circ\text{C}) = 6.462\tfrac{\text{kJ}}{\text{kg K}}\;, \\
h_6 &= h\,(10\,\text{bar}, s_5) = 2722\tfrac{\text{kJ}}{\text{kg}} &&, \; s_6 = s_5\;, \\
h_7 &= h\,(10\,\text{bar}, 500\,{}^\circ\text{C}) = 3479\tfrac{\text{kJ}}{\text{kg}} &&, \; s_7 = s\,(10\,\text{bar}, 500\,{}^\circ\text{C}) = 7.762\tfrac{\text{kJ}}{\text{kg K}}\;, \\
h_8 &= h\,(7.5\,\text{kPa}, s_5) = 2421\tfrac{\text{kJ}}{\text{kg}} &&, \; s_8 = s_7\;.
\end{aligned}
$$

The mass flow ratio is adjusted such that the mixed state (state 3) is saturated liquid. From applying the first law to the open feedwater heater, we find the mass flow ratio

$$
y = \frac{\dot{m}_B}{\dot{m}_A + \dot{m}_B} = \frac{h_3 - h_2}{h_6 - h_2} = 0.23\;.
$$

Net work and heat in per unit mass are

$$
\frac{\dot{W}}{\dot{m}_A + \dot{m}_B} = (1 - y)\,(h_1 - h_2 + h_7 - h_8) + h_3 - h_4 + h_5 - h_6 = 1418.4\frac{\text{kJ}}{\text{kg}}\;,
$$

$$
\frac{\dot{Q}_{in}}{\dot{m}_A + \dot{m}_B} = h_5 - h_4 + (1 - y)\,(h_7 - h_6) = 3147.3\frac{\text{kJ}}{\text{kg}}\;,
$$

so that the thermal efficiency of the cycle is

$$
\eta_{ofwh} = \frac{\dot{W}}{\dot{Q}_{in}} = 0.451\;.
$$

Compared to the standard reheat cycle, the single open feedwater heater improves thermal efficiency by 2.6%. The following calculation shows that this improvement is due to decreased irreversible losses.

The entropy generation rates for the two heat transfer steps in the boiler follow from the second law as

$$
\dot{S}_{gen,45} = (\dot{m}_A + \dot{m}_B)\left[s_5 - s_4 - \frac{h_5 - h_4}{T_B}\right]\;,
$$

$$
\dot{S}_{gen,67} = \dot{m}_A\left[s_7 - s_6 - \frac{h_7 - h_6}{T_B}\right]\;,
$$

and the entropy generation in the condenser is

$$\dot{S}_{gen,81} = \dot{m}_A \left[s_1 - s_8 - \frac{h_1 - h_8}{T_0} \right] .$$

Moreover, entropy is generated in the adiabatic feedwater heater through mixing:

$$\dot{S}_{gen,ofwh} = \dot{m}_A (s_3 - s_2) + \dot{m}_B (s_3 - s_6) .$$

The total entropy generation for the system is

$$\dot{S}_{gen} = \dot{S}_{gen,45} + \dot{S}_{gen,67} + \dot{S}_{gen,81} + \dot{S}_{gen,ofwh} ,$$

with a value per unit mass of

$$\frac{\dot{S}_{gen}}{\dot{m}_A + \dot{m}_B} = 2.3 \frac{\text{kJ}}{\text{kg K}} .$$

Now, the standard work loss $T_0 \dot{S}_{gen}$ relative to the power output is

$$\frac{\dot{W}_{\text{loss}}}{\dot{W}} = \frac{T_0 \dot{S}_{gen}}{\dot{W}} = 48.4\% .$$

12.3.3 Closed Feedwater Heater (with Pump)

Now we examine the reheat cycle with closed feedwater heater and pump as shown in Fig. 12.5, where

$$\begin{aligned}
h_1 &= h_f (7.5\,\text{kPa}) = 168.8 \tfrac{\text{kJ}}{\text{kg}} &, \quad s_1 &= s_f (7.5\,\text{kPa}) = 0.5764 \tfrac{\text{kJ}}{\text{kg K}} , \\
h_2 &= h (125\,\text{bar}, s_1) = 181.4 \tfrac{\text{kJ}}{\text{kg}} &, \quad s_2 &= s_1 , \\
h_3 &\simeq h_f (T_9) = 762.8 \tfrac{\text{kJ}}{\text{kg}} &, \quad s_3 &\simeq s_f (T_9) = 2.139 \tfrac{\text{kJ}}{\text{kg K}} , \\
h_4 &= 765.8 \tfrac{\text{kJ}}{\text{kg}} &, \quad s_4 &\simeq s_f (T_4) = 2.145 \tfrac{\text{kJ}}{\text{kg K}} , \\
h_5 &= h (125\,\text{bar}, 500\,^\circ\text{C}) = 3342 \tfrac{\text{kJ}}{\text{kg}} &, \quad s_5 &= s (125\,\text{bar}, 500\,^\circ\text{C}) = 6.462 \tfrac{\text{kJ}}{\text{kg K}} ,
\end{aligned}$$

$$\begin{aligned}
h_6 &= h (10\,\text{bar}, s_5) = 2722 \tfrac{\text{kJ}}{\text{kg}} &, \quad s_6 &= s_5 , \\
h_7 &= h (10\,\text{bar}, 500\,^\circ\text{C}) = 3479 \tfrac{\text{kJ}}{\text{kg}} &, \quad s_7 &= s (10\,\text{bar}, 500\,^\circ\text{C}) = 7.762 \tfrac{\text{kJ}}{\text{kg K}} , \\
h_8 &= h (7.5\,\text{kPa}, s_7) = 2421 \tfrac{\text{kJ}}{\text{kg}} &, \quad s_8 &= s_7 , \\
h_9 &= h_f (10\,\text{bar}) = 762.8 \tfrac{\text{kJ}}{\text{kg}} &, \quad s_9 &= s_f (10\,\text{bar}) = 2.139 \tfrac{\text{kJ}}{\text{kg K}} , \\
h_{10} &= h (125\,\text{bar}, s_9) = 775.8 \tfrac{\text{kJ}}{\text{kg}} &, \quad s_{10} &= s_9 .
\end{aligned}$$

State 3 is chosen by assuming perfect heat exchange in the feedwater heater, so that $T_3 = T_9 = T_{sat}(10\,\text{bar}) = 179.9\,^\circ\text{C}$. With that, the mass flow ratio is

$$y = \frac{\dot{m}_B}{\dot{m}_A + \dot{m}_B} = \frac{h_3 - h_2}{h_6 - h_9 + h_3 - h_2} = 0.229 .$$

The boiler feed state (state 4) has the enthalpy

$$h_4 = (1-y)\, h_3 + y h_{10} \,.$$

Net work and heat in per unit mass are

$$\frac{\dot{W}}{\dot{m}_A + \dot{m}_B} = (1-y)\,(h_1 - h_2 + h_7 - h_8) + y\,(h_9 - h_{10}) + (h_5 - h_6)$$

$$= 1423.2 \frac{\text{kJ}}{\text{kg}} \,,$$

$$\frac{\dot{Q}_{in}}{\dot{m}_A + \dot{m}_B} = h_5 - h_4 + (1-y)\,(h_7 - h_6) = 3160.0 \frac{\text{kJ}}{\text{kg}} \,,$$

so that the thermal efficiency of the cycle is

$$\eta_{cfwh\_p} = \frac{\dot{W}}{\dot{Q}_{in}} = 0.450 \,.$$

Compared to the standard reheat cycle, the single closed feedwater heater improves thermal efficiency by 2.5%, slightly below the cycle with open feedwater heater.

The entropy generation rates for the two heat transfer steps in the boiler follow from the second law as

$$\dot{S}_{gen,45} = (\dot{m}_A + \dot{m}_B) \left[s_5 - s_4 - \frac{h_5 - h_4}{T_B} \right] \,,$$

$$\dot{S}_{gen,67} = \dot{m}_A \left[s_7 - s_6 - \frac{h_7 - h_6}{T_B} \right] \,,$$

and the entropy generation in the condenser is

$$\dot{S}_{gen,81} = \dot{m}_A \left[s_1 - s_8 - \frac{h_1 - h_8}{T_0} \right] \,.$$

Moreover, entropy is generated in the feedwater heater and mixing chamber:

$$\dot{S}_{gen,cfwh} = \dot{m}_A\,(s_4 - s_2) + \dot{m}_B\,(s_4 - s_6) \,.$$

The total entropy generation for the system is

$$\dot{S}_{gen} = \dot{S}_{gen,45} + \dot{S}_{gen,67} + \dot{S}_{gen,81} + \dot{S}_{gen,cfwh}$$

and its value per unit mass is

$$\frac{\dot{S}_{gen}}{\dot{m}_A + \dot{m}_B} = 2.32 \frac{\text{kJ}}{\text{kg}\,\text{K}} \,.$$

As a measure for the associated work loss, we consider the standard work loss $T_0\dot{S}_{gen}$ relative to the power output:

$$\frac{\dot{W}_{loss}}{\dot{W}} = \frac{T_0\dot{S}_{gen}}{\dot{W}} = 48.5\% \,.$$

12.3.4 Closed Feedwater Heater (with Throttle)

Finally we examine the reheat cycle with closed feedwater heater and throttling into the condenser as shown in Fig. 12.6, for which we find

$$\begin{aligned}
h_1 &= h_f\,(7.5\,\text{kPa}) = 168.8\tfrac{\text{kJ}}{\text{kg}} &, \quad s_1 &= s_f\,(7.5\,\text{kPa}) = 0.5764\tfrac{\text{kJ}}{\text{kg K}}\,, \\
h_2 &= h\,(125\,\text{bar}, s_1) = 181.4\tfrac{\text{kJ}}{\text{kg}} &, \quad s_2 &= s_1\,, \\
h_3 &\simeq h_f\,(T_8) = 762.8\tfrac{\text{kJ}}{\text{kg}} &, \quad s_3 &\simeq s_f\,(T_8) = 2.139\tfrac{\text{kJ}}{\text{kg K}}\,, \\
h_4 &= h\,(125\,\text{bar}, 500\,°\text{C}) = 3342\tfrac{\text{kJ}}{\text{kg}} &, \quad s_4 &= s\,(125\,\text{bar}, 500\,°\text{C}) = 6.462\tfrac{\text{kJ}}{\text{kg K}}\,, \\
h_5 &= h\,(10\,\text{bar}, s_4) = 2722\tfrac{\text{kJ}}{\text{kg}} &, \quad s_5 &= s_4\,, \\
h_6 &= h\,(10\,\text{bar}, 500\,°\text{C}) = 3479\tfrac{\text{kJ}}{\text{kg}} &, \quad s_6 &= s\,(10\,\text{bar}, 500\,°\text{C}) = 7.762\tfrac{\text{kJ}}{\text{kg K}}\,, \\
h_7 &= h\,(7.5\,\text{kPa}, s_6) = 2421\tfrac{\text{kJ}}{\text{kg}} &, \quad s_7 &= s_6\,, \\
h_8 &= h_f\,(10\,\text{bar}) = 762.8\tfrac{\text{kJ}}{\text{kg}} &, \quad s_8 &= s_f\,(10\,\text{bar}) = 2.139\tfrac{\text{kJ}}{\text{kg K}}\,, \\
h_9 &= h_8 &, \quad s_9 &= s\,(7.5\,\text{kPa}, h_9) = 2.471\tfrac{\text{kJ}}{\text{kg K}}\,.
\end{aligned}$$

State 3 is chosen by assuming perfect heat exchange in the feedwater heater, so that $T_3 = T_8 = T_{sat}\,(10\,\text{bar}) = 179.9\,°\text{C}$. The energy balance for the feedwater heater now is

$$(\dot{m}_A + \dot{m}_B)\,(h_3 - h_2) = \dot{m}_B\,(h_5 - h_8)\,\,.$$

With that, the mass flow ratio is

$$y = \frac{\dot{m}_B}{\dot{m}_A + \dot{m}_B} = \frac{h_3 - h_2}{h_5 - h_8} = 0.297\,.$$

Net work and heat in per unit mass are

$$\begin{aligned}
\frac{\dot{W}}{\dot{m}_A + \dot{m}_B} &= h_1 - h_2 + h_4 - h_5 + (1-y)\,(h_6 - h_7) = 1351.4\frac{\text{kJ}}{\text{kg}}\,, \\
\frac{\dot{Q}_{in}}{\dot{m}_A + \dot{m}_B} &= h_4 - h_3 + (1-y)\,(h_6 - h_5) = 3111.6\frac{\text{kJ}}{\text{kg}}\,,
\end{aligned}$$

so that the thermal efficiency of the cycle is

$$\eta_{cwfh\_th} = \frac{\dot{W}}{\dot{Q}_{in}} = 43.4\% \,.$$

Compared to the standard reheat cycle, the single closed feedwater with throttling into the condenser improves thermal efficiency by only $\sim 1\%$. The improvement is significantly lower than for the case where the bled-off flow is pumped to boiler pressure, since a considerable amount of work is lost in the irreversible expansion through the throttling valve.

The entropy generation rates for the two heat transfer steps in the boiler follow from the second law as

$$\dot{S}_{gen,34} = (\dot{m}_A + \dot{m}_B)\left[s_4 - s_3 - \frac{h_4 - h_3}{T_B}\right] ,$$
$$\dot{S}_{gen,56} = \dot{m}_A\left[s_6 - s_5 - \frac{h_6 - h_5}{T_B}\right] ,$$

and the entropy generation in the condenser is

$$\dot{S}_{gen,cond} = \dot{m}_A\left[s_1 - s_7 - \frac{h_1 - h_7}{T_0}\right] + \dot{m}_B\left[s_1 - s_9 - \frac{h_1 - h_9}{T_0}\right] .$$

Moreover, entropy is generated in the feedwater heater and in throttling:

$$\dot{S}_{gen,cfwh} = (\dot{m}_A + \dot{m}_B)(s_3 - s_2) + \dot{m}_B(s_8 - s_5) ,$$
$$\dot{S}_{gen,th} = \dot{m}_B(s_9 - s_8) .$$

The total entropy generation is

$$\dot{S}_{gen} = \dot{S}_{gen,34} + \dot{S}_{gen,45} + \dot{S}_{gen,cond} + \dot{S}_{gen,cfwh} + \dot{S}_{gen,th}$$

and its value per unit mass is

$$\frac{\dot{S}_{gen}}{\dot{m}_A + \dot{m}_B} = 2.45\frac{\mathrm{kJ}}{\mathrm{kg\,K}} .$$

As a measure for the associated work loss, we consider the standard work loss $T_0\dot{S}_{gen}$ relative to the power output:

$$\frac{\dot{W}_{loss}}{\dot{W}} = \frac{T_0\dot{S}_{gen}}{\dot{W}} = 54.0\% .$$

12.3.5 Summary

The analysis of the four configurations, one without and three with feedwater heaters, shows that incorporation of feedwater heaters improves thermal efficiency of steam cycles. Due to the feedwater heaters, the average temperature for heat transfer between the working fluid and the boiler is smaller, hence there is less irreversibility for the boiler processes. All feedwater heaters are associated with irreversibilities through heat exchange over finite temperature difference, mixing, or throttling. The reduction of irreversibility in the boiler processes is larger than the additional irreversibility associated with the feedwater heaters, so that overall the irreversibility is reduced, and the process performance is improved. For this example, the improvement through open feedwater heater and closed feedwater heater with pump is similar. The system with open feedwater heater and throttling into the condenser yields a smaller improvement, due to the additional irreversibility in the throttling process; however, this system is cheaper to build and maintain. With more feedwater heaters employed, the temperature differences for heat transfer, and the pressure differences for throttling decrease, and thus irreversible losses are reduced.

Finally, it must be noted that losses occurring in irreversible pumps and turbines were ignored, to not overwhelm the computations with more detail.

12.4 Cogeneration Plants

12.4.1 Process Heat

Many industries require process heat and electrical power, for instance the chemical industry, pulp and paper plants, refineries, textile production, etc. Cogeneration power plant are variants of the Rankine cycle, which produce power and provide heat at the required temperature level. Adjustments in the process allow to provide variable amounts of process heat, depending on demand. If excess electrical power is produced, this can be sold to the grid, if more electrical power is needed, it must be purchased from the grid.

As an example we study a cogeneration plant that produces power and steam at intermediate pressures and temperatures (say, $5 - 7\,\mathrm{bar}$, $150 - 200\,^\circ\mathrm{C}$). Depending on the demand for process heat, steam can be routed into the process heater either directly from the steam generator (with throttling to lower the pressure), or it can be drawn out of the turbine at intermediate pressure, see Fig. 12.8.

In normal mode, all generated steam is supplied to the turbine, and only a portion of the steam is extracted from the turbine, that is

$$\dot{m}_6 = 0 \ , \ \dot{m}_5 \neq 0 \ , \ \dot{m}_4 \neq 0 \, .$$

If the demand for process heat is larger, all steam is taken from the turbine, so that

$$\dot{m}_6 = 0 \ , \ \dot{m}_5 \neq 0 \ , \ \dot{m}_4 = 0 \, .$$

This decreases the power generation. Only in case of extremely high demand one will bypass the turbine entirely, so that

$$\dot{m}_6 \neq 0 \ , \ \dot{m}_5 = 0 \ , \ \dot{m}_4 = 0 \, .$$

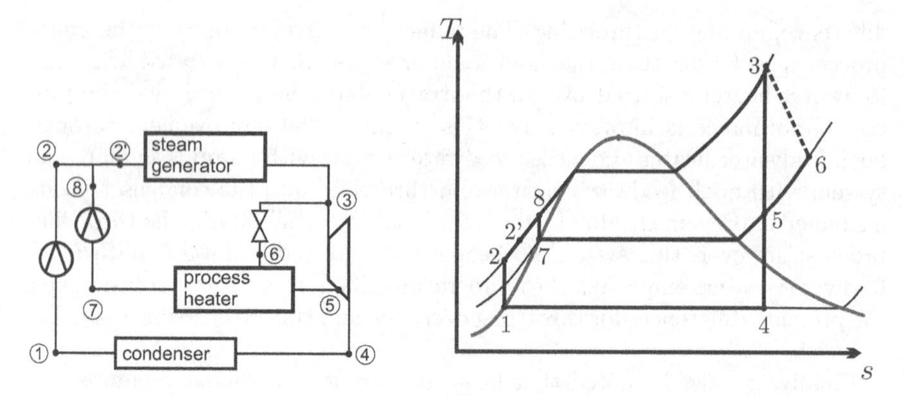

Fig. 12.8 Cogeneration power plant. The plant generates power and process heat. Depending on demand, steam for the process heater is supplied directly from the turbine feed flow (State 3), or extracted from the turbine (State 5).

In this case, only process heat is provided, no electrical power is generated.

From the diagram, we read off the pump and turbine work as

$$\dot{W}_P = \dot{m}_1 (h_1 - h_2) + (\dot{m}_5 + \dot{m}_6)(h_7 - h_8) \ ,$$
$$\dot{W}_T = \dot{m}_1 (h_3 - h_4) + \dot{m}_5 (h_3 - h_5) \ ,$$

and the net work of the plant is

$$\dot{W}_{net} = \dot{W}_P + \dot{W}_T \ .$$

The process heat is

$$\left|\dot{Q}_{proc}\right| = \dot{m}_5 (h_5 - h_7) + \dot{m}_6 (h_6 - h_7) \ ,$$

and the heat addition in the steam generator is

$$\dot{Q}_{in} = \dot{m}_1 (h_3 - h_2) + (\dot{m}_5 + \dot{m}_6)(h_3 - h_8) = (\dot{m}_1 + \dot{m}_5 + \dot{m}_6)(h_3 - h_{2'}) \ .$$

The so-called utilization factor of the plant is

$$\varepsilon_u = \frac{\dot{W}_{net} + \left|\dot{Q}_{proc}\right|}{\dot{Q}_{in}} \ .$$

Note that the utilization factor favors use of process heat: For the case where all heat is used in the process ($\dot{m}_5 = \dot{m}_4 = 0$), the utilization factor is unity. However, for the case where no process heat is used ($\dot{m}_5 = \dot{m}_6 = 0$), the utilization factor is equal to the thermal efficiency of the plant. Hence, as with all efficiency measures, one has to be careful in the interpretation of the utilization factor.

12.4.2 District Heating

Since heat cannot be completely converted to work, a power plant has to reject heat, typically to the environment. To maximize the work output, the temperature for heat rejection should be as close as possible to the local environmental temperature T_0. Heat transferred at environmental temperature has no further use, and thus has no value. For space heating in winter, one needs heat at somewhat elevated temperature. This heat can be provided from the heat rejection of power plants in district heating systems. For this, at times when space heating is required, the plant is operated at elevated condenser temperature. The condenser heat is rejected into a water circuit that then is used to bring the heat into buildings. In summer, when no heating is required, the condenser operates at lower temperatures, and the heat is rejected directly to the environment. District heating works best when the distance for heat transmission is short, so that little heat is lost to the environment in transmission. Hence, for combined heat and power plants, its is best to build smaller local power stations, which provide power and heat for the closer neighborhood, rather than large plants far from consumers. Utilization factors for combined heat and power plants can be defined similar to those for process heating.

12.5 Refrigeration Systems

We return to the discussion of vapor refrigeration and heat pump cycles. The basic cycle, consisting of compressor, condenser, throttle, and evaporator was discussed in 10.3. Refrigerators and heat pumps draw heat from a cold environment (T_L) and reject heat into a warm environment (T_H). This transfer of heat from cold to warm requires work, which in vapor and gas refrigeration systems is supplied to the compressor. Efficient refrigeration and heat pump cycles requires: (a) small temperature differences between the working fluid and the respective environments, and (b) efficient compressors.

Typically, compressors are adiabatic, which leads to relatively high temperature of the compressed vapor, and therefore to large temperature differences between the working gas and the warm environment. This implies large entropy generation in heat transfer, hence more work is required to overcome irreversibilities. Intercooling during compression lowers the work requirement for compressors (see also Sec. 13.3). Intercooling also leads to lower temperatures after compression, which reduces the work loss to irreversibilities. There are several ways to incorporate intercooling into refrigeration cycles, which we now discuss.

Multi-stage refrigeration systems are particularly useful when one has to bridge a large temperature difference between the cold and warm environment, as, e.g., in the production of frozen food. Heat pumps normally operate on smaller temperature differences, and multi-stage compression is not used.

12.5.1 Cascade Refrigeration System

In a cascade refrigeration system two or more refrigeration cycles operate on top of each other. Figure 12.9 shows an example with two cycles A and B connected by a closed heat exchanger, which serves as evaporator for cycle A, and as condenser for cycle B.

With the closed heat exchanger, the working fluids of the two cycles remain separated. Thus, the two cycles can employ different cooling fluids, and operate at different pressures; cooling fluids can be chosen with the best temperature-pressure characteristics in the respective temperature ranges. For simplicity, the T-s-diagram is drawn for the case that both cycles use the same working fluid, and for the case of reversible compressors.

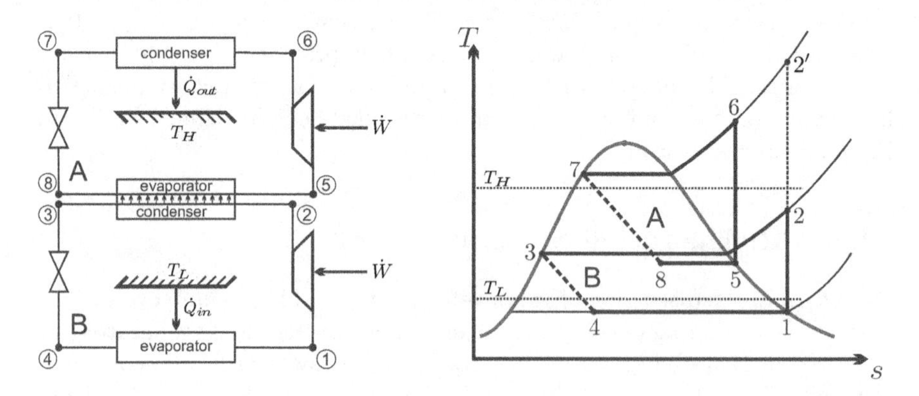

Fig. 12.9 Schematic and T-s-diagram for a cascade refrigeration system: A closed heat exchanger between two cycles serves as condensator for the low pressure cycle and as evaporator for the high pressure cycle

The energy balance for the heat exchanger relates the mass flows of both cycles,

$$\dot{m}_A (h_5 - h_8) = \dot{m}_B (h_2 - h_3) \implies \frac{\dot{m}_A}{\dot{m}_B} = \frac{h_2 - h_3}{h_5 - h_8} \,. \tag{12.18}$$

The cooling power of the cycle is

$$\dot{Q}_{in} = \dot{m}_B (h_1 - h_4) \,, \tag{12.19}$$

and the two compressors consume the power

$$\dot{W} = \dot{m}_B (h_1 - h_2) + \dot{m}_A (h_5 - h_6) \ , \tag{12.20}$$

so that the cycle has the coefficient of performance

$$\mathrm{COP_R} = \frac{\dot{Q}_{in}}{\left|\dot{W}\right|} = \frac{h_1 - h_4}{h_2 - h_1 + \frac{h_2 - h_3}{h_5 - h_8} (h_6 - h_5)} \ . \tag{12.21}$$

The T-s-diagram in Fig. 12.9 also indicates state 2′, which would be the compressor exit state in a single stage system. The corresponding temperature $T_{2'}$ is significantly above the temperature T_H of the heat receiving environment. The two stage system reaches the maximum temperature T_6, which is not as high, and has smaller temperature differences in all heat transfer processes, and thus lower irreversibility, and better COP.

12.5.2 Refrigeration with Flash Chamber

If only a single refrigerant is to be used, there are alternatives to design multi-stage refrigeration cycles that do not need closed heat exchangers, and use simpler open heat exchangers. Figure 12.10 shows a two-stage refrigeration system which uses a flash chamber at intermediate pressure to divide the exit flow of the upper throttle into saturated liquid and saturated vapor. The saturated liquid (state 7) is throttled further into the evaporator. The saturated vapor (state 9) is mixed with the vapor leaving the low pressure compressor (state 2); the mixture (state 3) is fed into the high pressure compressor. Effectively, this cycle compresses with intercooling to temperature T_3.

The mass flow $\dot{m}_A$ goes through the cycle 1-2-3-4-5-6-7-8-1, and the mass flow $\dot{m}_B$ goes through the cycle 3-4-5-6-3. Both mass flows are related through the quality x_6 of state 6,

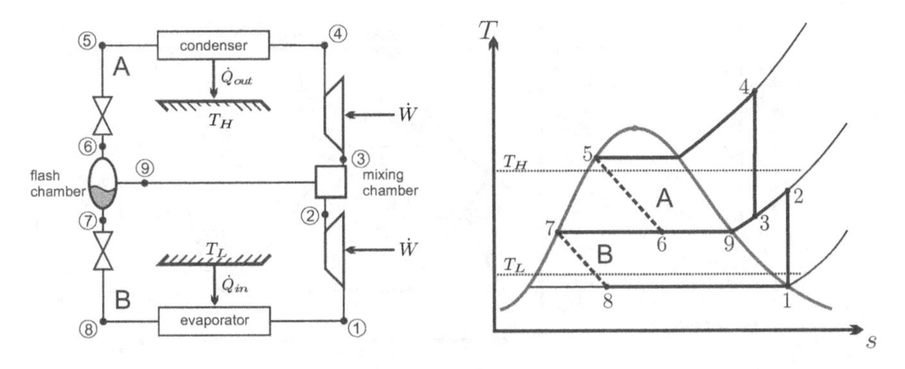

Fig. 12.10 Two-stage refrigeration cycle with flash chamber: Schematic and T-s-diagram

$$\dot{m}_A = (1 - x_6)(\dot{m}_A + \dot{m}_B) \quad , \quad \dot{m}_B = x_6(\dot{m}_A + \dot{m}_B) \ . \tag{12.22}$$

Cooling power and compressor work can be read from the diagram as

$$\dot{Q}_{in} = \dot{m}_A (h_1 - h_8) \ , \tag{12.23}$$

$$\dot{W} = \dot{m}_A (h_1 - h_2) + (\dot{m}_A + \dot{m}_B)(h_3 - h_4) \ , \tag{12.24}$$

and the coefficient of performance is

$$\mathrm{COP_R} = \frac{\dot{Q}_{in}}{\left|\dot{W}\right|} = \frac{h_1 - h_8}{h_2 - h_1 + \frac{h_4 - h_3}{1 - x_6}} \ . \tag{12.25}$$

12.6 Linde Method for Gas Liquefaction

Liquefaction of air, and other gases, provides low temperatures, and offers a method to separate gas mixtures by distillation. For instance, steel mills have their own air liquefaction plants to separate oxygen from air. The oxygen is needed to remove carbon from the pig iron that leaves the blast furnace.

The Linde method is a classical approach to achieve this goal, by throttling pressurized gas. For gases the critical point is relatively low, e.g., for air $T_{cr} = 132.5\,\mathrm{K}$, $p_{cr} = 37.7\,\mathrm{bar}$, and thus the process must include a pre-cooling stage.

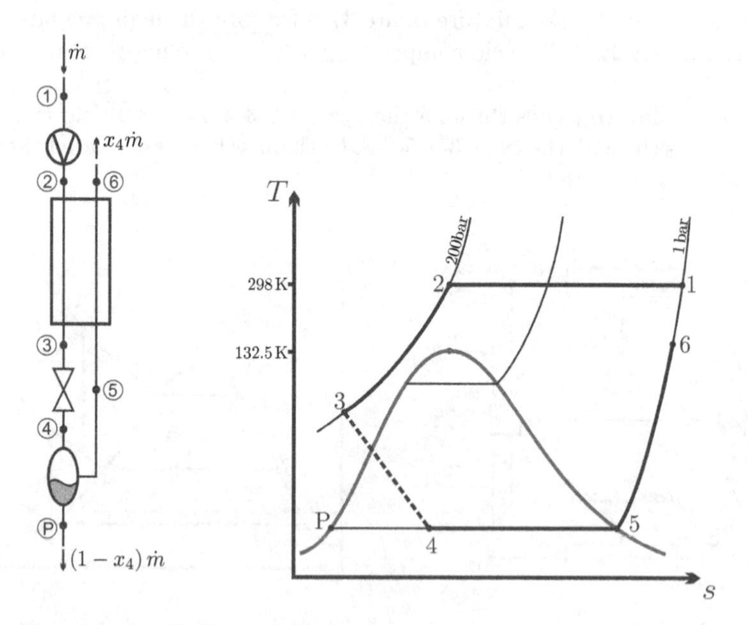

Fig. 12.11 Linde process for gas liquefaction

For air liquefaction, the process runs as shown in Fig. 12.11: Air at environmental conditions ($T_1 = 298\,\mathrm{K}$, $p_1 = 1\,\mathrm{bar}$) is compressed in a multi-stage compressor with intercooling to a high pressure $p_2 \simeq 200\,\mathrm{bar}$. Intercooling is necessary to reduce the work for compression as much as possible; for simplicity the T-s-diagram in the figure shows isothermal compression at $T_2 = T_1$. Next the compressed air is pre-cooled isobarically in the regenerator to a temperature T_3, and then throttled to p_1. The temperature T_3 must be sufficiently low, so that state 4 after throttling is in the two-phase region. In a flash chamber the flow is divided into saturated liquid—the desired product P—and saturated vapor (state 5) which is routed through the pre-cooler. Obviously, the conditions for this process are such that air cannot be described as an ideal gas; proper tables for air as superheated vapor and saturated liquid-vapor mix are required.

Due to the use of throttling, the Linde process is inherently irreversible, and thus requires more work than a reversible process for the same task. As we have seen, all refrigeration systems suffer from irreversibilities, and there is little alternative to this process for large scale gas liquefaction.

Problems

12.1. Regenerative Boiler

The boiler for a power plant is fitted with a regenerator of effectiveness 81.2% to preheat the incoming air before it is heated by burning of coal. Specifically, the system draws environmental air at $T_0 = 7\,^\circ\mathrm{C}$, the flame temperature is $827\,^\circ\mathrm{C}$, and the boiler exhaust is at $667\,^\circ\mathrm{C}$.

1. Determine the preheat temperature, the final exhaust temperature, the work potential of the boiler exhaust, and the work potential of the final exhaust.
2. Determine the entropy generation in the regenerator, and the corresponding work loss.

12.2. Mixing Chamber

Steam at 100 bar, $600\,^\circ\mathrm{C}$ is throttled to 10 bar, and fed into an adiabatic mixing chamber where it is mixed with compressed liquid water at 10 bar, $50\,^\circ\mathrm{C}$. The exiting mass flow is $100\frac{\mathrm{kg}}{\mathrm{s}}$ of saturated liquid at 10 bar. For steady state operation, determine

1. The mass flows of steam and liquid that enter.
2. The rate of entropy generation due to throttling, and the rate of entropy generation due to mixing.
3. The associated work loss.

As always: draw a sketch and a T-s-diagram.

12.3. Closed Feedwater Heater
In a closed feedwater heater, a mass flow of 200t/h of compressed liquid water at 80 bar, 40 °C is heated by heat exchange with a stream of saturated liquid-vapor mix at $x = 0.95$, $p = 10$ bar.

For the case that both streams leave the heat exchanger in saturated liquid state determine the mass flow of liquid-vapor mix, the entropy generation rate, and the work loss.

12.4. Steam Power Plant with Regeneration and Reheat
A steam power plant operates on a reheat-regenerative Rankine cycle with an open feedwater heater. Steam enters the high-pressure turbine at 100 bar, 550 °C, and leaves at a pressure of 8 bar as saturated vapor. Some steam is extracted at this pressure to heat the feedwater in an open feedwater heater which provides saturated liquid. The rest of the steam is reheated to 500 °C and then expanded in the low pressure turbine to the condenser pressure of 10 kPa.

The isentropic efficiency of the low pressure turbine is 0.95; all pumps can be considered to operate reversibly.

1. Draw a schematic and a T-s-diagram of the process, numerate corresponding points in schematic and diagram, and name the different devices (pump, turbine, etc.)
2. Make a list with the values of enthalpy at all relevant points of the process.
3. Compute the ratio of mass flow diverted to the feedwater heater after the first turbine.
4. Compute the thermal efficiency of the cycle.
5. The net power output of the plant is 100 MW. Determine the mass flow through the high pressure turbine.
6. Determine total entropy generation rate and work loss of the cycle.

12.5. Vapor Power Plant with Regeneration
A power plant operates on a regenerative vapor power cycle with one closed feedwater heater according to the following process:

1-2: Isentropic compression of saturated water from condenser pressure 0.04 bar to 60 bar.

2-3: Isobaric heating in the closed feedwater heater to 141.3 °C.

3-4: Isobaric heating in the steam generator to 60 bar, 550 °C.

4-5: Isentropic expansion into the condenser.

Some steam is extracted from the turbine at 4 bar to heat feedwater in a closed feedwater heater. This part of the steam undergoes the following two processes:

6-7: Isobaric cooling and condensation at 4 bar of diverted steam to saturated liquid state.

7-8: Throttling of condensate exiting the feedwater heater into the condenser.

1. Draw a schematic and a T-s-diagram of the process.
2. Make a list of the values of enthalpies at the points 1 to 8.
3. Compute the percentage of mass flow diverted into the feedwater heater at point 6.
4. Determine the thermal efficiency of the cycle.
5. Determine the mass flow rate into the turbine, if the net power developed is 320 MW.
6. Determine total entropy generation rate and work loss of the cycle.

12.6. Steam Power Plant with Two Feedwater Heaters, One Open, One Closed

Consider an ideal steam regenerative Rankine cycle with one open and one closed feedwater heater. Steam enters the turbine at 12.5 MPa, 550 °C, the condenser pressure is 10 kPa. Steam for the closed feedwater heater is extracted from the turbine at 0.8 MPa and for the open feedwater heater at 0.3 MPa. The feedwater is heated to the condensation temperature of the stream for the closed feedwater heater. The extracted steam leaves the closed feedwater heater at saturated state and is throttled into the open feedwater heater.

1. Draw a schematic of the process, and the corresponding T-s-diagram.
2. For a power output of 250 MW determine the mass flow rate through the steam generator, and the mass flows into the feedwater heaters.
3. Determine the thermal efficiency of the cycle.
4. Determine the entropy generation in the throttle, and estimate the corresponding work loss.

12.7. Steam Power Plant with Reheat and Two Feedwater Heaters, One Closed, One Open

The boiler pressure in a reheat steam power plant is 150 bar, the reheat pressure is 14 bar, and the condenser pressure is 10 kPa. For both turbines, the inlet temperature is 500 °C. After the high pressure turbine, some steam is bled-off and routed to the closed feedwater heater where it is fully condensed, and then pumped into the boiler feedwater. The remaining steam is reheated, and then runs through the low pressure turbine. Part of the flow is bled-off from the turbine at a pressure of 4 bar, while the main flow expands into the condenser. The diverted flow is mixed in the open feedwater heater with the flow that is pumped in from the condenser. The resulting mixture, which is in the saturated liquid state, is then pumped to boiler pressure before it enters the closed feedwater heater.

1. Draw a schematic of the process. Use the following numbering of processes: 1-2: Low pressure feedwater pump (from condenser). 3-4: Second feedwater

pump (after open feedwater heater). 6-7: High pressure steam generator. 7-8: High pressure turbine. 8-9: Reheat. 9-11: Low pressure turbine. 10: Bled-off for open feedwater heater. 11-1: Condenser. 12-13: Third feedwater pump.
2. Draw the corresponding T-s-diagram. Use different colors (or different line styles) to show the process curves for the main flow and the two bled-off flows.
3. Determine the enthalpies at all relevant states, based on the following assumptions: Reversible pumps and turbines, exit of open feedwater heater is saturated liquid (state 3), perfect heat exchange in closed feedwater heater, so that $T_5 = T_{12}$.
4. Determine the thermal efficiency of the plant.
5. Determine the three mass flows when the plant delivers a power of 500 MW.
6. Determine the overall entropy generation of the system, and the work loss to irreversibilities.
7. Determine the thermal efficiency of a standard reheat plant with the same pressures and turbines. Explain why the feedwater heaters improve efficiency.

12.8. Steam Power Plant with Reheat and Two Feedwater Heaters, One Closed, One Open

Repeat the previous problem, now considering irreversible pumps (isentropic efficiency $\eta_P = 0.85$) and turbines (isentropic efficiency $\eta_T = 0.92$).

12.9. Steam Power Plant with Reheat and Two Feedwater Heaters, One Open, One Closed

A reheat steam power plant has one closed feedwater heater (c.f.w.h.) and one open feedwater heater (o.f.w.h.). The boiler pressure is 150 bar, the reheat pressure is 15 bar, and the condenser pressure is 10 kPa. For both turbines, the inlet temperature is 500 °C. After the high pressure turbine, some steam is bled-off and routed to the o.f.w.h. The remaining steam is reheated, and then runs through the low pressure turbine. Part of the flow is bled-off from the turbine at a pressure of 5 bar. This flow is further routed through the c.f.w.h., where it fully condenses, and is then pumped into the o.f.w.h. The main flow expands into the condenser, which it leaves as saturated liquid. This flow is pumped to the o.f.w.h. pressure, heated in the c.f.w.h., and then mixed with the other flows in the o.f.w.h. The resulting mixture in the o.f.w.h., which is in the saturated liquid state, is then pumped to boiler pressure.

1. Draw a schematic of the process. Use the following numbering of processes: 1-2: Low pressure feedwater pump (from condenser). 2-3 and 9-11: Closed f.w.h. 4-5: Second feedwater pump (after o.f.w.h.). 5-6: High pressure steam generator. 6-7: High pressure turbine. 7-8: Reheat. 8-10: Low pressure turbine. 9: Bled-off for closed f.w.h. 10-1: Condenser. 11-12: Third feedwater pump.

2. Draw the corresponding T-s-diagram. Use different colors (or different line styles) to show the process curves for the main flow and the two bled-off flows.
3. Determine the enthalpies at all relevant states, based on the following assumptions: Reversible pumps and turbines, exit of o.f.w.h. is saturated liquid (state 4), perfect heat exchange in closed feedwater heater, so that $T_3 = T_{11}$.
4. Determine the relative amounts of the relevant mass flows.
5. Determine the thermal efficiency of the plant.
6. Determine the three mass flows when the plant delivers a power of 500 MW.
7. Determine the overall entropy generation of the system, and the work loss to irreversibilities.
8. Determine the thermal efficiency of a standard reheat plant with the same pressures and turbines. Explain why the feedwater heaters improve efficiency.

12.10. Steam Power Plant with Reheat and Two Feedwater Heaters, One Open, One Closed
Repeat the previous problem, now considering irreversible pumps (isentropic efficiency $\eta_P = 0.85$) and turbines (isentropic efficiency $\eta_T = 0.92$).

12.11. Cogeneration Power Plant
A cogeneration power plant with reheat produces 3 MW of power and supplies 7 MW of process heat. Steam enters the isentropic high-pressure turbine at 8 MPa and 500 °C and expands to a pressure of 1 MPa. At this pressure, part of the steam is extracted from the turbine and routed to the process heater; this stream leaves the process heater as compressed liquid at 120 °C. The remaining steam is reheated to 500 °C and then expanded in the isentropic low-pressure turbine to the condenser pressure of 15 kPa. The condensate is pumped to 1 MPa and then mixed with the stream of compressed liquid that comes from the process heater. The mixture is then pumped to the boiler pressure.

1. Make a schematic of the cycle, and draw the corresponding T-s-diagram.
2. Determine the heat input, the relative amount of steam running through the process heater, and the utilization factor.

12.12. Cogeneration Steam Power Plant with Regeneration
A small power plant that produces 30 MW of power operates on a regenerative vapor power cycle with one closed feedwater heater according to the following process:

Steam of 125 bar, 550 °C (state 1) enters the high pressure turbine where it is expanded isentropically to 10 bar (state 2). 50% of this steam are reheated to 500 °C (state 3) and then expanded in the low pressure turbine to the condenser pressure 0.075 bar (state 4). After condensation to saturated liquid state (state 5) this stream is pumped isentropically to 10 bar (state 6) and routed into the open feedwater heater. Part of the steam extracted after the high pressure turbine (state 2) is used for process heating. For this, the steam passes through a heat exchanger which it leaves as compressed liquid at 60 °C (state 7) that is fed into the open feedwater heater. The rest of the extracted steam of state 2 is directly routed into the feedwater heater. The water leaving the feedwater heater is in the saturated liquid state (state 8); an isentropic pump increases its pressure to the boiler pressure (state 9).

1. Draw a schematic and a T-s-diagram of the process.
2. Make a table with enthalpies at the relevant states of the process.
3. Determine the mass flows through the boiler and the process heater.
4. Determine the utilization factor of the plant.

12.13. District Heating
A 40 MW power plant is to be build to supply electrical power to a small town in the North where, due to the low average temperature, a large amount of space heating is required. One proposal suggest to set the condenser pressure a bit higher, so that the condenser is at temperature T_{C1}, and to use the removed heat for district heating. An alternative proposal suggests to set the condenser to the lower temperature T_{C2} so that the turbine delivers more work, which can then be used to run heat pumps between T_{C1} and T_{C2}. Discuss these proposals and make a recommendation to town council. Use thermodynamic arguments (of course!), it might be helpful to draw pictures with energy flows and temperatures.

12.14. Standard Vapor Cooling Cycle with Ammonia
A standard vapor refrigeration cycle operates with ammonia as cooling fluid. The maximum and minimum pressures reached are 1.5 atm and 10 atm, respectively. The adiabatic compressor draws saturated vapor, and has an isentropic efficiency of 0.9. The ammonia vapor leaving the compressor is cooled, condensed and further cooled to 20 °C before it enters the throttling device.

Draw the process into the log p-h diagram for ammonia, and find the enthalpies and temperatures at all principal points. For a cooling power of 2 kW, determine the power consumption and the COP.

12.15. Two-Stage Refrigeration Cycle

1. Draw a schematic, and the corresponding T-s-diagram for a two-stage refrigeration cycle with an open heat exchanger.
2. Indicate all principal points in both diagrams by numbers, and indicate the different elements by name (compressor, throttle, etc.).
3. Compute the mass flow ratio between upper and lower cycle in terms of enthalpies.
4. Give the expression for the COP of the system in terms of enthalpies and mass flow ratio.

Note: The next three problems compare cooling cycles running between the same upper and lower pressures. There is some data overlap, and to simplify proceedings an irreversible compressor is considered only in the first cycle. For all three, start with drawing a sketch, and the T-s-diagram

12.16. Standard Vapor Cooling Cycle with R134a

A standard vapor refrigeration cycles operates with R134a as cooling fluid between the pressures 1.2 MPa and 0.1 MPa, respectively. The adiabatic compressor draws saturated vapor, and has an isentropic efficiency of 0.9. The vapor leaving the compressor is cooled and fully condensed before it enters the throttling device. Draw a schematic and the T-s-diagram, and then determine:

1. The COP for the cycle with irreversible and with reversible compressor.
2. The mass flow rate and the power consumption for a cooling power of 200 kW.
3. Determine entropy generation rates for each process, the overall entropy generation rate, and work loss of the cycle. Assume $T_H = 30\,°\mathrm{C}$ and $T_L = -20\,°\mathrm{C}$.

12.17. Two-Stage Refrigeration Cycle with Flash Chamber

A two-stage compression refrigeration system operates with R134a between the pressures 1.2 MPa and 0.1 MPa. The refrigerant leaves the condenser as saturated liquid and is throttled to a flash chamber operating at 0.4 MPa. The refrigerant leaving the low-pressure compressor at 0.4 MPa is also routed to the flash chamber.

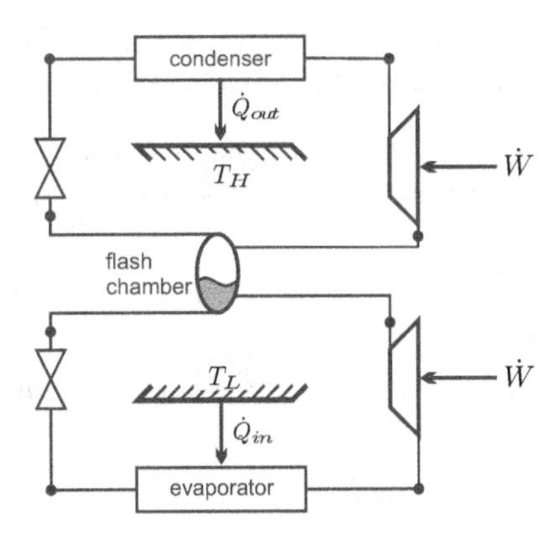

The saturated vapor leaving the flash chamber is compressed to the condenser pressure by the high-pressure compressor, while the saturated liquid leaving the flash chamber is throttled to the evaporator pressure. The refrigerant leaves the evaporator as saturated vapor and both compressors are isentropic. Draw a schematic and the T-s-diagram, and then determine:

1. The fraction of mass flows running through the two compressors.
2. The COP, and compare to that of the previous problem.
3. The two mass flow rates and the power consumption for a cooling power of 200 kW.
4. Determine entropy generation rates for each process, the overall entropy generation rate, and work loss of the cycle. Assume $T_H = 30\,°\mathrm{C}$ and $T_L = -20\,°\mathrm{C}$.

12.18. Two-Stage Refrigeration Cycle with Heat Exchanger
A two-stage cascade refrigeration system operates with R134a between the pressures 1.2 MPa and 0.1 MPa. Heat exchange between the two cycles takes place in an adiabatic counter-flow heat exchanger where the pressures are 0.32 and 0.4 MPa, respectively. In both cycles, the refrigerant is in saturated liquid state at the condenser exit and at saturated vapor state at the compressor inlet. Draw a schematic and the T-s-diagram, and then determine:

1. The fraction of mass flows running through the two compressors.
2. The coefficient of performance.
3. The two mass flow rates, and the power consumption for a cooling power of 200 kW.
4. The total entropy generation rate and work loss of the cycle. Assume $T_H = 30\,°\mathrm{C}$ and $T_L = -20\,°\mathrm{C}$.

12.19. Two-Stage Refrigeration Cycles
Repeat the previous two problems for the case where the compressors have an isentropic efficiency of 0.85.

12.20. Refrigeration Cycle with Intercooling
A vapor-compression refrigeration cycle operates at steady state with ammonia as working fluid according to the following cycle:

1-2: Adiabatic irreversible compression of saturated vapor at $p_1 = 1.75\,\mathrm{bar}$ to $p_2 = 5\,\mathrm{bar}$, isentropic compressor efficieny is $\eta_C = 0.8$.

2-3: Isobaric cooling to $20\,°\mathrm{C}$.

3-4: Adiabatic irreversible compression to $p_4 = 12\,\mathrm{bar}$, isentropic compressor efficieny is $\eta_C = 0.8$.

4-5: Isobaric heat rejection in condenser; state 5 is saturated liquid.

5-6: Throttling into the evaporator, $p_6 = p_1$.

6-1: Isobaric evaporation to state 1.

1. Draw a schematic and plot the process in a T-s-diagram.
2. Find the enthalpies at points 1-6.
3. Determine the coefficient of performance.

12.21. Advanced Cooling Cycle
In a hot climate, a two-stage cascade refrigeration system operates with refrigerant R134a. The evaporator temperature of the low pressure stage is $-20\,°\mathrm{C}$,

and the condenser temperature of the high pressure stage is 50 °C. Heat exchange between the cycles takes place in a counter-flow heat exchanger where the pressures are 0.4 and 0.5 MPa, respectively. In both cycles, the refrigerant is in saturated liquid state at the condenser exit, and in saturated vapor state at the compressor inlet. The isentropic efficiency of both compressors is 0.8.

1. Draw a T-s-diagram of the cycle with respect to saturation lines.
2. Make a list of the enthalpies and entropies at states 1 through 8.
3. Determine the ratio of mass flows entering the upper and the lower compressor.
4. Determine the COP of the cycle, and the power requirement for a cooling power of 120 kW.
5. Determine entropy generation rates for all processes, the overall entropy generation rate, and the work loss of the cycle. Assume $T_H = 40\,°C$ and $T_L = -10\,°C$.

12.22. Regenerative Gas Cooling System

A regenerative gas refrigeration cycle uses helium as working fluid. The helium enters the compressor at 100 kPa and −10 °C and is compressed to 300 kPa. Then, it is cooled to 20 °C by heat exchange with a cooling water flow. Next, the helium enters the regenerator where it is cooled further before it enters the turbine. Helium leaves the refrigerated space at −25 °C and enters the regenerator. Assume isentropic operation of turbine and regenerator, and determine

1. The temperature at the turbine inlet.
2. The COP of the cycle.
3. The net power input required for a mass flow rate of $0.45\,\frac{\text{kg}}{\text{s}}$.

Helium behaves as an ideal gas; it is monatomic, and thus has constant specific heat, $c_p = \frac{5}{2}R$.

Chapter 13
Gas Engines

13.1 Stirling Cycle

13.1.1 The Ideal Stirling Cycle

We have pointed out again and again that effective energy conversion requires the reduction of internal and external irreversibilities as much as possible. An important cause of external irreversibility is heat transfer between the system and its heat sources and sinks, e.g., a stream of combustion gas, or the environment. The related loss can be reduced if heat that is rejected in one process of a cycle can be added elsewhere within the system. This simultaneously reduces the heat rejection and the heat supply from the exterior, thus leading to efficiency improvements. Indeed, *regeneration*, i.e., internal exchange of heat within a system, is the most important tool to reduce external irreversibilities, and increase efficiency. Historically, the first engine which used a regenerator was the Stirling engine.

The idealized Stirling cycle consists of two isothermal and two isochoric processes, taking place at temperatures T_H and T_L, and volumes V_1, V_2, respectively. The working medium is an ideal gas, e.g., air or helium, which is confined permanently in a cylinder. The T-S- and p-V-diagrams are depicted in Fig. 13.1.

We ignore all internal irreversibilities, so that work and heat for the reversible processes within the ideal Stirling cycle are

$$\begin{aligned}
&\text{1-2 isothermal: } w_{12} = RT_H \ln \tfrac{V_2}{V_1} > 0 \,, && q_{12} = RT_H \ln \tfrac{V_2}{V_1} > 0 \,, \\
&\text{2-3 isochoric: } w_{23} = 0 && , \; q_{23} = u\left(T_L\right) - u\left(T_H\right) < 0 \,, \\
&\text{3-4 isothermal: } w_{34} = RT_L \ln \tfrac{V_1}{V_2} < 0 \,, && q_{34} = RT_L \ln \tfrac{V_1}{V_2} < 0 \,, \\
&\text{4-1 isochoric: } w_{41} = 0 && , \; q_{41} = u\left(T_H\right) - u\left(T_L\right) < 0 \,.
\end{aligned} \tag{13.1}$$

H. Struchtrup, *Thermodynamics and Energy Conversion*,
DOI: 10.1007/978-3-662-43715-5_13,

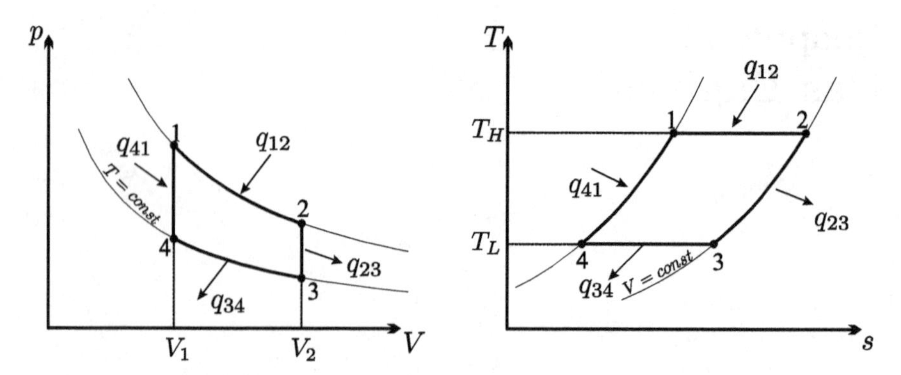

Fig. 13.1 Idealized Stirling power cycle in p-V- and T–s-diagrams

Since internal energy of the ideal gas depends only on temperature, $u = u(T)$, it turns out that the heats exchanged during the isochoric processes are equal, but of opposite sign,[1]

$$q_{23} = u(T_L) - u(T_H) = -q_{41} . \tag{13.2}$$

With this, the heat q_{23}, which is rejected during the isochoric cooling process, can be used for the isochoric heating q_{41}. In the Stirling engine, this is done by means of the regenerator, which allows for internal heat exchange. The working principle of the regenerator will be discussed further below.

When a regenerator is employed, the heat q_{41} is exchanged internally, and only the heat q_{12} must be provided from the outside, e.g., by burning a fuel. The thermal efficiency is obtained, with (13.1), as

$$\eta_{\text{St}} = \frac{w_\odot}{q_{in}} = \frac{w_{12} + w_{34}}{q_{12}} = 1 - \frac{T_L}{T_H} . \tag{13.3}$$

In case that T_L and T_H are the temperatures of reservoirs with which the cycle exchanges heat, the above is just the Carnot efficiency, that is the maximum efficiency for a process operating between reservoirs at temperatures T_H and T_L. It follows that the ideal Stirling process with regenerator is a realization of a Carnot engine, as long as the heat transfer with the reservoirs takes places at infinitesimal temperature difference.

We note that the amounts of heat exchanged during the isochoric processes, q_{23} and q_{41}, can only be equal in size for an ideal gas (with variable or constant specific heats), for which internal energy does not depend on specific volume, $u = u(T)$ and not $u = u(T, v)$. This is different for the Carnot cycle—another realization of a Carnot engine—which has the same efficiency independent of the working medium.

[1] This implies that in the T-s-diagram the areas below the curves 2-3 and 4-1 are equal.

Although the efficiency is independent of the type of ideal gas used, most Stirling engines use helium or hydrogen as working medium. The high heat conductivity of gases with low molecular masses leads to a faster heat exchange and thus allows to operate the engine at a higher frequency.

The power output of the engine depends on the mass m enclosed in the cylinder, and the rotation speed $\dot{n}$ of the engine,

$$\dot{W} = \dot{n} m w_{\odot} \,. \tag{13.4}$$

Fast changes in power demand, as they are necessary for use in cars, can be achieved by changing the amount of working gas in the cylinder, i.e., by pumping additional mass in or out. This, of course, adds to the complexity of the process. It is therefore no surprise that most of today's applications of the Stirling engine consider systems which run at constant load, and generate electricity, in particular with heat supply from solar collectors.

13.1.2 Working Principle of a Stirling Engine

It is difficult, if not impossible, to build an engine that operates on the ideal Stirling cycle. All real Stirling engines approximate the ideal cycle to some extent. There are many different working principles for Stirling engines, and here we discuss the operating principle of a Leybold Stirling engine for use in teaching laboratories, which operates in the same way as the original Stirling engine.[2]

The lab engine, sketched in Fig. 13.2, consists of a glass cylinder in which two pistons—the working piston and the displacement piston—move vertically with a phase shift of 90°. Mounted on top of the cylinder is the heating coil (electrical heating), which maintains the upper part of the engine at high temperature (T_H). The lower part of the cylinder is encased by a second glass cylinder, with cooling water flowing between the two cylinders and through the bottom of the displacement piston to maintain the lower part of the engine at low temperature (T_L).

The displacement piston shifts the gas between the upper high-temperature part of the engine and the lower low-temperature part. The movement of the displacement piston forces the gas through a cylindrical hole in the displacement piston that is filled with copper wool which acts as the regenerator. As the gas passes from the hot part of the engine to the cold part, the gas cools gradually by giving up heat to the copper wool. On the way back the gas takes heat from the regenerator and is thus gradually heated. The working piston seals the gas against the environment and serves to compress or expand it while exchanging work with the environment.

The actual thermodynamic cycle of the Stirling engine differs somewhat from the idealized Stirling cycle, see Fig. 13.3 for a qualitative comparison

[2] This engine is used in the teaching laboratory at the University of Victoria.

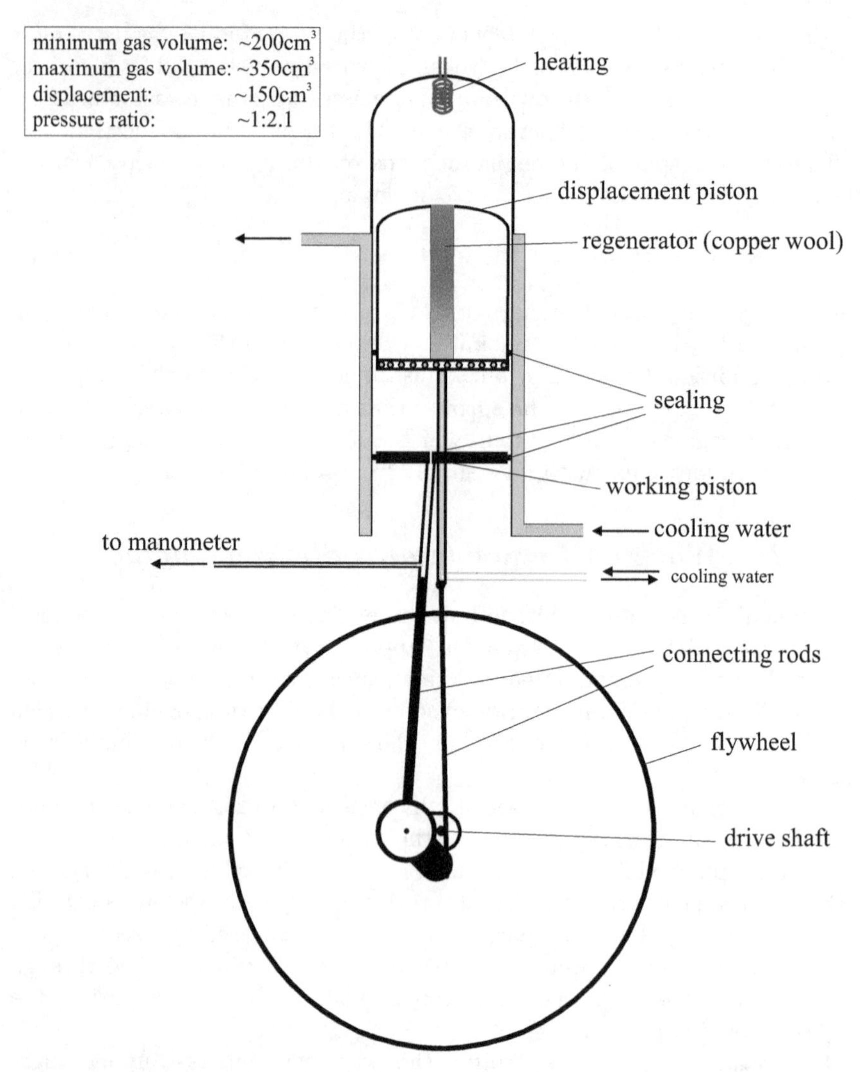

Fig. 13.2 Setup of Leybold-Stirling engine

of the idealized and the real cycle. In order to understand how the Stirling engine approximates the ideal cycle, it is best to study the p-V-diagram in Fig. 13.3, which also shows the displacement of the two pistons as function of the shaft angle.

I. *Isothermal expansion:* The displacement piston is at bottom dead center, almost at rest, so that most of the working gas is in the upper hot zone. The working piston moves downward and the gas expands, the heat supplied is transferred to work.

II. *Isochoric cooling:* The working piston is at bottom dead center, so that the total gas volume is (almost) fixed. The displacement piston is moving upwards and the working gas streams into the lower cold part of the engine. While flowing through the regenerator (copper wool), the gas transfers heat to the regenerator.
III. *Isothermal compression:* The displacement piston is at top dead center, and the working gas is in the lower cool part of the cylinder. The working piston moves upwards compressing the gas. The gas releases heat to the cooling water so that the gas temperature remains nearly constant.
IV. *Isochoric heating:* The working piston is at top dead center, while the displacement piston is moving downwards. The cool gas is streaming upwards through the regenerator, where it receives the energy which was stored in Step II.

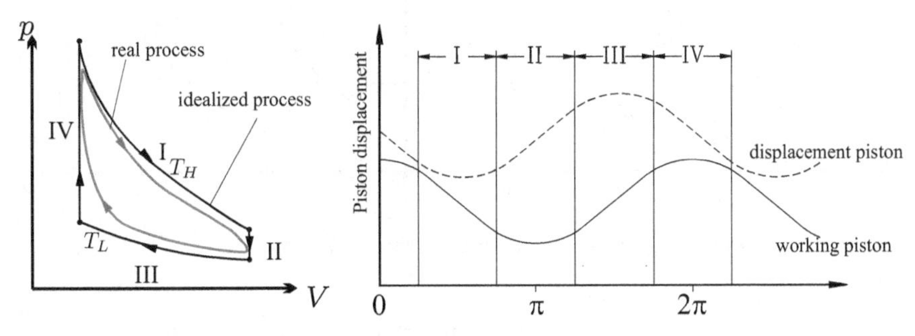

Fig. 13.3 Left: Ideal and real Stirling cycles in p-V-diagram. Right: Piston displacement as function of shaft angle.

The main reasons why the measured p-V-diagram deviates from the ideal one, and assumes the more oval form shown in Fig. 13.3 are:

1. Truly isochoric processes require that the working piston is at rest. However, since it is driven by a crankshaft, the working piston moves sinusoidally.
2. Compression and expansion (I and III) are fast, and do not take place isothermally.
3. The heating coil releases heat into the gas at all times, not only during step I .
4. Part of the working gas remains in the cool part of the engine at all times.
5. The regenerator is not 100% efficient.
6. Heat losses to the environment and friction dissipate energy.
7. Insufficient sealing leads to exchange of gas with the outside.

13.1.3 The Reverse Stirling Cycle

A Stirling engine can also operate as a refrigeration engine or a heat pump. In both cases the engine is driven by a motor, and the process curve in the p-V-diagram runs counter-clockwise, see Fig. 13.4 for the ideal process curve. During the isothermal processes (1-2, 3-4) heat is exchanged with the environment of the engine, while the heat transfer during the isochoric processes (2-3, 4-1) is an internal heat exchange by means of the regenerator.

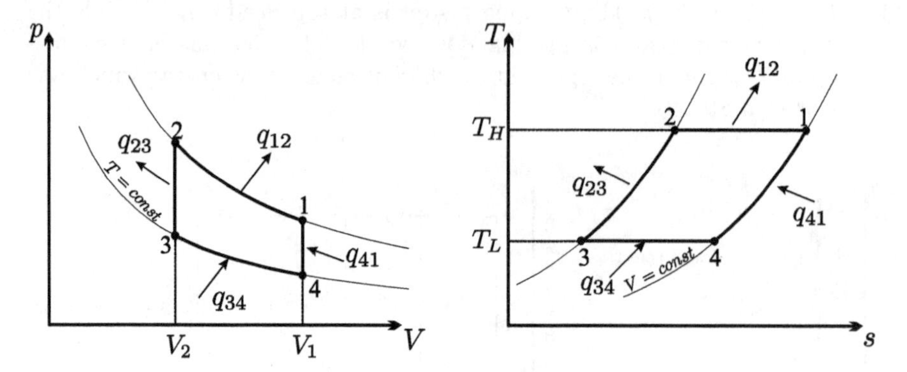

Fig. 13.4 Stirling refrigeration cycle in p-V- and T–s-diagrams

We find for the branches of this cycle the following exchange of heat and work with the environment:

$$\begin{array}{lll} \text{1-2 isothermal:} & w_{12} = RT_H \ln \frac{V_2}{V_1} < 0 \,, & q_{12} = RT_H \ln \frac{V_2}{V_1} < 0 \,, \\ \text{2-3 isochoric:} & w_{23} = 0 & , \; q_{23} = u(T_L) - u(T_H) < 0 \,, \\ \text{3-4 isothermal:} & w_{34} = RT_L \ln \frac{V_1}{V_2} > 0 \,, & q_{34} = RT_L \ln \frac{V_1}{V_2} > 0 \,, \\ \text{4-1 isochoric:} & w_{41} = 0 & , \; q_{41} = u(T_H) - u(T_L) > 0 \,. \end{array} \tag{13.5}$$

As for the Stirling heat engine, the heats for the isochoric processes have the same absolute value, $q_{23} = -q_{41}$, and the regenerator ensures internal heat exchange.

The coefficient of performance for the inverse cycle is

$$\mathrm{COP_R} = \frac{q_{in}}{|w_\odot|} = \frac{q_{34}}{|w_{12} + w_{34}|} = \frac{1}{\frac{T_H}{T_L} - 1} \tag{13.6}$$

for a refrigeration engine, and

$$\mathrm{COP_{HP}} = \frac{|q_{out}|}{|w_\odot|} = \frac{|q_{12}|}{|w_{12} + w_{34}|} = \frac{1}{1 - \frac{T_L}{T_H}} \tag{13.7}$$

for a heat pump, respectively. Again, these are the COP's of the respective Carnot engines, i.e., the maximum COP's that can be reached between the temperatures T_H, T_L.

The lab engine can run as both, heat pump and refrigerator. If run as a heat pump, the lower, water cooled, part of the engine is the cold part of the engine (at T_L) and heat is pumped to the upper part of the engine which becomes hot (T_H). When the operating direction of the engine is reversed, the lower, water cooled, part of the engine becomes the hot part of the engine (at T_H) and heat is pumped away from the upper part of the engine which becomes cold (T_L), the engine operates as a refrigerator.

13.1.4 Stirling Engines Then and Now

The Stirling engine was patented in 1816 by Robert Stirling (1790-1878), a minister of the Church of Scotland. At that time, steam engines had a rather low efficiency (2-10%), and were quite unsafe. Boilers exploded often, and the high pressure steam that was released had scalding effects. The Stirling hot-air engine was promising to overcome both problems: The regenerator gave it a good efficiency, and in the unlikely case of a bursting engine, only hot air was released so that the consequences of an accident were far less severe. However, the Stirling engine could never live up to the expectations. While the steam turbine was improved more and more to today's efficiencies of over 45%, and internal combustion engines, i.e., Otto and Diesel engines, prevailed for the use in cars and trucks, Stirling engines almost vanished from the scene.

Unlike an internal combustion engine, a Stirling engine does not exchange the working gas in each cycle, but contains the gas permanently. The heat is supplied outside the engine, so that any heat source is suitable to power a Stirling engine. Thus, a Stirling engine can be driven by carbon fuels (coal, natural gas, gasoline, Diesel oil), hydrogen, solar radiation, nuclear power, waste heat of industrial processes, etc. If a fuel is used, it is burned continuously, with lower emissions than in a reciprocating internal combustion engine.

Stirling refrigeration engines can give very low temperatures, and are widely used for small scale cryogenic cooling. A promising application for Stirling heat engines is the conversion of solar energy into electricity by means of parabolic mirror dishes with Stirling engines in the focus. While these devices can only operate under direct sunlight, they can be far more efficient than photovoltaic cells.

At present it is not possible to build a high efficiency Stirling engine at a competitive price. In order to have a high specific power (kW per litre of engine capacity), the working gas must be highly pressurized (goal: up to 190 bar), causing problems of sealing against the environment and lubrication. It is at the seals where a large portion of mechanical losses occur. The efficiency increases with the temperature difference between the cold and the hot part of the engine. Thus, highly efficient engines require non-standard materials that can operate at temperatures of 750 °C and more.

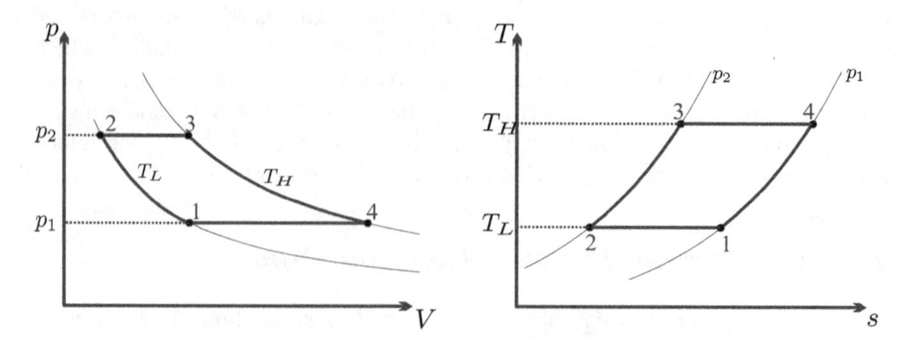

Fig. 13.5 p-V- and T-s-diagrams for the Ericsson cycle

13.2 Ericsson Cycle

The Stirling cycle operates in piston-cylinder assemblies, in closed systems. The Ericsson cycle is very similar, only that it consists of open system devices, namely isothermal turbine and compressor, and isobaric heat exchangers; again, the working fluid is an ideal gas. Figure 13.5 shows the corresponding p-V- and T-s-diagrams. Since we consider open system devices for the realization of the cycle, the corresponding work and heat contributions are

$$\begin{aligned}
&\text{1-2 compressor at } T_L: && w_{12} = -RT_L \ln \tfrac{p_2}{p_1} < 0 \,, && q_{12} = -RT_L \ln \tfrac{p_2}{p_1} < 0 \,, \\
&\text{2-3 heating at } p_2: && w_{23} = 0 && , q_{23} = h\left(T_H\right) - h\left(T_L\right) > 0 \,, \\
&\text{3-4 turbine at } T_H: && w_{34} = -RT_H \ln \tfrac{p_1}{p_2} > 0 \,, && q_{34} = -RT_H \ln \tfrac{p_1}{p_2} > 0 \,, \\
&\text{4-1 cooling at } p_2: && w_{41} = 0 && , q_{41} = h\left(T_L\right) - h\left(T_H\right) < 0 \,.
\end{aligned}$$

Since the enthalpy of an ideal gas depends only on temperature, but not on pressure, the amounts of heat exchanged on the isobaric legs are equal in magnitude with opposite signs: a regenerator can be used for internal exchange of heat. In fact, the regenerator must be a counter-flow heat exchanger, which leads to the schematic shown in Fig. 13.6.

With the use of the regenerator, only the heat q_{34} must be supplied from the outside, and the thermal efficiency of the ideal cycle becomes

$$\eta_{\mathrm{Er}} = \frac{w_{\odot}}{q_{in}} = \frac{w_{12} + w_{34}}{q_{34}} = 1 - \frac{T_L}{T_H} \,. \tag{13.8}$$

This, again, is the Carnot efficiency, that is the best possible efficiency obtainable from a heat engine operating between reservoirs at temperatures T_H, T_L.

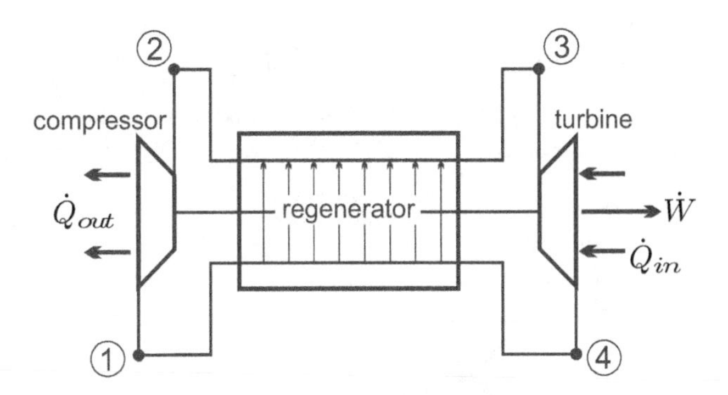

Fig. 13.6 Schematic of Ericsson engine

As the Stirling engine, the Ericsson engine uses external heat supply, which, in principle, allows the use of any heat source.

The Ericsson cycle relies on isothermal turbine and compressor, which require heat exchange during compression and expansion. Since heat transfer is slow, these are difficult to realize: Fast compressors and turbines normally are adiabatic, since the mass flows through too fast so that there is no time for significant heat transfer. Adiabatic compression with intercooling, and adiabatic expansion with reheat, together with regeneration, as discussed in the next sections, are means to bring gas turbine cycles closer to the Ericsson cycle, and thus improve their efficiency.

13.3 Compression with Intercooling

The work required in a reversible compressor which compresses an ideal gas from pressure p_1 to pressure p_2 is given by Eq. (9.25),

$$w_C = -\int_1^2 v dp \,. \tag{13.9}$$

Thus, as discussed earlier, the work is the area to the left of the process curve in the p-v-diagram, see Sec. 9.6; less work is required for smaller specific volume of the compressed substance.

Figure 13.7 shows the curves for an isothermal and an adiabatic compressor in the p-v-diagram, in the reversible case. The adiabatic curve is steeper than the isothermal curve (Sec. 7.7), since in the isothermal process the cooling during compression ensures a smaller specific volume. Therefore the isothermal compressor requires less work than the adiabatic one.

For an isothermal compressor we have $v = RT_1/p$, and the work is $w_C = -RT_1 \ln \frac{p_2}{p_1}$, while for an adiabatic compressor, assuming constant specific heats, we have $v = v_1 (p_1/p)^{1/k}$ and $w_C = c_p T_1 \left(1 - \left(\frac{p_2}{p_1}\right)^{\frac{k-1}{k}}\right)$. The difference in work requirement is indicated by the shaded area in the figure.

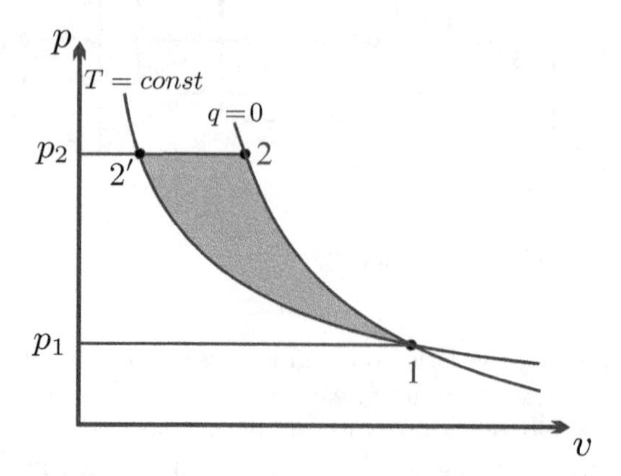

Fig. 13.7 Adiabatic (1-2) and isothermal (1-2') compressor in p-V-diagram. The grey area is the difference in the compressor work.

Obviously, to reduce the work requirement one would aim at isothermal compression. However, isothermal compression requires cooling during compression which is impossible to achieve for the large throughputs required, and cannot be used in practice.

An alternative is offered by multi-stage compressors with intercooling, in which the gas is compressed adiabatically in each stage, and then isobarically cooled to the environmental temperature T_1 before it enters the next stage. Intercooling reduces the gas volume, and thus the work requirement is reduced. As an example, Fig. 13.8 shows a three stage compressor with intercooling. As more stages are used, the process curve approaches the isothermal curve. It should be noted that construction is more costly than for a single stage compressor.

The work savings depend on the pressures chosen for intercooling, which must be optimized. We consider an n-stage compressor which takes in gas at p_1, T_1, and compresses it to the final pressure p_e; between each stage the gas is cooled back to T_1. Stage i compresses from T_1, p_i to p_{i+1}, and requires the work

$$w_{C_i} = \frac{h(T_1) - h(T_{i+1})}{\eta_{C_i}}, \tag{13.10}$$

where η_{C_i} is the isentropic efficiency of stage i. The temperature at the exit of an adiabatic reversible compressor between the same pressures, T_{i+1}, is obtained from the relation $\frac{p_r(T_{i+1})}{p_r(T_1)} = \frac{p_{i+1}}{p_i}$. Accordingly, the reversible work $w_{C,rev}(P_i) = h(T_1) - h(T_{i+1})$ is a function of the pressure ratio $P = \frac{p_{i+1}}{p_i}$ for the stage, and we can write

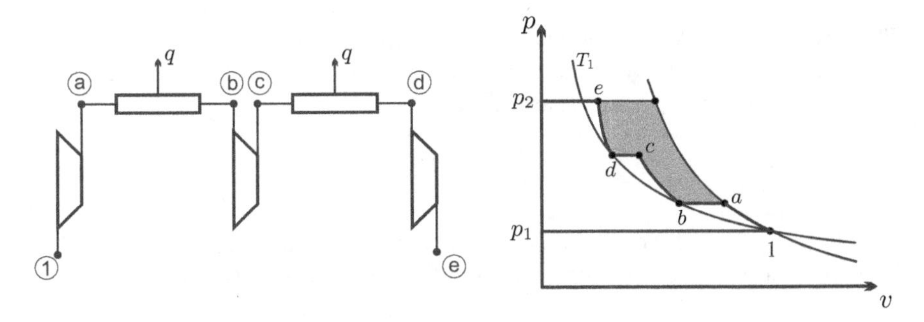

Fig. 13.8 Schematic and p-v-diagram for a compressor with three adiabatic stages and intercooling. The grey area is the amount of work saved in comparison to a single adiabatic compressor.

$$w_{C_i} = \frac{w_{C,rev}\left(\frac{p_{i+1}}{p_i}\right)}{\eta_{C_i}} . \tag{13.11}$$

The total work is just the sum over the individual compressors,

$$w_C = \sum_{i=1}^{n} \frac{w_{C,rev}\left(\frac{p_{i+1}}{p_i}\right)}{\eta_{C_i}} . \tag{13.12}$$

The minimum compression work is obtained from setting $\frac{\partial w_C}{\partial p_j} = 0$ for all intermediate pressures ($j = 1, 2, \ldots, n$). The derivative is evaluated as follows:

$$\begin{aligned}\frac{\partial w_C}{\partial p_j} &= \frac{\partial}{\partial p_j}\left[\frac{w_{C,rev}\left(\frac{p_j}{p_{j-1}}\right)}{\eta_{C_{j-1}}} + \frac{w_{C,rev}\left(\frac{p_{j+1}}{p_j}\right)}{\eta_{C_j}}\right] \\ &= \frac{w'_{C,rev}\left(\frac{p_j}{p_{j-1}}\right)}{\eta_{C_{j-1}}}\frac{1}{p_{j-1}} - \frac{w'_{C,rev}\left(\frac{p_{j+1}}{p_j}\right)}{\eta_{C_j}}\frac{p_{j+1}}{p_j^2} ,\end{aligned} \tag{13.13}$$

where $w'_{C,rev}(P)$ indicates the derivative of $w_{C,rev}(P)$ with respect to the pressure ratio. Setting the above to zero gives the condition for minimum work requirement

$$\frac{p_j}{p_{j-1}}\frac{w'_{C,rev}\left(\frac{p_j}{p_{j-1}}\right)}{\eta_{C_{j-1}}} = \frac{p_{j+1}}{p_j}\frac{w'_{C,rev}\left(\frac{p_{j+1}}{p_j}\right)}{\eta_{C_j}} . \tag{13.14}$$

For the further evaluation we consider only the case where all stages have the same isentropic efficiency, $\eta_{C_j} = \eta_C$, so that

$$\frac{p_j}{p_{j-1}} w'_{C,rev}\left(\frac{p_j}{p_{j-1}}\right) = \frac{p_{j+1}}{p_j} w'_{C,rev}\left(\frac{p_{j+1}}{p_j}\right) . \tag{13.15}$$

The reversible work is a monotonous function of the pressure ratio. It follows that the multi-stage compressor requires minimum work when all stages operate at the same pressure ratio P, i.e.,

$$\frac{p_{i+1}}{p_i} = P\,. \tag{13.16}$$

This implies that all stages consume the same work per unit mass, $w_C(P)$. Multiplication of the pressure ratios of all stages gives, with $p_{n+1} = p_e$,

$$\prod_{i=1}^{n} \frac{p_{i+1}}{p_i} = \frac{p_e}{p_1} = P^n\,, \tag{13.17}$$

so that

$$P = \left(\frac{p_e}{p_1}\right)^{\frac{1}{n}} \quad \text{and} \quad p_{i+1} = p_1 P^i = \sqrt[n]{p_1^{n-i}\, p_e^i}\,. \tag{13.18}$$

Special cases are a two stage compressor, which consumes minimum work for the intermediate pressure $p_m = \sqrt{p_1 p_e}$, and the three stage compressor in the figure, for which the optimum intermediate pressures are obtained as $p_b = \sqrt[3]{p_1^2 p_e}$, $p_d = \sqrt[3]{p_1 p_e^2}$.

In case that the isentropic compressor efficiency depends on the pressure within the compressor, one has to evaluate (13.14). We note that the above argument is valid also for polytropic compressors, as long as the polytropic exponent for all compressors is the same.

13.4 Gas Turbine Cycles with Regeneration and Reheat

13.4.1 Regenerative Brayton Cycle

We return to the discussion of the Brayton cycle, which was introduced in Sec. 10.5. As is evident from the discussion there, in particular from the T-s-diagram, the Brayton cycle expels rather warm exhaust into the environment. Since the exhaust is warmer than the environment, it has a work potential as was shown in Sec. 11.6. If the exhaust is just blown into the environment, this work potential remains unused. The ensuing equilibration between exhaust and environment is an irreversible process—it is an external irreversibility for the gas turbine.

To recover at least a portion of the exhaust work potential, the exhaust can be lead through a regenerator to heat the compressed air before it enters the combustion chamber. With this, less heat must be supplied from the outside, and the efficiency is increased.

Figure 13.9 shows schematic and T-s-diagram for a Brayton gas turbine cycle with regenerator, which is a counter-flow heat exchanger. Since heat

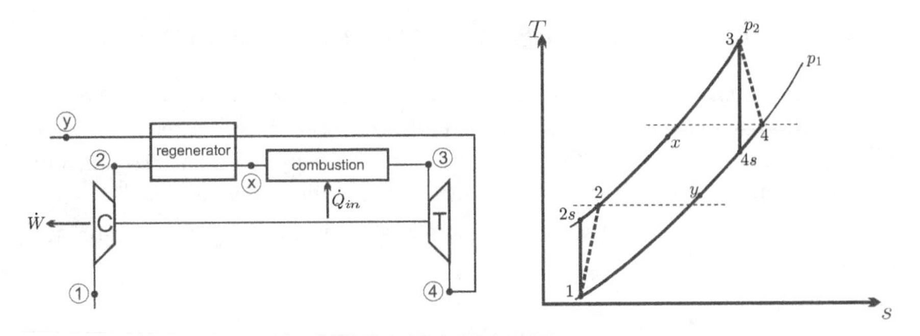

Fig. 13.9 Schematic and T-s-diagram for a Brayton cycle with regeneration

goes from warm to cold, the preheat temperature T_x cannot be larger than the turbine exhaust temperature T_4, while the final exhaust temperature T_y cannot be smaller than T_2. Thus, the use of a regenerator makes only sense when the turbine exhaust temperature T_4 is larger than the temperature after compression, T_2.

A perfect heat exchanger would yield $T_x = T_4$ and this is used to define the regenerator effectiveness as[3]

$$\eta_{reg} = \frac{h_x - h_2}{h_4 - h_2} \,. \tag{13.19}$$

A 100% effective regenerator would transfer heat at infinitesimal temperature differences, a realistic regenerator operates with finite temperature differences and around 80% effectiveness. The exhaust temperature T_y follows from the energy balance over the regenerator, assuming that no heat is lost to the exterior, as

$$h\left(T_y\right) = h_y = h_4 - h_x + h_2 \,. \tag{13.20}$$

The regenerator reduces the amount of heat that must be supplied from the fuel, which is $q_{x3} = h_3 - h_x$. Accordingly, the thermal efficiency for the depicted cycle $1-2-x-3-4-y$ is given by

$$\eta_{B,reg} = \frac{h_1 - h_2 + h_3 - h_4}{h_3 - h_x} = 1 - \frac{h_4 - h_1 - \eta_{reg}\left(h_4 - h_2\right)}{h_3 - h_2 - \eta_{reg}\left(h_4 - h_2\right)} \,. \tag{13.21}$$

For $\eta_{reg} = 0$, this reduces to the thermal efficiency of the standard Brayton cycle, $\eta_B = 1 - \frac{h_4 - h_1}{h_3 - h_2}$. For non-zero effectiveness, the efficiency is larger than η_B. This follows from the fact that, because $0 < \eta_B < 1$, $h_4 - h_1 < h_3 - h_2$, which implies that with growing regenerator effectiveness η_{reg} the thermal efficiency $\eta_{B,reg}$ grows as well.

The actual improvement depends on the detailed data of the process.

[3] Note that the working fluid is an ideal gas, where enthalpy is a function of temperature only, $h = h\left(T\right)$.

13.4.2 Example: Brayton Cycle with Regenerator

The impact of the regenerator is best studied by means of examples. We consider a Brayton cycle with compressor inlet temperature $T_1 = 290\,\mathrm{K}$, turbine inlet temperature $T_3 = 1500K$, and pressure ratio $p_2/p_1 = 10$. To simplify the computation, we rely on the cold-air approximation with $k = \frac{c_p}{c_v} = 1.4$, $c_p = \frac{k}{k-1}R$, which gives the temperatures after isentropic compressor and turbine as

$$T_{2s} = T_1 \left(\frac{p_2}{p_1}\right)^{\frac{k-1}{k}} = 560\,\mathrm{K} \;, \; T_{4s} = T_3 \left(\frac{p_1}{p_2}\right)^{\frac{k-1}{k}} = 777\,\mathrm{K} \,.$$

First, we consider the fully reversible cycle, with 100% effective regenerator. For this, according to (13.19, 13.20), the preheat temperature after the regenerator is $T_x = T_{4s}$, and the exhaust temperature is $T_y = T_{2s}$. With the computed temperatures, we find the specific work of the reversible cycle as

$$w_{\odot} = c_p \left(T_1 - T_{2s} + T_3 - T_{4s}\right) = 454.8 \frac{\mathrm{kJ}}{\mathrm{kg}} \,,$$

and the thermal efficiencies of the cycle without and with regenerator are

$$\eta_B = \frac{T_1 - T_{2s} + T_3 - T_{4s}}{T_3 - T_{2s}} = 48.2\% \,,$$
$$\eta_{B,reg} = \frac{T_1 - T_{2s} + T_3 - T_{4s}}{T_3 - T_x} = 62.7\% \,.$$

We see that a regenerator can give substantial improvement for cycle efficiency.

With the regenerator, the external loss is reduced, since the exhaust temperature, and thus the external irreversibility, is lowered considerably. Indeed, the exhaust of the cycle without regenerator (temperature $T_x = T_{4s}$) has the work potential

$$w_{ex} = c_p \left(T_{4s} - T_1 - T_1 \ln \frac{T_{4s}}{T_1}\right) = 201.9 \frac{\mathrm{kJ}}{\mathrm{kg}} \,,$$

while exhaust of the cycle with regenerator (temperature $T_y = T_{2s}$) has the work potential

$$w_{ex,reg} = c_p \left(T_y - T_1 - T_1 \ln \frac{T_y}{T_1}\right) = 79.5 \frac{\mathrm{kJ}}{\mathrm{kg}} \,.$$

Recall that the work potential of the exhaust is lost, since the exhaust is dumped into the environment. For this example, the regenerator reduces the exhaust loss by about 60%.

All efficiency values in the above example are relatively high, since no internal irreversibilities are accounted for. To study how internal irreversibilities affect the results, we now assume isentropic efficiencies for compressor and turbine of $\eta_T = \eta_C = 0.85$, and a regenerator effectiveness of $\eta_{reg} = 0.8$.

Then, we find the temperatures after compressor and turbine as

$$T_2 = T_1 + \frac{T_{2s} - T_1}{\eta_C} = 618\,\mathrm{K} \quad , \quad T_4 = T_3 + \eta_T \left(T_{4s} - T_3 \right) = 885\,\mathrm{K} \, ,$$

and the temperatures after heat exchange in the regenerator, from (13.19, 13.20), as

$$T_x = T_2 + \eta_{reg} \left(T_4 - T_2 \right) = 832\,\mathrm{K} \quad , \quad T_y = T_4 - T_x + T_2 = 671\,\mathrm{K} \, .$$

The specific work for the cycle is now

$$w_{\odot} = c_p \left(T_1 - T_2 + T_3 - T_4 \right) = 288 \frac{\mathrm{kJ}}{\mathrm{kg}} \, ,$$

and the thermal efficiencies for the cycle with and without regenerator are

$$\eta_B = \frac{T_1 - T_2 + T_3 - T_4}{T_3 - T_2} = 32.5\% \, ,$$

$$\eta_{B,reg} = \frac{T_1 - T_2 + T_3 - T_4}{T_3 - T_x} = 43\% \, .$$

The corresponding work potentials of the dumped exhaust are $w_{ex} = 273 \frac{\mathrm{kJ}}{\mathrm{kg}}$ and $w_{ex,reg} = 138 \frac{\mathrm{kJ}}{\mathrm{kg}}$, i.e., the regenerator reduces the exhaust loss by 50%. Nevertheless, the exhaust still has significant work potential, which is about 50% of the work actually delivered by the system. Another heat engine can be used to produce work from the exhaust—see the discussion of the combined cycle further below.

For motivation of the next section we compute the ratio between compressor and turbine work, i.e., the back work ratio, for this cycle as

$$\mathrm{bwr} = \frac{|w_C|}{w_T} = \frac{h_2 - h_1}{h_3 - h_4} = \frac{T_2 - T_1}{T_3 - T_4} = 53.3\% \, .$$

13.5 Brayton Cycle with Intercooling and Reheat

Our discussion of compressors has shown that multi-stage compression with intercooling reduces the work required for compression. Applying this idea in a gas turbine cycle reduces the back work ratio, and also the temperature T_2 behind the compressor. When a regenerator is used, the exhaust temperature T_y is limited by the temperature T_2 after compression, which is lower with

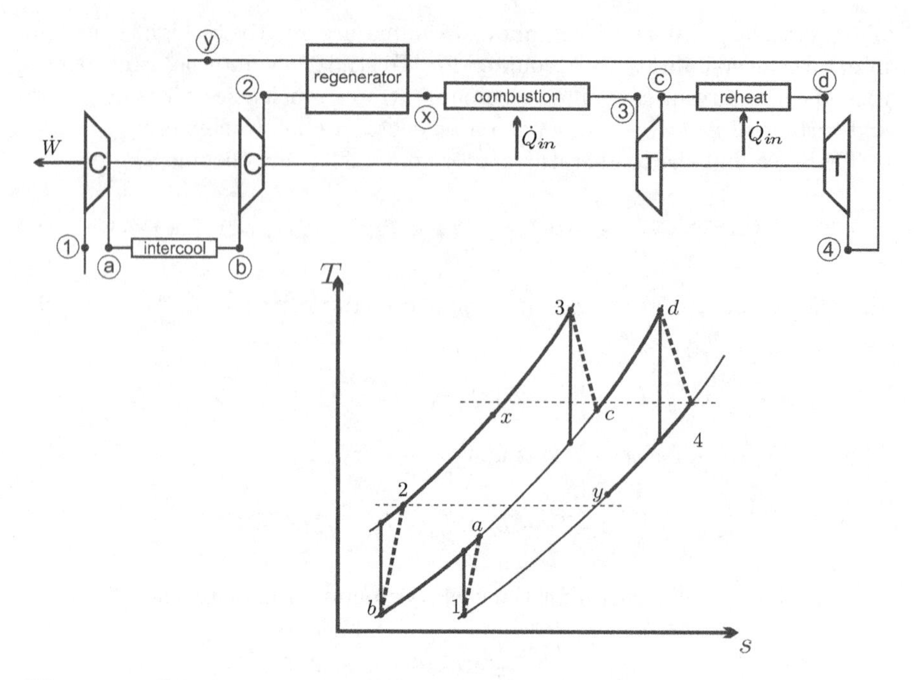

Fig. 13.10 Schematic and T-s-diagram for gasturbine cycle with two-stage compressor with intercooling, two-stage turbine with reheat, and regenerator

multi-stage compression. Lower exhaust temperature T_y lowers the external irreversibility, and thus gives better efficiency.

In short, further efficiency gain can be obtained by using multi-stage compression with intercooling together with regeneration. Figure 13.10 shows schematic and T-s-diagram for a system with two-stage compression and regenerator that also includes two turbines stages with intermediate reheat.

Reheat increases the average temperature for heating, and thus the efficiency. The optimum reheat pressure can be determined by maximizing work, similar to the discussion is Sec. 13.3. If the turbines have the same inlet temperature, and the same isentropic efficiency, the maximum work is obtained when they have the same pressure ratio.

Since reheat increases the turbine exit temperature, reheat will increase the thermal efficiency only if accompanied by regeneration. The thermal efficiency of this cycle is obtained as

$$\eta = \frac{w_{C1} + w_{C2} + w_{T1} + w_{T2}}{q_{comb} + q_{reheat}} = 1 - \frac{h_a - h_1 + h_2 - h_b + h_4 - h_x}{(h_3 - h_x) + (h_d - h_c)} \,. \tag{13.22}$$

With intercooling and reheat, a larger portion of the heat can be exchanged in the regenerator, which reduces the exhaust temperature T_y and the corresponding exhaust loss, thus increasing the thermal efficiency. With more and more intercooling and reheat stages, the process becomes more similar to the Ericsson process.

13.6 Combined Cycle

As we have seen, even in a gas turbine with regeneration there are significant exhaust losses. An alternative to regeneration is using the gas turbine exhaust to provide heat for another heat engine. In the *combined cycle* the gas turbine exhaust is used in a *heat recovery steam generator* (HSRG) to provide heat for a steam power cycle. Figure 13.11 shows schematic and energy flow diagram for the combination of a standard Brayton cycle with a standard Rankine cycle. Real power plants use state of the art regenerative steam cycles (see Section 12.2).[4]

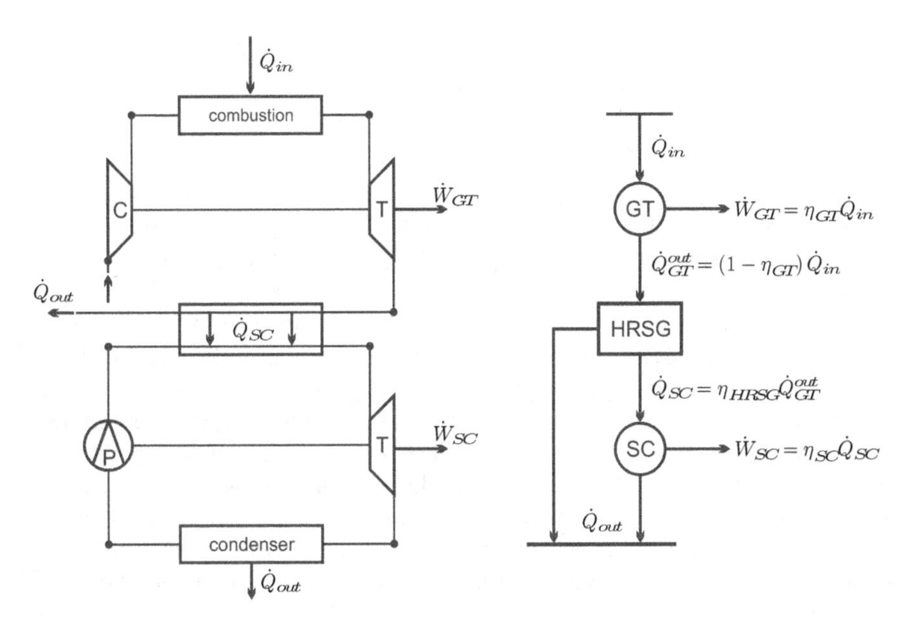

Fig. 13.11 Schematic and energy flows in a combined cycle

In order to evaluate the combined cycle, we consider a gas turbine cycle with thermal efficiency η_{GT} and a steam cycle with thermal efficiency η_{SC}, which are connected by an HSRG with effectiveness η_{HSRG}.

The heat into the system from combustion in the gas turbine is $\dot{Q}_{in}$. The gas turbine cycle produces the work

[4] In case that the gas exhaust from the HRSG still has marked work potential, a regenerator can be added to the gas cycle for preheating of the combustion air (not shown in Figure).

$$\dot{W}_{GT} = \eta_{GT} \dot{Q}_{in} \; ,$$

and rejects the heat

$$\dot{Q}_{GT}^{out} = (1 - \eta_{GT}) \dot{Q}_{in}$$

to the HSRG. The latter delivers the heat

$$\dot{Q}_{SC} = \eta_{HSRG} \dot{Q}_{GT}^{out} = \eta_{HSRG} (1 - \eta_{GT}) \dot{Q}_{in}$$

to the steam cycle, which thus produces the power

$$\dot{W}_{SC} = \eta_{SC} \dot{Q}_{SC} = \eta_{SC} \eta_{HSRG} (1 - \eta_{GT}) \dot{Q}_{in} \; .$$

The combined power output of both cycles is

$$\dot{W} = \dot{W}_{GT} + \dot{W}_{SC} = (\eta_{GT} + \eta_{SC} \eta_{HSRG} (1 - \eta_{GT})) \dot{Q}_{in} \; ,$$

which gives the thermal efficiency of the combined cycle as

$$\eta = \frac{\dot{W}}{\dot{Q}_{in}} = \eta_{GT} + \eta_{SC} \eta_{HSRG} (1 - \eta_{GT}) \; .$$

The efficiency of the combined cycle is always greater than the efficiency of the gas turbine alone (unless $\eta_{HSRG} = 0$), and is also larger than the efficiency of the steam cycle alone as long as η_{HSRG} is large enough (in particular $\eta > \eta_{SC}$ if $\eta_{HRSG} = 1$).

A gas turbine with $\eta_{GT} = 0.31$ combined with a steam cycle of $\eta_{SC} = 0.45$ by means of a HRSG with $\eta_{HRSG} = 0.9$ has an overall efficiency of 60%. Indeed, combined cycle power plants have the highest available efficiencies among all combustion driven power plants. They allow large upper temperatures in the gas turbines, and reject heat at relatively low temperatures. The only drawback to their use is that they require gaseous or liquid fuels, and cannot be fed directly with coal. Recall that the average efficiency of the World's combustion power plants is about 35% or less. Much better use of fossil fuels could be made by using coal gasification or liquefaction and combined cycle plants.

13.7 The Solar Tower

We mentioned solar power conversion as an application for Stirling engines. Here, we discuss an interesting application for solar power conversion, which relies on the chimney effect, which, in turn, relies on the variation of air pressure with height as expressed in the barometric formula (2.25).

The solar tower, or solar chimney, sketched in Fig. 13.12, works as follows: Solar radiation provides heat $\dot{Q}_{\odot}$ which passes through a glass roof, is absorbed by black mats on the ground, and the warm mats heat up air. The

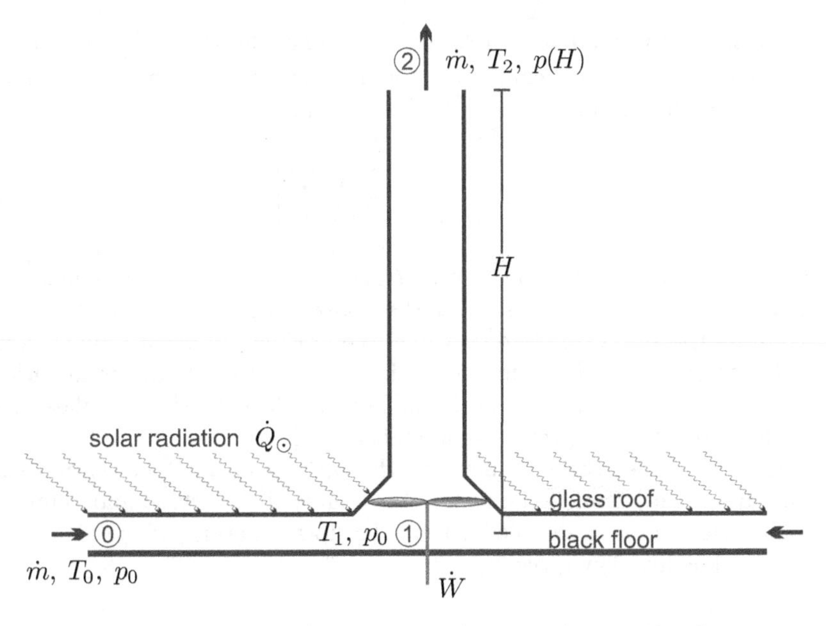

Fig. 13.12 Solar Tower. The sketch shows a cut trough a circular device.

warm air rises through the chimney and drives turbines which are connected to a generator to produce the power $\dot{W}$. When a layer of water is placed below the black mats (e.g., one might use black sacks filled with water), the heat provided from the sun goes partly into air and partly into the water. When the solar heat supply stops after sunset, the warm water heats the air, and the tower still produces electricity until the water has cooled down. An experimental plant with a 180 m tower was build some years ago in Spain, plans to build a tower with a height of 1000 m and a diameter of the glass roof of 6 km in Australia are presently on hold.

We aim at a thermodynamic evaluation of the solar tower, and ask in particular for its thermal efficiency. The temperatures involved are rather low, and thus we can use the cold air approximation, i.e., we assume the specific heats of air to be constants. The exterior air is assumed to have constant temperature T_0, and the pressure depends on height z according to the barometric formula, $p(z) = p_0 \exp\left[-\frac{gz}{RT}\right]$ with p_0 the pressure at the ground.

The incoming air is at $\{T_0, p_0\}$, and as it flows towards the turbines it is heated isobarically until it reaches the temperature T_1 just before the turbines. To compute this temperature, we apply the first law for open systems between the outer rim and the point just before the turbine. Since the radius is large, the flow velocity $\mathcal{V}_0$ at the outer rim can be neglected, and the first law gives

$$\dot{m}\left[h_1 - h_0 + \frac{1}{2}\mathcal{V}_1^2\right] = \dot{Q}_\odot\ .$$

The velocity at turbine inlet is related to mass flow by $\mathcal{V}_1 = \frac{\dot{m}}{\rho_1 A_1}$, where A_1 is the cross section of the chimney at point 1, and the pressure of the flow remains constant, so that $\rho_1 = \rho_0 \frac{T_0}{T_1}$. Thus we find

$$\dot{m}\left[c_p\left(T_1 - T_0\right) + \frac{1}{2}\left(\frac{\dot{m}T_1}{\rho_0 T_0 A_1}\right)^2\right] = \dot{Q}_\odot \,. \tag{13.23}$$

To avoid a detailed discussion of heat transfer mechanisms, we assume the temperature T_1 to be given, so that the above is an equation for the heat supply from the sun, $\dot{Q}_\odot$. Note that due to emission and absorption of radiation the temperature T_1 is limited, as in a greenhouse. A proper radiation heat transfer analysis must be performed to establish the size the glass roof must have, so that the specified temperature T_1 is reached.

Next, we consider the flow between the turbine inlet (Point 1) and the exhaust from the chimney (Point 2). We assume that turbine and chimney are adiabatic, and, for simplicity, that the flow is reversible, so that it is isentropic. The first law gives

$$\dot{m}\left[h_2 - h_1 + \frac{1}{2}\left(\mathcal{V}_2^2 - \mathcal{V}_1^2\right) + gH\right] = -\dot{W} \,.$$

The pressure at H follows from the barometric law, and thus the adiabatic relations give, with $p_1 = p_0$,

$$\frac{T_2}{T_1} = \left(\frac{p_2}{p_1}\right)^{\frac{k-1}{k}} = \exp\left[-\frac{gH}{RT_0}\frac{k-1}{k}\right] \simeq 1 - \frac{gH}{RT_0}\frac{k-1}{k} \,,$$

$$\frac{\rho_2}{\rho_1} = \left(\frac{p_2}{p_1}\right)^{\frac{1}{k}} = \exp\left[-\frac{gH}{RT_0}\frac{1}{k}\right] \simeq 1 - \frac{gH}{RT_0}\frac{1}{k} \,;$$

the Taylor expansions of the exponentials are well justified for $H = 1000\,\mathrm{m}$, $T_0 = 298\,\mathrm{K}$.

We use all this, and $\mathcal{V}_2 = \frac{\dot{m}}{\rho_2 A_2}$, $c_p = \frac{k}{k-1}R$, to find the power produced as

$$\dot{W} = \dot{m}\left[\left(\frac{T_1}{T_0} - 1\right)gH - \frac{\dot{m}^2}{2}\left(\frac{RT_1}{p_0 A_2}\right)^2\left(1 + \frac{2gH}{RT_0}\frac{1}{k} - \left(\frac{A_2}{A_1}\right)^2\right)\right] . \tag{13.24}$$

This equations gives power $\dot{W}$ in dependence of mass flow $\dot{m}$, all other quantities are given by material and construction.

Before we study the full result, we have a look at a further simplification, where the contribution of kinetic energies, that is all terms with factor $\dot{m}^2$, are ignored. In this case (13.24) and (13.23) reduce to

$$\dot{W} = \dot{m}\left(\frac{T_1}{T_0} - 1\right)gH \quad , \quad \dot{Q}_\odot = \dot{m}c_p T_0\left(\frac{T_1}{T_0} - 1\right) ,$$

which yields a thermal efficiency of

$$\eta = \frac{\dot{W}}{\dot{Q}_{\odot}} = \frac{gH}{c_p T_0} \,.$$

Accordingly, it is beneficial to build the tower as high as possible. For a height of 200 m the efficiency is $\eta = 0.65\%$ and this increases to $\eta = 3.25\%$ when the height is raised to 1000 m. Note that the thermal efficiency is very low nevertheless. Here it must be considered that the main investment is to build the plant, while the energy source—solar radiation—is available for free, as long as no clouds are present, which is, of course, why one would build such a solar tower power plant in a sunny country.

Since the investment costs are high, one will aim to harvest as much power as possible, which will be achieved by optimizing the operating conditions. We return to (13.24) and determine the optimum mass flow to maximize power from the condition $d\dot{W}/d\dot{m} = 0$ as

$$\dot{m}_{\max} = \frac{p_0 A_2}{R T_1} \sqrt{\frac{\frac{2}{3}\left(\frac{T_1}{T_0} - 1\right) gH}{1 + \frac{2gH}{RT_0}\frac{1}{k} - \left(\frac{A_2}{A_1}\right)^2}} \,.$$

The corresponding power output is

$$\dot{W}_{\max} = \dot{m}_{\max} \frac{2}{3}\left(\frac{T_1}{T_0} - 1\right) gH \,.$$

Figure 13.13 shows power output and thermal efficiency as a function of mass flow for the following data: $g = 9.81 \frac{\mathrm{m}}{\mathrm{s}^2}$, $H = 1000\,\mathrm{m}$, $T_1 = 345\,\mathrm{K}$, $T_0 = 300\,\mathrm{K}$, $R = 0.287 \frac{\mathrm{kJ}}{\mathrm{kg\,K}}$, $c_p = 1.004 \frac{\mathrm{kJ}}{\mathrm{kg\,K}}$, $k = 1.4$, $A_1 = 2A_2$, $A_2 = \pi r^2$ with $r = 45\,\mathrm{m}$. These are curves for reversible operation. As for all processes, irreversible processes, mainly in the turbines, will reduce power generation and efficiency. The power curve exhibits the maximum computed above, and the thermal efficiency drops with increasing mass flow. The maximum power output is $\dot{W}_{\max} = 206\,\mathrm{MW}$, where the efficiency is $\eta\left(\dot{m}_{\max}\right) = \frac{2}{3}\frac{gH}{c_p T_0} = 2.2\%$. The corresponding heat intake is $\dot{Q}_{\odot}\left(\dot{m}_{\max}\right) = 9521\,\mathrm{MW}$. For an average absorbed irradiation of $I = 340 \frac{\mathrm{W}}{\mathrm{m}^2}$, the glass roof must have a surface of $A_{roof} = \pi r_{roof}^2 = \frac{\dot{Q}_{\odot}(\dot{m}_{\max})}{I} = 28\,\mathrm{km}^2$, which corresponds to a radius of 3 km.

13.8 Simple Chimney

In a simple chimney, no work is extracted, $\dot{W} = 0$, the chimney serves to drive an air flow. Equation (13.24), with $\dot{W} = 0$, gives a relation between mass flow and chimney height,

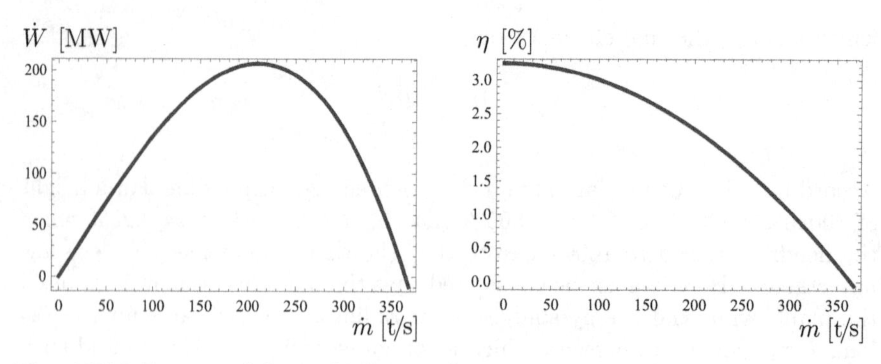

Fig. 13.13 Power and thermal efficiency for a solar tower with height $H = 1000\,\text{m}$ and chimney radius $r = 45\,\text{m}$

$$\dot{m} = \frac{p_0 A_2}{RT_1}\sqrt{\frac{2\left(\frac{T_1}{T_0}-1\right)gH}{1+\frac{2gH}{RT_0}\frac{1}{k}-\left(\frac{A_2}{A_1}\right)^2}}\,.$$

We restrict the attention to cases where $A_2 \ll A_1$ and low enough heights, so that $\frac{2gH}{RT_0}\frac{1}{k} \ll 1$ and

$$\dot{m} = \rho_0 A_2 \sqrt{2gH}\frac{T_0}{T_1}\sqrt{\frac{T_1}{T_0}-1}\,.$$

The draught of the chimney depends on the square root of the height H. The mass flow also has a non-linear dependence on the ratio between the temperatures outside (T_0) and at the foot of the chimney (T_1) with a maximum for $T_1/T_0 = 2$.

In the past, the chimney effect was used for instance to drive the combustion air for a coal power plant. As the above analysis shows, a well working natural draught chimney requires a relatively high temperature ($T_1 = 2T_0$), which implies discharge of large amounts of warm gas, and correspondingly high entropy generation, and work loss. In modern power plants, and other applications, discharge is effected by fans, which allow low exhaust temperatures. High chimneys are still build today, not to increase draught, but to expel the exhaust into higher layers of the atmosphere for better dispersion of the exhaust.

13.9 Aircraft Engines

13.9.1 Thrust and Propulsive Power

While stationary turbines drive generators to produce electrical power, aircraft turbines accelerate the incoming air flow to produce thrust, which is the force F to push the airplane. The engines are powered through the combustion of a fuel.

We consider an airplane moving with the velocity $\mathcal{V}_A$ through environmental air. Due to the motion of the airplane, air is swept into the engine with velocity $\mathcal{V}_A$ and the engine accelerates the flow to the exhaust velocity $\mathcal{V}_E$ measured with respect to the engine. The temperature of the incoming air is T_A and the exhaust is at T_E; inlet and exit pressure are just the environmental pressure, but higher pressures occur inside the engine.

A sketch of the flow through the engine is shown in Fig. 13.14, which also indicates the fuel that is required to drive the engine; the working principle of the engine is discussed later.

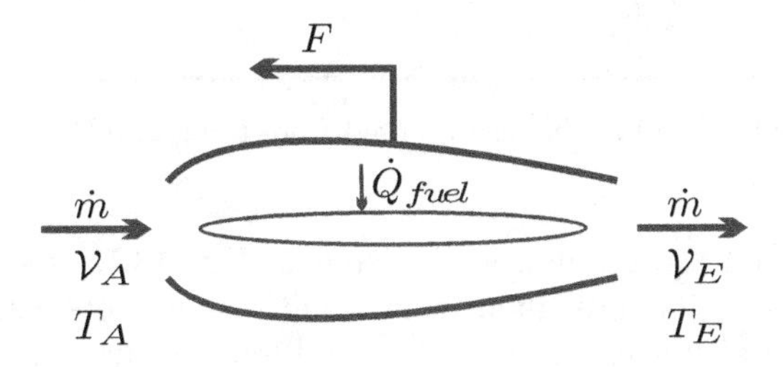

Fig. 13.14 Acceleration of air and thrust in an air engine

According to Newton's second law, the thrust is the rate of change of momentum of the air that passes the engine,

$$F = \dot{m}\left(\mathcal{V}_E - \mathcal{V}_A\right) . \tag{13.25}$$

Power is force times velocity, and thus the propulsive power provided by the engine is the product of the force acting on the airplane, i.e., the thrust, and the airplane velocity, that is

$$\dot{W}_P = F\mathcal{V}_A = \dot{m}\left(\mathcal{V}_E - \mathcal{V}_A\right)\mathcal{V}_A . \tag{13.26}$$

Before we look inside the engine to discuss its working principles, we take a look at the engine as a whole from different points of view, in order to find criteria for engine performance. Indeed, depending on the point of view of the observer, the engine seems to be performing different tasks:

For an observer resting with the engine, e.g., the pilot or a passenger, the engine consumes fuel, and accelerates and heats air that passes through. This is the frame of reference used in Fig. 13.14. For this observer the first law reads

$$\dot{m}\left[h_E - h_A + \frac{1}{2}\mathcal{V}_E^2 - \frac{1}{2}\mathcal{V}_A^2\right] = \dot{Q}_{fuel} . \tag{13.27}$$

Note, that this observer does not notice thrust and propulsive power.

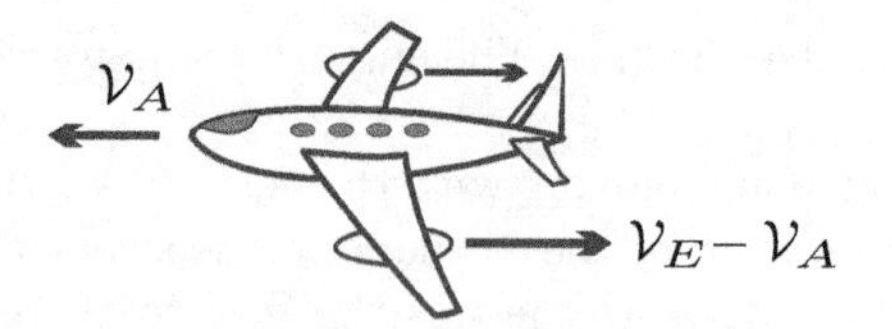

Fig. 13.15 Airplane and engine as observed from the ground

An observer on the ground, as depicted in Fig. 13.15, sees the airplane flying with velocity $\mathcal{V}_A$ through air which is at rest, and observes air expelled from the engine with velocity $(\mathcal{V}_E - \mathcal{V}_A)$ and temperature T_E. To obtain the appropriate form of the first law for this observer we use the identity

$$\frac{1}{2}\left(\mathcal{V}_E^2 - \mathcal{V}_A^2\right) = \frac{1}{2}\left(\mathcal{V}_E - \mathcal{V}_A\right)^2 + \mathcal{V}_A\left(\mathcal{V}_E - \mathcal{V}_A\right) , \tag{13.28}$$

which, when inserted into (13.27), gives the energy balance as

$$\dot{m}\left(h_E - h_A\right) + \frac{1}{2}\dot{m}\left(\mathcal{V}_E - \mathcal{V}_A\right)^2 + \dot{m}\left(\mathcal{V}_E - \mathcal{V}_A\right)\mathcal{V}_A = \dot{Q}_{fuel} . \tag{13.29}$$

We introduce the abbreviations

$$\dot{E}_{kin}^{ex} = \frac{1}{2}\dot{m}\left(\mathcal{V}_E - \mathcal{V}_A\right)^2 ,$$

for the exhaust flow of kinetic energy, and

$$\dot{Q}_H = \dot{m}\left(h_E - h_A\right) \tag{13.30}$$

for the heating rate of the air passing through the engine, that is the amount of heat required to heat the air isobarically from T_A to T_E. With these, and the definition of propulsive power (13.26), the first law (13.29) assumes the compact form

$$\dot{Q}_H + \dot{E}_{kin}^{ex} + \dot{W}_P = \dot{Q}_{fuel} . \tag{13.31}$$

This is the first law for the observer on the ground, who understands that the heat $\dot{Q}_{fuel}$ supplied through combustion of the fuel leads to three effects:

(a) Propulsive power $\dot{W}_P$; (b) Acceleration of environmental air, so that the kinetic energy of exhaust is $\dot{E}^{ex}_{kin}$; (c) Heating $\dot{Q}_H$ of the air to temperature T_E.

Of course, both forms (13.27, 13.31) of the first law are equivalent, they just differ by the point of view of the observer. However, the latter form is better suited for evaluation of engine performance.

Subsonic flight and supersonic flight have different aerodynamics, with more propulsive power required in supersonic flows. To optimize speed and efficiency, commercial airliners fly at about 80-90% of the speed of sound. For military applications speed is essential, and many fighter planes fly at supersonic speeds.

13.9.2 Air Engine Efficiency

The engine is build to deliver the propulsive power $\dot{W}_P$. Heating $\dot{Q}_H$ and acceleration $\dot{E}^{ex}_{kin}$ are side effects which must be considered as losses. The exhaust leaving the engine at T_E, $\mathcal{V}_E - \mathcal{V}_A$ has work potential (exergy) against the environment as discussed in Sec. 11.6. Since there is no way to put the exhaust to use after it is expelled from an engine in flight, it just equilibrates with the surrounding air—this is an irreversible loss, i.e., an external irreversibility associated with the process. For efficient use of the fuel one must diminish external losses, that is aim for processes with low exhaust exergy, which means low exhaust temperature T_E and low exhaust velocity $\mathcal{V}_E - \mathcal{V}_A$. We also note that low exit velocities diminish engine noise significantly.

The obvious measure for engine performance is the thermal efficiency for propulsion, defined as

$$\eta_P = \frac{\dot{W}_P}{\dot{Q}_{fuel}} = \frac{\dot{W}_P}{\dot{Q}_H + \dot{E}^{ex}_{kin} + \dot{W}_P} , \tag{13.32}$$

where the first law (13.29) was used. For fixed value of $\dot{W}_P$ the thermal efficiency of propulsion grows, when heating $\dot{Q}_H$ and exhaust kinetic energy $\dot{E}^{ex}_{kin}$ become smaller.

A common measure from fluid dynamics for propulsive efficiency is the Froude propulsive efficiency (William Froude, 1810-1879) η_F which asks how much of the gain in kinetic energy produced in the engine, as seen from the observer resting with the engine, is actually converted to propulsive power, that is

$$\eta_F = \frac{\dot{W}_P}{\dot{m}\left(\frac{1}{2}\mathcal{V}_E^2 - \frac{1}{2}\mathcal{V}_A^2\right)} = \frac{\dot{W}_P}{\dot{W}_P + E^{ex}_{kin}} = \frac{2\mathcal{V}_A}{\mathcal{V}_A + \mathcal{V}_E} . \tag{13.33}$$

The Froude efficiency is a purely mechanical measure, other than the thermal efficiency η_P it does not account for the loss through expulsion of hot exhaust. The Froude efficiency approaches unity when the outflow velocity $\mathcal{V}_E$ approaches the inflow velocity $\mathcal{V}_A$.

For a certain airplane, the required propulsive power $\dot{W}_P$ and flight velocity $\mathcal{V}_A$ are given, while inflow temperature T_A and inflow pressure p_A depend on the local condition of the air the airplane is flying through. Both efficiencies, η_F and η_P, show that an efficient engine for the airplane will have a small increase in velocity $(\mathcal{V}_E - \mathcal{V}_A)$, but a large mass flow $\dot{m}$, so that $\dot{W}_P = \dot{m}(\mathcal{V}_E - \mathcal{V}_A)\mathcal{V}_A$ has the required value. Moreover, to keep the thermal loss small, the exhaust temperature should be as low as possible.

13.9.3 Turbojet Engine

Standard air turbines operate similar to stationary gas turbines for power generation. While in the latter the turbine serves to drive the compressor and the generator, an air engine has a smaller turbine, which only serves to drive the compressor. After the turbine, the still hot and compressed air is expanded in a nozzle to accelerate the flow to $\mathcal{V}_E$.

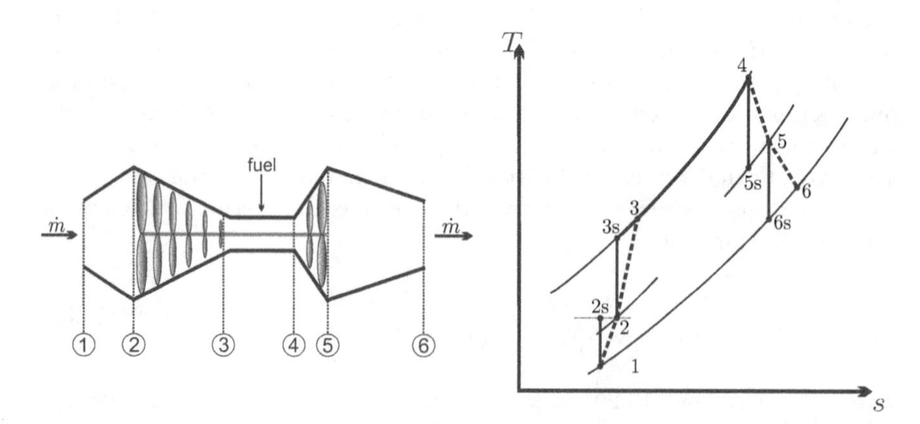

Fig. 13.16 Schematic and T-s-diagram for a standard air engine with diffuser (1-2), compressor (2-3), combustion chamber (3-4), turbine (4-5) and nozzle (5-6)

Figure 13.16 shows a schematic and the T-s-diagram for a simple standard air turbine, i.e., a turbojet engine, consisting of diffuser, compressor, combustion chamber, turbine and nozzle. The diffuser decelerates the inflow to increase the pressure, thus lowering the work required for compression. The turbine is used solely to drive the compressor, the work for both is equal, with different sign. The hot pressurized combustion product is expanded and accelerated in the nozzle. Since the throughput is fast, there is no time to exchange heat, and diffuser, compressor, turbine and nozzle are considered to be adiabatic.

For the discussion of air engines we shall rely, again, on the air standard analysis, that is we ignore any composition changes and treat the working fluid as air. The standard air turbine operates on the following cycle

$$\begin{array}{lll} \text{1-2 adiabatic diffuser:} & h_2 = h_1 + \frac{1}{2}\mathcal{V}_A^2 \,, \\ \text{2-3 adiabatic compressor:} & w_C = h_2 - h_3 \,, \\ \text{3-4 isobaric heating (combustion):} & q_{in} = h_4 - h_3 \,, \\ \text{4-5 adiabatic turbine:} & w_T = -w_C = h_4 - h_5 \,, \\ \text{5-6 adiabatic nozzle:} & \mathcal{V}_E = \sqrt{2\left(h_5 - h_6\right)} \,, \\ \text{6-1 equilibration with environment.} & \end{array} \tag{13.34}$$

Again, we consider the ideal process under cold-air approximation (constant specific heat) to get insight into the parameters that determine performance. Inlet conditions T_1, p_1, $\mathcal{V}_A$ are given; compressor pressure ratio $P_C = \frac{p_3}{p_2}$ and turbine inlet temperature T_4 are design parameters for the engine. For reversible processes, we find the following relations between the properties at the corner points of the process:

$$\begin{array}{ll} T_2 = T_1 + \frac{\mathcal{V}_A^2}{2c_p} \,, & p_2 = p_1 \left(\frac{T_2}{T_1}\right)^{\frac{k}{k-1}} \,, \\ T_3 = T_2 P_C^{\frac{k-1}{k}} \,, & p_3 = p_4 = p_2 P_C \,, \\ T_5 = T_4 + T_2 - T_3 \,, & p_5 = p_4 \left(\frac{T_5}{T_4}\right)^{\frac{k}{k-1}} \,, \\ T_6 = T_5 \left(\frac{p_1}{p_5}\right)^{\frac{k-1}{k}} \,, & p_6 = p_1 \,. \end{array} \tag{13.35}$$

Combining all results yields the exhaust velocity, specific propulsive power, exhaust temperature and heat supply as

$$\begin{aligned} \mathcal{V}_E &= \sqrt{2c_p \left[T_4 \left(1 - P^{\frac{1-k}{k}}\right) + T_1 \left(1 - P^{\frac{k-1}{k}}\right)\right] + \mathcal{V}_A^2} \,, \\ w_P &= \frac{\dot{W}_P}{\dot{m}} = \left(\mathcal{V}_E - \mathcal{V}_A\right) \mathcal{V}_A \,, \\ T_6 &= T_4 P^{\frac{1-k}{k}} \,, \\ q_{in} &= c_p \left[T_4 - T_1 P^{\frac{k-1}{k}}\right] \,; \end{aligned} \tag{13.36}$$

here, $P = \frac{p_2}{p_1} P_C$ is the overall pressure ratio of the engine. For given turbine inlet temperature T_4, the propulsive power has a maximum at pressure ratio $P_{\max} = \left(\frac{T_4}{T_1}\right)^{\frac{k}{2k-2}}$, while propulsive efficiency grows with P.

For a fixed pressure ratio P, exhaust temperature T_6 and heat supply q_{in} grow linearly with the turbine inlet temperature T_4, while exhaust speed $\mathcal{V}_E$ and propulsive power w_P grow slower. Thus, increase of the turbine inlet temperature increases propulsive power, but reduces the propulsive thermal efficiency $\eta_P = w_p/q_{in}$, due to larger external irreversibilities.

From the above follows that engines with high pressure ratio and large turbine inlet temperatures provide large propulsive power. If one is interested mainly in power, and efficiency has lower importance, one will run a engine under these conditions. However, fuel cost is significant, and directly related to efficiency. Most of today's jet engines are high bypass turbofan engines, which are far more efficient, as will be discussed in Sec. 13.9.5.

Standard turbojet engines, or turbofan engines with low bypass ratio, are employed for supersonic propulsion, mostly for military applications. In afterburner engines additional boost is obtained by injecting fuel into the hot nozzle flow, where it burns, and further heats the flow, and thus gives even higher nozzle exit velocities.

13.9.4 Example: Turbojet Engine

An airplane cruises with a velocity $\mathcal{V}_A = 300 \frac{\text{m}}{\text{s}}$ at about 9000 m altitude where the local pressure and temperature are $p_1 = 32\,\text{kPa}$ and $T_1 = 241\,\text{K}$. The compressor pressure ratio is $P = 12$, and the turbine inlet temperature is 1400 K. We assume isentropic efficiencies of 80% for compressor and turbine, and 95% for diffuser and nozzle, and compute the velocity and temperature of the exhaust gas, propulsive power for a mass flow of $\dot{m} = 50 \frac{\text{kg}}{\text{s}}$, thermal and Froude efficiencies, and the work potential of the exhaust. The working fluid is air, as ideal gas with variable specific heats. The process is as shown in Fig. 13.16, with irreversible subprocesses. We go through the processes step by step, all numerical values will be entered into a table that is found further below.

Diffuser (1-2): The first law gives $h_2 = h(T_2) = h_1 + \frac{1}{2}\mathcal{V}_A^2$ which determines T_2. The pressure after an isentropic diffuser that reaches T_2 is obtained as $p_{2s} = p_1 \frac{p_r(T_2)}{p_r(T_1)}$; however, this is not the pressure after the irreversible diffuser. The isentropic efficiency of the diffuser $\eta_D = \frac{h_x - h_1}{h_2 - h_1}$ defines the fictitious temperature T_x that would be obtained by isentropic compression to the actual pressure p_2 for the irreversible compressor. We find $h_x = h(T_x) = h_1 + \eta_D(h_2 - h_1)$, and hence T_x. With that, the pressure after the diffuser is $p_2 = p_x = p_1 \frac{p_r(T_x)}{p_r(T_1)}$.

Compressor (2-3): The exit pressure of the compressor is $p_3 = p_2 P$. Moreover, we find T_{3s} from $p_r(T_3) = p_r(T_2) P$, and then h_3 and T_3 follow from the isentropic compressor efficiency as $h_3 = h(T_3) = h_2 + (h_{3s} - h_2)/\eta_C$.

Heating (3-4): As always, the combustion chamber is assumed to be isobaric, $p_4 = p_3$; the temperature T_4 is given.

Turbine (4-5): The turbine is required to drive the compressor, that is $w_T = -w_C$, or $h_4 - h_5 = h_3 - h_2$ from which we find h_5 and then T_5. A reversible turbine expanding to the same pressure p_5 would end with enthalpy $h_{5s} = h_4 + (h_5 - h_4)/\eta_T$ which follows from the definition of the isentropic turbine efficiency η_T. The pressure finally is obtained from the isentropic relation between points 4 and 5s, $p_5 = p_{5s} = p_4 \frac{p_r(T_{5s})}{p_r(T_4)}$.

Nozzle (5-6): The nozzle expands the turbine exhaust to the environmental pressure $p_6 = p_1$. An isentropic nozzle would give the turbine exit temperature T_{6s}, and thus the enthalpy $h_{6s} = h(T_{6s})$, which follow from $p_r(T_{6s}) = p_r(T_5) \frac{p_6}{p_5}$. Exit enthalpy, and temperature, follow from the isentropic nozzle efficiency η_N, one finds $h_6 = h_5 + (h_{6s} - h_5)\eta_N$. The nozzle exit velocity is $\mathcal{V}_6 = \sqrt{2(h_5 - h_6)}$.

The complete data is shown in the following table, which also contains entropy values for inlet and exhaust:

	p/ kPa	T/ K	$h/\frac{\text{kJ}}{\text{kg}}$	$p_r(T)$	$\mathcal{V}/\frac{\text{m}}{\text{s}}$	$s^0(T)/\frac{\text{kJ}}{\text{kg K}}$
1	32.0	241.0	241.9	0.476	300	6.917
2s	58.2				0	
x	56.5	283.7	284.7	0.841	0	
2	56.5	286.0	286.9	0.865	0	
3s	678	576.6	583.3	10.38	0	
3	678	647.0	657.4		0	
4	678	1400	1516	332.2	0	
5s	175	1005	1052	85.77	0	
5	175	1086	1145	116.8	0	
6s	32	701.1	715.3	21.36	0	
6	32	721.1	736.8		904	8.040

With all data known, we can compute propulsive power and heat supply,

$$\dot{W}_P = \dot{m}(\mathcal{V}_E - \mathcal{V}_A)\mathcal{V}_A = 9.0\,\text{MW}\,,$$
$$\dot{Q}_{in} = \dot{m}(h_4 - h_3) = 42.9\,\text{MW}\,.$$

Thermal propulsive efficiency and Froude efficiency thus are

$$\eta_P = \frac{\dot{W}_P}{\dot{Q}_{in}} = 21.1\% \quad \text{and} \quad \eta_F = \frac{2\mathcal{V}_A}{\mathcal{V}_A + \mathcal{V}_E} = 49.9\%\,.$$

The efficiencies are low, because the exhaust has considerable work potential against the environment, which is not used. Substituting the appropriate values into Eq. 11.18, the work potential of the exhaust, which is just the work lost to external irreversibilities, is[5]

$$\dot{W}_{rev} = \dot{m}\left[h_6 - h_1 - T_1\left[s^0(T_6) - s^0(T_1)\right] + \frac{1}{2}(\mathcal{V}_E - \mathcal{V}_A)^2\right] = 20.3\,\text{MW}\,.$$

[5] Note that inflow (state 1) and outflow (state 6) are at the same pressure; therefore there is no pressure contribution to the entropy difference.

Almost half of the loss (9.12 MW) is due to kinetic energy losses, the remainder is due to thermal losses. Altogether, the external work loss is more than twice the actual propulsive power produced. Both must be reduced to improve engine efficiency.

13.9.5 Bypass Turbofan Engine

The general discussion of air engine efficiency has shown that engines with relatively slow and cold exhaust are more efficient. Since propulsive power is given as $\dot{W}_P = \dot{m}\,(\mathcal{V}_E - \mathcal{V}_A)\,\mathcal{V}_A$, lower exit velocity $\mathcal{V}_E$ must be compensated by larger mass flow, to generate the desired power. Bypass turbofan engines have smaller exit velocities, increased mass flow, and also smaller (average) exit temperature, and thus have small external losses, and high efficiency.

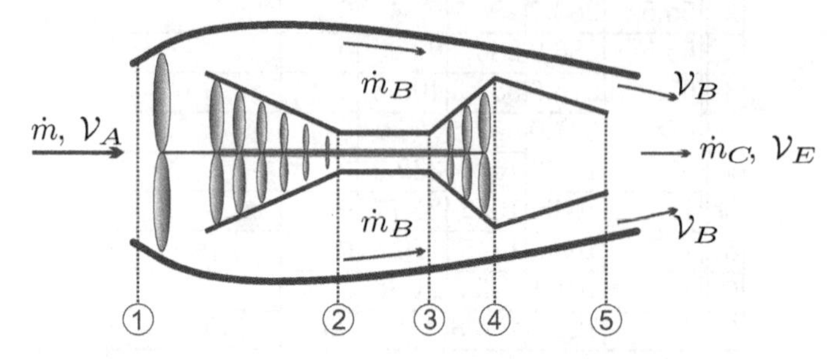

Fig. 13.17 Bypass turbofan engine with fan, compressor, combustion chamber, turbine and nozzle. The fan forces air through the bypass, where it is accelerated

In a turbofan engine, sketched in Fig. 13.17, an additional turbine stage is added to drive the fan. The fan forces air through a duct outside the engine core, called the bypass, where the air is accelerated to velocity $\mathcal{V}_B$. The incoming mass flux $\dot{m}$ is split into two streams, the bypass stream $\dot{m}_B$ and the core stream $\dot{m}_C$. Only the core stream runs through compressor, combustion chamber, turbine and nozzle. The bypass flow is not heated; this lowers the exhaust temperature and improves efficiency.

This arrangement increases the mass flow through the engine and decreases the outflow velocity and temperature, and thus allows to produce propulsive power at higher efficiency. The bypass ratio is defined as the ratio of the inlet cross sections of core and bypass, $M_B = \frac{A_B}{A_C} = \frac{\dot{m}_B}{\dot{m}_C}$, and we expect better efficiency for larger bypass ratio. The fan is particularly important at take-off, where the power demand is high, but the airplane speed $\mathcal{V}_A$ is low.

Again, we study the ideal process under cold-air standard conditions. The power demand for the adiabatic fan follows from the first law. For the ideal process, we have isentropic compression by the fan followed by isentropic expansion in the duct. Inlet and exit pressures agree, as do the respective entropies, and thus the exit temperature equals the inlet temperature. Hence, the fan work is just the difference in kinetic energy for the bypass flow,

$$\dot{W}_F = \dot{m}_B \left[\frac{1}{2}\mathcal{V}_A^2 - \frac{1}{2}\mathcal{V}_B^2 \right] . \tag{13.37}$$

For irreversible fan and duct, the exit temperature would be slightly higher, and enthalpy terms would appear in the first law.

The processes in the core are the same as for the standard turbine, but now we use the numbering as in Fig. 13.17. As long as all processes are ideal, the evaluation is easiest when we begin with the first law balanced over the total core, between states 1 and 5. The turbine work is used to drive the compressor and the fan, but since turbine and compressor are within the control volume, only the fan work appears in the first law, which reads[6]

$$\dot{m}_C \left[h_5 - h_1 + \frac{1}{2}\mathcal{V}_E^2 - \frac{1}{2}\mathcal{V}_A^2 \right] = \dot{Q}_{in} - \left(-\dot{W}_F \right) . \tag{13.38}$$

The heat supply to the compressed air is

$$\dot{Q}_{in} = \dot{m}_C \left(h_3 - h_2 \right) = \dot{m}_C c_p \left[T_3 - T_1 P^{\frac{k-1}{k}} \right] , \tag{13.39}$$

where T_3 is the turbine inlet temperature, and $T_1 P^{\frac{k-1}{k}} = T_2$ is the temperature after the compressor; here, P is the overall pressure ratio of the engine.

The compressed gas at state 3 is expanded reversibly, first in the adiabatic turbine to state 4, and then in the adiabatic nozzle to exhaust state 5, so that the exit temperature is

$$T_5 = T_3 P^{\frac{1-k}{k}} . \tag{13.40}$$

For simplicity, we assume that core and bypass exit velocities are equal, $\mathcal{V}_B = \mathcal{V}_E$. Then, combining the above equations and solving for the exit velocity gives

$$\mathcal{V}_E = \sqrt{\frac{1}{1+M_B} 2c_p \left[T_3 \left(1 - P^{\frac{1-k}{k}} \right) + T_1 \left(1 - P^{\frac{k-1}{k}} \right) \right] + \mathcal{V}_A^2} . \tag{13.41}$$

This result differs from the result for the standard turbine (13.36) only in that the factor $\frac{1}{1+M_B} = \frac{\dot{m}_C}{\dot{m}_C + \dot{m}_B}$ appears in the first term; the previous result is found for $M_B = 0$.

For given turbine inlet temperature T_3, the exit velocity $\mathcal{V}_E$ and the propulsive power $\dot{W}_P = (\dot{m}_B + \dot{m}_C)(\mathcal{V}_E - \mathcal{V}_A)\mathcal{V}_A$ both have a maximum for the

[6] $\dot{W}_F < 0$ is defined as the work consumed by the fan, the work delivered from the core to drive the fan is $\left(-\dot{W}_F \right)$.

overall pressure ratio $P_{\max} = \left(\frac{T_3}{T_1}\right)^{\frac{k}{2k-2}}$. In modern turbines some of the compressed air is forced through small ducts in the turbine blades for efficient blade cooling. This allows very high turbine inlet temperatures of up to $T_3 = 1700\,\mathrm{K}$, with corresponding pressure ratios of $P_{\max} \simeq 30$ (for $T_1 = 240\,\mathrm{K}$).

At the optimal pressure ratio $P_{\max}$, exit velocity, propulsive power, and heat supply become

$$\mathcal{V}_E^{\max} = \sqrt{\frac{1}{1+M_B} 2c_p \left(\sqrt{T_3} - \sqrt{T_1}\right)^2 + \mathcal{V}_A^2}\,, \tag{13.42}$$

$$\dot{W}_P^{\max} = \dot{m}_C \left(1 + M_B\right) \left(\mathcal{V}_E^{\max} - \mathcal{V}_A\right) \mathcal{V}_A\,, \tag{13.43}$$

$$\dot{Q}_{in}^{\max} = \dot{m}_C c_p \sqrt{T_3} \left(\sqrt{T_3} - \sqrt{T_1}\right)\,, \tag{13.44}$$

so that the propulsive thermal efficiency becomes

$$\eta_{\max} = \frac{\dot{W}_P^{\max}}{\dot{Q}_{in}^{\max}} = \frac{\left[\sqrt{\frac{2c_p}{1+M_B}\left(\sqrt{T_3} - \sqrt{T_1}\right)^2 + \mathcal{V}_A^2} - \mathcal{V}_A\right] \mathcal{V}_A}{\frac{c_p}{1+M_B}\sqrt{T_3}\left[\sqrt{T_3} - \sqrt{T_1}\right]}\,. \tag{13.45}$$

This efficiency depends on turbine inlet temperature T_3 and bypass ratio M_B. Figure 13.18 shows the thermal efficiency $\eta_{\max}$ as function of M_B, T_3. For given turbine inlet temperature T_3, the propulsive efficiency grows with the bypass ratio, while for given bypass ratio there is a optimum for the temperature.

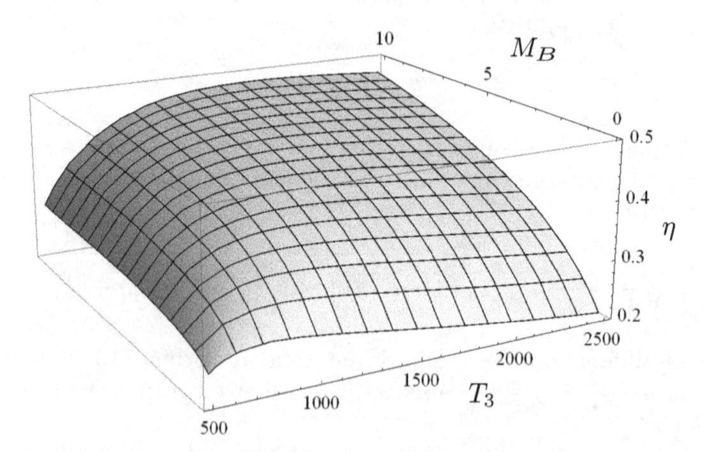

Fig. 13.18 Efficiency $\eta_{\max}$ of bypass turbofan engine over bypass ratio M_B and turbine inlet temperature T_3 (for optimal pressure ratio $P_{\max}$, $\mathcal{V}_A = 280\,\frac{\mathrm{m}}{\mathrm{s}}$)

The bypass ratio is limited, due to size and weight limitations of the engine, and we ask for the optimum turbine inlet temperature T_3 for given bypass ratio M_B, which follows from the condition

$$\left(\frac{\partial\eta_{\max}}{\partial T_3}\right)_{M_B} = 0\,.$$

The resulting equation between the optimal temperature $T_3^{\rm opt}$ and M_B is best solved for the bypass ratio,

$$M_B = \frac{c_p T_1}{\frac{1}{2}\mathcal{V}_A^2}\frac{\left(\sqrt{\frac{T_3^{\rm opt}}{T_1}}-1\right)^4}{2\sqrt{\frac{T_3^{\rm opt}}{T_1}}-1} - 1\,. \tag{13.46}$$

For an aircraft travelling at speed $\mathcal{V}_A = 280\frac{m}{s}$ through air ($c_p = 1.004\frac{\rm kJ}{\rm kg\,K}$) at $T_1 = 240\,{\rm K}$, the optimal turbine inlet temperature has the values $T_3^{\rm opt} = \{1300\,{\rm K}, 1500\,{\rm K}, 1700\,{\rm K}\}$ for bypass ratios $M_B = \{4.22, 6.78, 9.83\}$, the outflow velocity is $449\frac{\rm m}{\rm s}$.

The optimum efficiency for a bypass turbojet engine with given turbine inlet temperature is obtained from inserting (13.46) into (13.45) as

$$\eta_{\max}^{\rm opt} = \frac{\dot{W}_P^{\max}}{\dot{Q}_{in}^{\max}} = \frac{\left(1-\sqrt{\frac{T_1}{T_3}}\right)^2}{1-\frac{1}{2}\sqrt{\frac{T_1}{T_3}}}\,. \tag{13.47}$$

Optimum propulsive efficiency grows with increasing turbine inlet temperature. With the same data as before ($T_1 = 240\,{\rm K}$, $T_3 = 1700\,{\rm K}$) the optimized reversible bypass turbojet engine reaches a thermal propulsive efficiency of 48%.

The optimized values computed above from simplifying assumptions (cold air, air standard, reversible processes, same exit velocity for bypass and core flows) are not too far from those encountered in state-of the art turbofan engines. These engines are quite complex, with 2 or 3 turbine-compressor and turbine-fan pairs running on concentric shafts, and sometimes gears, so that high and low pressure turbines and compressors, and the fan run at optimal rotational speeds.

Noise reduction is an important task in commercial aircraft. Engine noise is high when engine exit flows are supersonic, hence real-life engines must be constructed to have subsonic exit velocities. From the above discussion it is evident that turbofans allow low exit velocities, and hence relatively silent operation.[7]

[7] The above optimization of the bypass engine relied on several simplifying assumptions (cold air standard, all processes reversible) which leads to somewhat inaccurate results. Hence, there is no value in including the requirement for subsonic outflow into the arguments.

In propjet engines, fan and bypass are replaced by a propeller, which provides all thrust, the turbine expands to the environmental pressure and provides the work to run the propeller. Propeller engines are efficient, but for aerodynamic reasons are limited to lower velocities of not more than $600\,\frac{\text{m}}{\text{s}}$, while turbofan engines can operate at subsonic and supersonic flight speeds.

Problems

13.1. Stirling Cycle

A Stirling cycle with 5 g of air as working fluid is heated by solar radiation and rejects heat to the environment. The highest and lowest temperatures reached in the cycle are 1000 K and 300 K, respectively, and the maximum pressure ratio is 10. Draw the process in a T-s-diagram and in a p-v-diagram, then determine the thermal efficiency and the power produced when the engine runs at 400 rpm. Determine the entropy changes, heat and work per unit mass for all four processes.

13.2. Stirling Cycle for Refrigeration

A Stirling engine with helium as working gas is considered for refrigeration purposes. The goal is to withdraw heat at a temperature of $T_L = 150\,\text{K}$ and reject it to the environment at $T_H = 300\,\text{K}$. Helium is a monatomic gas, which is well described as an ideal gas with constant specific heats.

1. Draw T-s-diagram and p-v-diagram for the cycle.
2. For a volume ratio between largest and smallest volume of 3, compute heat and work per unit mass for all four processes, and the coefficient of performance.
3. The computation of specific heat and work is independent of the pressure. Discuss the role of the pressure in the performance of the engine, why is a high pressure desirable?
4. Determine the pressure at all corner points when the highest pressure in the engine is 10 bar.

13.3. Stirling Cycle for Refrigeration

A Stirling engine with argon as working fluid is used for refrigeration purposes. Heat exchange with the cold and warm surroundings takes place at $T_H = 27\,°\text{C}$ and $T_L = -73\,°\text{C}$, respectively. The highest pressure in the cycle is $p_H = 12\,\text{bar}$ and the smallest volume is one third of the largest volume.

1. Plot the Stirling cycle in a T-s-diagram, and in a p-v-diagram, number the corner points.
2. Compute the pressures and specific volumes on all corner points
3. Discuss the regenerator: show that the amounts of heat rejection and heat supply in the two isochoric processes have the same absolute value, but different signs.
4. Compute the coefficient of performance of the cycle.

5. Assume the cylinder of the engine contains an air mass of 40 g, and the engine runs at 300 rpm – what is its refrigeration capacity?

13.4. Compression Modes
Air (ideal gas with variable specific heats) is compressed in a compressor from $p_1 = 1.2\,\text{bar}$, $T_1 = 280\,\text{K}$ to $p_2 = 12\,\text{bar}$. The incoming volume flow is $1\frac{\text{m}^3}{\text{s}}$. Determine the power consumption for the following cases:

1. Isothermal reversible compression.
2. Isentropic compression.
3. Polytropic reversible compression with $n = 1.2$.
4. Compression in two isentropic stages with intercooling to T_1 at $p_m = \sqrt{p_1 p_2}$.

Draw a p-v- and a T-s-diagram which shows the four process curves. Hint: For computation of isothermal and polytropic case use that $w_{12} = -\int_1^2 v dp$.

13.5. Two Stage Compressor with Irreversibilities
A two stage compression system with intercooling is used to increase the pressure of an ideal gas. Specifically, the gas enters the system at p_1, T_1, and leaves the first compressor (isentropic efficiency η_{C1}) at pressure p_2. It is then isobarically intercooled to T_1, and compressed to p_4 in the second compressor (isentropic efficiency η_{C2}). Assume constant specific heats, and determine the pressure p_2 that should be chosen to minimize the work requirement of the system.

13.6. Gas Turbine Cycle with Regeneration

1. Draw a schematic for a gas turbine system for electricity generation with irreversible single stage compression, two stages of irreversible expansion with reheat, and a regenerator. Enumerate the relevant corner points of the process.
2. Draw the corresponding T-s-diagram.
3. Express the thermal efficiency in terms of enthalpies.

13.7. Brayton Cycle with Regeneration
A gas turbine running on the Brayton cycle has an efficiency of 35.9%, at pressure ratio 14.7. The turbine inlet temperature is 1288 °C, and the air entering the engine is at 1 bar, 20 °C. The engine produces a net power of 174.9 MW and the mass throughput is $1690\frac{\text{t}}{\text{h}}$.

1. Determine the isentropic efficiencies of turbine and compressor.
2. Determine the thermal efficiency for this gas turbine for the case that a regenerator with 80% effectiveness is added to the cycle.

13.8. Optimal Reheat Pressure
Prove the following statement from the text for a reheat turbine with n-stages: If the turbines have the same inlet temperature, and the same isentropic efficiency, the maximum work is obtained when they have the same pressure ratio.

13.9. Brayton Cycle with Intercooling, Reheat and Regeneration
A regenerative gas turbine cycle uses two stage of compression with intercooling, and two stages of expansion with reheating. The pressure ratio for each stage is 3.5, the turbine inlet temperature is 1400 K for both turbines, and between the compressors the air is cooled back to the environmental temperature of 290 K. The isentropic efficiencies of the compressors and turbines are 80% and 85%, respectively, and the regenerator effectiveness is 80%. Determine:

1. The enthalpies at all principal states.
2. The net work and the back work ratio.
3. The thermal efficiency for the system as described, and for the case that no regenerator is present.
4. The work potential of the turbine exhaust, and of the final exhaust.

As usual: draw schematic and diagrams.

13.10. Gas Turbine with Regenerator
A gas turbine with air (non-constant specific heats) as working fluid operates according to the following cycle:

1-2: Adiabatic compression of air at $T_1 = 300\,\mathrm{K}$, $p_1 = 1\,\mathrm{bar}$ to $T_2 = 620\,\mathrm{K}$, $p_2 = 9.74\,\mathrm{bar}$.

2-3: Isobaric heating of the working fluid in the regenerator, the temperature T_3 is 40 K below the temperature of the turbine exhaust, T_5.

3-4: Further isobaric heating in the combustion chamber to $T_4 = 1300\,\mathrm{K}$.

4-5: Adiabatic expansion in turbine to pressure $p_5 = p_1$, with isentropic efficiency of 92%.

5-6: Isobaric cooling in the regenerator.

1. Draw a schematic, and a T-s-diagram.
2. Make a table with pressures, temperatures and enthalpies at the points 1 to 6.
3. Determine the thermal efficiency and the back-work-ratio of the cycle:
   a) when it operates with regenerator
   b) when it operates without regenerator
4. Compute the isentropic efficiency of the compressor.

13.11. Combined Cycle: Gas Turbine and Steam Power Plant
A combined cycle power plant consists of a gas turbine cycle (thermal efficiency 28%), and a steam power plant (thermal efficiency 46%). The exhaust of the gas turbine is used to provide the heat for generating steam in a heat recovery steam generator. Assume that the HRSG has an efficiency of 92%, and compute the overall efficiency of the system.

13.12. Turbojet Engine
A turbojet engine drives an airplane traveling with velocity $290\,\frac{\mathrm{m}}{\mathrm{s}}$ at a height where the pressure is 28 kPa, and the temperature is $-40\,^{\circ}\mathrm{C}$. The compressor

pressure ratio is 11, and the turbine inlet temperature is 1300 K. The mass flow through the engine is $60\frac{\text{kg}}{\text{s}}$.

Assume isentropic efficiencies of 82% for compressor and turbine, 95% for the nozzle, and 100% for the diffuser. Determine the velocity of the exhaust gas, the propulsive power, the rate of fuel consumption when the heating value of the fuel is $42000\frac{\text{kJ}}{\text{kg}}$, the thermal efficiency, and the Froude propulsive efficiency.

13.13. Air Engine

An airplane propelled by a standard turbo-jet engine flies at Mach number $M = 0.9$ in an environment where the pressure is 40 kPa and the temperature is 240 K. The heat added to the air flowing through the engine is $q = 550\frac{\text{kJ}}{\text{kg}}$ and the hot air leaves the engine at 650 K. The engine inlet has a diameter of 1 m. Assume that the working fluid is air as ideal gas.

Determine outflow velocity, thrust, propulsive power, thermal efficiency, and Froude efficiency of the engine.

13.14. Air Engine

Air at 25 kPa, 225 K enters a turbojet engine in flight at an altitude of 10 000 m, the flight velocity is $290\frac{\text{m}}{\text{s}}$. The pressure ratio across the compressor is 10. The turbine inlet temperature is 1300 K, and the pressure at the nozzle exit is 25 kPa again. The diffuser and nozzle processes are isentropic, compressor and turbine have isentropic efficiencies of 90% and 95%, respectively, and there is no pressure drop for flow through the combustor.

Consider air as ideal gas with constant specific heats, $R = 0.287\frac{\text{kJ}}{\text{kg K}}$, $c_p = 1.004\frac{\text{kJ}}{\text{kg K}}$.

Neglect kinetic energy except at the diffuser inlet and the nozzle exit.

1. Draw a schematic of the engine, and the corresponding T-s-diagram.
2. Make a table with the values of pressure and temperature at each principal state.
3. Compute the velocity at the nozzle exit.
4. Compute the thrust of the engine and the propulsive power for a mass flow rate of $80\frac{\text{kg}}{\text{s}}$.
5. Determine thermal efficiency and Froude efficiency.

13.15. Bypass Turbofan Engine

A bypass turbo fan engine has a bypass ratio of 5.5 (the mass flow through the bypass is 5.5 times the mass flow through the gas turbine), and propels an aircraft cruising at $250\frac{\text{m}}{\text{s}}$ in high altitude where the pressure is 30 kPa and the temperature is 230 K. The mass flow through the gas turbine core is $30\frac{\text{kg}}{\text{s}}$. Assume variable specific heats.

The flow through the bypass consists of isentropic diffuser, fan, nozzle.

The gas turbine process is as follows:

1-2: Compression in isentropic diffuser.

2-3: Isentropic compressor, pressure ratio $p_3/p_2 = 10$.

3-4: Isobaric heating in combustion chamber to 1300 K.
4-5: Turbine TC to drive the compressor.
5-6: Turbine TF to drive the fan.
6-7: Isentropic expansion in nozzle.

1. Make a sketch of the engine, and draw the corresponding T-s-diagram.
2. Determine the power required to drive the fan, when the bypass outflow velocity is $420\frac{\mathrm{m}}{\mathrm{s}}$.
   Hint: Balance the complete bypass. Pressures at inlet and outlet are equal to environmental pressure. Then, for isentropic operation, the outlet temperature is equal to the inlet temperature (show that!)
3. Determine temperature, pressure, relative pressure, enthalpy, and outflow velocity at all 7 points. Provide a table with the values.
4. Compute the propulsive power of the engine, its thermal efficiency, its Froude efficiency, and the power that could be generated from the exhaust by equilibrating it to the environment.

Chapter 14
Compressible Flow: Nozzles and Diffusers

14.1 Sub- and Supersonic Flows

Gas flows through nozzles and diffusers show an interesting behavior when their speed reaches, or is above, the speed of sound of the gas. This behavior must be well understood for the proper design of devices and engines. In this section we study the laws that govern the transition from sub- to supersonic flow and vice-versa, and draw relevant conclusions for the design of nozzles and diffusers, and for rocket motors. We will see that subsonic nozzles have a converging cross section, while supersonic nozzles have a converging-diverging cross section.

14.2 Speed of Sound

The ear reacts to small pressure oscillations,[1] that is sound is a pressure wave. The speed of sound is the speed with which such a small pressure disturbance travels through a medium. To compute its value, we consider a simple experiment, shown in Fig. 14.1 (left): In a long pipe, a wave is created by a small push of a piston. This wave travels through the pipe with the speed of sound, a.

To analyze the wave, and to compute the speed of sound, it is best to consider the wavefront from a co-moving frame, as depicted on the right of the figure. The wave travels into undisturbed fluid with properties p, ρ, h, s, and the observer travelling with the wave sees fluid coming in at the speed of sound a. The fluid left behind by the wave has slightly altered properties $p + dp, \rho + d\rho, h + dh, s + ds$, and the observer on the wave sees it leaving with speed $a - d\mathcal{V}$.

Since the wave travels fast, there is no time for heat exchange, thus we can assume that the wave is adiabatic. Moreover, we shall ignore irreversibilities through friction, that is we consider reversible wave propagation.

[1] Large pressure difference can damage the ear!

H. Struchtrup, *Thermodynamics and Energy Conversion*,
DOI: 10.1007/978-3-662-43715-5_14, © Springer-Verlag Berlin Heidelberg 2014

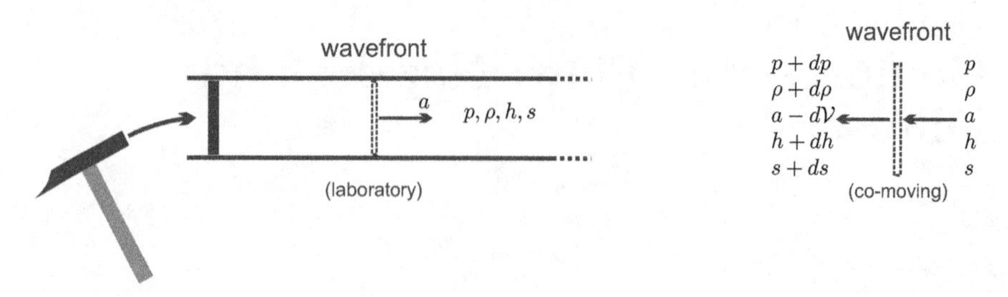

Fig. 14.1 A soundwave travelling through a pipe. Left: In the laboratory frame. Right: In a co-moving frame.

Indeed, sound waves in air are only very weakly damped. That we hear noises from far away only weakly is due to the spherical propagation of sound waves, where the sound energy is distributed over shells of surface $4\pi R^2$ when R is the distance from the sound source. Thus local sound energy, which is what we hear, is inversely proportional to the square of the distance from the source, that is proportional to $1/R^2$. In a pipe, however, the propagation is one-dimensional, waves travel as plane waves with unchanged wave surface area, which is equal to the pipe cross section.

Applying the mass balance to the wave front yields

$$\dot{m} = \rho A \mathcal{V} = const. \tag{14.1}$$

Since the cross section A does not change, this yields

$$(\rho + d\rho)(a - d\mathcal{V}) = \rho a . \tag{14.2}$$

The changes $d\mathcal{V}$ and $d\rho$ are so small that their product can be ignored, so that the above reduces to

$$\frac{1}{\rho} d\rho = \frac{1}{a} d\mathcal{V} . \tag{14.3}$$

Since the wave is adiabatic, and does not exchange any work, the first law simply gives

$$\dot{m}\left(h + \frac{1}{2}\mathcal{V}^2\right) = const. , \tag{14.4}$$

which, with $\dot{m} = const.$, reduces to

$$h + dh + \frac{1}{2}(a - d\mathcal{V})^2 = h + \frac{1}{2}a^2 , \tag{14.5}$$

so that, with (14.3)

$$dh = a d\mathcal{V} = \frac{a^2}{\rho} d\rho . \tag{14.6}$$

Since the wave is adiabatic and reversible, the second law gives

$$\dot{m}s = const. \tag{14.7}$$

or, since $\dot{m} = const.$,

$$s = const. \text{ or } \quad ds = 0\,; \tag{14.8}$$

the entropy remains unchanged, the flow is isentropic.

We insert the above equations (14.3, 14.6, 14.8) into the Gibbs equation $Tds = dh - \frac{1}{\rho}dp$ to find

$$0 = a^2 \frac{1}{\rho} d\rho - \frac{1}{\rho} dp\,. \tag{14.9}$$

Solving for a we find

$$a = \sqrt{\left(\frac{\partial p}{\partial \rho}\right)_s}\,, \tag{14.10}$$

where the subscript s indicates that the derivative must be taken at constant entropy.

14.3 Speed of Sound in an Ideal Gas

The speed of sound depends on the material, and for its computation from (14.10) one needs to know the thermal and caloric equations of state. Here, we determine the speed of sound for an ideal gas. Starting point is, again, the Gibbs equation, together with the caloric equation of state, $h = h(T) = \int c_p dT$,

$$Tds = dh - \frac{1}{\rho}dp = c_p dT - \frac{1}{\rho}dp\,. \tag{14.11}$$

With $ds = 0$ and the ideal gas law written as $T = \frac{p}{\rho R}$, we obtain

$$0 = c_p d\left(\frac{p}{\rho R}\right) - \frac{1}{\rho}dp = \frac{c_p}{\rho R}dp - \frac{c_p p}{\rho^2 R}d\rho - \frac{1}{\rho}dp\,. \tag{14.12}$$

With $c_v = c_p - R$ this yields

$$\frac{c_v}{\rho R}dp = \frac{c_p p}{\rho^2 R}d\rho\,, \tag{14.13}$$

or

$$\left(\frac{\partial p}{\partial \rho}\right)_s = \frac{c_p}{c_v}\frac{p}{\rho} = kRT\,. \tag{14.14}$$

Thus, the speed of sound in an ideal gas depends on temperature as

$$a = \sqrt{kRT}\,. \tag{14.15}$$

Keep in mind that k depends on temperature: For air ($R = 0.287\frac{\text{kJ}}{\text{kg K}}$) at $T = 298$ we have $k = 1.4$ and thus $a = 346\frac{\text{m}}{\text{s}}$, while at $T = 1200\,\text{K}$ we have $k = 1.314$ and find $a = 672.8\frac{\text{m}}{\text{s}}$.

The speed of sound depends on the type of gas through the ratio of specific heats, k, and through the gas constant, $R = \frac{\bar{R}}{M}$. Gases with smaller molar mass M have larger speed of sound. As example we consider helium ($M_{\text{He}} = 4\frac{\text{kg}}{\text{kmol}}, k_{\text{He}} = \frac{5}{3}$), argon ($M_{\text{Ar}} = 39.95\frac{\text{kg}}{\text{kmol}}, k_{\text{Ar}} = \frac{5}{3}$), and carbon dioxide ($M_{\text{CO}_2} = 44\frac{\text{kg}}{\text{kmol}}, k_{\text{CO}_2} = \frac{4}{3}$) at 298 K, for which we find $a_{\text{He}} = 1016\frac{\text{m}}{\text{s}}, a_{\text{Ar}} = 321.5\frac{\text{m}}{\text{s}}$ and $a_{\text{CO}_2} = 274\frac{\text{m}}{\text{s}}$.

14.4 Area-Velocity Relation

The main design parameter for nozzles and diffusers is the change of cross section, and we ask how flow properties, in particular velocity and pressure, change with the cross section. Figure 14.2 shows an adiabatic and reversible, i.e., isentropic, flow through a duct with varying cross section. We consider a small slice of the duct of width dx, and apply the balances of mass, energy and entropy, similar to what we did for the wave in the pipe above.

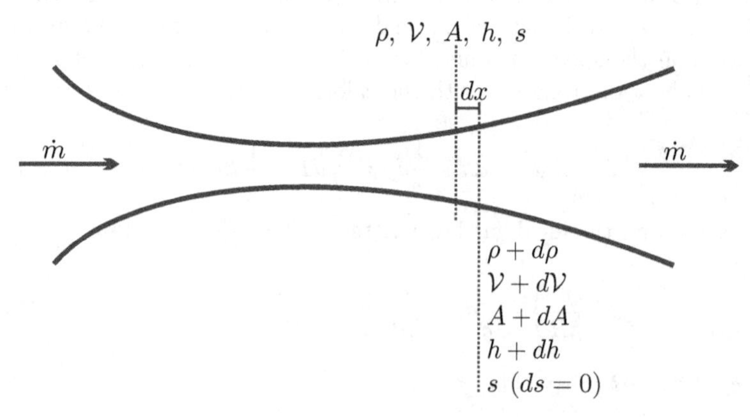

Fig. 14.2 Adiabatic reversible, i.e., isentropic, flow through a duct with changing cross section

The mass balance yields

$$\dot{m} = \rho \mathcal{V} A = const. \tag{14.16}$$

and thus

$$\frac{d\dot{m}}{\dot{m}} = \frac{dA}{A} + \frac{d\mathcal{V}}{\mathcal{V}} + \frac{d\rho}{\rho} = 0\,. \tag{14.17}$$

The first law gives at first

$$h + \frac{1}{2}\mathcal{V}^2 = h_0 = const. \tag{14.18}$$

and thus

$$dh + \mathcal{V}d\mathcal{V} = 0 \,. \tag{14.19}$$

Here, h_0 is the stagnation enthalpy, defined as the enthalpy the flow would obtain when brought to rest adiabatically. Since the flow is isentropic, we have $ds = 0$, and the Gibbs equation gives

$$Tds = dh - \frac{1}{\rho}dp = 0 \,. \tag{14.20}$$

Elimination of enthalpy between the last two equations yields the relation between pressure and velocity changes in isentropic nozzles and diffusers,

$$\frac{1}{\rho}dp = -\mathcal{V}d\mathcal{V} \,, \tag{14.21}$$

which we discussed already in Sec. 9.11. We use this to eliminate $\mathcal{V}$ from the mass balance to find

$$\frac{dA}{A} = \frac{1}{\mathcal{V}^2}\frac{1}{\rho}dp - \frac{d\rho}{\rho} = \frac{1}{\rho}dp\left(\frac{1}{\mathcal{V}^2} - \frac{1}{\left(\frac{\partial p}{\partial \rho}\right)_s}\right) \,, \tag{14.22}$$

where the subscript indicates isentropic flow.

To proceed, we introduce the Mach number

$$\mathrm{Ma} = \frac{\mathcal{V}}{a} \,,$$

which compares flow velocity to speed of sound: $\mathrm{Ma} < 1$ for subsonic flows, $\mathrm{Ma} > 1$ for supersonic flows, and $\mathrm{Ma} = 1$ for sonic flows. Flows with $\mathrm{Ma} \gg 1$ are called hypersonic and flows with $\mathrm{Ma} \simeq 1$ are called transonic.

With the definition (14.10) of the speed of sound we thus can write the relation (14.22) as

$$\frac{dA}{A} = \frac{1}{\rho\mathcal{V}^2}dp\left(1 - \mathrm{Ma}^2\right) \,, \tag{14.23}$$

or, by eliminating pressure,

$$\frac{dA}{A} = -\frac{d\mathcal{V}}{\mathcal{V}}\left(1 - \mathrm{Ma}^2\right) \,. \tag{14.24}$$

Equations (14.23) and (14.24) are the area-pressure relation and the area-velocity relation for isentropic duct flows. Both relations carry the factor $\left(1 - \mathrm{Ma}^2\right)$ which has different sign for subsonic and supersonic flows.

Accordingly, a change of cross section has different effect when applied to sub- and supersonic flows.

Subsonic Flows (Ma<1): For a converging duct we have from (14.23, 14.24)

$$dA < 0 \quad \Longrightarrow \quad dp < 0\,, \;\; d\mathcal{V} > 0\,;$$

the flow is accelerated while pressure drops; this is a nozzle.

For a diverging duct, on the other hand, we have the opposite signs,

$$dA > 0 \quad \Longrightarrow \quad dp > 0\,, \;\; d\mathcal{V} < 0\,;$$

pressure grows, and the flow decelerates; this is a diffuser.

Supersonic Flows (Ma>1): For a converging duct we have from (14.23, 14.24)

$$dA < 0 \quad \Longrightarrow \quad dp > 0\,, \;\; d\mathcal{V} < 0\,;$$

pressure grows, and the flow decelerates; this is a diffuser.

For a diverging duct, on the other hand, we have the opposite signs,

$$dA > 0 \quad \Longrightarrow \quad dp < 0\,, \;\; d\mathcal{V} > 0\,;$$

the flow is accelerated while pressure drops; this is a nozzle.

In other words, a converging duct acts as a nozzle in subsonic flow, but as a diffuser in supersonic flow. A diverging duct acts as a diffuser in subsonic flow, but as a nozzle in supersonic flow. Figure 14.3 shows a summary.

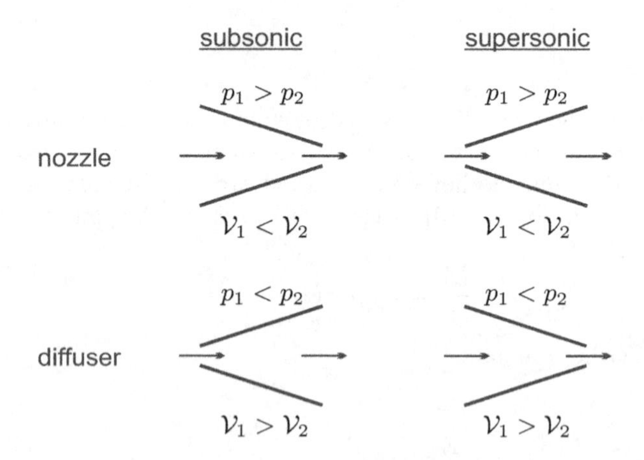

Fig. 14.3 Nozzles and diffuser cross section variation in sub- and supersonic flow

14.5 Nozzle Flows

Rocket motors and some jet engines expel supersonic flows for propulsion, and thus need appropriate nozzle geometries. To accelerate a subsonic flow to supersonic speed requires a converging-diverging nozzle, where the flow is accelerated to sonic speed in the converging part, and then to supersonic speed in the diverging part. After its inventor Gustaf de Laval (1845-1913), such a nozzle is called Laval nozzle.

We will almost exclusively deal with isentropic flows. So that we can perform analytical calculations, we restrict the treatment to ideal gases with constant specific heats. We shall discuss flows through purely converging nozzles, and through converging-diverging nozzles. In both cases, the balances of mass, energy and entropy reduce to

$$\begin{aligned}\dot{m} &= \rho \mathcal{V} A = const.\,,\\ h + \frac{1}{2}\mathcal{V}^2 &= h_0 = const.\,,\\ \frac{T}{p^{\frac{k-1}{k}}} &= \frac{T_0}{p_0^{\frac{k-1}{k}}}\,,\quad \frac{p}{\rho^k} = \frac{p_0}{\rho_0^k}\,,\end{aligned} \tag{14.25}$$

where $\rho, \mathcal{V}, T, h$ are the properties at a given cross section of the nozzle, and T_0, p_0, h_0 are stagnation properties. The stagnation state is defined as the hypothetical state that is reached by bringing the flow to rest isentropically.

With $h - h_0 = c_p (T - T_0)$ and $c_p = \frac{k}{k-1} R$ we find from the above the local velocity as

$$\mathcal{V} = \sqrt{2(h_0 - h)} = \sqrt{\frac{2kRT_0}{k-1}\left(1 - \frac{T}{T_0}\right)} = \sqrt{\frac{2kRT_0}{k-1}}\sqrt{1 - \left(\frac{p}{p_0}\right)^{\frac{k-1}{k}}}\,. \tag{14.26}$$

With the isentropic relation for density, we can write the mass flow through the nozzle as

$$\dot{m} = \rho_0 \sqrt{\frac{2kRT_0}{k-1}} A \left[\left(\frac{p}{p_0}\right)^{\frac{1}{k}} \sqrt{1 - \left(\frac{p}{p_0}\right)^{\frac{k-1}{k}}}\right] = const. \tag{14.27}$$

The mass flow is a product of three factors: The constant $\rho_0 \sqrt{\frac{2kRT_0}{k-1}}$ which is fixed by the stagnation state (ρ_0, T_0), the cross section A, and the flow function ψ, which we define as

$$\psi\left(\frac{p}{p_0}\right) = \left(\frac{p}{p_0}\right)^{\frac{1}{k}} \sqrt{1 - \left(\frac{p}{p_0}\right)^{\frac{k-1}{k}}}\,. \tag{14.28}$$

Figure 14.4 shows the flow function for $k = 1.33, 1.4, 1.67$.

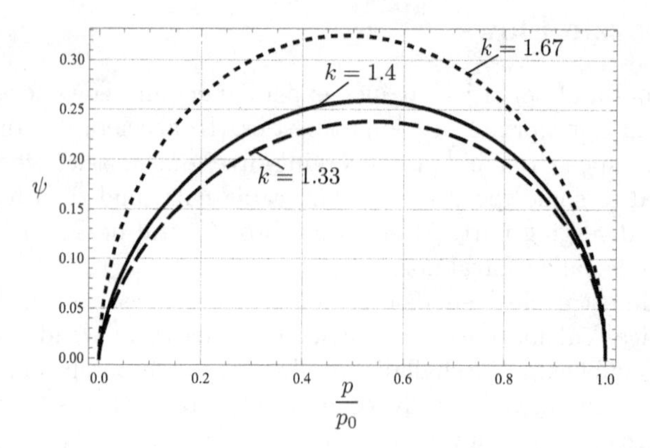

Fig. 14.4 Flow function $\psi\,(p/p_0)$ for $k = 1.33, 1.4, 1.67$. The function has a maximum at $\frac{p^*}{p_0} = \left(\frac{2}{k+1}\right)^{\frac{k}{k-1}} = 0.540, 0.528, 0.488$.

The argument $\left(\frac{p}{p_0}\right)$ of the flow function assumes values between 0 and 1, and the curve exhibits a maximum with the critical values

$$\frac{p^*}{p_0} = \left(\frac{2}{k+1}\right)^{\frac{k}{k-1}} \quad , \quad \psi^* = \left(\frac{2}{k+1}\right)^{\frac{1}{k-1}} \sqrt{\frac{k-1}{k+1}}\,; \tag{14.29}$$

for the k values in the figure, the critical pressure assumes the values $\frac{p^*}{p_0} = 0.534, 0.528, 0.487$.

The condition of constant mass flow is equivalent to

$$A\,\psi\left(\frac{p}{p_0}\right) = const.\,, \tag{14.30}$$

and this relation will be used now to understand converging and converging-diverging nozzle flows. For this, we study the outflow from a large container into a nozzle, where the gas in the container is in the constant stagnation state (T_0, p_0). The flow is driven by the difference between the back pressure outside the nozzle, p_b, and the stagnation pressure p_0, as indicated in Fig. 14.5. No flow occurs when $p_b = p_0$, and we now study what happens when p_b is lowered gradually.

14.6 Converging Nozzle

As the back pressure p_b is lowered a bit, the flow develops. The cross section A is decreasing along the nozzle coordinate x, see Fig. 14.6. According to Fig. 14.4, the flow function is hill-shaped, and the nozzle feed state is on the right foot of that hill, at $p = p_0$, $\psi = 0$. Since $A\psi\left(\frac{p}{p_0}\right)$ is constant, for

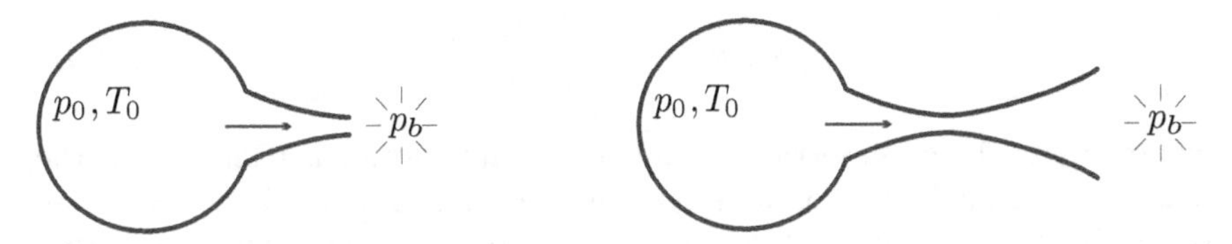

Fig. 14.5 Nozzle flows from stagnation. Left: Converging nozzle. Right: Converging-diverging nozzle (Laval nozzle).

decreasing cross section A the flow function must grow, i.e., go uphill. The flow function grows along x, and reaches its largest value in the smallest cross section A_{th}, the throat of the nozzle, which is at the end of the converging nozzle. As ψ grows along the nozzle coordinate, the pressure decreases, until it assumes the pressure $p_e = p_b$ in the end cross section of the nozzle. Further decrease of the back pressure leads to lower pressures along the nozzle, and larger values of the flow function.

When the back pressure p_b assumes the critical value p^*, the flow function in the exit is at its maximum ψ^*. No further growth of ψ is possible when the back pressure is lowered further. Thus, the exit state remains at $\psi_e = \psi^*$ and $p_e = p^*$, even when the back pressure is lowered below p^*. Figure 14.6 visualizes this behavior.

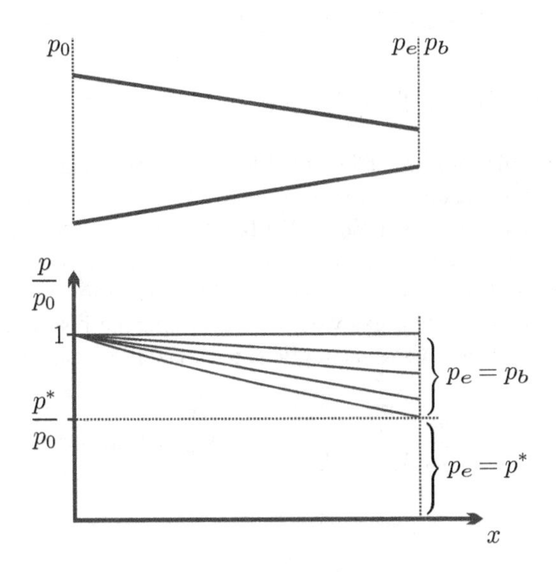

Fig. 14.6 Converging nozzle: Geometry and pressure profile

When the pressure p^* is reached in the nozzle throat, one speaks of choked flow. Indeed, under this condition, the mass flow obtains a maximum value,

$$\dot{m}^* = \rho_0 \sqrt{kRT_0} A_{th} \left(\frac{2}{k+1} \right)^{\frac{k+1}{2k-2}} , \tag{14.31}$$

where A_{th} is the cross section at the nozzle end. No further increase of the mass flow through the nozzle is possible. To understand this behavior, we determine temperature and velocity in the exit at choked conditions. From the adiabatic relation and (14.26, 14.29) we find

$$T^* = T_0 \left(\frac{p^*}{p_0} \right)^{\frac{k-1}{k}} = \frac{2T_0}{k+1} \quad \text{and} \quad \mathcal{V}^* = \sqrt{\frac{2kRT_0}{k+1}} = \sqrt{kRT^*} = a^* . \tag{14.32}$$

Thus, for choked flow, the exit speed is just the local speed of sound, a^*. We recall that the speed of sound is the velocity of a pressure disturbance. When the back pressure is lowered, the information on the pressure change travels with the speed of sound relative to the gas. Since the gas leaves with just the same speed, the information on pressure change is not transmitted into the nozzle, and no changes can occur inside, the exit velocity and the mass flow are limited.

With $p_b < p^*$, a pressure discontinuity occurs at the nozzle exit, which contributes to thrust for the airplane or rocket. The exhaust expands outside the nozzle to the back pressure, and accelerates, and this expansion is somewhat irreversible, more so with bigger pressure differences.[2] Most commercial airplanes have converging nozzles and subsonic outflow, to reduce noise.

14.7 Example: Safety Valve

The mass flow limitation must be considered for the design of safety valves. As an example we consider a steam boiler that produces 10t/h of saturated vapor at $p_0 = 15\,\text{bar}$ (so that $T_0 = 471\,\text{K}$). In case of emergency, all steam produced must be discharged through a safety valve. We compute the minimum cross section the valve can have under the assumption that the steam can be described as an ideal gas with constant specific heats and $k = 1.135$. Solving (14.31) for the cross section gives, with the ideal gas law $p_0 = \rho_0 R T_0$,

$$A_{th} = \frac{\dot{m}^*}{p_0} \sqrt{\frac{RT_0}{k}} \left(\frac{k+1}{2} \right)^{\frac{k+1}{2k-2}} = 13.5\,\text{cm}^2 .$$

[2] In the discussion of jet engines we have considered adiabatic expansion to the back pressure, which, for $p_b < p^*$, is only possible in Laval nozzles as discussed below. This simplification ignores the external expansion; however, the associated irrversibilities can be included into the nozzle efficiency. A full discussion of the exterior expansion, and the related thrust is beyond the scope of this book.

14.8 Laval Nozzle

To reach exit velocities above the speed of sound requires converging-diverging nozzles, which will be discussed next. The cross section A is first decreasing until is reaches its smallest value A_{th} in the throat of the nozzle, and then it is increasing to the exit cross section A_e. Again we consider the pressure distribution in the nozzle which results from lowering the back pressure p_b, see Fig. 14.7.

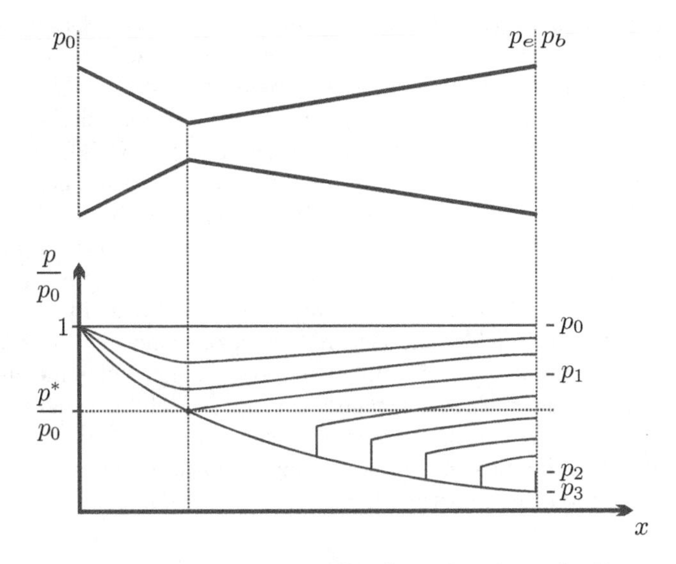

Fig. 14.7 Geometry and pressure profiles for a Laval nozzle. Pressure profiles depend on the back pressure as discussed in the text.

When $p_b = p_0$, there is no flow and homogenous pressure p_0 throughout the nozzle. When the back pressure is lowered somewhat, a flow occurs. According to Fig. 14.4, the flow function is hill-shaped, and the nozzle feed state is on the right foot of that hill, at $p = p_0$, $\psi = 0$. Since $A\psi\left(\frac{p}{p_0}\right)$ is constant, the flow function must grow—go uphill—in the converging part of the nozzle, and it must decrease—go downhill—in the diverging part.

We first consider relatively large back pressures in the pressure range $p_1 < p_b < p_0$ in the figure. As the nozzle converges, the flow state climbs uphill until a value $p_{th} > p^*$, $\psi_{th} < \psi^*$ is reached in the throat. This point is to the right of the maximum. As the cross section grows in the diverging part, the flow function must decrease, and this is only possible by returning to higher pressures, that is by going back downhill towards the right. The flow is accelerated in the converging part of the nozzle, and decelerated in the diverging part, the outflow velocity is relatively low, and subsonic. The extreme case of this flow type is reached for $p_b = p_1$, when the air is in the critical state in the throat—on top of the hill with sonic speed—but then is decelerated again.

When the flow reaches the critical state in the throat, the flow function can decrease by going down the left side of the hill, towards lower pressures, and higher, i.e., supersonic, velocities. As indicated in the figure, this requires low back pressures $p_2 < p_b \leq 0$. In this range the flow is isentropic inside the nozzle. If the back pressure is just at p_3, the end pressure is equal to the back pressure, and no external irreversibilities occur. If the back pressure is in the range $p_2 < p_b < p_3$, the end pressure is below the back pressure and pressure is equilibrated through oblique shocks outside the nozzle. For lower back pressures $p_b < p_3$, the end pressure is above the back pressure and pressure is equilibrated through external expansion waves.

In the range $p_2 < p_b < p_1$, isentropic flow inside the nozzle is not possible. The flow will follow isentropic flow conditions for a while, accelerating to supersonic flow behind the throat, and then a irreversible normal shock will occur, that is a sudden jump from low pressure supersonic flow to high pressure subsonic flow. Behind the shock, the flow will be isentropic again, and the gas will expand to the back pressure.

We shall not further discuss normal and oblique shocks, but stress that they are strongly irreversible, and thus reduce nozzle efficiency substantially. In a shock, the flow changes from supersonic to subsonic, which means significant reduction of thrust. Nozzle geometry must be carefully designed so that the flow conditions are optimal. Some supersonic aircraft have nozzles with variable geometry, to adjust for the wide range of back pressures encountered between take-off and high altitude flight.

14.9 Rockets, Ramjet and Scramjet

Jet engines are air-breathing, that is they carry the fuel on board, and burn it with oxygen from the ambient air that passes through the engine. Rockets carry the oxygen on board, either as liquefied oxygen, or in the form of a compound. Thus, rockets are independent of ambient air, and can fly at extremely high altitude, and in space. In a rocket motor, fuel and oxidizer are burnt at high pressure in the combustion chamber and then expand through a Laval nozzle, which provides large supersonic exit velocities, and thus large thrust, see Fig. 14.8.

The oxygen a rocket has to carry on board increases the take-off weight, and reduces the payload that can be carried along. Some military applications require fast transport of payload through the atmosphere,[3] and rockets are used because they offer extremely high velocities.

Ramjet and scramjet are conceptually simple air-breathing engines for supersonic flight, hence they do not require to carry the heavy oxidizer on board.

[3] ... It is left to the reader to fathom what kind of delivery would be that urgent ...

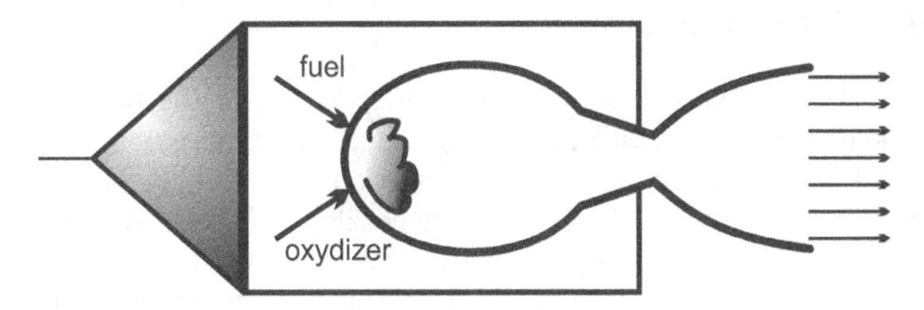

Fig. 14.8 A rocket

In both pressure build-up in the engine is affected only through diffusers, see Fig. 14.9 for sketches.

In a ramjet, the incoming air is slowed down to subsonic speed and high pressure by means of a converging-diverging diffuser, which operates on the same principles as a Laval nozzle, only inversely. Fuel is injected into the compressed air and burned, the hot combustion product then expands through a Laval nozzle to high velocities. Ramjets can be used for supersonic flight with speeds up to Ma = 6.

In a scramjet (supersonic combustion ramjet), the flow stays supersonic at all times. A supersonic diffuser slows down the flow and pressure increases. Fuel is injected into the flow and burned, and the hot pressurized combustion product leaves through a supersonic nozzle. Flight speeds could be up to Ma = 15 or so. Due to the high air velocity at the burner, it is quite difficult to maintain stable combustion of the fuel. To get the scramjet engine started it must be accelerated to supersonic speed first, so that the converging diffuser leads to pressure build-up.

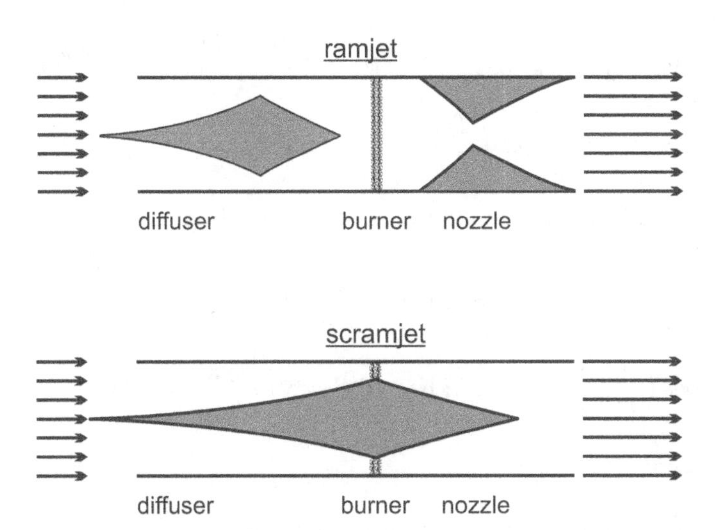

Fig. 14.9 Ramjet and scramjet

14.10 Example: Ramjet

A ramjet is to fly at $\mathrm{Ma}_I = 3$ in high altitude, where the temperature is $T_I = 200\,\mathrm{K}$ and the pressure is $p_I = 0.3\,\mathrm{bar}$. The diffuser inlet, diffuser outlet, and nozzle inlet all have the cross section $A_R = 0.1\,\mathrm{m}^2$. We determine the throat cross sections of diffuser and nozzle, and the nozzle exit cross section for the case that the nozzle expands isentropically to the outside pressure. Moreover, we will compute exit velocity, thrust, and propulsive power. To simplify proceedings, we consider the working gas air as ideal gas with constant specific heats, $R = 0.287 \frac{\mathrm{kJ}}{\mathrm{kg\,K}}$, $k = 1.4$, $c_p = \frac{k}{k-1}R$.

Diffuser: The incoming flow is at T_I, p_I and Ma_I. The inlet density and velocity are

$$\rho_I = \frac{p_I}{RT_I} = 0.523 \frac{\mathrm{kg}}{\mathrm{m}^3}$$

$$\mathcal{V}_I = \mathrm{Ma}_I\, a_I = \mathrm{Ma}_I\, \sqrt{kRT_I} = 850.4 \frac{\mathrm{m}}{\mathrm{s}}\,.$$

This gives the mass flow

$$\dot{m} = \rho_I \mathcal{V}_I A_R = 44.45 \frac{\mathrm{kg}}{\mathrm{s}}\,.$$

To be able to use the equations for nozzles, which also hold for diffusers, we first need to determine the stagnation state for the diffuser. The stagnation temperature T_{D0} follows from the first law (adiabatic deceleration), and the corresponding stagnation pressure p_{D0} follows from the isentropic relation (with $h(T) = c_p (T - T_0)$):

$$h_{in} + \frac{1}{2}\mathcal{V}_{in}^2 = h_{D0} \quad \Longrightarrow \quad T_{D0} = \frac{c_p T_{in} + \frac{1}{2}\mathcal{V}_{in}^2}{c_p} = 560.2\,\mathrm{K}\,,$$

$$p_{D0} = p_{in} \left(\frac{T_{D0}}{T_{in}}\right)^{\frac{k}{k-1}} = 11.03\,\mathrm{bar}\,.$$

The critical state for the diffuser is determined from (14.29, 14.32) as

$$p^{D*} = p_{D0}\left(\frac{2}{k+1}\right)^{\frac{k}{k-1}} = 5.82\,\mathrm{bar}\,,\quad \psi^{D*} = \left(\frac{2}{k+1}\right)^{\frac{1}{k-1}} \sqrt{\frac{k-1}{k+1}} = 0.2588\,,$$

$$T^{D*} = T_{D0}\left(\frac{p_{D0}}{p^{D*}}\right)^{\frac{k-1}{k}} = 466.7\,\mathrm{K}\,,\quad \mathcal{V}^{D*} = \sqrt{kRT^{D*}} = 433.0 \frac{\mathrm{m}}{\mathrm{s}}\,,$$

$$\rho^{D*} = \frac{p^{D*}}{RT^{D*}} = 4.35 \frac{\mathrm{kg}}{\mathrm{m}^3}\,.$$

The throat cross section follows from the mass flow,

$$A^{D*} = \frac{\dot{m}}{\rho^{D*}\mathcal{V}^{D*}} = 0.0236\,\mathrm{m}^2 \ .$$

The diffuser exit area is given, and the corresponding exit pressure p_D follows from the constant mass flow relation $A\psi = const.$ by first determining the exit flow function ψ_D and then finding the corresponding pressure; this yields

$$\psi_D = \psi^{D*}\frac{A^{D*}}{A_R} = 0.611 \quad \Longrightarrow \quad p_D = 0.987 p_{D0} = 10.87\,\mathrm{bar} \ .$$

The pressure can either be found from the plot of the flow function (Fig. 14.4), or by numerical solution of (14.28); note that there are two solutions, the larger one refers to subsonic flow. Temperature, density and velocity at the diffuser outlet are

$$T_D = T^{D*}\left(\frac{p_D}{p^{D*}}\right)^{\frac{k-1}{k}} = 557.9\,\mathrm{K} \ , \ \rho_D = \frac{p_D}{RT_D} = 6.79\frac{\mathrm{kg}}{\mathrm{m}^3} \ ,$$

$$\mathcal{V}_D = \frac{\dot{m}}{\rho_D A_R} = 65.4\frac{\mathrm{m}}{\mathrm{s}} \ .$$

Burner: In the burner, fuel is injected into the compressed air and isobarically burned. As always, we ignore the mass of the fuel added, and treat the combustion product as air. With the temperature after the burner at $T_B = 1300\,\mathrm{K}$ and the pressure $p_B = p_D$, we find density and velocity as

$$\rho_B = \frac{p_D}{RT_D} = 2.91\frac{\mathrm{kg}}{\mathrm{s}} \ , \ \mathcal{V}_B = \frac{\dot{m}}{\rho_B A_R} = 152.5\frac{\mathrm{m}}{\mathrm{s}} \ .$$

The total heat supplied is

$$\dot{Q}_B = \dot{m}c_p\left(T_B - T_D\right) = 33.1\,\mathrm{MW} \ .$$

Nozzle: The inlet state for the nozzle is the exit state of the burner $(\rho_B, T_B, \mathcal{V}_B)$. For the computation of the nozzle, we must first determine its stagnation state:

$$T_{N0} = \frac{c_p T_B + \frac{1}{2}\mathcal{V}_B^2}{c_p} = 1311.6\,\mathrm{K} \ , \ p_{N0} = p_B\left(\frac{T_{N0}}{T_B}\right)^{\frac{k}{k-1}} = 11.22\,\mathrm{bar} \ .$$

Critical data in the nozzle throat are

$$p^{N*} = p_{N0}\left(\frac{2}{k+1}\right)^{\frac{k}{k-1}} = 5.92\,\text{bar} \ , \ \psi^{N*} = 0.2588 \, ,$$

$$T^{N*} = T_{N0}\left(\frac{p_{N0}}{p^{N*}}\right)^{\frac{k-1}{k}} = 1093.0\,\text{K} \ , \ \mathcal{V}^{N*} = \sqrt{kRT^{N*}} = 622.7\frac{\text{m}}{\text{s}} \, ,$$

$$\rho^{N*} = \frac{p^{N*}}{RT^{N*}} = 1.89\frac{\text{kg}}{\text{m}^3} \ , \ A^{N*} = \frac{\dot{m}}{\rho^{N*}\mathcal{V}^{N*}} = 0.0355\,\text{m}^2 \ .$$

The computation of the throat cross section from critical data and the mass flow is already included in the above list.

The last process to consider is the isentropic expansion in the diverging part of the nozzle to the outside pressure $p_I = 0.3\,\text{bar}$. We find the following data for the nozzle exit:

$$\psi_E = \left(\frac{p_I}{p_{N0}}\right)^{\frac{1}{k}}\sqrt{1-\left(\frac{p_I}{p_{N0}}\right)^{\frac{k-1}{k}}} = 0.0604 \ , \ A_E = A^{N*}\frac{\psi^{N*}}{\psi_E} = 0.152\,\text{m}^2 \, ,$$

$$T_E = T^{N*}\left(\frac{p_I}{p^{N*}}\right)^{\frac{k-1}{k}} = 466.1\,\text{K} \ , \ \rho_E = \frac{p_I}{RT_E} = 0.224\frac{\text{kg}}{\text{m}^3} \, .$$

Exit velocity and Mach number are

$$\mathcal{V}_E = \frac{\dot{m}}{\rho_E A_E} = 1303.3\frac{\text{m}}{\text{s}} \ , \ \text{Ma}_E = \frac{\mathcal{V}_E}{\sqrt{kRT_E}} = 3.01 \, .$$

Power and Efficiency: All cross sections and all property data for reversible operation were computed above. From the given data we find thrust and propulsive power as

$$F = \dot{m}\left(\mathcal{V}_E - \mathcal{V}_{in}\right) = 201.12\,\text{kN} \ , \ \dot{W}_P = F\mathcal{V}_{in} = 17.12\,\text{MW} \, .$$

This corresponds to the thermal propulsive efficiency and the Froude efficiency

$$\eta_P = \frac{\dot{W}_P}{\dot{Q}_B} = 0.517 \ , \ \eta_F = \frac{2\mathcal{V}_{in}}{\mathcal{V}_{in} + \mathcal{V}_E} = 0.79 \, .$$

Problems

14.1. Speed of Sound

Determine the speed of sound in helium (based on constant specific heats) and air (based on variable specific heats, tables) at 300 K and 1500 K. Compute the corresponding Mach numbers for a velocity of $290\frac{\text{m}}{\text{s}}$.

14.2. Speed of Sound in R-134a

Determine the speed of sound in refrigerant R-134a at 1 MPa, 60 °C. Use table data!

14.3. Security Valve
A steam boiler produces saturated vapor at 17.5 bar. The security valve has a smallest free area of $20\,\text{cm}^2$. Determine the maximum mass flow that can be produced so that no pressure is building up when the valve is open.

14.4. Laval Nozzle
A Laval nozzle is to be designed such that it delivers $4\frac{\text{kg}}{\text{s}}$ of air at $10\,°\text{C}$ and 1 bar at twice the speed of sound. The air that expands in the nozzle is delivered by an isentropic compressor that draws air at 1 bar and $10\,°\text{C}$.

Consider air as ideal gas with constant specific heats, $R = 0.287\frac{\text{kJ}}{\text{kg K}}$, $c_p = 1.004\frac{\text{kJ}}{\text{kg K}}$.

1. Determine the cross section at the end of the nozzle.
2. Determine temperature, pressure, mass density and velocity in the throat.
3. Determine stagnation pressure and stagnation temperature for the nozzle flow.
4. Determine the power consumed by the compressor.
5. Determine the heat that must be withdrawn from the flow between compressor and nozzle.

14.5. Nozzle Flow
Consider a Laval nozzle for rocket propulsion. Pressure and temperature in the combustion chamber are 10 bar and 2500 K, respectively. The mass flow through the nozzle is $30\frac{\text{kg}}{\text{s}}$ of combustion product (ideal gas, constant specific heats, $R = 0.287\frac{\text{kJ}}{\text{kg K}}$, $k = 1.4$), the cross section at the end of the nozzle is $A_e = 700\,\text{cm}^2$ and the flow is isentropic throughout the nozzle. The environmental pressure is 0.9 bar.

1. Do you expect supersonic or subsonic flow at the outlet? Why?
2. Compute the area of the throat of the nozzle.
3. Find pressures and gas velocities at throat and end.
4. Discuss the flow behind the nozzle

14.6. Rocket Engine
A converging-diverging nozzle is fed from a combustion chamber at temperature $T_0 = 2200\,\text{K}$. The flow through the nozzle is isentropic, and the outflow is supersonic with the velocity $v = 1400\frac{\text{m}}{\text{s}}$. The pressure in the throat is measured as $p^* = 4\,\text{bar}$.

The gas flowing through the nozzle can be considered as an ideal gas with constant specific heats: $c_p = 0.98\frac{\text{kJ}}{\text{kg K}}$, $k = c_p/c_v = 1.4$.

1. Determine the pressure in the combustion chamber.
2. Determine temperature and pressure at the nozzle exit.
3. Determine the speed of sound at the nozzle exit.

Chapter 15
Transient and Inhomogeneous Processes in Open Systems

15.1 Introduction

So far, we have considered steady state processes in open systems, and time dependent processes in closed systems. In this chapter, to widen the scope a bit, we show some simple applications of space dependent and time dependent open systems.

The full discussion of inhomogeneous processes, steady state or transient, requires the solution of the partial differential equations of hydrodynamics, e.g., the Navier-Stokes equations and Fourier's law of heat conduction. The derivation of the transport equations is the subject of *Non-equilibrium Thermodynamics*, and their solution is a question of mathematics and numerical methods. *Fluid Dynamics*, and *Heat and Mass Transfer* are disciplines which rely heavily on the study of solutions of the appropriate transport equations.

Below, we first discuss one-dimensional co- and counter-flow heat exchangers as a simple application of inhomogeneous systems. Then, as a relatively easy application of open time-dependent systems, we consider filling and discharge processes, e.g., of gas bottles, rooms, or cavities, as long as the content can be assumed to have homogeneous properties.

15.2 Heat Exchangers

15.2.1 Basic Equations

We discuss the principles of simple heat exchange between two flows, which are either running in the same direction (co-flow), or in opposite directions (counter-flow), see Fig. 15.1 for a basic sketch of the set-up. Heat exchange is assumed to be a one-dimensional process, where the temperatures $T_A(x)$ and $T_B(x)$ of the two flows depend only on the space coordinate in flow direction,

H. Struchtrup, *Thermodynamics and Energy Conversion*,
DOI: 10.1007/978-3-662-43715-5_15,

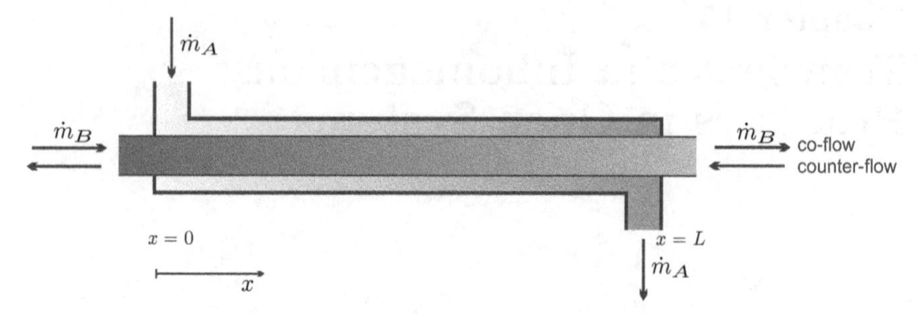

Fig. 15.1 Heat exchange between two flows A and B. Co- and counter flow settings are indicated by the arrows that show the direction of flows.

x. In particular, temperature profiles perpendicular to the flow are ignored, that is the given temperatures are cross-sectional averages.

Heat is transferred between the two flows due to different temperatures $T_A(x)$ and $T_B(x)$. Typically, the inflow temperatures of the flows are given, and one aims at determining the outflow temperatures.

Since the temperatures depend on the location x, we need to balance energy for each location. Figure 15.2 shows small elements of infinitesimal width dx and the corresponding flows and properties for co- and counter-flow settings.

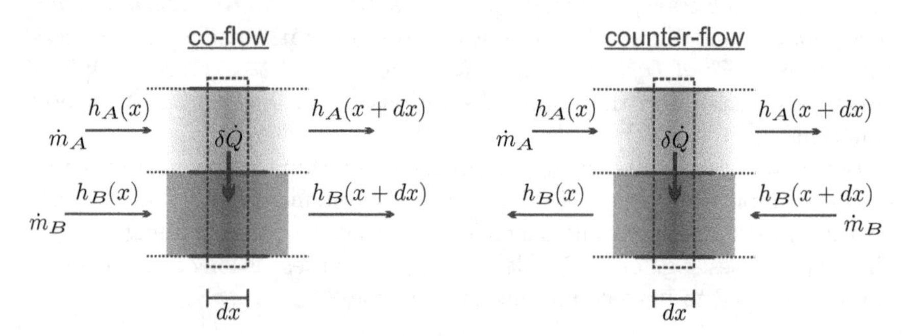

Fig. 15.2 A volume element dx of the heat exchanger, and the corresponding fluxes for co- and counter-flow heat exchangers

The task at hand is to determine the temperature curves in both flows. Applying the first law to each flow in each element gives

$$\begin{aligned} \dot{m}_A \left[h_A(x+dx) - h_A(x)\right] &= -\delta\dot{Q} \,, \\ \pm\dot{m}_B \left[h_B(x+dx) - h_B(x)\right] &= \delta\dot{Q} \,, \end{aligned} \tag{15.1}$$

where in the second equation the upper sign refers to co-flow, and the lower sign refers to counter-flow. The heat exchange between the flows within the element dx is denoted by $\delta\dot{Q}$, and is given by Newton's law of cooling (4.35), which we write here as

$$\delta\dot{Q} = \alpha\left(T_A - T_B\right)dx\ . \tag{15.2}$$

Heat transfer is proportional to the contact area bdx. In the above, the width b is absorbed into the coefficient α and only the length dx of the element is made explicit. The heat transfer coefficient α will be assumed to be constant.

Taylor expansion allows to relate the enthalpy differences to the specific heats and the temperature gradients as

$$\begin{aligned} h_A\left(x+dx\right) - h_A\left(x\right) &= c_p^A \frac{dT_A}{dx} dx\ , \\ h_B\left(x+dx\right) - h_B\left(x\right) &= c_p^B \frac{dT_B}{dx} dx\ ; \end{aligned} \tag{15.3}$$

for simplicity we shall assume that the specific heats are independent of temperature.

Combining all of the above yields two coupled differential equations for the temperatures,

$$\frac{dT_A}{dx} = \hat{\alpha}_A\left(T_B - T_A\right) \quad , \quad \frac{dT_B}{dx} = \mp\hat{\alpha}_B\left(T_B - T_A\right)\ , \tag{15.4}$$

with the abbreviations

$$\hat{\alpha}_A = \frac{\alpha}{\dot{m}_A c_p^A} \quad , \quad \hat{\alpha}_B = \frac{\alpha}{\dot{m}_B c_p^B}\ . \tag{15.5}$$

The coupled equations (15.4) can be integrated easily,[1] and the solutions read

$$T_A\left(x\right) = K_2 \exp\left[-\left(\hat{\alpha}_A \pm \hat{\alpha}_B\right)x\right] + \frac{\hat{\alpha}_A}{\hat{\alpha}_A \pm \hat{\alpha}_B} K_1\ , \tag{15.6a}$$

$$T_B\left(x\right) = \mp\frac{\hat{\alpha}_B}{\hat{\alpha}_A} K_2 \exp\left[-\left(\hat{\alpha}_A \pm \hat{\alpha}_B\right)x\right] + \frac{\hat{\alpha}_A}{\hat{\alpha}_A \pm \hat{\alpha}_B} K_1\ , \tag{15.6b}$$

where K_1 and K_2 are integrating constants, and, as in all equations in this section, the upper sign is for co-flow, and the lower sign is for counter-flow exchangers.

---

[1] Take the difference of both to get an equation for $(T_B - T_A)$ that can be integrated. Then use the result to eliminate T_B in the equation for T_A, and solve for T_A.

15.2.2 Co-flow Heat Exchangers

We consider co-flow heat exchangers (upper sign) first. The known inflow conditions are the temperatures $T_A(0)$ and $T_B(0)$, for which we find from (15.6)

$$T_A(0) = K_2 + \frac{\hat{\alpha}_A}{\hat{\alpha}_A + \hat{\alpha}_B} K_1 \quad , \quad T_B(0) = -\frac{\hat{\alpha}_B}{\hat{\alpha}_A} K_2 + \frac{\hat{\alpha}_A}{\hat{\alpha}_A + \hat{\alpha}_B} K_1 \,. \tag{15.7}$$

Solving this for the constants K_1 and K_2 and inserting these into (15.6) gives the temperature curves for the co-flow heat exchanger as

$$\begin{aligned} T_A(x) &= \frac{T_A(0) + \frac{\hat{\alpha}_A}{\hat{\alpha}_B} T_B(0)}{1 + \frac{\hat{\alpha}_A}{\hat{\alpha}_B}} + \frac{T_A(0) - T_B(0)}{1 + \frac{\hat{\alpha}_B}{\hat{\alpha}_A}} \exp\left[-\left(\hat{\alpha}_A + \hat{\alpha}_B\right) x\right] \,, \\ T_B(x) &= \frac{T_A(0) + \frac{\hat{\alpha}_A}{\hat{\alpha}_B} T_B(0)}{1 + \frac{\hat{\alpha}_A}{\hat{\alpha}_B}} - \frac{T_A(0) - T_B(0)}{1 + \frac{\hat{\alpha}_A}{\hat{\alpha}_B}} \exp\left[-\left(\hat{\alpha}_A + \hat{\alpha}_B\right) x\right] \,. \end{aligned} \tag{15.8}$$

According to these equations, the two temperatures approach a common value exponentially. The common value is obtained in the limit $x \to \infty$ as

$$T_A(\infty) = T_B(\infty) = \frac{T_A(0)}{1 + \frac{\hat{\alpha}_A}{\hat{\alpha}_B}} + \frac{T_B(0)}{1 + \frac{\hat{\alpha}_B}{\hat{\alpha}_A}} \,. \tag{15.9}$$

In a heat exchanger of finite length L, the exit temperatures $T_A(L)$ and $T_B(L)$ differ from this value, see Fig. 15.3.

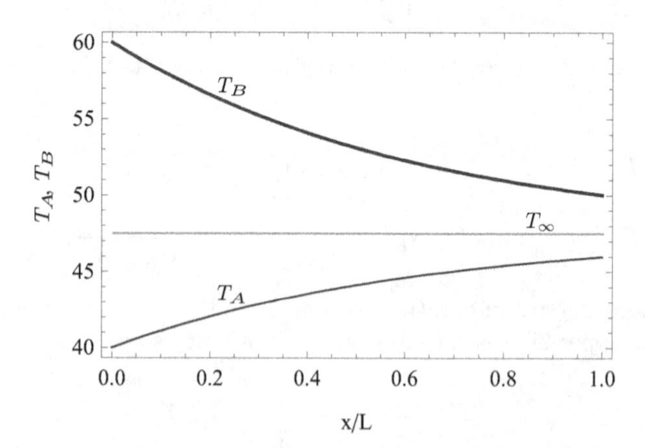

Fig. 15.3 Temperatures T_A and T_B in a co-flow heat exchanger with inflow temperatures $T_A(0) = 40\,°\text{C}$ and $T_B(0) = 60\,°\text{C}$, for $\hat{\alpha}_A/L = 0.7$ and $\hat{\alpha}_B/L = 1$. The asymptotic value $T_A(\infty) = T_B(\infty) = 47.5\,°\text{C}$ is shown as well.

The entropy generation is a useful measure for the work loss associated with the heat exchanger. Since we consider the heat exchanger to be adiabatic to the outside, the entropy generation is just the difference between entropy flowing in and out; for simple incompressible fluids it is computed as

$$\dot{S}_{gen} = \dot{m}_A c_p^A \ln \frac{T_A(L)}{T_A(0)} + \dot{m}_B c_p^B \ln \frac{T_B(L)}{T_B(0)} . \tag{15.10}$$

In a co-flow heat exchanger, all heat transfer takes places over finite temperature differences and the entropy generation is always finite. Therefore co-flow heat exchangers always have an associated work loss.

15.2.3 Counter-Flow Heat Exchangers

Now we consider counter-flow heat exchangers (lower sign). The known inflow conditions are the temperatures $T_A(0)$ and $T_B(L)$, for which we find from (15.6)

$$\begin{aligned} T_A(0) &= K_2 + \frac{\hat{\alpha}_A}{\hat{\alpha}_A - \hat{\alpha}_B} K_1 , \\ T_B(L) &= \frac{\hat{\alpha}_B}{\hat{\alpha}_A} K_2 \exp\left[-\left(\hat{\alpha}_A - \hat{\alpha}_B\right) L\right] + \frac{\hat{\alpha}_A}{\hat{\alpha}_A - \hat{\alpha}_B} K_1 . \end{aligned} \tag{15.11}$$

Solving this for the constants K_1 and K_2 and inserting these into (15.6) gives the temperature curves for the counter-flow heat exchanger as

$$\begin{aligned} T_A(x) &= T_A(0) + \left[T_B(L) - T_A(0)\right] \frac{\exp\left[\left(\hat{\alpha}_B - \hat{\alpha}_A\right) x\right] - 1}{\frac{\hat{\alpha}_B}{\hat{\alpha}_A} \exp\left[\left(\hat{\alpha}_B - \hat{\alpha}_A\right) L\right] - 1} , \\ T_B(x) &= T_B(L) + \left[T_B(L) - T_A(0)\right] \frac{\exp\left[\left(\hat{\alpha}_B - \hat{\alpha}_A\right) x\right] - \exp\left[\left(\hat{\alpha}_B - \hat{\alpha}_A\right) L\right]}{\exp\left[\left(\hat{\alpha}_B - \hat{\alpha}_A\right) L\right] - \frac{\hat{\alpha}_A}{\hat{\alpha}_B}} . \end{aligned} \tag{15.12}$$

This solution becomes singular for the special case $\hat{\alpha}_A = \hat{\alpha}_B = \hat{\alpha}$. L'Hôpital's rule must be used to find the temperature curves for this case as

$$T_A(x) = T_A(0) + \left[T_B(L) - T_A(0)\right] \frac{\hat{\alpha} x}{1 + \hat{\alpha} L} , \tag{15.13}$$

$$T_B(x) = T_B(L) + \left[T_B(L) - T_A(0)\right] \frac{\hat{\alpha}(x - L)}{1 + \hat{\alpha} L} . \tag{15.14}$$

Thus, in general, we will observe exponential curves for the temperatures, but straight lines in the case that $\hat{\alpha}_A = \hat{\alpha}_B$. Figure 15.4 shows the temperature curves for three cases with different or equal values of $\hat{\alpha}_A$ and $\hat{\alpha}_B$.

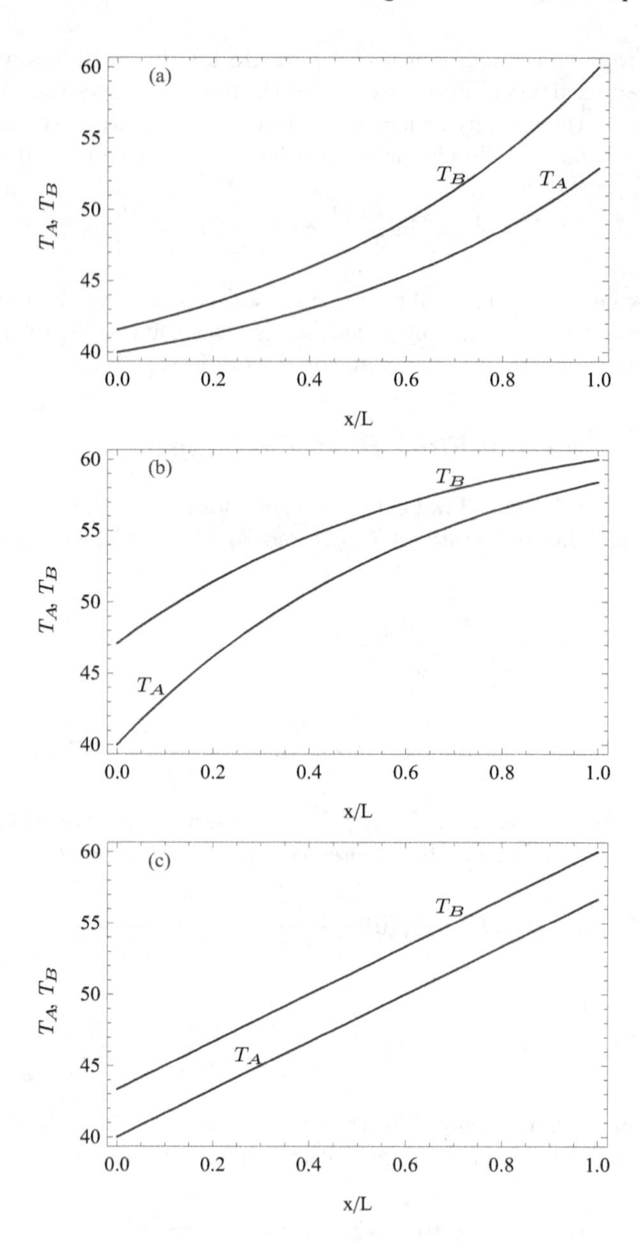

Fig. 15.4 Temperature curves for a counter-flow heat exchanger with $T_A(0) = 40\,^\circ\mathrm{C}$ and $T_B(0) = 60\,^\circ\mathrm{C}$, for the three cases (a) $\hat{\alpha}_A/L = 3.5$ and $\hat{\alpha}_B/L = 5$; (b) $\hat{\alpha}_A/L = 5$ and $\hat{\alpha}_B/L = 3.5$; (c) $\hat{\alpha}_A/L = \hat{\alpha}_B/L = 5$

In particular we note that the exit temperatures are not limited by a common mean value as for co-flow exchange, but can be quite close to the inlet temperature of the other stream. For an exchanger of length L, the exit temperatures of the two streams are

$$T_A(L) = T_A(0) + [T_B(L) - T_A(0)] \frac{\exp[(\hat{\alpha}_B - \hat{\alpha}_A) L] - 1}{\frac{\hat{\alpha}_B}{\hat{\alpha}_A} \exp[(\hat{\alpha}_B - \hat{\alpha}_A) L] - 1},$$

$$T_B(0) = T_B(L) + [T_B(L) - T_A(0)] \frac{1 - \exp[(\hat{\alpha}_B - \hat{\alpha}_A) L]}{\exp[(\hat{\alpha}_B - \hat{\alpha}_A) L] - \frac{\hat{\alpha}_A}{\hat{\alpha}_B}}. \tag{15.15}$$

We limit our attention to a case where $\hat{\alpha}_B < \hat{\alpha}_A$, and ask for the limiting exit temperatures for infinite length, $L \to \infty$, which are

$$T_A(L_\infty) = T_B(L_\infty), \tag{15.16}$$

$$T_B(0) = T_A(0) \frac{\hat{\alpha}_B}{\hat{\alpha}_A} + \left(1 - \frac{\hat{\alpha}_B}{\hat{\alpha}_A}\right) T_B(L_\infty). \tag{15.17}$$

Thus, stream A exits in equilibrium with stream B at $x = L_\infty$, but stream B cannot achieve equilibrium with the incoming stream A at $x = 0$. When $\hat{\alpha}_B > \hat{\alpha}$, the behavior is opposite. The only case where both exiting streams are in equilibrium with the incoming streams, in the case $L \to \infty$, is for $\hat{\alpha}_B = \hat{\alpha}_A = \hat{\alpha}$.

The above discussion already gives indication that a counter-flow heat exchanger works particularly well when $\hat{\alpha}_B = \hat{\alpha}_A = \hat{\alpha}$, which is the case when the mass flows are matched such that $\dot{m}_A c_p^A = \dot{m}_B c_p^B$, see (15.5). The discussion of the entropy generation rate of the heat exchanger sheds more light on this. Again ignoring heat loss to the exterior, the entropy generation is

$$\dot{S}_{gen} = \dot{m}_A c_p^A \ln \frac{T_B(L)}{T_B(0)} + \dot{m}_B c_p^B \ln \frac{T_B(0)}{T_B(L)}, \tag{15.18}$$

and Fig. 15.5 shows the reduced entropy generation rate

$$\frac{\dot{S}_{gen}}{\sqrt{\dot{m}_A c_p^A \dot{m}_B c_p^B}} = \ln \left[\left(\frac{T_B(L)}{T_B(0)} \right)^{\sqrt{\frac{\hat{\alpha}_B}{\hat{\alpha}_A}}} \left(\frac{T_B(0)}{T_B(L)} \right)^{\sqrt{\frac{\hat{\alpha}_A}{\hat{\alpha}_B}}} \right], \tag{15.19}$$

as a function of the ratio $\frac{\hat{\alpha}_A}{\hat{\alpha}_B} = \frac{\dot{m}_B c_p^B}{\dot{m}_A c_p^A}$ for various total exchanger lengths L. If the heat exchanger is sufficiently long, the entropy generation develops a minimum for $\frac{\hat{\alpha}_A}{\hat{\alpha}_B} = 1$, which therefore is the optimum condition for running counter-flow heat exchangers.

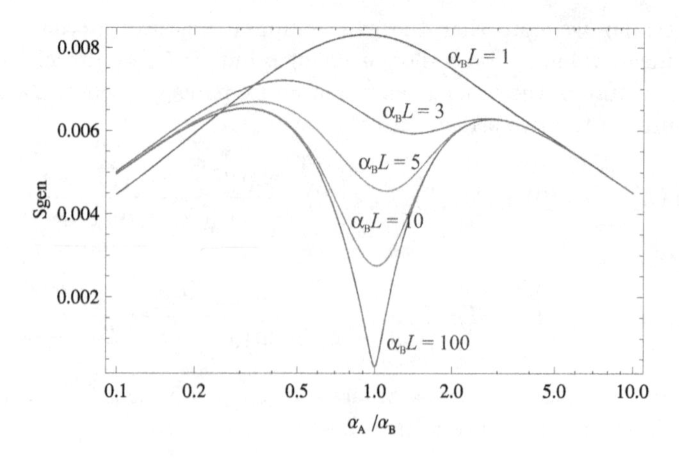

Fig. 15.5 Counter-flow heat exchanger: Reduced entropy generation rate over the ratio $\frac{\hat{\alpha}_A}{\hat{\alpha}_B} = \frac{\dot{m}_B c_p^B}{\dot{m}_A c_p^A}$ for various values of $\hat{\alpha}_B L$; note the logarithmic scale

15.2.4 Summary

In an infinitely long counter-flow heat exchanger running at optimum condition $\frac{\hat{\alpha}_A}{\hat{\alpha}_B} = 1$, both streams have the same temperatures at all locations (with an infinitesimal difference for heat transfer, of course), and no entropy is generated. In co-flow heat exchangers, the entropy generation is always non-zero, since the temperature difference at the common inlet is given by the temperatures of the incoming streams.

Realistic heat exchangers have finite length, and thus some generation of entropy, which is always less in counter-flow heat exchangers as compared to co-flow systems. In the latter, both flows approach an intermediate temperature, while in the counter-flow case, the flows approach an exchange of their temperatures: the exit temperature of one flow is close to the inlet temperature of the other. This certainly is the more desirable outcome.

Entropy generation arguments should be used in the design of heat exchangers, which should be build such that the overall entropy generation in the heat exchanger is as small as possible.

15.3 Heating of a House

As first example for transient processes in open systems, we study the heating of a house in a simplified model. The house of volume V_0 is initially in equilibrium with the outside environment at p_0, T_0. To elevate the temperature in the house to a pleasant value T_H, the house is heated at constant rate $\dot{Q}_{heat}$. Heat transfer through the walls leads to heat loss $\dot{Q}_{loss}$ to the outside. Moreover, the house exchanges air through gaps in doors, windows and walls, and we assume that the pressure stays constant at all times. We ask for the amount of heat that must be supplied to keep the house at the inside temperature T_H, and for the time required to heat the house to that temperature.

The house consists of the mass m_S of structural material (wood, stone, concrete, plaster, ...), which has the average specific heat c_S. To simplify the calculation, we shall assume that the whole structure is at the temperature $T(t)$, that is we ignore any inhomogeneous temperature distribution in the outside walls, which, in reality, would be colder further outside. The mass of air in the house is denoted as $\dot{m}_A(t)$ and is subject to change due to inflow or outflow through gaps.

The mass and energy balances for the house read

$$\frac{dm_A}{dt} = -\dot{m}_A \quad , \quad \frac{dU}{dt} = \dot{Q}_{heat} - \alpha A (T - T_0) - \dot{m}_A h_A \, , \tag{15.20}$$

where α is the heat transfer coefficient for the house, U is the total internal energy (air and structure) and A is the outside surface. Kinetic and potential energies are ignored, and the house and air have the homogeneous temperature T. The last term in the energy balance, $\dot{m}_A h_A$, is the convective outflow of energy. As long as the house is heated, we expect only outflow, due to isobaric expansion of the inside air. In cooling of the house, say at night when the heating is switched off, the air contracts, and outside air (at h_0) flows in.

Before we consider the time dependent process, we have a look at the final steady state, for which $\frac{dm_A}{dt} = \frac{dU}{dt} = 0$ and $T = T_H$, so that the final house temperature is obtained as

$$T_H = T_0 + \frac{\dot{Q}_{heat}}{\alpha A} \, . \tag{15.21}$$

This relation shows that the house temperature is controlled through adjustment of the heating rate $\dot{Q}_{heat}$. Improved insulation reduces the heat transfer coefficient α and thus the heat requirement $\dot{Q}_{heat}$.

With suitable choice of the energy constants, the internal energy of the house and the enthalpy of the air are

$$U = (m_S c_S + m_A c_v) T \quad , \quad h = u + RT = c_p T \, , \tag{15.22}$$

where c_v and $c_p = c_v + R$ are the specific heats of air, assumed to be constant. The mass of air in the house follows from the ideal gas law as

$$m_A = \frac{p_0 V_0}{RT} \, . \tag{15.23}$$

Combining the above, including eliminating the air mass by means of (15.23), yields a differential equation for temperature, which after some simplifications reads

$$\left(m_S c_S + c_p \frac{p_0 V_0}{RT}\right) \frac{dT}{dt} = \dot{Q}_{heat} - \alpha A \left(T - T_0\right) . \tag{15.24}$$

Separation of variables gives

$$\frac{m_S c_S + c_p \frac{p_0 V_0}{RT}}{\dot{Q}_{heat} - \alpha A \left(T - T_0\right)} dT = dt , \tag{15.25}$$

and integration between the initial state (T_0, t_0) and the final state (T, t) gives the solution[2]

$$\frac{p_0 V_0}{R} c_p \frac{\ln\left[\frac{T}{T_0} \frac{\dot{Q}_{heat}}{\dot{Q}_{heat} - \alpha A (T - T_0)}\right]}{\dot{Q}_{heat} + T_0 \alpha A} + \frac{c_S m_S}{\alpha A} \ln\left[\frac{\dot{Q}_{heat}}{\dot{Q}_{heat} - \alpha A \left(T - T_0\right)}\right] = (t - t_0) . \tag{15.26}$$

We can use the equilibrium condition (15.21) to write this in a more compact form,

$$m_H c_p \ln \frac{\frac{T_H}{T_0} - 1}{\frac{T_H}{T} - 1} + m_S c_S \ln\left[\frac{T_H - T_0}{T_H - T}\right] = \alpha A \left(t - t_0\right) , \tag{15.27}$$

where $m_H = \frac{p_0 V_0}{R T_H}$ is the final mass of air in the house. This equation describes the evolution of the inside temperature T towards T_H over time t, in an implicit way. The first term on the left hand side describes the heating of the air, and the second term describes the heating of the structure.

We consider a small house with a footprint of $10\,\mathrm{m} \times 10\,\mathrm{m}$, a height of $3\,\mathrm{m}$ and a flat roof. The air volume of the house is approximately $300\,\mathrm{m}^3$ which at final conditions (T_H, p_0) corresponds to a mass of $m_H = \frac{p_0 V_0}{R T_H} = 350\,\mathrm{kg}$. Clearly, this mass, and the corresponding heat capacity $m_H c_p$, is much less than mass m_S and heat capacity $m_S c_S$ of the structure.

If we completely ignore the contribution to air heating, we obtain for the temperature an exponential relation,

$$T = T_0 + (T_H - T_0)\left(1 - \exp\left[-\frac{\alpha A}{c_S m_S}\left(t - t_0\right)\right]\right) , \tag{15.28}$$

where we have used (15.21).

In a house, normally first the air is heated, e.g., by radiators or forced air heating, and then heat is transferred from the warm air to the structure. If heat transfer to the structure is slow, the structure heating can be ignored, and the air temperature approaches the final temperature according to

$$T = \frac{T_H}{1 + \left(\frac{T_H}{T_0} - 1\right) \exp\left[-\frac{\alpha A}{c_p m_A}\left(t - t_0\right)\right]} , \tag{15.29}$$

[2] If you cannot do the integral with pencil and paper, you can use an integration table, or a mathematical software package like Mathematica.

where α now is the heat transfer coefficient between inside air (at T) and structure (at T_0).

The approach to the final temperature is determined by the time constants $\frac{\alpha A}{c_S m_S}$ and $\frac{\alpha A}{c_p m_A}$, respectively. Due to the large differences in mass between air and structure, the air heats up much faster than the structure.

15.4 Reversible Filling of an Adiabatic Container

When one inflates a bicycle tire with a hand pump, tire and pump become warm. While some of the heat comes from friction in the pump—no seal is friction free—a large part of the heat is due to the rise of temperature of the air being compressed.

To fix ideas we consider the simplest possible case, namely the filling of an adiabatic container with an ideal gas under the assumption that all processes are fully reversible. The system to be considered consists of the adiabatic container of volume V_0 and an adiabatic reversible compressor which draws outside air at T_0, p_0 and compresses it to the pressure p inside the container, which grows over time. Figure 15.6 gives a sketch of the system considered.

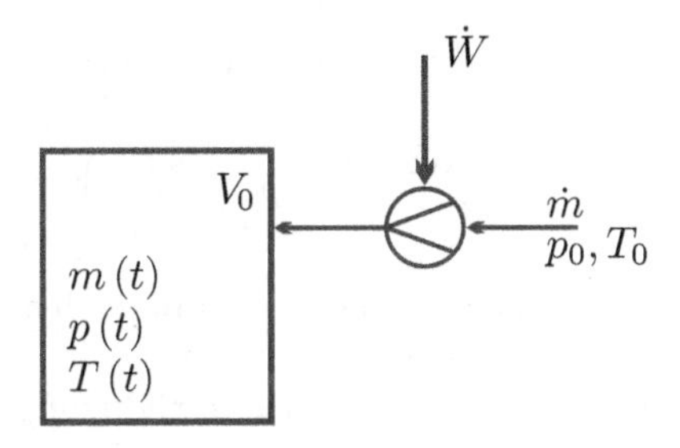

Fig. 15.6 Filling of a container with air

When all processes are reversible, it is convenient to set the system boundary such that the system contains container and compressor. Then, the balances for mass, energy and entropy become

$$\frac{dm}{dt} = \dot{m} \ , \quad \frac{dU}{dt} = -\dot{W} + \dot{m}h_0 \ , \quad \frac{dS}{dt} = \dot{m}s_0 \ , \tag{15.30}$$

where $\dot{W}$ is the power to run the compressor, and $\dot{m}$ is the mass flow pushed into the container. There is no generation of entropy due to the assumption of reversibility. We assume that initially the container is in equilibrium with the environment, so that it contains the mass $m_0 = \frac{p_0 V_0}{RT_0}$, and that the mass flow has the same value at all times, i.e., $\dot{m} = const$. Integration of the mass balance gives the mass in the container at time t as

$$m(t) = \dot{m}t + m_0 . \tag{15.31}$$

The entropy balance can be integrated easily, to give

$$S = \dot{m}s_0 t + S_0 , \tag{15.32}$$

where $S_0 = m_0 s_0$ is the entropy of the initial filling, and $S = ms$ is the entropy at time t. Combining the last two equations we find

$$S = ms = \dot{m}s_0 t + m_0 s_0 = ms_0 . \tag{15.33}$$

Therefore the specific entropy of the gas stays constant,

$$s = \frac{S}{m} = s_0 . \tag{15.34}$$

With the entropy of the ideal gas, $s - s_0 = c_p \ln \frac{T}{T_0} - R \ln \frac{p}{p_0}$, and the ideal gas law $p = \frac{mRT}{V_0}$, follows a relation between temperature and mass ($k = c_p/c_v$),

$$T(t) = T_0 \left(\frac{m(t)}{m_0}\right)^{k-1} . \tag{15.35}$$

As air mass is added to the container, the pressure rises to

$$p(t) = \frac{m(t) RT(t)}{V_0} = p_0 \left(\frac{m(t)}{m_0}\right)^k . \tag{15.36}$$

The air inside the container is compressed adiabatically, as is the intake air when it passes through the compressor.

Since the pressure rises, the power to drive the compressor is increasing over time. From the first law, with $U = mu = mc_v T$, $h_0 = c_p T_0$, we find

$$\dot{W} = \dot{m}h_0 - \frac{dU}{dt} = -\dot{m}c_p T_0 \left[\left(\frac{m(t)}{m_0}\right)^{k-1} - 1\right] . \tag{15.37}$$

We ask for the total work required for an n-fold increase of mass, so that $m = nm_0$. The filling time is

$$t_n = \frac{(n-1) m_0}{\dot{m}} , \tag{15.38}$$

and the corresponding work is

$$W_n = \int_0^{t_n} \dot{W} dt = -m_0 c_v T_0 \left[n^k - 1 - k(n-1)\right] . \tag{15.39}$$

15.5 Reversible Discharge from an Adiabatic Container

In the previous section we filled a container with compressed air. Expansion of the compressed air yields work. Thus, compressing air into a container can be used as a means to store energy. The reversible work potential of the air in the filled container is the closed system exergy of the compressed air (11.31),

$$\Xi = m\left(u - u_E\right) + p_E\left(V - V_E\right) - T_E m\left(s - s_E\right) . \tag{15.40}$$

For the evaluation we must consider the appearing quantities with some care: V is the system volume, which for the present notation is the container volume V_0, while $V_E = \frac{mRT_E}{p_E}$ is the volume of the system when expanded to environmental pressure and temperature $p_E = p_0$, $T_E = T_0$. Note that, due to the reversible adiabatic filling process considered, $s = s_0 = s_E$. With this, and the filling state $m = nm_0$, T, p, as given above, the exergy of the compressed air becomes

$$\Xi_n = m_0 c_v T_0 \left[n^k - 1 - k\left(n - 1\right)\right] = -W_n . \tag{15.41}$$

So, not surprisingly, the work potential of the compressed air equals the work that is required for its reversible filling (with opposite sign). Thus, if all processes are fully reversible, compressed air storage gives a 100% efficient means of energy storage. All work required to fill the container could be taken out to produce work again, e.g. as electricity.

15.6 Reversible Discharge after Cooling

Real life compressed air storage will have a storage efficiency below 100% due to irreversible losses in compressor and turbine during filling and discharge, and due to energy loss by heat transfer from the hot compressed gas to the environment. To get some insight, we assume that, while filling and discharge happen adiabatically and reversibly, the container loses some heat during storage, so that the temperature drops to T_C. In this case, the state of the air in the container is

$$T = T_C \quad , \quad m_C = nm_0 \quad , \quad p_C = \frac{m_C R T_C}{V_0} , \tag{15.42}$$

with the reversible work potential

$$\begin{aligned} \Xi_C &= mc_v\left(T_C - T_0\right) + p_0\left(V_0 - \frac{mRT_0}{p_0}\right) - T_0 m\left(c_p \ln\frac{T_C}{T_0} - R\ln\frac{p_C}{p_0}\right) \\ &= m_0 c_v T_0\left[\frac{p_C}{p_0} - n\ln\frac{p_C}{p_0} + n\ln n^k - 1 - \left(n-1\right)k\right] . \end{aligned} \tag{15.43}$$

Since pressure and temperature are lower than directly after filling, the work potential has dropped, $\Xi_C < \Xi_n$.

Realizing the work potential is another question. We study the inversion of the filling process, that is a turbine for the reversible discharge of the air into the environment. Again, we assume an adiabatic process with constant mass flow. During discharge, the state of the air in the container changes due to expansion. For this computation it is best to consider container and turbine separately. With the present state in the container denoted by T, p, u, s and so on, the discharge from the adiabatic container is given by mass and energy balance,

$$\frac{dm}{dt} = -\dot{m} \ , \quad \frac{dU}{dt} = -\dot{m}h \ . \tag{15.44}$$

Integration of the mass balance with the initial mass $m_C = nm_0$ gives the mass at time t as

$$m(t) = m_C - \dot{m}t \ . \tag{15.45}$$

With $U = mc_vT$ and $h = c_pT$, the first law gives the temperature of the air remaining in the container as

$$\frac{d\ln T}{dt} = -(1-k)\frac{d\ln m}{dt} \implies T = T_C\left(\frac{m}{m_C}\right)^{k-1} \ . \tag{15.46}$$

The temperature drops as mass is discharged, and the same is true for pressure,

$$p = \frac{mRT}{V_0} = \frac{m_C R T_C}{V_0}\left(\frac{m}{m_C}\right)^k = p_C\left(\frac{m}{m_C}\right)^k \ . \tag{15.47}$$

The discharged gas is then expanded through the turbine from p to p_0. For the reversible adiabatic turbine the entropy is constant, so that the turbine exit temperature is

$$T_e = T\left(\frac{p_0}{p}\right)^{\frac{k-1}{k}} = T_C\left(\frac{p_0}{p_C}\right)^{\frac{k-1}{k}} \ . \tag{15.48}$$

Interestingly, while the pressure ratio changes throughout the process, the turbine exit temperature remains at the same value.

The power provided by the turbine is

$$\dot{W} = \dot{m}(h - h_e) = \dot{m}c_p(T - T_e) = \dot{m}c_pT_C\left[\left(\frac{m}{m_C}\right)^{k-1} - \left(\frac{p_0}{p_C}\right)^{\frac{k-1}{k}}\right] \ . \tag{15.49}$$

The discharge will be finished when $p = p_0$, that is

$$\left(\frac{p_0}{p_C}\right)^{\frac{1}{k}} = \frac{m_C - \dot{m}t_{end}}{m_C} \quad \Longrightarrow \quad t_{end} = \frac{m_C}{\dot{m}}\left[1 - \left(\frac{p_0}{p_C}\right)^{\frac{1}{k}}\right] . \tag{15.50}$$

Then, the mass left in the container, and the temperature, are

$$m_{end} = m_C\left(\frac{p_0}{p_C}\right)^{\frac{1}{k}} , \quad T_{end} = T_C\left(\frac{m_{end}}{m_C}\right)^{k-1} = T_C\left(\frac{p_0}{p_C}\right)^{\frac{k-1}{k}} = T_e . \tag{15.51}$$

The total work delivered is

$$\begin{aligned} W_C &= \int_0^{t_{end}} \dot{W} dt = \dot{m}c_p T_C\left[\int_0^{t_{end}}\left(\frac{m}{m_C}\right)^{k-1} dt - \left(\frac{p_0}{p_C}\right)^{\frac{k-1}{k}} t_{end}\right] \\ &= \cdots = m_0 c_v T_0\left[\frac{p_C}{p_0} - k\left(\frac{p_C}{p_0}\right)^{\frac{1}{k}} + k - 1\right] . \end{aligned} \tag{15.52}$$

The pressure p_C assumes values between the reversible filling pressure $p_n = p_0 n^k$ and np_0, which is the pressure when the compressed air is thermally equilibrated with the environment at T_0.

When no heat loss occurs ($T_C = T_n$), the turbine exit temperature is just the environmental temperature, $T_e = T_n\left(\frac{p_0}{p_n}\right)^{\frac{k-1}{k}} = T_0$, and it lies below T_0, when the air lost energy to the environment. In the extreme case that $T_C = T_0$, the turbine exit temperature is $T_0 n^{\frac{1-k}{k}} < T_0$. The actual work delivered is below the work potential (15.43), since the air leaving the turbine is colder than the environmental air, and thus there is work potential due to temperature difference between exhaust and environment. When the cold exhaust just mixes with the environmental air, entropy is produced, and this work potential is lost.

The second law efficiency for the discharge process alone is the ratio between the work produced and the initial work potential,

$$\eta_{II} = \frac{W_C}{\Xi_C} = \frac{\frac{p_C}{p_0} - 1 - k\left(\left(\frac{p_C}{p_0}\right)^{\frac{1}{k}} - 1\right)}{\frac{p_C}{p_0} - 1 - k\,(n-1) + n^k \ln\left[n\left(\frac{p_0}{p_C}\right)^{\frac{1}{k}}\right]} . \tag{15.53}$$

The second law efficiency is $\eta_{II} = 1$ for the case without heat loss in storage ($T_C = T_n$ and $T_e = T_0$) and $\eta_{II} = \left[\frac{k}{k-1}\left(n - n^{\frac{1}{k}}\right) - n + \right] / \left[\frac{n^k}{k}\ln n - n + 1\right]$ for complete heat loss ($T_C = T_0$ and $T_e = T_0 n^{-\frac{k-1}{k}}$). This efficiency does only account for the work loss in discharge, but not for the loss associated to the heat loss during storage.

The storage efficiency for the complete filling and discharge process is the ratio between the work produced in discharge, and the work required for filling,

$$\eta_{st} = \left|\frac{W_C}{W_n}\right| = \frac{\frac{p_C}{p_0} - 1 - k\left(\left(\frac{p_C}{p_0}\right)^{\frac{1}{k}} - 1\right)}{n^k - 1 - k(n-1)} . \tag{15.54}$$

The storage efficiency is $\eta_{st} = 1$ for the case without heat loss in storage ($T_C = T_n$ and $T_e = T_0$), and $\eta_{st} = \left[n - 1 - k\left(n^{\frac{1}{k}} - 1\right)\right] / \left[n^k - 1 - k(n-1)\right]$ for complete heat loss ($T_C = T_0$ and $T_e = T_0 n^{\frac{k-1}{k}}$). This efficiency accounts for the work lost in storage and discharge.

The values on both efficiencies depend on the total mass exchange, nm_0, e.g. for $n = 20$, we find for the second law efficiency values in $[0.16, 1]$ and for the storage efficiency values in $[0.22, 1]$.

15.7 Reversible Filling of a Gas Container with Heat Exchange

In the previous sections we assumed that no heat is exchanged during filling and discharge, which would be the case when the processes are rather fast. For a lower filling rate, since real containers are not adiabatic, there will be heat exchange during filling, and here we study a container with heat exchange. We assume that the container is in contact with the environment at T_0, and use Newton's law of cooling. The balances for mass, energy and entropy become

$$\frac{dm}{dt} = \dot{m} , \quad \frac{dU}{dt} = \dot{Q} - \dot{W} + \dot{m}h_0 , \quad \frac{dS}{dt} - \frac{\dot{Q}}{T} = \dot{m}s_0 , \tag{15.55}$$

where $\dot{Q} = \alpha(T_0 - T)$ is the heat exchange between the compressed gas at T and the environment. As before, we assume that initially the container is in equilibrium with the environment, so that it contains the mass $m_0 = \frac{p_0 V_0}{RT_0}$, and that the mass flow is constant, $\dot{m} = const$. Integration of the mass balance gives the mass in the container at time t as

$$m(t) = \dot{m}t + m_0 .$$

With $S = ms$ and the mass balance, the entropy balance assumes the form

$$m\frac{ds}{dt} = \frac{\alpha(T_0 - T)}{T} - \dot{m}(s - s_0) . \tag{15.56}$$

The entropy of the ideal gas is

$$s - s_0 = c_v \ln\frac{T}{T_0} + R\ln\frac{V_0/m}{V_0/m_0} = c_v \ln\frac{T}{T_0} - R\ln\frac{m}{m_0} . \tag{15.57}$$

After some manipulation, the second law reduces to a differential equation for the gas temperature,

$$\frac{d\ln\frac{T}{T_0}}{dt} = -\frac{\alpha}{mc_v}\left(1-\frac{T_0}{T}\right) - \frac{\dot{m}}{m}\left(\ln\frac{T}{T_0} - (k-1)\left(\ln\frac{m}{m_0}+1\right)\right) . \quad (15.58)$$

Solving the first law for power, and use of the appropriate constitutive equation and the differential equation for T results in

$$\dot{W} = \dot{m}c_v T\left[\ln\frac{T}{T_0} - (k-1)\ln\frac{m}{m_0} + k\left(\frac{T_0}{T}-1\right)\right] . \quad (15.59)$$

The filling pressure follows from the ideal gas equation, $p = \frac{mRT}{V_0}$. Of course, for the adiabatic case ($\alpha = 0$) the solution of Sec. 15.4 fulfills both equations.

For the non-adiabatic case, the equation for temperature cannot be solved with pencil and paper, but must be solved on a computer. The numerical solution $T(t)$ can then be used to determine the compressor power (15.59) and the total work $W = \int \dot{W}dt$. Figure 15.7 shows, in dimensionless form, mass, temperature, pressure and compressor work as functions of time, comparing the adiabatic case ($\alpha = 0$, dashed), and the case with heat loss to the environment ($\alpha \neq 0$, continuous). Less work is required in the latter case, since, due to the constant heat loss, the temperature, and thus the pressure is lower. For the data given with the figure, the total work requirement for filling (dimensionless) is, $\frac{W}{m_0RT_0} = -34.26$ for the adiabatic case, and $\frac{W}{m_0RT_0} = -22.28$ for the non-adiabatic case.

While non-adiabatic filling requires less work, the work potential of the filling (same mass) is less as well, due to lower pressure. The heat loss to the environment will continue after filling is completed. Accordingly, the filling pressure, and the work potential will drop after filling is completed. We consider the work potential (exergy) in case that the filling temperature has dropped to the environmental temperature T_0. Then the state of the air in the cavern is

$$\frac{m}{m_0} = n \ , \quad T = T_0 \ , \quad p = \frac{mRT_0}{V_0} \ , \quad s = c_p\ln\frac{T_0}{T_0} - R\ln\frac{p}{p_0} + s_0 \quad (15.60)$$

with the work potential (exergy)

$$\Xi_0 = m_0c_vT_0(k-1)(n\ln n - n + 1) . \quad (15.61)$$

The non-dimensional value corresponding to $n = 11$ (the final state in Fig. 15.7) is $\frac{\Xi_0}{m_0RT_0} = 16.38$. Thus, if the container is filled adiabatically and then cools to environmental temperature, the work potential is only half of the filling work. Of course, irreversibilities during filling will increase the work requirement, and irreversibilities during discharge will reduce the work output below Ξ_0 so that in the end the storage efficiency is relatively low.

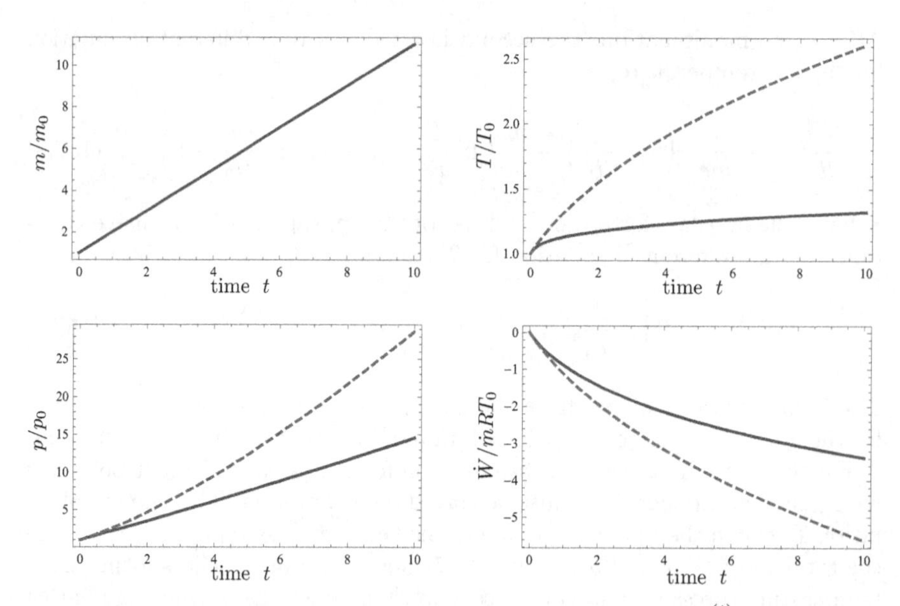

Fig. 15.7 Reversible filling of a container: Dimensionless mass $\frac{m(t)}{m_0}$, temperature $\frac{T(t)}{T_0}$, pressure $\frac{p(t)}{p_0}$, and compressor work $\frac{\dot{W}(t)}{\dot{m}RT_0}$ as functions of dimensionless time $\frac{\dot{m}}{m_0}t$ for: (a) Adiabatic filling ($\alpha = 0$, dashed curves), and (b) Filling with heat loss ($\frac{\alpha}{\dot{m}c_v} = 10$, continous curves)

15.8 CAES: Compressed Air Energy Storage

Renewable energy resources such as wind power or solar energy suffer from the fact that one cannot align the times of energy generation and energy demand. Large scale use of renewable energy therefore requires efficient storage mechanisms for the energy. Compressed air energy storage (CAES) systems are suggested as one means to this end. The idea is to use renewable energy at the time of production to fill a big cavern, e.g. an abandoned salt mine, with compressed air. Later, when energy demand arises, the compressed air in the cavity can be expanded through a turbine to generate electricity.

The filling and discharge problems discussed above can be considered as prototypes of the processes in a compressed air storage system. But, of course, what happens in realistic systems differs from the idealized description considered so far: First of all, the compressor used to fill the cavern, and the turbine used to expand the air, will not be reversible, but both will loose some work to friction. Typically, the temperature of the gas fed into the cavern by the compressor differs from the temperature of the gas in the cavern, entropy is generated as the two mix. As the air in the cavern heats up due to compression there will be heat flow from the hot gas to the colder cavern wall, entropy is generated in heat transfer over finite temperature differences. Moreover, when simply a turbine is used to expand the air leaving the cavern, the exhaust temperature will typically lie below the environmental temperature. More entropy is generated as the cold turbine exhaust mixes with the environmental air. Since entropy generation means work loss, the storage efficiency will always be below 100%.

A simplified CAES system is depicted in Fig. 15.8: An adiabatic irreversible compressor (1-2) pushes air into the cavern, the compressor power is provided by renewable energy, indicated in the figure by a wind turbine. After filling is completed, the air sits a while in the cavern (state C). Cavern temperature $T_C(t)$ and pressure $p_C(t)$ change over time. Since the simple expansion of the air through a turbine leads to cooling and external loss, one typically reheats the air that leaves the cavern. In the system studied here, the reheat occurs by means of a regenerator (C-3) and a combustion chamber (3-4), that is a fuel is used to add heat to the system. The compressed reheated air is then expanded in the turbine (4-5) and the turbine exhaust is run through the regenerator (5-6) before it is discharged into the environment.

With this set-up, a CAES system is not a pure energy storage system but a mix between a storage system and a conventional gas turbine. In a conventional gas turbine the compressor work is provided by the turbine, and thus comes from the fuel. In a CAES system, the compressor work is provided by an external source, either a regenerative source like wind or solar, or surplus electricity from conventional power plants at times of weak demand.

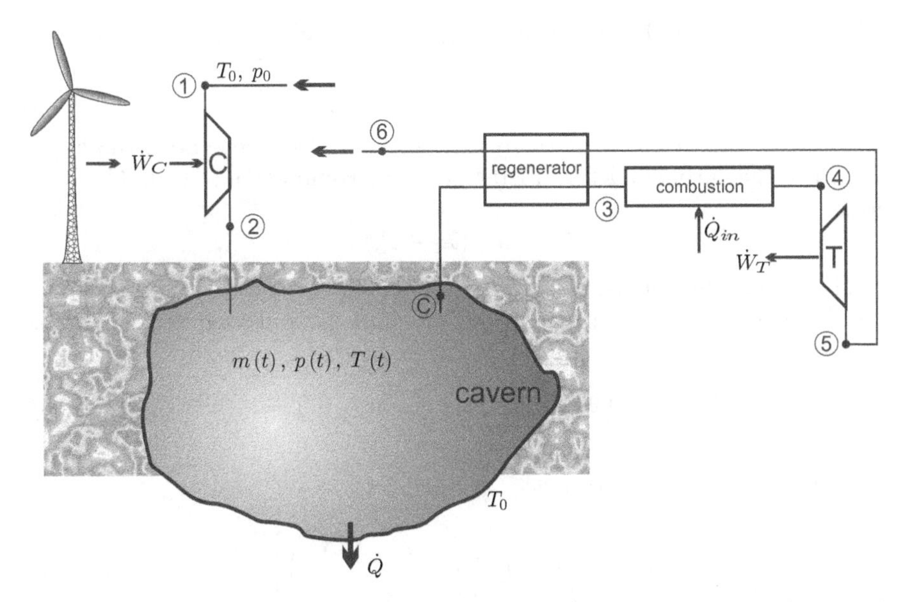

Fig. 15.8 A simple compressed air energy storage system, with reheat and regenerator

In a realistic system, the compressor power depends on the amount of renewable energy available for storage. To simplify the problem we consider constant mass flow $\dot{m}$ and treat air as ideal gas with constant specific heats. Typically, the cavern is not emptied to environmental pressure, but is maintained in a window, $p_{\min} \leq p_C(t) \leq p_{\max}$, to ensure sufficient pressure ratio in expansion and even power generation.

For compressor, regenerator and turbine, we can use the usual expressions for exit properties, work, and heat. The only difference to steady state operation is that due to the changing state in the cavern all temperatures and pressures change over time.

The adiabatic irreversible compressor draws air at environmental state $p_1 = p_0$, $T_1 = T_0$ and compresses it to the cavern pressure $p_C(t)$. The compressor power is given by

$$\dot{W}_C = \dot{m}\frac{1}{\eta_C}c_pT_0\left(1-\left(\frac{p_C}{p_0}\right)^{\frac{k-1}{k}}\right) = \dot{m}c_p\left(T_0 - T_2\right) , \qquad (15.62)$$

where η_C is the isentropic efficiency. The compressor exit temperature follows from the second equation as

$$T_2 = T_0\left[1-\frac{1}{\eta_C}\left(1-\left(\frac{p_C}{p_0}\right)^{\frac{k-1}{k}}\right)\right] . \qquad (15.63)$$

The regenerator with effectiveness η_{reg} heats the pressurized air to the temperature

$$T_3 = T_C + \eta_{reg}\left(T_5 - T_C\right) , \qquad (15.64)$$

where T_5 is the turbine exit temperature. After the regenerator, the air is heated to the turbine inlet temperature T_4 by combustion of fuel. The heat added is

$$\dot{Q} = \dot{m}c_p\left(T_4 - T_3\right) .$$

The irreversible adiabatic turbine (isentropic efficiency η_T) expands the heated cavern air to the environmental pressure p_0, and produces the power

$$\dot{W}_T = \dot{m}\eta_Tc_pT\left(1-\left(\frac{p_0}{p_C}\right)^{\frac{k-1}{k}}\right) = \dot{m}c_p\left(T_4 - T_5\right) . \qquad (15.65)$$

The turbine exit temperature is

$$T_5 = T_4\left[1-\eta_T\left(1-\left(\frac{p_0}{p_C}\right)^{\frac{k-1}{k}}\right)\right] . \qquad (15.66)$$

For the description of the cavern state we need the time dependent first law to describe the change. We must distinguish between the periods of filling, resting, and discharge. During filling the mass balance reads

$$\frac{dm}{dt} = \dot{m} \,, \tag{15.67}$$

and the first law reduces to

$$\frac{dU}{dt} = \alpha \left(T_0 - T_C\right) + \dot{m} h_2 \,, \tag{15.68}$$

where we have employed Newton's law of cooling to describe the heat loss to the cavern wall. With the ideal gas law, $T_C = \frac{p_C V_C}{mR}$, the internal energy can be written as

$$U = m c_v T_C = \frac{c_v}{R} m R T_C = \frac{1}{k-1} p_C V_C \,, \tag{15.69}$$

where V_C is the constant cavern volume. With this, the ideal gas law, and the above result for T_2, the first law becomes a differential equation for cavern pressure,

$$\frac{dp}{dt} = (k-1) \frac{\alpha}{V_C} \left(T_0 - \frac{p_C V_C}{mR}\right) + k \frac{\dot{m} R T_0}{V_C} \left[1 - \frac{1}{\eta_C} \left(1 - \left(\frac{p_C}{p_0}\right)^{\frac{k-1}{k}}\right)\right] . \tag{15.70}$$

While the air sits in the closed cavern, it cools down, and the pressure drops. In this case the second term is absent, and pressure changes according to

$$\frac{dp_C}{dt} = (k-1) \frac{\alpha}{V_C} \left(T_0 - \frac{p_C V_C}{mR}\right) . \tag{15.71}$$

During discharge, mass balance and first law become

$$\frac{dm}{dt} = -\dot{m} \,, \quad \frac{dU}{dt} = \alpha \left(T_0 - T_C\right) - \dot{m} h_C \,, \tag{15.72}$$

where h_C is the enthalpy of the air in the cavern. Applying the ideal gas law gives an equation for the change of pressure during discharge,

$$\frac{dp_C}{dt} = \frac{\alpha}{\frac{c_v}{R} V_C} \left(T_0 - \frac{p_C V_C}{mR}\right) - k \frac{\dot{m}}{m} p_C \,. \tag{15.73}$$

The above equations can be easily solved on the computer. To get some idea of the solution behavior, we assume a regular 24 hour cycle of four 6 hour long segments of filling–resting–discharge–resting. The cavern volume is $V_C = 250000\,\mathrm{m}^3$. All other parameters are set somewhat arbitrarily, so that the basic characteristics of the process become evident: The coefficient for Newton's law was chosen as $\alpha = 1000 \frac{\mathrm{kW}}{\mathrm{K}}$, which is high enough to almost equilibrate air and cavern wall over the resting period. The mass in the cavern, and the mass flow $\dot{m}$, are adjusted so that the cavern pressure oscillates between 20 and 50 bar. The turbine inlet temperature, which is controlled by fuel addition, is set to $T_4 = 1000\,\mathrm{K}$. The equation presented above can be solved with an easy numerical stepping system. To have no influence of initial conditions, the equations must be solved over several cycles. Figure 15.8 shows, for one cycle, mass, pressure, temperatures, work and heat for the cavern and the devices.

We point to some distinctive features: During filling, cavern pressure and temperature go up, they reach their maximum when filling stops. Subsequently, both drop, due to isochoric cooling by heat transfer to the cavern walls. In discharge both drop further due to expansion. When discharge is finished, temperature is slightly below the wall temperature, hence it increases during the low pressure rest period. Compressor power and exit temperature increase during the filling period, since the compressor has to work over an increasing pressure ratio. Turbine power decreases during discharge, since the pressure ratio goes down.

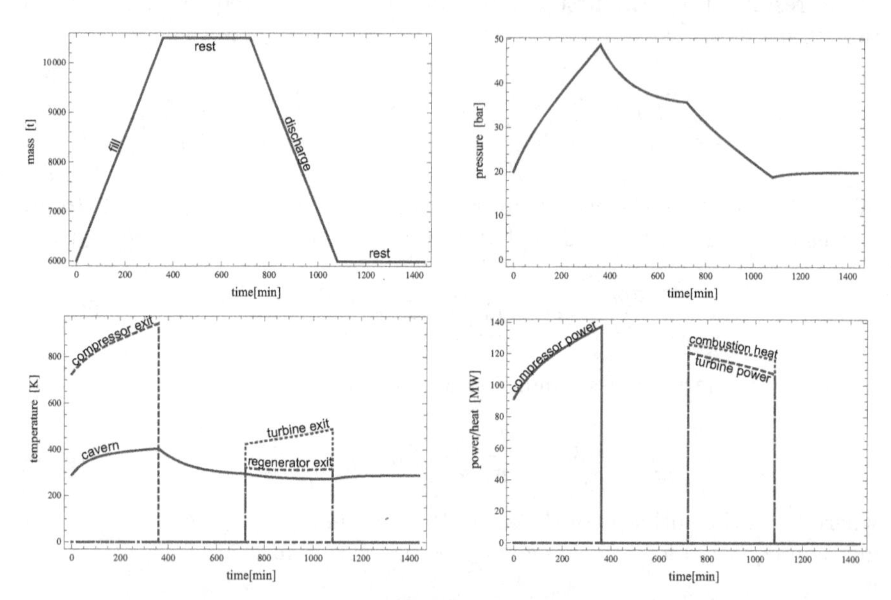

Fig. 15.9 Numerical solution for CAES plant (see text). Upper figures show the mass in the cavern, $m(t)$, and the pressure, $p_C(t)$. Left lower figure shows exit temperatures of compressor, regenerator and turbine, right lower figure shows compressor and turbine power, and heating rate.

There are several efficiency measures that can be used to evaluate the performance of the system. The simplest is the ratio between overall work produced by the turbine and energy input from renewable energy and fuel,

$$\eta_{overall} = \frac{W_T}{Q + |W_C|} = 0.475 \,. \tag{15.74}$$

Here, W_T, W_C, Q are the time integrals of $\dot{W}_T, \dot{W}_C, \dot{Q}$. This measure assigns the same importance to the renewable energy used to drive the compressor and the heat from the fuel. For the given data, less than half of the energy input is converted into work.

When the interest is mainly about the fuel usage, or the accompanying release of CO_2 into the atmosphere, which both are proportional to the heat required to produce the work output, one will use the thermal efficiency

$$\eta_{thermal} = \frac{W_T}{Q} = 0.940 \,. \tag{15.75}$$

Since the compressor work must not be provided by the turbine, the back work ratio is zero, and the thermal efficiency is about twice that of a conventional Brayton cycle with regenerator. This efficiency assigns no cost to the renewable energy.

The storage efficiency of the system is the percentage of compressor work (input) that is recovered in the expansion (output). The output comes from the fuel and the storage, and it is a question of interpretation how much can be attributed to the fuel. It seems sensible to compare the CAES with a heat engine with a thermal efficiency η_{HE}, so that its net work is $W_{HE} = \eta_{HE} Q$. The difference between the CAES turbine work, W_T and the net work W_{HE} is due to the storage, and therefore we define the storage efficiency as

$$\eta_{storage} = \frac{W_T - \eta_{HE} Q}{W_C} \,. \tag{15.76}$$

Clearly, the efficiency value depends on the type of heat engine one considers for comparison. State of the art combined cycle power plants reach thermal efficiencies of 60%, which gives a rather low storage efficiency $\eta_{storage} = 0.347$. However, the comparison between CAES and combined cycle is somewhat unfair, since both will be used for different purposes. The steam engine in the combined cycle has large start-up and shut-down times, so that the cycle must run at steady states for long periods to provide base power. By construction, the CAES only operates from time to time, and one will run it at peak power demand. Gas turbine systems have short start-up times, and therefore are often used to provide power at times of large demand. Therefore, it is more reasonable to compare CAES to a typical Brayton cycle, which, with regenerator, can have efficiencies of up to $\eta_{HE} = 42\%$. The corresponding storage efficiency is $\eta_{storage} = 0.531$.

Problems

15.1. Heating of a Room

A room of volume $75\,\mathrm{m}^3$ contains air, initially at $T_0 = 273\,\mathrm{K}$. A heater supplies heat at a rate of 2000 W. Since air can leave or enter the room through small gaps in windows and doors, the pressure in the room is equal to the outside pressure of 1 bar at all times. Assume that no air *enters* the room.

Consider air as an ideal gas with constant specific heats, and assume that all properties inside the room are homogeneous.

1. Assuming that there are no heat losses to the environment, compute mass and temperature of the air in the room as a function of time and plot the result.
2. Consider the same problem with heat losses. For this, assume that the heat losses are proportional to the difference between the temperatures inside and outside (assume $T_{outside} = T_0 = 273\,\mathrm{K}$), that is $\dot{Q}_{loss} = \alpha A (T_0 - T)$. Here, $\alpha = 90 \frac{\mathrm{kJ}}{\mathrm{m}^2\,\mathrm{h}\,\mathrm{K}}$, is the heat transfer coefficient, and $A = 60\,\mathrm{m}^2$ is the wall surface. Plot the resulting curves.

15.2. Heat Transfer Loss in Isobaric Pipe Flow

Consider a mass flow $\dot{m}$ through a pipe of length L at steady state without pressure loss; the fluid enters the pipe at temperature T_1. The heat exchange of the pipe section of length dx with the environment at T_0 can be described by Newton's law as $\delta\dot{Q} = \alpha (T_0 - T)\, dx$. For the following, assume that the fluid has constant specific heat c_p and ignore kinetic and potential energies.

1. Compute the temperature of the fluid, $T(x)$.
2. Consider a control volume just around the fluid in the pipe, and compute entropy generation and work loss in that volume. Discuss your findings.
3. Consider a control volume around the pipe whose boundaries are at the environmental state and compute entropy generation and work loss. Show that the entropy generation is positive (for any choice of temperatures).

15.3. Friction Loss in Pipe Flow (Incompressible Fluid)

Consider a mass flow $\dot{m}$ through an adiabatic pipe of length L at steady state with pressure loss. The fluid enters the pipe at temperature T_1 and pressure p_1. Due to friction, the pressure drops as $\delta p = -\beta dx$ along the distance dx. Assume that the fluid is incompressible and has constant specific heat c (note: this means $du = cdT$, but not $dh = cdT$!). Ignore potential and kinetic energies.

1. Compute the temperature of the fluid, $T(x)$.
2. Consider a control volume just around the fluid in the pipe, and compute entropy generation and work loss in that volume.

15.4. Friction Loss in Pipe Flow (Ideal Gas)

Consider a mass flow $\dot{m}$ through a pipe of length L at steady state with pressure loss; the fluid enters the pipe at temperature T_1 and pressure p_1.

Due to friction, the pressure drops as $\delta p = -\beta dx$ along the distance dx; the pipe is adiabatic. Assume that the fluid is an ideal gas with constant specific heats. Ignore potential and kinetic energies.

1. Compute the temperature of the gas, $T(x)$.
2. Consider a control volume just around the gas in the pipe, and compute entropy generation and work loss in that volume.
3. Compare the results for a gas with those obtained for the incompressible fluid in the previous problem (discuss!).

Chapter 16
More on Property Relations

16.1 Measurability of Properties

Some properties are easy to measure, and thus quite intuitive, e.g., pressure p, temperature T and specific volume v. Accordingly, the thermal equation of state, $p(T, v)$ can be measured with relative ease. Other properties cannot be measured directly, for instance internal energy u or enthalpy h, which must be determined by means of applying the first law to a calorimeter, or entropy s, which must be determined from other properties by integration of the Gibbs equation.

The Gibbs equation

$$Tds = du + pdv \tag{16.1}$$

gives a differential relation between properties for any simple substance. Its analysis with the tools of multivariable calculus, as presented below, shows that specific internal energy $u = U/m$, specific enthalpy $h = H/m$, specific Helmholtz free energy $f = u - Ts$, and specific Gibbs free energy $g = h - Ts$ are potentials when considered as functions of particular variables. The evaluation of the potentials leads to a rich variety of relations between thermodynamic properties which will be derived and explored in this chapter. In particular, these relate properties that are more difficult, or even impossible, to measure to those that are more easy to measure, and thus reduce the necessary measurements to determine data for all properties.

Later, in Chapter 17, it will be seen that the thermodynamic potentials play an important role in finding the equilibrium states of a system.

16.2 Thermodynamic Potentials and Maxwell Relations

We rewrite the Gibbs equation as

$$du = Tds - pdv \ . \tag{16.2}$$

H. Struchtrup, *Thermodynamics and Energy Conversion*,
DOI: 10.1007/978-3-662-43715-5_16,

This equation suggests to consider the internal energy u as a function of entropy s and volume v, that is $u(s,v)$. Its exact differential (u is a state function) is given by

$$du = \left(\frac{\partial u}{\partial s}\right)_v ds + \left(\frac{\partial u}{\partial v}\right)_s dv \,, \tag{16.3}$$

and by comparison with (16.2) we identify

$$T = \left(\frac{\partial u}{\partial s}\right)_v \quad , \quad -p = \left(\frac{\partial u}{\partial v}\right)_s \,. \tag{16.4}$$

Thus, the internal energy $u(s,v)$ is a potential, in the sense that temperature $T(s,v)$ and pressure $p(s,v)$ are obtained from derivatives of u.

The order of second derivatives can be exchanged,

$$\frac{\partial^2 u}{\partial v \partial s} = \frac{\partial^2 u}{\partial s \partial v} \,, \tag{16.5}$$

and since u is a potential for T and p, we find a relation between the derivatives of T and p, one of the so-called Maxwell relations,

$$\left(\frac{\partial T}{\partial v}\right)_s = -\left(\frac{\partial p}{\partial s}\right)_v \,. \tag{16.6}$$

For the novice, expressions like (16.6) seem to be just mathematical symbols. However, all expressions have a clear meaning which derives from the properties they entail. For instance, the expression $\left(\frac{\partial T}{\partial v}\right)_s$ gives the change of temperature with volume at constant entropy. It can be measured, in principle, by varying the volume of the system by a small amount dv while the entropy s stays constant, and measuring the associated change in temperature dT. Similarly, the expression $\left(\frac{\partial p}{\partial s}\right)_v$ describes the change of pressure with entropy at constant volume. It can be measured, in principle, in a system at constant volume v by varying the entropy by ds and recording the associated pressure change dp. Similar interpretations apply to the expressions in (16.4).

The above partial derivatives are, in fact, not accessible to measurements since one of the variables is entropy: since there is no direct measurement of entropy, it is very difficult, if not impossible, to conduct an experiment in which entropy is changed by a given amount ds, or fixed at a given value s.[1]

To find more relations of the same kind, we use that $-pdv = -d(pv) + vdp$, which leads to a shift in variables from v to p, a so-called Legendre transform

[1] Earlier we learned that an adiabatic and reversible process is isentropic. Thus, a process at constant entropy s can be achieved in a very slow experiment in an isolated container. However, the actual value of entropy for the given experiment cannot be determined through measurement.

(Adrien-Marie Legendre, 1752-1833). With this, the Gibbs equation (16.2) becomes $du = Tds + vdp - d(pv)$, and by introducing enthalpy as

$$h = u + pv \,, \tag{16.7}$$

the Gibbs equation assumes the alternative form

$$dh = Tds + vdp \,. \tag{16.8}$$

This equation points to considering enthalpy h as a function of entropy s and pressure p, where now $h(s,p)$ is a potential, with

$$T = \left(\frac{\partial h}{\partial s}\right)_p \,, \quad v = \left(\frac{\partial h}{\partial p}\right)_s \,, \tag{16.9}$$
$$\left(\frac{\partial T}{\partial p}\right)_s = \left(\frac{\partial v}{\partial s}\right)_p \,.$$

The last equation is the Maxwell equation for this potential, it results from exchanging the order of derivatives, $\frac{\partial^2 h}{\partial p \partial s} = \frac{\partial^2 h}{\partial s \partial p}$. Again all partial derivatives contain the entropy s as one of the variables, and therefore they are not accessible to measurements.

Similarly, we can apply a Legendre transform to the term $Tds = d(Ts) - sdT$ in the Gibbs equation. This transformation exchanges the variable s, which cannot be measured directly, by the variable T, which can be measured. As a result, we find new potentials, the free energies.

The Helmholtz free energy is defined as

$$f = u - Ts \,, \tag{16.10}$$

and with this the Gibbs equation can be written in the alternative form

$$df = -sdT - pdv \,.$$

We recognize that $f(T,v)$ is a potential, with

$$-s = \left(\frac{\partial f}{\partial T}\right)_v \,, \quad -p = \left(\frac{\partial f}{\partial v}\right)_T \,, \tag{16.11}$$
$$\left(\frac{\partial s}{\partial v}\right)_T = \left(\frac{\partial p}{\partial T}\right)_v \,.$$

The last equation is the Maxwell equation for this potential, it results from exchanging the order of derivatives, $\frac{\partial^2 f}{\partial v \partial T} = \frac{\partial^2 f}{\partial T \partial v}$. Most remarkably, the Maxwell relation $(16.11)_3$ contains the expression $\left(\frac{\partial p}{\partial T}\right)_v$, which describes the change of pressure p with temperature T in an experiment at constant volume v. Since p, T and v can be measured, this expression can be found experimentally. In fact, measurement of $\{p,T,v\}$ gives the thermal equation of state $p(T,v)$, and we can say that $\left(\frac{\partial p}{\partial T}\right)_v$ can be determined from the thermal equation of state. The other expression, $\left(\frac{\partial s}{\partial v}\right)_T$, cannot be measured by itself, since it contains entropy s. Hence, with the Maxwell relation the expression $\left(\frac{\partial s}{\partial v}\right)_T$ can be measured through measurement of the thermal equation of state.

The Gibbs free energy is defined as

$$g = h - Ts = f + pv \,, \tag{16.12}$$

and with another Legendre transform, the Gibbs equation becomes

$$dg = -sdT + vdp \,.$$

We recognize that $g(T,p)$ is a potential, with

$$-s = \left(\frac{\partial g}{\partial T}\right)_p \,, \quad v = \left(\frac{\partial g}{\partial p}\right)_T \,, \tag{16.13}$$
$$\left(\frac{\partial s}{\partial p}\right)_T = -\left(\frac{\partial v}{\partial T}\right)_p \,.$$

The last equation is the Maxwell equation for this potential, it results from exchanging the order of derivatives, $\frac{\partial^2 g}{\partial p \partial T} = \frac{\partial^2 g}{\partial T \partial p}$. Also this Maxwell relation contains one expression that can be determined from the thermal equation of state, namely $\left(\frac{\partial v}{\partial T}\right)_p$, relating it to an expression that is not accessible to direct measurement, namely $\left(\frac{\partial s}{\partial p}\right)_T$. The usefulness of the above differential relations, in particular of those that involve expressions that can be measured, will become evident in the subsequent sections.

It is important to note that the thermodynamic properties u, h, f, g are only potentials when considered as functions of the given variables. That is only $u(s,v)$, $h(s,p)$, $f(T,v)$, $g(T,p)$ are potentials! One can use property relations to change the variables, for instance with the thermal equation of state $p(T,v)$ one obtains $g(T,p) = g(T,p(T,v)) = g(T,v)$—however, as function of T and v the Gibbs free energy is *not* a potential.

16.3 Two Useful Relations

Gibbs equation and Maxwell relations can be used to obtain additional relations between properties which will form the center of the following section on measurement of thermodynamic properties. To proceed, we consider energy and entropy in the Gibbs equation (16.2) as functions of temperature and volume, $u(T,v)$, $s(T,v)$, and evaluate their differentials as

$$du = \left(\frac{\partial u}{\partial T}\right)_v dT + \left(\frac{\partial u}{\partial v}\right)_T dv \quad , \quad ds = \left(\frac{\partial s}{\partial T}\right)_v dT + \left(\frac{\partial s}{\partial v}\right)_T dv \,. \quad (16.14)$$

After some reordering, the Gibbs equation (16.2) yields

$$\left[\left(\frac{\partial u}{\partial T}\right)_v - T\left(\frac{\partial s}{\partial T}\right)_v\right] dT + \left[\left(\frac{\partial u}{\partial v}\right)_T - T\left(\frac{\partial s}{\partial v}\right)_T + p\right] dv = 0 \,. \quad (16.15)$$

Two properties can always be controlled independently, that is in the above dT and dv must be independent. This implies that their factors, in square brackets, must vanish,

$$\left(\frac{\partial u}{\partial v}\right)_T = T\left(\frac{\partial s}{\partial v}\right)_T - p \quad , \quad \left(\frac{\partial u}{\partial T}\right)_v = T\left(\frac{\partial s}{\partial T}\right)_v \,. \quad (16.16)$$

A faster approach to the first relation is to take the partial derivative of the Gibbs equation (16.2) with respect to v while keeping T constant. Similarly the second relation follows from taking the partial derivative of the Gibbs equation (16.2) with respect to T while keeping v constant.

With the Maxwell relation $(16.11)_3$ to replace the entropy derivative $\left(\frac{\partial s}{\partial v}\right)_T$ in $(16.16)_1$, we find an equation for the volume dependence of internal energy that is entirely determined by the thermal equation of state $p\,(T, v)$,

$$\left(\frac{\partial u}{\partial v}\right)_T = T\left(\frac{\partial p}{\partial T}\right)_v - p \,. \quad (16.17)$$

Since internal energy cannot be measured directly, the left hand side cannot be determined experimentally. However, the right hand side is determined by the thermal equation of state, $p\,(T, v)$, which is easy to measure. The equation states that the volume dependence of the internal energy is known from measurement of the thermal equation of state.

For instance for the ideal gas $p\,(T, v) = RT/v$ and thus $\left(\frac{\partial u}{\partial v}\right)_T = 0$— the internal energy of the ideal gas is independent of volume, and therefore depends only on temperature, $u = u\,(T)$. While we have used this from the beginning as an experimental fact, we see here that it is a direct consequence of the Gibbs equation and the thermal equation of state.

To obtain a similar relation for enthalpy, we take the partial derivative of the Gibbs equation $(16.9)_1$ with respect to p while keeping T constant, to obtain

$$\left(\frac{\partial h}{\partial p}\right)_T = T\left(\frac{\partial s}{\partial p}\right)_T + v \,. \quad (16.18)$$

With the Maxwell relation $(16.13)_3$ to replace the entropy derivative $\left(\frac{\partial s}{\partial p}\right)_T$, this gives an equation for the pressure dependence of enthalpy, which is entirely determined by the thermal equation of state $p\,(T, v)$, or rather $v\,(T, p)$,

$$\left(\frac{\partial h}{\partial p}\right)_T = -T\left(\frac{\partial v}{\partial T}\right)_p + v \,. \tag{16.19}$$

For instance for the ideal gas $v(T,p) = RT/p$ and thus $\left(\frac{\partial h}{\partial p}\right)_T = 0$—the enthalpy of the ideal gas is independent of pressure, and depends only on temperature, $h = h(T)$.

16.4 Relation between Specific Heats

Our first use of (16.17) is to derive a relation between the specific heats at constant volume and constant pressure,

$$c_v = \left(\frac{\partial u}{\partial T}\right)_v \quad \text{and} \quad c_p = \left(\frac{\partial h}{\partial T}\right)_p \,. \tag{16.20}$$

It will be seen that, as long as the thermal equation of state is known, it suffices to measure one of the specific heats, the other can then be determined from the relation to be derived.

We start with the first law for reversible processes in differential form, $du = \delta q - pdv$, and insert $u(T,v)$, so that with the above definition of c_v

$$\delta q = du + pdv = c_v dT + \left[\left(\frac{\partial u}{\partial v}\right)_T + p\right] dv \,. \tag{16.21}$$

With the just found relation (16.17) this simplifies to

$$\delta q = c_v dT + T\left(\frac{\partial p}{\partial T}\right)_v dv \,. \tag{16.22}$$

For a constant volume process we have $dv = 0$ and thus $\delta q = c_v dT$ which relates the specific heat c_v to the heat that must be added in an isochoric process to raise the temperature by dT; in fact this is the definition of c_v as a measurable quantity. Measurement of the specific heat $c_v(T,v)$ is done in calorimeters, where a substance in a rigid container ($dv = 0$) originally at temperature T is carefully heated by a known amount δq, e.g., by means of a electrical resistor, and the corresponding temperature change dT is recorded. Great care must be taken to control heat leaks, so that the amount of heat added to the material is known as accurately as possible.

To introduce the specific heat at constant pressure, c_p, we proceed as follows: In the last equation, we consider volume v as function of T and p by means of the thermal equation of state $v(T,p)$, so that $dv = \left(\frac{\partial v}{\partial T}\right)_p dT + \left(\frac{\partial v}{\partial p}\right)_T dp$, to obtain

$$\delta q = \left[c_v + T \left(\frac{\partial p}{\partial T} \right)_v \left(\frac{\partial v}{\partial T} \right)_p \right] dT + T \left(\frac{\partial p}{\partial T} \right)_v \left(\frac{\partial v}{\partial p} \right)_T dp \, . \tag{16.23}$$

On the other hand, Legendre transform in (16.21) gives, with (16.19),

$$\delta q = dh - vdp = c_p dT + \left[\left(\frac{\partial h}{\partial p} \right)_T - v \right] dp = c_p dT - T \left(\frac{\partial v}{\partial T} \right)_p dp \, . \tag{16.24}$$

By comparison we find from the last two equations

$$c_v + T \left(\frac{\partial p}{\partial T} \right)_v \left(\frac{\partial v}{\partial T} \right)_p = c_p \quad \text{and} \quad \left(\frac{\partial p}{\partial T} \right)_v \left(\frac{\partial v}{\partial p} \right)_T = - \left(\frac{\partial v}{\partial T} \right)_p \, . \tag{16.25}$$

Thus, the two specific heats are related as

$$c_p - c_v = T \left(\frac{\partial p}{\partial T} \right)_v \left(\frac{\partial v}{\partial T} \right)_p = -T \left(\frac{\partial p}{\partial v} \right)_T \left[\left(\frac{\partial v}{\partial T} \right)_p \right]^2 \, . \tag{16.26}$$

The right hand side is known when the thermal equation of state, $p(T, v)$, is known. Thus, if one of the specific heats and the thermal equation of state are measured, the other specific heat is known. It is easy to show from the above that for the ideal gas $c_p - c_v = R$.

Finally, we note that straightforward application of the Gibbs equation in (16.20) yields the equivalent expressions for the specific heats

$$c_v = T \left(\frac{\partial s}{\partial T} \right)_v \quad \text{and} \quad c_p = T \left(\frac{\partial s}{\partial T} \right)_p \, , \tag{16.27}$$

see the above derivation of (16.16) for details.

16.5 Measurement of Properties

Only few thermodynamic properties can be measured easily, namely temperature T, pressure p, and volume v. These are related by the thermal equation of state $p(T, v)$ which is therefore relatively easy to measure.

The specific heats (16.20) can be measured in careful measurements where, because of (16.26), it suffices to measure either c_v or c_p. These calorimetric measurements employ the first law, where the change in temperature in response to the heat added to the system is measured.

Other important quantities, however, e.g., u, h, f, g, s, cannot be measured directly. In the following we shall study how they can be related to measurable quantities, i.e., T, p, v, and c_v by means of the Gibbs equation and the differential relations derived above.

We first consider the measurement of internal energy. The differential of $u(T,v)$ is

$$du = c_v dT + \left(\frac{\partial u}{\partial v}\right)_T dv \,. \tag{16.28}$$

Therefore, the internal energy $u(T,v)$ can be determined by integration when c_v and $\left(\frac{\partial u}{\partial v}\right)_T$ are known from measurements. By (16.17) the term $\left(\frac{\partial u}{\partial v}\right)_T$ is known through measurement of the thermal equation of state, and we can write

$$du = c_v dT + \left[T\left(\frac{\partial p}{\partial T}\right)_v - p\right] dv \,. \tag{16.29}$$

Thus, only the specific heat $c_v(T,v)$ must be measured when the thermal equation of state $p(T,v)$ is already known.

To determine what measurements must be taken to determine the specific heat $c_v(T,v) = \left(\frac{\partial u}{\partial T}\right)_v$, we consider its differential,

$$dc_v = \left(\frac{\partial c_v}{\partial T}\right)_v dT + \left(\frac{\partial c_v}{\partial v}\right)_T dv \,. \tag{16.30}$$

From the definition of c_v follows, with (16.17),

$$\begin{aligned}\left(\frac{\partial c_v}{\partial v}\right)_T &= \frac{\partial^2 u}{\partial v \partial T} = \frac{\partial^2 u}{\partial T \partial v} = \frac{\partial}{\partial T}\left(\frac{\partial u}{\partial v}\right)_T \\ &= \frac{\partial}{\partial T}\left[T\left(\frac{\partial p}{\partial T}\right)_v - p\right] = T\left(\frac{\partial^2 p}{\partial T^2}\right)_v \,. \end{aligned} \tag{16.31}$$

Thus, the volume dependence of c_v follows from measurement of the thermal equation of state.

Accordingly, in order to determine the specific heat $c_v(T,v)$ for all T and v it is sufficient to measure the thermal equation of state $p(T,v)$ for all (T,v) and the specific heat $c_v(T,v_0)$ for all temperatures T but only one volume v_0. Then, $c_v(T,v)$ follows from integration of (16.30). Finally, integration of (16.28) gives the internal energy $u(T,v)$.

Integration is performed from a reference state (T_0,v_0) to the actual state (T,v). Since internal energy is a point function, its differential is exact, and the integration is independent of the path chosen. The easiest integration is in two steps, first at constant volume v_0 from (T_0,v_0) to (T,v_0), then at constant temperature T from (T,v_0) to (T,v). The integration results in

$$\begin{aligned} u(T,v) - u(T_0,v_0) &= \\ &= \int_{T_0}^{T} c_v(T',v_0)\,dT' + \int_{v_0}^{v}\left[T\left(\frac{\partial p}{\partial T}\right)_{v'} - p(T,v')\right] dv' + \sum_i \Delta u_i \,. \end{aligned} \tag{16.32}$$

The internal energy can only be determined apart from a reference value $u(T_0, v_0)$. As long as no chemical reactions occur, the energy constant $u(T_0, v_0)$ can be arbitrarily chosen; see Chapter 23 on chemical reactions for additional discussion. When phase changes are involved, the respective energies Δu_i have to be added.

Enthalpy can be obtained from integration of its differential

$$dh = c_p dT + \left(\frac{\partial h}{\partial p}\right)_T dp = c_p dT + \left[v - T\left(\frac{\partial v}{\partial T}\right)_p\right] dp \, , \qquad (16.33)$$

where (16.19) was used. Here, integration is performed from (T_0, p_0) to (T, p). When we integrate in two steps, first at constant pressure p_0 from (T_0, p_0) to (T, p_0), then at constant temperature T from (T, p_0) to (T, p), the integration gives

$$h(T,p) - h(T_0, p_0) =$$
$$= \int_{T_0}^{T} c_p(T', p_0)\, dT' + \int_{p_0}^{p} \left[v(T, p') - T\left(\frac{\partial v}{\partial T}\right)_{p'}\right] dp' + \sum_i \Delta h_i \, . \qquad (16.34)$$

Here we have explicitly introduced the heats of phase change Δh_i which must be added whenever the line of integration crosses a saturation curve in the p-T-diagram. The reference enthalpy $h(T_0, p_0)$ can be chosen arbitrarily as long as no chemical reactions occur. In case of chemical reactions, it should be chosen as the enthalpy of formation, see the discussion in the chapter on chemical reactions.

Entropy $s(T, p)$ follows by integration of the Gibbs equation, e.g., in the form

$$ds = \frac{1}{T} dh - \frac{v}{T} dp = \frac{c_p}{T} dT - \left(\frac{\partial v}{\partial T}\right)_p dp \, ,$$

as

$$s(T,p) - s(T_0, p_0) = \int_{T_0}^{T} \frac{c_p(T', p_0)}{T'} dT' - \int_{p_0}^{p} \left(\frac{\partial v}{\partial T}\right)_{p'} dp' + \sum_i \frac{\Delta h_i}{T_i} \, ; \qquad (16.35)$$

Also entropy can be determined only apart from a reference value $s(T_0, v_0)$ which only plays a role when chemical reactions occur; see Chapter 23. When the line of integration crosses saturation lines in the p-T-diagram, the corresponding entropy changes $\Delta s_i = \frac{\Delta h_i}{T_i}$ must be included. This can be seen as follows: At an equilibrium phase interface, temperature T and (saturation) pressure $p_{\text{sat}}(T)$ are continuous. Integration of the Gibbs equation $Tds = dh - vdp$ across the phase interface yields $T\Delta s = \Delta h$.

For the ideal gas, where $\left(\frac{\partial v}{\partial T}\right)_p = \frac{R}{p}$ and the specific heat depends on T only, enthalpy and entropy assume the familiar forms (with suitable choice of integration constants)

$$h(T) = \int_{T_0}^{T} c_p(T')\,dT' ,$$

$$s(T,p) = s^0(T) - R\ln\frac{p}{p_0} \quad \text{with} \quad s^0(T) = \int_{T_0}^{T} \frac{c_p(T',p_0)}{T'}dT' . \tag{16.36}$$

After u, h and s are determined, Helmholtz free energy f and Gibbs free energy g simply follow by means of their definitions (16.10, 16.12). Thus the measurement of *all* thermodynamic quantities requires only the measurement of the thermal equation of state $p(T,v)$ for all (T,v) and the measurement of the specific heat at constant volume $c_v(T,v_0)$ for all temperatures, but only one volume, e.g., in a constant volume calorimeter.[2] All other quantities follow from differential relations that are based on the Gibbs equation, and integration.

Above we have outlined the necessary measurements to fully determine all relevant thermodynamic properties. We close this section by pointing out that all properties can be determined if just one of the thermodynamic potentials is known, this is shown in the next example. Since all properties can be derived from the potential, the expression for the potential is sometimes called the *fundamental relation.*

16.6 Example: Gibbs Free Energy as Potential

In this example we consider a particular function for the Gibbs free energy $g(T,p)$, to show that knowledge of one potential allows to determine all relevant property relations, including all other potentials.

We consider the fundamental relation (A is a constant with the appropriate dimensions)

$$g(T,p) = -A\frac{T^4}{p} .$$

We first use that $g(T,p)$ is a potential (16.13), which gives the entropy and the specific volume as derivatives,

$$s(T,p) = -\left(\frac{\partial g}{\partial T}\right)_p = 4A\frac{T^3}{p} ,$$

$$v(T,p) = \left(\frac{\partial g}{\partial p}\right)_T = A\frac{T^4}{p^2} .$$

The caloric equations of state then follow from the definition of $g = h - Ts = u + pv - Ts = f - pv$ as

[2] Or, alternatively, the measurement of the specific heat $c_p(T,p_0)$ at all temperatures but only one pressure p_0.

$$h(T,p) = g + Ts = g - T\left(\frac{\partial g}{\partial T}\right)_p = 3A\frac{T^4}{p}\,,$$

$$u(T,p) = h - pv = h - p\left(\frac{\partial g}{\partial p}\right)_T = 2A\frac{T^4}{p}\,,$$

$$f(T,p) = g - pv = g - p\left(\frac{\partial g}{\partial p}\right)_T = -2A\frac{T^4}{p}\,.$$

A switch of variables is obtained by solving the thermal equation of state for pressure,

$$p(T,v) = \sqrt{A}\frac{T^2}{\sqrt{v}}\,,$$

for which we find

$$\begin{aligned} s(T,v) &= 4\sqrt{A}T\sqrt{v}\,, \\ u(T,v) &= 2\sqrt{A}T^2\sqrt{v}\,, \\ h(T,v) &= 3\sqrt{A}T^2\sqrt{v}\,, \\ f(T,v) &= -2\sqrt{A}T^2\sqrt{v}\,, \\ g(T,v) &= -\sqrt{A}T^2\sqrt{v}\,. \end{aligned}$$

For the variables (T,v), the Helmholtz free energy $f(T,v)$ is a potential. It is easy to verify that the above expression for f fulfills

$$s = -\left(\frac{\partial f}{\partial T}\right)_v\,, \quad p = \left(\frac{\partial f}{\partial v}\right)_T\,.$$

Temperature $T(s,v)$ and the potentials $u(s,v)$ and $h(s,p)$ follow as

$$T(s,v) = \frac{s}{4\sqrt{A}\sqrt{v}}\,, \quad u(s,v) = \frac{1}{8\sqrt{A}}\frac{s^2}{\sqrt{v}}\,, \quad h(s,p) = \frac{3}{4^{\frac{4}{3}}}\frac{1}{A^{\frac{1}{3}}}s^{\frac{4}{3}}p^{\frac{1}{3}}\,;$$

the further evaluation of these potentials is left to the reader.

To determine the specific heat at constant volume we have to consider energy as function of temperature and volume, $u(T,v)$

$$c_v(T,v) = \left(\frac{\partial u}{\partial T}\right)_v = \frac{\partial u(T,v)}{\partial T} = 4\sqrt{A}T\sqrt{v}\,.$$

A shift in variables gives, e.g., $c_v(T,p) = 4A\frac{T^3}{p}$ or $c_v(s,p) = s$.

16.7 Compressibility, Thermal Expansion

The isothermal compressibility gives information about the volume change of a substance when pressure is changed isothermally, it is defined as

$$\kappa_T = -\frac{1}{v}\left(\frac{\partial v}{\partial p}\right)_T ; \tag{16.37}$$

the minus sign is convention, and guarantees a positive value of the compressibility.

The coefficient of thermal expansion gives information about the volume change with temperature when the pressure is kept constant, it is defined as

$$\alpha = \frac{1}{v}\left(\frac{\partial v}{\partial T}\right)_p . \tag{16.38}$$

We also define the coefficient

$$\beta = \frac{1}{p}\left(\frac{\partial p}{\partial T}\right)_v , \tag{16.39}$$

which describes the increase of pressure with temperature in an isochoric process.

These and similar quantities are important for the design of thermal devices, e.g. for load calculations etc. Obviously, they can be determined from the measurement of the thermal equation of state $p(v,T)$. We shall show next that they are not independent.

For this, we begin with a mathematical exercise: Consider a function $z(x,y)$ and its differential

$$dz = \left(\frac{\partial z}{\partial x}\right)_y dx + \left(\frac{\partial z}{\partial y}\right)_x dy . \tag{16.40}$$

We also have, by inversion, $x = x(y,z)$ and the corresponding differential

$$dx = \left(\frac{\partial x}{\partial y}\right)_z dy + \left(\frac{\partial x}{\partial z}\right)_y dz . \tag{16.41}$$

Eliminating dx between the two equations gives

$$\left[1 - \left(\frac{\partial z}{\partial x}\right)_y \left(\frac{\partial x}{\partial z}\right)_y\right] dz = \left[\left(\frac{\partial z}{\partial x}\right)_y \left(\frac{\partial x}{\partial y}\right)_z + \left(\frac{\partial z}{\partial y}\right)_x\right] dy . \tag{16.42}$$

Since z and y can be varied independently, the factors in square brackets must vanish, and thus we have

$$\left(\frac{\partial z}{\partial x}\right)_y = \frac{1}{\left(\frac{\partial x}{\partial z}\right)_y} , \quad \left(\frac{\partial z}{\partial x}\right)_y \left(\frac{\partial x}{\partial y}\right)_z \left(\frac{\partial y}{\partial z}\right)_x = -1 . \tag{16.43}$$

These two equations hold for any choice of the functions[3] x, y, z; e.g., Eq. $(16.25)_2$ is a special case of $(16.43)_2$.

A special choice is $x = T$, $y = v$, $z = p$ so that

$$1 = -\left(\frac{\partial p}{\partial T}\right)_v \left(\frac{\partial T}{\partial v}\right)_p \left(\frac{\partial v}{\partial p}\right)_T = \frac{p\beta\kappa_T}{\alpha} ; \tag{16.44}$$

this shows that the coefficients introduced above are dependent.

As an example we consider the ideal gas where $pv = RT$. We find

$$\kappa_T = \frac{1}{p} , \quad \alpha = \frac{1}{T} , \quad \beta = \frac{1}{T} . \tag{16.45}$$

From (16.26) we find

$$c_p - c_v = \alpha\beta vpT = \kappa_T \beta^2 vp^2 T . \tag{16.46}$$

Below we shall see that thermodynamic stability implies $\kappa_T \geq 0$, and thus the above implies $c_p \geq c_v$. For incompressible substances the isothermal compressibility vanishes, $\kappa_T = 0$, and thus the specific heats at constant pressure and volume agree, $c_p = c_v = c$.

A relation between the isentropic and the isothermal compressibilities can be found by the following chain of arguments, which uses (16.37, 16.27, 16.43),

$$\begin{aligned}
\kappa_s &= -\frac{1}{v}\left(\frac{\partial v}{\partial p}\right)_s = -\frac{1}{v}\left(\frac{\partial v}{\partial T}\right)_s \left(\frac{\partial T}{\partial p}\right)_s \\
&= -\frac{1}{v}\left[-\left(\frac{\partial s}{\partial T}\right)_v \left(\frac{\partial v}{\partial s}\right)_T\right]\left(\frac{\partial T}{\partial p}\right)_s \\
&= \frac{c_v}{Tv}\left[\left(\frac{\partial v}{\partial p}\right)_T \left(\frac{\partial p}{\partial s}\right)_T\right]\left(\frac{\partial T}{\partial p}\right)_s \\
&= -\frac{c_v}{T}\kappa_T\left[-\left(\frac{\partial T}{\partial s}\right)_p\right] = \frac{c_v}{c_p}\kappa_T .
\end{aligned} \tag{16.47}$$

16.8 Example: Van der Waals Gas

The van der Waals equation (6.29) was developed to describe non-ideal gases, it reads

$$p = \frac{RT}{v-b} - \frac{a}{v^2} . \tag{16.48}$$

Here, the constant b accounts for the reduction of the volume accessible for a gas particle due to the finite size of the other molecules, and the constant

---

[3] In particular the first equation implies $\left(\frac{\partial z}{\partial y}\right)_x = 1/\left(\frac{\partial y}{\partial z}\right)_x$ and $\left(\frac{\partial x}{\partial y}\right)_z = 1/\left(\frac{\partial y}{\partial x}\right)_z$ as well, which were used to find the second equation.

a accounts for the reduction in pressure due to attractive forces between the particles. In ideal gases, the specific volume is relatively high, so that $b \ll v$ and $a \ll v^2$, in which case the equation reduces to the ideal gas law $p = \frac{RT}{v}$.

While the van der Waals equation offers intuitive insight into the deviation from ideal gas behavior, it only offers a qualitative description of real gas behavior, including, as will be seen later, phase changes. Thus, despite its quantitative inaccuracy, it serves as an important example to study thermodynamic principles and methods.

16.8.1 Determination of Constants a, b

The constants a and b are related to microscopic quantities, namely the eigenvolume of gas molecules, and the interaction potential between gas molecules. Their values a, b can be determined from property data at the critical point.

The critical isotherm has a horizontal inflection point at the critical point, so that

$$\left(\frac{\partial p}{\partial v}\right)_{T,cr} = \left(\frac{\partial^2 p}{\partial v^2}\right)_{T,cr} = 0 . \tag{16.49}$$

Thus, together with the van der Waals equation (16.48) itself evaluated at the critical point, we have three conditions to be fulfilled at the critical point, which gives, after a brief calculation,

$$p_{cr} = \frac{1}{27}\frac{a}{b^2} \quad , \quad v_{cr} = 3b \quad , \quad RT_{cr} = \frac{8}{27}\frac{a}{b} . \tag{16.50}$$

At the critical point helium and water have the following properties:

$$\begin{array}{rl} \text{helium:} & T_{cr} = 5.3\,\mathrm{K} \;\; , \; p_{cr} = 0.23\,\mathrm{MPa} \; , \; v_{cr} = 1.445 \times 10^{-2}\,\frac{\mathrm{m}^3}{\mathrm{kg}} \; , \\ \text{water:} & T_{cr} = 647.3\,\mathrm{K} \; , \; p_{cr} = 22.09\,\mathrm{MPa} \; , \; v_{cr} = 3.156 \times 10^{-3}\,\frac{\mathrm{m}^3}{\mathrm{kg}} \; , \end{array} \tag{16.51}$$

from which one finds the constants as

$$\begin{array}{rl} \text{helium:} & a = 144.07\frac{\mathrm{m}^5}{\mathrm{s}^2\,\mathrm{kg}} \; , \; b = 4.817 \times 10^{-3}\,\frac{\mathrm{m}^3}{\mathrm{kg}} \; , \; R = 1672\frac{\mathrm{J}}{\mathrm{kg\,K}} \; , \\ \text{water:} & a = 660.1\frac{\mathrm{m}^5}{\mathrm{s}^2\,\mathrm{kg}} \; , \; b = 1.0526 \times 10^{-3}\,\frac{\mathrm{m}^3}{\mathrm{kg}} \; , \; R = 287\frac{\mathrm{J}}{\mathrm{kg\,K}} \; . \end{array} \tag{16.52}$$

The actual values for the gas constants of helium and water are $2079\frac{\mathrm{J}}{\mathrm{kg\,K}}$ and $462\frac{\mathrm{J}}{\mathrm{kg\,K}}$, respectively, and we see that the van der Waals equation predicts wrong values of the gas constants, with an error of about 20% for helium and 38% for water. The bigger error for water can be attributed to the more complex character of the dipole water molecules and their interaction among themselves, which makes them, other than the monatomic "spherical" helium atoms, not well suited for the arguments on interaction and eigenvolume that lead to the van der Waals equation.

The derivation of the van der Waals equation shows that the eigenvolume of a particle v_0 is related to the constant b as $v_0 = \frac{1}{2} bM/N_A$ where M is the molar mass and $N_A = 6.022 \times 10^{23} \frac{1}{\text{mol}}$ is the Avogadro constant. Assuming spherical particles, the corresponding molecule diameter is

$$d_0 = \sqrt[3]{\frac{6}{\pi} v_0} = \sqrt[3]{\frac{3}{\pi} \frac{bM}{N_A}} , \tag{16.53}$$

which gives $d_0^{\text{He}} = 3.126 \times 10^{-10}\,\text{m}$ and $d_0^{H_2O} = 3.11 \times 10^{-10}\,\text{m}$. While certainly not exact, this numbers give a good indication of the molecule size.

16.8.2 Isotherms

When dimensionless pressure, volume and temperature are introduced by means of critical point data as

$$\pi = \frac{p}{p_{cr}} \quad , \quad \upsilon = \frac{v}{v_{cr}} \quad , \quad \tau = \frac{T}{T_{cr}} , \tag{16.54}$$

use of (16.50) gives the dimensionless van der Waals equation

$$\pi = \frac{8\tau}{3\upsilon - 1} - \frac{3}{\upsilon^2} . \tag{16.55}$$

In this dimensionless form, the equation is independent of the type of gas, all factors are independent of the type of gas. This somewhat surprising finding is known as the principle of corresponding states.

Figure 16.1 shows isothermal lines in the π-υ-diagram (i.e., the dimensionless p-v-diagram). The critical isotherm ($\tau = 1$) exhibits the horizontal inflection point at the critical point ($\pi = \upsilon = 1$). At supercritical temperatures ($\tau > 1$), the isotherms are monotonically decreasing with volume, and for large temperatures they agree with the curves from the ideal gas law.

For sub-critical temperatures, however, the curves are non-monotonic; in particular there is a portion of the curves where $\left(\frac{\partial v}{\partial p}\right)_T \geq 0$, which means that the isothermal compressibility is negative, $\kappa_T = -\frac{1}{v}\left(\frac{\partial v}{\partial p}\right)_T < 0$. In the next chapter we will see that thermodynamic stability requires positive compressibility, $\kappa_T \geq 0$. Thus, these portions of the curves are unstable and cannot be attained. In Section 17.8 it will be seen how these unstable states are bridged by splitting of the van der Waals fluid into liquid and vapor phases.

For sufficiently small temperatures the isothermal curves predict negative pressure. States of negative pressure are unstable, but can be reached by very careful experiments in which they appear to be metastable.

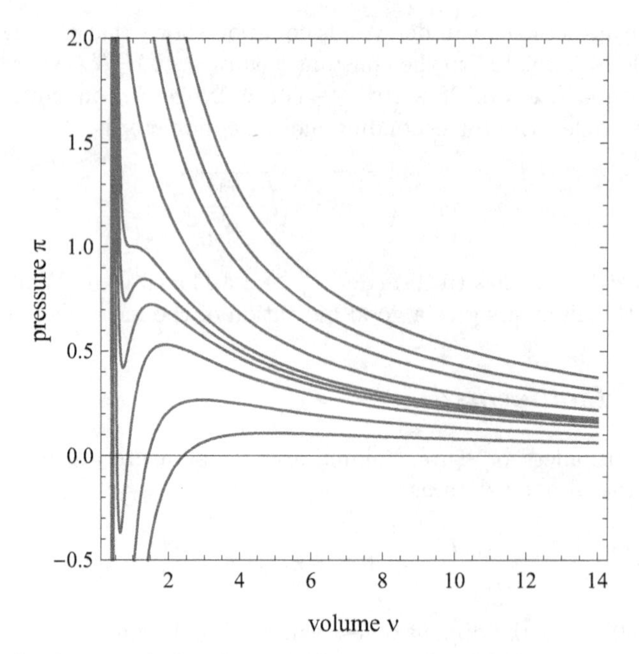

Fig. 16.1 Isotherms of the dimensionless van der Waals equation, for $\tau = 0.4, 0.6, 0.8, 0.9, 1.0, 1.2, 1.5, 1.7, 2.0$

16.8.3 Internal Energy and Entropy

Next we determine property relations for energy and entropy of the van der Waals gas. By inserting the van der Waals equation (16.48) into (16.17), we find

$$\left(\frac{\partial u}{\partial v}\right)_T = T\left(\frac{\partial p}{\partial T}\right)_v - p = \frac{a}{v^2}\,, \tag{16.56}$$

and from (16.31) we find

$$\left(\frac{\partial c_v}{\partial v}\right)_T = \frac{\partial}{\partial T}\left(\frac{\partial u}{\partial v}\right) = T\left(\frac{\partial^2 p}{\partial T^2}\right)_v = 0\,. \tag{16.57}$$

This implies that for a van der Waals gas, as for the ideal gas, the specific heat depends only on temperature, but not on volume. Energy itself, other than for the ideal gas, has a dependence on volume. Insertion of the above into (16.32) and explicit integration over volume then gives the internal energy of the van der Waals gas as

$$u\,(T,v) - u\,(T_0,v_0) = \int_{T_0}^{T} c_v\,(T')\,dT' - a\left(\frac{1}{v} - \frac{1}{v_0}\right)\,. \tag{16.58}$$

We recall that the coefficient a describes the influence of long range interaction between the particles. The above shows the explicit contribution of these interactions on the energy of the gas. For large volume, the average particle distance is very large, and the average particle-particle interaction energy vanishes.

To compute the entropy $s(T,v)$ we use the Gibbs equation,

$$Tds = du + pdv = \left(\frac{\partial u}{\partial T}\right)_v dT + \left[\left(\frac{\partial u}{\partial v}\right)_T + p\right] dv = c_v dT + \frac{RT}{v-b} dv\,, \tag{16.59}$$

where for the last equation the equation of state and the above result for internal energy were used. Integration gives

$$s(T,v) - s(T_0,v_0) = \int_{T_0}^{T} \frac{c_v(T')}{T'} dT' + R\ln\frac{v-b}{v_0-b}\,. \tag{16.60}$$

As we have seen earlier, entropy measures the number of possible microscopic realizations of the macroscopic state of a gas. In the van der Waals gas, the volume accessible to a particle is reduced by the presence of the other particles (measured by the coefficient b), and this reduces the number of possible configurations, and thus the entropic contribution from volume. For large volumes, $v - b \simeq v$, i.e., the contribution of particle volume can be ignored.

16.9 Joule-Thomson Coefficient

In a throttling process, the pressure drops while the enthalpy stays constant, the process is isenthalpic and irreversible. The temperature, however, may rise, fall, or stay constant. The Joule-Thomson coefficient $(\partial T/\partial p)_h$ describes the change of temperature with pressure in an isenthalpic process.

With the choice $x = h$, $y = p$, $z = T$, the relation $(16.43)_2$ gives an expression for the coefficient,

$$\left(\frac{\partial T}{\partial p}\right)_h = -\frac{1}{c_p}\left(\frac{\partial h}{\partial p}\right)_T\,. \tag{16.61}$$

Use of the relation (16.19) and the definition of the coefficient of thermal expansion α (16.38) gives

$$\left(\frac{\partial T}{\partial p}\right)_h = \frac{vT}{c_p}\left(\alpha - \frac{1}{T}\right)\,. \tag{16.62}$$

For an ideal gas where $\alpha = \frac{1}{T}$ we obtain $\left(\frac{\partial T}{\partial p}\right)_h = 0$; this reflects that enthalpy is a function only of temperature, which therefore must be constant in an isenthalpic process.

For a real gas, the curve where

$$\left(\frac{\partial T}{\partial p}\right)_h = 0 \quad \text{or} \quad \alpha = \frac{1}{T}\,, \tag{16.63}$$

is the inversion curve, it separates states where the temperature increases or decreases in throttling processes.

16.10 Example: Inversion Curve for the Van der Waals Gas

We compute the inversion curve for the van der Waals gas in the dimensionless variables π, v, τ, where the condition assumes the form $\alpha = \frac{1}{v}\left(\frac{\partial v}{\partial \tau}\right)_\pi = \frac{1}{\tau}$, or

$$v\left(\frac{\partial \tau}{\partial v}\right)_\pi = \tau\,.$$

By solving the dimensionless van der Waals equation (16.55) for τ, inserting it into the above on both sides, and performing the derivatives, we find the relation between pressure and volume on the inversion curve,

$$\pi_{\text{inv}} = \frac{18}{v} - \frac{9}{v^2}\,.$$

Eliminating π with the van der Waals equation gives the inversion relation between τ and v,

$$\tau_{\text{inv}} = \frac{3}{4}\left(3 - \frac{1}{v}\right)^2 \quad \text{or} \quad v_{\text{inv}} = \frac{1}{3 - 2\sqrt{\frac{\tau}{3}}}\,.$$

The above can be combined to

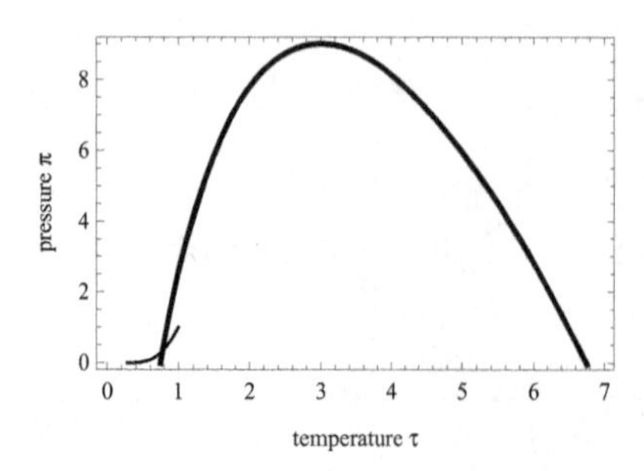

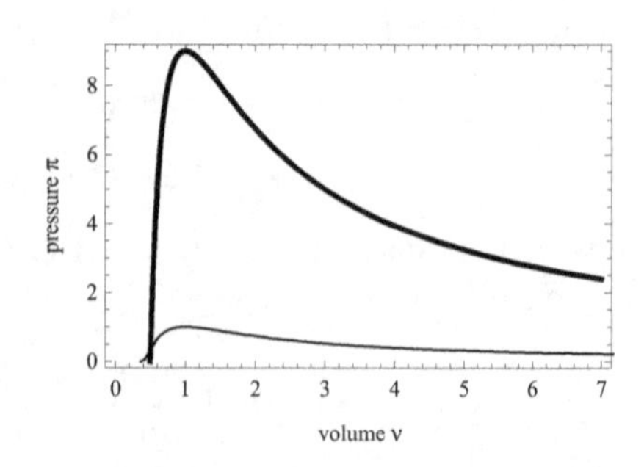

Fig. 16.2 Inversion curves for the van der Waals gas in π-τ- and π–v-diagrams. The π-τ-diagram also shows the saturation line, and the π–v-diagram shows the two phase region (thinner lines).

$$\pi_{\text{inv}} = 24\sqrt{3\tau} - 12\tau - 27 \, .$$

The corresponding curves in the π-τ- and π-υ-diagrams are shown in Fig. 16.2, with respect to saturation lines, which will be determined in Sec. 17.9. For states below the curve the Joule-Thomson coefficient $\left(\frac{\partial T}{\partial p}\right)_h$ is positive, so that pressure reduction leads to cooling.

Problems

16.1. Fundamental Relation for Ideal Gas

Internal energy and entropy of an ideal gas with constant specific heats are $u = c_v T$, $s = c_v \ln \frac{T}{T_0} + R \ln \frac{v}{v_0}$. Determine the fundamental relations $u\,(s, v)$, $h\,(s, p)$, $f\,(T, v)$ and $g\,(T, p)$. Take the appropriate first derivatives of all to verify that they are potentials.

16.2. Fundamental Relation for Ideal Incompressible Liquid

Internal energy and entropy of an ideal incompressible liquid with constant specific heat are $u = cT$, $s = c \ln \frac{T}{T_0}$. Determine the fundamental relations $u\,(s, v)$, $h\,(s, p)$, $f\,(T, v)$ and $g\,(T, p)$. Take the appropriate first derivatives of all to verify that they are potentials.

16.3. Ideal Gas Table (Air)

The specific heat of air is in good accuracy given as

$$\bar{c}_p\,(T) = a + bT + cT^2 + dT^3 + \frac{e}{T^2} \, ,$$

with the constants:
for $T < 1000\,\mathrm{K}$:

$$a = 30.0051 \frac{\mathrm{kJ}}{\mathrm{kmol\,K}} \, , \quad b = -8.86766 \times 10^{-3} \frac{\mathrm{kJ}}{\mathrm{kmol\,K^2}} \, ,$$
$$c = 2.212730 \times 10^{-5} \frac{\mathrm{kJ}}{\mathrm{kmol\,K^3}} \, , \quad d = -1.02450 \times 10^{-8} \frac{\mathrm{kJ}}{\mathrm{kmol\,K^4}} \, ,$$
$$e = 8.38737 \times 10^{2} \frac{\mathrm{kJ\,K}}{\mathrm{kmol}} \, ;$$

for $1000\,\mathrm{K} < T < 2200\,\mathrm{K}$:

$$a = 36.7781 \frac{\mathrm{kJ}}{\mathrm{kmol\,K}} \, , \quad b = -3.90661 \times 10^{-3} \frac{\mathrm{kJ}}{\mathrm{kmol\,K^2}} \, ,$$
$$c = 3.46633 \times 10^{-6} \frac{\mathrm{kJ}}{\mathrm{kmol\,K^3}} \, , \quad d = -7.46611 \times 10^{-10} \frac{\mathrm{kJ}}{\mathrm{kmol\,K^4}} \, ,$$
$$e = -2.52571 \times 10^{6} \frac{\mathrm{kJ\,K}}{\mathrm{kmol}} \, .$$

1. Use the above data to prepare a table with the values of $\bar{u}(T)$, $\bar{h}(T)$, $\bar{s}^0(T)$ for air in the temperature range $230 - 2200\,\mathrm{K}$. Adjust enthalpy and entropy so that $\bar{h}(300\,\mathrm{K}) = 8693.5\frac{\mathrm{kJ}}{\mathrm{kmol}}$ and $\bar{s}(T_0 = 298.15\,\mathrm{K}, p_0 = 1\,\mathrm{bar}) = 206.565\frac{\mathrm{kJ}}{\mathrm{kmol\,K}}$.
2. For easy handling of isentropic processes, it is convenient to have the relative pressure p_r and the relative volume v_r in the tables, see Sec. 7.5. Review their definition, and add them to your table as well.

16.4. Ideal Gas Table (H_2O)

When water vapor can be considered as ideal gas, its specific heat is in good accuracy given in polynomial form as

$$\bar{c}_p(T) = a + bT + cT^2 + dT^3 \,,$$

with the constants

$$a = 32.24\frac{\mathrm{kJ}}{\mathrm{kmol\,K}} \,, \quad b = 0.1923 \times 10^{-2}\frac{\mathrm{kJ}}{\mathrm{kmol\,K^2}} \,,$$
$$c = 1.055 \times 10^{-5}\frac{\mathrm{kJ}}{\mathrm{kmol\,K^3}} \,, \quad d = -3.595 \times 10^{-9}\frac{\mathrm{kJ}}{\mathrm{kmol\,K^4}} \,.$$

1. Use this data to prepare a table with the values of $\bar{u}(T)$, $\bar{h}(T)$, $\bar{s}^0(T)$ for water vapor in the temperature range $273 - 1800\,\mathrm{K}$. Chose the energy and entropy constants such that you have agreement (as close as possible) with the table for water vapor as an ideal gas.
2. Next make tables for $u(T)$, $h(T)$, $s(T,p)$, for various values of p. Readjust the integration constants thus that your data matches the tables for superheated steam at $T = 50\,°\mathrm{C}$, $p = 0.01\,\mathrm{MPa}$. Make sets of tables at different pressures to compare with actual steam tables. Discuss the validity of the ideal gas assumption for vapor for high and low pressures, and high and low temperatures.

16.5. Thermodynamic Potential for a Gas

The Gibbs free energy of a gas is given as (a, b, c are constants with appropriate units)

$$g(T,p) = a\left(T - T\ln\frac{T}{T_0}\right) - \frac{b}{2}T^2 - \frac{c}{6}T^3 + RT\ln\frac{p}{p_0} \,.$$

1. Determine the equations of state for entropy $s(T,p)$, specific volume $v(T,p)$, enthalpy $h(T,p)$, internal energy $u(T,p)$, Helmholtz free energy $f(T,p)$.
2. Determine the equations of state for entropy $s(T,v)$, specific volume $v(T,v)$, enthalpy $h(T,v)$, internal energy $u(T,v)$, Helmholtz free energy $f(T,v)$.
3. Determine the specific heat at constant pressure, $c_p(T,p)$.
4. Determine the specific heat at constant volume, $c_v(T,v)$.

16.6. Thermodynamic Potential
The Helmholtz free energy of a substance is given as (A is a positive constant–determine its unit!)

$$f(T, v) = -AT^{\alpha}v^{\beta} .$$

1. Determine the equations of state for entropy $s(T, v)$, pressure $p(T, v)$, internal energy $u(T, v)$, enthalpy $h(T, v)$, Gibbs free energy $g(T, v)$.
2. Thermodynamic stability requires positive specific heat $c_v \geq 0$, and positive isothermal compressibility $\kappa_T \geq 0$. Use these requirements to identify the possible ranges of the exponents α and β.

16.7. Isothermal Compressibility and Thermal Expansion
Use tabulated data for superheated water vapor to estimate the isothermal compressibility κ_T and the coefficient of thermal expansion α at 600 °C and 7.0 MPa. Compare to the ideal gas values of $\kappa_{T,id.gas} = 1/p$ and $\alpha = 1/T$. Also determine the factor β from tabulated values, and test how well your approximations fulfill the relation $\frac{p\beta\kappa_T}{\alpha} = 1$.

16.8. Isothermal Compressibility
Use tabulated data for superheated vapor of R134a to estimate the isothermal compressibility κ_T at 60 °C and 1.4 MPa. Compare to the ideal gas value of $\kappa_{T,\text{id.gas}} = 1/p$. Is it easier to compress ideal gas or R134a (at this state)? Why is that?

16.9. Coefficient of Thermal Expansion and Joule-Thomson Coefficient
Use tabulated data for superheated water vapor to estimate the specific heat at constant pressure, c_p, and the coefficient of thermal expansion, α, at 550 °C and 20 MPa. Use your results to determine the Joule-Thomson coefficient at the same state. When the vapor is throttled, will the temperature go up or down?

16.10. Measuring the Coefficient of Thermal Expansion
In the temperature range between 0 °C and 50 °C the coefficient of volume expansion of a liquid L is measured as $\alpha_L = 1.2 \times 10^{-3} \frac{1}{\text{K}}$. To measure the coefficient of volume expansion for a solid S, the following experiment is conducted: A cylinder made of S is immersed in L and the percentage of immersed volume of solid is measured at 0 °C and 50 °C as 82.1% and 86.6%, respectively. Use Archimedes' principle to determine the coefficient α_S from this data.

16.11. Thermosyphon
In warm countries, one finds often a simple device for heating of water, the thermosyphon.

Solar radiation provides a heat flux $\dot{Q}$ which heats water. Since warm water has a smaller mass density than cold water, the heated water will rise. The goal is to compute the mass flow $\dot{m}$ and the temperature difference $T_t - T_b$ that will be observed.

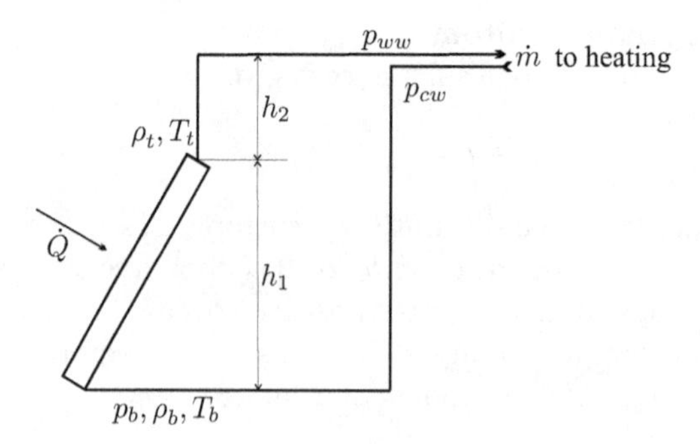

Fig. 16.3 Thermosyphon

1. Give arguments that $\frac{\rho_b - \rho_t}{\rho_b} = \alpha (T_t - T_b)$ where α is the thermal expansion coefficient of water.
2. Assume that the density is a linear function of height between 0 and h_1 and show that the difference between the pressures in warm and cold water is given by $\Delta p = p_{ww} - p_{cw} = (\rho_b - \rho_t) g \left(\frac{1}{2} h_1 + h_2\right)$. This refers to the pressure difference at rest, when $\dot{m} = 0$.
3. When the water is allowed to flow, the pressure difference determined above is consumed by friction. The law of Hagen-Poiseuille relates the volume flow and the pressure drop in a pipe of radius R and length L as $\dot{V} = \frac{\pi R^4}{8 \eta L} \Delta p$, where η is the viscosity. The first law for the collector gives a relation between mass flow and temperature difference, $\dot{Q} = \dot{m} c_w (T_t - T_b)$. Combine all results to obtain equations for the temperature difference and the mass flow. Discuss what parameters must be increased to raise mass flow or temperature difference.
4. The maximum of solar radiation is $1300 \frac{\mathrm{W}}{\mathrm{m}^2}$. Consider a collector with an area of $1\,\mathrm{m}^2$, and the data below to compute the temperature difference and mass flow.

$$R = 1\,\mathrm{cm}\,, \ L = 10\,\mathrm{m}\,, \ h_1 = 1/\sqrt{2}\,\mathrm{m}\,, \ h_2 = 0.5\,\mathrm{m}\,,$$

$$c_w = 4.18 \frac{\mathrm{kJ}}{\mathrm{kg\,K}}\,, \ \eta = 10^{-3} \frac{\mathrm{kg}}{\mathrm{m\,s}}\,, \ \alpha = 2 \times 10^{-4} \frac{1}{\mathrm{K}}\,.$$

16.12. Van der Waals Equation
Go through the arguments of Sec. 16.8 step by step to derive Eq. (16.50). Use critical point data for argon, oxygen, nitrogen to compute their van der Waals constants. Compare the values for R with their actual values, and discuss. Plot isotherms in a p-v-diagram. Also follow step by step through the arguments of Sec. 16.10 to determine and plot the inversion curve.

Chapter 17
Thermodynamic Equilibrium

17.1 Equilibrium Conditions

We introduced the second law of thermodynamics to formalize the statement that a system which is left to itself will approach a final stable equilibrium state. A system left completely to itself is isolated, and does not exchange heat, work or mass with its surroundings, therefore $\dot{Q} = \dot{W} = 0$; for such a system, the second law states that in equilibrium entropy will assume a maximum. While the initial state of a system typically is inhomogeneous, in equilibrium we expect homogeneous temperatures and zero velocity, since internal heat transfer will equilibrate temperature, and internal friction will dissipate all kinetic energy. If gravity can be ignored, pressure and density (in a single phase system) are homogeneous as well, else they might be inhomogeneous, as, e.g., in the barometric formula.

Below, we shall confirm these expectations by evaluating the second law for isolated systems. Thereafter, we generalize the discussion to thermodynamic equilibria of closed systems with various boundary conditions. The state of the system can be controlled from the surroundings of the system in a number of ways. When the system is in thermal contact with a temperature reservoir it will assume the temperature of the reservoir, and thus the system temperature is controlled. The system volume can be controlled by confining material into a closed box. The system pressure can be controlled by exerting a constant force on a piston that closes the system. The system's energy E is controlled when heat and work balance, i.e., $\dot{Q} = \dot{W}$, so that $\frac{dE}{dt} = 0$. Systems at controlled temperatures or pressures will exchange heat or work, and change their volume, as they approach their equilibrium state.

We shall see that, depending on the boundary conditions, different thermodynamic properties will attain a minimum or a maximum in equilibrium. However, the resulting equilibria share the same characteristics.

H. Struchtrup, *Thermodynamics and Energy Conversion*,
DOI: 10.1007/978-3-662-43715-5_17,

17.2 Equilibrium in Isolated Systems

An isolated system does not exchange energy or mass with its surroundings. Accordingly, first and second law reduce to

$$\frac{dE}{dt} = 0 \quad , \quad \frac{dS}{dt} = \dot{S}_{gen} \geq 0 \, , \tag{17.1}$$

with a constant mass m in the system. Since no work is exchanged, the volume V must be constant as well. According to the second law, the state of the system will change until the entropy has reached a maximum. However, since mass and energy of the system do not change over time, at all times the process is restricted by having the initial mass and energy enclosed in the system. The approach to equilibrium is a reorganization of the local properties of the system towards the final equilibrium state.

We study the approach to equilibrium for a single phase system; heterogeneous systems will be discussed later, in Sec. 17.7. For this we have to consider the total mass, energy and entropy by integration over the full system,

$$m = \int_V \rho dV \quad , \quad E = \int_V \rho \left(u + \frac{1}{2}\mathcal{V}^2 + \gamma z \right) dV \quad , \quad S = \int_V \rho s dV \, . \tag{17.2}$$

In order to avoid confusion with the Gibbs free energy, the gravitational acceleration is denoted by γ. Here, ρ, T, $\mathcal{V}$, and $u(\rho, T)$, $s(\rho, T)$ are the *local* values of the thermodynamic properties, that is, $\rho = \rho(\vec{r})$, $T = T(\vec{r})$ etc., where $\vec{r}$ is the location in the volume V of the system, see Sec. 2.7.

Before we proceed, we need to state the momentum vector $\vec{M}$ of the system. Typically, we are interested in systems that are globally at rest, where the overall momentum vanishes, but we might consider also systems moving with a constant velocity $\vec{v}$, so that $\vec{M} = m\vec{v}$. Since all elements of the system have their own velocity $\vec{\mathcal{V}}(\vec{r})$, we find the total momentum by summing over the system,

$$\vec{M} = m\vec{v} = \int_V \rho \vec{\mathcal{V}} dV \, ; \tag{17.3}$$

here $\vec{\mathcal{V}}$ is the local velocity vector with $\mathcal{V} = \sqrt{\vec{\mathcal{V}} \cdot \vec{\mathcal{V}}}$. As long as no forces act on the system, its momentum will be constant; total momentum vanishes for a system at rest in the observer frame, $\vec{M} = 0$.

The equilibrium state is the maximum of entropy S under the constraints of given mass m, momentum $\vec{M}$, and energy E. The best way to account for the constraints is the use of Lagrange multipliers Λ_ρ, $\vec{\Lambda}_M$ and Λ_E to incorporate the constraints and maximize not S but

$$\Phi = \int_V \rho s dV - \Lambda_\rho \left(\int_V \rho dV - m\right) - \vec{\Lambda}_M \cdot \left(\int_V \rho \vec{\mathcal{V}} dV - \vec{M}\right)$$
$$- \Lambda_E \left(\int_V \rho \left(u + \frac{1}{2}\mathcal{V}^2 + \gamma z\right) dV - E\right) . \quad (17.4)$$

The maximization of Φ will give the local values of the thermodynamic equilibrium properties $\{\rho, T, \mathcal{V}\}$ in terms of the Lagrange multipliers, which then must be determined from the given values of $\left\{m, \vec{M}, E\right\}$.

For the solution of this problem, we employ some rules of variational calculus. The condition for an extremum of the integral $\int_{x_0}^{x_1} X(x, y, y')\, dx$ with $y = y(x)$ and $y' = dy/dx$, where $X(x, y, y')$ is known, is that the first variation of the integral vanishes. This requirement results in Euler's differential equation of variational calculus, $\frac{d}{dx}\frac{\partial X}{\partial y'} - \frac{\partial X}{\partial y} = 0$ (Leonhard Euler, 1707 - 1783). The solution of Euler's equation yields the desired function $y(x)$ that maximizes the integral. Euler's equation holds also when x and y are vectors. In our case we identify $x = \vec{r}$, $y = \left\{\rho, \vec{\mathcal{V}}, T\right\}$ and

$$X = \rho \left[s - \Lambda_\mu - \vec{\Lambda}_M \cdot \vec{\mathcal{V}} - \Lambda_E \left(u + \frac{1}{2}\mathcal{V}^2 + \gamma z\right)\right] . \quad (17.5)$$

In this particular case the integrand X is independent of $y' = \left\{\frac{d\rho}{d\mathbf{r}}, \frac{d\vec{\mathcal{V}}}{d\mathbf{r}}, \frac{dT}{d\mathbf{r}}\right\}$, so that Euler's equation reduces to

$$\frac{\partial X}{\partial y} = \left\{\frac{\partial X}{\partial \rho}, \frac{\partial X}{\partial \vec{\mathcal{V}}}, \frac{\partial X}{\partial T}\right\} = 0 , \quad (17.6)$$

or, in detail,

$$\frac{\partial X}{\partial \rho} = \left[s - \Lambda_\rho - \mathbf{\Lambda}_M \cdot \vec{\mathcal{V}} - \Lambda_E \left(u + \frac{1}{2}\mathcal{V}^2 + \gamma z\right)\right]$$
$$+ \rho \left[\left(\frac{\partial s}{\partial \rho}\right)_T - \Lambda_E \left(\frac{\partial u}{\partial \rho}\right)_T\right] = 0 , \quad (17.7)$$

$$\frac{\partial X}{\partial \vec{\mathcal{V}}} = \rho \left[-\vec{\Lambda}_M - \Lambda_E \vec{\mathcal{V}}\right] = 0 . \quad (17.8)$$

$$\frac{\partial X}{\partial T} = \rho \left[\left(\frac{\partial s}{\partial T}\right)_\rho - \Lambda_E \left(\frac{\partial u}{\partial T}\right)_\rho\right] = 0 , \quad (17.9)$$

We proceed with evaluating these three conditions to find the equilibrium state. For convenience, we begin with the middle equation, (17.8), which gives immediately that the velocity is homogeneous in equilibrium,

$$\vec{\mathcal{V}} = -\frac{\vec{\Lambda}_M}{\Lambda_E} . \tag{17.10}$$

For the case of a system at rest, where

$$0 = \vec{M} = \int_V \rho \vec{\mathcal{V}} dV = -\frac{\vec{\Lambda}_M}{\Lambda_E} \int_V \rho dV = -\frac{\vec{\Lambda}_M}{\Lambda_E} m , \tag{17.11}$$

this implies that in equilibrium all local elements are at rest,

$$\vec{\mathcal{V}} = \vec{\Lambda}_M = \vec{M} = 0 . \tag{17.12}$$

To evaluate the last condition, (17.9), we recall that the Gibbs equation $Tds = du - \frac{p}{\rho^2} d\rho$ gives $\left(\frac{\partial s}{\partial T}\right)_\rho = \frac{1}{T} \left(\frac{\partial u}{\partial T}\right)_\rho$. Hence, the condition becomes

$$\left(\frac{\partial s}{\partial T}\right)_\rho - \Lambda_E \left(\frac{\partial u}{\partial T}\right)_\rho = \frac{1}{T} \left(\frac{\partial u}{\partial T}\right)_\rho - \Lambda_E \left(\frac{\partial u}{\partial T}\right)_\rho = 0 . \tag{17.13}$$

It follows that in equilibrium the temperature is homogeneous, and equal to the inverse Lagrange multiplier,

$$T = \frac{1}{\Lambda_E} . \tag{17.14}$$

To evaluate the first condition, (17.7), we insert the above results for Λ_E, $\mathbf{\Lambda}_M$, $\vec{\mathcal{V}}$ and use again the Gibbs equation, which gives $\left(\frac{\partial s}{\partial \rho}\right)_T - \frac{1}{T} \left(\frac{\partial u}{\partial \rho}\right)_T = -\frac{p}{T\rho^2}$. After some reordering, we find

$$g = u - Ts + \frac{p}{\rho} = -T\Lambda_\rho - \gamma z , \tag{17.15}$$

where g is the Gibbs free energy, and γ is gravitational acceleration. With the temperature homogeneous, and the constant Lagrange multiplier Λ_ρ, this is an implicit equation for the equilibrium density, which appears as an argument in the Gibbs free energy $g(\rho, T)$, or, alternatively, it is an equation for pressure p, if we write $g(p, T)$. Often we consider systems in which the potential energy can be ignored. For such systems, the Gibbs free energy is homogeneous, $g(\rho, T) = -T\Lambda_\rho$. Homogeneous Gibbs free energy and temperature implies that density and pressure are homogeneous as well. Phase equilibrium will be discussed in Sec. 17.9.

In summary, maximizing entropy in the isolated system yields that the system is fully at rest, $\mathcal{V} = 0$, has homogeneous temperature, $T = 1/\Lambda_E$, and, in the gravitational field, has inhomogeneous density and pressure, given implicitly by $g(T, \rho) = -T\Lambda_\rho - \gamma z$. What remains is to determine the Lagrange multipliers $\Lambda_E = 1/T$ and Λ_ρ, which follow from the given values of mass

$m = \int_V \rho dV$ and energy $E = \int_V \rho \left[u(T,\rho) + \gamma z\right] dV$ in the system. Their detailed values depend on the size and geometry of the system.

17.3 Barometric and Hydrostatic Formulas

To gain insight into the influence of potential energy, we evaluate (17.15) for ideal gases and incompressible fluids. For an ideal gas, the Gibbs free energy is $g(\rho,T) = h(T) - T\left(s^0(T) - R\ln\frac{\rho RT}{p_0}\right)$. Using this in (17.15) and solving for density gives the barometric formula,

$$\rho = \rho^0 \exp\left[-\frac{\gamma z}{RT}\right] , \qquad (17.16)$$

where $\rho^0 = \frac{p_0}{RT}\exp\left[-\frac{\Lambda_\rho}{R} - \frac{h(T)-Ts^0(T)}{RT}\right]$ is the density at reference height $z = 0$. The ideal gas law gives the corresponding expression for pressure as $p = p^0 \exp\left[-\frac{\gamma z}{RT}\right]$, where $p^0 = \rho^0 RT$ is the pressure at $z = 0$.

For incompressible fluids, $\rho = const.$, and internal energy and entropy depend only on temperature, so that the Gibbs free energy is $g(T,p) = u(T) + \frac{p}{\rho} - Ts(T)$. Using this in (17.15) and solving for pressure gives the hydrostatic pressure formula,

$$p = p^0 - \rho\gamma z , \qquad (17.17)$$

where $p^0 = \rho T\left[s(T) - u(T)/T - \Lambda_\rho\right]$ is the pressure at reference height $z = 0$.

17.4 Thermodynamic Stability

The equilibrium state determined in the previous sections should be stable, which means that, indeed, it should be a maximum of the integral Φ as defined in (17.4). This requires that the second variation of Φ must be negative. In our case, where the integrand X depends only on y, this requires negative values for the second derivatives $\partial^2 X/\partial y^2$ at the location of the maximum. With the help of the Gibbs equation, the second derivatives can be written as

$$\frac{\partial X}{\partial\rho\partial\rho} = \left[\frac{1}{T} - \Lambda_E\right]\left[2\left(\frac{\partial u}{\partial\rho}\right)_T + \rho\left(\frac{\partial^2 u}{\partial\rho^2}\right)_T\right] - \frac{1}{\rho T}\left(\frac{\partial p}{\partial\rho}\right)_T ,$$

$$\frac{\partial^2 X}{\partial\vec{\mathcal{V}}^2} = -\rho\Lambda_E ,$$

$$\begin{aligned}
\frac{\partial^2 X}{\partial T^2} &= \rho \left[\frac{1}{T} - \Lambda_E \right] \left(\frac{\partial^2 u}{\partial T^2} \right)_\rho - \frac{\rho}{T^2} \left(\frac{\partial u}{\partial T} \right)_\rho , \\
\frac{\partial^2 X}{\partial \rho \partial T} &= \frac{\partial^2 X}{\partial T \partial \rho} = \left[\frac{1}{T} - \Lambda_E \right] \left(\frac{\partial u}{\partial T} \right)_\rho , \\
\frac{\partial X}{\partial \rho \partial \vec{\mathcal{V}}} &= \frac{\partial X}{\partial \vec{\mathcal{V}} \partial \rho} = -\vec{\Lambda}_M - \Lambda_E \vec{\mathcal{V}} , \\
\frac{\partial X}{\partial T \partial \vec{\mathcal{V}}} &= \frac{\partial X}{\partial \vec{\mathcal{V}} \partial T} = 0 .
\end{aligned} \tag{17.18}$$

These must now be evaluated at the equilibrium state, $T = 1/\Lambda_E$ and $\vec{\mathcal{V}} = -\vec{\Lambda}_M/\Lambda_E$, where they must be negative. With the definitions of isothermal compressibility κ_T (16.37) and the specific heat at constant volume c_v (16.20), the resulting conditions can be written as

$$\begin{aligned}
\frac{\partial X}{\partial \rho \partial \rho}_{|eq} &= -\frac{1}{\rho T} \left(\frac{\partial p}{\partial \rho} \right)_T = -\frac{1}{\rho^2 T \kappa_T} < 0 , \\
\frac{\partial^2 X}{\partial \vec{\mathcal{V}}^2}_{|eq} &= -\frac{\rho}{T} < 0 , \\
\frac{\partial^2 X}{\partial T^2}_{|eq} &= -\frac{\rho}{T^2} \left(\frac{\partial u}{\partial T} \right)_\rho = -\frac{\rho}{T^2} c_v < 0 ;
\end{aligned} \tag{17.19}$$

all mixed derivatives vanish in equilibrium. With the mass density being positive, thermodynamic stability thus requires that isothermal compressibility, specific heat, and thermodynamic temperature are positive,

$$\kappa_T > 0 \quad , \quad c_v > 0 \quad , \quad T \geq 0 . \tag{17.20}$$

These conditions imply that the volume decreases when pressure is increased isothermally, and that the temperature rises when heat is added to the system. While this matches our daily experience, it is nevertheless remarkable that it is guaranteed by the second law as a universal principle, valid for all materials.

17.5 Equilibrium in Non-isolated Systems

Non-isolated systems exchange work or heat with their surroundings. For the study of their equilibria, we use the first and second law in their global forms,

$$\frac{d}{dt} \left(U + E_{pot} \right) = \dot{Q} - p_B \frac{dV}{dt} \quad , \quad \frac{dS}{dt} - \frac{\dot{Q}}{T_B} = \dot{S}_{gen} \geq 0 , \tag{17.21}$$

which are valid when the system exchanges work only via a piston. Here, p_B is the pressure at the piston boundary, and the system exchanges heat only at boundary temperature T_B; we shall consider only cases with homogeneous pressure and temperature at the system boundary. For simplicity, we ignore kinetic energy, which can be incorporated as in the previous sections, with the same result that all elements of the system will be at rest in equilibrium.

For all systems discussed below, if single phase systems are considered, the respective maximization or minimization requirements are mathematically very similar to the maximization of entropy as discussed above.

In cases where the homogeneous boundary temperature T_B is prescribed, the role of the Lagrange multiplier Λ_E is assumed by the boundary temperature T_B, and thus the homogeneous equilibrium temperature of the system is $T = T_B$.

In cases where the piston pressure is prescribed, the pressure condition $g(p,T) = -T\Lambda_\rho - \gamma z$ must be compatible with the pressure prescribed at the piston. If gravity can be ignored, this gives $g(p,T) = g(p_B,T) = -T\Lambda_\rho$, hence homogeneous pressure $p = p_B$. In cases with gravity, since we have assumed homogeneous piston pressure, this implies horizontal piston and $g(p,T) = g(p_B,T) - \gamma(z - z_B)$, where z_B is the height of the piston.

17.5.1 Adiabatic and Isochoric System

For an adiabatic system, we have $\dot{Q} = 0$ and thus

$$\frac{d}{dt}(U + E_{pot}) = -p_B \frac{dV}{dt} \quad , \quad \frac{dS}{dt} \geq 0 \, . \tag{17.22}$$

Entropy grows in an adiabatic process, until it reaches a maximum in equilibrium. We note that for an isochoric process, where $V = const.$, or $\frac{dV}{dt} = 0$, the total energy $E = U + E_{pot}$ stays constant as well. Thus, we have in particular

$$S \Longrightarrow \text{Maximum and } U + E_{pot} = const. \ \text{ for } \ \dot{Q} = 0, \ V = const. \tag{17.23}$$

Indeed, this is the case of a fully isolated system as discussed above, which does not exchange heat and work with its surroundings. The equilibrium state for this case follows from maximizing entropy under constraints of given values for mass m and energy $U + E_{pot}$.

17.5.2 Adiabatic and Isobaric System

A Legendre transform gives an alternative form of the first law,

$$\frac{d}{dt}(U + p_B V + E_{pot}) = V \frac{dp_B}{dt} \, , \tag{17.24}$$

and we conclude that

$$S \Longrightarrow \text{Maximum and } U + p_B V + E_{pot} = const. \text{ for } \dot{Q} = 0,\ p_B = const. \tag{17.25}$$

Note that $H = U + pV$ is the enthalpy. The equilibrium state for this case follows from maximizing entropy under constraints of given values for mass m and $U + p_B V + E_{pot}$.

17.5.3 Isentropic and Isochoric System

For the discussion of non-adiabatic systems, we eliminate the heat $\dot{Q}$ between the first and the second law, to find

$$\frac{d}{dt}\left(U + E_{pot}\right) - T_B \frac{dS}{dt} + p_B \frac{dV}{dt} = -T_B \dot{S}_{gen} \leq 0\,. \tag{17.26}$$

It follows that in a process with constant entropy and constant volume, where $\frac{dS}{dt} = \frac{dV}{dt} = 0$, the total energy will assume a minimum in equilibrium,

$$E = U + E_{pot} \Longrightarrow \text{Minimum for } S = const.,\ V = const. \tag{17.27}$$

The equilibrium state for this case follows from minimizing energy $U + E_{pot}$ under constraints of given values for mass m and entropy S. Note that entropy is difficult to control, and thus this case is typically not encountered in applications.

17.5.4 Isothermal and Isochoric System

By means of a Legendre transform, (17.26) can be rewritten as

$$\frac{d}{dt}\left(U - T_B S + E_{pot}\right) + S \frac{dT_B}{dt} + p_B \frac{dV}{dt} = -T_B \dot{S}_{gen} \leq 0\,. \tag{17.28}$$

It follows that in a process with constant boundary temperature and volume, where $\frac{dT_B}{dt} = \frac{dV}{dt} = 0$, the combination $E - T_B S$ assumes a minimum in equilibrium,

$$U - T_B S + E_{pot} \Longrightarrow \text{Minimum for } T_B = const.,\ V = const. \tag{17.29}$$

Recall that $U - TS = F$ is the Helmholtz free energy. The equilibrium state for this case follows from minimizing $U - T_B S + E_{pot}$ under constraint of given value for mass m.

17.5.5 Isothermal and Isobaric System

Another Legendre transform shows that for a process with constant boundary pressure and temperature ($\frac{dT_B}{dt} = \frac{dp_B}{dt} = 0$) the combination $E + p_B V - T_B S$ assumes a minimum,

$$U + p_B V - T_B S + E_{pot} \Longrightarrow \text{Minimum for } T_B = const.,\ p_B = const. \quad (17.30)$$

Recall that $U + pV - TS = H - TS = G$ is the Gibbs free energy. The equilibrium state for this case follows from minimizing $U + p_B V - T_B S + E_{pot}$ under constraint of given value for mass m.

17.5.6 Energy vs. Entropy

Temperature, volume and pressure are far easier to control than energy and entropy, and thus one normally encounters the last two cases for the computation of equilibria. For simple one-phase systems the results are straightforward: homogeneous temperature T, and, if gravity is ignored, homogeneous pressures p. More complex systems, in particular systems in several phases, and reacting and inert mixtures of several components have additional degrees of freedom that approach equilibrium values, and it is convenient to determine these equilibrium values under the assumption that thermal and mechanical equilibrium, i.e., homogeneous temperature and pressure, are established already. Then, the computation of equilibrium states typically entails to find minima of free energies, either of the Helmholtz free energy $F = U - TS$, or of the Gibbs free energy $G = H - TS$.

The free energies describe the competition between energy and entropy, with the temperature as factor to determine their relative importance. We take a look at this for the Helmholtz free energy, $F = U - TS$. The Helmholtz free energy can attain a minimum state either by making the energy U small, or by making the entropic term TS large. At low temperatures, the product TS is relatively small, thus the entropic term does not matter much, and energy is more important; states of low energies are assumed, for instance the liquid state, which is due to the attractive potential between molecules. For high temperatures, however, the entropic term TS dominates, and states of large entropy are assumed, e.g., the vapor state. For intermediate temperatures, energy and entropy find a compromise, e.g., the coexistence of vapor, which has large entropy, and liquid, which has low energy, in phase equilibrium.

17.6 Interpretation of the Barometric Formula

We discuss the barometric formula (17.16) in the context of the competition between energy and entropy, where the temperature is the deciding factor.

The barometric formula is quite interesting as a rough indicator on the behavior of planetary atmospheres. For an exact discussion, however, one should account for temperature variances within the atmosphere, and for the spherical geometry of the planets.

We consider a column of atmosphere of base area A. The number of moles in a layer of the atmosphere at height z is $dn = \frac{\rho(z)}{M} A dz$, while the total number of moles in the column is $N = m/M$, with $m = \int \rho A dz$ being the total mass in the column. The probability to find a particle in the layer at z is given by

$$\pi dz = \frac{dn}{N} = \frac{\rho(z)\, A dz}{m} = \frac{\gamma}{RT} \exp\left[-\frac{\gamma z}{RT}\right] dz \,. \tag{17.31}$$

$\pi(z)$ as defined here is a probability density, which fulfills $\int_0^\infty \pi(z)\, dz = 1$. Mean value and variance of the height of a particle are

$$\bar{z} = \int_0^\infty z\pi(z)\, dz = \frac{\bar{R}T}{M\gamma} \quad , \quad \sigma = \sqrt{\int_0^\infty (z-\bar{z})^2\, \pi(z)\, dz} = \frac{\bar{R}T}{M\gamma} \,. \tag{17.32}$$

For large values of $\bar{z}$ and σ, gases are more likely to escape a planet. Obviously, $\bar{z}$ and σ grow with temperature, which explains why hot planets, e.g., Mercury, have lost their atmosphere. Moreover, $\bar{z}$ and σ are smaller for larger gravitation γ, which explains why heavier planets have more stable atmospheres: Jupiter, for instance, is a heavy gas planet. Finally, $\bar{z}$ and σ grow with decreasing molar mass M which explains why light elements are more likely to escape from the atmosphere of a planet. Indeed, there is only little helium left in Earth's atmosphere, although helium is one of the most abundant elements in the universe. A good source for helium is natural gas which was formed long ago, when Earth's atmosphere was richer in helium.

The above discussion can be seen in the context of competition between energy and entropy. When the temperature is low, the entropy is less important, and the equilibrium state has a low potential energy, $\bar{z}$ is small, and $\bar{z} = 0$ for $T = 0$. But when the temperature is high, entropy is more important, and tries to establish a state of even distribution within the accessible volume. The actual state, with exponential decay, is a compromise between the two opposing tendencies. We shall explore this competition more as we proceed.

17.7 Equilibrium in Heterogeneous Systems

The thermodynamic equilibrium conditions, e.g., system entropy assumes a maximum in isolated systems (17.23), or Gibbs free energy assumes a minimum when pressure and temperature are prescribed at the boundary (17.30), are universally valid. In this section, we evaluate the equilibrium state for a heterogeneous system, which consists of two parts in thermal and mechani-

U_1	U_2
V_1	V_2
$S(U_1, V_1)$	$S(U_2, V_2)$

Fig. 17.1 An externally adiabatic system at constant volume, containing two different materials or phases

cal contact. To be specific, we consider an adiabatically enclosed system at constant volume that is divided into two parts as depicted in Fig. 17.1. The two parts may contain different substances, or the same substance, and they might contain different phases. The divider between the two parts can move freely, and is diathermal, i.e., heat can pass, potential and kinetic energies are ignored. Due to the boundary conditions for the system, total energy, $U = U_1 + U_2$, and total volume, $V = V_1 + V_2$ are constants, but energy and volume of the parts might change. We assume the system is in thermal equilibrium and consider small perturbations from the equilibrium state such that the energies and volumes of the two parts are

$$U_1 + \delta U \quad , \quad U_2 - \delta U \quad \text{and} \quad V_1 + \delta V \quad , \quad V_2 - \delta V \, . \tag{17.33}$$

This perturbation yields a change in entropy, so that the entropy of the perturbed state is $S + \delta S$. Since the perturbed state is an equilibrium state, the entropy $S = S_1(U_1, V_1) + S_2(U_2, V_2)$ is a maximum; accordingly, the perturbation in entropy must be negative, $\delta S < 0$. We have

$$S + \delta S = S_1(U_1 + \delta U, V_1 + \delta V) + S_2(U_2 - \delta U, V_2 - \delta V) \, , \tag{17.34}$$

and from Taylor expansion to first order we find

$$\delta S = \left(\frac{\partial S_1}{\partial U_1}\right)_{V_1} \delta U + \left(\frac{\partial S_1}{\partial V_1}\right)_{U_1} \delta V - \left(\frac{\partial S_2}{\partial U_2}\right)_{V_2} \delta U - \left(\frac{\partial S_2}{\partial V_2}\right)_{U_2} \delta V \, . \tag{17.35}$$

From the Gibbs equation $TdS = dU + pdV$ we identify $(\partial S/\partial U)_V = 1/T$ and $(\partial S/\partial V)_U = p/T$, and thus the above can be rewritten (with some reordering) as

$$0 > \delta S = \left[\frac{1}{T_1} - \frac{1}{T_2}\right] \delta U + \left[\frac{p_1}{T_1} - \frac{p_2}{T_2}\right] \delta V \, . \tag{17.36}$$

Since δU and δV can have arbitrary positive or negative values, the sign condition on δS can only be fulfilled when both terms vanish. This gives the expected equilibrium conditions for the two parts, namely that they have the same temperatures and pressures

$$T_1 = T_2 \quad \text{and} \quad p_1 = p_2 \,. \tag{17.37}$$

The above discussion can be performed for any splitting of the system, and for different substances in the subsystem. It follows that all possible subsystems have the same temperature and pressure, that is pressure and temperature are homogeneous within the system.

If potential energy, e.g., gravitation, plays a role, pressure is not homogeneous, see Sec. 17.3. Pressure distribution within one substance or phase is then given by $g(p,T) = -T\Lambda_\rho - \gamma z$, while the pressure is continuous at the interface between two substances or phases.

17.8 Phase Equilibrium

A particular class of equilibrium states concerns equilibria between different phases of the same substance, e.g., liquid-vapor equilibria.

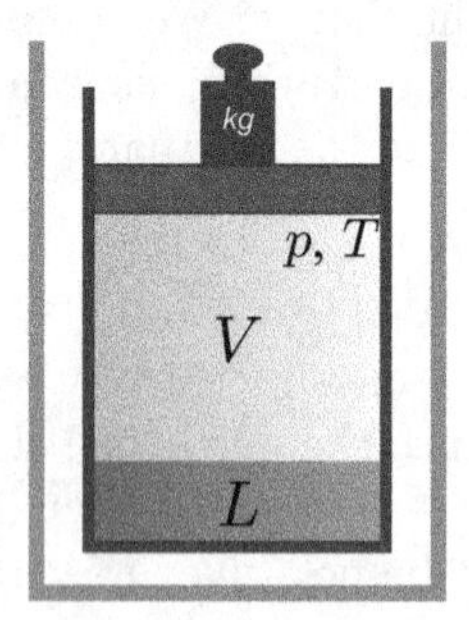

Fig. 17.2 Liquid (L) and vapor (V) phase in equilibrium at given pressure p and temperature T

Figure 17.2 shows liquid and vapor in equilibrium in a system where pressure p and temperature T are fixed at the boundaries by the given mass of the piston, and exposure to a large reservoir at T. According to (16.1) the equilibrium state of this system is determined by a minimum of the Gibbs free energy G, which is just the sum of the Gibbs free energies of the two phases. The mass $m = m_L + m_V$ within the system is constant, and thus we have

$$G = m_V g_V(T,p) + m_L g_L(T,p) = m_V g_V(T,p) + (m - m_V) g_L(T,p) \,. \tag{17.38}$$

The specific free energies of the individual phases, g_L and g_V, depend only on the intensive variables p and T. When thermal and mechanical equilibrium are established, T and p are homogeneous throughout both phases, and the vapor mass m_V is the only variable. The chemical equilibrium is assumed when G becomes a minimum, that is for $dG/dm_V = 0$, which gives

$$g_V(T,p) = g_L(T,p) \ . \tag{17.39}$$

Hence, in a two phase system in equilibrium, pressure, temperature *and* Gibbs free energies are homogeneous. It follows that both phases can coexist only at values for pressure and temperature (T,p) that fulfill the above condition. Solving for p gives the saturation pressure $p_{\text{sat}}(T)$, with the well known value of $p_{\text{sat}}(100\,°\text{C}) = 1\,\text{atm}$ for water. Solving for T gives the saturation temperature, $T_{\text{sat}}(p)$.

In case that temperature and pressure are chosen such that the Gibbs free energies of liquid and vapor are different, the Gibbs free energy (17.38) assumes a boundary minimum with either $m_L = m$, $m_V = 0$ (compressed liquid) or $m_V = m$, $m_L = 0$ (superheated vapor). In detail we have for a specified pressure p:

$$\begin{aligned} T < T_{\text{sat}}(p) \quad &\Longrightarrow \quad g_L(T,p) < g_V(T,p) \quad \Longrightarrow \quad m_L = m,\ m_V = 0\ , \\ T > T_{\text{sat}}(p) \quad &\Longrightarrow \quad g_L(T,p) > g_V(T,p) \quad \Longrightarrow \quad m_V = m,\ m_L = 0\ . \end{aligned}$$

The phase change can be understood as a competition between energy and entropy. Recall that Gibbs free energy is $g = h - Ts$. For small temperatures, the entropic term $(-Ts)$ is relatively small, and energetic effects dominate. Then the Gibbs free energy is small for the liquid, where the potential energy between particles due to the molecular interaction is at a minimum, the particles are close to each other, and the volume is small. For larger temperatures, the entropic contribution becomes more important, and the Gibbs free energy becomes small for large entropies. Since vapor entropy grows with volume,[1] the vapor state prevails and the volume is large. At saturation, energetic and entropic contributions are of comparable size, and both phases coexist.

Alternatively, we have for a specified temperature T:

$$\begin{aligned} p > p_{\text{sat}}(T) \quad &\Longrightarrow \quad g_L(T,p) < g_V(T,p) \quad \Longrightarrow \quad m_L = m,\ m_V = 0\ , \\ p < p_{\text{sat}}(T) \quad &\Longrightarrow \quad g_L(T,p) > g_V(T,p) \quad \Longrightarrow \quad m_V = m,\ m_L = 0\ . \end{aligned}$$

Since vapor entropy grows with lower pressure[2], the entropic term will dominate even at low temperatures, if only the pressure is sufficiently small. Thus, exposing a substance to low pressure might induce phase change.

While we used liquid and vapor as example, the above derivation is not restricted to any particular phases. For any two phases to be in equilibrium, their Gibbs free energies must agree. For an example, revisit Fig. 6.4 in Chapter 6 which shows the saturation lines for water as ice, liquid, and vapor.

At the triple point, all three phases coexist in equilibrium, and their free energies must agree (S stands for solid),

$$g_V(T,p) = g_L(T,p) = g_S(T,p) \ . \tag{17.40}$$

---

[1] This can be seen from the ideal gas entropy in the form $s - s_0 = c_v \ln \frac{T}{T_0} + R \ln \frac{v}{v_0}$.
[2] This can be seen from the ideal gas entropy in the form $s - s_0 = c_p \ln \frac{T}{T_0} - R \ln \frac{p}{p_0}$.

These are two conditions for T, p and thus there is only one pair of values T_{tr}, p_{tr} at which three phases can coexist, the triple point (e.g., for water: $T_{tr} = 0.01\,^\circ\mathrm{C}$, $p_{tr} = 611\,\mathrm{Pa}$).

The conditions derived above describe the thermodynamic equilibrium of two phases, which is not always attained. Some substances can exist for very long periods in metastable states, outside of equilibrium. A typical example is tin, which below 13.2 °C is stable as a semiconductor phase, and is metallic above. However, the phase transition does only occur at much lower temperatures. Another example is carbon, for which the stable phase at room temperature is graphite, while diamond is metastable, which obviously does not diminish its value, both as a gem, and for toolmaking.

17.9 Example: Phase Equilibrium for the Van der Waals Gas

We consider the van der Waals equation (16.48) for its ability to describe phase equilibrium.

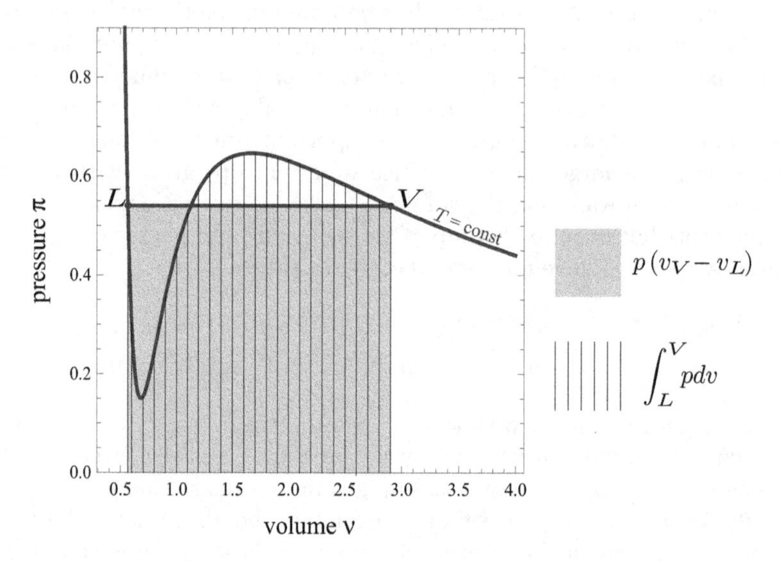

Fig. 17.3 Phase equilibrium for van der Waals gas

Figure 17.3 shows the sketch of an undercritical isotherm for the van der Waals gas, which is not monotonous, but exhibits an unstable region where $\kappa_T = -\frac{1}{v}\left(\frac{\partial v}{\partial p}\right)_T \leq 0$ (so that the isotherm has a positive slope). Instead of following the curve, the gas can split into two phases: L (liquid) and V (vapor) as indicated. The question is, where the connecting line which determines the saturation pressure should be drawn. The answer comes from evaluation of (17.39) which can be written as

$$g_L = g_V \quad \text{or, with } g = f + pv, \text{ as} \quad f_L - f_V = p\left(v_V - v_L\right) , \tag{17.41}$$

where p is the common pressure of the two phases, i.e., the saturation pressure.

The difference in Helmholtz free energy can be reformulated as

$$f_L - f_V = -\underset{\text{isotherm}}{\int_L^V} \left(\frac{\partial f}{\partial v}\right)_T dv = \underset{\text{isotherm}}{\int_L^V} pdv , \tag{17.42}$$

where we used that $f\left(T, v\right)$ is a potential, see (16.11). Thus the condition for the saturation pressure reads

$$\underset{\text{isotherm}}{\int_L^V} pdv = p\left(v_V - v_L\right) . \tag{17.43}$$

This is Maxwell's equal area rule, which states that the areas below the $\mathcal{S}$-shaped van der Waals curve and the straight line connecting vapor and liquid, as shown in the figure, must be equal.

The evaluation of this condition leads to transcendental equations which must be solved numerically. Figure 17.4 shows, in dimensionless variables, some isotherms and the computed vapor dome in the p-v-diagram.

17.10 Clapeyron Equation

The Clapeyron equation describes the slope of the saturation curve. The two phases, and the corresponding properties are denoted as phase $'$ and phase $''$, respectively. To find the Clapeyron equation, we consider a point $\{p_{\text{sat}}, T\}$ on the saturation curve, and an adjacent point on the curve $\{p_{\text{sat}} + dp_{\text{sat}}, T + dT\}$. For the Gibbs free energy of the latter, for one phase, we find by Taylor expansion

$$\begin{aligned} g'\left(p_{\text{sat}} + dp_{\text{sat}}, T + dT\right) &= g'\left(p_{\text{sat}}, T\right) + \left(\frac{\partial g'}{\partial p_{\text{sat}}}\right)_T dp_{\text{sat}} + \left(\frac{\partial g'}{\partial T}\right)_{p_{\text{sat}}} dT \\ &= g'\left(p_{\text{sat}}, T\right) + v' dp_{\text{sat}} - s' dT , \end{aligned} \tag{17.44}$$

where we have used (16.13). We consider the corresponding equation for phase $''$ as well,

$$g''\left(p_{\text{sat}} + dp_{\text{sat}}, T + dT\right) = g''\left(p_{\text{sat}}, T\right) + v'' dp_{\text{sat}} - s'' dT \tag{17.45}$$

and take the difference,

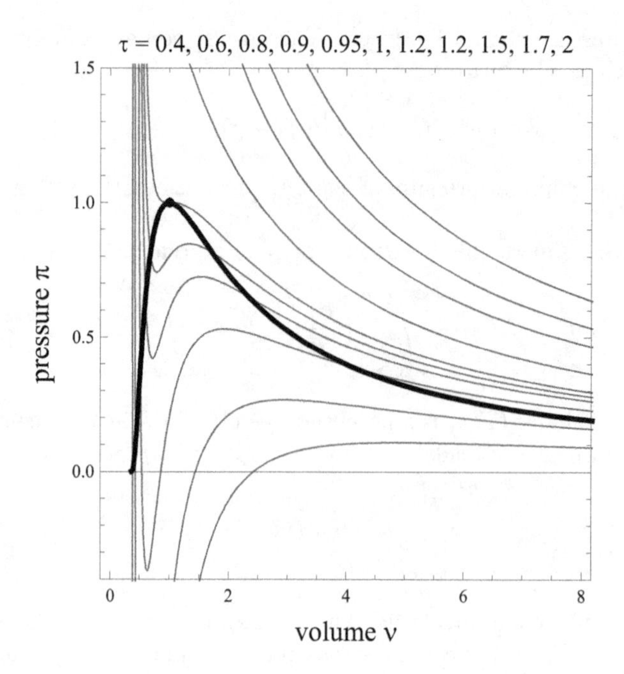

Fig. 17.4 Van der Waals isotherms and two phase region in the $\pi - \upsilon$-diagram

$$
\begin{aligned}
g''\left(p_{\text{sat}} + dp_{\text{sat}}, T + dT\right) - g'\left(p_{\text{sat}} + dp_{\text{sat}}, T + dT\right) \\
= g''\left(p_{\text{sat}}, T\right) - g'\left(p_{\text{sat}}, T\right) + \left(\upsilon'' - \upsilon'\right) dp_{\text{sat}} - \left(s'' - s'\right) dT \,.
\end{aligned}
$$

Since both points are on the saturation curve, the differences of the free energies vanish on both sides, and we find the Clapeyron equation[3]

$$
\frac{dp_{\text{sat}}}{dT} = \frac{s'' - s'}{\upsilon'' - \upsilon'} = \frac{1}{T} \frac{h'' - h'}{\upsilon'' - \upsilon'} \,. \tag{17.46}
$$

$h'' - h'$ is the heat of phase change, for instance the heat of evaporation or the heat of melting.

17.11 Example: Estimate of Heat of Evaporation

The Clapeyron equation can be used to find an estimate for the vapor pressure curve. We consider liquid-vapor equilibrium, where the Clapeyron equations reads

$$
\frac{dp_{\text{sat}}}{dT} = \frac{1}{T} \frac{h_{LV}}{\upsilon_V - \upsilon_L} \,, \tag{17.47}
$$

[3] Note that $g' = g''$ implies $h'' - h' = T\left(s'' - s'\right)$.

with the heat of evaporation $h_{LV} = h_V - h_L$. To proceed, we assume that the specific volume of the vapor phase (V) is far larger than the volume of the liquid phase (L), so that $v_V - v_L \simeq v_V$. Moreover, we assume that the vapor can be described by the ideal gas law, so that $v_V = RT/p$. Both assumptions are reasonably accurate at pressures not too far from the triple point. The equation then reduces to the Clausius-Clapeyron equation

$$\frac{1}{p_{\text{sat}}} \frac{dp_{sat}}{dT} = \frac{h_{LV}}{RT^2} ,$$

which can be integrated with the further assumption that the heat of evaporation h_{LV} is constant, to give

$$\ln \frac{p_{\text{sat}}(T_2)}{p_{\text{sat}}(T_1)} = \frac{h_{LV}}{R} \left(\frac{1}{T_1} - \frac{1}{T_2} \right) . \tag{17.48}$$

We use this equation to estimate the heat of evaporation at 7.5 °C from measurements of saturation pressures and volumes. From steam tables we find the following data:

$$T_1 = 278.15\,\text{K} , \quad p_{\text{sat}}(T_1) = 0.8721\,\text{kPa} , \quad v_V(T_1) = 147.12 \frac{\text{m}^3}{\text{kg}} ,$$
$$T_2 = 283.15\,\text{K} , \quad p_{\text{sat}}(T_2) = 1.2276\,\text{kPa} , \quad v_V(T_2) = 106.38 \frac{\text{m}^3}{\text{kg}} .$$

With $R = 0.462 \frac{\text{kJ}}{\text{kg K}}$ we find $h_{LV}(7.5\,°\text{C}) = 2488.18 \frac{\text{kJ}}{\text{kg}}$, which is reasonably close to the exact value of $2515 \frac{\text{kJ}}{\text{kg}}$.

17.12 Example: Ice Skating

Another interesting application is the discussion of ice flows and ice skating. We consider the solid-liquid equilibrium at temperatures not too far from $0\,°\text{C} = 273.15\,\text{K}$, where $h_{SL} = 333.7 \frac{\text{kJ}}{\text{kg}}$, $v_L = 0.001 \frac{\text{m}^3}{\text{kg}}$ and $v_S = 0.001085 \frac{\text{m}^3}{\text{kg}}$. Since the solid, i.e., the ice, has a larger volume than the liquid, the melting curve has a negative slope,

$$\frac{dp_{\text{sat}}}{dT} = \frac{1}{T} \frac{h_{SL}}{v_L - v_S} = -142 \frac{\text{bar}}{\text{K}} . \tag{17.49}$$

Due to the density anomaly of water, ice swims on water. Would this not be so, ponds and lakes would freeze completely in cold climates (ice would form on the top, and sink), and no life could survive. Thus, the anomaly is of some importance to our ecosystem. Note also that, considering its low molar mass, water condenses at rather high temperatures (100 °C at $p = 1\,\text{atm}$; compare to saturation temperature of oxygen (O_2) of −218.8 °C); this is due to the strong attractive intermolecular forces between water molecules (hydrogen bonds).

Coming back to (17.49), we conclude that in order to lower the melting temperature by 1 °C, the pressure must be increased by 142 bar. In other words, to melt ice of −10 °C the pressure must be increased by 1420 bar. Due to the weight of the ice, the pressure inside glaciers is quite high, and the slow flow of glaciers might be attributed to this effect, but also to plastic deformation.

Quite often, however, melting under pressure is used to explain the physics of ice skating, but this is not a valid explanation. Indeed, high pressures would break the ice before it could melt. The contact area for the skates would have to be very small, to bring these high pressures on the ice. As most people know, ice is very slippery even with street shoes, which have a large contact area. Also one would expect a strong dependence of the skating ability on temperature, since higher pressures are required for melting at lower temperatures. Moreover, since no liquid water exist below −23 °C, skating would not be possible at temperatures below that. Canadian students report that they skated without problems at temperatures below −30 °C.

Today, it is believed that the slipperiness of ice is due to a molecular layer of water molecules at the surface which are not fixated in the lattice structure of the ice. There seems to be no stable configuration with an energetic minimum for the surface molecules, and they dangle about. This molecular layer behaves almost like liquid water, and thus ice is slippery.[4]

Problems

17.1. Barometric Formula

Atmospheric air can be considered as an ideal gas with $R = 0.287 \frac{\text{kJ}}{\text{kg K}}$.

1. Balance the forces on a layer of the atmosphere of thickness dz to show that $\frac{dp}{dz} = -\rho g$.
2. Consider an isothermal atmosphere and compute $p(z)$ with $p(z=0) = p_0 = 1 bar$.
3. In reality, the temperature of the atmosphere is decreasing with height according to

$$T(z) = T_0 \left(1 - \frac{\gamma z}{T_0}\right) \quad \text{where} \quad \gamma = \frac{0.65\,\text{K}}{100\,\text{m}}, \; T_0 = 288\,\text{K},$$

   (valid for $0 \leq z \leq 10,000\,\text{m}$). Compute $p(z)$ for this case (again with $p(z=0) = p_0$).
4. In your last result, consider the limit $\gamma \longrightarrow 0$ and show that you obtain the same result as in ii.).

---

[4] See: S.C. Colbeck, Pressure melting and ice skating, Am. J. Phys. **63**, 888-890 (1995). Wettlaufer & Dash, Melting Below Zero, Scientific American Magazine, February 2000.

5. Plot the two curves, and discuss—what pressures are predicted for Mt. Baker, Mt. Everest?

17.2. Three-Phase Equilibrium
Consider a phase mixture of solid, liquid and vapor in equilibrium, in a closed system at constant pressure and temperature. Minimize the Gibbs free energy to find the equilibrium condition (17.40).

17.3. Solid Carbon
Pure carbon in solid form can appear in different crystal lattices, as graphite ($\bar{h}_f^0 = 0\frac{\text{kJ}}{\text{mol}}$, $\bar{s}_f^0 = 5.75\frac{\text{kJ}}{\text{kmol K}}$) or as diamond ($\bar{h}_f^0 = 1.89\frac{\text{kJ}}{\text{mol}}$, $\bar{s}_f^0 = 2.38\frac{\text{kJ}}{\text{kmol K}}$). Here, $\bar{h}_f^0$, $\bar{s}_f^0$ denote enthalpy and entropy of formation, that is the values of enthalpy and entropy at standard conditions ($T_0 = 298\,\text{K}$, $p_0 = 1\,\text{atm}$).

1. Experience shows that both forms exist at standard conditions, but thermodynamically only one is stable—is it graphite or diamond? State your argument.
2. The mass density of graphite is $2.25\frac{\text{kg}}{\text{litre}}$ and that of diamond is $3.52\frac{\text{kg}}{\text{litre}}$. When the pressure is increased to 1.6×10^4 bar, is your answer to the previous question the same? Explain.
3. At which pressure are graphite and diamond in equilibrium (at T_0)?

17.4. Stirred Water
A rigid adiabatic container contains 1 litre of water at 20 °C. The water was briefly stirred with a propeller, so that the average velocity is $15\frac{\text{m}}{\text{s}}$. Consider only the time after stirring (but not the water motion) has stopped. Under the conditions of this process, water can be described as an ideal incompressible liquid $\left(v = v_0 = 0.001\frac{\text{m}^3}{\text{kg}}\right)$ with constant specific heat $c_p = c_w = 4.18\frac{\text{kJ}}{\text{kg K}}$.

1. Show that incompressibility implies $c_p = c_v = c_w$.
2. Combine the 1st and 2nd law of thermodynamics to show that, while the water still moves after stirring stops, it comes to rest over time, and the rest state is the equilibrium state, that is the kinetic energy will go to zero. Hint: Use the Gibbs equation, and account for adiabatic process and incompressibility.
3. Determine the increase of temperature between the stirred state and the final rest state when the system is adiabatic.

17.5. Approximate Equation for Saturation Pressure
In order to find an approximate equation for the saturation pressure of a substance, assume that the liquid can be considered as an incompressible liquid, and the vapor as an ideal gas.

1. For the liquid (f) show first that incompressibility implies that the specific heats at constant pressure and constant volume are the same. Next, assume

constant specific heat c_f and incompressibility and find internal energy, enthalpy and entropy by integration. Show that

$$u_f = c_f \left(T - T_0\right) \quad , \quad h_f = c_f \left(T - T_0\right) \quad , \quad s_f = c_f \ln \frac{T}{T_0} \; .$$

The assumption of incompressibility is not sufficient to obtain the relation for h_f. What contribution is missing, and why (or when) can it be ignored?

2. Consider vapor (g) as an ideal gas with constant specific heats, and show that specific enthalpy and entropy are given by

$$h_g = c_p \left(T - T_0\right) + h_{fg} \left(T_0\right) \quad , \quad s_g = c_p \ln \frac{T}{T_0} - R \ln \frac{p}{p_{\text{sat}} \left(T_0\right)} + \frac{h_{fg} \left(T_0\right)}{T_0} \; ;$$

c_p is the specific heat of the vapor, $h_{fg}\left(T_0\right)$ is the specific heat of evaporation at reference temperature T_0, and $p_{\text{sat}}\left(T_0\right)$ is the saturation pressure at T_0. Discuss the choice of integrating constants and give clear arguments why $h_{fg}\left(T_0\right)$ appears in both relations.

3. Find an expression for the heat of evaporation $h_{fg}\left(T\right)$.
4. Use the condition for phase equilibrium $g_f\left(T, p_{\text{sat}}\left(T\right)\right) = g_g\left(T, p_{\text{sat}}\left(T\right)\right)$ to find an equation for the saturation pressure $p_{\text{sat}}\left(T\right)$.
5. Use data for water and chose $T_0 = 273.15\,\text{K}$ to make a table with values of saturation pressure and heat of evaporation for several temperatures. Compare to tabled data: When an error of 5% is acceptable, what is the maximum temperature for which the approximation can be used?
6. An equation that is regularly used, and gives a better fit, is the Antoine equation

$$\log p_{\text{sat}}\left(T\right) = A - \frac{B}{C + T} \; .$$

Use data for water at 0.01 °C, 50 °C, and 100 °C to obtain the constants A, B, C. You may use a computer program to find the constants. Plot the saturation pressure as function of T, and make a list of values for temperatures up to the critical temperature 374.14 °C. Compare with tabulated data.

17.6. Heat of Evaporation

Use the Clausius-Clapeyron equation together with the Antoine equation for water (see previous problem) to find a relation for the heat of vaporization, $h_{fg}\left(T\right)$. Assume that the liquid volume can be ignored against the vapor volume, and that the vapor can be described as an ideal gas. Plot $h_{fg}\left(T\right)$ over T, and compare with tabulated data. Discuss the result.

17.7. Property Data: Interpolation

For a thermodynamic computation you need the Gibbs free energy, the entropy, and the enthalpy of liquid water at a temperature of 97.5 °C and a pressure of 1 atm. In an old and incomplete table you find the following data:

compressed liquid:

$$g(95\,^\circ\mathrm{C}, 1\,\mathrm{atm}) = -62\frac{\mathrm{kJ}}{\mathrm{kg}}\,,$$

saturated vapor:

$$g_g(100\,^\circ\mathrm{C}) = -68.5\frac{\mathrm{kJ}}{\mathrm{kg}}\,,\quad p_{sat}(100\,^\circ\mathrm{C}) = 1\,\mathrm{atm}\,.$$

Use only this data to determine (a) $g(97.5\,^\circ\mathrm{C}, 1\,\mathrm{atm})$, (b) $s(97.5\,^\circ\mathrm{C}, 1\,\mathrm{atm})$ and (c) $h(97.5\,^\circ\mathrm{C}, 1\,\mathrm{atm})$.

17.8. Phase Equilibrium of a Van der Waals Gas
Use Maxwell's equal area rule to construct the two phase region for the dimensionless van der Waals equation. For given temperature τ you need to determine saturation pressure $\pi_{sat}(\tau)$ and saturation volumes $\upsilon_f(\tau)$ and $\upsilon_g(\tau)$. It is best to prescribe a value for υ_f and then find the corresponding values for τ, π_{sat}, and υ_g.

1. Write down the equations you need to solve the problem.
2. The equations are transcendental and thus must be solved numerically. Use one of the convenient mathematics programs like Mathematica, Maple, Matlab, etc. to solve the problem.
3. Plot the two-phase region and some isothermal curves in a p-v-diagram.
4. Plot the saturation curve in the p-T-diagram.

17.9. Homogeneity
Solve problems 4.15 and 4.16.

Chapter 18
Mixtures

18.1 Introduction

Many applications of thermodynamics involve not single substances but mixtures. The challenge is to track mixture composition, and to find the property data for the given composition. As long as mixture composition does not change, one can deal with mixtures the same way as with simple substances, including tabulating their properties; our treatment of air is the prime example of this.

Mixture composition can change through mixing or separation processes, through phase changes when the components have different vapor pressures, and through chemical reactions.

There is a vast array of applications for mixture theory, in particular in chemical engineering. Applications to be discussed include desalination of seawater, osmotic power plants, phase equilibrium and distillation processes, chemical equilibrium and NH_3 production, and combustion.

In this and the following chapters we shall provide the tools to properly describe and evaluate these processes. The present chapter introduces additional properties to account for mixture composition, and relations between properties of components and the mixture as a whole.

18.2 Mixture Composition

We consider mixtures of ν components, indicated by greek subscripts $\alpha = 1, 2, \ldots, \nu$. The present chapter deals with non-reacting mixtures, reacting mixtures will be discussed later.

Throughout the following we assume that all components have the same temperature T. The mixture is contained in the volume V, and the mixing state is homogeneous, so that each component is equally distributed in V.

The composition of the mixture can either be described through the masses m_α of the components contained in the volume V, or by their amount in

H. Struchtrup, *Thermodynamics and Energy Conversion*,
DOI: 10.1007/978-3-662-43715-5_18, © Springer-Verlag Berlin Heidelberg 2014

molecule numbers N_α. Rather than tracking actual particle numbers, one uses the mole as a unit for counting particles, with the mole number defined as

$$n_\alpha = \frac{N_\alpha}{A} = \frac{m_\alpha}{M_\alpha} \,. \tag{18.1}$$

Here $A = 6.022 \times 10^{23} \frac{1}{\text{mol}}$ is Avogadro's number, which defines the number of particles in one mole, and M_α is the molar mass, i.e., the mass of 1 mol of particles of type α.

The total mass m and the total mole number n of the mixture are obtained by summation over all components,

$$m = \sum_{\alpha=1}^{\nu} m_\alpha \;, \quad n = \sum_{\alpha=1}^{\nu} n_\alpha = \sum_{\alpha=1}^{\nu} \frac{m_\alpha}{M_\alpha} = \frac{m}{M} \,. \tag{18.2}$$

The last equation defines the average molar mass M of the mixture.

Often we will not be interested in the absolute amounts of the components, but in the relative amounts. Mass fraction c_α (sometimes denoted as "mass concentration") and mole fraction X_α are defined as

$$c_\alpha = \frac{m_\alpha}{m} \;, \quad X_\alpha = \frac{n_\alpha}{n} = \frac{N_\alpha}{N} \;; \tag{18.3}$$

according to their definitions we have

$$\sum_{\alpha=1}^{\nu} X_\alpha = \sum_{\alpha=1}^{\nu} c_\alpha = 1 \,. \tag{18.4}$$

18.3 Example: Composition and Molar Mass of Air

The average molar mass of a mixture is given by

$$M = \frac{m}{n} = \frac{1}{n} \sum_{\alpha=1}^{\nu} m_\alpha = \frac{1}{n} \sum_{\alpha=1}^{\nu} M_\alpha n_\alpha = \sum_{\alpha=1}^{\nu} M_\alpha X_\alpha \,. \tag{18.5}$$

Air is a mixture of several gases, the main components and their mole fractions and molar masses are

$$\begin{array}{lll}
\text{nitrogen:} & X_{N_2} = 0.7808 \;, & M_{N_2} = 28.02 \frac{\text{kg}}{\text{kmol}} \;, \\
\text{oxygen:} & X_{O_2} = 0.2095 \;, & M_{O_2} = 32 \frac{\text{kg}}{\text{kmol}} \;, \\
\text{argon:} & X_{Ar} = 0.0093 \;, & M_{Ar} = 39.94 \frac{\text{kg}}{\text{kmol}} \;, \\
\text{carbon dioxide:} & X_{CO_2} = 0.000397 \;, & M_{CO_2} = 44.01 \frac{\text{kg}}{\text{kmol}} \,.
\end{array}$$

Accordingly, the average molar mass of air is $M_{\text{air}} = 28.97\frac{\text{kg}}{\text{kmol}}$. The corresponding mass fractions are

$$c_\alpha = \frac{m_\alpha}{m} = \frac{n_\alpha M_\alpha}{nM} = X_\alpha \frac{M_\alpha}{M} \,,$$

so that

$$c_{\text{N}_2} = 0.755,\; c_{\text{O}_2} = 0.231,\; c_{\text{Ar}} = 0.013,\; c_{\text{CO}_2} = 0.000455\,.$$

18.4 Mixture Properties

In previous chapters, we have mainly used specific properties, that is properties per unit mass which are denoted as, e.g., $v_\alpha, u_\alpha, h_\alpha,\, s_\alpha$. For mixtures it is often more convenient do refer to particle numbers, and thus we will often use mole based properties, denoted as, e.g., $\bar{v}_\alpha, \bar{u}_\alpha, \bar{h}_\alpha,\, \bar{s}_\alpha$.

Mole and mass based quantities are related through the molar mass M_α, in particular we have

mass/mole density: $\rho_a = \frac{m_\alpha}{V} \;,\; \bar{\rho}_a = \frac{n_\alpha}{V} \,,$

specific/molar volume: $v_\alpha = \frac{V}{m_\alpha} \;,\; \bar{v}_\alpha = \frac{V}{n_\alpha} = v_\alpha M_\alpha \,,$

specific/molar internal energy: $u_\alpha \;,\; \bar{u}_\alpha = u_\alpha M_\alpha \,,$

specific/molar enthalpy: $h_\alpha \;,\; \bar{h}_\alpha = h_\alpha M_\alpha \,,$

specific/molar entropy: $s_\alpha \;,\; \bar{s}_\alpha = s_\alpha M_\alpha \,.$

Properties of the mixture are obtained as weighted sums over the properties of the individual components. We study this for the total internal energy, for which we have

$$U = mu = \sum_{\alpha=1}^{\nu} m_\alpha u_\alpha = n\bar{u} = \sum_{\alpha=1}^{\nu} n_\alpha \bar{u}_\alpha \,. \tag{18.6}$$

The specific internal energy, and the molar internal energy of the mixture are obtained by division with m or n, as

$$u = \frac{U}{m} = \sum_{\alpha=1}^{\nu} c_\alpha u_\alpha \;,\quad \bar{u} = \frac{U}{n} = \sum_{\alpha=1}^{\nu} X_\alpha \bar{u}_\alpha \,. \tag{18.7}$$

Enthalpy and entropy of the mixture are obtained in the same way:

$$h = \sum_{\alpha=1}^{\nu} c_\alpha h_\alpha \;,\quad \bar{h} = \sum_{\alpha=1}^{\nu} X_\alpha \bar{h}_\alpha \,, \tag{18.8}$$

$$s = \sum_{\alpha=1}^{\nu} c_\alpha s_\alpha \;,\quad \bar{s} = \sum_{\alpha=1}^{\nu} X_\alpha \bar{s}_\alpha \,. \tag{18.9}$$

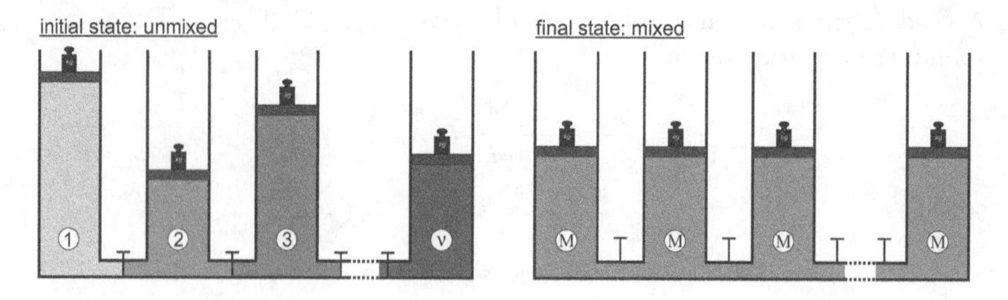

Fig. 18.1 Mixing of components at constant T and p

Above, we have not indicated the dependencies between properties. In general, the properties of one component will depend on the presence of all other components. For instance, the internal energy of component α will depend on temperature T and total pressure p of the mixture, and on all mole fractions X_β, $\beta = 1, \ldots, \nu$, that is $\bar{u}_\alpha = \bar{u}_\alpha (T, p, X_\beta)$. Therefore tabulated data for single components (where $X_\alpha = 1$ and $X_\beta = 0$ for $\beta \neq \alpha$) normally cannot be used. As will be seen, tabulated data for pure components can only be used for ideal gas mixtures, and ideal mixtures.

While all components have the same temperature T, they contribute to pressure differently. The partial pressure p_α is the contribution of component α to total pressure p, with

$$p = \sum_{\alpha=1}^{\nu} p_\alpha \,. \tag{18.10}$$

Note that, in general, $p_\alpha = p_\alpha (T, p, X_\beta)$, that is the partial pressure of a component will depend on the state and composition of the mixture.

18.5 Mixing Volume, Heat of Mixing and Entropy of Mixing

As components are mixed at constant temperature and pressure the extensive properties might change. To properly account for the change, we consider ν components in an initial unmixed state (I) where each component is at the same temperature, T, and pressure, p. The components are mixed while keeping temperature and pressure constant, the final state (E) is a homogeneous mixture, see Fig. 18.1. We ask for the corresponding changes in total volume, total enthalpy, and total entropy.

The volume change between initial and final state is computed as[1]

[1] This and the other equations in this section are written with molar quantities and mole numbers. It is straightforward to write all in specific properties and masses, e.g. $V_{\text{mix}} = m_\gamma v_\gamma (T, p, c_\beta) - \sum_{\alpha=1}^{\nu} m_\alpha v_\alpha (T, p)$.

$$V_{\text{mix}} = V_E - V_I = n_\gamma \bar{v}_\gamma (T, p, X_\beta) - \sum_{\alpha=1}^{\nu} n_\alpha \bar{v}_\alpha (T, p) \ , \tag{18.11}$$

where $\bar{v}_\alpha (T, p)$ denotes the specific volume of component α alone at (T, p) and $\bar{v}_\gamma (T, p, X_\beta)$ denotes the specific volume of *any* component γ in the mixture of composition X_β $(\beta = 1, \ldots, \nu)$ at (T, p). Note that in the mixed state all components are distributed over the volume of the mixture, V_E.

The change of volume is due to spatial hindrances or advantages on the molecular scale. For instance a mixture of 1 litre of water with 1 litre of ethanol (C_2H_5OH) yields a mixing volume of 1.93 litres.

When volume ratios are used to define the composition of a mixture, it must be clarified whether the volume of the component is related to the volume of the mixture, V_E, or the to the total volume of the components *before* mixing, V_I. A widely used measure for the alcohol content of beverages is "percent of alcohol by volume", defined as volume of the ethanol component *alone* over total volume of the *mixture*,

$$\%\text{ABV} = \frac{m_{alc} v_{alc} (T, p)}{V_E} \ .$$

Isothermal mixing of components might release or require heat, which must be transferred. The first law applied to the isothermal and isobaric mixing process yields

$$H_{\text{mix}} = H_E - H_I = \sum_{\alpha=1}^{\nu} n_\alpha \left[\bar{h}_\alpha (T, p, X_\beta) - \bar{h}_\alpha (T, p) \right] \ , \tag{18.12}$$

where H_{mix} is the heat that must be exchanged in order to keep the temperature T constant for the mixing process. Here, $\bar{h}_\alpha (T, p)$ is the molar enthalpy of component α alone at (T, p) and $\bar{h}_\alpha (T, p, X_\beta)$ is the molar enthalpy of the component α in a mixture of composition X_β $(\beta = 1, \ldots, \nu)$ at (T, p).

The enthalpy and internal energy are influenced by the interaction potential between molecules. In a pure substance, particles of type α interact only with particles of the same type. In a mixture, however, particles of type α are surrounded by different types of particles β $(\beta = 1, \ldots, \nu)$, which leads to different molecular interaction potentials, and thus a change in internal energy $\bar{u}_\alpha$ and enthalpy $\bar{h}_\alpha$ for the particles of type α as compared to the pure substance. A more detailed exploration of this will come in Sec. 22.9.

The entropy of mixing is computed in the same way, as

$$S_{\text{mix}} = S_E - S_I = \sum_{\alpha=1}^{\nu} n_\alpha \left[\bar{s}_\alpha (T, p, X_\beta) - \bar{s}_\alpha (T, p) \right] \ . \tag{18.13}$$

We shall discuss the enthalpy and entropy of mixing as we proceed. To simplify the discussion we shall ignore volume changes from now on.

18.6 Ideal Gas Mixtures

We first consider ideal gas mixtures, which are particularly simple. In ideal gases, due to the large average distance between particles, the potential energies between particles can be ignored against their microscopic kinetic energies. Then, the individual components are not affected by the presence of other components, there is no enthalpy of mixing, and the ideal gas law holds for the individual components and the mixture.[2]

The partial pressure of one component distributed over the mixture volume V is given by the ideal gas law

$$p_\alpha = \frac{m_\alpha}{V} R_\alpha T = \frac{n_\alpha}{V} \bar{R} T \,, \tag{18.14}$$

where $\bar{R} = 8.314 \frac{\text{kJ}}{\text{kmol K}}$ is the universal gas constant, and $R_\alpha = \bar{R}/M_\alpha$. The second form of the ideal gas law shows that the behavior of all ideal gases depends only on mole number, temperature and volume, but not on the type of gas.

The total pressure is just the sum of the partial pressures. For ideal gases, where the partial pressures are unaffected by the presence of other molecules, this is known as Dalton's law (John Dalton, 1766-1844),

$$p = \sum_{\alpha=1}^{\nu} p_\alpha = \sum_{\alpha=1}^{\nu} \frac{n_\alpha}{V} \bar{R} T = \frac{n}{V} \bar{R} T = \frac{m}{V} R T \,. \tag{18.15}$$

Here $R = \bar{R}/M$ is the gas constant for the mixture and M is the mixture's mean molar mass (18.5). Thus, with the proper molar masses, the components and the mixture obey the ideal gas law.

Division of the ideal gas laws for the component and the mixture shows that for ideal gas mixtures the pressure ratio equals the mole ratio,

$$\frac{p_\alpha}{p} = \frac{n_\alpha}{n} = X_\alpha \,. \tag{18.16}$$

In case that the ideal gas mixture is separated, so that each individual component is at the mixture pressure p and temperature T in its own volume V_α, the ideal gas law for the components reads

$$V_\alpha = \frac{n_\alpha \bar{R} T}{p} \quad \text{with} \quad \sum_{\alpha=1}^{\nu} V_\alpha = \sum_{\alpha=1}^{\nu} \frac{n_\alpha \bar{R} T}{p} = \frac{n \bar{R} T}{p} = V \,. \tag{18.17}$$

Here, V is the volume of the mixture in the mixed state. This is Amagat's law (Émile Amagat 1841-1915) which states that there will be no volume

[2] Recall the van der Waals equation $p = \frac{RT}{v-b} - \frac{a}{v^2}$, where the second term accounts for attractive forces between the molecules; for large specifc volume $\frac{a}{v^2} \to 0$ and $v - b \to v$.

change when ideal gases are mixed, $V_{\text{mix}} = 0$, as long as the pressures and temperatures before and after mixing are the same.

18.7 Energy, Enthalpy and Specific Heats for Ideal Gases

For ideal gases, all energies and enthalpies depend only on the temperature T. The potential energy between particles is not relevant due to their large average distance, and thus there is no energy of mixing ($U_{\text{mix}} = 0$), and no enthalpy of mixing, $H_{\text{mix}} = U_{\text{mix}} + pV_{\text{mix}} = 0$. With that, the energies and enthalpies of the components have the same temperature dependence in the mixture and in the pure state,

$$u(T, c_\alpha) = \sum_{\alpha=1}^{\nu} c_\alpha u_\alpha(T) \quad , \quad \bar{u}(T, X_\alpha) = \sum_{\alpha=1}^{\nu} X_\alpha \bar{u}_\alpha(T) \; , \tag{18.18}$$

$$h(T, c_\alpha) = \sum_{\alpha=1}^{\nu} c_\alpha h_\alpha(T) \quad , \quad \bar{h}(T, X_\alpha) = \sum_{\alpha=1}^{\nu} X_\alpha \bar{h}_\alpha(T) \; . \tag{18.19}$$

Specific heats for the components and the mixture follow from differentiation with respect to temperature,

$$c_v = \left(\frac{\partial u}{\partial T}\right)_{v,c_\beta} = \sum_{\alpha=1}^{\nu} c_\alpha c_{v,\alpha} \quad , \quad \bar{c}_v = \left(\frac{\partial \bar{u}}{\partial T}\right)_{v,X_\beta} = \sum_{\alpha=1}^{\nu} X_\alpha \bar{c}_{v,\alpha} \tag{18.20}$$

$$c_p = \left(\frac{\partial h}{\partial T}\right)_{p,c_\beta} = \sum_{\alpha=1}^{\nu} c_\alpha c_{p,\alpha} \quad , \quad \bar{c}_p = \left(\frac{\partial \bar{h}}{\partial T}\right)_{p,X_\beta} = \sum_{\alpha=1}^{\nu} X_\alpha \bar{c}_{p,\alpha} \tag{18.21}$$

Here, $c_{v,\alpha} = \frac{du_\alpha(T)}{dT}$ etc. are the specific heats for the pure components.

18.8 Entropy of Mixing for Ideal Gas

While volume, energy and enthalpy do not change when originally separated ideal gases at (T, p) (state I) are mixed, so that the mixture is at (T, p) (state II), the entropy does change. We compute the entropy for the two cases.

Component α fills the mixture volume V at temperature T and is at its partial pressure p_α. Its entropy is not affected by the presence of other components, it reads just as for a single component,

$$\bar{s}_\alpha(T, p_\alpha) = \bar{s}_\alpha^0(T) - \bar{R} \ln \frac{p_\alpha}{p_0} \; , \tag{18.22}$$

where $\bar{s}_\alpha^0(T)$ denotes the entropy of α alone at (T, p_0) and is tabulated.

With $X_\alpha = \frac{p_\alpha}{p}$ the molar entropy of one component in the mixture can be rewritten as

$$\bar{s}_\alpha(T, p, X_\beta) = \bar{s}^0_\alpha(T) - \bar{R}\ln X_\alpha - \bar{R}\ln\frac{p}{p_0}\,. \tag{18.23}$$

Accordingly, the entropy of component α depends only on the relative amount of α, while the composition of other components X_β ($\beta \neq \alpha$) does not play a role.

In the unmixed state, all components are at (T, p) (i.e., $X_\alpha = 1$), and in the mixed state the components are at (T, p_α). The corresponding total entropies follow from summation over all components as

$$S_{\mathrm{I}} = \sum_{\alpha=1}^{\nu} n_\alpha \left[\bar{s}^0_\alpha(T) - \bar{R}\ln\frac{p}{p_0}\right], \tag{18.24}$$

$$S_{\mathrm{II}} = \sum_{\alpha=1}^{\nu} n_\alpha \left[\bar{s}^0_\alpha(T) - \bar{R}\ln X_\alpha - \bar{R}\ln\frac{p}{p_0}\right]. \tag{18.25}$$

The entropy of mixing is just the difference,

$$S_{\mathrm{mix}} = S_{\mathrm{II}} - S_{\mathrm{I}} = -\bar{R}\sum_{\alpha=1}^{\nu} n_\alpha \ln\frac{p_\alpha}{p} = -\bar{R}\sum_{\alpha=1}^{\nu} n_\alpha \ln X_\alpha > 0\,. \tag{18.26}$$

The entropy of mixing is positive, since $X_\alpha \leq 1$. Thus, entropy grows in mixing, and we conclude that mixing is an irreversible process, with a work loss. We shall later discuss how work could be obtained by reversible mixing.

The molar entropy of a mixture follows from (18.25) by division with the mole number n as

$$\bar{s} = \sum X_\alpha \bar{s}_\alpha(T, p, X_\beta) = \sum X_\alpha\left(\bar{s}^0_\alpha(T) - \bar{R}\ln X_\alpha - \bar{R}\ln\frac{p}{p_0}\right). \tag{18.27}$$

18.9 Gibbs Paradox

We consider two different ideal gases in a container of Volume V, halved by a slider. Both gases are at the same temperature and pressure (T, p), and their mole numbers are equal $n_1 = n_2 = \frac{n}{2}$. When the slider is pulled out, the two gases mix while pressure and temperature remain unchanged. The partial pressures for both gases are

$$p_1 = p_2 = \frac{n_1 RT}{V} = \frac{n}{2}\frac{RT}{V} = \frac{p}{2}\,. \tag{18.28}$$

In the final equilibrium state, both gases are evenly distributed throughout the container. Entropy has grown, and the entropy of mixing for this process is

$$S_{\text{mix}} = -\bar{R}\sum_{\alpha=1}^{2} n_\alpha \ln \frac{p_\alpha}{p} = n\bar{R}\ln 2 > 0\,. \tag{18.29}$$

When the slider is pushed back to separate the two parts of the container, the state is quite different from the initial state, since both parts now contain mixture. The difference between the two states becomes manifest in the positive value of the entropy of mixing.

Let us now consider that both gases are identical. The slider can be pulled out and pushed back at any time, without any change in the state of the gas, which is always the same gas at (T, p) in both compartments. Yet, the expression for the entropy of mixing, Eq. (18.29), predicts an entropy increase. This increase should not be present for a single gas, and thus the prediction of a positive entropy of mixing is considered to be a paradox, known as Gibbs paradox.

The paradox can be resolved by noting that the particles of just one type of gas cannot be distinguished. In particular, we cannot recognize whether an arbitrary particle picked from the gas was originally in a particular partition of the container. Indeed, could we mark the particles in one of the partitions at the initial time (before the slider is pulled out), the final state would show that particles that were on one side initially are now distributed over the whole container. In short, the entropy of mixing can only be computed for the mixing of distinguishable species.

18.10 Example: Isentropic Expansion through a Nozzle

We consider the isentropic expansion of a mixture of oxygen ($X_{O_2} = 0.2$) and carbon dioxide ($X_{CO_2} = 1 - X_{O_2} = 0.8$) entering a nozzle at $\mathcal{V}_1 = 25\frac{\text{m}}{\text{s}}$, $p_1 = 6\,\text{bar}$, $T_1 = 1200\,\text{K}$ and leaving at $p_2 = 1\,\text{bar}$. We ask for the temperature T_2 and the velocity $\mathcal{V}_2$ at the nozzle exit.

The molar mass of the mixture is $M = X_{O_2}M_{O_2} + X_{CO_2}M_{CO_2} = 41.6\frac{\text{kg}}{\text{kmol}}$. Since the process is reversible and adiabatic, it is isentropic, that is the molar entropy (18.27) of the mixture stays constant,

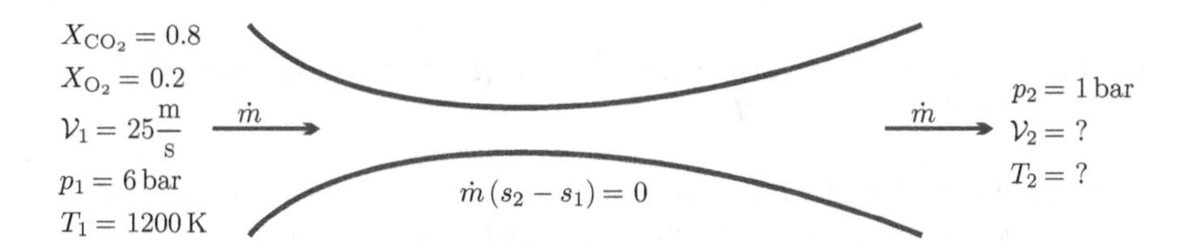

Fig. 18.2 Isentropic expansion in nozzle

$$\bar{s}_1 = \bar{s}_2 \, .$$

Since the mixing state does not change, this results in

$$X_{O_2}\left[\bar{s}^0_{O_2}(T_1) - \bar{R}\ln X_{O_2} - \bar{R}\ln\frac{p_1}{p_0}\right] + X_{CO_2}\left[\bar{s}^0_{CO_2}(T_1) - \bar{R}\ln X_{CO_2} - \bar{R}\ln\frac{p_1}{p_0}\right] = X_{O_2}\left[\bar{s}^0_{O_2}(T_2) - \bar{R}\ln X_{O_2} - \bar{R}\ln\frac{p_2}{p_0}\right] + X_{CO_2}\left[\bar{s}^0_{CO_2}(T_2) - \bar{R}\ln X_{CO_2} - \bar{R}\ln\frac{p_2}{p_0}\right]$$

or, after simplifying,

$$X_{O_2}\bar{s}^0_{O_2}(T_2) + X_{CO_2}\bar{s}^0_{CO_2}(T_2) = X_{O_2}\bar{s}^0_{O_2}(T_1) + X_{CO_2}\bar{s}^0_{CO_2}(T_1) + \bar{R}\ln\frac{p_2}{p_1} \, .$$

The right hand side can be computed from tabulated data as

$$A(T_2) = X_{O_2}\bar{s}^0_{O_2}(T_2) + X_{CO_2}\bar{s}^0_{CO_2}(T_2) = 259.20\frac{\text{kJ}}{\text{kmol K}} \, .$$

The temperature T_2 must be found from trial and error, and interpolation. We find $A(890\,\text{K}) = 258.90\frac{\text{kJ}}{\text{kmol K}}$ and $A(900\,\text{K}) = 259.46\frac{\text{kJ}}{\text{kmol K}}$ and conclude from interpolation that $T_2 = 895.4\,\text{K}$.

The velocity follows from the first law for the nozzle as

$$\mathcal{V}_2 = \sqrt{2(h_1 - h_2) + \mathcal{V}_1^2} \, .$$

We have

$$h_1 - h_2 = \frac{1}{M}\left[X_{O_2}\left(\bar{h}_{O_2}(T_1) - \bar{h}_{O_2}(T_2)\right) + X_{CO_2}\left(\bar{h}_{CO_2}(T_1) - \bar{h}_{CO_2}(T_2)\right)\right] ,$$

and thus $h_1 - h_2 = 372.7\frac{\text{kJ}}{\text{kg}}$ and $\mathcal{V}_2 = 611\frac{\text{m}}{\text{s}}$.

18.11 Example: Isochoric Mixing of Two Gases at Different p, T

We consider an adiabatic rigid container of volume V which is separated by a membrane. The two parts of the container hold different ideal gases, with the given data $\left(m_1, T_1^I, p_1^I, M_1, c_{v,1}\right)$ and $\left(m_2, T_2^I, p_2^I, M_2, c_{v,2}\right)$, the superscript I indicates the initial state, the final state will be denoted with superscript E. To simplify the problem we assume the specific heats to be constant.

The dividing membrane is removed, and we ask for the final pressure and temperature, and for the amount of entropy generated.

The volumes of the compartments and of the container are

$$V_1 = \frac{n_1 \bar{R} T_1^I}{p_1^I} \quad , \quad V_2 = \frac{n_2 \bar{R} T_2^I}{p_2^I} \quad , \quad V = V_1 + V_2 \; .$$

Since the container is adiabatic and rigid, $V = const.$, the first law applied to the volume V gives that the total internal energy is constant, $U^I = U^E$, so that the final temperature is

$$T^E = \frac{m_1 c_{v,1} T_1^I + m_2 c_{v,2} T_2^I}{m_1 c_{v,1} + m_2 c_{v,2}} = \frac{X_1 \bar{c}_{v,1} T_1^I + X_2 \bar{c}_{v,2} T_2^I}{X_1 \bar{c}_{v,1} + X_2 \bar{c}_{v,2}} \; .$$

The final pressure follows from the ideal gas law for the mixture as

$$p^E = \frac{n \bar{R} T^E}{V} = \frac{1}{\frac{X_1 T_1^I}{p_1^I T^E} + \frac{X_2 T_2^I}{p_2^I T^E}} \; .$$

The change in entropy of the two gases is

$$\Delta S = n_1 \left[\bar{c}_{p,1} \ln \frac{T^E}{T_1^I} - \bar{R} \ln \frac{X_1 p^E}{p_1^I} \right] + n_2 \left[\bar{c}_{p,2} \ln \frac{T^E}{T_2^I} - \bar{R} \ln \frac{X_2 p^E}{p_2^I} \right] ,$$

which can be split into three contributions that refer to the entropy change due to heat exchange, pressure equilibration and mixing,

$$\begin{aligned} \Delta S = n &\left[X_1 \bar{c}_{p,1} \ln \frac{T^E}{T_1^I} + X_2 \bar{c}_{p,2} \ln \frac{T^E}{T_2^I} \right] \\ &+ n\bar{R} \left[X_1 \ln \frac{p_1^I}{p^E} + X_2 \ln \frac{p_2^I}{p^E} \right] - n\bar{R} \left[X_1 \ln X_1 + X_2 \ln X_2 \right] \; . \end{aligned}$$

18.12 Ideal Mixtures

Ideal mixtures are defined as mixtures with vanishing volume and heat of mixing, and an entropy of mixing just as that of the ideal gas, i.e.,

$$V_{\text{mix}} = H_{\text{mix}} = 0 \quad , \quad S_{\text{mix}} = -\sum_\alpha n_\alpha \bar{R} \ln X_\alpha \; . \qquad (18.30)$$

Ideal gas mixtures are a special case of ideal mixtures. In particular, the theory of ideal mixtures can be applied to dilute liquid solutions.

As for the ideal gas, the first two conditions $(18.30)_{1,2}$ for ideal mixtures state that mixing does not affect energy. This would be the case if intermolecular potential of the various components are equal, or at least rather similar, so that the potential energy of a pair of identical molecules (α-α-pair) is similar to that of a dissimilar pair (α-β-pair).

The entropy of mixing warrants more detailed discussion. It is best motivated through Boltzmann's microscopic interpretation of entropy as discussed earlier, in Sec. 4.14. The microscopic definition of entropy reads

$$S = k \ln \Omega \,, \tag{18.31}$$

where $k = \bar{R}/A = 1.3806 \times 10^{-23} \frac{\mathrm{J}}{\mathrm{K}}$ is Boltzmann's constant, and Ω is the number of microscopic realizations of a given macroscopic state. For instance, for an ideal gas, simply speaking, a macroscopic state is given by the values of temperature, pressure, and velocity of the gas, while a microscopic state is given by the location and velocities of the gas particles. The entropy can be computed from (18.31) for rather complex systems, e.g., polymers, but for many systems the evaluation of the equation becomes too cumbersome to be done analytically.

A macroscopic state is for instance given by mass, volume, and temperature of a sample of substance. While the macroscopic state is maintained, due to the thermal motion of the particles the system runs through a succession of its accessible microscopic states. At standard condition, the average speed of a gas molecule is of the order of the speed of sound ($\sim 350 \frac{\mathrm{m}}{\mathrm{s}}$), and it undergoes about 10^{10} collisions per second. Accordingly, the system goes through a vast number of microstates per second.

While the full evaluation of $S = k \ln \Omega$ requires consideration of the full microstate, i.e., locations and velocities of particles, we can consider the locations alone to compute the entropy of mixing. We study a container with $N = \sum_\alpha N_\alpha$ particles of different types $\alpha = 1, \ldots, \nu$. We divide the container into N cells of equal size, and assume that there can be only one particle per cell. Since we cannot distinguish between different particles of the same kind, the number of possibilities to distribute the $\sum_\alpha N_\alpha$ particles over the N cells is

$$\Omega = \frac{N!}{\prod N_\alpha!} \,. \tag{18.32}$$

The corresponding macrostate for a mixed state is simply to have $N = \sum_\alpha N_\alpha$ particles in the container.

Before we further evaluate this expression, we study the example of just four particles of two different types, two of each type. In the state before mixing, a wall separates the different components into the configuration $\left[1\;1 \;||\; 2\;2\right]$. There is only one microscopic realization of this configuration, since the particles cannot pass the wall. As soon as the wall is removed, the particles can exchange positions due to thermal motion, and thus access a larger number of microstates. Equation (18.32) gives $\Omega = 6$; the states are

$$\begin{bmatrix} 1\,1\,2\,2 \\ 2\,2\,1\,1 \\ 1\,2\,1\,2 \\ 2\,1\,1\,2 \\ 1\,2\,2\,1 \\ 2\,1\,2\,1 \end{bmatrix} .$$

Thus, the mixed state, where the wall is removed has more microscopic realizations than the unmixed state, and the higher entropy.

Note that the original configuration forms one of the accessible microstates of the unrestrained system as well. The probability that this state is $1/\Omega$, which is indeed the probability that *any* particular microstate is assumed. For real systems, the particle numbers are huge, and the number of possible configurations is enormous. The probability to find the original configuration, which was maintained by the wall before it was removed, is negligible, and most microstates will be mixed—thus mixed states will be observed.

We proceed to evaluate (18.31) with (18.32) for large numbers N_α. By means of Stirling's formula $\ln N! \simeq N \ln N - N$, which holds for large N, we find

$$S_{\text{mix}} = k \ln \frac{N!}{\prod N_\alpha!} = k \left(N \ln N - \sum_\alpha N_\alpha \ln N_\alpha \right)$$
$$= k \sum_\alpha \left(N_\alpha \ln N - N_\alpha \ln N_\alpha \right) . \qquad (18.33)$$

When we introduce the mole number $n_\alpha = N_\alpha / A$, the mole fraction $X_\alpha = \frac{n_\alpha}{n} = \frac{N_\alpha}{N}$, and $\bar{R} = kA$, we recover $(18.30)_3$,

$$S_{\text{mix}} = - \sum_\alpha n_\alpha \bar{R} \ln X_\alpha . \qquad (18.34)$$

In thermodynamic systems the number of particles is normally very large, with 6.022×10^{23} particles per mole. For example in an equimolar binary mixture with $N_1 = N_2 = 10^{23}$ one finds $\ln \Omega = 1.386 \times 10^{23}$, $S_{\text{mix}} = 1.9139 \frac{\text{J}}{\text{K}}$. The probability to recover the initial unmixed state as one of the microstates is incredibly small at $\frac{1}{\Omega} = 4^{-10^{23}}$: spontaneous unmixing is not impossible, but incredibly unlikely, and cannot be expected to be observed in the lifetime of the universe. Return to Sec. 4.14 for additional discussion on the microscopic interpretation of entropy.

The computation of the entropy of mixing in this section relied on the assumption that there are as many location cells as particles, which is appropriate for simple liquids. The reader might wonder how the entropy of mixing for an ideal gas comes about, for which there are far more cells than particles. We briefly run through the necessary arguments: To deal with an ideal gas mixture, we consider empty cells as an additional species with count

N_e. With $N = N_g + N_e$ as the total number of cells and $N_g = \sum_\alpha N_\alpha$ as number of gas particles, we have one additional term:

$$\begin{aligned} S_{\text{mix,id.gas}} &= k \ln \frac{N!}{N_e! \prod N_\alpha!} \\ &= k \left[\sum_\alpha \left(N_\alpha \ln N - N_\alpha \ln N_\alpha \right) + \left(N_e \ln N - N_e \ln N_e \right) \right] \end{aligned} \tag{18.35}$$

With the mole fraction of gas species $X_\alpha = N_\alpha / N$, this can be simplified to

$$S_{\text{mix,id.gas}} = -k \sum_\alpha N_\alpha \ln X_\alpha + k N_g \left[\ln \frac{N}{N_g} - \frac{N - N_g}{N_g} \ln \frac{N - N_g}{N} \right] . \tag{18.36}$$

The first term is the entropy of mixing as discussed before. We proceed with the discussion of the second term. The total volume filled by the gas is $V = N v_c$, where v_c is the cell volume. Therefore $\ln \frac{N}{N_g} = \ln \frac{\bar{v}}{\bar{v}_0}$, where $\bar{v}$ is mole volume and $\bar{v}_0 = v_c A$ is a reference mole volume. For the ideal gas, there are far more cells than gas particles, so that $N \gg N_g$, and in the limit we find $\lim_{\frac{N_g}{N} \to 0} \frac{N - N_g}{N_g} \ln \frac{N - N_g}{N} = -1$. Hence, with $k = \bar{R}/A$, we find three contributions to ideal gas entropy, namely entropy of mixing, the well-known volume dependence of entropy, and a constant,

$$S_{\text{id.gas}} = \bar{R} n_g \left[- \sum_\alpha X_\alpha \ln X_\alpha + \ln \frac{\bar{v}}{\bar{v}_0} + 1 \right] . \tag{18.37}$$

18.13 Entropy of Mixing and Separation Work

We consider mixing and separation of ideal mixtures at constant temperature T. The combined first and second law for a closed system at constant temperature T reads

$$\dot{W} = -T \dot{S}_{gen} - \frac{dU - TS}{dt} , \tag{18.38}$$

or, after integration over the duration of the process,

$$W_{12} + T S_{gen} = - \left(U_2 - U_1 \right) + T \left(S_2 - S_1 \right) , \tag{18.39}$$

where $S_{gen} = \int_{t_1}^{t_2} \dot{S}_{gen} dt$ is the total entropy generation for the process.

We consider mixing first, where state 1 is the unmixed state, and state 2 is the mixed state. Then, since $U_{\text{mix}} = 0$ for an ideal mixture,

$$W_{12} + T S_{gen} = T S_{\text{mix}} - U_{\text{mix}} = T S_{\text{mix}} > 0 . \tag{18.40}$$

When the mixing is fully irreversible, no work is drawn, $W_{12} = 0$, and the entropy generation is just the entropy of mixing, $S_{gen} = S_{\text{mix}}$. However, the equation shows that it is possible to generate the work

$$W_{12} = TS_{\text{mix}} - TS_{gen} > 0 \,. \tag{18.41}$$

In a real device, the irreversibilities of the process diminish the work by TS_{gen}. The maximum work is obtained from a fully reversible process as

$$W_{\text{mix}}^{rev} = TS_{\text{mix}} \,. \tag{18.42}$$

Note that more work can be produced at higher temperature.

Now we consider separation, where state 1 is the mixed state, and state 2 is the unmixed state. Then, again with $U_{\text{mix}} = 0$,

$$W_{12} = -TS_{gen} - TS_{\text{mix}} < 0 \,. \tag{18.43}$$

Work is required for separation ($TS_{\text{mix}} > 0$) and to overcome irreversibilities in the device ($TS_{gen} > 0$). The minimum separation work is obtained or a fully reversible process, where $S_{gen} = 0$, as

$$W_{\text{sep}}^{rev} = -TS_{\text{mix}} \,. \tag{18.44}$$

The separation work is directly proportional to the entropy of mixing, less work is required for separation at lower temperatures.

Section 21 will present a closer look at desalination plants, which separate salt from water, and osmotic power plants, which use mixing for power generation. 21.5.

18.14 Non-ideal Mixtures

The equilibrium condition for a mixture at given pressure p and temperature T is that the Gibbs free energy assumes a minimum. We denote the Gibbs free energies of the unmixed and the mixed states by G_{unmixed}, G_{mixed}, respectively. Both are related as

$$G_{\text{mixed}} = G_{\text{unmixed}} + G_{\text{mix}} \;\text{ with }\; G_{\text{mix}} = H_{\text{mix}} - TS_{\text{mix}} \,. \tag{18.45}$$

The equilibrium state will be the mixed state for negative Gibbs free energy of mixing, $G_{\text{mix}} < 0$, so that $G_{\text{mixed}} < G_{\text{unmixed}}$, but it will be the unmixed state, if $G_{\text{mix}} > 0$, so that $G_{\text{unmixed}} < G_{\text{mixed}}$.

For an ideal mixture we have $H_{\text{mix}} = 0$, hence $G_{\text{mix}} = -TS_{\text{mix}} < 0$. Accordingly, ideal mixtures will assume a mixed equilibrium state.

The Gibbs free energy of mixing will only be positive if the enthalpy of mixing is large, that is for $H_{\text{mix}} > TS_{\text{mix}}$. Large enthalpy of mixing is observed when the intermolecular forces between like particles are much larger than those between unlike particles.

Due to its dipole structure, water has strong polar bonds between its molecules, while oil is non-polar. Breaking up the water bonds to form water-oil pairs instead of water-water pairs requires energy, and hence the enthalpy of mixing, H_{mix}, is positive. From the fact that oil separates from water, we can conclude that the energetic effect exceeds the entropic effect, that is $H_{\text{mix}} > TS_{\text{mix}}$.

Table salt (sodium chloride, NaCl) dissociates in water into charged ions, Na^+ and Cl^-, which have energetic bonds with water. Nevertheless, for water-salt solutions, the enthalpy of mixing is positive, but smaller than the entropic term TS_{mix}. In dilute solutions, the energetic interaction between salt ions and water molecules can be ignored, and the solution can be approximated as an ideal mixture. Dissociation requires shielding of salt ions by the polar water molecules, where several water molecules are shielding one salt ion. If all water molecules are used for the hydrogen shells, no additional salt ions can be dissolved, the solution becomes saturated.

Problems

18.1. Mixture Properties

An ideal gas mixture consists of 6 kg of O_2, 5 kg of N_2 and 12 kg of CO_2.

1. Determine the mass and mole fraction of each component.
2. Determine the average molar mass and the gas constant of the mixture.
3. Compute the partial pressures of all components when the total pressure is 2 bar.
4. Compute the entropy of mixing between mixed and unmixed state.

18.2. Heating of Mixture

A piston-cylinder device contains a mixture of 1 kg of H_2 and 2 kg of N_2, initially at 200 kPa and 280 K. The mixture is heated at constant pressure until the volume is three times the initial volume. Determine the temperature of the final state, the total heat transferred, and the change of entropy of the mixture. Use tabulated property data.

18.3. Isobaric Cooling of Mixture

A piston-cylinder device contains a mixture of 0.75 kg of N_2 and 2 kg of CO at 300 kPa and 860 K. Heat is now transferred from the mixture at constant pressure until the volume is one third of the initial volume. Determine the heat transfer, the work done, and the change of entropy. Use tabulated property data.

18.4. Mixing of H_2 and CO_2

An adiabatic rigid tank is divided into two parts. One part contains 4.4 kg of CO_2 at 25 °C and 200 kPa, and the other part contains 1 kg of H_2 at 80 °C and 400 kPa. After the divider is removed, the gases mix and the mixture assume a final equilibrium state.

1. Determine the equilibrium temperature and the equilibrium pressure.
2. Compute the entropy generated in the process.

Assume constant specific heats at 300 K for both gases.

18.5. Compressor
A compressor draws an ideal gas mixture (molar composition 50% CO_2, 33.3% CO, 16.7% O_2) at 37 °C, 1 bar, 50 $\frac{m}{s}$. The mass flow rate is 5 $\frac{kg}{s}$, and the exit state is 237 °C, 80 $\frac{m}{s}$. The compressor is not adiabatic, heat is lost to the surroundings at a rate of 3 $\frac{kJ}{kg}$ of mixture flowing.

1. Determine the power to run the compressor.
2. Assume that the compression is polytropic, and determine the polytropic exponent. Then find the exit pressure.

18.6. Adiabatic Turbine
The combustion product in a gas turbine system consists of nitrogen, oxygen, carbon dioxide and water with the following mole flow rates: $\dot{n}_{N_2} = 1.7\frac{kmol}{s}$, $\dot{n}_{O_2} = 0.15\frac{kmol}{s}$, $\dot{n}_{H_2O} = 0.25\frac{kmol}{s}$, $\dot{n}_{CO_2} = 0.2\frac{kmol}{s}$.

At turbine inlet, the pressure is 12 bar, and the temperature is 1500 K. Determine the power production of the turbine in isentropic expansion to an external pressure of 1 bar.

Hint: To estimate exit temperature for trial and error, you might want to look at expansion of air first.

18.7. Mixing and Separation
1 kg of argon at 10 bar, 400 K and 0.5 kg of xenon at 10 bar, 1000 K are isobarically and adiabatically mixed in a closed piston-cylinder system.

1. Determine the equilibrium temperature of the mixture.
2. Determine the initial system volume, and the volume change between initial and final state. Explain the result.
3. Find the reversible work required for separation of the mixture.

Remark: Both gases are monatomic, with $M_{Ar} = 39.95\frac{kg}{kmol}$, and $M_{Xe} = 131.3\frac{kg}{kmol}$.

18.8. Mixing of Argon and Helium
An adiabatic cylinder is closed by a moveable piston. The cylinder contains 4litres of argon at 150 kPa and a rubber balloon which contains 1litre of helium at 3 bar. Both gases are initially at a temperature of 25 °C, and the piston rests due to its own weight. Assume that the pressure volume characteristic of the balloon is of the form $\Delta p = a\left(V_B - V_0\right)^2$ where Δp is the pressure difference between inside and outside, V_B is the actual volume, the reference volume V_0 is 0.25litres, and a is a material constant. This implies

that the balloon shell stores some energy when stretched. As soon as the balloon has reached the volume V_0, there are no further stresses in the balloon, and the balloon shell will just collapse. For simplicity, ignore the thermal mass of the balloon.

A small hole opens in the balloon through which all helium escapes; in the final equilibrium state the gases are mixed.

1. Determine the pressure, temperature and cylinder volume in the final equilibrium state.
2. Compute the entropy changes for both gases, and the total entropy generated in the process. Compare to the entropy of mixing, S_{mix}, and discuss the difference.

Chapter 19
Psychrometrics

19.1 Characterization of Moist Air

Psychros and *metro* are Greek words meaning *cold* and *measure*, respectively, and psychrometrics describes moist air: mixtures of air and water vapor with possibly some liquid water present as well. Psychrometrics is most important for designing proper air conditioning systems for buildings, where the air should be not too dry or moist, to make the environment comfortable; moreover, moisture buildup at (or in!) walls must be prevented.

This chapter describes how to characterize and analyze moist air mixtures, and discusses basic processes for moisturizing and dehumidification in HVAC systems (Heating-Ventilating-Air-Conditioning).

We consider air-vapor mixtures at temperature T and pressure $p = p_\mathrm{a} + p_\mathrm{v}$. Air behaves as an ideal gas, hence the partial pressure of air, p_a, follows the ideal gas law. At the relevant temperatures, the partial pressure of the vapor in moist air, p_v, is so low that the vapor can be described as an ideal gas as well.

The vapor pressure cannot exceed the saturation pressure $p_\mathrm{sat}(T)$. If $p_\mathrm{v} < p_\mathrm{sat}(T)$ there is no liquid water present, but if $p_\mathrm{v} = p_\mathrm{sat}(T)$ some liquid water will be present either in form of droplets (fog), or as an larger amount on the bottom. Figure 19.1 illustrates undersaturated moist air as a mixture of air and vapor, and saturated moist air as a mixture of air, vapor, and liquid water.

Since the saturation pressure $p_\mathrm{sat}(T)$ increases with temperature, warm air can hold more water vapor than cold air. Cooling of moist air can lead to condensation of water, e.g., on cold bottles, or on eyeglasses when one enters the warm and humid air of a house coming in from a cold winter environment.

The humidity ratio ω, also known as specific humidity, is defined as the ratio of vapor and air mass in a sample of moist air of the volume V,

$$\omega = \frac{m_\mathrm{v}}{m_\mathrm{a}} = \frac{\frac{p_\mathrm{v} V}{R_\mathrm{v} T}}{\frac{p_\mathrm{a} V}{R_\mathrm{a} T}} = \frac{M_\mathrm{v}}{M_\mathrm{a}} \frac{p_\mathrm{v}}{p_\mathrm{a}} = 0.622 \frac{p_\mathrm{v}}{p - p_\mathrm{v}} , \tag{19.1}$$

H. Struchtrup, *Thermodynamics and Energy Conversion*,
DOI: 10.1007/978-3-662-43715-5_19,

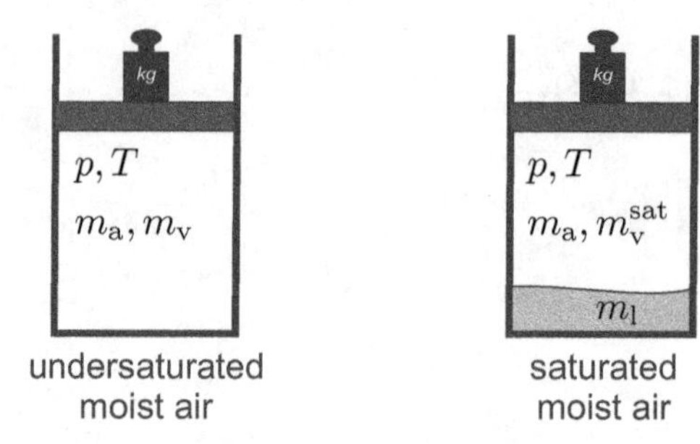

Fig. 19.1 Undersaturated and saturated moist air

where we have used that $\frac{M_v}{M_a} = 0.622$. The humidity ratio for the saturated state, where $p_v = p_{sat}(T)$, is a function of temperature and pressure

$$\omega_{sat}(T,p) = 0.622 \frac{p_{sat}(T)}{p - p_{sat}(T)} . \tag{19.2}$$

The relative humidity ϕ is defined as the ratio between the actual mole fraction of vapor in the sample, and the vapor mole fraction in the saturated state,

$$\phi = \frac{X_v}{X_{sat}(T)} = \frac{p_v}{p_{sat}(T)} , \tag{19.3}$$

where it was used that, for ideal gas mixtures, $X_\alpha = \frac{p_\alpha}{p}$. Whether we perceive moist air as comfortable or not depends on the temperature and the relative humidity. In a dry environment, the human body loses a lot of moisture to evaporation from the skin and in breathing; one must drink a lot to replenish the moisture. In deserts, it helps to cover the body loosely with cloth, to prevent exposure of skin to the dry air, thus limiting evaporation from the skin. In high humidity, the air cannot accept more vapor, and thus sweat does not evaporate which results in difficulty to regulate the body temperature. For buildings, a relative humidity of $\phi \simeq 0.6$ is providing the most pleasant environment.

The enthalpy of a moist air sample is

$$H = H_a + H_v = m_a h_a + m_v h_v , \tag{19.4}$$

where $h_a(T)$ and $h_v = h_g(T)$ are the specific enthalpies of air and water vapor. Since the amount of vapor changes due to evaporation and condensation, it is convenient to base the specific enthalpy of moist air on the dry air mass and we write[1]

[1] The subscript $1+\omega$ serves to distinguish a specific property per unit mass of dry air, which corresponds to $1+\omega$ unit masses of moist air. This notation is uncommon in the North-American literature, but it is useful to avoid confusion with proper specific enthalpies.

$$h_{1+\omega}(T,\omega) = \frac{H}{m_a} = h_a(T) + \omega h_v(T) \ . \tag{19.5}$$

Since at these low pressures air and vapor are ideal gases, their enthalpies depend only on temperature T, while the enthalpy of moist air, $h_{1+\omega}$, depends also on the humidity ratio ω.

The specific volume of moist air per unit mass of dry air can be computed from the Amagat model as, again with $\frac{M_v}{M_a} = 0.622$,

$$v_{1+\omega} = \frac{V_a + V_v}{m_a} = \frac{\frac{m_a R_a T}{p} + \frac{m_v R_v T}{p}}{m_a} = \left(1 + \frac{\omega}{0.622}\right) \frac{R_a T}{p} \ . \tag{19.6}$$

19.2 Dewpoint

In isobaric cooling, see Fig. 19.2, the partial pressures of air and vapor stay constant as long as no water condenses. The dewpoint temperature T_d is defined as the temperature at which vapor starts to condense when moist air is isobarically cooled,[2]

$$p_{sat}(T_d) = p_v \quad \text{or} \quad T_d = T_{sat}(p_v) \ . \tag{19.7}$$

Figure 19.3 illustrates the cooling and condensation process for water in air in a T-s-diagram. Initially, the vapor is at state 1. No water condenses as the vapor is cooled until it reaches the dewpoint (state d). In the final state, the air is mixed with saturated vapor (state 2) and saturated liquid (state 3).

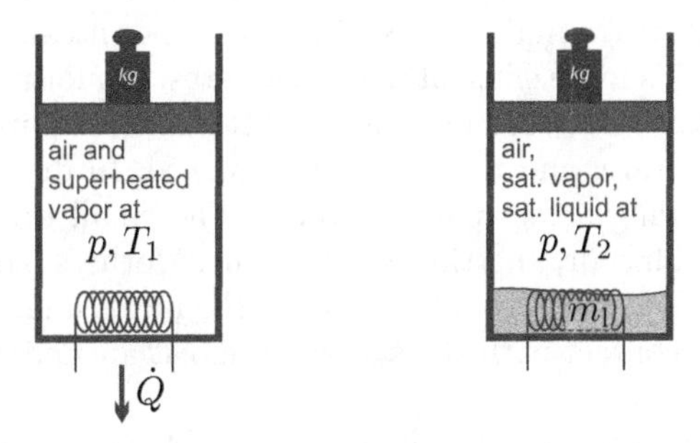

Fig. 19.2 Isobaric cooling of moist air

[2] Due to the presence of air, the saturation pressure will be slightly different for vapor in air as compared to water alone (see the discussion in later chapters). The difference is small, however, and can be ignored.

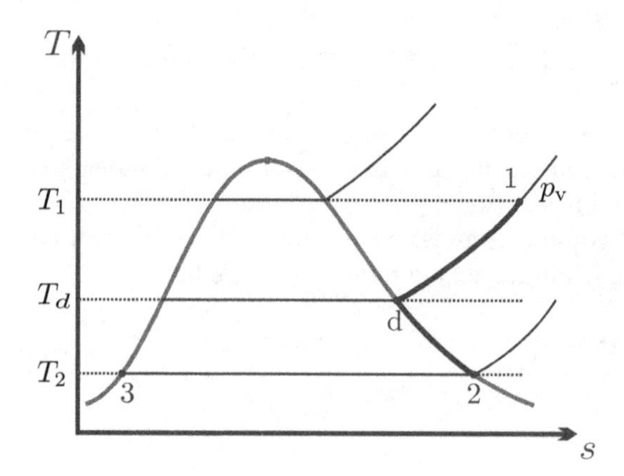

Fig. 19.3 T-s-diagram of water for the isobaric cooling of moist air

19.3 Adiabatic Saturation and Wet-Bulb Temperature

When moist air of temperature T and humidity ratio $\omega < \omega_{\text{sat}}$ flows over a surface of water, some water will evaporate and the humidity of the air will increase. Evaporation requires the heat of evaporation h_{fg}, which is drawn from the air and the liquid, which therefore cool down as liquid evaporates. If there is sufficient contact between air and water, water will evaporate until the air will finally be saturated.

This effect is an important part of our life: When the body gets hot, humans sweat, the sweat evaporates by drawing the heat of vaporization from the body. The dryer the surrounding air is, the more vapor it can accept, and thus cooling by sweating is more efficient in dry climates. In moist climates, e.g., in tropical rainforests, the air can only accept little or no additional vapor, the sweat cannot evaporate, and no cooling is achieved. In dry climates, patios are cooled by spraying a fine mist of water. The small droplets evaporate immediately in the dry air, and this cools the air. Mothers blow air over their babies' food to cool it, good restaurants serve the meals under covers, so that the food is only in contact with the saturated moist air under the cover, and remains hot.

We study the system depicted in Fig. 19.4. Moist air at (T, p, ω) flows through a wetted porous material, which provides a large contact surface between air and liquid water. Pressure losses in the flow through the porous material are ignored in the following. The air leaves in saturated state at the so-called wet-bulb temperature T_{wb}. We assume a steady state process under adiabatic conditions, that is no heat is added to the air flow from the exterior, and we assume that the make-up water flow from the reservoir is at the wet-bulb temperature T_{wb}. The first law for the system then reduces to the equality between incoming and outgoing enthalpy flows, $\sum_{in} \dot{m}_\alpha h_\alpha =$

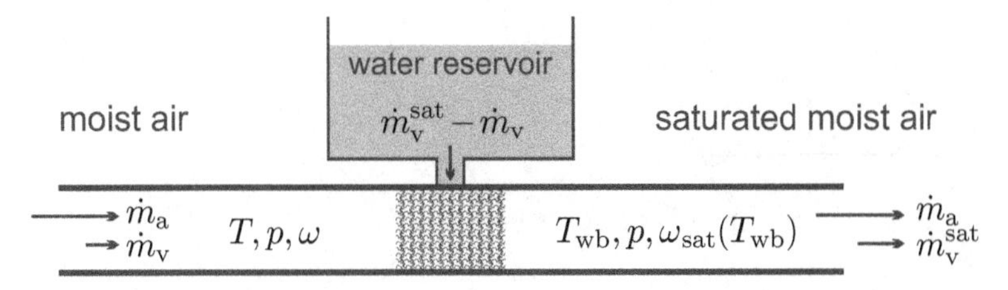

Fig. 19.4 Saturation of a moist air flow

$\sum_{out} \dot{m}_\alpha h_\alpha$, or, in detail,

$$\dot{m}_a h_a(T) + \dot{m}_v h_v(T) + \left(\dot{m}_v^{sat} - \dot{m}_v\right) h_f(T_{wb}) = \dot{m}_a h_a(T_{wb}) + \dot{m}_v^{sat} h_v(T_{wb}) \ . \tag{19.8}$$

With $\omega = \frac{\dot{m}_v}{\dot{m}_a}$ and $\omega_{sat}(T_{wb}, p) = \frac{\dot{m}_v^{sat}}{\dot{m}_a} = 0.622 \frac{p_{sat}(T_{wb})}{p - p_{sat}(T_{wb})}$ we find an equation for the humidity ratio of the incoming moist air,

$$\omega(T, T_{wb}, p) = \frac{h_a(T_{wb}) - h_a(T) + \omega_{sat}(T_{wb}, p)\left[h_v(T_{wb}) - h_f(T_{wb})\right]}{h_v(T) - h_f(T_{wb})} \ . \tag{19.9}$$

The temperatures of the incoming air, T, and of the wet bulb, T_{wb}, can be measured easily, and this measurement allows to determine the humidity ratio ω from the above equation.

Indeed, for the measurement of the wet-bulb temperature it is sufficient to cover a thermometer with a wet cloth, and expose it to air flow. After a while (before the cloth has dried, of course) a steady state is reached, and the thermometer shows the wet-bulb temperature of the air flow. For the measurement in standing air (e.g. in a room), the wet thermometer has to be moved, and one uses a sling psychrometer: two thermometers, one dry, one wet, on a handle are rotated in the air.

19.4 Psychrometric Chart

From (19.9) it is evident that determining humidity from the measurement of dry- and wet-bulb temperatures involves some work for finding property data. Instead of working with property tables it is practical to use a psychrometric chart, from which all interesting data for moist air can be extracted. Through the humidity ratio, the equations depend on the total pressure, p, and one should take care to use the proper chart. Small daily pressure changes at a location do not affect the results much, but one will have to account for changes of environmental pressure with height. Figure 19.5 shows a chart for $p = 0.8$ bar which would be appropriate for a location at a height of about 2000 m above sea level.

The psychrometric chart has the dry-bulb temperature T on the abscissa and the humidity ratio ω on the ordinate. The diagram shows lines of constant

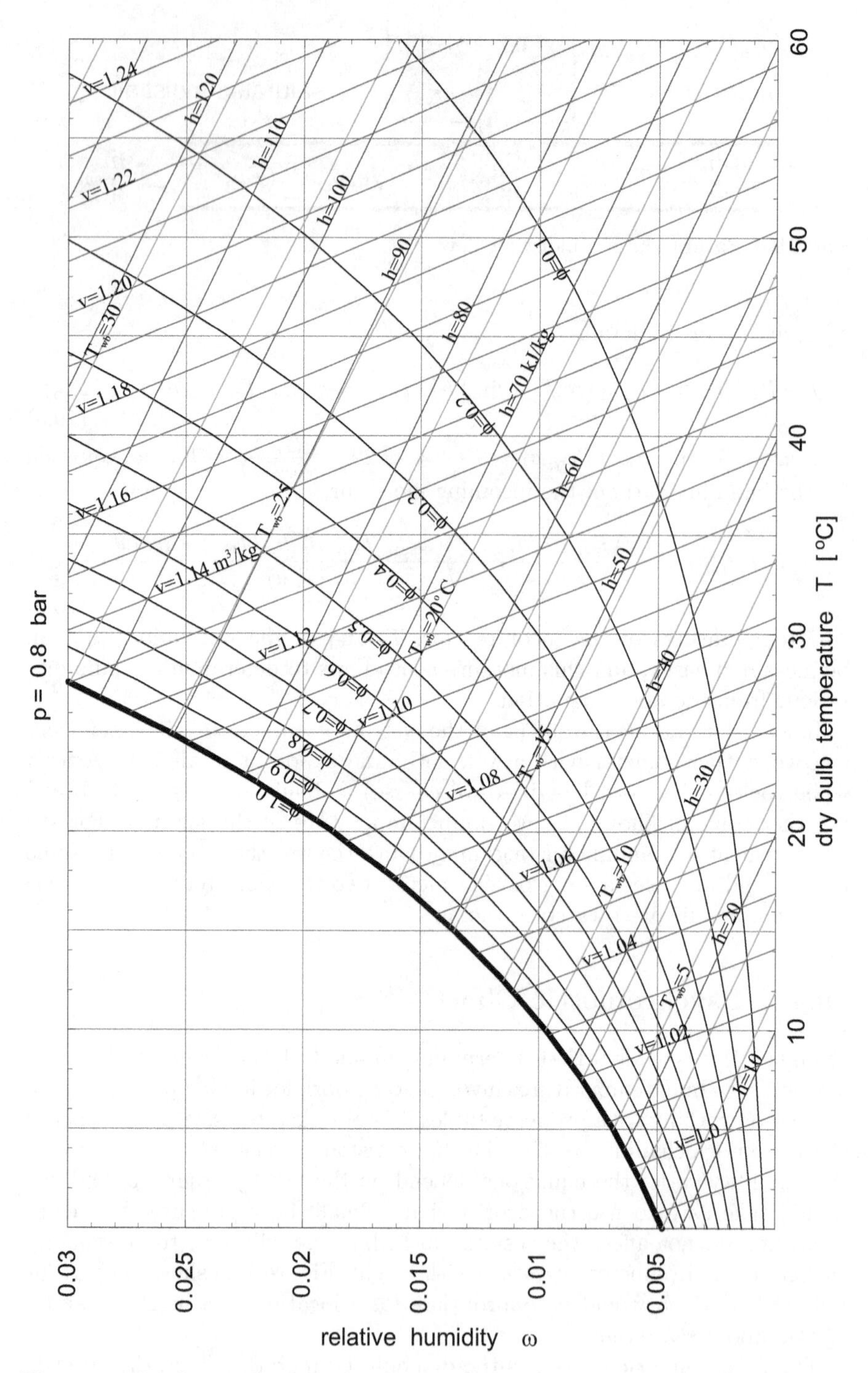

Fig. 19.5 Psychrometric chart for $p = 0.8\text{bar}$

relative humidity ϕ, constant wet-bulb temperature T_{wb}, constant enthalpy $h_{1+\omega}$, and constant specific volume $v_{1+\omega}$. When two of these six quantities are known, all others can be easily read of the diagram. The chart will be used in subsequent sections.

The construction of the psychrometric chart is not difficult. For typical HVAC applications, vapor and dry air can be described as ideal gases with constant specific heats, and liquid water can be described as incompressible liquid with constant specific heat, so that

$$h_{\text{a}}(T) = c_p^{\text{a}}(T - T_R) \ , \ \ h_{\text{v}}(T) = c_p^{\text{v}}(T - T_R) + h_{fg}(T_R) \ , \ \ h_f(T) = c_f(T - T_R) \ , \tag{19.10}$$

where $T_R = 273.15\,\text{K}$, $c_p^{\text{a}} = 1.005\frac{\text{kJ}}{\text{kg K}}$, $c_p^{\text{v}} = 1.88\frac{\text{kJ}}{\text{kg K}}$, $c_f = 4.18\frac{\text{kJ}}{\text{kg K}}$, and $h_{fg}(T_R) = 2500\frac{\text{kJ}}{\text{kg}}$. This choice of reference values for enthalpies and temperature (T_R) ensures that the psychrometric chart is compatible with standard steam tables, for which the triple point enthalpy of the saturated liquid is typically set to zero.

At low pressures, the saturation pressure $p_{\text{sat}}(T)$ of water can be described by the Antoine equation

$$p_{\text{sat}}(T) = p_{tr} \exp\left[17.0361 - \frac{3974.54}{T/\,^\circ\text{C} + 233.290}\right] , \tag{19.11}$$

where T is the temperature in °C, and $p_{tr} = 0.611\,\text{kPa}$ is the triple point pressure.

To plot the lines of constant ϕ, T_{wb}, $h_{1+\omega}$ and $v_{1+\omega}$, we need, for the given total pressure p, the humidity ratio ω as a function of the dry-bulb temperature T, and the quantity in question, i.e., relative humidity, enthalpy, etc. Lines of constant wet-bulb temperature T_{wb} follow immediately by plotting (19.9), which gives $\omega(T, T_{\text{wb}}, p)$, for fixed values of T_{wb} and p. Equations (19.1) and (19.3) can be combined into

$$\omega(T, \phi, p) = \frac{0.622}{\frac{1}{\phi}\frac{p}{p_{\text{sat}}(T)} - 1} \tag{19.12}$$

which, when plotted for fixed ϕ and p, gives the lines of constant relative humidity. Solving (19.5) for ω yields

$$\omega(T, h_{1+\omega}) = \frac{h_{1+\omega} - h_{\text{a}}(T)}{h_{\text{v}}(T)} , \tag{19.13}$$

which gives the lines of constant enthalpy per unit mass of dry air, $h_{1+\omega}$. Finally, lines of constant specific volume per unit mass of dry air follow from (19.6) as

$$\omega(T, v_{1+\omega}, p) = \left(\frac{p v_{1+\omega}}{R_{\text{a}} T} - 1\right) 0.622 \, . \tag{19.14}$$

The psychrometric chart for $p = 0.8\,\text{bar}$ of Fig. 19.5 was obtained from the above equations. The chart exhibits lines of constant enthalpy ($h_{1+\omega} = 10, 20...120\frac{\text{kJ}}{\text{kg}}$), constant relative humidity ($\phi = 0.1, 0.2, ..., 1$), constant wet-bulb temperature ($T_{\text{wb}} = 0, 2.5, 5, 7.5, ...32.5\,°\text{C}$), and of constant specific volume ($v_{1+\omega} = 0.99, 1.00, ..., 1.24$). Lines for enthalpy, volume and wet-bulb temperature have no meaning for $\phi > 1$ and are not drawn. As can be seen from the equations above, the lines of constant wet-bulb temperature, constant enthalpy $h_{1+\omega}$ and constant volume $v_{1+\omega}$ are not straight. Their curvature is so small, however, that in the chart they appear to be straight lines. Also note that enthalpy and wet-bulb temperature lines are not parallel.

Differences in enthalpy values between diagrams obtained from the equations above and those commonly distributed could arise due to different reference states for enthalpies, while enthalpy differences will agree. Figure 19.6 shows a standard ASHRAE[3] chart for $p = 1.01325\,\text{bar}$; the process depicted in the chart will be discussed in the next section.

19.5 Dehumidification

When moist air is cooled below its dewpoint, water condenses, and the humidity ratio ω drops. This process forms the basis for dehumidification systems, in which moist air is cooled below the dewpoint, some water condenses, and then the air is reheated, so that a desired final state is reached, Figure 19.7 shows a schematic for such a process.

We study the process by means of an example, that also shows how to use the psychrometric chart. A mass flow $\dot{m}_a = 10\frac{\text{kg}}{\text{s}}$ of outside air of state 1 ($T_1 = 30\,°\text{C}$, $T_{\text{wb}}^1 = 22.5\,°\text{C}$, $p = 1\,\text{atm}$) is to be dehumidified and cooled so that the final state 4 is $T_4 = 20\,°\text{C}$, $\phi_4 = 0.7$. To achieve this state, the flow is first cooled isobarically. As long as the temperature is above the dewpoint, the humidity ratio does not change. At state 2 ($\phi_2 = 1$), water starts to condense. The moist air is cooled further to state 3, while some water condenses. State 3 has the same humidity ratio as the desired final state 4, which is finally obtained by isobaric heating. The process curve is shown in the psychrometric chart, Fig. 19.6.

From the diagram, we read the following data for the process

$$
\begin{array}{llll}
\omega_1 = 0.0145 & , \omega_2 = \omega_1 & , \omega_3 = \omega_4 & , \omega_4 = 0.0105 \,, \\
\phi_1 = 0.55 & , \phi_2 = 1 & , \phi_3 = 1 & , \phi_4 = 0.7 \,, \\
T_1 = 30\,°\text{C} & , T_2 = 19\,°\text{C} & , T_3 = 14.5\,°\text{C} & , T_4 = 20\,°\text{C} \,, \\
T_{\text{wb}}^1 = 22.5\,°\text{C} & , T_{\text{wb}}^2 = T_2 & , T_{\text{wb}}^3 = T_3 & , T_{\text{wb}}^4 = 16.5\,°\text{C} \,, \\
h_{1+\omega}^1 = 66\frac{\text{kJ}}{\text{kg}} & , h_{1+\omega}^2 = 55\frac{\text{kJ}}{\text{kg}} & , h_{1+\omega}^3 = 41\frac{\text{kJ}}{\text{kg}} & , h_{1+\omega}^4 = 47\frac{\text{kJ}}{\text{kg}} \,, \\
v_{1+\omega}^1 = 0.878\frac{\text{m}^3}{\text{kg}} & , v_{1+\omega}^2 = 0.847\frac{\text{m}^3}{\text{kg}} & , v_{1+\omega}^3 = 0.829\frac{\text{m}^3}{\text{kg}} & , v_{1+\omega}^4 = 0.844\frac{\text{m}^3}{\text{kg}} \,.
\end{array}
$$

[3] American Society of Heating, Refrigerating and Air Conditioning Engineers.

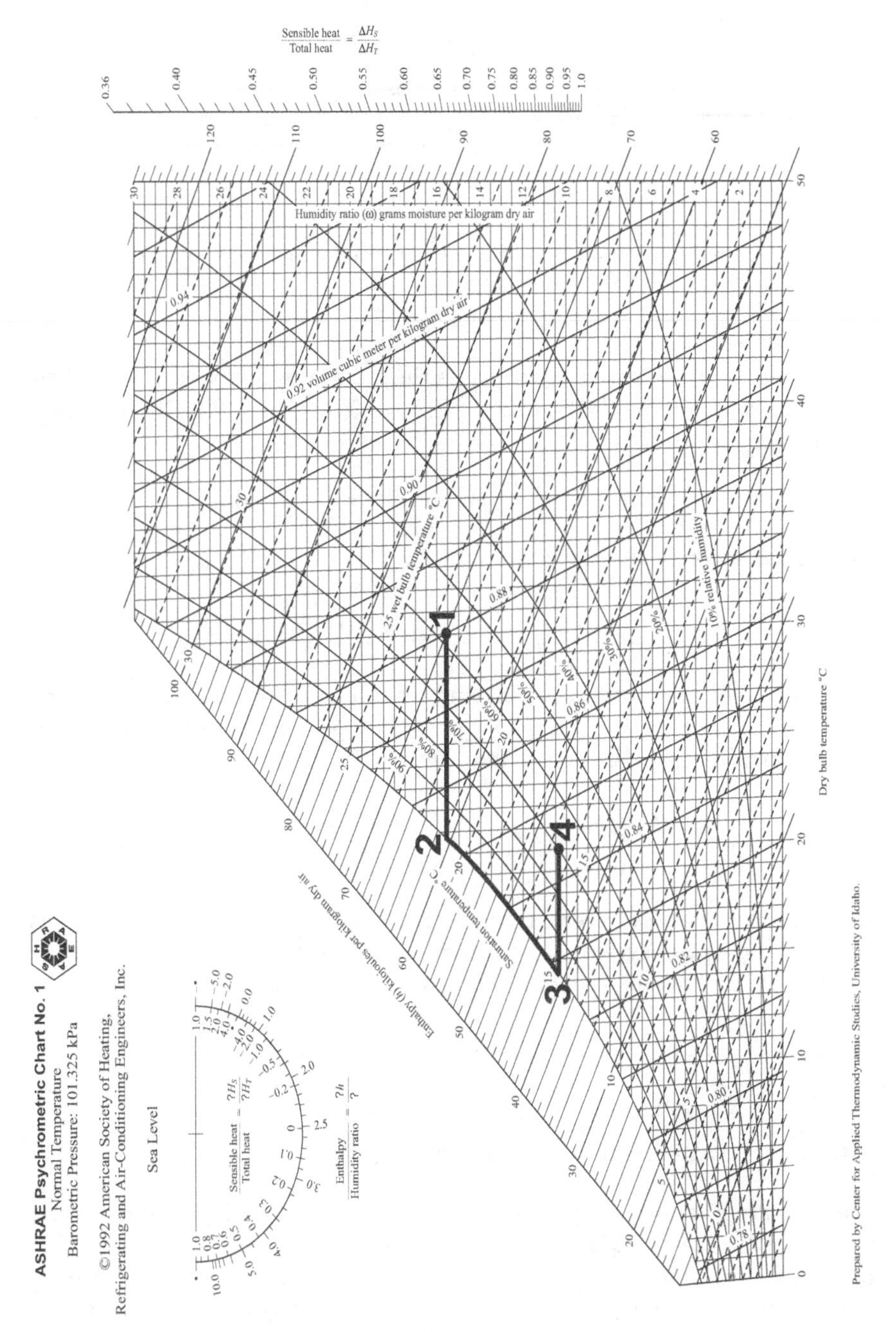

Fig. 19.6 ASHRAE psychrometric chart for standard atmospheric pressure $p_0 = 1\,\text{atm} = 1.01325\,\text{bar}$. The line 1-2-3-4 depicts the dehumidification process of Example 19.5.

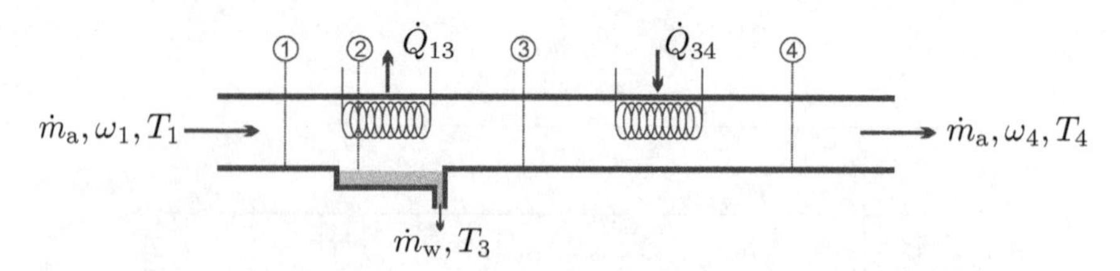

Fig. 19.7 Dehumidification by cooling (1-2), condensation (2-3), and reheating (3-4)

We assume that the liquid water leaves the system at temperature T_3 where it has the enthalpy $h_f(T_3) = c_f(T_3 - T_0) = 60.61\frac{\text{kJ}}{\text{kg}}$, see (19.10). The mass balance for water reads

$$\dot{m}_\text{v}^1 = \dot{m}_\text{w} + \dot{m}_\text{v}^4 \quad \Longrightarrow \quad \omega_1 \dot{m}_\text{a} = \dot{m}_\text{w} + \omega_4 \dot{m}_\text{a} \; ,$$

so that the amount of water removed is,

$$\dot{m}_\text{w} = (\omega_1 - \omega_4)\,\dot{m}_\text{a} = 0.04\frac{\text{kg}}{\text{s}} \; .$$

The heat exchange rates for cooling and reheating are obtained from the first law, which here reduces to $\dot{Q} = \sum_{out} \dot{m}_\alpha h_\alpha - \sum_{in} \dot{m}_\alpha h_\alpha$ (no work, kinetic and potential energies ignored), and hence

$$\begin{aligned}\dot{Q}_{13} &= \dot{m}_\text{a} h^3_{1+\omega} + \dot{m}_\text{w} h_f\,(T_3) - \dot{m}_\text{a} h^1_{1+\omega} = -248\,\text{kW} \; , \\ \dot{Q}_{34} &= \dot{m}_\text{a}\left(h^4_{1+\omega} - h^3_{1+\omega}\right) = 60\,\text{kW} \; .\end{aligned}$$

The cooling process requires a refrigeration system, some of the heat rejected by the refrigeration system can be used for reheating.

The volume flows entering and leaving the system are

$$\dot{V}_1 = v^1_{1+\omega}\dot{m}_\text{a} = 8.78\frac{\text{m}^3}{\text{s}} \quad , \quad \dot{V}_4 = v^4_{1+\omega}\dot{m}_\text{a} = 8.44\frac{\text{m}^3}{\text{s}} \; .$$

19.6 Humidification with Steam

The humidity ratio of dry or moist air can be increased by adding water either as steam or as liquid at pressure p. Steam injection increases humidity ratio and temperature, as long as the steam temperature is above the air temperature. Injection of liquid water, e.g., by spraying of fine mist, leads to cooling of the air, due to evaporation. We study both processes by means of examples, beginning with steam.

We study a steam injection process as depicted in Fig. 19.8. A volume flow $\dot{V}_1 = 10\frac{\text{m}^3}{\text{s}}$ of moist air with dry- and wet-bulb temperatures $T_1 = 14\,°\text{C}$,

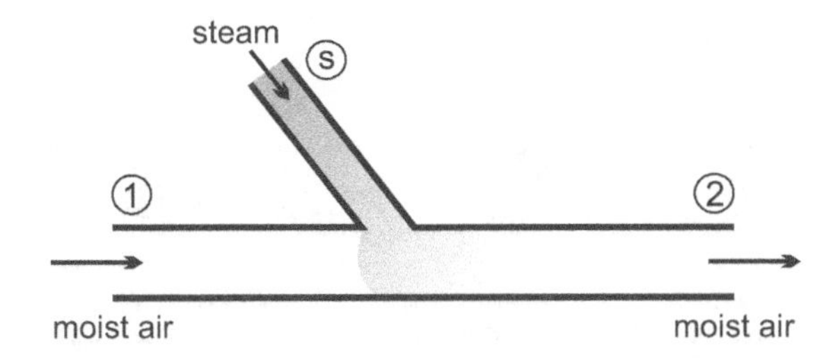

Fig. 19.8 Humidification of moist air by addition of steam

$T_{\mathrm{wb},1} = 5\,°\mathrm{C}$ flows at a pressure of $p = 0.8\,\mathrm{bar}$. Superheated steam at $T_\mathrm{s} = 211\,°\mathrm{C}$, $p_\mathrm{s} = 0.8\,\mathrm{bar}$ is injected at a rate of $\dot{m}_\mathrm{s} = 306\frac{\mathrm{kg}}{\mathrm{h}}$. We ask for the final state (state 2) of the air.

From the psychrometric chart Fig. 19.5 we find the properties of state 1 as

$$\omega_1 = 0.0032\,,\; \phi_1 = 0.26\,,\; h^1_{1+\omega} = 22.5\frac{\mathrm{kJ}}{\mathrm{kg}}\,,\; v^1_{1+\omega} = 1.035\frac{\mathrm{m}^3}{\mathrm{kg}}\,,$$

and the enthalpy of the injected steam follows from (19.10) as $h_\mathrm{s} = 2897\frac{\mathrm{kJ}}{\mathrm{kg}}$. The mass flow of dry air is $\dot{m}_\mathrm{a} = \dot{V}_1/v^1_{1+\omega} = 9.66\frac{\mathrm{kg}}{\mathrm{s}}$.

For this continuous flow process, the balances for vapor mass and energy read

$$\omega_1\dot{m}_\mathrm{a} + \dot{m}_\mathrm{s} = \omega_2\dot{m}_\mathrm{a}\,,$$
$$\dot{m}_\mathrm{a}h^1_{1+\omega} + \dot{m}_\mathrm{s}h_s = \dot{m}_\mathrm{a}h^2_{1+\omega}\,.$$

From these, we find

$$\omega_2 = \omega_1 + \frac{\dot{m}_\mathrm{s}}{\dot{m}_\mathrm{a}} = 0.012\,,$$
$$h^2_{1+\omega} = h^1_{1+\omega} + (\omega_2 - \omega_1)\,h_s = 48\frac{\mathrm{kJ}}{\mathrm{kg}}\,.$$

With the above values for ω_2 and $h^2_{1+\omega}$ state 2 can be localized in the chart, and we find the following other properties: $\phi_2 = 0.6$, $T_2 = 20\,°\mathrm{C}$, $T^2_\mathrm{wb} = 14.5\,°\mathrm{C}$, $v^2_{1+\omega} = 1.07\frac{\mathrm{m}^3}{\mathrm{kg}}$.

19.7 Evaporative Cooling

Next we consider cooling and moisturizing of air by addition of liquid water, again by means of an example, see Fig. 19.9. The initial state is relatively dry air at a pressure of $p = 1\,\mathrm{bar}$, relative humidity $\phi_1 = 0.2$, and temperature $T_1 = 40\,°\mathrm{C}$, so that $\omega_1 = 0.009$. Liquid water at $T_\mathrm{w} = 20\,°\mathrm{C}$ is sprayed into

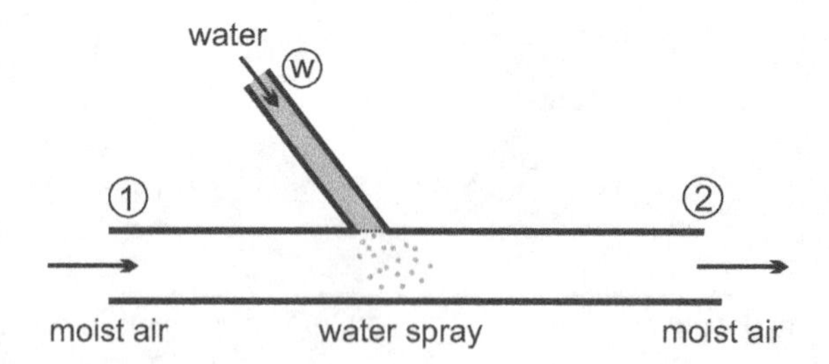

Fig. 19.9 Evaporative cooling: Water is sprayed into moist air and evaporates

the air, and we ask how much liquid must be added per kg of air in order to lower the temperature to $T_2 = 30\,^\circ\text{C}$. The conservation laws for water mass and energy for this process read

$$\begin{aligned} \omega_1 \dot{m}_\text{a} + \dot{m}_\text{w} &= \omega_2 \dot{m}_\text{a} \,, \\ \dot{m}_\text{a} h_{1+\omega}\left(T_1, \omega_1\right) + \dot{m}_\text{w} h_f\left(T_\text{w}\right) &= \dot{m}_\text{a} h_{1+\omega}\left(T_2, \omega_2\right) . \end{aligned} \tag{19.15}$$

First, we solve the problem analytically. With the above relations for enthalpies (19.5, 19.10) we find

$$\begin{aligned} \omega_2 - \omega_1 &= \frac{h_\text{a}\left(T_1\right) - h_\text{a}\left(T_2\right) + \omega_1\left[h_\text{v}\left(T_1\right) - h_\text{v}\left(T_2\right)\right]}{h_\text{v}\left(T_2\right) - h_f\left(T_\text{w}\right)} \qquad (19.16) \\ &= \frac{\left[c_p^\text{a} + \omega_1 c_p^\text{v}\right]\left(T_1 - T_2\right)}{c_p^\text{v}(T_2 - T_R) + h_{fg}\left(T_R\right) - c_f(T_\text{w} - T_R)} = 4.13\frac{\text{g}}{\text{kg}} = 0.00413 \,, \end{aligned}$$

and thus $\omega_2 = 0.0131$.

When one plots this process into the psychrometric chart, one notices that the wet-bulb temperatures of both states are very close. To understand this behavior, we consider two states $\alpha = 1, 2$ with the same wet-bulb temperature T_wb so that, from (19.9),

$$h_\text{a}\left(T_\text{wb}\right) + \omega_\text{sat}\left(T_\text{wb}\right)\left[h_\text{v}\left(T_\text{wb}\right) - h_f\left(T_\text{wb}\right)\right] = h_\text{a}\left(T_\alpha\right) + \omega_\alpha\left[h_\text{v}\left(T_\alpha\right) - h_f\left(T_\text{wb}\right)\right] . \tag{19.17}$$

By taking the difference of this equation for $\alpha = 1$ and $\alpha = 2$ we find

$$\omega_2 - \omega_1 = \frac{h_\text{a}\left(T_1\right) - h_\text{a}\left(T_2\right) + \omega_1\left[h_\text{v}\left(T_1\right) - h_\text{v}\left(T_2\right)\right]}{h_\text{v}\left(T_2\right) - h_f\left(T_\text{wb}\right)} . \tag{19.18}$$

This almost agrees with the expression (19.16), the only difference is the value of the temperature of the added water in the denominator. If the added liquid water in (19.16) is at the wet-bulb temperature of state 1, then the wet-bulb temperature will stay constant. Since under HVAC conditions the enthalpy of the vapor exceeds the enthalpy of the added liquid by far, the denominators in (19.16) and (19.18) will be very close, and both equations will give almost the same result.

In short, evaporative cooling, that is injection of liquid water into moist air, can be well approximated as a process of constant wet-bulb temperature. The psychrometric chart shows lines of constant $T_{\rm wb}$ and can be used to evaluate these processes. Moreover, since the lines of constant enthalpy $h_{1+\omega}$ are almost parallel to the lines of constant wet-bulb temperature, some authors suggest to describe evaporative cooling as a constant enthalpy process.

19.8 Adiabatic Mixing

We consider the adiabatic and isobaric mixing of two moist air streams of states 1 and 2. Mass and energy balances relate the final state 3 to the incoming streams as

$$\begin{aligned} \dot{m}_{\rm a}^1 + \dot{m}_{\rm a}^2 &= \dot{m}_{\rm a}^3 \,, \\ \omega_1 \dot{m}_{\rm a}^1 + \omega_2 \dot{m}_{\rm a}^2 &= \omega_3 \dot{m}_{\rm a}^3 \,, \\ \dot{m}_{\rm a}^1 h_{1+\omega}^1 + \dot{m}_{\rm a}^2 h_{1+\omega}^2 &= \dot{m}_{\rm a}^3 h_{1+\omega}^3 \,. \end{aligned} \tag{19.19}$$

Elimination of $\dot{m}_{\rm a}^3$ gives

$$\frac{\dot{m}_{\rm a}^1}{\dot{m}_{\rm a}^2} = \frac{\omega_3 - \omega_2}{\omega_1 - \omega_3} = \frac{h_{1+\omega}^3 - h_{1+\omega}^2}{h_{1+\omega}^1 - h_{1+\omega}^3} \,, \tag{19.20}$$

which implies that in the psychrometric chart the mixed state 3 lies on the line connecting states 1 and 2, see Fig. 19.10 for illustration.

As an example we consider mixing of two streams at $p = 1\,\text{atm}$ and

$$\begin{aligned} T_1 &= 15\,^\circ\text{C} \,, \quad \dot{V}_1 = 30\,\tfrac{\text{m}^3}{\text{min}} \,, \quad \phi = 1 \,, \\ T_2 &= 30\,^\circ\text{C} \,, \quad \dot{V}_2 = 40\,\tfrac{\text{m}^3}{\text{min}} \,, \quad \phi = 0.5 \,. \end{aligned}$$

From the psychrometric chart we read

$$\begin{aligned} h_{1+\omega}^1 &= 42\,\tfrac{\text{kJ}}{\text{kg}} \,, \quad \omega_1 = 0.011 \,, \quad v_{1+\omega}^1 = 0.830\,\tfrac{\text{m}^3}{\text{kg}} \,, \\ h_{1+\omega}^2 &= 63\,\tfrac{\text{kJ}}{\text{kg}} \,, \quad \omega_2 = 0.013 \,, \quad v_{1+\omega}^1 = 0.876\,\tfrac{\text{m}^3}{\text{kg}} \,. \end{aligned}$$

The corresponding mass flows of dry air are

$$\dot{m}_{\rm a}^1 = \frac{\dot{V}_1}{v_{1+\omega}^1} = 36.15\frac{\text{kg}}{\text{min}} \quad , \quad \dot{m}_{\rm a}^2 = \frac{\dot{V}_2}{v_{1+\omega}^2} = 45.66\frac{\text{kg}}{\text{min}} \,,$$

and the final state is

$$\omega_3 = \frac{\dot{m}_{\rm a}^1 \omega_1 + \dot{m}_{\rm a}^2 \omega_2}{\dot{m}_{\rm a}^1 + \dot{m}_{\rm a}^2} = 0.012 \quad , \quad h_{1+\omega}^3 = \frac{\dot{m}_{\rm a}^1 h_{1+\omega}^1 + \dot{m}_{\rm a}^2 h_{1+\omega}^2}{\dot{m}_{\rm a}^1 + \dot{m}_{\rm a}^2} = 54\frac{\text{kJ}}{\text{kg}} \,.$$

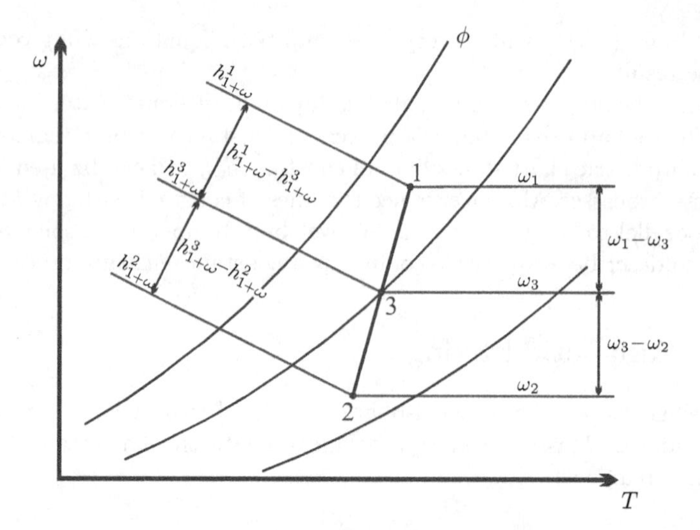

Fig. 19.10 Adiabatic mixing of moist air streams: In the psychrometric chart, the mixed state 3 is on the line connecting the initial states 1 and 2

A special situation may arise due to the convexity of the saturation line ($\phi = 1$). It can happen that the line connecting the two initial states lies outside the accessible region of the diagram. Figure 19.11 shows this for the special case of mixing of two saturated states. In these cases, some liquid water will fall-out as fog, and the mixture will be in the saturated state. Obviously, formation of fog must be avoided in HVAC applications. The relevant equations are again the conservation laws for air and vapor mass, and for energy, which now read

$$\begin{aligned} \dot{m}_{\mathrm{a}}^1 + \dot{m}_{\mathrm{a}}^2 &= \dot{m}_{\mathrm{a}}^3 \,, \\ \omega_1 \dot{m}_{\mathrm{a}}^1 + \omega_2 \dot{m}_{\mathrm{a}}^2 &= \omega_{\mathrm{sat}}\left(T_3\right) \dot{m}_{\mathrm{a}}^3 + \dot{m}_{\mathrm{w}} \,, \\ \dot{m}_{\mathrm{a}}^1 h_{1+\omega}^1 + \dot{m}_{\mathrm{a}}^2 h_{1+\omega}^2 &= \dot{m}_{\mathrm{a}}^3 h_{1+\omega}^3\left(T_3, \omega_{\mathrm{sat}}\right) + \dot{m}_{\mathrm{w}} h_{\mathrm{w}}\left(T_3\right) \,. \end{aligned}$$

Due to the occurrence of $\omega_{\mathrm{sat}}\left(T_3\right)$, these are three non-linear equations for the three unknowns T_3, $\dot{m}_{\mathrm{w}}$, $\dot{m}_{\mathrm{a}}^3$, which are best solved numerically.

The air on top of water bodies normally is saturated. When two streams of water at different temperatures meet, fog will occur as a result of the mixing of the two accompanying air flows.

19.9 Cooling Towers

Evaporate cooling is used in cooling towers for steam power plants, which require a large amount of heat rejection in the condenser. Figure 19.12 shows

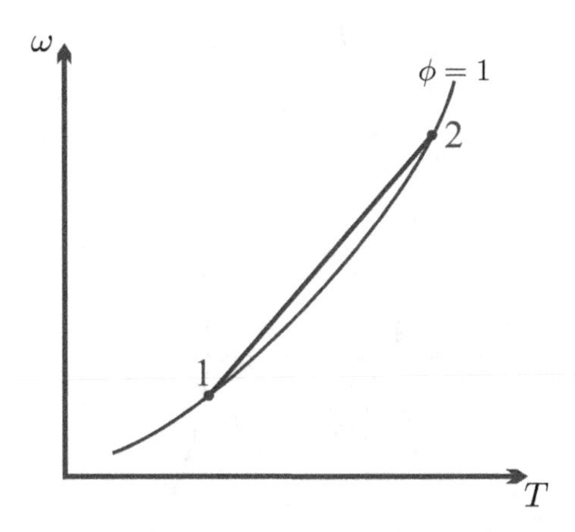

Fig. 19.11 Adiabatic mixing with fall-out of liquid water

a schematic for a natural draft cooling tower, the hyperbolic shape is chosen for structural strength and low material use.

The cooling water flow $\dot{m}_{\text{cw}}$ comes from the condenser of the power plant, where it was heated to T_1 while the water circulating in the steam cycle was condensed. The incoming cooling water is sprayed into the cooling tower, where some of it evaporates, which leads to cooling of the liquid. Since moist air is lighter than dry air—the low molar mass of vapor lowers the average molar mass—moist air rises and leaves the tower, while fresh environmental air at (T_3, ω_3) is drawn in at the bottom. Make-up water at $\dot{m}_{\text{m}}, T_5$ is added to compensate the loss of evaporated water and the mass flow $\dot{m}_{\text{cw}}$ leaves the cooling tower towards the condenser at T_2. Normally, the make-up water is drawn from rivers or lakes, and that is why power plants are build close to these. As the rising moist air equilibrates with the environment, some of the added water might condense, which leads to clouds that normally can be seen above cooling towers.

The balances for air and water mass, and for energy, read

$$\begin{aligned} \dot{m}_{\text{a}} &= const.\,, \\ \dot{m}_{\text{cw}} + \dot{m}_{\text{m}} + \dot{m}_{\text{a}}\omega_3 &= \dot{m}_{\text{cw}} + \dot{m}_{\text{a}}\omega_4\,, \\ \dot{m}_{\text{cw}} h_f(T_1) + \dot{m}_{\text{m}} h_f(T_5) + \dot{m}_{\text{a}} h_{1+\omega}^3 &= \dot{m}_{\text{cw}} h_f(T_2) + \dot{m}_{\text{a}} h_{1+\omega}^4\,. \end{aligned} \tag{19.21}$$

19.10 Example: Cooling Tower

As an example we study the cooling tower for a $\dot{W} = 300\,\text{MW}$ power plant with a thermal efficiency $\eta = 0.4$. In the condenser, the cooling water is heated from $T_2 = 30\,°\text{C}$ to $T_1 = 40\,°\text{C}$ (the numbers refer to Fig. 19.12), thus the mass flow of cooling water is

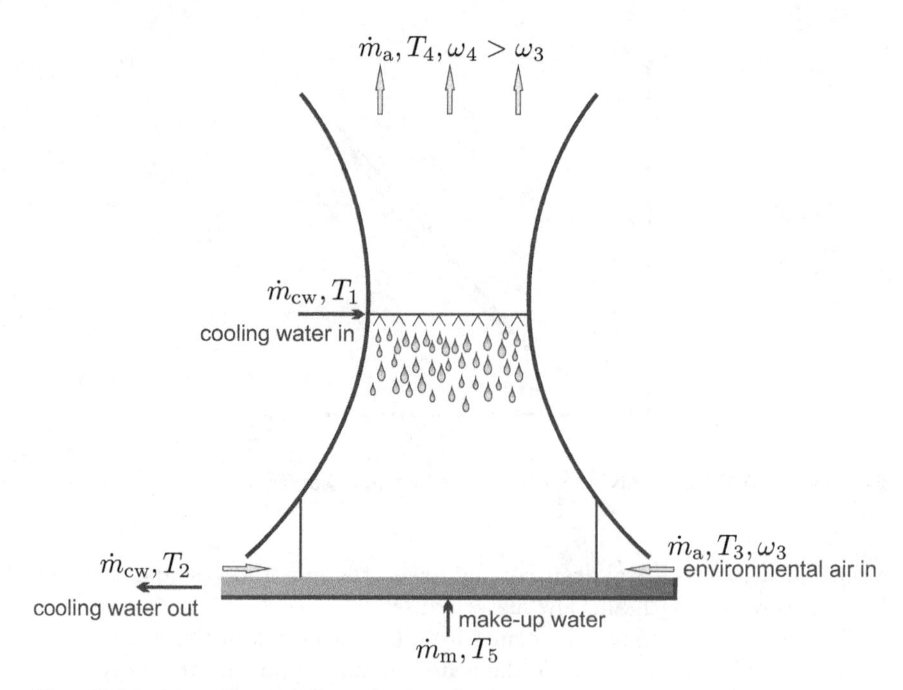

Fig. 19.12 Air and water flows in a cooling tower

$$\dot{m}_{cw} = \frac{\dot{Q}_{cw}}{h_1 - h_2} = \frac{\dot{Q}_{cw}}{c_f \left(T_1 - T_2\right)} = \frac{1-\eta}{\eta} \frac{\dot{W}}{c_f \left(T_2 - T_1\right)} = 10.77 \frac{\mathrm{t}}{\mathrm{s}} .$$

We ask for the required flows of air and make-up water, which both depend on the state of the incoming and exiting moist air. For further computation we assume that the incoming air is at $T_3 = 25\,^{\circ}\mathrm{C}$, $\phi_3 = 0.5$, so that $\omega_3 = 0.01$, $h^3_{1+\omega} = 51 \frac{\mathrm{kJ}}{\mathrm{kg}}$, and that the make-up water is at $T_5 = 25\,^{\circ}\mathrm{C}$, so that, with (19.10), $h_f \left(T_5\right) = 104.5 \frac{\mathrm{kJ}}{\mathrm{kg}}$, and $h_f \left(T_1\right) = 167.2 \frac{\mathrm{kJ}}{\mathrm{kg}}$, $h_f \left(T_2\right) = 125.4 \frac{\mathrm{kJ}}{\mathrm{kg}}$. Furthermore, we assume that the exiting air is saturated, so that $\omega_4 = \omega_{\mathrm{sat}} \left(T_4\right)$.

The remaining unknowns in this problem are the air mass flow $\dot{m}_a$, the make-up water flow $\dot{m}_m$ and the exit temperature T_4. This problem differs from evaporative cooling as discussed above, due to the large amount of warm water sprayed into the air. The heat transfer between air droplets and air could only be described by a detailed heat transfer analysis. To simplify the problem, we assume $T_4 = 30\,^{\circ}\mathrm{C}$ which implies $h^4_{1+\omega} = 100 \frac{\mathrm{kJ}}{\mathrm{kg}}$ and $\omega_4 = \omega_{\mathrm{sat}} \left(T_4\right) = 0.0272$. Then we find from the conservation laws

$$\dot{m}_a = \frac{h_f \left(T_1\right) - h_f \left(T_2\right)}{h^4_{1+\omega} - h^3_{1+\omega} - \left[\omega_{\mathrm{sat}} \left(T_4\right) - \omega_3\right] h_f \left(T_5\right)} \dot{m}_{cw} = 9.61 \frac{\mathrm{t}}{\mathrm{s}} ,$$

$$\dot{m}_m = \dot{m}_a \left[\omega_{\mathrm{sat}} \left(T_4\right) - \omega_3\right] = 0.165 \frac{\mathrm{t}}{\mathrm{s}} .$$

The river that provides the make-up water should have a sufficiently large mass flow rate, so that the removal of the make-up water will not disturb the ecological equilibrium of the river. An alternative to cooling towers is the direct use of river or lake water as cooling water. In this case, the heat rejected by the power plant is added to the river or lake. The related increase in water temperature changes the chemical environment, e.g., the amount of oxygen dissolved decreases with increasing temperature (Henry's law, Sec. 22.10), which might disturb the ecological equilibrium more than the removal of some water for use in cooling towers.

Problems

19.1. Compression of Moist Air

Air initially at 1 atm, 25 °C and relative humidity of 60% is compressed isothermally until condensation of water occurs. Determine the pressure at the onset of condensation. Draw the process for the vapor into a T-s-diagram.

19.2. Compressed Air

To avoid condensation of water in compressed air lines, it might be necessary to dehumidify the compressed air. To study this, consider a compressor that draws outside air at 93 kPa, 14 °C and relative humidity of 40%, and compresses it to 800 kPa. After compression, the air flows through ducts for distribution, where it is cooled to the workshop temperature of 22 °C. Determine the dewpoint of the compressed air–will there be condensation in the pipes?

19.3. Air Conditioning

An air conditioning system provides a volume flow of $3\frac{\text{m}^3}{\text{s}}$ of moist air at 1 atm, 22 °C and 50% relative humidity by conditioning outside air at 34 °C and 50% relative humidity. For this, the outside air is first cooled and dehumidified, and then heated to the final temperature. Assume that the condensate leaves the system at 10 °C and determine the temperature after dehumidification is completed, the amount of heat that must be withdrawn in the cooling process, and the heat added in the heating process per unit mass of dry air.

19.4. Air Conditioning

An air conditioning system provides air at 1 atm, 20 °C and 60% relative humidity which is obtained from outside air at 38 °C and 70% relative humidity as follows: The outside air is first cooled and dehumidified, and then heated to the desired final temperature. The pressure stays constant throughout the process. Determine the temperature after dehumidification is completed, and assume that the condensate leaves the system at this temperature. Next, determine the heat that must be withdrawn in the cooling process, and the heat added in the heating process, both per unit mass of dry air. Finally, determine the mass flow and the volume flow of the air delivered, when the cooling power of the system is 150 kW.

19.5. Humidification

An air conditioning system draws $22\frac{\text{m}^3}{\text{min}}$ outside air at 1 atm, 10 °C and 40% relative humidity. The air is first heated to 22 °C, and then humidified by injection of steam. The air leaves the system at 25 °C and 55% relative humidity. Determine the rate of heat supply in the heating section, the mass flow rate of steam required, and the temperature of the steam.

19.6. Humidification

At an elevated location, an air conditioning system draws $20\frac{\text{m}^3}{\text{min}}$ outside air at 0.8 bar. With a psychrometer it is determined that the dry- and wet-bulb temperatures of the incoming air are 10 °C and 2.5 °C, respectively. To reach the desired state, the air is first heated to a temperature T_2, and then humidified by injection of superheated steam at 0.8 bar, 150 °C, where the enthalpy is $2777.84\frac{\text{kJ}}{\text{kg}}$. The air leaves the system at 22 °C and 55% relative humidity. Determine the mass flow rate of steam required, the rate of heat supply in the heating section, and the temperature of the air before steam is injected.

19.7. Dehumidification and Mixing

The outside air of a building is at 26 °C, and 90% relative humidity, the pressure is 1 atm. The air conditioning system of the building is required to provide air at 22 °C and 50% relative humidity. To reach that state, the flow of incoming outside air is split into two streams.

One stream is dehumidified by cooling to 5 °C so that liquid water condenses, and subsequent reheating. The cooling system removes 6 kW from this flow.

Then, the dehumidified stream is mixed with the other stream, so that the desired state is reached.

For the solution use the psychrometric chart.

1. Determine the dry air mass flows of the two streams.
2. Determine the heat required for the reheating of the dehumidified stream.
3. The final air flow should not be faster than $3\frac{\text{m}}{\text{s}}$, determine the cross section of the duct.

19.8. Air Conditioning

An air conditioning system draws a volume flow of $20\frac{\text{m}^3}{\text{min}}$ of outside air at 30 °C and 90% relative humidity (state 1). The air flow is divided in to two streams, stream A and stream B. Stream A is first dehumidified by cooling to 5 °C (state 2), and then heated to state 3. Stream A and stream B are then mixed adiabatically. The mixture has a dry-bulb temperature of 20 °C at 50% relative humidity (state 4). The pressure is constant at 1 atm throughout the process.

1. Indicate the states 1,2,3,4 in the psychrometric chart.
2. Compute the ratios of dry air mass flows , and the values of the two mass flows.
3. Compute the heat to be removed from (1-2) and added to (2-3) stream A.
4. Is it feasible to do both, heating and cooling with a single refrigeration cycle? Discuss?

19.9. Evaporative Cooling
To provide air at a desired state, a volume flow of $10\frac{\text{m}^3}{\text{s}}$ outside air (dry-bulb temperature 15 °C, wet-bulb temperature 10.8 °C) is first heated to 30 °C and then cooled and humidified by spraying of liquid water. The final temperature is 25 °C. Determine the relative humidity at the exit, the mass flow of water added, and the heating rate required. Use the psychrometric chart.

19.10. Evaporative Cooling
An air conditioning system draws a volume flow of $50\frac{\text{m}^3}{\text{min}}$ of outside air at 40 °C and 10% relative humidity (state 1). To produce moist air of pleasant conditions, this air is first cooled by evaporative cooling to state 2, and then by heat exchange with a cooling system to the final state, with dry-bulb temperature $T_3 = 20\,°\text{C}$ at $\phi_3 = 60\%$ relative humidity. The pressure is constant at 1 atm throughout the process.

1. Indicate the states 1,2,3 in the attached psychrometric chart.
2. Determine the mass flow of water required for evaporative cooling (1-2).
3. Determine the heat to be removed, $\dot{Q}_{23}$, in kilowatts.
4. A leak occurs in the system, and moist air at state 3 is mixed with outside air. Determine the dry air mass flow of leaked air when the mixture has a dry-bulb temperature of $T_4 = 22\,°\text{C}$.

19.11. Mixing of Two Moist Air Streams
Consider the adiabatic mixing of two streams of moist air at $p = 1\,\text{atm}$. Stream 1 is saturated moist air of 20 °C at a volumetric flow rate of $60\frac{\text{m}^3}{\text{min}}$ and stream 2 is moist air of 34 °C, 20% relative humidity. The relative humidity after mixing is 60%. Mark all relevant points on the psychrometric chart.

1. For the incoming flows and for the mixture, determine the values for enthalpy, temperature, relative humidity, humidity ratio, and specific volume.
2. Determine the dewpoint temperature and the wet-bulb temperature of the mixture
3. Compute the volumetric flow of stream 2.

19.12. Mixing of Air Streams
Two streams of moist air are mixed adiabatically at 1 atm. One stream has a dry-bulb temperature of 40 °C and a wet-bulb temperature of 32 °C, and the mass flow rate is $8\frac{\text{kg}}{\text{s}}$. The other stream is saturated air at 18 °C with a mass flow rate of $6\frac{\text{kg}}{\text{s}}$. Determine the state of the mixture (temperature, specific humidity, relative humidity, enthalpy, volume flow).

19.13. Air Conditioning in the Desert

In the desert: The outside air of a building is at 40°C, and 10% relative humidity, the pressure is 1 atm. The air conditioning system of the building is required to provide air at 20 °C and 50% relative humidity. To reach that state, the flow of incoming outside air is split into two streams, A and B:

Stream A is spray-cooled by injection of liquid water to a relative humidity of 100%, the mass flow of water (at 20 °C) injected is $10\frac{\text{kg}}{\text{h}}$.

Stream B is cooled to temperature T_B by a standard refrigeration cycle with COP of 3. Then, the spray-cooled stream A is mixed with stream B, so that the desired end state is reached.

1. Make a sketch of the process, and enter the relevant points in the psychrometric chart.
2. Determine the temperature T_B.
3. Determine the dry air mass flows of both streams.
4. Determine the heat removed from stream B, and the power requirement of the refrigerator.

19.14. Cooling Tower

In a 500 MW steam power plant, the condenser is cooled by a cooling water flow that enters the condenser at 26 °C, and leaves at 40 °C. The cooling water is cooled back to 26 °C in a natural-draft cooling tower which draws environmental air at 1 atm with dry- and wet-bulb temperatures of 23 °C and 18 °C, respectively, and discharges saturated air at 37 °C. The thermal efficiency of the power plant is 43.5%. Determine mass flow of cooling water, volume flows of air into and out of the cooling tower, and the required mass flow of makeup water.

19.15. Clouds

Cumulus clouds are formed when air at the ground is heated, takes up moisture, and then rises due to its buoyancy. While rising, the moist air expands, more or less adiabatically, since the pressure decreases with height. During expansion the temperature of the rising air is decreasing. When the temperature reaches the dew point temperature, water vapor condenses, and a cloud is formed.

In order to compute the height of the clouds, assume that the pressure in the atmosphere is given by the barometric formula 2.26.

Consider a fixed mass of moist air, that occupies a volume V, and has enthalpy H. Consider the moist air as an ideal gas, so that its enthalpy and volume are given as

$$H = m_{\text{a}} \left[c_p^{\text{a}} \left(T_{\text{a}} - T_R \right) + \omega \left(h_{fg} \left(T_R \right) + c_p^{\text{v}} \left(T_{\text{a}} - T_R \right) \right) \right] ,$$
$$V = m_{\text{a}} \left(R_{\text{a}} + \omega R_{\text{v}} \right) \frac{T_{\text{a}}}{p} .$$

Here, m_{a} is the mass of dry air, ω is the humidity ratio, c_p^{a}, c_p^{v} and R_{a}, R_{v} are the specific heats and gas constants of dry air and vapor, $h_{fg}\left(T_R\right)$ is the heat of evaporation of water at $T_R = 273.15\,\mathrm{K}$, T_a is the temperature of the rising moist air, and p is the local pressure.

1. Discuss the assumptions behind the equations for enthalpy and volume.
2. Show that the first law for an adiabatic process for the moist air gives $dH = Vdp$.
3. Show that the temperature of the rising moist air is given by

$$T_\mathrm{a}(z) = T_M \left(\frac{p(z)}{p_0}\right)^{\frac{R_\mathrm{a}+\omega R_\mathrm{v}}{c_p^\mathrm{a}+\omega c_p^\mathrm{v}}},$$

   where T_M is the temperature of the moist air at the ground, just before rising (at $z = 0$).
4. Employ the ideal gas law for the vapor to find its partial pressure in the moist air as

$$p_v(z) = \frac{\omega}{\omega + \frac{R_\mathrm{a}}{R_\mathrm{v}}} p_0 \left[1 - \frac{\alpha}{T_0} z\right]^{\frac{g}{\alpha R_\mathrm{a}}}.$$

5. The saturation pressure for the vapor is given by $p_\mathrm{sat}(T_\mathrm{a}(z))$, and water will condense, when $p_v(z) > p_\mathrm{sat}(T_\mathrm{a}(z))$. Set $T_M = 298\,\mathrm{K}$, $\omega = 0.01$ (or other values), and find the height z_Cloud, where clouds begin to form.

Chapter 20
The Chemical Potential

20.1 Definition and Interpretation

Changes in the composition of the mixture, either by addition or removal of components or by reaction, change the properties of the mixture, in particular the Gibbs free energy. A straightforward extension of the Gibbs equation which accounts for the change of Gibbs free energy with varying composition is

$$dG = -SdT + Vdp + \sum_{\alpha=1}^{\nu} \bar{\mu}_\alpha dn_\alpha \, , \tag{20.1}$$

where the new quantity $\bar{\mu}_\alpha$ is the *chemical potential.* We have by definition,

$$\bar{\mu}_\gamma = \left(\frac{\partial G}{\partial n_\gamma} \right)_{T,p,n_\alpha(\alpha\neq\gamma)} \, . \tag{20.2}$$

The chemical potential is of fundamental importance in the thermodynamics of inert and reacting mixtures. To understand its physical meaning, we consider two mixtures of different composition and different pressures but equal temperatures T, which are divided by a semi-permeable membrane that only allows component γ to pass, as depicted in Fig. 20.1. Since pressure and temperature are controlled, we know that in equilibrium the Gibbs free energy of the system must assume a minimum,

$$G \longrightarrow \text{Minimum} \, . \tag{20.3}$$

The total Gibbs free energy of the system is the sum of the Gibbs free energies of the two parts I and II of the system, which depend only on the pressure, temperature and mole numbers within their portion of the system,

$$G = G^{\mathrm{I}} \left(T, p^{\mathrm{I}}, n_1^{\mathrm{I}}, \ldots, n_\gamma^{\mathrm{I}}, \ldots, n_{\nu_{\mathrm{I}}}^{\mathrm{I}} \right) + G^{\mathrm{II}} \left(T, p^{\mathrm{II}}, n_1^{\mathrm{II}}, \ldots, n_\gamma^{\mathrm{II}}, \ldots, n_{\nu_{\mathrm{II}}}^{\mathrm{II}} \right) \, . \tag{20.4}$$

H. Struchtrup, *Thermodynamics and Energy Conversion*,
DOI: 10.1007/978-3-662-43715-5_20,

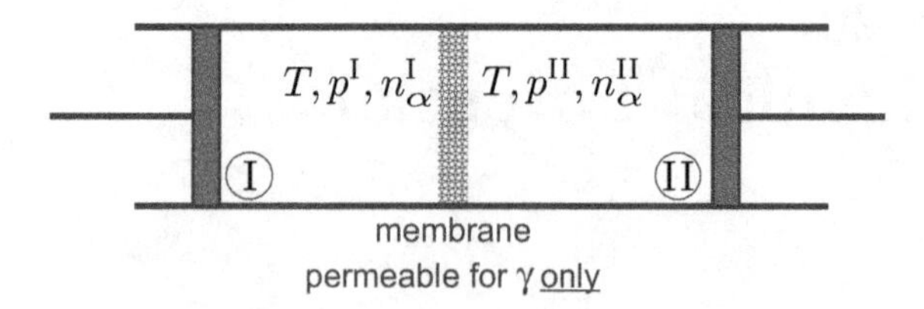

Fig. 20.1 System divided by semi-permeable membrane

Only the mole numbers n^{I}_{γ} and $n^{\mathrm{II}}_{\gamma} = n_{\gamma} - n^{\mathrm{I}}_{\gamma}$ can change due to migration through the semi-permeable membrane. All other mole numbers $(n^{\mathrm{I}}_{\alpha}, n^{\mathrm{II}}_{\alpha})$, the pressures $(p^{\mathrm{I}}, p^{\mathrm{II}})$ and the temperature T are fixed. Thus, the condition for equilibrium of the system, where G is at a minimum, is

$$\left(\frac{\partial G^{\mathrm{I}}}{\partial n^{\mathrm{I}}_{\gamma}}\right)_{T,p^{\mathrm{I}},n^{\mathrm{I}}_{\alpha}(\alpha\neq\gamma)} + \left(\frac{\partial G^{\mathrm{II}}}{\partial n^{\mathrm{I}}_{\gamma}}\right)_{T,p^{\mathrm{II}},n^{\mathrm{II}}_{\alpha}(\alpha\neq\gamma)} = 0 \tag{20.5}$$

or, since $n^{\mathrm{I}}_{\gamma} + n^{\mathrm{II}}_{\gamma} = n_{\gamma} = const.$ and thus $\frac{\partial}{\partial n^{\mathrm{I}}_{\gamma}} = -\frac{\partial}{\partial n^{\mathrm{II}}_{\gamma}}$,

$$\bar{\mu}^{\mathrm{I}}_{\gamma} = \left(\frac{\partial G^{\mathrm{I}}}{\partial n^{\mathrm{I}}_{\gamma}}\right)_{T,p^{\mathrm{I}},n^{\mathrm{I}}_{\alpha}(\alpha\neq\gamma)} = \left(\frac{\partial G^{\mathrm{II}}}{\partial n^{\mathrm{II}}_{\gamma}}\right)_{T,p^{\mathrm{II}},n^{\mathrm{II}}_{\alpha}(\alpha\neq\gamma)} = \bar{\mu}^{\mathrm{II}}_{\gamma} \ . \tag{20.6}$$

In short, in thermodynamic equilibrium, the chemical potentials of the substance γ that can pass through a semipermeable membrane are equal,

$$\bar{\mu}_{\gamma}\left(T, p^{\mathrm{I}}, n^{\mathrm{I}}_{\alpha}\right) = \bar{\mu}_{\gamma}\left(T, p^{\mathrm{II}}, n^{\mathrm{II}}_{\alpha}\right) \ . \tag{20.7}$$

Just as the continuity of temperature at diathermic walls allows us to measure temperature, the continuity of the chemical potential at semi-permeable walls allows us its measurement, at least in principle. And just as temperature differences lead to heat flow and allow for power generation, differences in chemical potential cause particle flow and allow for power generation, as will be discussed in Section 20.7.

Nevertheless, the chemical potential is a rather abstract quantity, and we need to study it further, and relate it to quantities we are more familiar with.

20.2 Properties of the Chemical Potential

The Gibbs free energy $G(T, p, n_{\beta})$ is an extensive quantity, which implies that it is a homogeneous function of mole number,

$$zG(T, p, n_{\beta}) = G(T, p, zn_{\beta}) \ . \tag{20.8}$$

We take the derivative of the above with respect to z, to obtain

$$\begin{aligned} G(T,p,n_\beta) &= \sum_{\alpha=1}^{\nu} \frac{\partial G(T,p,zn_\beta)}{\partial (zn_\alpha)} \frac{d(zn_\alpha)}{dz} \\ &= \sum_{\alpha=1}^{\nu} \frac{\partial G(T,p,zn_\beta)}{\partial (zn_\alpha)} n_\alpha = \sum_{\alpha=1}^{\nu} \bar{\mu}_\alpha (T,p,zn_\beta)\, n_\alpha \,. \end{aligned} \tag{20.9}$$

This must hold for all z. Since the left hand side is independent of z, the right hand side should not depend on z as well. This is so, when the chemical potential does not depend on all mole numbers n_β, but only on quotients like the mole ratio X_β, which is independent of z, since $X_\beta(n_\gamma) = \frac{n_\beta}{n}$, so that $X_\beta(zn_\gamma) = X_\beta(n_\gamma)$. Thus we have

$$\bar{\mu}_\alpha = \bar{\mu}_\alpha (T,p,X_\beta) \,. \tag{20.10}$$

With this we obtain from (20.9)

$$G(T,p,n_\beta) = \sum_{\alpha=1}^{\nu} n_\alpha \bar{\mu}_\alpha (T,p,X_\beta) \,. \tag{20.11}$$

According to (20.11), in a mixture the total Gibbs free energy is the sum of the component mole numbers times their chemical potentials, $G = \sum n_\alpha \bar{\mu}_\alpha$. Since

$$G = H - TS = \sum_{\alpha=1}^{\nu} n_\alpha \bar{h}_\alpha - T \sum_{\alpha=1}^{\nu} n_\alpha \bar{s}_\alpha = \sum_{\alpha=1}^{\nu} n_\alpha \left(\bar{h}_\alpha - T\bar{s}_\alpha\right) \,, \tag{20.12}$$

this implies that the chemical potential is the specific Gibbs free energy of the component in the mixture, in the sense that

$$\bar{\mu}_\alpha (T,p,X_\beta) = \bar{h}_\alpha (T,p,X_\beta) - T\bar{s}_\alpha (T,p,X_\beta) \,. \tag{20.13}$$

Here, $\bar{h}_\alpha (T,p,X_\beta)$ and $\bar{s}_\alpha (T,p,X_\beta)$ are the specific enthalpy and entropy of component α in the mixture at temperature T, pressure p, and composition X_β.

The molar Gibbs free energy of the mixture is defined as

$$\bar{g} = \frac{G}{n} = \sum_{\alpha=1}^{\nu} X_\alpha \bar{\mu}_\alpha (T,p,X_\beta) \,. \tag{20.14}$$

For a single substance the chemical potential is equal to the molar Gibbs free energy,

$$G = n\bar{g} = n\bar{\mu} \quad \text{so that} \quad \bar{\mu}(T,p) = \bar{g}(T,p) \,. \tag{20.15}$$

For this reason, one sometimes finds the specific Gibbs free energy denoted as the chemical potential. For the description of the component, it is useful to distinguish between the Gibbs free energy $\bar{g}_\alpha(T,p)$ that describes the component α alone at (T,p) and the chemical potential $\bar{\mu}_\alpha(T,p,X_\beta)$ that describes the component α in a mixture at (T,p) with mole fractions X_β, $\beta=1,\ldots,\nu$.

Since the order of derivatives can be exchanged, we have the symmetry property

$$\frac{\partial\bar{\mu}_\beta}{\partial n_\gamma}=\frac{\partial^2 G}{\partial n_\gamma\partial n_\beta}=\frac{\partial^2 G}{\partial n_\beta\partial n_\gamma}=\frac{\partial\bar{\mu}_\gamma}{\partial n_\beta}\,. \tag{20.16}$$

20.3 Gibbs and Gibbs-Duhem Equations

We revisit the Gibbs equation for the mixture (20.1),

$$dG=-SdT+Vdp+\sum_{\alpha=1}^{\nu}\bar{\mu}_\alpha dn_\alpha\,. \tag{20.17}$$

The Gibbs equation for molar quantities of the mixture follows by setting $G=n\bar{g}$, $S=n\bar{s}$, $V=n\bar{v}$ and $n_\alpha=nX_\alpha$ as

$$d\bar{g}=-\bar{s}dT+\bar{v}dp+\sum_{\alpha=1}^{\nu}\bar{\mu}_\alpha dX_\alpha\,. \tag{20.18}$$

Alternative forms of the Gibbs equation can be obtained through Legendre transforms, which yield, e.g.,

$$Td\bar{s}=d\bar{u}+pd\bar{v}-\sum_{\alpha=1}^{\nu}\bar{\mu}_\alpha dX_\alpha\,. \tag{20.19}$$

As discussed in Sec. 16.2, the Gibbs equation leads to Maxwell relations, such as

$$\left(\frac{\partial\mu_\alpha}{\partial T}\right)_{p,X_\beta}=-\left(\frac{\partial\bar{s}}{\partial X_\alpha}\right)_{T,p,X_\beta}\,,\quad\left(\frac{\partial\mu_\alpha}{\partial p}\right)_{T,X_\beta}=\left(\frac{\partial\bar{v}}{\partial X_\alpha}\right)_{T,p,X_\beta}\,. \tag{20.20}$$

With another Legendre transform (20.17) can be rewritten as

$$dG=-SdT+Vdp+\sum_{\alpha=1}^{\nu}d\left(\bar{\mu}_\alpha n_\alpha\right)-\sum_{\alpha=1}^{\nu}n_\alpha d\bar{\mu}_\alpha\,, \tag{20.21}$$

which, together with (20.11), yields the Gibbs-Duhem relation

$$0 = -SdT + Vdp - \sum_{\alpha=1}^{\nu} n_\alpha d\bar{\mu}_\alpha \,. \tag{20.22}$$

Taking the derivative of the Gibbs-Duhem relation (20.22) with respect to a mole number n_γ yields, with the symmetry property (20.16),

$$\sum_{\alpha=1}^{\nu} n_\alpha \left(\frac{\partial \bar{\mu}_\alpha}{\partial n_\gamma}\right)_{T,p,n_{\beta(\beta\neq\gamma)}} = \sum_{\alpha=1}^{\nu} n_\alpha \left(\frac{\partial \bar{\mu}_\gamma}{\partial n_\alpha}\right)_{T,p,n_{\beta(\beta\neq\gamma)}} = 0 \,. \tag{20.23}$$

20.4 Mass Based Chemical Potential

It is an easy exercise to translate the above mole based relations into mass based relations, which read

$$dG = -SdT + Vdp + \sum_{\alpha=1}^{\nu} \mu_\alpha dm_\alpha \,, \tag{20.24}$$

$$dg = -sdT + vdp + \sum_{\alpha=1}^{\nu} \mu_\alpha dc_\alpha \,, \tag{20.25}$$

$$\mu_\alpha = \left(\frac{\partial G}{\partial m_\gamma}\right)_{T,p,m_\alpha(\alpha\neq\gamma)} \,, \tag{20.26}$$

$$\mu_\alpha = \mu_\alpha\left(T,p,c_\beta\right) \,, \tag{20.27}$$

$$G\left(T,p,m_\beta\right) = \sum_{\alpha=1}^{\nu} m_\alpha \mu_\alpha\left(T,p,c_\beta\right) \,, \tag{20.28}$$

$$\mu_\alpha\left(T,p,c_\beta\right) = h_\alpha\left(T,p,c_\beta\right) - Ts_\alpha\left(T,p,c_\beta\right) \,, \tag{20.29}$$

$$\mu\left(T,p\right) = g\left(T,p\right) \,, \tag{20.30}$$

$$\frac{\partial \mu_\beta}{\partial m_\gamma} = \frac{\partial \mu_\gamma}{\partial m_\beta} \,, \tag{20.31}$$

$$0 = -SdT + Vdp - \sum_{\alpha=1}^{\nu} m_\alpha d\mu_\alpha \,, \tag{20.32}$$

$$\sum_{\alpha=1}^{\nu} m_\alpha \left(\frac{\partial \mu_\alpha}{\partial m_\gamma}\right)_{T,p,m_{\beta(\beta\neq\gamma)}} = \sum_{\alpha=1}^{\nu} m_\alpha \left(\frac{\partial \mu_\gamma}{\partial m_\alpha}\right)_{T,p,m_{\beta(\beta\neq\gamma)}} = 0 \,. \tag{20.33}$$

The verification of the above is left to the reader.

20.5 The Chemical Potential for an Ideal Mixture

The Gibbs free energy of an ideal mixture at (T, p) is given by

$$G = \sum_{\alpha=1}^{\nu} n_\alpha \bar{\mu}_\alpha = \sum_{\alpha=1}^{\nu} n_\alpha \bar{g}_\alpha + H_{\text{mix}} - T S_{\text{mix}} \,, \tag{20.34}$$

where $\sum_\alpha n_\alpha \bar{g}_\alpha$ is the Gibbs free energy of the unmixed state, and $H_{\text{mix}} - TS_{\text{mix}}$ is the Gibbs free energy of mixing. For the ideal mixture $H_{\text{mix}} = 0$, and S_{mix} is given by (18.26) so that

$$G = \sum_{\alpha=1}^{\nu} n_\alpha \bar{\mu}_\alpha = \sum_{\alpha=1}^{\nu} n_\alpha \bar{g}_\alpha + \bar{R}T \sum_{\alpha=1}^{\nu} n_\alpha \ln X_\alpha \,. \tag{20.35}$$

Thus the chemical potential of the ideal mixture is the sum of the Gibbs free energy of the component alone at mixing conditions (T, p) plus a contribution from the entropy of mixing,

$$\bar{\mu}_\alpha \left(T, p, X_\beta\right) = \bar{g}_\alpha \left(T, p\right) + \bar{R}T \ln X_\alpha \,. \tag{20.36}$$

20.6 The Chemical Potential for an Ideal Gas Mixture

Using that $g = h - Ts$ and the property relations for the ideal gas, the chemical potential of an ideal gas is

$$\bar{\mu}_\alpha \left(T, p, X_\beta\right) = \bar{h}_\alpha \left(T\right) - T \bar{s}_\alpha^0 \left(T\right) + \bar{R}T \ln \frac{X_\alpha p}{p_0} \,. \tag{20.37}$$

As an example we study two ideal gas mixtures at $\left(T, p^{\text{I}}\right)$ and $\left(T, p^{\text{II}}\right)$, separated by a semi-permeable membrane that only allows component ν to pass. The equilibrium condition (20.7) is

$$\bar{\mu}_\nu \left(T, p^{\text{I}}, X_\beta^I\right) = \bar{\mu}_\nu \left(T, p^{\text{II}}, X_\beta^{\text{II}}\right) \,, \tag{20.38}$$

and evaluation with (20.37) gives

$$X_\nu^{\text{I}} p^{\text{I}} = X_\nu^{\text{II}} p^{\text{II}} \,. \tag{20.39}$$

Thus, we find the intuitive result that for an ideal gas the partial pressure $p_\nu = X_\nu p$ is continuous at the ideal semipermeable membrane. This means, that the component that can pass the membrane behaves as if the membrane is not present, it is homogeneously distributed over the entire accessible volume.

20.7 The Chemical Potential as Driving Force for Mass Transfer

We consider the set-up shown in Fig. 20.2: Two reservoirs I and II contain mixtures of different temperature T, pressure p, and composition X_α, which remain constant at all times. Through semipermeable membranes the two reservoirs are connected to a heat and mass conducting duct. Due to the non-equilibrium between the two reservoirs we expect flows of mass and heat. The duct might contain devices to extract work from these flows.

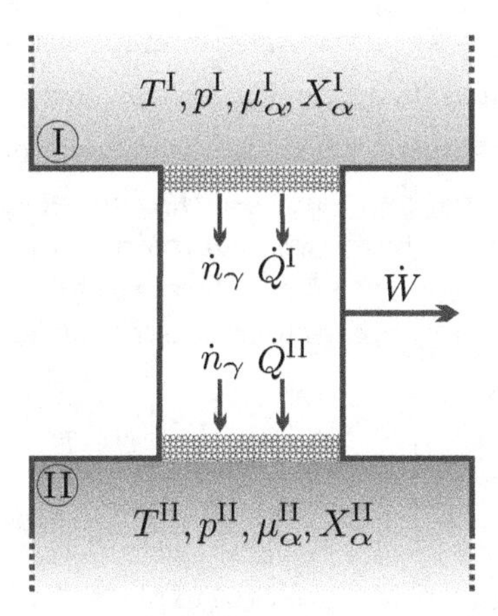

Fig. 20.2 Chemical potential difference as driving force

The membranes let only component γ pass and we consider only steady state processes. Since only component γ passes through membranes and duct, the first and second law for the duct read

$$\dot{n}_\gamma\left(\bar{h}_\gamma^{\mathrm{II}}-\bar{h}_\gamma^{\mathrm{I}}\right)=\dot{Q}^{\mathrm{I}}+\dot{Q}^{\mathrm{II}}-\dot{W} \quad , \quad \dot{n}_\gamma\left(\bar{s}_\gamma^{\mathrm{II}}-\bar{s}_\gamma^{\mathrm{I}}\right)=\frac{\dot{Q}^{\mathrm{II}}}{T^{\mathrm{II}}}+\frac{\dot{Q}^{\mathrm{I}}}{T^{\mathrm{I}}}+\dot{S}_{gen} \,. \quad (20.40)$$

Elimination of the heat rejected to reservoir II gives, with $\bar{s}_\gamma=\frac{\bar{h}_\gamma-\bar{\mu}_\gamma}{T}$,

$$\dot{W}+T^{\mathrm{II}}\dot{S}_{gen}=\dot{n}_\gamma T^{\mathrm{II}}\left[\frac{\bar{\mu}_\gamma^{\mathrm{I}}}{T^{\mathrm{I}}}-\frac{\bar{\mu}_\gamma^{\mathrm{II}}}{T^{\mathrm{II}}}\right]+\left(\dot{n}_\gamma\bar{h}_\gamma^{\mathrm{I}}+\dot{Q}^{\mathrm{I}}\right)\left[1-\frac{T^{\mathrm{II}}}{T^{\mathrm{I}}}\right] \,. \quad (20.41)$$

The right hand side of this equation vanishes in the case of thermal and chemical equilibrium, where $T^{\mathrm{I}}=T^{\mathrm{II}}$ and $\mu_\gamma^{\mathrm{I}}=\mu_\gamma^{\mathrm{II}}$.

Power can be generated from the differences in temperature and chemical potential. We have studied heat engines, which are driven by temperature differences, in great detail in previous chapters. The generation of power from differences in the chemical potential, known as osmotic power generation, will be discussed further below. The maximum possible power is generated when transfer through the duct occurs fully reversibly, so that $\dot{S}_{gen} = 0$.

When the duct is fully irreversible, so that no power is generated, $\dot{W} = 0$, the amount of entropy produced (which must be non-negative) is

$$\dot{S}_{gen} = \dot{n}_\gamma \left[\frac{\bar{\mu}_\gamma^{\mathrm{I}}}{T^{\mathrm{I}}} - \frac{\bar{\mu}_\gamma^{\mathrm{II}}}{T^{\mathrm{II}}} \right] + \left(\dot{n}_\gamma \bar{h}_\gamma^{\mathrm{I}} + \dot{Q}^{\mathrm{I}} \right) \left[\frac{1}{T^{\mathrm{II}}} - \frac{1}{T^{\mathrm{I}}} \right] \geq 0 \,. \tag{20.42}$$

As before, in Sec. 4.10, we interpret the entropy generation in terms of thermodynamic forces and fluxes. The forces are the differences in reduced chemical potential, $\left[\frac{\bar{\mu}_\gamma^{\mathrm{I}}}{T^{\mathrm{I}}} - \frac{\bar{\mu}_\gamma^{\mathrm{II}}}{T^{\mathrm{II}}} \right]$, and inverse temperature, $\left[\frac{1}{T^{\mathrm{II}}} - \frac{1}{T^{\mathrm{I}}} \right]$, between the reservoirs. The related fluxes are the matter flow $\dot{n}_\gamma$ and the total energy flow $\left(\dot{n}_\gamma \bar{h}_\gamma^{\mathrm{I}} + \dot{Q}^{\mathrm{I}} \right)$. Previously, when we discussed Newton's law of cooling, we had only one flux-force pair, and set the flux proportional to the force. For the present setting, we have two such pairs, which offers a richer set of transport laws. Indeed, we allow for *both* forces to affect *both* fluxes, and write fluxes and forces as vectors, with a matrix of coefficients α_{ij},

$$\begin{bmatrix} \dot{n}_\gamma \\ \\ \dot{n}_\gamma \bar{h}_\gamma^{\mathrm{I}} + \dot{Q}^{\mathrm{I}} \end{bmatrix} = A \begin{bmatrix} \alpha_{11} & \alpha_{12} \\ \\ \alpha_{21} & \alpha_{22} \end{bmatrix} \cdot \begin{bmatrix} \frac{\bar{\mu}_\gamma^{\mathrm{I}}}{T^{\mathrm{I}}} - \frac{\bar{\mu}_\gamma^{\mathrm{II}}}{T^{\mathrm{II}}} \\ \\ \frac{1}{T^{\mathrm{II}}} - \frac{1}{T^{\mathrm{I}}} \end{bmatrix} . \tag{20.43}$$

As always, we made the transfer surface A explicit. To ensure positive entropy generation, the matrix of transport coefficients α_{ij} must be positive definite, which is the case for $\alpha_{11} > 0$ and $\alpha_{11}\alpha_{22} - \alpha_{12}\alpha_{21} > 0$.

We note that transfer of matter does not only occur due to differences in pressure (typical flow), and composition (diffusion), but also in response to a temperature gradient (thermodiffusion, or Soret effect, Charles Soret 1854-1904)). Heat transfer does not only occur due to a temperature difference, but also due to differences in pressure or composition (Dufour effect, Louis Dufour 1832-1892).

A deep discussion of laws of this type reveals the Onsager reciprocity relations (Lars Onsager, 1903-1976), which for this case state that the matrix must be symmetric, $\alpha_{21} = \alpha_{21}$. Careful experiments show that the so-called cross effects, which are described by the off-diagonal terms in the matrix, are less important than the direct effects, which are described by the diagonal terms. In engineering applications the cross effects are often ignored.[1]

---

[1] The arguments we use here are typical in the field of Non-equilibrium Thermodynamics. A thorough yet accessible introduction into this topic is presented in: S. Kjelstrup, D. Bedeaux, E. Johannessen, J. Gross: *Non-equilibrium Thermodynamics for Engineers*, World Scientific, Singapore 2010.

When the flow $\dot{n}_\gamma$ goes from I to II as shown, the reduced chemical potential $\frac{\bar{\mu}_\gamma^{\mathrm{I}}}{T^{\mathrm{I}}}$ must be larger than $\frac{\bar{\mu}_\gamma^{\mathrm{II}}}{T^{\mathrm{II}}}$. In other words, a difference in chemical potential causes a flow against the gradient of the (reduced) chemical potential.

When we ignore temperature differences ($T^{\mathrm{I}} = T^{\mathrm{II}} = T$), we can expect a linear law of the form

$$\dot{n}_\nu = \kappa A \left[\bar{\mu}_\gamma^{\mathrm{I}} - \bar{\mu}_\gamma^{\mathrm{II}} \right] , \tag{20.44}$$

where $\kappa = \alpha_{11}/T$ is a positive constitutive coefficient, and A is the cross section of the duct.

In the special case when both mixtures are ideal, and have the same pressures and temperatures, we have $\bar{\mu}_\gamma^{\mathrm{I}} - \bar{\mu}_\gamma^{\mathrm{II}} = \bar{R} T \ln \frac{X_\gamma^{\mathrm{I}}}{X_\gamma^{\mathrm{II}}}$: there will be a particle flux unless both mole fractions are equilibrated.

In short, the desire to equilibrate the chemical potential leads to a flux of that component that can pass the duct. The flux direction is against the gradient, which reduces the gradient. According to Eq. (20.41), this flux can be used to produce power $\dot{W} > 0$. Inversion of the flux, that is forcing flow in the direction of the gradient, requires power input, $\dot{W} < 0$. This is fully equivalent to what we have seen on temperature differences: Heat will flow from hot to cold—against the gradient—by itself, and this heat flux can be used to produce power in a heat engine. Transferring heat from cold to hot—in the direction of the gradient—requires work input, i.e., a heat pump.

Problems

20.1. Symmetry Property for Ideal Mixture
For an ideal mixture, prove the symmetry property (20.16), as well as (20.23).

20.2. Chemical Potential for Ideal Mixture
Determine the chemical potential of an ideal mixture (20.36) by taking the derivative of (20.35), $\bar{\mu}_\alpha = \left(\frac{\partial G}{\partial n_\alpha} \right)_{T,p,n_\gamma}$.

20.3. Mass Based Chemical Potential
Derive the equations for the mass based chemical potential in Sec. 20.4.

20.4. Mass Based Chemical Potential for Ideal Mixture
Rewrite the Gibbs free energy of an ideal mixture (20.35) in terms of mass based quantities, $G(T, p, m_\beta)$ and determine the mass based chemical potential from the derivative, $\mu_\alpha = \left(\frac{\partial G}{\partial m_\alpha} \right)_{T,p,m_\gamma}$. Show that $\bar{\mu}_\alpha = M_\alpha \mu_\alpha$.

20.5. Gas Mixture
A semipermeable membrane which allows only hydrogen to pass divides two containers. In thermodynamic equilibrium container A holds 800 g of nitrogen and 150 g of hydrogen, while container B holds 1800 g of carbon dioxide and 100 g of hydrogen. Container A is kept at a pressure of 20 bar and the temperature of both containers is 300 K.

1. Determine the pressure in container B.
2. Determine the minimum work to separate the hydrogen out of container A.

20.6. Gas Mixture

A semipermeable membrane which allows only hydrogen to pass divides two containers which are kept at 300 K. Initially, container A holds 1000 g of CO_2 at 10 bar, and container B holds 300 g of O_2 at 20 bar. Now 400 g of H_2 are added to the system, such that the pressures of A and B do not change.

1. Determine the amounts of H_2 in containers A and B.
2. Determine the minimum work to separate the H_2 out of container A.

20.7. Gas Mixture

A semipermeable membrane which allows only N_2 to pass divides two containers which are kept at 350 K. Initially, container A holds 2 kg of CO_2 at 10 bar, and container B holds 3 kg of Xe at 20 bar. Now an unknown amount of nitrogen is added to the system, such that the pressures of A and B do not change. After thermodynamic equilibrium is established, it is observed that the volume of container A has increased by 25%.

1. Determine the amount of N_2 added to container A.
2. Determine the amount of N_2 added to container B.
3. Determine the minimum work to separate the nitrogen out of container A.

20.8. Gas Separation

A binary mixture of ideal gases (α, β) is to be partly separated. The mixture is at temperature $T = 300\,\mathrm{K}$ and the initial mole fraction is X_α^I. The goal is to separate half of the initial amount of α from the rest.

1. Determine the mole fraction of α in the mixture remaining after separation.
2. Show that the minimum work for separation is the difference in the entropy of mixing of the final and initial mixtures, times the temperature of the mixture.
3. Determine the minimum work for the required separation, per mole of separated component α.
4. Suppose a semipermeable membrane exists that allows only α to pass. Determine the pressure to which the mixture has to be compressed in order to achieve the desired separation; assume the separated gas is at pressure p_0.
5. Specify the results for minimum separation work and pressure for $X_\alpha^I = 0.5, 0.1, 0.01, 0.001, 0.0001$ and discuss the results. Would you think membrane separation makes sense? What would be the alternatives?

20.9. Ideal Gas in Gravitational Field

Show that in a mixture of ideal gases at temperature T_0 in the gravitational field ($\gamma = 9.81\,\frac{\mathrm{m}}{\mathrm{s}^2}$) the individual components obey the barometric formula. Hint: Minimize free energy to show first that $\bar{\mu}_\beta - \bar{R}T_0 + M_\beta \gamma z = const.$, and then evaluate the chemical potential for ideal gases.

20.10. Binary Mixture in Gravitational Field
An ideal liquid mixture of incompressible components $\alpha = 1, 2$ is enclosed in a piston cylinder system in the gravitational field. The pressure at the piston is p, and the system is kept at constant temperature T. The total mole numbers of the components are n_α, and their local mole densities are denoted as $\nu_\alpha(z)$ with $n_\alpha = \int_0^H \nu_\alpha(z)\, A dz$, where A is the constant cross section of the cylinder, z is the height coordinate, and H is actual height. To simplify the computation, assume that the total mole density is constant, $\nu_1(z) + \nu_2(z) = \nu_0$. By minimizing $G + E_{pot}$ (see Chapter 17) under constraints of given mole numbers n_α, find an equation for the mole fraction as function of height, $X_1(z) = \frac{\nu_1(z)}{\nu_0}$. Show that the mole fraction depends on the difference between molar masses M_α. Simplify for the case of a dilute solution, where $X_1 \ll 1$.

20.11. Gibbs Equation
Show that the Gibbs equation (20.25) leads to the alternative forms

$$Tds = du + pdv - \sum_\alpha \mu_\alpha dc_\alpha \ , \quad df = -sdT - pdv + \sum_\alpha \mu_\alpha dc_\alpha \ .$$

20.12. Equilibrium in a Mixture
Consider a mixture in equilibrium under controlled volume, energy and partial masses. Maximize entropy

$$S = \int \rho s dV \ .$$

under the constraints

$$U = \int \rho u dV \ , \quad m_\alpha = \int \rho c_\alpha dV \ ,$$

and show that in equilibrium pressure p, temperature T and chemical potential μ_α of all components are homogeneous. Use the Gibbs equation in the form

$$Tds = du + pdv - \sum_\alpha \mu_\alpha dc_\alpha \ .$$

Remark: s, u, v, μ_α are specific quantities (per unit mass of mixture); dm is the mass element of mixture. Use Lagrange multipliers.

Chapter 21
Mixing and Separation

21.1 Osmosis and Osmotic Pressure

The word "osmosis" comes from the greek word for pushing and refers to the passing of a substance through a semi-permeable membrane. Applications are, e.g., cell membranes in the human body, or membranes for desalination. For the discussion of osmotic phenomena we shall assume ideal mixtures only.

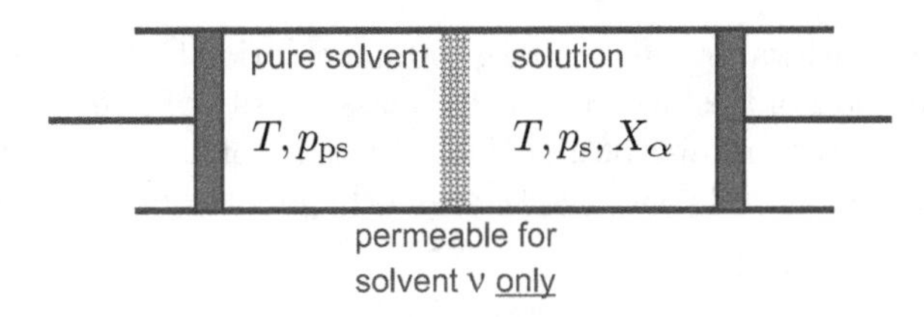

Fig. 21.1 Pressure difference at a membrane

The osmotic pressure is the pressure on the membrane due to those components of the mixture that cannot pass. Specifically, we consider a membrane which separates the pure solvent ν at pressure p_{ps} from an ideal mixture at pressure p_s, see Fig. 21.1. The osmotic pressure is defined as the pressure difference over the membrane in equilibrium,

$$p_{osm} = p_s - p_{ps} \ . \tag{21.1}$$

The equilibrium condition for the membrane is

$$\bar{\mu}_\nu \left(T, p_{ps}\right) = \bar{\mu}_\nu \left(T, p_s, X_\alpha\right) \ . \tag{21.2}$$

For the pure solvent, the chemical potential is just the specific Gibbs free energy, and we find by Taylor expansion

H. Struchtrup, *Thermodynamics and Energy Conversion*,
DOI: 10.1007/978-3-662-43715-5_21,

$$\bar{\mu}_\nu(T,p_{\text{ps}}) = \bar{g}_\nu(T,p_{\text{ps}}) = \bar{g}_\nu(T,p_{\text{s}}) + \left(\frac{\partial \bar{g}_v}{\partial p}\right)_{T,p=p_{\text{s}}} (p_{\text{ps}} - p_{\text{s}}) + \frac{1}{2}\left(\frac{\partial^2 \bar{g}_v}{\partial p^2}\right)_{T,p=p_{\text{s}}} (p_{\text{ps}} - p_{\text{s}})^2 + \ldots \quad (21.3)$$

From now on, we assume that both, solution and solvent, are incompressible liquids. In the equation above, $\left(\frac{\partial \bar{g}_v}{\partial p}\right)_T = \bar{v}_\nu$ is the molar volume of the solvent alone which is assumed to be constant (incompressible!), so that $\left(\frac{\partial^n \bar{g}_v}{\partial p^n}\right)_T = \left(\frac{\partial^{n-1} \bar{v}_v}{\partial p^{n-1}}\right)_T = 0 \ (n \geq 2)$, and

$$\bar{\mu}_\nu(T,p_{\text{ps}}) = \bar{g}_\nu(T,p_{\text{s}}) - \bar{v}_\nu p_{\text{osm}} \,. \quad (21.4)$$

The chemical potential of the solvent in the ideal mixture is

$$\bar{\mu}_\nu(T,p_{\text{s}},X_\alpha) = \bar{g}_\nu(T,p_{\text{s}}) + \bar{R}T \ln X_\nu \,, \quad (21.5)$$

and the equilibrium condition (21.2) reduces to

$$p_{\text{osm}} = -\frac{\bar{R}T}{\bar{v}_\nu} \ln X_\nu \,. \quad (21.6)$$

One kilogram of seawater contains 35 g sodium chloride (NaCl, with $M_{\text{NaCl}} = 58.5\frac{\text{g}}{\text{mol}}$) and 965 g of freshwater. It has a mass density of about $1029\frac{\text{kg}}{\text{m}^3}$, i.e., one litre of seawater contains 36 g of NaCl and 993 g of freshwater. Since NaCl dissociates into Na^+ and Cl^- ions, the osmotic pressure of seawater at 277 K (4 °C) is

$$p_{\text{osm}}^{sw} = -\frac{\bar{R}T}{M_w v_w} \ln \frac{n_w}{n_w + n_{\text{Na}^+} + n_{\text{Cl}^-}} = -\frac{\bar{R}T}{M_w v_w} \ln \frac{1}{1 + 2\frac{M_w m_{\text{NaCl}}}{m_w M_{\text{NaCl}}}} = 28.2\,\text{bar}. \quad (21.7)$$

Recall that solutions of salt in water are not ideal mixtures, unless they are diluted, see Sec. 18.14.

21.2 Osmotic Pressure for Dilute Solutions

The osmotic pressure assumes a rather interesting form in the limit of dilute solutions, where most of the solution is solvent ν, that is $n_\nu \gg n_\alpha$ holds. Then we can expand

$$-\ln X_\nu = -\ln\frac{n_\nu}{n} = -\ln\frac{n_\nu}{n_\nu + \sum_{\alpha=1}^{\nu-1} n_\alpha} = \ln\left(1 + \sum_{\alpha=1}^{\nu-1}\frac{n_\alpha}{n_\nu}\right) \simeq \sum_{\alpha=1}^{\nu-1}\frac{n_\alpha}{n_\nu} \tag{21.8}$$

so that, when we also use that $n_\nu \bar{v}_\nu = V$ is the volume of the mixture, the osmotic pressure is

$$p_{\mathrm{osm}} = \sum_{\alpha=1}^{\nu-1}\frac{n_\alpha \bar{R}T}{V} \,. \tag{21.9}$$

The dissolved substances in a dilute solution exert an ideal gas pressure on the membrane; the osmotic pressure is the sum of these ideal gas pressures.

21.3 Example: Pfeffer Tube

An instructive example for the strength of the osmotic forces is the Pfeffer tube (Wilhelm Pfeffer, 1845-1920), depicted in Fig. 21.2. The solvent is water. A tube is closed by a membrane that only lets the solvent ν pass, and is set vertically into a bath of solvent. Some solvent will pass the membrane, so that the solvent inside and outside the tube are at the same level. Then salt is added to the tube. This leads to an additional amount of solvent drawn from the bath into the tube; in the final equilibrium, the osmotic pressure and the hydrostatic pressure are balanced.

Since this is a system in which temperature and (environmental) pressure are controlled, the total Gibbs free energy $G = H - TS + E_{pot}$ of the system will minimize. Drawing solvent into the tube increases the entropy of the

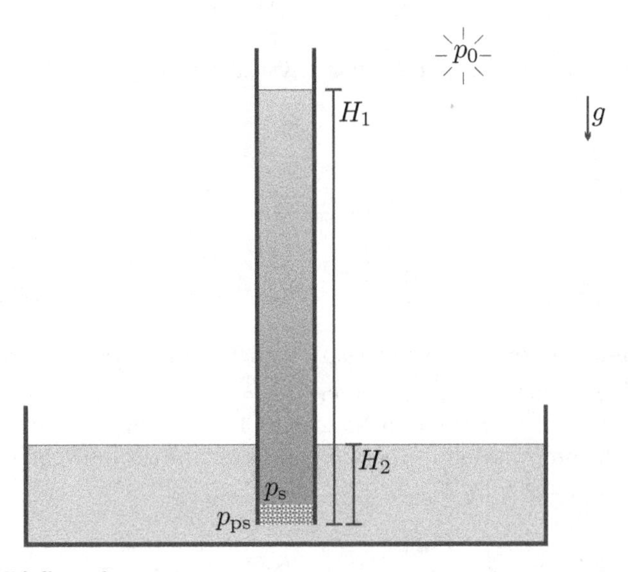

Fig. 21.2 Pfeffer tube

dissolved particles, which now have a greater volume to access, but also increases the potential energy. The final equilibrium state, a balance of height increasing entropic force and height reducing energetic force, will be calculated next under some simplifying assumptions. The goal is the computation of the height H_1 in the tube.

Directly at the membrane, the hydrostatic pressures of the solution in the tube and solvent in the bath are

$$p_{\mathrm{s}} = p_0 + \rho_{\mathrm{s}} g H_1 \quad , \quad p_{\mathrm{ps}} = p_0 + \rho_\nu g H_2 \; ,$$

and these are balanced by the osmotic pressure, so that

$$p_{\mathrm{osm}} = p_{\mathrm{s}} - p_{\mathrm{ps}} = \rho_{\mathrm{s}} g H_1 - \rho_\nu g H_2 \; .$$

With the simplified expression (21.9) we find at first.

$$\frac{\bar{R}T}{Vg} \sum_{\alpha=1}^{\nu-1} n_\alpha = \rho_{\mathrm{s}} H_1 - \rho_{\mathrm{ps}} H_2 \; .$$

$V = AH_1$ is the volume of solution in the tube of cross section A. The density of the solution is $\rho_{\mathrm{s}} = \rho_\nu + \sum_{\alpha=1}^{\nu-1} \rho_\alpha \simeq \rho_{\mathrm{ps}} + \sum_{\alpha=1}^{\nu-1} \frac{m_\alpha}{V}$, where we approximated the density of the solvent in solution, ρ_ν, by the density of the pure solvent, ρ_{ps}. After multiplication with H_1/ρ_{ps} we obtain a quadratic equation for H_1,

$$H_1^2 + H_1 \left(\frac{1}{\rho_{\mathrm{ps}} A} \sum_{\alpha=1}^{\nu-1} m_\alpha - H_2 \right) - \frac{\bar{R}T}{A \rho_{\mathrm{ps}} g} \sum_{\alpha=1}^{\nu-1} n_\alpha = 0 \; .$$

Only the positive solution for H_1 has physical meaning,

$$H_1 = \frac{1}{2} \left(H_2 - \frac{1}{\rho_{\mathrm{ps}} A} \sum_{\alpha=1}^{\nu-1} m_\alpha \right) + \sqrt{\frac{1}{4} \left(H_2 - \frac{1}{\rho_{\mathrm{ps}} A} \sum_{\alpha=1}^{\nu-1} m_\alpha \right)^2 + \frac{\bar{R}T}{A \rho_{\mathrm{ps}} g} \sum_{\alpha=1}^{\nu-1} n_\alpha} \; .$$

As example we consider water ($\rho_{\mathrm{ps}} = 1000 \frac{\mathrm{kg}}{\mathrm{m}^3}$) in a tube of cross section $A = 1\,\mathrm{cm}^2$ in which $\sum_{\alpha=1}^{\nu-1} m_\alpha = m_{\mathrm{NaCl}} = 1\,\mathrm{g}$ of cooking salt (NaCl) is dissolved. The salt dissolves into Na^+ and Cl^- ions, and thus we have to be careful in the computation of the mole number of dissolved particles, which is (with $M_{\mathrm{NaCl}} = 58.5 \frac{\mathrm{g}}{\mathrm{mol}}$)

$$\sum_{\alpha=1}^{\nu-1} n_\alpha = n_{\mathrm{Na}^+} + n_{\mathrm{Cl}^-} = 2 \frac{m_{\mathrm{NaCl}}}{M_{\mathrm{NaCl}}} = 3.419\, 10^{-2}\,\mathrm{mol} \; .$$

When the tube is immersed into the pure solvent by $H_2 = 1\,\mathrm{cm}$, the solution in the tube will reach a height of $H_1 = 9.29\,\mathrm{m}$, which corresponds to about 1 litre of water in the tube! This shows the enormous forces that are present in osmosis, which are due to the desire of the salt to increase the entropy of mixing by making the accessible volume (i.e., the volume of the solution), as large as possible.

Due to the difference in density between salt solution and pure water, the height H_1 becomes smaller as the tube is pushed deeper into the solvent (i.e., for larger H_2), as the problems show, it might even happen that $H_2 > H_1$.

The strong desire of the salt to draw water can be used (and was widely used in the past) to cure meat: While water can pass cell membranes, salt cannot, and thus a piece of meat or fish immersed in salt will be depleted from some water. The same is true for bacteria on the meat which will die from dehydration, and thus cannot spoil the meat.

Since seawater contains more salt than the cells of the human body, drinking seawater is deadly: The seawater will draw water from the cells, which will be damaged by dehydration. In fact, after drinking seawater one will be more thirsty than before.

Another example for osmotic forces is putting sugar on strawberries, which will draw water (juice) out of the strawberries that mixes with the sugar. The opposite can be seen when raisins are put into water: the sugar inside the raisin draws water in, and the raisin swells.

21.4 Desalination in a Continuous Process

In the previous section we have seen that salt can draw water through a semipermeable membrane. In the inverse process, fresh water is obtained from saltwater, by pressing the saltwater against a semipermeable membrane, which only allows freshwater to pass, e.g., think of increasing the pressure on the saltwater column in the Pfeffer tube. We ask for the minimum work required for desalination, that is for the work required in a reversible process.

Figure 21.3 shows the general set-up for a continuous desalination plant, without specifying how desalination is taking place inside the plant, which is drawn as a grey box. A mole flow $\dot{n}_{sw}$ of salt water with salt mole fraction X_{sw} enters the plant at environmental pressure and temperature (p_0, T_0). Work $\dot{W}$ is supplied to the plant, which also exchanges heat $\dot{Q}$ with the environment. Two streams leave the plant at (p_0, T_0), a stream of freshwater $\dot{n}_{fw}$, and the brine stream $\dot{n}_b$ which contains all salt and has a salt mole fraction $X_b > X_{sw}$.

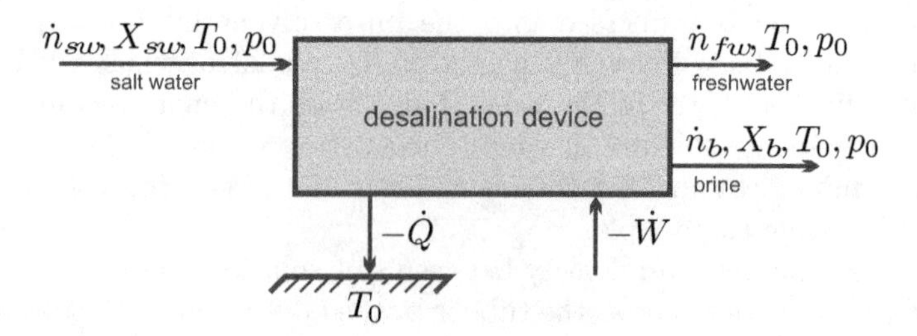

Fig. 21.3 Continuous desalination

This is a best case scenario, where the streams entering and leaving the plant are in thermal (T_0) and mechanical (p_0) equilibrium with the environment. The following calculation will assume reversible processes, and thus the actual work that a real plant requires will be larger. It is the task of the engineer to design a process that uses as little work as possible, and the calculation of the minimum work will serve as an important guideline on the quality of the actual process devised.

The overall mole balance and the mole balance for salt read

$$\dot{n}_{sw} = \dot{n}_{fw} + \dot{n}_b \quad , \quad \dot{n}_{sw} X_{sw} = \dot{n}_b X_b \, . \tag{21.10}$$

With the freshwater/seawater ratio $y = \dot{n}_{fw}/\dot{n}_{sw}$ we have

$$X_b = \frac{X_{sw}}{1-y} \, ; \tag{21.11}$$

note that y reaches its maximum $y_{\max} = 1 - X_{sw}$ for full desalination, where the brine would be pure salt ($X_b = 1$).

First and second law for the plant, which exchanges heat only at T_0, read

$$\sum_{out} \dot{n}_\alpha \bar{h}_\alpha - \sum_{in} \dot{n}_\alpha \bar{h}_\alpha = \dot{Q} - \dot{W} \quad , \quad \sum_{out} \dot{n}_\alpha \bar{s}_\alpha - \sum_{in} \dot{n}_\alpha \bar{s}_\alpha = \frac{\dot{Q}}{T_0} + \dot{S}_{gen} \, , \tag{21.12}$$

all properties are taken at environmental conditions (p_0, T_0). Elimination of $\dot{Q}$ gives

$$\dot{W} = -T_0 \dot{S}_{gen} - \Delta \dot{G}^0 \, , \tag{21.13}$$

where

$$\begin{aligned} \Delta \dot{G}^0 &= \sum_{out} \dot{n}_\alpha \left(\bar{h}_\alpha - T_0 \bar{s}_\alpha \right) - \sum_{in} \dot{n}_\alpha \left(\bar{h}_\alpha - T_0 \bar{s}_\alpha \right) \\ &= \sum_{out} \dot{n}_\alpha \bar{\mu}_\alpha \left(T_0, p_0, X_\beta \right) - \sum_{in} \dot{n}_\alpha \bar{\mu}_\alpha \left(T_0, p_0, X_\beta \right) \, . \end{aligned} \tag{21.14}$$

Since all flows are at (p_0, T_0), $\Delta\dot{G}^0$ is the difference in Gibbs free energy at environmental conditions flowing through per unit time.

According to (21.4), for reversible operation, $\dot{S}_{gen} = 0$, the work required for desalination is given by $-\Delta\dot{G}^0$. To simplify the problem we assume ideal mixtures, where $\bar{\mu}_\alpha(T_0, p_0, X_\beta) = \bar{g}_\alpha(T_0, p_0) + \bar{R}T_0 \ln X_\alpha$, and find the minimum work as

$$\dot{W}_{rev} = -\Delta\dot{G}^0 = T_0\Delta\dot{S}_0 = \bar{R}T_0\left[\sum_{in}\dot{n}_\alpha \ln X_\alpha - \sum_{out}\dot{n}_\alpha \ln X_\alpha\right] . \quad (21.15)$$

Inserting the proper flows and mole numbers yields at first

$$\begin{aligned}\dot{W}_{rev} = \bar{R}T_0 &\left[\dot{n}_{sw}X_{sw}\ln X_{sw} + \dot{n}_{sw}(1 - X_{sw})\ln(1 - X_{sw})\right. \\ &\left.-\left(\dot{n}_{fw}\ln X_{fw} + \dot{n}_b X_b \ln X_b + \dot{n}_b(1 - X_b)\ln(1 - X_b)\right)\right] ;. \quad (21.16)\end{aligned}$$

note that the freshwater outflow is unmixed, i.e., $X_{fw} = 1$. After some algebra, this finally assumes the form

$$\begin{aligned}\dot{W}_{rev} = -p_{\text{osm}}\dot{V}_{fw}&\left[\frac{(1-y)\ln(1-y) + (1-X_{sw})\ln(1-X_{sw})}{y\ln(1-X_{sw})}\right. \\ &\left.-\frac{(1-y-X_{sw})\ln(1-y-X_{sw})}{y\ln(1-X_{sw})}\right] , \quad (21.17)\end{aligned}$$

where $\dot{V}_{fw} = \bar{v}_{fw}\dot{n}_{fw}$ is the volume flow of freshwater produced, and $p_{\text{osm}} = -\frac{\bar{R}T}{\bar{v}_{fw}}\ln(1 - X_{sw})$ is the osmotic pressure of the incoming seawater.

The minimum work required per liter of seawater, $\dot{W}_{rev}/\dot{V}_{fw}$, depends on the salt concentration X_{sw} in the incoming seawater, and on the extraction ratio y. In the limit that only a small amount of freshwater is extracted, $y \ll 1$, the work is proportional to the osmotic pressure of the seawater,

$$\dot{W}_{rev}(y = 0) = -p_{\text{osm}}\dot{V}_{fw} . \quad (21.18)$$

For the case of complete desalination, $y \to y_{\max} = (1 - X_{sw})$, the work required is larger,

$$\dot{W}_{rev}(y_{\max}) = -p_{\text{osm}}\dot{V}_{fw}\left(1 + \frac{X_{sw}\ln X_{sw}}{(1-X_{sw})\ln(1-X_{sw})}\right) = -\dot{n}_{sw}T\Delta\bar{s}_{mix} , \quad (21.19)$$

where $\Delta\bar{s}_{mix}$ is the entropy of mixing for mixing brine and freshwater per mole of mixture, i.e., for one mole of seawater. It should be noted that the assumption of ideal mixture will not hold for large salt content, thus this is a hypothetical value.

Figure 21.4 shows the work required as function of the freshwater ratio y. For smaller values of y the increase in work requirement is not too large, and one might consider to go to values of $y = 0.5$, or so. Note that larger values of y reduce the amount of seawater drawn for a given amount of freshwater produced, and thus the overall size of the device or plant. As y approaches the maximum value, we observe fast increase of the work requirement.

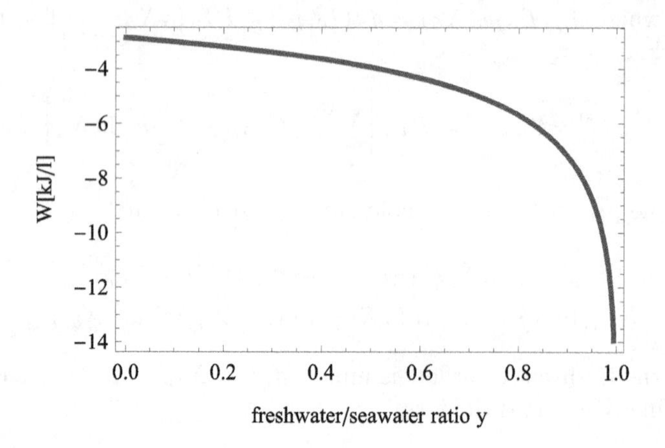

Fig. 21.4 Work required per litre of seawater over freshwater/seawater ratio y

The work requirement will be larger in actual processes, where additional work input is needed to overcome irreversible losses. For a membrane-based desalination system there are losses to friction in the pipes, the pump, in filters, and as water is pressed through the membrane, and losses due to non-uniform salt concentration: salt might accumulate in front of the membrane and this *concentration polarization* increases the local osmotic pressure and thus the work requirement. Note that less work is needed when reverse osmosis takes place at lower temperatures, where the osmotic pressure is lower. Typical values for separation work in commercial membrane desalination systems are about 10 kJ per litre of freshwater.

21.5 Reversible Mixing: Osmotic Power Generation

Rivers transport freshwater into the sea, where freshwater and saltwater mix. The difference in salt content can be used to drive osmotic power plants.

An intuitive basic set-up is to enclose saltwater in a piston-cylinder system and bring it into contact with the freshwater via a semipermeable membrane. The saltwater will draw freshwater into the cylinder to increase its entropy, the piston will be pushed and work is done. Note that, in principle, the salt inside the cylinder can draw an infinite amount of freshwater! Alternatively, one can enclose the freshwater in the cylinder and bring it into contact with the seawater via a semipermeable membrane. The saltwater will draw freshwater out of the cylinder to increase its entropy, the piston will be pulled in and work is done.

Osmotic power generation is in development, with one trial plant operating in Norway. A main bottleneck for large scale commercial application is the development of suitable membranes.

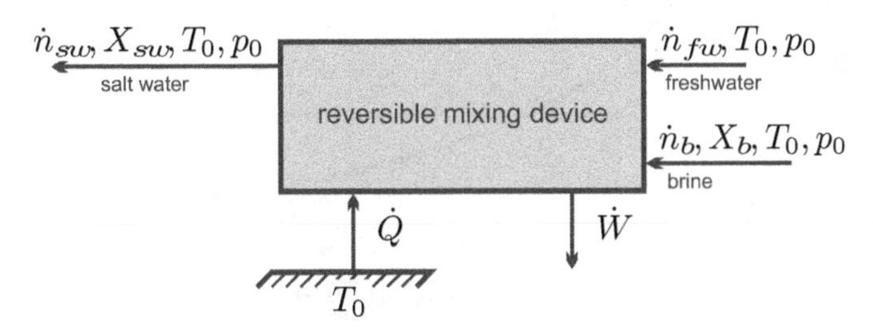

Fig. 21.5 Mole flows, work and heat for reversible mixing

The amount of work that could be obtained can be determined from inverting the continuous desalination process discussed in the previous section, see Fig. 21.5. To be able to use the same notation, we consider the reversible mixing of brine ($\dot{n}_b$, X_b) and freshwater ($\dot{n}_{fw}, X_{fw} = 1$) to saltwater ($\dot{n}_{sw}$, X_{sw}); for mixing of seawater and river water, X_b is the composition of the *seawater*.

We introduce the freshwater/brine ratio as $x = \frac{\dot{n}_{fw}}{\dot{n}_b}$, for which we have

$$1 - y = \frac{\dot{n}_{sw} - \dot{n}_{fw}}{\dot{n}_{sw}} = \frac{\dot{n}_b}{\dot{n}_b + \dot{n}_{fw}} = \frac{1}{1+x} \quad , \quad X_{sw} = \frac{X_b}{1+x} \, . \tag{21.20}$$

The work obtained by reversible mixing per litre of fresh water follows from (21.17) as (note that all signs must be inverted for the inverted process)

$$\begin{aligned}\frac{\dot{W}_{rev}}{\dot{V}_{fw}} = \frac{\bar{R}T}{\bar{v}_{fw}} \frac{1}{x} \big[&(1+x)\ln(1+x) + (1-X_b)\ln(1-X_b) \\ &- (1+x-X_b)\ln(1+x-X_b) \big] \, .\end{aligned} \tag{21.21}$$

Here, $\dot{V}_{fw} = \bar{v}_{fw}\dot{n}_{fw}$ is the volume flow of freshwater. It is assumed that the two streams have the same temperature.

In the limit $x \to 0$, where the freshwater (fw) mixes with an infinite amount of seawater (b), the work is proportional to the osmotic pressure of the seawater, i.e.,

$$\dot{W}_{rev} = -\frac{\bar{R}T}{\bar{v}_{fw}} \dot{V}_{fw} \ln(1 - X_b) = p_{\mathrm{osm}} \dot{V}_{fw} \, . \tag{21.22}$$

For the opposite case, where salty brine (b) mixes with a large amount of freshwater (fw), so that $x \gg 1$, the work can also be expressed with the volume flow of brine,

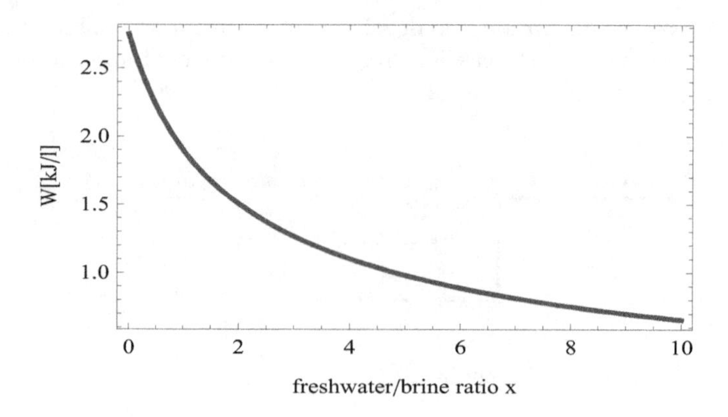

Fig. 21.6 Work produced per litre of freshwater over freshwater/saltwater ratio x

$$\dot{W}_{rev} = \dot{V}_{fw} \frac{\bar{R}T}{\bar{v}_{fw}} X_b \frac{\ln x}{x} = \dot{V}_b \frac{\bar{R}T}{\bar{v}_b} X_b \ln x \,. \tag{21.23}$$

Figure 21.6 shows the work that could be produced in the best, i.e., reversible, case per litre of freshwater for $X_b = 0.0218$ (which is the value for seawater) at $T = 5\,°\mathrm{C}$. The maximum work ($2.8 \frac{\mathrm{kJ}}{\mathrm{litre}}$) could be obtained by reversibly mixing the freshwater with an infinite amount of seawater (limit $x \to 0$). When even amounts of salt- and freshwater are mixed ($x = 1$), one still could produce $1.9 \frac{\mathrm{kJ}}{\mathrm{litre}}$. The river Rhine, Germany's largest river, discharges in average a volume flow of $2200 \frac{\mathrm{m}^3}{\mathrm{s}}$. The power potential of an osmotic power plant that uses 1/10 of the volume flow and operates at $x = 1$ is $\dot{W}_{pot} = 418\,\mathrm{MW}$. There are many rivers with large[1] or small discharge, which in principle offer a tremendous potential for regenerative energy production (driven by weather processes, i.e., the sun). Not all locations will be suitable, and a feasible technology that operates on large and small scales would be helpful.

Particular care must be taken in keeping the osmotic pressure of the incoming seawater high. Normally the freshwater discharged from a river into the ocean will mix with the local seawater, and the mixture in front of the river mouth will have a lower salt content, and therefore lower osmotic pressure, than the ocean far away. This implies that the mixture that leaves the osmotic power plant must be discharged at some distance from the origin of the saltwater entering the plant.

[1] Amazon: $219,000 \frac{\mathrm{m}^3}{\mathrm{s}}$, Congo $41,800 \frac{\mathrm{m}^3}{\mathrm{s}}$, Mississippi/Missouri: $16,200 \frac{\mathrm{m}^3}{\mathrm{s}}$, Mackenzie/Peace/Finlay: $9,910 \frac{\mathrm{m}^3}{\mathrm{s}}$.

The Norwegian company Statkraft (www.statkraft.com) is developing a power plant based on the concept of pressure retarded osmosis, see Fig. 21.7. The process relies on the fact that seawater pressurized to a pressure p_{sw} between the environmental and the osmotic pressure of the seawater, $(p_{sw} - p_0) < p_{\text{osm}}$, draws freshwater at the environmental pressure p_0 through a semipermeable membrane. This increases the flow of pressurized brackish (lower salt content than seawater) water, which is split into two streams. The first stream runs through a turbine to produce power, while the other stream is used to pressurize the incoming seawater in a pressure exchanger. The relative amount of freshwater, i.e., x in the above calculation, depends on the pressure p_{sw}. In reversible operation, as soon as the osmotic pressure of the brackish water, which decreases due to dilution, has reached the value $(p_{sw} - p_0)$, no further freshwater is drawn in. Irreversible pressure losses in pipes and pressure exchanger will reduce the amount of power produced.

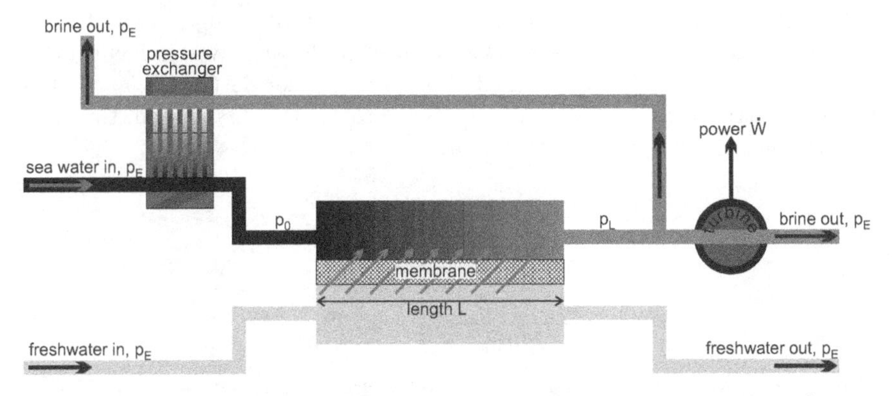

Fig. 21.7 Pressure retarded osmosis power plant (after www.statkraft.com)

21.6 Example: Desalination in Piston-Cylinder Device

A reciprocating desalination device as depicted in Fig. 21.8 operates according to the following cycle:

1-2: Intake of seawater at constant pressure p_0, until the intake valve closes at volume V_2.

2-3: Compression of seawater up to p_3, which is the pressure at which freshwater starts to pass a semipermeable membrane. The seawater can be considered as incompressible.

3-4: Further increase of pressure forces freshwater through the membrane, until a maximum pressure p_4 is reached.

4-5: The exit valve opens, and the pressure drops to p_0.

5-6: The brine is pushed out at constant pressure p_0.

6-1: Exit valve closes, inlet valve opens.

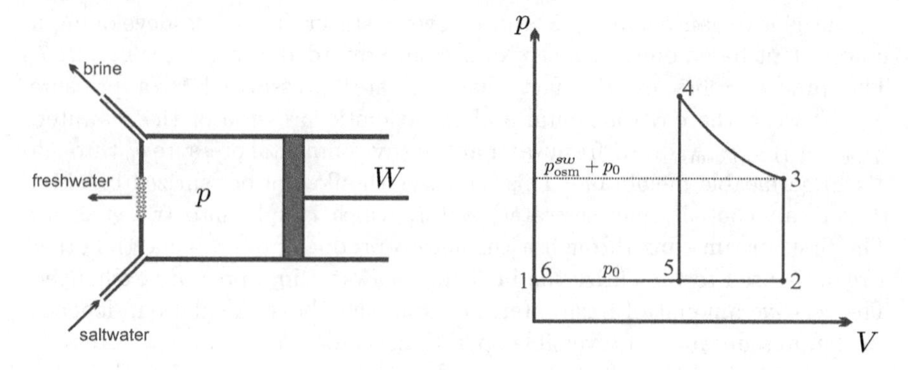

Fig. 21.8 Reciprocating desalination cycle

During the process the temperature remains constant at 15 °C.

We ask for the work to produce 1 litre of freshwater ($\rho_{H_2O} = 1000\frac{g}{litre}$) in relation to the freshwater-seawater ratio. The local seawater has a mass density $\rho_{sw} = 1030\frac{kg}{m^3}$, it contains 35 g sodium chloride (NaCl) and 995 g water per litre, so that $\rho_{NaCl}^{sw} = 35\frac{g}{litre}$. Since the salt is dissociated, the mole numbers of salt and water at state 2 are

$$n_{NaCl} = 2\frac{\rho_{NaCl}^{sw}}{M_{NaCl}}V_2 = 1.197\frac{mol}{litre}V_2$$

and

$$n_{H_2O}^{sw} = \frac{\rho_{sw} - \rho_{NaCl}^{sw}}{M_{H_2O}}V_2 = 55.28\frac{mol}{litre}V_2 \, .$$

Thus, the mole fraction of salt in the seawater is

$$X_{sw} = \frac{n_{NaCl}}{n_{NaCl} + n_{H_2O}^{sw}} = 0.021 \, ,$$

and its osmotic pressure is

$$p_{osm}^{sw} = p_3 - p_0 = -\rho_{H_2O}R_{H_2O}T \ln\left[1 - X_{sw}\right] = 28.3\,\text{bar} \, .$$

Next we ask for the relation between pressure and volume during the desalination in step 3-4. The volume of fresh water produced is $V_{fw} = V_2 - V$ which corresponds to the mole number

$$n_{fw} = \frac{\rho_{H_2O}}{M_{H_2O}}\left(V_2 - V\right) \, .$$

The mole number of water molecules remaining in the volume V is

$$n_r = n_{H_2O}^{sw} - n_{fw} = \frac{\rho_{sw} - \rho_{NaCl}^{sw}}{M_{H_2O}}V_2 - \frac{\rho_{H_2O}}{M_{H_2O}}\left(V_2 - V\right) = \frac{\rho_{H_2O}}{M_{H_2O}}\left(V - V_s\right) \, ,$$

where

$$V_s = \frac{\rho^{sw}_{\mathrm{NaCl}} - \rho_{sw} + \rho_{\mathrm{H_2O}}}{\rho_{\mathrm{H_2O}}} V_2$$

is the volume of the salt;[2] for the given data $V_s/V_2 = 0.005$. Since all salt remains in V, the mole fraction of salt in V is

$$X_r = \frac{n_{\mathrm{NaCl}}}{n_{\mathrm{NaCl}} + n_r} = \left[1 + \frac{1}{2}\frac{M_{\mathrm{NaCl}}}{M_{\mathrm{H_2O}}}\frac{\rho_{\mathrm{H_2O}}}{\rho^{sw}_{\mathrm{NaCl}}}\frac{V - V_s}{V_2}\right]^{-1} = \left[1 + \frac{1}{\gamma}\frac{V - V_s}{V_2}\right]^{-1} ;$$

for the given data $\gamma = 0.02154$. Then the pressure p in the cylinder is $p(V) = p_0 + p^r_{\mathrm{osm}}(V)$ with

$$p^r_{\mathrm{osm}}(V) = -\rho_{\mathrm{H_2O}} R_{\mathrm{H_2O}} T \ln\left[1 - X_r(V)\right] .$$

The work for the cycle is

$$W = \oint p dV = \int_{V_3}^{V_4} p^r_{\mathrm{osm}}(V)\, dV .$$

and with the abbreviations $\beta = \frac{V_{fw}}{V_2}$, $\alpha = 1 - \frac{V_s}{V_2} = 0.995$, we obtain for the work per litre of freshwater produced

$$\frac{W}{V_{fw}} = \rho_{\mathrm{H_20}} R_{\mathrm{H_2O}} T \frac{1}{\beta}\left[(\gamma + \alpha - \beta)\ln(\gamma + \alpha - \beta) + \alpha \ln \alpha \right.$$
$$\left. - (\alpha - \beta)\ln(\alpha - \beta) - (\gamma + \alpha)\ln(\gamma + \alpha)\right] .$$

When plotted as a function of the freshwater-seawater ratio by volume, β, the curve is very similar to the one shown in Fig. 21.4. Indeed, it is an easy, but somewhat cumbersome, exercise to show that the above agrees with (21.17).

21.7 Example: Removal of CO_2

The effect of greenhouse gases such as CO_2 on Earth's climate is widely discussed. One of the solutions to at least reduce the impact of burning carbon fuels is to remove CO_2 from the combustion product, compress it, and and store it in depleted gas or oil reservoirs below ground (CCS - *carbon capture and storage*). As an alternative, it is sometimes suggested to remove CO_2 from the atmosphere. This is a less viable alternative, since the concentration of CO_2 in the atmosphere is very low, which makes it more costly to remove it, as will be seen below.

[2] Assuming ideal mixture. As said before, this assumption is only valid for low salt concentrations. The computation of V_s is an extrapolation, and V_s should be seen as a useful abbreviation.

In the oxy-fuel process, oxygen is separated from air, then mixed with fuel and a portion of the power plant's exhaust, then the mixture is burned to provide heat for power production. The exhaust is only CO_2 and water, which can be separated rather easily by condensation of water. Since the concentration of oxygen in the air is far higher than the concentration of CO_2 in air, the separation work for O_2 is lower in comparison.

To estimate the work requirements, we consider air as a mixture of "gas" (no need to specify the composition), and carbon dioxide or oxygen with mole fraction X and ask for the work required per mole of CO_2 or O_2 to reduce the mole fraction in the air to $\hat{X} = \alpha X$. For the computation we consider a given amount n_{air} of air at (T, p), and a fully reversible separation process. The total amount of air is

$$n_{air} = n_{gas} + n_s ,$$

where n_s is the number of moles of the component to be separated (CO_2 or O_2) in the air, and n_{gas} is the number of moles of all other gases in the air.

Since all contributing gases are ideal, and since the temperature remains constant, the first law just gives that the work for separation is equal to the heat that must be removed,

$$W_{12} = Q_{12} .$$

The second law for the reversible process gives

$$S_2 - S_1 = \frac{Q_{12}}{T} ;$$

thus the separation work is

$$W_{12} = T (S_2 - S_1) .$$

Since total pressure and temperature remain unchanged, only the mixing contributions to entropy are relevant.

In state 1, before separation, we have n_{gas} moles of "gas" mixed with $n_s = \frac{X}{1-X} n_{gas}$ moles of CO_2 (or O_2). In state 2 we have n_{gas} moles of "gas" mixed with $\hat{n}_s = \frac{\hat{X}}{1-\hat{X}} n_{gas}$ moles of CO_2 (or O_2), and $[n_s - \hat{n}_s]$ moles of pure separated CO_2 (or O_2). The corresponding mixing contributions to entropy are

$$S_1 = -\bar{R} \sum n_\alpha \ln X_\alpha = -\bar{R} \left[n_{gas} \ln (1 - X) + n_s \ln X \right] ,$$
$$S_2 = -\bar{R} \sum n_\alpha \ln X_\alpha = -\bar{R} \left[n_{gas} \ln \left(1 - \hat{X}\right) + \hat{n}_s \ln \hat{X} \right] .$$

We obtain the work per mole of separated CO_2 (or O_2) as

$$\bar{w} = \frac{W_{12}}{n_s - \hat{n}_s} = -\bar{R}T \frac{\ln(1-\alpha X) + \frac{\alpha X}{1-\alpha X}\ln(\alpha X) - \ln(1-X) - \frac{X}{1-X}\ln X}{\frac{X}{1-X} - \frac{\alpha X}{1-\alpha X}} .$$

The absolute work is larger for smaller original mole fraction. Figure 21.9 shows $\bar{w}$ for $\alpha = \{0.9, 0.5, 0.1\}$ as function of the original mole fraction X. The figure indicates that removal from carbon rich exhaust streams, where X is larger, is cheaper.

It is worthwhile to look at some numbers: Removal of CO_2 from atmospheric air, where[3] $X_{CO_2} = 400\,\text{ppm}$, requires a reversible separation work of $\{19.6,\ 20.2,\ 21.3\}\,\frac{\text{kJ}}{\text{mol}}$ for $\alpha = \{0.9, 0.5, 0.1\}$. On the other hand, removal of O_2 from atmospheric air, where $X_{O_2} = 0.21 = 210000\,\text{ppm}$, requires a reversible separation work of approximately $\{3.99,\ 4.59,\ 5.54\}\,\frac{\text{kJ}}{\text{mol}}$, that is just a quarter of the work required for CO_2.

The Virgin Earth Challenge[4] asks for the removal of *one billion metric tons* of carbon dioxide per year, that is $2.273 \times 10^{10}\,\frac{\text{kmol}}{\text{year}}$. This would require—in form of work!— $4.5 \times 10^{14}\,\frac{\text{kJ}}{\text{year}}$ at least (in a fully reversible process), that is about 0.1 percent of the world's energy consumption of $16\,\text{TW} = 5 \times 10^{17}\,\frac{\text{kJ}}{\text{year}}$, and about 0.6 percent of the world's power generation of $2.3\,\text{TW} = 7.2 \times 10^{16}\,\frac{\text{kJ}}{\text{year}}$. The world's yearly emissions of CO_2 are ca. $30 \times 10^{12}\,\frac{\text{kg}}{\text{year}}$ or $6.8 \times 10^{11}\,\frac{\text{kmol}}{\text{year}}$. The minimum (!) work requirement for the removal of this amount directly from the atmosphere is $1.45 \times 10^{16}\,\frac{\text{kJ}}{\text{year}}$—about 2.9% of the world's energy consumption, and 20% of the world's power generation.

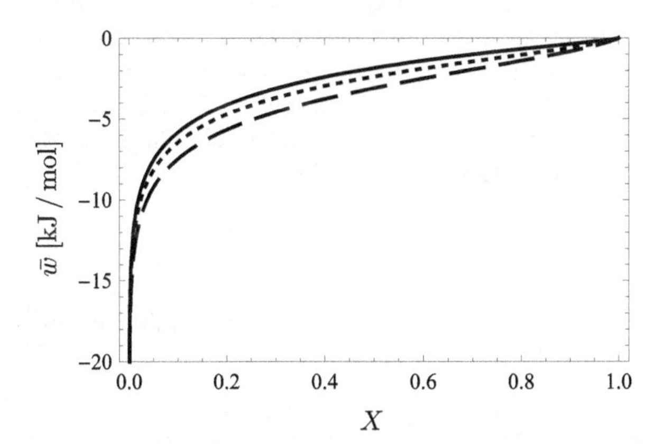

Fig. 21.9 Work $\bar{w}$ to remove 1 mole of gas from air with initial gas mole fraction X, so that remaining air has gas content αX for $\alpha = 0.9$ (continuous), $\alpha = 0.5$ (short dashes), $\alpha = 0.1$ (long dashes)

[3] Value for January 2014. In average, the yearly increase of the CO_2 content is about 2 ppm. This is about half of the newly emitted CO_2, the remainder is absorbed into the oceans.

[4] Launched in 2007, `http://www.virginearth.com`

The reversible separation of oxygen ($X_{O_2} = 0.21$) requires only about $5\frac{\text{kJ}}{\text{mol}}$, and thus the cost of oxygen separation is lower by approximately a factor of four. Note that one oxygen molecule is required to form one CO_2 molecule in combustion.

Obviously, additional cost occurs for compressing, pumping and storing of the removed carbon dioxide, and, most importantly, for overcoming the irreversibilities in the separation process. Thus, it might be that the separation cost is only a small part of the bill—only a careful cost analysis can show the relative importance of the various cost factors.[5]

Indeed, an important cost factor that must not be forgotten is the occurrence of *irreversible* losses in the separation process. After all, the above calculation only considers the reversible limit for separation work. The actual amount of losses will depend strongly on the process chosen for separation, and we can only make some general estimates.

It is a reasonable assumption that the work loss for separation of one mole of CO_2 (or of O_2, in the case of the oxy-fuel process) is proportional to the amount of air that must be moved through the separation plant and to the work $\bar{w}_\text{p}(X,\alpha)$ required to move and process the incoming air in the plant.

To proceed, we need to estimate $\bar{w}_\text{p}(X,\alpha)$. The process through the separation plant is a flow process and we can estimate the processing work as

$$\bar{w}_\text{p} = \bar{v}_0 \Delta p \,,$$

where $\bar{v}_0$ is the molar volume of the air at environmental conditions. The overall pressure loss Δp is associated with the actual transport and separation process; examples are the pressure difference required to press the air through a semipermeable membrane as given by Darcy's law, or the pressure difference required to transport air over larger distances, which is necessary to remove the depleted air far from the plant, so that re-circulation of depleted air is minimized. With the ideal gas law we obtain

$$\bar{w}_\text{p} = \bar{R}T_0 \frac{\Delta p}{p_0} \,.$$

Based on the above, the work to overcome irreversible losses required per mole of gas separated can be estimated as

$$\bar{w}_\text{irr} = \frac{n_{gas} + n_s}{n_s - \hat{n}_s} \bar{w}_\text{p}(X,\alpha) = \frac{1-\alpha X}{(1-\alpha)X} \bar{R}T_0 \frac{\Delta p(X,\alpha)}{p_0} \,.$$

Due to the low CO_2 content of air, one needs to move about 550 times more air through a separation plant to remove one mole of CO_2, than one would need to remove one mole of O_2 for the oxy-fuel process. For $\alpha = 0.5$ the value for the work to overcome irreversible losses is $12990 \frac{\Delta p(X,\alpha)}{p_0} \frac{\text{kJ}}{\text{mol}}$ for separation

[5] See D.W. Keith, M. Ha-Duong, J. K. Stolaroff, Climate Strategy with CO_2 Capture from the Air, Climatic Change **74**(1-3), pp. 17-45 (2006)]

of CO_2 from air, while it is only $21.26 \frac{\Delta p(X,\alpha)}{p_0} \frac{\text{kJ}}{\text{mol}}$ for the separation of O_2 from air.

Figure 21.10 compares $\bar{w}_{\text{irr}}$ for $X = 390\,\text{ppm}$ (the CO_2 fraction in air in 2010) and $X = 0.21$ (the O_2 fraction in air) as function of $\alpha = \hat{X}/X$ (for small α a larger amount of CO_2 or O_2 is removed), for $\frac{\Delta p(X,\alpha)}{p_0} = 0.01$. Already for this small pressure loss, the loss to irreversible work for taking CO_2 directly from the air is about 5 times the reversible work for separation, so that the overall work would be about 6 times the reversible work! Presently, no technology exists that would even reach this value.

Obviously, the actual values for $\bar{w}_{\text{irr}}$ will depend on the pressure losses $\Delta p\,(X,\alpha)$ for the actual process devised. One will expect higher pressure losses for smaller gas content X and larger extraction ratio, i.e. smaller α.

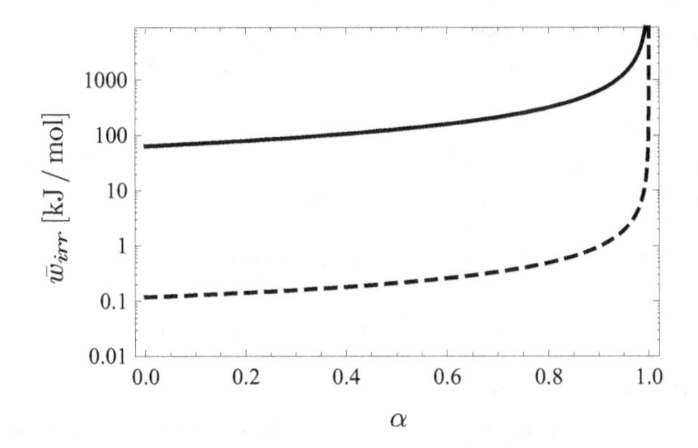

Fig. 21.10 Irreversible contribution to separation work for $X_{CO2} = 384\text{ppm}$ (continuous) and $X_{CO_2} = 0.21$ (dashed) as function of $\alpha = \bar{X}/X$, for $\frac{\Delta p}{p_0} = 0.01$.

The separation processes discussed earlier in this chapter rely on membrane separation. For a membrane that allows only one component to pass, we can see from (20.39) that the pressure ratio over the membrane must be ($X_\nu^{\text{II}} = 1$ for the pure separated gas)

$$\frac{p^{\text{I}}}{p^{\text{II}}} > \frac{1}{X_\nu^{\text{I}}} \,.$$

With $X_{CO_2} = 400\,\text{ppm}$ this implies a pressure ratio larger than 2500, while for O_2 separation, the pressure ratio must be only above 5. Clearly, while membrane separation might be feasible for O_2, it is not at all feasible for CO_2 removal from air. Not only would the large pressure ratio require extremely sturdy membrane materials, but the creation of huge pressure ratios will lead to huge irreversible losses.

Thus, chemical processes must be considered, where the air passes over a substance A that is keen to react with the CO_2 to form A_n-CO_2 compounds.

In a subsequent step, the reaction product must be split into A and CO_2 to dispose the CO_2 and re-use the A.

Another alternative is separation by distillation as discussed in the next chapter. This requires cooling and partial liquefaction of air. All known large-scale cooling processes involve substantial irreversibilities in compressors, throttling valves, heat exchangers, also through heat leaks, and, again, the actual work required will be substantially bigger than the reversible work.

Problems

21.1. Physiological Solution for Use in Hospitals

The osmotic pressure in bodily fluids of mammals is 7.7 atm at 36 °C. Compute the amount of salt (NaCl) that must be added to 1 litre water to give a solution with the same osmotic pressure.

$M_{\text{NaCl}} = 58.5 \frac{\text{kg}}{\text{kmol}}$. Note: NaCl dissociates in solution.

21.2. Salt Water

One kilogram of water and 80 grams of NaCl are mixed, the mixture has a temperature of 80 °C. Assume the mixture is ideal.

1. Compute the osmotic pressure of the solution.
2. Compute the entropy of mixing.
3. Compute the minimum work required for separation of salt and water.

21.3. Osmotic Equilibrium I

A semipermeable membrane which allows only water to pass, divides two containers. There is 1 litre of water in total, and 10 g of NaCl in each container. One container is kept at a pressure of 10 bar and the other at 20 bar; the temperature of both is 300 K.

1. Show that in equilibrium the pressure difference between the two containers equals the difference in osmotic pressures.
2. Set up the equation needed to determine how water is distributed between the two containers, and determine the water masses in both containers.
3. Will the equilibrium change when the temperature is lowered to 20 °C? If so, in which direction does water move, low to high pressure, or high to low?

21.4. Osmotic Equilibrium II

A semipermeable membrane which allows only water to pass, divides two containers. In thermodynamic equilibrium container A holds 400 g of water and 20 g salt, container B holds 600 g of water and 20 g of salt. Moreover, container A is kept at a pressure of 20 bar; the temperature of both is 300 K. Determine the pressure in container B.

21.5. Osmotic Equilibrium with Temperature Difference I
Two piston-cylinder systems are connected by a semipermeable membrane that allows only water to pass. Both cylinders contain water and NaCl. The left cylinder is pressurized to 15 bar and its temperature is maintained at 320 K . The right cylinder is maintained at a temperature of 325 K.

Determine the pressure that must be exerted on the right cylinder so that in chemical equilibrium the mole fraction of salt in both containers is $X_{s,L} = X_{s,R} = 0.05$.

Hint: Careful, the temperature difference affects the chemical potentials!

21.6. Osmotic Equilibrium with Temperature Difference II
Two piston-cylinder systems are connected by a semipermeable membrane that allows only water to pass. Both cylinders contain water and NaCl. The left cylinder is pressurized to 10 bar and its temperature is maintained at 300 K. The right cylinder is pressurized to 20 bar and its temperature is maintained at 305 K. The mole fraction of salt in the left container is measured as $X_{s,L} = 0.05$.

Determine the mole fraction of salt $X_{s,R}$ in the right cylinder for the case of chemical equilibrium.

Hint: Careful, the temperature difference affects the chemical potentials!

21.7. Partial Separation of a Binary Gas Mixture I
Some helium is to be separated from an equimolar mixture of argon and helium. For this, the mixture is pressurized to a pressure p_M, and then flows past a membrane, which allows only helium to pass. The helium pressure on the back of the membrane is 4 bar. Determine the pressure p_M that is necessary to remove 50% of the helium, in the best case.

21.8. Cooling Fluid
A cooling fluid consists of a mixture of water and ethylene glycol ($C_2H_6O_2$, $M = 62\frac{\text{g}}{\text{mol}}$). The glycol mass fraction is $c_g = 0.3$. Assume the mixture is ideal.

1. Determine the mean molar mass of the mixture.
2. The mole volume of glycol is $\bar{v}_g = 0.056\frac{\text{m}^3}{\text{kmol}}$. Compute the specific volume of the mixture, and its mole volume.
3. Determine the entropy of mixing for 1 kg of mixture.
4. What is the minimum work required for complete separation at 0 °C, per kilogram of mixture?
5. The above cooling fluid is brought into contact with a semipermeable membrane that allows only water to pass. On the other side of the membrane is a salt (NaCl) solution. In equilibrium at 300 K the cooling fluid is at pressure $p_c = 45$ bar and the salt solution is at pressure $p_s = 5$ bar. Determine the mole fraction X_s of salt in the salt solution.

21.9. Minimum Work for Reverse Osmosis

We computed the work loss in irreversible mixing as TS_{mix} . This is also the minimum work required for separation.

1. Discuss the above statement.
2. Compute the minimum work for the separation of $1\,\text{m}^3$ of salt water at $35\,°\text{C}$. The saltwater contains 75 g sodium chloride (NaCl, $M_{\text{NaCl}} = 58.5\frac{\text{kg}}{\text{kmol}}$) per litre and has a mass density of $1060\frac{\text{kg}}{\text{m}^3}$. Remember that NaCl dissociates into Na^+ and Cl^- ions. Note: You will get different results if you consider splitting into H_2O, Na^+, Cl^-, than if you consider splitting into H_2O and NaCl. The difference is the entropy of mixing between sodium and chlorine.

21.10. Reversible Mixing

A fabrication process produces a salty waste flow (density: $1040\frac{\text{g}}{\text{litre}}$, 50 g of salt per litre). How much work could be obtained by mixing 1 litre of seawater (density: $1025\frac{\text{g}}{\text{litre}}$, 32 g of salt per litre) with 1 litre of the waste? Assume all flows are at $8\,°\text{C}$.

21.11. Desalination in Piston-Cylinder Device

2 litres of saltwater are enclosed in a piston cylinder device. The saltwater is compressed up to the pressure p_2, which is the pressure at which freshwater just starts to pass a semipermeable membrane. Further increase of pressure forces freshwater through the membrane, until one litre of saltwater remains in the cylinder.

The saltwater contains 50 g sodium chloride (NaCl) per litre and has a mass density of $1040\frac{\text{kg}}{\text{m}^3}$; it can be considered as incompressible ideal mixture, the temperature remains constant at $15\,°\text{C}$.

Assume that the osmotic pressure can be computed from the approximation $p_{\text{osm}} = -\frac{R_W T}{v_W} \ln X_W \simeq \frac{R_W T}{v_W}(1 - X_W)$.

1. Compute the pressure p_2.
2. Find the relation between the volume remaining in the cylinder and the pressure.
3. Compute the work required for the process.

21.12. Reverse Osmosis

Consider the continuous desalination device depicted below. Fresh seawater at $15\,°\text{C}$, 1 bar is pumped isothermally to a pressure of 35 bar, and then flows past a semipermeable membrane, which allows only fresh water to pass. The exiting brine drives a turbine, and leaves the system at 1 bar. Assume that the temperature remains constant throughout the device, and that all processes are reversible.

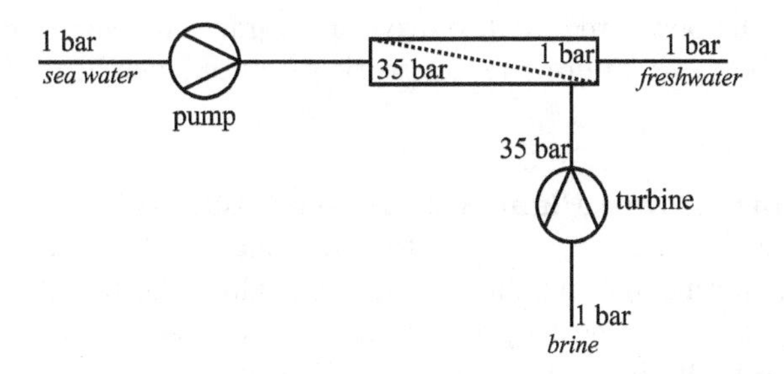

The mass densities of sea water, freshwater and brine are given as: $\rho_{sw} = 1030\frac{\text{kg}}{\text{m}^3}$, $\rho_{fw} = 1000\frac{\text{kg}}{\text{m}^3}$, $\rho_b = 1040\frac{\text{kg}}{\text{m}^3}$, and the seawater contains $35\frac{\text{g}}{\text{litre}}$ of sodium chloride ($M_{\text{NaCl}} = 58.5\frac{\text{kg}}{\text{kmol}}$).

1. Determine the volume flow of freshwater produced per volume flow of sea water.
2. Determine the work required to drive the pump, and the work that can be recovered by the turbine. Compute the net work required per litre of freshwater.
3. Assume now that pump and turbine are irreversible, with efficiencies of 85%, and determine the net work.

21.13. Osmotic Power Plant

Saltwater at 15 °C, 1 bar is pumped to a pressure of 20 bar and then flows along a semi-permeable membrane through which freshwater enters and dilutes the saltwater. The exiting solution (diluted saltwater) drives a turbine, and leaves the system at 1 bar. Assume that the temperature remains constant throughout the device, and that all processes are reversible.

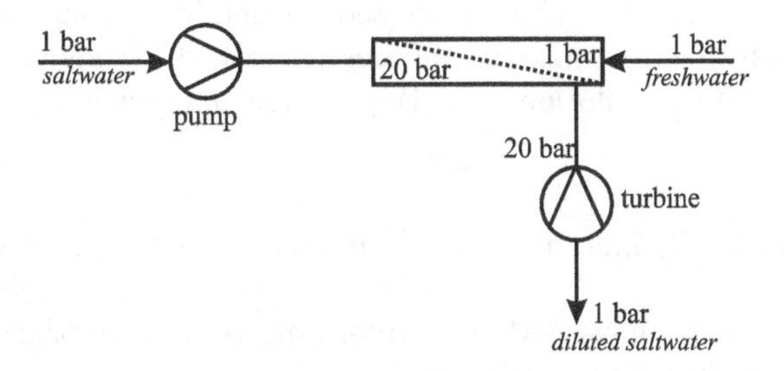

The incoming saltwater contains 45 g of sodium chloride ($M_{\text{NaCl}} = 58.5\frac{\text{kg}}{\text{kmol}}$) per litre. The mass densities of saltwater and freshwater are given as: $\rho_{sw} = 1035\frac{\text{kg}}{\text{m}^3}$, $\rho_{fw} = 1000\frac{\text{kg}}{\text{m}^3}$, and there is no mixing volume.

1. Determine the osmotic pressure of the incoming saltwater.
2. The volume flow of saltwater is $1000\frac{\text{litres}}{\text{min}}$. Determine the volume flows of the freshwater drawn in, and of the diluted saltwater that enters the turbine.

3. Determine the power required to drive the pump, and the power produced by the turbine. Compute the net work produced per litre of freshwater drawn in.

21.14. Osmotic Power Plant with Irreversibilities

Saltwater at 25 °C, 1 bar is pumped to a pressure of 22 bar, and then flows along a semi- permeable membrane, through which fresh water enters and dilutes the saltwater. Due to pressure loss in the desalination system, the exiting solution (diluted saltwater) leaves the desalinator at a pressure of 19 bar. This flow drives a turbine, and leaves the system at 1 bar. Assume that the temperature remains constant throughout the device, and that pump and turbine are irreversible with isentropic efficiencies of 0.85.

The incoming saltwater contains 42 g of sodium chloride ($M_{\mathrm{NaCl}} = 58.5\frac{\mathrm{kg}}{\mathrm{kmol}}$) per litre. The mass densities of saltwater and freshwater are given as: $\rho_{sw} = 1033\frac{\mathrm{kg}}{\mathrm{m}^3}$, $\rho_{fw} = 1000\frac{\mathrm{kg}}{\mathrm{m}^3}$, and there is no mixing volume.

1. Determine the osmotic pressure of the incoming saltwater.
2. Determine the mole fraction of water in the exiting diluted water.
3. The volume flow of saltwater is 1000 litres/minute. Determine the volume flows of the freshwater drawn in, and of the diluted saltwater that enters the turbine
4. Determine the power required to drive the pump, and the power produced by the turbine. Compute the net work produced per litre of freshwater drawn in.

21.15. Desalination I

A pipe is closed at one end with a semipermeable membrane which only allows water to pass. The pipe is pressed vertically into an ocean, see the following sketch. The temperature of the ocean is 4 °C, and the salt water contains 35 g sodium chloride (NaCl) per litre, moreover $\rho_{sw} = 1030\frac{\mathrm{kg}}{\mathrm{m}^3}$, $\rho_{fw} = 1000\frac{kg}{m^3}$

1. To what height H_1 must the pipe be immersed, before fresh water passes the membrane?
2. At what height H_2 have both, sea water and fresh water, the same height, $h = H$? Explain why this is possible.
3. When $h > H_2$ one can run a water wheel with the fresh water leaving the pipe. Of course, this setup is not a perpetual motion engine. Why? (for some discussion see: Scientific American, June 1971, p. 124-125, and April 1972, p110-111). The above calculation assumed constant density of the seawater, and a homogeneous salt content. Can that be expected?

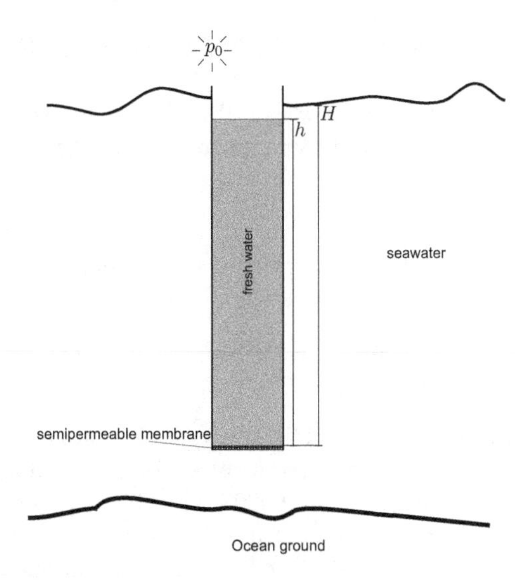

21.16. Desalination II

The solution of the previous problem assumed constant density of the seawater. Consider the same problem for the case that the salt concentration follows the barometric law, so that

$$\rho_{sw} = \rho_{fw} + \rho_{s,0} \exp\left[-\frac{gz}{2R_sT}\right] ,$$

where R_s is the gas constant for the salt (factor 2 accounts for dissociation) and $\rho_{s,0}$ is a constant specified by the density at the surface (at $z = 0$). Note that z points upwards, so that the density is larger at greater depth.

1. Discuss the salt concentration profile in the oceans. Under what circumstances would you expect the exponential and the constant profile, respectively?
2. Compute the difference $h - H$ for the exponential profile.
3. Discuss your findings.

Remark: Not all information on the web that you might find on this issue is correct (the same is true for any other topic, of course).

21.17. Desalination III

Assume that the salt concentration in a salt water lake follows the barometric law $\rho_s = \rho_{s,0} \exp\left[-\frac{gz}{2R_sT}\right]$ (salt density increasing with depth).

A pipe is closed at one end with a semipermeable membrane which only allows water to pass. The pipe is pressed vertically into the lake. The temperature of the lake is 4°C, and the salt water at the surface contains 20 g sodium chloride (NaCl) per litre. For the following, assume that the density of salt water is given by $\rho_{sw} = \rho_{fw} + \rho_s$ where $\rho_w = 1000 \frac{\text{kg}}{\text{m}^3}$.

1. To what height H_1 must the pipe be immersed, before fresh water passes the membrane?
2. Compute the freshwater height h as a function of sea water height H.
3. Is it possible, that both, sea water and fresh water have the same height, $h = H$?

21.18. Separation
Consider a mixture of three ideal gases, say oxygen, nitrogen and carbon dioxide. Compute the minimum work required to

1. Separate the carbon dioxide from the two other gases
2. Separate all three components.

21.19. Partial Separation of a Binary Gas Mixture II
Some carbon dioxide is to be separated from a mixture of nitrogen and carbon dioxide with mole fraction $X_{CO_2} = 0.0205$. For this, the mixture is pressurized to a pressure of 50 bar, and then flows past a membrane, which allows only CO_2 to pass. The CO_2 pressure on the back of the membrane is 1 bar. Determine:

1. The mole fraction of CO_2 in the exiting mixture, in the best case.
2. The percentage of CO_2 separated from the mixture.
3. The change of the entropy of mixing.
4. The minimum separation work per mole of CO_2, when the environment temperature is $T_0 = 300\,\mathrm{K}$.

21.20. Removal of Carbon Dioxide from the Atmosphere
From http://www.virginearth.com/ (2007): *The Virgin Earth Challenge is a prize of $25m for whoever can demonstrate to the judges' satisfaction a commercially viable design which results in the removal of anthropogenic, atmospheric greenhouse gases so as to contribute materially to the stability of Earth's climate.* From Wikipedia: *The prize will be awarded to the first scheme that is capable of removing one billion metric tons of carbon dioxide from the atmosphere per year for 10 years.*

Let's evaluate the goal thermodynamically, by computing how much work is necessary.

1. The Earth radius is 6370 km, the pressure at ground level is 1.01325 bar and the gravitational acceleration is $9.81\frac{\mathrm{m}}{\mathrm{s}^2}$. Estimate mass and mole number of Earth's atmosphere.
2. In January 2014, the mole fraction of CO_2 in the atmosphere was 397.80 ppm, up from 393.14 ppm in January 2012 (data from http://co2now.org/). The pre-industrial level was 284 ppm. Determine mole number and mass of CO_2 in the atmosphere, the amount added in the past year, and the amount added since industrialization began.
3. The yearly emissions from fossil fuels and cement production are 33.5 Gt of CO_2. Compare this number to the amount added to the atmosphere computed above. Where is the remaining CO_2 going?

4. Assume that there is a winner of the competition, the device is built, and 1 billion tons of CO_2 are removed in one year. Determine the new mole fraction of CO_2 after the year.
5. Compute the minimum work required to remove 10 billion tons from the atmosphere per year (assume $T = 290\,\mathrm{K}$ and January 2014 composition), and compare to the world energy consumption of about 16 TW, and the world generation of electric power of about 2.5 TW. Also compute the minimum work required to remove the amount added to the atmosphere per year. To simplify, consider a binary mixture of 'air molecules' and CO_2 molecules.
6. Discuss all results and also the question of irreversibilities in the processes.

Chapter 22
Phase Equilibrium in Mixtures

22.1 Phase Mixtures

In this chapter we discuss equilibria between mixtures in different phases, e.g. liquid-vapor equilibria, or liquid-solid equilibria. Phase mixtures normally are characterized by different mole fractions (or concentrations) in the different phases. We shall discuss phase diagrams, changes of melting and evaporation temperatures in solutions, distillation processes, and gas solubility.

22.2 Gibbs' Phase Rule

In extension to our previous notation we introduce Latin superscripts to denote the phases, and we assume that we have f different phases. For instance X_α^j ($\alpha = 1, \ldots, \nu$, $j = 1, \ldots, f$) denotes the mole fraction of component α in phase j. Also, $\mu_\alpha^j\left(T, p, X_\beta^j\right)$ is the chemical potential of component α in phase j; note that the chemical potential depends only on the mole fractions of the same phase.

For each phase, the mole fractions sum to unity, that is

$$\sum_{\beta=1}^{\nu} X_\beta^j = 1 \quad \text{for } j = 1, 2, \ldots, f \,. \tag{22.1}$$

It follows that a mixture with ν components in f phases is characterized by $(\nu - 1)\, f$ independent mole fractions. In addition, pressure p and temperature T are variables, so that the mixture is characterized by $(\nu - 1)\, f + 2$ intensive variables, $\{p, T, X_\alpha^j\}$.

Phase boundaries are permeable interfaces between phases that allow all components to pass, and thus in equilibrium the chemical potentials of all components must be continuous between any two phases,

$$\bar{\mu}_\alpha^1 = \bar{\mu}_\alpha^2 = \cdots = \bar{\mu}_\alpha^f \quad \text{for } \alpha = 1, \ldots, \nu \,. \tag{22.2}$$

H. Struchtrup, *Thermodynamics and Energy Conversion*,
DOI: 10.1007/978-3-662-43715-5_22, © Springer-Verlag Berlin Heidelberg 2014

These equations are known as Gibbs' phase rule, and they give $(f-1)\,\nu$ conditions[1] that restrict the $(\nu-1)\,f+2$ variables. The difference between these numbers gives the number of degrees of freedom for the system,

$$\mathrm{F}=[(\nu-1)\,f+2]-[(f-1)\,\nu]=2+\nu-f\,. \tag{22.3}$$

That is, we can freely chose F properties of the mixture, while the remaining $(f-1)\,\nu$ properties are then fixed through Gibbs' phase rule.

We consider this rule for the simple example of a single component, where $\nu=1$. The variables are (p,T) and we find $\mathrm{F}=3-f$.

In a single phase, we have $f=1$, so that there are $\mathrm{F}=2$ degrees of freedom: p and T can be chosen independently.

When there are two phases, either liquid-vapor, liquid-solid, or vapor-solid, we have $f=2$, so that there is only one degree of freedom, $\mathrm{F}=1$: pressure and temperature cannot be chosen independently, but are dependent. The sole equilibrium condition $\mu^1\,(p,T)=\mu^2\,(p,T)$ defines the saturation pressure $p_{\mathrm{sat}}\,(T)$.

For three phases to coexist, we have $f=3$, so that $\mathrm{F}=0$: this is possible only at one pair of values $(p,T)_{tr}$—the triple point.

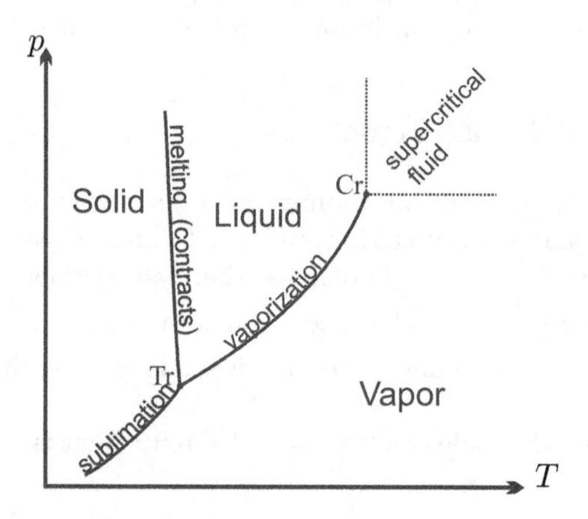

Fig. 22.1 p-T-diagram of a single substance

Figure 22.1 shows again a sketch of the p-T-diagram of a single substance (for water, actually) with the saturation curves and the triple point.

22.3 Liquid-Vapor-Mixtures: Idealized Raoult's Law

We study vapor-liquid equilibrium in mixtures where we denote the liquid phase by $'$ and the vapor phase by $''$. The Gibbs phase rule for this case reads

[1] Just count the equal signs in each of the equations (22.2).

$$\bar{\mu}'_\alpha\left(T,p,X'_\beta\right)=\bar{\mu}''_\alpha\left(T,p,X''_\beta\right)\quad,\quad\alpha=1,\ldots,\nu\,.\tag{22.4}$$

Our goal is to simplify the phase rule by using constitutive equations for incompressible liquids and ideal gases, respectively. Deviations from this idealized behavior will be studied later. In particular, we make the following assumptions:

1. The liquids are ideal mixtures, so that

$$\bar{\mu}'_\alpha\left(T,p,X'_\beta\right)=\bar{g}'_\alpha\left(T,p\right)+\bar{R}T\ln X'_\alpha\,;\tag{22.5}$$

$\bar{g}'_\alpha\left(T,p\right)$ denotes the Gibbs free energy of α alone at (T,p) in the liquid state.

2. The pure liquids are incompressible, so that

$$\bar{g}'_\alpha\left(T,p\right)=\bar{g}'_\alpha\left(T,p^{\mathrm{sat}}_\alpha\left(T\right)\right)+\bar{v}'_\alpha\left(p-p^{\mathrm{sat}}_\alpha\left(T\right)\right)\,.\tag{22.6}$$

The above results from Taylor expansion in pressure, with $\left(\frac{\partial\bar{g}}{\partial p}\right)_T=\bar{v}=$ *const.*; $p^{\mathrm{sat}}_\alpha\left(T\right)$ is the saturation pressure for component α alone.

3. The vapor can be described as a mixture of ideal gases, so that, since $\bar{g}''_\alpha\left(T,p\right)=\bar{h}''_\alpha\left(T\right)-T\left[\bar{s}^{0\prime\prime}_\alpha\left(T\right)-\bar{R}\ln\frac{p}{p_0}\right]$,

$$\begin{aligned}\bar{\mu}''_\alpha\left(T,p,X''_\beta\right)&=\bar{g}''_\alpha\left(T,p\right)+\bar{R}T\ln X''_\alpha\\&=\bar{g}''_\alpha\left(T,p^{\mathrm{sat}}_\alpha\left(T\right)\right)+\bar{R}T\ln\frac{p}{p^{\mathrm{sat}}_\alpha\left(T\right)}+\bar{R}T\ln X''_\alpha\,.\end{aligned}\tag{22.7}$$

4. The mixture is sufficiently far away from the critical points of all components, so that for the components alone

$$\bar{v}'_\alpha\ll\bar{v}''_\alpha=\frac{\bar{R}T}{p^{\mathrm{sat}}_\alpha}\,.\tag{22.8}$$

When we use the four assumptions in (22.4) together with the definition of the saturation pressure of the single components, $\bar{g}'_\alpha\left(T,p^{\mathrm{sat}}_\alpha\left(T\right)\right)=\bar{g}''_\alpha\left(T,p^{\mathrm{sat}}_\alpha\left(T\right)\right)$, we find Raoult's law for ideal mixtures (François-Marie Raoult, 1830-1901)

$$X'_\alpha p^{\mathrm{sat}}_\alpha\left(T\right)=X''_\alpha p\quad,\quad\alpha=1,\ldots,\nu\,.\tag{22.9}$$

The partial pressure of α in the vapor is given by $p_\alpha=X''_\alpha p$. Raoult's law states that the amount of component α in the liquid is proportional to its partial pressure in the vapor.

22.4 Phase Diagrams for Binary Mixtures

For the case of binary mixtures in liquid-vapor equilibrium, we have $\nu=2$ and Raoult's law gives the equations

$$X_1' p_1^{\text{sat}}(T) = X_1'' p \quad , \quad (1 - X_1')\, p_2^{\text{sat}}(T) = (1 - X_1'')\, p\,. \tag{22.10}$$

The variables are (T, p, X_1', X_1''), and according to Gibbs' phase rule there are two degrees of freedom, F = 2, so that when two variables are prescribed, the others are fixed.

In particular, when pressure and temperature are prescribed, the values for the mole fractions in both phases are computed as

$$X_1' = \frac{p_2^{\text{sat}}(T) - p}{p_2^{\text{sat}}(T) - p_1^{\text{sat}}(T)} \quad , \quad X_1'' = \frac{p_1^{\text{sat}}(T)}{p} \frac{p_2^{\text{sat}}(T) - p}{p_2^{\text{sat}}(T) - p_1^{\text{sat}}(T)}\,. \tag{22.11}$$

For given temperature T, we can draw the two curves into a p-X-diagram. Solving for pressure gives the saturated liquid line,

$$p(T, X') = p_2^{\text{sat}}(T) - \left(p_2^{\text{sat}}(T) - p_1^{\text{sat}}(T)\right) X_1'\,, \tag{22.12}$$

and the saturated vapor line,

$$p(T, X'') = \frac{p_1^{\text{sat}}(T)\, p_2^{\text{sat}}(T)}{(1 - X_1'')\, p_1^{\text{sat}}(T) + X_1'' p_2^{\text{sat}}(T)}\,. \tag{22.13}$$

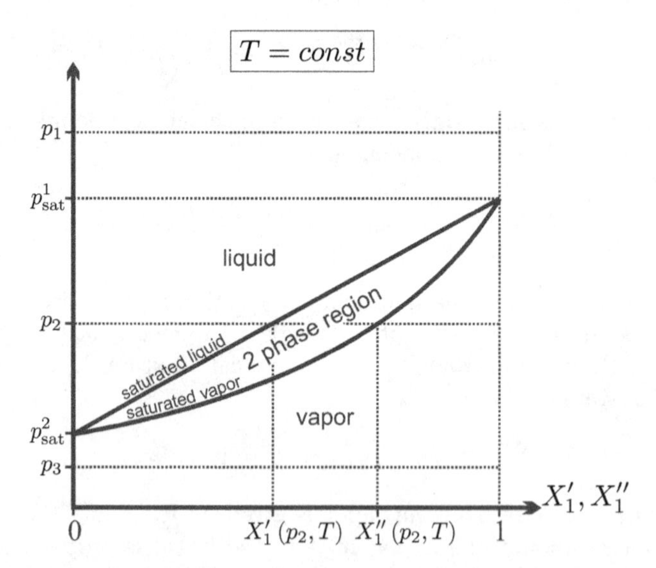

Fig. 22.2 p-X-diagram for a binary mixture at temperature T, according to Raoult's law in the idealized case

Figure 22.2 shows a p-X-diagram for fixed T with the two curves that meet for $X_1' = X_1'' = 0$ and $X_1' = X_1'' = 1$ at the respective saturation pressures. For the assumptions used, the saturated liquid curve is a straight line.

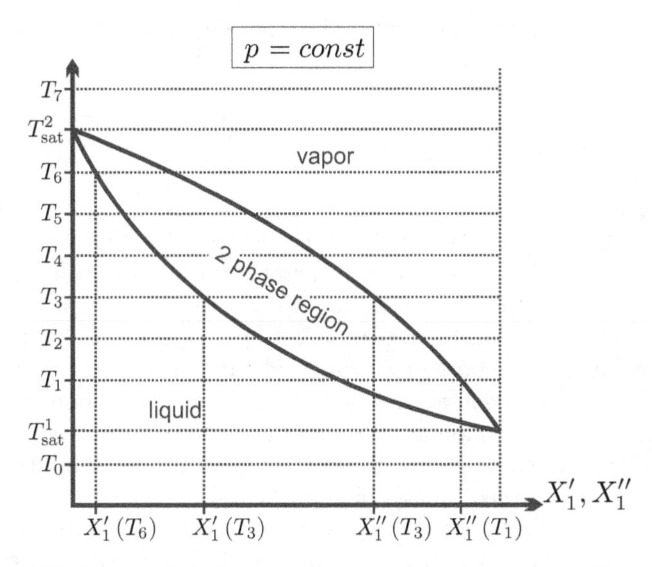

Fig. 22.3 T-X-diagram for a binary mixture at pressure p, according to Raoult's law in the idealized case

Also indicated in the figure are three pressures. At p_1 the pressure is higher than both saturation pressures, and the there is only one single liquid phase. At p_3 the pressure is lower than both saturation pressures, and there is only one single vapor phase. At the intermediate pressure p_2, which lies between the two saturation pressures, the mixture can be found either as pure liquid or pure vapor phase, or it can split into two phases. What happens depends on the overall mole fraction of the mixture $X_1 = \frac{n_1'+n_1''}{n_1'+n_1''+n_2'+n_2''}$. If $X_1 \leq X_1'(p_2, T)$, there is only a single liquid phase, and for $X_1 \geq X_1''(p_2, T)$, there is only a single vapor phase. If $X_1'(p_2, T) < X_1 < X_1''(p_2, T)$, the mixture splits into two phases, the liquid phase with mole fraction $X_1'(p_2, T)$ and the vapor phase with mole fraction $X_1''(p_2, T)$. The component with the larger vapor pressure (here component 1) is more volatile, i.e. more keen to evaporate, and the vapor is richer in the more volatile component while the liquid is depleted of it.

The equations (22.12, 22.13) can be numerically solved for T; this requires equations or tables for the saturation pressures $p_\alpha^{sat}(T)$. The resulting T-X-diagram for constant pressure is sketched in Fig. 22.3. The interpretation of the diagram follows the same lines as for the p-X-diagram: For temperatures above both saturation temperatures $T_\alpha^{sat}(p)$, e.g., T_7, the mixture will be pure vapor for all values X_1. The mixture will be pure liquid for temperatures below both saturation temperatures, e.g., T_0 in the figure. For temperatures between the two saturation temperatures (T_1 to T_6), the mixture will split into two phases at $X_1'(T, p)$ and $X_1''(T, p)$, if the overall mole fraction X_1 lies between these values; else the mixture is either liquid (for $X_1 \leq X_1'(T,p)$) or vapor (for $X_1 \geq X_1''(T,p)$).

22.5 Distillation

Distillation is a separation procedure based on the different compositions of vapor and liquid mixture in equilibrium. Figure 22.4 shows a sketch for a bubble tray column for distillation that matches the T-X-diagram of Fig. 22.3. A temperature gradient is imposed by heating at the bottom and cooling at the top of the column. Several bubble trays are inserted in the column. Industrial columns have up to 100 trays and can reach heights of 60 metres. Heating at the bottom generates vapor that rises in the column, while liquid is generated by condensation at the cooled top and drips down.

On the trays, which all have different temperatures following the temperature gradient established in the column, the rising vapor passes as bubbles through the liquid. This contact between vapor and liquid establishes thermodynamic equilibrium on the tray, determined by the pressure in the column and the temperature on the tray. Consider the tray at T_3: The vapor leaving the tray upwards has the mole fraction $X_1''(T_3)$ while the liquid dripping down has the mole fraction $X_1'(T_3)$. Figure 22.5 illustrates how liquid and vapor pass through the tray: The vapor is forced to bubble through the liquid, and this leads to exchange of components and energy.

For the mixture shown in Figs. 22.3 and 22.4, component 1 is more volatile, so that the vapor becomes richer in component 1 as it ascends, while the liquid becomes richer in component 2 as it descends.

The design and dimensioning of distillation columns is the task of chemical engineers and will not be discussed further. Obviously, distillation becomes more complex when multicomponent mixtures are involved, when saturation temperatures of components are close, and when the mixtures exhibit non-ideal behavior, e.g., azeotropes, see Sec. 22.11.

22.6 Saturation Pressure and Temperature of a Solvent

We consider solutions of low volatility components, e.g., salts, in a more volatile solvent, e.g., water, in liquid-vapor equilibrium. The mole fraction of solvent vapor is

$$X_\nu'' = 1 - \sum_{\alpha=1}^{\nu-1} X_\alpha'' = 1 - \sum_{\alpha=1}^{\nu-1} X_\alpha' \frac{p_\alpha^{\text{sat}}(T)}{p} , \tag{22.14}$$

where we used Raoult's law for the mole fractions of the dissolved components, X_α''. For the non-volatile substances the saturation pressures $p_\alpha^{\text{sat}}(T)$

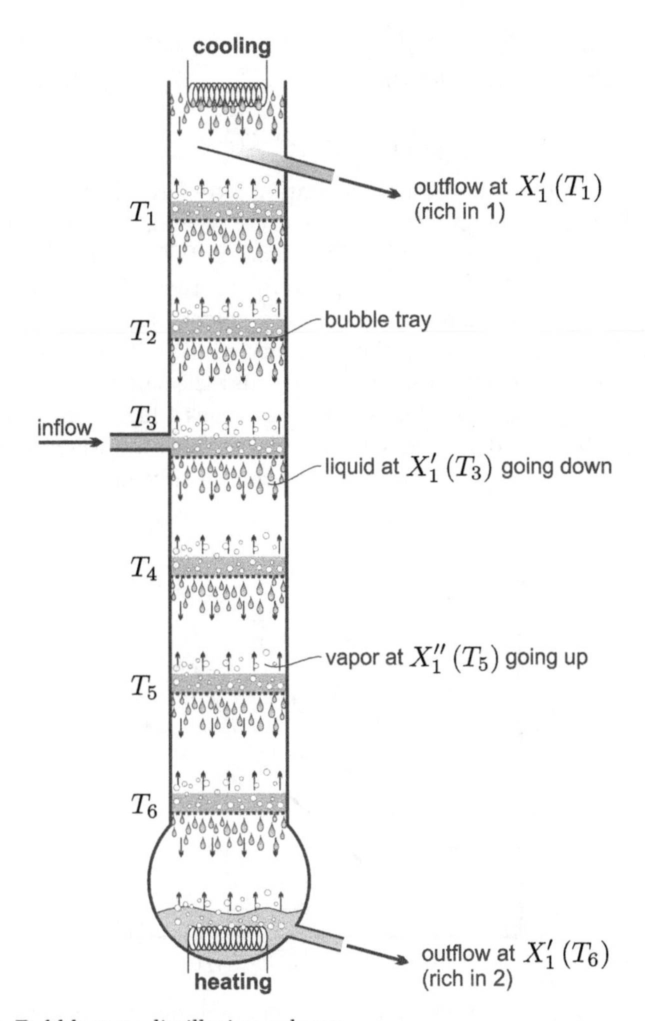

Fig. 22.4 Bubble-tray distillation column

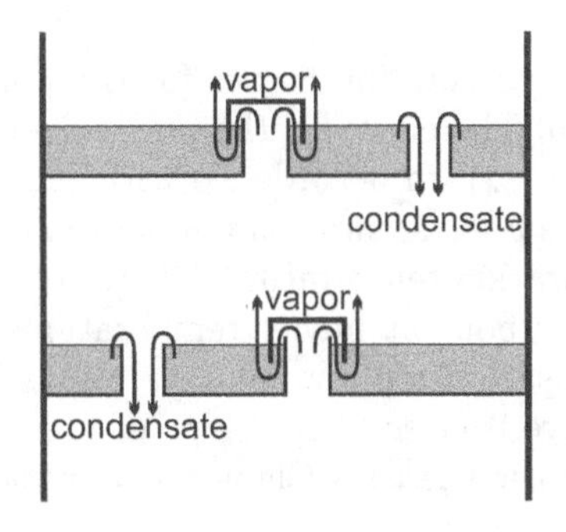

Fig. 22.5 Schematic close-up of bubble trays

are very low, so that $\sum_{\alpha=1}^{\nu-1} X'_\alpha \frac{p_\alpha^{\text{sat}}(T)}{p} \ll 1$; accordingly we can approximate $X''_\nu \simeq 1$. Raoult's law (22.9) for the solvent, $X''_\nu p = X'_\nu p_\nu^{\text{sat}}(T)$, then simplifies to

$$p_{\text{sol}}(T) = X'_\nu p_\nu^{\text{sat}}(T) . \tag{22.15}$$

Here, $p_{\text{sol}}(T)$ denotes the actual saturation pressure of the solvent vapor over the solution at temperature T. In other words, the ideal mixture of composition X'_ν at temperature T will boil when the pressure is $p_{\text{sol}}(T)$.

According to (22.15) the actual pressure in the solvent vapor, p_{sol}, is smaller than the saturation pressure of the solvent alone, $p_\nu^{\text{sat}}(T)$, that is the dissolved substances reduce the volatility of the solvent. The reason for this is a competition between solvent vapor and the salt dissolved in the liquid to increase their entropy: The solvent vapor has a larger entropy than the solvent liquid, but the dissolved salt has a larger entropy when it can access a larger volume, that is when there is more liquid solvent.

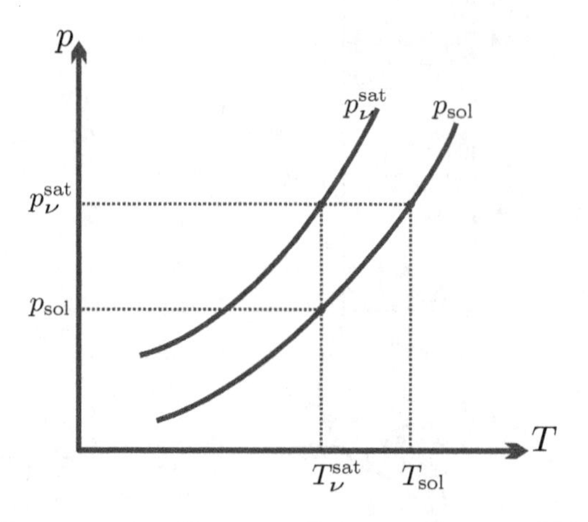

Fig. 22.6 p-T-diagram: saturation curve for pure solvent and for solution

Figure 22.6 shows the saturation curves for the pure solvent and the solution in a p-T-diagram. The curve for the pure solvent (p_ν^{sat}) lies above the curve for the solution (p_{sol}), in accordance with (22.15). The figure shows that for a given pressure p the saturation temperature $T_{\text{sol}}(p)$ of the solution is higher than the saturation temperature $T_\nu^{\text{sat}}(p)$ of the solvent alone: at a given pressure saltwater boils at higher temperatures than pure water. To estimate the change in saturation temperature, we assume that the slope of the new saturation curve is close to the slope of the saturation curve of the solvent as described by the Clausius-Clapeyron equation (17.46), that is

$$\frac{T_{\text{sol}} - T_\nu^{\text{sat}}}{p_\nu^{\text{sat}} - p_{\text{sol}}} \simeq \frac{dT_\nu^{sat}}{dp} = T_\nu^{sat} \frac{v''_\nu - v'_\nu}{h_{LV}} . \tag{22.16}$$

With the assumptions that the liquid volume can be neglected against the vapor volume, and that the vapor follows the ideal gas law, and with (22.15), this yields

$$T_{\rm sol} - T_\nu^{\rm sat}(p) = \frac{R_\nu \left[T_\nu^{sat}(p)\right]^2}{h_{LV}(p)} \left(1 - X'_\nu\right) . \tag{22.17}$$

$T_{\rm sol}$ is the temperature at which an ideal mixture of composition X'_ν at pressure p will boil.

22.7 Freezing of a Liquid Solution

Ice crystals cannot easily accept salt molecules into their lattice, and thus when a salt solution is cooled, first water will freeze out, while the salt will remain in the liquid as long as possible. Sea-ice, that is ice that freezes out of oceans in cold climates, contains no salt. As sea-ice forms, the salt content of the water just below the ice increases. This salty water has increased density, and will sink towards the bottom of the ocean, thus driving ocean currents. Completely frozen salt-water solution is a mixture of pure ice and regions of salt-water crystals, e.g., $NaCl\text{-}2H_2O$ crystals.

The dissolved salt ions move in the liquid, they have a larger entropy when they can move in a larger volume, that is when more liquid water is present. So in order to gain entropy, the dissolved salt prevents water from freezing, which is observed as a drop in the temperature at which water will freeze. To estimate this drop, we consider solid and liquid as incompressible ideal mixtures, so that we can approximate the chemical potentials of the solvent in liquid solution ($'$) and ice ($''$) as

$$\mu'_\nu = g'_\nu\left(T, p_\nu^{\rm sat}(T)\right) + v'_\nu\left(p_{\rm melt} - p_\nu^{\rm sat}(T)\right) + R_\nu T \ln X'_\nu \, , \tag{22.18}$$

$$\mu''_\nu = g''_\nu\left(T, p_\nu^{\rm sat}(T)\right) + v''_\nu\left(p_{\rm melt} - p_\nu^{\rm sat}(T)\right) + R_\nu T \ln X''_\nu \, . \tag{22.19}$$

With no salt in the ice, we have $X''_\nu = 1$, and equating the chemical potentials, $\mu'_\nu = \mu''_\nu$, gives

$$p_{\rm melt}(T) - p_\nu^{\rm sat}(T) = -\frac{R_\nu T}{v'_\nu - v''_\nu} \ln X'_\nu \, . \tag{22.20}$$

For water, the volume of ice is larger than the volume of liquid, $v'_\nu - v''_\nu < 0$, and hence the melting pressure of the solution, $p_{\rm melt}(T)$, is lower than that of pure water ($p_\nu^{\rm sat}(T)$).

To compute the change in melting temperature, we use Fig. (22.7) and the Clausius-Clapeyron equation to estimate

$$\frac{T_v^{\rm sat}(p) - T_{\rm melt}(p)}{p_{\rm melt}\left(T_v^{\rm sat}\right) - p_\nu^{\rm sat}\left(T_v^{\rm sat}\right)} \simeq \frac{dT_v^{\rm sat}}{dp} = T_v^{\rm sat}\frac{v''_\nu - v'_\nu}{h''_\nu - h'_\nu} \, , \tag{22.21}$$

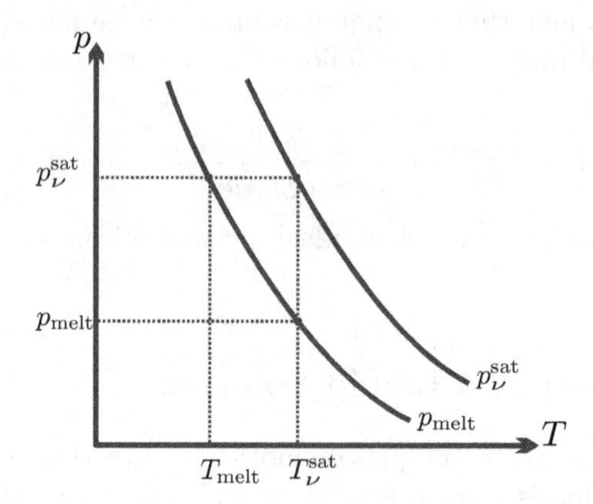

Fig. 22.7 p-T-diagram: melting curve for pure solvent and for solution

so that, with (22.20),

$$T_{melt}(p) - T_v^{sat}(p) = \frac{R_\nu \left[T_v^{sat}\right]^2}{h'_\nu - h''_\nu} \ln X'_\nu < 0\,. \tag{22.22}$$

Here, $h'_\nu - h''_\nu$ is the heat of melting, e.g., $h'_\nu - h''_\nu = 333.7\frac{\mathrm{kJ}}{\mathrm{kg}}$ for water at $0\,°\mathrm{C}$.

22.8 Non-ideal Mixtures: Activity and Fugacity

So far, we considered only ideal mixtures. In this section, we study how non-ideal effects are incorporated into the description.

The activity coefficient $\gamma_\alpha(T,p,X_\beta)$ and the activity $a_\alpha(T,p,X_\beta) = \gamma_\alpha(T,p,X_\beta)\,X_\alpha$ are defined such that the chemical potential reads

$$\bar{\mu}_\alpha(T,p,X_\beta) = \bar{g}_\alpha(T,p) + \bar{R}T\ln a_\alpha = \bar{g}_\alpha(T,p) + \bar{R}T\ln\left[\gamma_\alpha X_\alpha\right]\,. \tag{22.23}$$

As always, $\bar{g}_\alpha(T,p)$ denotes the Gibbs free energy of component α alone, under the same pressure and temperature as the mixture. For ideal mixtures or ideal gases, the activity coefficient reduces to $\gamma_\alpha = 1$, and the activity becomes $a_\alpha = X_\alpha$.

For the description of vapors, one uses typically the fugacity coefficient $\varphi_\alpha(T,p,X_\beta)$ and the fugacity $f_\alpha(T,p,X_\beta) = \varphi_\alpha X_\alpha \frac{p}{p_\alpha^{sat}(T)}$, which are defined such that

$$\begin{aligned}\bar{\mu}_\alpha(T,p,X_\beta) &= \bar{g}_\alpha\left(T,p_\alpha^{sat}(T)\right) + \bar{R}T\ln f_\alpha \\ &= \bar{g}_\alpha\left(T,p_\alpha^{sat}(T)\right) + \bar{R}T\ln\left[\varphi_\alpha X_\alpha \frac{p}{p_\alpha^{sat}(T)}\right]\,.\end{aligned} \tag{22.24}$$

Here, $\bar{g}_\alpha\left(T, p_\alpha^{\mathrm{sat}}(T)\right)$ denotes the Gibbs free energy of component α alone at saturation under the same temperature as the mixture. When the vapor can be described as an ideal gas, the fugacity coefficient reduces to $\varphi_\alpha = 1$.

Just as the chemical potential, activity and fugacity depend on the detailed composition, X_β, of the respective phase. Thus, on the first glance, it might seem that the introduction of activity and fugacity and their coefficients has no advantage, since one replaces one unknown function, the chemical potential, with another, activity or fugacity. The latter, however, are accessible to measurements, and this is why they are used in chemical engineering.

To understand how fugacity can be measured, we consider the Gibbs equation (20.1) and take mixed derivatives with respect to p and n_α, to find

$$\frac{\partial^2 G}{\partial p \partial n_\alpha} = \left(\frac{\partial \bar{\mu}_\alpha}{\partial p}\right)_{T, n_\alpha, n_\beta} = \left(\frac{\partial V}{\partial n_\alpha}\right)_{T, p, n_\beta} = \frac{\partial^2 G}{\partial n_\alpha \partial p} \,. \tag{22.25}$$

Here, we insert the chemical potential in the form (22.24) and separate the term containing the fugacity φ_α. The result is an expression for the change of φ_α with pressure,

$$\left(\frac{\partial \ln \varphi_\alpha}{\partial p}\right)_{T, n_\alpha, n_\beta} = \frac{1}{\bar{R}T}\left(\frac{\partial V}{\partial n_\alpha}\right)_{T, p, n_\beta} - \frac{1}{p} \,. \tag{22.26}$$

In the limit $p \to 0$ the vapor will behave as an ideal gas, and therefore will have the fugacity coefficient $\varphi_\alpha\left(T, p = 0, X_\beta\right) = 1$. Integration at constant temperature between $p = 0$ and the actual pressure p thus gives the fugacity coefficient as

$$\ln \varphi_\alpha\left(T, p, X_\beta\right) = \frac{1}{\bar{R}T} \int_{p=0}^{p} \left[\left(\frac{\partial V}{\partial n_\alpha}\right)_{T, p', n_\beta} - \frac{\bar{R}T}{p'}\right] dp' \,. \tag{22.27}$$

The function under the integral can be measured by systematically adding component α, and determining the resulting volume change; this must be repeated for many pressures. Thus, the fugacity can be measured in the vapor phase.

Measurements of activity and activity coefficients rely on Raoult's law. We reconsider Raoult's law for the equilibrium between a liquid and a vapor mixture, where we now use activity and fugacity coefficients to describe the liquid and the vapor, respectively. Gibbs' phase rule $\bar{\mu}'_\alpha\left(T, p, X'_\beta\right) = \bar{\mu}''_\alpha\left(T, p, X''_\beta\right)$ becomes

$$\bar{g}'_\alpha(T, p) + \bar{R}T \ln a'_\alpha = \bar{g}''_\alpha\left(T, p_\alpha^{\mathrm{sat}}(T)\right) + \bar{R}T \ln f''_\alpha \,. \tag{22.28}$$

When we consider the liquid as almost incompressible, and assume that its molar volume can be ignored compared to the molar volume of the vapor,[2]

[2] The same assumptions were used in Sec. 22.3.

this reduces to

$$a'_\alpha = f''_\alpha \quad \text{or} \quad \gamma'_\alpha X'_\alpha = \varphi''_\alpha X''_\alpha \frac{p}{p_\alpha^{\text{sat}}(T)} \,. \tag{22.29}$$

It follows that activity can be measured when fugacity is known.

22.9 A Simple Model for Heat of Mixing and Activity

Ideal mixtures have zero enthalpy of mixing H_{mix}, and their activity coefficients are unity. We shall now develop a simple model for H_{mix} to describe non-ideal behavior. We consider binary mixtures only.

Part of the internal energy of a substance comes from the interaction potential between neighboring molecules. For a component alone, this contribution to energy is included in enthalpy $\bar{h}_\alpha(T,p)$ or free energy $\bar{g}_\alpha(T,p)$, and we can consider identical neighboring particles as energetically neutral.

In a mixture between components 1 and 2, there will be neighboring pairs of the same type (1-1, 2-2) and pairs of different type (1-2). While the former are energetically neutral, the formation of the latter can either release or require energy. When the attractive force between different particles (1-2) is stronger than that between equal particles (1-1, 2-2), 1-2 pairs are energetically preferred over neutral pairs, due to negative interaction energy; then energy is released when a mixed pair (1-2) is formed. When the attractive force between different particles (1-2) is weaker than between equal particles (1-1, 2-2), neutral pairs are energetically preferred over (1-2) pairs, due to positive interaction potential; then energy is required to form a pair.

We consider a mixture of $N = N_1 + N_2$ particles of types 1, 2. The probability to find a particle of type α is $\frac{N_\alpha}{N}$ and thus the probability to find a pair (1-2) is proportional to $\frac{N_1 N_2}{N^2}$. The total number of pairs is of the order of the total number of particles N, and thus the number of (1-2) pairs is proportional to $N\frac{N_1 N_2}{N^2} = \frac{N_1 N_2}{N}$. When we introduce mole numbers, the enthalpy of mixing is

$$H_{\text{mix}} = \bar{\varepsilon} \frac{n_1 n_2}{n_1 + n_2} \quad , \quad \bar{\varepsilon} \gtrless 0 \,. \tag{22.30}$$

Depending on its sign, $\bar{\varepsilon}$ is the energy released or required to form one mole of (1-2) pairs.

With the entropy of mixing as before, the Gibbs free energy (20.34) of the binary mixture becomes

$$G = n_1 \bar{g}_1(T,p) + n_2 \bar{g}_2(T,p) + \bar{\varepsilon} \frac{n_1 n_2}{n_1 + n_2} + \bar{R}T \left(n_1 \ln \frac{n_1}{n_1 + n_2} + n_2 \ln \frac{n_2}{n_1 + n_2} \right) . \tag{22.31}$$

The chemical potential is the derivative of the total Gibbs free energy (20.2),

$$\bar{\mu}_\alpha = \left(\frac{\partial G}{\partial n_\alpha}\right)_{T,p,n_\beta} = \bar{g}_\alpha(T,p) + \bar{\varepsilon}(1 - X_\alpha)^2 + \bar{R}T\ln X_\alpha\,. \tag{22.32}$$

Comparison with (22.23) gives the activity coefficient as

$$\gamma_\alpha = \exp\left[\frac{\bar{\varepsilon}}{\bar{R}T}(1 - X_\alpha)^2\right]. \tag{22.33}$$

22.10 Gas Solubility: Henry's Law

We consider phase equilibrium between a liquid non-ideal mixture with the activity coefficient from the model of the previous section, and an ideal gas vapor phase (fugacity coefficient $\varphi''_\alpha = 1$). Then, with the activity coefficient (22.33), Raoult's law (22.29) assumes the form

$$\exp\left[\frac{\bar{\varepsilon}}{\bar{R}T}(1 - X'_\alpha)^2\right] X'_\alpha = \frac{p''_\alpha}{p^{\mathrm{sat}}_\alpha(T)}, \tag{22.34}$$

where $p''_\alpha = X''_\alpha p$ is the partial pressure of α in the vapor.

We consider a not too volatile liquid solvent, say water, under an atmosphere of volatile vapors, say air, where we can expect that the mole fractions of air components (oxygen, nitrogen, argon, ...) in the liquid are rather small. Then, in the above, we can set $(1 - X'_\alpha)^2 \simeq 1$, and find the mole fraction of gases dissolved in the liquid given by Henry's law (William Henry, 1774-1836),

$$X'_\alpha = \frac{p''_\alpha}{p^{\mathrm{sat}}_\alpha(T)\exp\left[\frac{\bar{\varepsilon}}{\bar{R}T}\right]} = \frac{p''_\alpha}{H_\alpha(T)}, \tag{22.35}$$

with Henry's constant $H_\alpha(T)$. The saturation pressure can be approximated by (17.48), so that Henry's constant becomes[3]

$$H_\alpha(T) = H_\alpha(T_0)\exp\left[\Delta H_\alpha\left(\frac{1}{T_0} - \frac{1}{T}\right)\right] \quad \text{with} \quad \Delta H_\alpha = \frac{\bar{h}^{LV}_\alpha - \bar{\varepsilon}}{\bar{R}}. \tag{22.36}$$

For oxygen, nitrogen, and carbon dioxide the following data can be found ($T_0 = 298\,\mathrm{K}$):

$$\begin{aligned}
H_{\mathrm{O_2}}(T_0) &= 43102\,\mathrm{bar} &, \quad \Delta H_{\mathrm{O_2}} &= 1700\,\mathrm{K}\,,\\
H_{\mathrm{N_2}}(T_0) &= 85590\,\mathrm{bar} &, \quad \Delta H_{\mathrm{N_2}} &= 1300\,\mathrm{K}\,,\\
H_{\mathrm{CO_2}}(T_0) &= 1648\,\mathrm{bar} &, \quad \Delta H_{\mathrm{CO_2}} &= 2400\,\mathrm{K}\,,\\
H_{\mathrm{He}}(T_0) &= 149700\,\mathrm{bar} &, \quad \Delta H_{\mathrm{He}} &= 230\,\mathrm{K}\,.
\end{aligned}$$

[3] Rolf Sander: Compilation of Henry's Law Constants for Inorganic and Organic Species of Potential Importance in Environmental Chemistry, `http://www.henrys-law.org/`

It should be noted that at environmental temperatures oxygen and nitrogen are well above their critical point, where no saturation pressure exists. Therefore, it is somewhat surprising that the above derivation gives a meaningful result for these temperatures.

In carbonated drinks, carbon dioxide dissolves in water under pressure (normally around 2 bar, depending on temperature). When the pressure is released by opening the bottle or can, the gas bubbles out, to establish the equilibrium for the partial pressure of CO_2 at the surface. From (22.36) we see that Henry's constant grows with temperature, so that less gas is dissolved in warmer water. This is the reason for the effervescence of a warm pop can as compared to a cold one.

Guinness beer is nitrogenated with a special tab. Due to the low dissolution of N_2, bottling it would give a very flat beer (under the same pressure one could dissolve 50 times more CO_2 than N_2), unless a "widget" is used, where N_2 is enclosed in a small capsule that releases the gas when the pressure in the can drops after it is opened.

Colder oceans are richer in oxygen, and thus are rich in marine life. Therefore big whales migrate between the polar waters where they find most food.

For the solvent, the mole fraction in the liquid is almost unity, $X'_\nu \simeq 1$, which implies that the activity coefficient is close to unity as well, $\gamma'_\nu = \exp\left[\frac{\bar{\varepsilon}}{\bar{R}T}\left(1 - X'_\nu\right)^2\right] \simeq 1$. Hence, evaluation of Raoult's law gives $p_\alpha^{\text{sat}}(T) = p''_\alpha$, i.e., the partial pressure of water in a saturated water-air mixture is, for all practical applications, equal to the saturation pressure of water, as used in the discussion of psychrometrics.

22.11 Phase Diagrams with Azeotropes

Next, we study phase diagrams for non-ideal mixtures, based on Raoult's law (22.29). To obtain interesting behavior, it suffices to consider non-ideality in the liquid phase, expressed through activity coefficients of the form (22.33), while the vapor is considered as a mixture of ideal gases where the fugacity coefficients are unity.

For a binary mixture, we obtain the two equations (with $X_2 = 1 - X_1$),

$$X''_1 p = p''_1\left(T, X'_1\right) = p_1^{\text{sat}}(T)\, X'_1 \exp\left[\frac{\bar{\varepsilon}}{\bar{R}T}\left(1 - X'_1\right)^2\right], \quad (22.37)$$

$$\left(1 - X''_1\right) p = p''_2\left(T, X'_1\right) = p_2^{\text{sat}}(T)\left(1 - X'_1\right) \exp\left[\frac{\bar{\varepsilon}}{\bar{R}T} X'^2_1\right]. \quad (22.38)$$

Note that, as indicated, the left hand sides are just the partial pressures in the vapor, expressed through the right hand sides as functions of temperature and liquid composition.

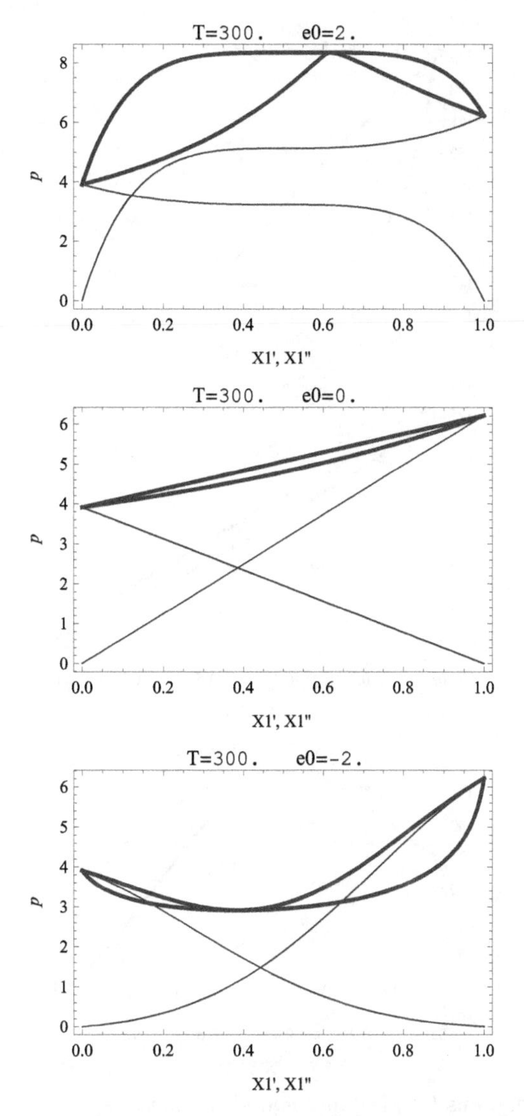

Fig. 22.8 p-X-digrams for ideal and non-ideal mixtures. Bold curves: saturated liquid and saturated vapor lines. Thin curves: partial vapor pressures $p''_\alpha(T, X'_1)$.

The saturated liquid curve is obtained from adding both equations to eliminate X''_1,

$$p(T, X'_1) = p_1^{\mathrm{sat}}(T)\, X'_1 \exp\left[\frac{\bar{\varepsilon}}{\bar{R}T}(1 - X'_1)^2\right] + p_2^{\mathrm{sat}}(T)\,(1 - X'_1) \exp\left[\frac{\bar{\varepsilon}}{\bar{R}T} X'^2_1\right] . \tag{22.39}$$

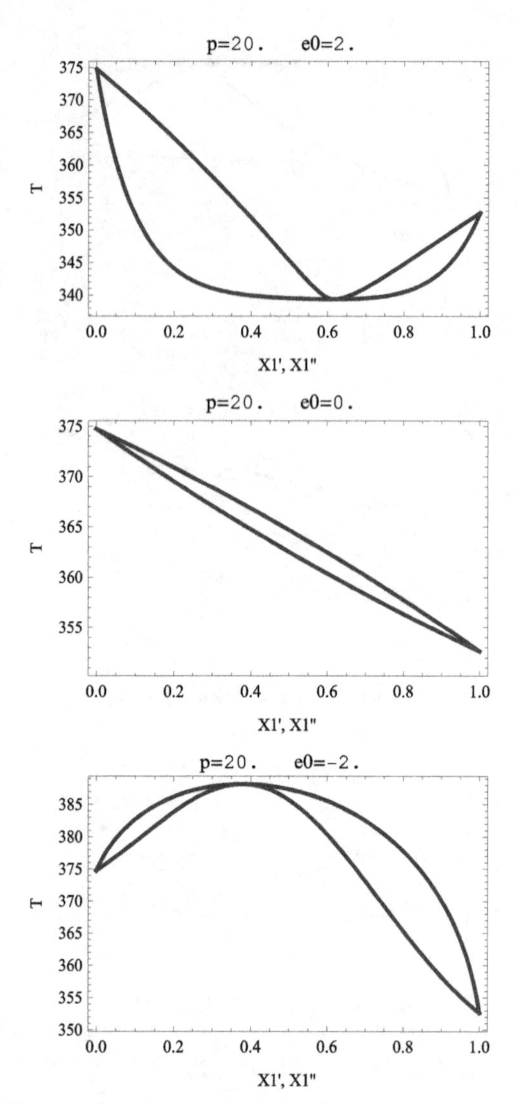

Fig. 22.9 T-X-digrams for ideal and non-ideal mixtures

The saturated vapor curve $p(T, X_1'')$ cannot be computed analytically, since it is not possible to analytically eliminate X_1' from the equations. Instead, we obtain it as follows: Division of the two equations for p_1'' and p_2'' gives

$$X_1''(T, X_1') = \frac{1}{1 + \frac{p_2^{sat}(T)}{p_1^{sat}(T)} \left(\frac{1}{X_1'} - 1\right) \exp\left[-\frac{\bar{\varepsilon}}{\bar{R}T}\left(1 - 2X_1'\right)\right]} . \tag{22.40}$$

With this, the saturated vapor curve can be expressed in parametric form as

$$\{p(T, X_1'), X_1''(T, X_1')\} \tag{22.41}$$

with X_1' as curve parameter.[4]

Figure 22.8 shows the phase diagrams together with the vapor pressures $p_\alpha''(T, X_1') = X_\alpha'' p$ for various values of $\bar{\varepsilon}$, based on data given in problem 22.25. For non-ideal mixtures, the saturated liquid and vapor lines meet not only for the pure substances ($X_1 = 1$ or $X_1 = 0$), but also in a point between, the so-called azeotrope.

The corresponding T-X-diagrams at constant p, which are of importance to understand distillation processes, can be constructed as follows: For the saturated liquid line one prescribes (X_1', p) in (22.39) and solves numerically for T to find one point (X_1', T) of the saturated liquid curve $T(p, X_1')$. The curve results from repetition for many values X_1'. For the saturated vapor curve one evaluates (22.40) for values on the saturated liquid curve to compute $X_1''(T(p, X_1'), X_1')$ and then uses the saturated liquid line to find points on the saturated liquid lines as $\{T(p, X_1'), X_1''(T(p, X_1'), X_1')\}$. Figure 22.9 shows the resulting T-X-diagrams corresponding to the p-X-diagrams above, they show azeotropes as well.

As compared to the ideal mixture, when the formation of mixed pairs (1-2) is not favored, which is the case for $\bar{\varepsilon} > 0$, the mixture becomes more volatile, as can be seen from the larger vapor pressures in Fig. 22.8 and the lower saturation temperatures in Fig. 22.9. When the formation of mixed pairs is favored, $\bar{\varepsilon} < 0$, the mixture is less volatile, vapor pressures are lower, and saturation temperatures are higher.

Azeotropic mixtures cannot be separated by distillation processes, as becomes obvious when one considers the distillation column discussed earlier in conjunction with an azeotropic T-X-diagram. The location of the azeotrope might change with pressure, and separation can be achieved by distillation at different pressures (for the model used to generate these figures, this effect is very weak, however). Other methods for separation of azeotropic mixtures are discussed in the chemical engineering literature.

Problems

22.1. Ammonia and Water

Absorption refrigeration systems often employ two phase equilibrium mixtures of ammonia (NH_3) and water (H_2O). For the following problems, assume an ideal mixture:

1. A liquid-vapor mixture of ammonia and water at 30 °C has a mole fraction of ammonia in the liquid of 60%. Determine the mole fractions of water

[4] In other words, for given T, for each value X_1' one finds a point $\{p(T, X_1'), X_1''(T, X_1')\}$ of the saturated vapor curve in the p-X-diagram.

and ammonia in the vapor phase, and the total pressure of the mixture. Saturation pressure of NH_3 at $p_{sat}(30\,°C) = 1116.5\,kPa$.
2. A two-phase mixture of ammonia and water is in equilibrium at 50 °C. The composition of the vapor phase is 99 percent NH_3 and 1 percent H_2O by mole numbers. Determine the composition of the liquid phase and the total pressure. Saturation pressure of NH_3 at 50 °C is 2033.5 kPa.
3. A liquid-vapor mixture of ammonia and water is at 40 °C and 1000 kPa. Determine the mole fractions of water and ammonia in the vapor and liquid phase. Saturation pressure of NH_3 at $p_{sat}(40\,°C) = 1554.33\,kPa$.
4. A liquid-vapor mixture of ammonia and water is in equilibrium at 10 °C and 500 kPa. Determine the mole fractions of water and ammonia in the vapor and liquid phase. Saturation pressure of NH_3: $p_{sat}(10\,°C) = 615.29\,kPa$.

22.2. Binary Mixture of Propane and 1-Butane I
Consider a mixture of $n_1 = 200\,kmol$ propane (C_3H_8) and $n_2 = 100\,kmol$ 1-butane (C_4H_{10}), with pressure $p = 10\,bar$ at a temperature of 320 K. The saturation pressures of the pure substances are $p_1^{sat}(320\,K) = 15.722\,bar$, and $p_2^{sat}(320\,K) = 4.537\,bar$.

1. Use the idealized Raoult law to find the mole fractions X_1', X_1''.
2. Determine the mole numbers n_1', n_2', n_1'', n_2'' .

22.3. Binary Mixture of Propane and 1-Butane II
Consider a mixture of $n_1 = 25\,kmol$ propane (C_3H_8) and $n_2 = 100\,kmol$ 1-butane (C_4H_{10}), at $T = 350\,K$ and $p = 12\,bar$. The saturation pressures of the pure substances can be found from the Antoine equation $\ln p_{sat}(T) = A - \frac{B}{C+T}$, where the pressure is measured in bars, the temperature in K, and the constants have the values

	A	B	C
C_3H_8	9.1058	1872.46	-25.16
C_4H_{10}	9.0580	2154.9	-34.42

1. Use the idealized Raoult law to find the mole fractions X_1', X_1''.
2. Determine the mole numbers n_1', n_2', n_1'', n_2'' .

22.4. Binary Mixture of Propane and 1-Butane III
Consider a two phase mixture of $n_1 = 150\,kmol$ propane (C_3H_8) and an unknown amount n_2 of butane (C_4H_{10}), at a temperature of 320 K.

At $p = 10\,bar$ the mole fraction of propane in the liquid phase is $X_1' = 0.488$, and the amount of propane in the liquid phase is 75 kmol.

The saturation pressure of propane is $p_{sat,1}(320\,K) = 15.7\,bar$.

1. Use the idealized Raoult law to find the mole fraction X_1''.
2. Determine the mole numbers n_2', n_1'', n_2''.
3. Determine the saturation pressure of butane at 320 K.

22.5. Binary Mixture of Propane and 1-Butane IV
Consider an ideal two phase mixture of propane (C_3H_8) and butane (C_4H_{10}), at temperature of 330 K. At this temperature, the saturation pressures of propane and butane are 19.4 bar and 5.86 bar, respectively. The mixture is equimolar, with 220 kmol of each component present, and the mole fraction of butane in the vapor phase is 30%.

1. Determine the mole fractions of butane and propane in both phases.
2. Determine the pressure of the mixture, p .
3. Determine the total mole number in the liquid.

22.6. Phase Diagrams for Propane and 1-Butane
Construct phase diagrams for a mixture of propane and 1-butane (ideal mixture). Use a computer with suitable software for plotting and evaluating of equations.

The saturation pressures of the pure substances can be found from the Antoine equation $\ln p_{\text{sat}}(T) = A - \frac{B}{C+T}$ with

	A	B	C
C_3H_8	9.1058	1872.46	-25.16
C_4H_{10}	9.0580	2154.9	-34.42

where the pressure is measured in bar, and the temperature in K.

1. Construct and plot the p-X phase diagram for a variety of temperatures (stay in between critical and triple points!).
2. Construct and plot the T-X diagram for a variety of pressures.

22.7. Water and R134a
Mixtures of water and refrigerant are used in self-foaming products. Consider a mixture of 120 g R134a (molar mass $102\frac{\text{g}}{\text{mol}}$) and 50 g water, initially at 20 °C and 2 bar, later at 20 °C and 1 bar.

1. Find the overal mole fraction of R134a in the mixture.
2. Find the mole fractions X'_R, X''_R, X'_W, X''_W for both pressures. Use the idealized Raoult law.
3. Draw a p-X-diagram for the mixture, and indicate the two states and the overall mole fraction in the diagram. Carefully consider whehter the mixture is all liquid, liquid-vapor mix, or all vapor.
4. Determine the mole numbers n'_R, n''_R, n'_W, n''_W for both states.
5. Assuming ideal mixtures for both phases, determine the volume change of the system.

22.8. Saturation Temperature and Pressure

1. Pure water at $p = 1$ bar boils at a temperature of 99.63 °C. How much salt must be added per kg of water to lower the saturation pressure for the same temperature by 1%?

2. For the same solution, find the boiling temperature for a pressure of 1.5 bar.

   Note: NaCl dissociates in solution.

22.9. Water and Salt
A solution of salt (NaCl, $M_{\mathrm{NaCl}} = 58.5 \frac{\mathrm{kg}}{\mathrm{kmol}}$) in water at 1.5 bar has a boiling temperature of 112 °C. How many grams of salt are dissolved in one litre of water?

Note: NaCl dissociates in solution.

22.10. Increase of Boiling Temperature
50 g sodium chloride are dissolved in 1 litre of water. Compute the boiling temperature at 1 bar.

22.11. Melting
In winter, after ice rain, a street is covered with a 1 cm thick layer of ice. What is the minimum mass of salt (NaCl) required per square meter to melt the ice when the temperature is −10 °C?

22.12. Cooling Liquid for a Car
The coolant of a car is required to freeze only below −20 °C. How many moles of ethylene glycol ($C_2H_6O_2$) or NaCl must be mixed to water, in order to lower the temperature of freezing to the required value? Why is glycol preferable above NaCl?

22.13. Cooking Pasta in the Mountains
A cylindrical pot (base area 250 cm^2, height 25 cm) contains 4 kg of pure water and 2 mol of NaCl. A lid which has a mass of 500 g rests freely on the top of the pot. The outside pressure is 0.8175 atm. At what temperature will the saltwater in the pot start to boil?

22.14. Boiling Point
100 g of glucose ($C_6H_{12}O_6$) and 20 g NaCl are dissolved in one litre of water, the mixture can be described as ideal mixture.

1. Find the boiling temperature of this mixture on a mountain, where the local pressure is 0.75 bar.
2. Determine the minimum work to remove only the salt, when the temperature is 23 °C.

22.15. A Lake
A lake in the mountains has a surface area of 1.2 km^2, its average depth is 24 m, and the water temperature is 17 °C. Air ($X_{N_2} = 0.79$, $X_{CO_2} = 0.0004$) at 0.8 bar stands over the water. Compute the masses of dissolved nitrogen and carbon dioxide.

22.16. Gas Mixture
Water is in contact with a gaseous mixture of nitrogen and carbon dioxide. The mole fractions of the gases in the liquid phase are measured as 1.1×10^{-4} for nitrogen and 0.006 for carbon dioxide. For a temperature of 25 °C, determine the mole fraction in the gas mixture, and the overall gas pressure.

22.17. Carbon Dioxide in Water

A piston cylinder system, which is maintained at a temperature of 298 K, contains 1 litre of liquid water, water vapor, and 5 g of carbon dioxide. The gas phase fills a volume of 1/10 litre.

1. Determine the pressure in the system.
2. Determine the percentage of CO_2 that is dissolved.

Hint: The solution becomes much easier with the assumption that in the liquid the mole amount of CO_2 is much smaller than the mole amount of water. If you use the assumption, verify it.

22.18. Henry's Law: Sparkling Water

A bottle of volume $V = 1.05$ litre contains 1 litre of liquid water, water vapor, and carbon dioxide. The pressure in the bottle is 2 bar at a temperature of 55 °C. Compute the masses of water and CO_2 in the bottle. Consider vapor and gaseous CO_2 as ideal gases. Ignore the volume (but not the amount) of dissolved CO_2 . Henry's law constant for CO_2: $H_{CO_2}(55\,°C) = 3200$ bar.

22.19. Henry's Law: Nitrogen in Water I

A bottle of volume $V = 0.55$ litre contains 0.5 litre of liquid water, water vapor, and nitrogen. The pressure in the bottle is 1.5 bar at a temperature of 35 °C. Compute the masses of water and N_2 in the bottle. Consider vapor and gaseous N_2 as ideal gases. Ignore the volume of dissolved N_2. Henry's law constant for N_2: $H_{N_2}(35\,°C) = 98600$ bar.

Remark: Guinness beer is nitrogenated with a special tab. This problem shows, why bottling it might give a very flat beer, unless a "widget" is used.

22.20. Henry's Law: Nitrogen in Water II

A bottle of total volume $V = 0.55$ litre contains 0.5 litre of liquid water, some water vapor, and a total of 0.006 mol of nitrogen. The temperature of the bottle and its contents is 35 °C.

1. Compute the total mass of water in the bottle (use tables).
2. Determine the amounts of nitrogen (in moles) dissolved in the liquid, and in the gas phase.
3. Determine the pressure in the bottle.

For the solution, ignore the volume (but not the amount) of dissolved N_2. Henry's law constant for N_2: $H_{N_2}\,(35\,°C) = 98600$ bar.

22.21. A Diver

A deep sea diver breathes a mixture of 21% O_2 and 79% He, which is at the local water pressure. The divers body contains 5 litre of blood; for simplicity, assume that blood behaves like water.

1. Assume the blood (i.e., water at 36°C) saturates with helium, and determine the amount (in moles) of helium dissolved in the blood at sea level and at a depth of 200 m.

2. For both cases determine also the volume (in litre) that the helium assumes at standard conditions.
3. Explain in thermodynamic terms why the driver needs to decompress slowly.

22.22. Henry's Law: Measurement of Oxygen Content of Lake Water

In order to retrieve water from deeper layers of a lake, the following simple device is suggested: A cylinder of volume $V = 0.11$ litre is filled with helium at 2 bar. The device is brought to the desired depth, and a valve is opened which allows water to enter the system, while the helium remains inside. Then the system is brought to the surface, and a sample of the water is taken to analyze the oxygen content.

The temperatures of all gases and liquids are assumed to be 9 °C, and the pressure on top of the lake is 0.9 bar. Density of water is 1 kg/litre. Henry's constant for O_2 in water at 9 °C is $H_{O_2}(T) = 31200$ bar.

1. 100 ml of water enter the system—estimate the depth at which the valve was opened.
2. It is measured that the mole fraction of oxygen in the water sample is 1.1×10^{-6}. Use this information to compute the mole fraction of oxygen in the water where the sample was taken.

Hint: Obviously, some of the oxygen will gas out and enter the gas volume spanned by the helium, and this leads to a lower gas content in the liquid.

22.23. Measurement of Henry's Constant

To measure Henry's constant for oxygen in water, the following device is suggested. A container is divided into two equal parts, one part is filled with highly purified water, the other part is filled with oxygen at 10 bar. Then, the division is removed, and oxygen and water assume an equilibrium state. The gas pressure is measured as 9.66 bar, and throughout the measurement the temperature is kept at 20 °C. Determine Henry's constant $H_{O_2}(T)$, and the relative amount of oxygen that enters the liquid.

Hint: Ignore the change in volume due to oxygen entering the liquid.

22.24. Henry's Law: Mass of CO_2 In Air and Oceans

1. On earth, the pressure at ground level is 1 atm = 1.01325 bar, the gravitational acceleration is $9.81\frac{\text{m}}{\text{s}^2}$, and the earth radius is 6300 km. Use this data to compute the mass of the atmosphere.
2. 0.04 mole% of the air is carbon dioxide (CO_2) and 20.95 mole% is oxygen (O_2). Compute the masses of CO_2 and O_2 in the atmosphere.
3. 70% of the earth is covered by oceans and the average depth of the oceans is 4000 m. Assuming the mass density of the oceans is $1000\frac{\text{kg}}{\text{m}^3}$, compute the mass of water in the oceans.

4. For an average water temperature of 4 °C, compute the masses of CO_2 and O_2 dissolved in the worlds oceans. Use Henry's law in the form

$$X'_\alpha = \frac{p''_\alpha}{H_\alpha(T)} ,$$

where $H_\alpha(T)$ is Henry's law constant.

Comment: This problem assumes equilibrium conditions and ignores chemical effects. The actual amount of CO_2 dissolved differs from the value obtained here, due to formation of carbonic acid and carbonates, as well as non-equilibrium conditions.

22.25. Azeotropic Curves

Consider a binary mixture in liquid-vapour equilibrium, for two substances whose vapor pressures follow the Antoine law $\ln p = A - \frac{B}{C+T}$, where p is pressure in bar, and T is the temperature in Kelvin, with

	A	B/K	C/K
component 1	9.1	2000	-25
component 2	9.0	2100	-25

Assume that the vapor can be described as an ideal gas, while the activity coefficients in the liquid phase are given as

$$\gamma_\alpha = \exp\left[\varepsilon_0 \frac{T_0}{T}\left(1 - X'_\alpha\right)^2\right] ,$$

where ε_0 is a measure for the energy of interaction between particles of different type and $T_0 = 300\,\mathrm{K}$.

1. Set up the two equations for Raoult's law for this case.
2. By eliminating X''_1, find the equation for the saturated liquid line, $p(X'_\alpha, T)$.
3. It is not possible to find the saturated vapor line analytically. Find a parameter form of the saturated vapor line $\{p(X'_\alpha, T), X''_\alpha(X'_\alpha, T)\}$, where X'_α plays the role of the curve parameter.
4. Plot the two lines, and the partial pressures in the vapor in a p-X-diagram for $\varepsilon_0 = -2, 0, 2$ (and other values of your choice) for $T = 300\,\mathrm{K}$ (and other values of your choice). Discuss the curves, in particular the azeotropic point.
5. Construct the T-X-diagram, and plot it for $p = 20\,\mathrm{bar}$.
6. Discuss distillation/rectification for a mixture with an azeotropic point. Is it possible to completely separate such a mixture by distillation?

Chapter 23
Reacting Mixtures

23.1 Stoichiometric Coefficients

In a chemical reaction, the composition of a mixture changes due to the formation of chemical compounds. The chemical changes are expressed in reaction equations, e.g., for the formation of water from hydrogen and oxygen, where two hydrogen molecules and one oxygen molecule react to form two water molecules,

$$2H_2 + 1O_2 \rightleftharpoons 2H_2O \tag{23.1}$$

or

$$-2H_2 - 1O_2 + 2H_2O = 0\ . \tag{23.2}$$

The first reaction equation uses arrows to indicate the possibility of forward and backward reactions, while the second is written as an actual equation. The coefficients $(-2,\ -1,\ +2)$ are the stoichiometric coefficients γ_α, which count how many particles of species α are involved in the reaction. Positive stoichiometric coefficients refer to products, negative coefficients refer to reactants.

Chemical reactions can occur forward and backward. In chemical equilibrium, the composition of the reacting mixture does not change and there are as many forward as backward reactions. Thus, the definition of "forward" and "backward" is somewhat arbitrary, which implies that the signs of the stoichiometric coefficients can be switched. Moreover, multiplication of all stoichiometric coefficients with the same factor is possible as well, e.g., an alternative chemical equation for the above reaction is (multiplication with $-\frac{1}{2}$)

$$H_2 + \frac{1}{2}O_2 - H_2O = 0 \tag{23.3}$$

with the stoichiometric coefficients $\left(1, \frac{1}{2}, -1\right)$.

H. Struchtrup, *Thermodynamics and Energy Conversion*,
DOI: 10.1007/978-3-662-43715-5_23, © Springer-Verlag Berlin Heidelberg 2014

23.2 Mass and Mole Balances

Due to chemical reactions, the mole numbers of all components involved in the reaction change. The reaction rate λ is defined as the number of net reactions (counted in moles) per unit time and volume. The reaction rate can be positive or negative, depending whether forward or backward reactions prevail. For each mole of reactions there are γ_α moles of component α produced or consumed. The rate of change of mole number for component α reads

$$\frac{dn_\alpha}{dt} = \gamma_\alpha V \lambda \, . \tag{23.4}$$

Multiplication with the molar mass M_α gives the corresponding rate of change for mass,

$$\frac{dm_\alpha}{dt} = \gamma_\alpha M_\alpha V \lambda \, . \tag{23.5}$$

The total mass, $m = \sum_\alpha m_\alpha$, is conserved,[1] that is

$$\frac{dm}{dt} = V\lambda \sum_\alpha \gamma_\alpha M_\alpha = 0 \, , \tag{23.6}$$

where the summation has to be taken over all substances involved in the reaction, that is products and reactants. Thus, the mass of the products is equal to the mass of the reactants,

$$\sum_\alpha \gamma_\alpha M_\alpha = 0 \, . \tag{23.7}$$

Division of (23.4) by the stoichiometric coefficient, and taking the difference of the result for different components gives a conservation law,

$$\frac{d}{dt}\left(\frac{n_\alpha}{\gamma_\alpha} - \frac{n_\beta}{\gamma_\beta}\right) = 0 \, . \tag{23.8}$$

This can be integrated to give

$$\frac{n_\alpha}{\gamma_\alpha} - \frac{n_\beta}{\gamma_\beta} = \frac{n_\alpha^0}{\gamma_\alpha} - \frac{n_\beta^0}{\gamma_\beta} \, , \tag{23.9}$$

where n_a^0 are the mole numbers at the beginning of the reaction. For the water reaction, for example, we find

$$\frac{n_{H_20}}{2} - \frac{n_{O_2}}{-1} = \frac{n_{H_20}^0}{2} - \frac{n_{O_2}^0}{-1} \, , \quad \frac{n_{H_2O}}{2} - \frac{n_{H_2}}{-2} = \frac{n_{H_20}^0}{2} - \frac{n_{H_2}^0}{-2} \tag{23.10}$$

or

[1] As long as we ignore the relativistic mass defect, which is extremely small.

$$\frac{n_{H_2O}}{2} + n_{O_2} = \frac{n^0_{H_2O}}{2} + n^0_{O_2} \quad , \quad n_{H_2O} + n_{H_2} = n^0_{H_20} + n^0_{H_2} \,. \tag{23.11}$$

A third equation, for O_2 and H_2, is not shown; this equation is a linear combination of those above, and does not give additional information.

Alternatively, one can balance the mole numbers of elements, since these must be conserved (nuclear reactions excluded). The only elements in the water reaction are oxygen and hydrogen, and the conservation of elements gives rise to the equations

$$\mathrm{O} : n_{H_2O} + 2n_{O_2} = n^0_{H_2O} + 2n^0_{O_2} \quad , \quad \mathrm{H} : 2n_{H_2O} + 2n_{H_2} = 2n^0_{H_2O} + 2n^0_{H_2} \,. \tag{23.12}$$

Obviously, this is equivalent to the previous equations. Balancing of elements must be used when multiple reactions occur.

23.3 Heat of Reaction

Chemical reactions are accompanied by changes in energy. We consider a mixture in a piston-cylinder system at (T, p), in which reactions take place at constant pressure and temperature. The heat of reaction is defined as the amount of heat that must be exchanged to keep temperature and pressure constant.

Since the system is isobaric, the first law assumes the form

$$H_2 - H_1 = Q_{12} \,. \tag{23.13}$$

The initial enthalpy is

$$H_1 = \sum_\alpha n^1_\alpha \bar{h}_\alpha \,, \tag{23.14}$$

and when a total of $\Lambda = \int_1^2 V\lambda dt$ moles of net reactions take place, the final enthalpy is

$$H_2 = \sum_\alpha \left(n^1_\alpha + \gamma_\alpha \Lambda\right) \bar{h}_\alpha \,. \tag{23.15}$$

Thus, the heat of reaction for one mole of reactions is[2]

$$\Delta\bar{h}_R\,(T,p) = \frac{Q_{12}}{\Lambda} = \sum_\alpha \gamma_\alpha \bar{h}_\alpha \,. \tag{23.16}$$

A reaction with $\Delta\bar{h}_R < 0$ is an exothermic reaction, heat must be withdrawn to keep the temperature constant. A reaction with $\Delta\bar{h}_R > 0$ is an endothermic reaction, heat must be added to keep the temperature constant.

---

[2] Here we have implicitely assumed an ideal mixture, where the enthalpy $\bar{h}_\alpha$ of the component is not affected by the change of composition, so that $\bar{h}^1_\alpha = \bar{h}^2_\alpha$.

The heat of reaction can be measured. Normally, one finds $\Delta\bar{h}_R$ tabulated at standard conditions, i.e. at $T_0 = 298.15\,\mathrm{K}$, $p_0 = 1\,\mathrm{bar}$. Some values are given in the following table:

$$
\begin{array}{lll}
\frac{1}{2}H_2 \rightleftharpoons H & : & \Delta\bar{h}_R = 218.0\frac{\mathrm{kJ}}{\mathrm{mol}} \\
\frac{1}{2}O_2 \rightleftharpoons O & : & \Delta\bar{h}_R = 249.2\frac{\mathrm{kJ}}{\mathrm{mol}} \\
H_2 + \frac{1}{2}O_2 \rightleftharpoons H_2O(l) & : & \Delta\bar{h}_R = -285.8\frac{\mathrm{kJ}}{\mathrm{mol}} \\
H_2 + \frac{1}{2}O_2 \rightleftharpoons H_2O(v) & : & \Delta\bar{h}_R = -241.8\frac{\mathrm{kJ}}{\mathrm{mol}} \\
C + \frac{1}{2}O_2 \rightleftharpoons CO & : & \Delta\bar{h}_R = -110.5\frac{\mathrm{kJ}}{\mathrm{mol}} \\
C + O_2 \rightleftharpoons CO_2 & : & \Delta\bar{h}_R = -393.5\frac{\mathrm{kJ}}{\mathrm{mol}} \\
\frac{3}{2}H_2 + \frac{1}{2}N_2 \rightleftharpoons NH_3 & : & \Delta\bar{h}_R = -46.2\frac{\mathrm{kJ}}{\mathrm{mol}} \\
\frac{1}{2}H_2 + \frac{1}{2}J_2 \rightleftharpoons HJ & : & \Delta\bar{h}_R = 25.9\frac{\mathrm{kJ}}{\mathrm{mol}} \\
CO_2 + H_2O \rightleftharpoons \frac{1}{6}C_6H_{12}O_6 + O_2 & : & \Delta\bar{h}_R = 466.3\frac{\mathrm{kJ}}{\mathrm{mol}} \\
CH_4 + 2O_2 \rightleftharpoons CO_2 + 2H_2O(l) & : & \Delta\bar{h}_R = -890.3\frac{\mathrm{kJ}}{\mathrm{mol}} \\
CH_4 + 2O_2 \rightleftharpoons CO_2 + 2H_2O(v) & : & \Delta\bar{h}_R = -802.3\frac{\mathrm{kJ}}{\mathrm{mol}}
\end{array}
$$

If one of the products is water, the heat of reaction depends on whether the product water is liquid (l) or vapor (v). Note that a change of sign or value of the stoichiometric coefficients changes sign and value of the heat of reaction; e.g., for the reaction $2H_2O\,(l) \rightleftharpoons 2H_2 + 1O_2$, we find $\Delta\bar{h}_R = 571.6\frac{\mathrm{kJ}}{\mathrm{mol}}$.

23.4 Heating Value

The heating value is defined as the energy released in the combustion of 1 kg of fuel at reference conditions. One distinguishes between the lower heating value (LHV), where the product water is vapor (v) and the higher heating value (HHV), where the product water is liquid (l). The difference is just the heat of evaporation of the the product water at standard reference temperature T_0 which is $2442\frac{\mathrm{kJ}}{\mathrm{kg}}$ or $43.96\frac{\mathrm{kJ}}{\mathrm{mol}}$.

For instance for the combustion of methane (CH_4), the higher and the lower heating values are

$$
\mathrm{HHV} = \frac{|\Delta\bar{h}_R\,(l)|}{M_{CH_4}} = 55643\frac{\mathrm{kJ}}{\mathrm{kg}} \quad , \quad \mathrm{LHV} = \frac{|\Delta\bar{h}_R\,(v)|}{M_{CH_4}} = 50150\frac{\mathrm{kJ}}{\mathrm{kg}} \; .
$$

23.5 Enthalpy of Formation

When we discussed the measurement of properties, it became clear that internal energy, enthalpy and entropy cannot be measured directly. What can be measured is, for instance, the specific heat at constant pressure, and the thermal equation of state. The enthalpy follows by integration, as shown in (16.34), which for the molar enthalpy of α assumes the form

$$\bar{h}_\alpha(T,p) - \bar{h}^0_\alpha =$$
$$= \int_{T_0}^{T} \bar{c}_{p,\alpha}(T',p_0)\,dT' + \int_{p_0}^{p}\left[\bar{v}_\alpha(T,p') - T\left(\frac{\partial v_\alpha}{\partial T}\right)_{p'}\right]dp' + \sum_i \Delta\bar{h}_{\alpha,i}\,. \tag{23.17}$$

When chemical reactions take place, the reference enthalpy $\bar{h}^0_\alpha = \bar{h}_\alpha(T_0,p_0)$ cannot be chosen arbitrarily. Indeed, for a reaction occurring at (T_0,p_0) the heat of reaction—which can be measured—is $\Delta\bar{h}_R(T_0,p_0) = \sum\gamma_\alpha\bar{h}^0_\alpha$. Thus, the reference enthalpies $\bar{h}^0_\alpha$ for different substances must be properly related, so that they give the proper, i.e., measured, heat of reaction—for all possible reactions.

To ensure this, the following convention is used to define the reference enthalpies as *enthalpies of formation* $\bar{h}^0_f$ at (T_0,p_0): Stable elements at (T_0,p_0), such as O_2, N_2, H_2, C are assigned values of zero enthalpy,

$$\bar{h}^0_{f,\alpha} = 0 \quad \text{(stable elements)}\,. \tag{23.18}$$

The values $\bar{h}^0_{f,\alpha}$ for compounds follows from measurement of $\Delta\bar{h}_R$. For instance for the water reaction

$$\Delta\bar{h}_R = \bar{h}^0_{f,H_2O} - \bar{h}^0_{f,H_2} - \frac{1}{2}\bar{h}^0_{f,O_2} = \bar{h}^0_{f,H_2O}\,. \tag{23.19}$$

While it would be convenient to use the $\bar{h}^0_{f,\alpha}$ as reference for thermodynamic property tables, this is most often not done. Most property tables that list enthalpy and internal energy as functions of temperature and pressure refer to other reference states. For ideal gas tables it is customary to scale such that the extrapolated enthalpy at $T = 0\,\mathrm{K}$ is zero. Vapor tables often have the internal energy or the enthalpy at the triple point set to zero. Thus, the tabulated values must be re-scaled before they are used for computations involving chemical reactions.

We discuss how the proper data can be found, when the molar enthalpy $\tilde{h}_\alpha(T,p)$ is tabulated. When no chemical reactions occur, the tabulated data can be used as is. However, when chemical reactions are considered, the tabulated values $\tilde{h}_\alpha(T,p)$ must be corrected to the proper reference. Since enthalpy differences are independent of the choice of the reference state, and since the enthalpy of formation is the enthalpy at (T_0,p_0), we have $\bar{h}_\alpha(T,p) - \bar{h}^0_{f,\alpha} = \tilde{h}_\alpha(T,p) - \tilde{h}_\alpha(T_0,p_0)$, or

$$\bar{h}_\alpha(T,p) = \bar{h}^0_{f,\alpha} + \left[\tilde{h}_\alpha(T,p) - \tilde{h}_\alpha(T_0,p_0)\right]\,. \tag{23.20}$$

Here, $\tilde{h}_\alpha(T,p)$ denotes tabulated enthalpy values, and $\bar{h}_\alpha(T,p)$ denotes the enthalpies with proper reference value that must be used for the discussion of chemical reactions.

Water is a common reaction product, in particular when combustion of fuels is considered. Pure water at (T_0, p_0) is liquid, with $\bar{h}^0_{f,H_2O}(l) = -285.83\frac{\text{kJ}}{\text{mol}}$, but when the product is a gas mixture some or all of the water can be in the vapor state, with $\bar{h}^0_{f,H_2O}(v) = -241.82\frac{\text{kJ}}{\text{mol}}$; see the section on psychrometrics for more discussion.

23.6 The Third Law of Thermodynamics

Also the entropy of a substance can be determined from integration based on measurements of specific heat and the thermal equation of state. Equation (16.35) written for the molar entropy of component α reads

$$\bar{s}_\alpha(T,p) - \bar{s}^0_{f,\alpha} = \int_{T_0}^{T} \frac{\bar{c}_{p,\alpha}(T', p_0)}{T'} dT' - \int_{p_0}^{p} \left(\frac{\partial \bar{v}_\alpha}{\partial T}\right)_{p'} dp' + \sum_i \frac{\Delta \bar{h}_{\alpha,i}}{T_i}, \tag{23.21}$$

with the entropies of formation $\bar{s}^0_{f,\alpha} = \bar{s}_\alpha(T_0, p_0)$.

The *Third Law of Thermodynamics*, formulated by Walther Nernst (1864-1941) based on experimental evidence, states that at absolute zero ($T = 0\,\text{K}$), the entropy of any crystalline substance is a constant, and independent of pressure and other properties (e.g. magnetization). Based on the microscopic definition of entropy, $S = k \ln \Omega$, Max Planck (1858-1947) found that the value of the constant depends on the quantum mechanical ground state, and it is zero for crystals that only have one ground state ($\Omega = 1$); then $s(T = 0\,\text{K}) = 0$. Systems with more ground states have a residual entropy.

The third law allows to determine the entropy of formation $\bar{s}^0_{f,\alpha} = \bar{s}_\alpha(T_0, p_0)$ from (23.21) by evaluating it at $T = 0\,\text{K}$ and $p = p_0$, which yields [3]

$$\bar{s}^0_{f,\alpha} = \bar{s}_\alpha(T_0, p_0) = \int_0^{T_0} \frac{\bar{c}_{p,\alpha}(T', p_0)}{T'} dT' + \sum_i \frac{\Delta \bar{h}_{\alpha,i}}{T_i}. \tag{23.22}$$

For substances with residual entropy, the latter must be added on the right hand side. Thus, the third law assigns absolute values to the entropy constants $\bar{s}^0_\alpha = \bar{s}_\alpha(T_0, p_0)$, i.e., the entropies of formation, which can be found in tables.

Most data tables for ideal gases show the entropy values $\bar{s}_\alpha(T, p_0)$ with respect to the proper reference, so that the tabulated value $\bar{s}_\alpha(T_0, p_0)$ is the entropy of formation of α at standard reference conditions. In other words, in contrast to enthalpy data, tabulated entropy data normally needs not to be corrected. A notebale exception are saturation tables, e.g., for water one

[3] Note that the sign of the enthalpies of phase change, $\Delta \bar{h}_{\alpha,i}$, depends on the direction of the phase change as the line of integration crosses a saturation curve in the p-T-plane. When the integration goes from (T_0, p_0) to $(0\text{K}, p_0)$, the $\Delta \bar{h}_{\alpha,i}$ are negative. For (23.22) the sign was switched, so that the $\Delta \bar{h}_{\alpha,i}$ appear with a plus sign.

often finds the reference set such that the entropy of liquid at the triple point is set to zero.

23.7 The Third Law and Absolute Zero

An interesting implication of the third law is that a temperature of absolute zero cannot be reached in a finite number of process steps. This can best be shown with the help of the T-s-diagram in Fig. 23.1, which shows lines of constant property ψ (e.g., pressure, magnetization, ...).

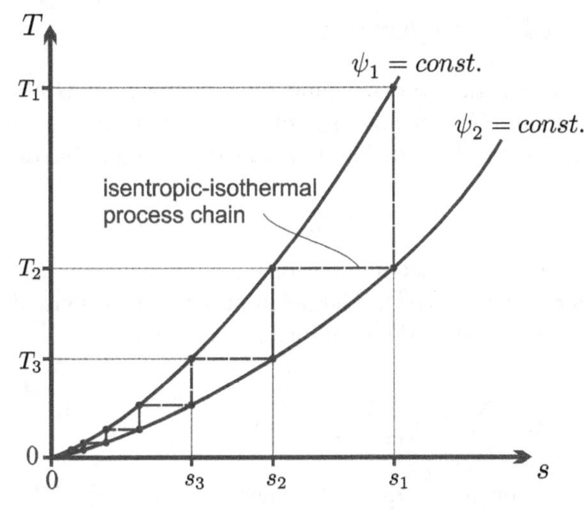

Fig. 23.1 Series of isothermal-isentropic processes between lines of constant ψ to reach low temperatures; ψ can be any property (except T, s)

Due to the third law, the lines of constant ψ all emerge from one point at $T = 0\,\text{K}$. To reach low temperatures, one can take a relatively large amount of substance at (T_1, ψ_1) and subject it to an adiabatic process ending at (T_2, ψ_2). Next, one takes a part of the substance at (T_2, ψ_2) and brings it to (T_2, ψ_1) isothermally, where the rest of substance at (T_2, ψ_2) serves as a temperature reservoir. Now one repeats the series of adiabatic and isothermal processes, with smaller and smaller amounts of substance. The resulting zigzag process is shown in the figure, where for simplicity the adiabatic processes are drawn as adiabatic reversible—i.e., isentropic—processes. Since the curves of constant ψ meet in the origin, the distance between the curves becomes smaller and smaller, and each step leads to a temperature reduction smaller than the previous. Accordingly, the heat $\delta q = Tds$ removed per unit mass in each isothermal step becomes smaller and smaller as well. Evidently, absolute zero cannot be reached with a finite number of steps.

Another conclusion from the third law refers to the specific heat: When phase changes are excluded, specific heat and entropy are related by

$$\bar{s}(T,p) = \int_0^T \frac{\bar{c}_p(T',p)}{T'} dT' . \tag{23.23}$$

Entropy must remain finite in the limit $T \to 0\,\mathrm{K}$, which implies that the specific heat must vanish in the limit,

$$\lim_{T \to 0\,\mathrm{K}} \bar{c}_p(T',p) = 0 . \tag{23.24}$$

23.8 Law of Mass Action

In equilibrium at constant pressure and temperature, the Gibbs free energy assumes a minimum, $G(T,p,n_\alpha) \longrightarrow \min$. In a chemical reaction, the mole numbers n_α are related by stoichiometry. From the mole balance (23.8) follows

$$n_\alpha = n_\alpha^0 + \gamma_\alpha \Lambda , \tag{23.25}$$

where $\Lambda = \int_1^2 V\lambda dt$ is the net number of reactions between the actual state (n_α) and a reference state (n_α^0). The net number of reactions, Λ, is the only variable in the system, thus the equilibrium condition is

$$\frac{dG}{d\Lambda} = \sum_{\alpha=1}^{\nu} \left(\frac{\partial G}{\partial n_\alpha}\right)_{p,T,n_\beta} \frac{\partial n_\alpha}{\partial \Lambda} = \sum_{\alpha=1}^{\nu} \bar{\mu}_\alpha \gamma_\alpha = 0 . \tag{23.26}$$

The resulting equilibrium condition for chemical reacting mixtures is known as the law of mass action,

$$\sum_{\alpha=1}^{\nu} \gamma_\alpha \bar{\mu}_\alpha = 0 . \tag{23.27}$$

The computation of reactive equilibria requires knowledge of the chemical potential. In the following we shall study equilibria in ideal mixtures.

Phase changes can be interpreted as chemical reactions, e.g., with stoichiometric coefficients $\gamma_V = 1$ and $\gamma_L = -1$ for evaporation. Since for single substances the chemical potentials are just the Gibbs free energies, the law of mass action gives the equilibrium condition, $\sum_{\alpha=1}^{\nu} \gamma_\alpha \bar{\mu}_\alpha = \bar{g}_V - \bar{g}_L = 0$.

23.9 Law of Mass Action for Ideal Mixtures and Ideal Gases

For ideal mixtures the chemical potential is given by $\bar{\mu}_\alpha = \bar{g}_\alpha(T,p) + \bar{R}T \ln X_\alpha$, and the law of mass action gives

$$\sum_{\alpha=1}^{\nu} \gamma_\alpha \ln X_\alpha = -\frac{\sum_{\alpha=1}^{\nu} \gamma_\alpha \bar{g}_\alpha (T,p)}{\bar{R}T} \ . \tag{23.28}$$

We define the Gibbs free energy of reaction as

$$\Delta \bar{g}_R (T,p) = \sum_{\alpha=1}^{\nu} \gamma_\alpha \bar{g}_\alpha (T,p) \ , \tag{23.29}$$

and rewrite the equilibrium conditions as

$$\prod_{\alpha=1}^{\nu} X_\alpha^{\gamma_\alpha} = \exp\left[-\frac{\Delta \bar{g}_R (T,p)}{\bar{R}T}\right] = K_X (T,p) \ . \tag{23.30}$$

The chemical equilibrium is determined through the equilibrium constant[4] $K_X (T,p)$ which, in ideal mixtures, can be determined from property data tables for single substances.

For ideal gases, the pressure dependence of the Gibbs free energy is explicitly known, we have $\bar{g}_\alpha (T,p) = \bar{g}_\alpha (T,p_0) + \bar{R}T \ln \frac{p}{p_0}$, where $\bar{g}_\alpha (T,p_0) = \bar{h}_\alpha (T) - T\bar{s}_\alpha^0 (T)$ is the temperature dependent part of the Gibbs free energy. Thus, the previous equation can be recast as

$$\prod_{\alpha=1}^{\nu} p_\alpha^{\gamma_\alpha} = K_p (T) \ , \tag{23.31}$$

where $K_p (T)$ is known as the chemical constant, and is given by

$$K_p (T) = \prod_{\alpha=1}^{\nu} p_0^{\gamma_\alpha} \exp\left[-\frac{\Delta \bar{g}_R (T,p_0)}{\bar{R}T}\right] \ . \tag{23.32}$$

Here, $\Delta \bar{g}_R (T,p_0) = \sum_{\alpha=1}^{\nu} \gamma_\alpha \bar{g}_\alpha (T,p_0)$ is the Gibbs free energy of reaction at standard pressure $p_0 = 1\,\text{bar}$.

Remarkably, the chemical constant $K_p (T)$ depends only on temperature, but not on pressure. Nevertheless, as we shall see, the chemical equilibrium is affected by total pressure $p = \sum_\alpha p_\alpha$.

As presented in (23.31) the chemical constant has the unit $[\text{bar}]^{\sum_\alpha \gamma_\alpha}$. Some authors prefer a dimensionless form, and write[5]

$$\prod_{\alpha=1}^{\nu} \left(\frac{p_\alpha}{p_0}\right)^{\gamma_\alpha} = \tilde{K}_p (T) \ . \tag{23.33}$$

[4] It might be a bit misleading to refer to $K_X (T,p)$ and $K_p (T)$ as "constants", since they are functions of temperature and pressure, but this is the name they have.

[5] Of course, they write without the tilde that was introduced to distinguish between the two definitions.

with

$$\tilde{K}_p(T) = \prod_{\alpha=1}^{\nu} \exp\left[-\frac{\Delta\bar{g}_R(T, p_0)}{\bar{R}T}\right] . \tag{23.34}$$

23.10 Example: NH_3 Production (Haber-Bosch Process)

Ammonia (NH_3) is one of the most important materials in chemical industry. It is used for the production of fertilizers and explosives, as cooling fluid in refrigeration systems, and as source material for many other chemical processes. The world ammonia production is about $130 \times 10^6 \frac{\mathrm{t}}{\mathrm{y}}$, and an ammonia plant that produces $1500 \frac{\mathrm{t}}{\mathrm{d}}$ consumes about 650 MW of energy; the energy required to produce one ton is ca. 35 GJ. Approximately 1% of the world's energy usage is devoted to the production of ammonia!

Before Haber and Bosch found out how to produce ammonia industrially, the supply came from guano fields off the cost of South America (guano is ... bird droppings). Here, we discuss the basic principles of the Haber-Bosch process which was developed by Fritz Haber (1868-1934) and brought to industrial production by Carl Bosch (1874-1940).

Ammonia is produced by combining hydrogen, normally obtained from natural gas by steam methane reforming, with nitrogen from air, where the oxygen is removed by reaction with carbon monoxide and hydrogen from the hydrogen production step. The chemical equation for ammonia synthesis reads

$$-N_2 - 3H_2 + 2NH_3 = 0 ,$$

and the law of mass action (23.31) for this reaction assumes the form

$$\frac{p_{NH_3}^2}{p_{N_2} p_{H_2}^3} = K_p(T) .$$

In order to determine the equilibrium compositions of ammonia, nitrogen and hydrogen at given (T, p), we require the balances of hydrogen and nitrogen mole numbers, and the total pressure,

$$n_{NH_3} + 2n_{N_2} = 2n_{N_2}^0 \quad , \quad 3n_{NH_3} + 2n_{H_2} = 2n_{H_2}^0 \quad , \quad p = p_{NH_3} + p_{N_2} + p_{H_2} .$$

Pressures and mole numbers are related via the ideal gas law

$$n_\alpha = \frac{p_\alpha V}{\bar{R}T} ,$$

where V is the reactor volume. Thus, we have four equations for the four unknowns $(p_{NH_3}, p_{N_2}, p_{H_2}, V)$,

$$\frac{p_{NH_3}^2}{p_{N_2} p_{H_2}^3} = K_p(T) \quad , \quad p = p_{NH_3} + p_{N_2} + p_{H_2} \, .$$

$$p_{NH_3} + 2p_{N_2} = 2n_{N_2}^0 \frac{\bar{R}T}{V} \quad , \quad 3p_{NH_3} + 2p_{H_2} = 2n_{H_2}^0 \frac{\bar{R}T}{V} \, .$$

We assume a stoichiometric mixture of the reactants, $n_{H_2}^0 = 3n_{N_2}^0$, to find

$$p_{H_2} = 3p_{N_2} \quad , \quad p_{N_2} = \frac{p - p_{NH_3}}{4} \quad , \quad V = \frac{4n_{N_2}^0}{1 + \frac{p_{NH_3}}{p}} \frac{\bar{R}T}{p} \, ,$$

and a quadratic equation for the equilibrium ammonia mole fraction $X_{NH_3} = \frac{p_{NH_3}}{p}$, with the solution

$$X_{NH_3} = 1 + \frac{8}{3p\sqrt{3K_p(T)}} - \sqrt{\left(1 + \frac{8}{3p\sqrt{3K_p(T)}}\right)^2 - 1} \, .$$

Tabulated values for the chemical constant are

$$K_p(300\,\mathrm{K}) = 4.67 \times 10^5 \frac{1}{\mathrm{bar}^2} \, ,$$
$$K_p(600\,\mathrm{K}) = 1.862 \times 10^{-3} \frac{1}{\mathrm{bar}^2} \, ,$$
$$K_p(773\,\mathrm{K}) = 1.585 \times 10^{-5} \frac{1}{\mathrm{bar}^2} \, .$$

Thus, a larger ammonia fraction is encountered for smaller temperatures, where $K_p(T)$ is larger, and for larger pressures. For instance, at $T = 300\,\mathrm{K}$ and $p = 1\,\mathrm{bar}$, one finds $X_{NH_3} = 0.935$ and this increases to $X_{NH_3} = 0.9970$ when the pressure is increased to 500 bar.

However, when one mixes hydrogen and nitrogen at $(300\,\mathrm{K}, 1\,\mathrm{bar})$, the mixture does not approach thermodynamic equilibrium, but remains in a metastable state; nothing happens, since the reaction rate is too slow. As Haber found out, iron catalysts are required to advance the reaction, but these work only at relatively high temperatures, where the ammonia yield is relatively small. In order to have a significant yield, the process must be performed under high pressures. For a reactor temperature of $T = 773\,\mathrm{K}$, the mole fraction is $X_{NH_3} = 0.309$ for $p = 500\,\mathrm{bar}$, but only $X_{NH_3} = 0.0013$ for $p = 1\,\mathrm{bar}$.

In continuous reactors, the product is cooled (at pressures below p_{sat}), the ammonia condenses and is removed, while the unused portions of hydrogen and nitrogen are fed back into the production process.

23.11 Le Chatelier Principle

The ammonia reaction shifts its equilibrium towards the desired product when the temperature is lower, and the pressure is higher. To study how changes of temperature and pressure affect the chemical equilibrium in other reactions (for ideal mixtures), we turn to (23.30) and compute the change of the equilibrium constant $K_X(T,p)$ with pressure and temperature. According to the definition, larger K_X implies more product.

We compute first the change of K_X with temperature, as

$$\left(\frac{\partial \ln K_X}{\partial T}\right)_p = -\frac{1}{\bar{R}T}\left[\left(\frac{\partial \Delta\bar{g}_R(T,p)}{\partial T}\right)_p - \frac{\Delta\bar{g}_R(T,p)}{T}\right] = \frac{\Delta\bar{h}_R(T,p)}{\bar{R}T^2}\,. \tag{23.35}$$

Here we have used $\left(\frac{\partial \bar{g}}{\partial T}\right)_p = -\bar{s}$, which implies

$$\left(\frac{\partial \Delta\bar{g}_R(T,p)}{\partial T}\right)_p = -\sum_\alpha \gamma_\alpha \bar{s}_\alpha = -\Delta\bar{s}_R\,. \tag{23.36}$$

It follows that an endothermic reaction ($\Delta\bar{h}_R > 0$) will advance further at larger temperatures, while an exothermic reaction ($\Delta\bar{h}_R < 0$) will advance further at lower temperatures. The ammonia reaction is exothermic, and thus temperature increase reduces the yield. Dissociation processes, e.g., $H_2 \rightleftharpoons 2H$, are endothermic and thus pronounced dissociation takes place at high temperatures.

For the change of K_X with pressure we find

$$\left(\frac{\partial \ln K_X}{\partial p}\right)_T = -\frac{1}{\bar{R}T}\left(\frac{\partial \Delta\bar{g}_R(T,p)}{\partial p}\right)_T = -\frac{\Delta\bar{v}_R(T,p)}{\bar{R}T}\,. \tag{23.37}$$

Here we have used $\left(\frac{\partial \bar{g}}{\partial p}\right)_T = \bar{v}$, which implies

$$\left(\frac{\partial \Delta\bar{g}_R(T,p)}{\partial p}\right)_T = \sum_\alpha \gamma_\alpha \bar{v}_\alpha(T,p) = \Delta\bar{v}_R(T,p)\;; \tag{23.38}$$

we might call $\Delta\bar{v}_R$ the volume of reaction. Note that $\bar{v}_\alpha(T,p)$ is the volume of component α alone at temperature and pressure of the mixture. We have $\Delta\bar{v}_R < 0$ when the volume of the product is less than the volume of the reactants, and $\Delta\bar{v}_R > 0$ when the volume of the product is larger than the volume of the reactants. The reaction will advance further at higher pressures for negative volumes of reaction, $\Delta\bar{v}_R < 0$.

For ideal gases, we have $\bar{v}(T,p) = \bar{R}T/p$, hence, when all components are ideal gases, the volume of reaction becomes

$$\Delta\bar{v}_R(T,p) = \sum_\alpha \gamma_\alpha \bar{v}_\alpha(T,p) = \frac{\bar{R}T}{p}\sum_\alpha \gamma_\alpha \,. \qquad (23.39)$$

In the ammonia reaction one mole of N_2 and three moles of H_2 combine to two moles of NH_3, hence $\sum_\alpha \gamma_\alpha = -2$. Thus, the product has half the volume of the reactants, the volume of reaction is negative, and pressure increase advances the reaction. In dissociation processes, e.g., $H_2 \to 2H$, the product has a larger volume, and pressure increase reduces the amount of dissociated gas.

The above statements are examples, for ideal mixtures, of Le Chatelier's Principle (Henry Le Châtelier, 1850-1936) which states that *A change in one of the variables (temperature, pressure, concentration, ...) that describe a system in equilibrium, produces a shift in the position of the equilibrium that counteracts the change.*

For instance, when a reactive mixture with exothermic reaction is heated to a higher temperature, backward reactions occur which consume energy and reduce the temperature. When the pressure is increased on a reactive mixture of ideal gases with negative reaction volume $\Delta\bar{v}_R$, reactions occur that reduce the overall particle number, and increase the mole volume $\bar{v}$, to reduce the pressure $p = \frac{RT}{\bar{v}}$.

23.12 Multiple Reactions

In many technical applications multiple reactions occur simultaneously. We denote the reaction rate densities for the various reactions as λ_a and the stoichiometric coefficient for component α in reaction a as γ_α^a, where the superscript $a = 1, \ldots, N$ indicates the reaction. The mole balance for component α then reads

$$\frac{dn_\alpha}{dt} = \sum_a \gamma_\alpha^a \lambda_a V \,. \qquad (23.40)$$

Integration upon time yields

$$n_\alpha = n_\alpha^0 + \sum_a \gamma_\alpha^a \Lambda_a \,, \qquad (23.41)$$

where Λ_a is the net number of reactions of type a. The Λ_a are the free parameters for the establishment of equilibrium, hence the equilibrium conditions are

$$\frac{dG}{d\Lambda_a} = \sum_\alpha \frac{\partial G}{\partial n_\alpha}\frac{\partial n_\alpha}{\partial \Lambda_a} = \sum_\alpha \bar{\mu}_\alpha \gamma_\alpha^a = 0 \quad \text{for } a = 1, \ldots, N \,. \qquad (23.42)$$

In other words, the law of mass action must hold for all reactions individually, that is

$$\sum_{\alpha} \bar{\mu}_\alpha \gamma_\alpha^a = 0 \quad \text{for } a = 1, \ldots, N . \tag{23.43}$$

For ideal mixtures and ideal gases we obtain generalizations of (23.30, 23.31),

$$\prod_{\alpha=1}^{\nu} X_\alpha^{\gamma_\alpha^a} = K_X^a (T, p) \quad \text{for } a = 1, \ldots, N. \tag{23.44}$$

and

$$\prod_{\alpha=1}^{\nu} p_\alpha^{\gamma_\alpha^a} = K_p^a (T) \quad \text{for } a = 1, \ldots, N. \tag{23.45}$$

Problems

23.1. Enthalpy of Formation and Heat of Reaction

Measurement of the heat of reaction at standard reference state T_0, p_0 for several reactions gave the following values:

$C + \frac{1}{2}O_2 = CO$: $\Delta \bar{h}_R^0 = -110 \frac{\text{kJ}}{\text{mol}}$,
$CO_2 = CO + \frac{1}{2}O_2$: $\Delta \bar{h}_R^0 = 280 \frac{\text{kJ}}{\text{mol}}$,
$H_2 + \frac{1}{2}O_2 = H_2O$: $\Delta \bar{h}_R^0 = -240 \frac{\text{kJ}}{\text{mol}}$,
$CH_4 + 2O_2 = CO_2 + 2H_2O$: $\Delta \bar{h}_R^0 = -800 \frac{\text{kJ}}{\text{mol}}$.

1. Use this data to determine the heat of formation $\bar{h}_f^0$ for H_2O, CO_2, CH_4. Do not read values from tables, but show how you use the data given above!
2. Measurements show that for temperatures below 800 K the specific heat of methane (CH_4) can be approximated as $\bar{c}_p = 20.1 \frac{\text{kJ}}{\text{kmol K}} + 0.053 T \frac{\text{kJ}}{\text{kmol K}^2}$. Use this and the data from above to determine the enthalpy of methane at 600 K.
3. Use the results from above and the gas tables for oxygen, carbon dioxide and water to compute the heat of reaction for $CH_4 + 2O_2 = CO_2 + 2H_2O$ at 600 K.

23.2. Dimethyl Ether (DME)

The heat of reaction at standard conditions for the combustion of gaseous dimethyl ether, C_2H_6O, is experimentally found as $-1460 \frac{\text{kJ}}{\text{mol}}$, when the product water is liquid.

1. Use this measurement and tabled data to determine the enthalpy of formation of DME.
2. Determine the heat of reaction when the product water is vapor.

23.3. Shifting the Chemical Equilibrium

Consider a mixture of CO_2, CO and O_2 in chemical equilibrium. Now the pressure is doubled. Will the number of moles of CO_2, CO and O_2 change? How? How does the equilibrium change when the temperature is increased?

23.4. Changes in Chemical Equilibrium
An equimolar mixture of CO and $H_2O(g)$ reacts to form an equilibrium mixture of CO_2, CO, H_2O and H_2 at 1727 °C, 1 atm.

1. Will decreasing the pressure while keeping the temperature constant increase or decrease the amount of H_2 present? Explain.
2. Will lowering the temperature increase or decrease the amount of H_2 present? Explain.

23.5. Shift in Chemical Equilibrium through Inert Addition
An equimolar mixture of O_2 and H_2 reacts to form an equilibrium mixture of O_2, H_2, and H_2O. After equilibrium is reached, N_2 is added to the mixture isobarically. As nitrogen is added, does the amount of water increase or decrease? Use Le Chatelier's principle for a first answer. Then perform a detailed analysis to find a relation between the amounts of N_2 added and H_2O present (assume stoichiometric mix of H_2 and O_2).

23.6. Shift in Chemical Equilibrium
Methane, CH_4, reacts with stoichiometric air to form an equilibrium mixture of CH_4, CO_2, H_2O, O_2, N_2. Will the equilibrium between CH_4, CO_2, H_2O, O_2 be different when the reaction takes place at the same temperature and pressure, but no nitrogen is present? State your arguments, e.g., consider the quotient p_{CH_4}/p_{CO_2}.

23.7. Dissociation of Oxygen
Measurement of oxygen at 3800 K shows equal mole fractions of $O_2$and O.

1. Determine the pressure.
2. Now the pressure is doubled. Determine the mole fractions of O_2 and O.

23.8. Law of Mass Action: Methanol Synthesis
Methanol (CH_3OH) is produced by catalytic hydrogenation of carbon monoxide according to the reaction

$$CO + 2H_2 - CH_3OH = 0 .$$

Assume that all partners in the reaction are ideal gases.

1. Carbon monoxide and hydrogen are mixed in stoichiometric ratio. Find an expression that relates the chemical constant $K_p(T)$, the methanol mol ratio X_{CH_3OH} and the total pressure p of the mixture.
2. A measurement at 600 K shows that $K_p(600\,\mathrm{K}) = 12000\,\mathrm{bar}^2$. For which total pressure p do we have a methanol ratio of 30%?
3. The reaction is exothermal. What does that mean? Would a further increase of temperature increase the methanol ratio?

23.9. Law of Mass Action: Formation of NO

Air (79% N_2 and 21% O_2) is heated to 2000 K at a constant pressure of 2 bar. Assume that the equilibrium mixture at this temperature consists of N_2, O_2, and NO.

For the reaction equation $\frac{1}{2}N_2 + \frac{1}{2}O_2 = NO$ one finds at this temperature that $\ln K_p(T) = -3.931$.

1. Compute the mole fractions of the three components in equilibrium.
2. Will the equilibrium composition change when the pressure is doubled?
3. For this reaction, $\ln K_p$ grows with temperature. Does that mean the reaction is exothermic or endothermic?

23.10. Law of Mass Action: N_2O_4

One kmol of N_2O_4 dissociates at 25 °C, 1 atm to form an equilibrium ideal gas mixture of N_2O_4 and NO_2, in which the amount of N_2O_4 is 0.8154 kmol.

1. Determine the mole number of NO_2 in equilibrium.
2. Determine the chemical constant $K_p(T)$ at 25 °C for the reaction.
3. Determine the amount of N_2O_4 that would be present if the pressure is 0.5 bar.

Hint: Determine first the absolute mole numbers of NO_2 and N_2O_4.

23.11. Chemical Reaction

Consider the combustion of hydrogen with oxygen according to $H_2O - H_2 - \frac{1}{2}O_2 = 0$, when the pressure in the combustion chamber is fixed at 2 bar. Hydrogen and oxygen enter the chamber in the stoichiometric ratio at $T_{in} = 298.15\,\mathrm{K}$, and the mass flow of hydrogen is $1\frac{\mathrm{kg}}{\mathrm{min}}$. The temperature in the chamber can be fixed by controlling the heat flux.

1. Compute the mass flow of oxygen.
2. The partial pressure of water vapor in the exhaust should be larger than 95% What is the maximum temperature for the combustion chamber, and what heat flux must be removed? Repeat for water conent of 98%.
3. Discuss the flame temperature under adiabatic conditions. Derive an equation that contains only T_{flame} as the unknown (or a set of equations, which will serve for the same purpose) – that equation will contain $K_p(T)$, $h_\alpha(T)$ for all components).

T/K	298.15	300	400	500	600	700	800	900	1000
$\mathrm{Log}[\frac{K_p(T)}{\sqrt{\mathrm{bar}}}]$	40.047	39.7868	29.2307	22.8855	18.6323	15.5832	13.2285	11.4978	10.0010

T/K	1100	1200	1300	1400	1500	1750	2000	2500	3000
$\mathrm{Log}[\frac{K_p(T)}{\sqrt{\mathrm{bar}}}]$	8.8830	7.8980	7.0637	6.3475	5.7254	4.4796	3.546	2.232	1.344

23.12. Steam Methane Reforming
Steam methane reforming is used to produce hydrogen from methane, in particular for further use in ammonia production. The first reaction step in steam methane reforming is the reaction

$$CH_4 + H_2O \leftrightarrow 3H_2 + CO \ ,$$

which is then followed by the water gas shift reaction discussed in the next two problems.

At 1173 K, the chemical constant for this reaction is $K_p\,(1173\,\mathrm{K}) = 1.43 \times 10^3\,\mathrm{bar}^2$. Methane and steam are mixed in the ratio 1 to α by mole, and react with help of catalysts at a pressure of 30 bar.

Find the equations to determine the partial pressures p_{CH_4}, p_{H_2O}, p_{H_2}, p_{CO}. Use a computer to determine the mole fractions of all components in chemical equilibrium at the given pressure and temperature for $\alpha = 1$ and $\alpha = 5$. Also determine as a measure for the relative conversion of methane into hydrogen the ratio p_{CH_4}/p_{H_2}. Interpret the result–can you explain it?

Assume that all components behave as ideal gases.

23.13. Property data: Water Gas Shift Reaction I
The water gas shift reaction is the second reaction for the production of hydrogen from methane. The reaction equation reads

$$CO + H_2O \leftrightarrow CO_2 + H_2 \ .$$

1. Assuming that all components are ideal gases, determine the heat of reaction for the temperature $T = 600\,\mathrm{K}$. Is the reaction exothermic or endothermic?
2. Determine the Gibbs free energy of reaction at $T = 600\,\mathrm{K}$ and $p = 1\,\mathrm{atm}$.
3. Compute the chemical constant $K_X(T,p)$. Note: this asks for K_X, not for K_p.
4. For this reaction, does the Gibbs free energy of reaction depend on pressure? Explain! Determine $K_p\,(T)$.

23.14. Law of Mass Action: Water Gas Shift Reaction II
Consider the chemical equilibrium of carbon monoxide, carbon dioxide, water and hydrogen through the water gas shift reaction, as above, at $T = 800\,\mathrm{K}$ and $p = 1\,\mathrm{atm}$. The chemical constant for this reaction is $K_X(800\,\mathrm{K}, 1\,\mathrm{atm}) = 16.424$.

1. Determine the equilibrium mole fractions for all components, when the initial state was a stoichiometric mixture of carbon monoxide and water vapor.
2. Can the equilibrium be shifted by a change of pressure? Explain!

23.15. Law of Mass Action: High Temperature Combustion
Reaction of a stoichiometric mixture of benzene (C_6H_6) and dry air at 200 kPa, with a small heat loss, results in a flame temperature of 2400 K. Assume that nitrogen, oxygen and water do *not* dissociate, and consider the equilibrium between CO_2, COand O_2. Determine the partial pressures of all components in the mixture.

23.16. Law of Mass Action: Incomplete Combustion
Octane (C_8H_{18}) is burned with 95% theoretical air. The resulting equilibrium mixture consists of CO_2, CO, H_2O, H_2 and N_2. Determine the mole fractions of all constituents when the mixture is at 1000 K and 2 bar.

23.17. Law of Mass Action: Methane
At 3000 K and 8 bar, methane (CH_4) reacts with the stoichiometric amount of pure oxygen according to $CH_4+2O_2 \leftrightarrow CO_2+2H_2O$. The chemical constants for the following reactions are given (all at 3000 K):

$$\begin{array}{ll} C + 2H_2 \leftrightarrow CH_4 & \text{with } K_p(T) = 9.685\frac{1}{\text{bar}^2} , \\ 2H_2 + O_2 \leftrightarrow 2H_2O & \text{with } K_p(T) = 484.9\frac{1}{\text{bar}} , \\ C + O_2 \leftrightarrow CO_2 & \text{with } K_p(T) = 15.87\frac{1}{\text{bar}} , \\ 2CO_2 \leftrightarrow 2CO + O_2 & \text{with } K_p(T) = 0.1089\,\text{bar} . \end{array}$$

Determine the mole fraction of CH_4 in equilibrium.

23.18. Law of Mass Action with Multiple Reactions
A mixture consists originally of 3 kmol of CO_2 and 1 kmol of O_2. Assume that in equilibrium at 3000 K and 3 bar only CO_2, CO, O_2 and O are present, and determine the equilibrium composition.

23.19. Law of Mass Action: Nitrogen and Oxygen
A gas mixture consisting of 1 kmol of NO, 10 kmol of O_2 and 40 kmol of N_2 reacts to form an equilibrium mixture of N_2, NO_2, NO, and O_2 at 500 K, 0.1 atm. Determine the composition of the equilibrium mixture.

1. Solve for the case that the molecular nitrogen stays inert.
2. Solve for the case that the molecular nitrogen reacts.

Chapter 24
Activation of Reactions

24.1 Approaching Chemical Equilibrium

So far, we have considered chemical equilibrium, which stands at the end of a reactive process, but not the details of reaching the equilibrium. In the discussion of the ammonia synthesis, we have mentioned the need for catalysts, which facilitate the reactions, and are required to reach the equilibrium state in reasonable time. In the present chapter, we shall discuss reaction rates, and the activation of reactions. Activation losses are an important cause of losses in fuel cells, and thus we will come back to this topic in the discussion of thermodynamics of fuel cells.

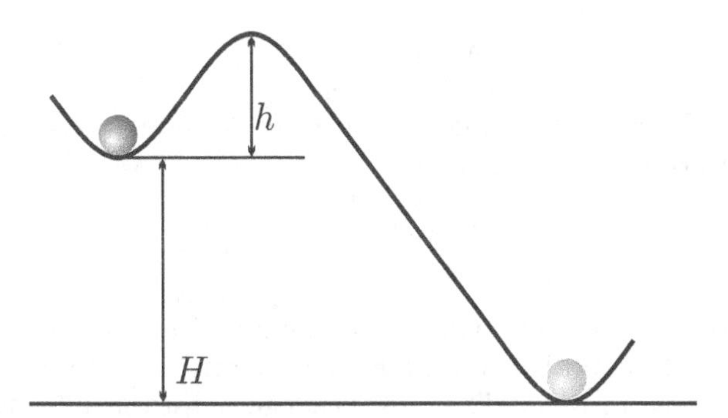

Fig. 24.1 Metastable state in mechanics

To introduce the problem, we consider a mechanical analogy, see Fig. 24.1. The stable state for the ball is the minimum at the bottom, but it is trapped in a metastable state at a local minimum at height H. In order to reach the absolute minimum, the ball has to overcome the well of height h, which requires energy.

For chemical reactions to occur, the reactant molecules must be split into parts that then can take part in forming the products. This requires energy,

H. Struchtrup, *Thermodynamics and Energy Conversion*,
DOI: 10.1007/978-3-662-43715-5_24, © Springer-Verlag Berlin Heidelberg 2014

which can come, e.g., from collisional impact. As an example we consider the formation of water from hydrogen and oxygen. The reactants are gases, and even at high pressures, most collisions involve only two particles, while three-particle collisions are very rare. In the following list X denotes a collision partner that provides or removes energy to or from the collisions, but is not involved in the reaction itself. The reactions that happen in the formation of water are

$$
\begin{aligned}
\mathrm{H_2 + X} &\rightleftharpoons \mathrm{2H + X} \\
\mathrm{O_2 + X} &\rightleftharpoons \mathrm{2O + X} \\
\mathrm{H + O_2} &\rightleftharpoons \mathrm{O + OH} \\
\mathrm{O + H_2} &\rightleftharpoons \mathrm{H + OH} \\
\mathrm{OH + H_2} &\rightleftharpoons \mathrm{H_2O + H} \\
\mathrm{OH + H + X} &\rightleftharpoons \mathrm{H_2O + X}
\end{aligned}
$$

From the statement that two-particle collisions are far more frequent than three-particle collisions one might conclude that the first two reactions will mainly happen in the forward direction (high energy impact of X on H_2 or O_2 splits these), while the last reaction will mainly happen in the backward direction (splitting of water by impact of X). The rate at which reactions take place will be proportional to the probability for a collision and to the probability that a reaction actually occurs when a collision takes place.

24.2 Reaction Rates and the Chemical Constant

We consider a simple reaction of the type

$$\mathrm{A + B \rightleftharpoons C + D} \,. \tag{24.1}$$

Accordingly, the rates of change of the mole densities $\nu_\alpha = n_\alpha / V$ of the involved species must be related as

$$\frac{d\nu_{\mathrm{A}}}{dt} = \frac{d\nu_{\mathrm{B}}}{dt} = -\frac{d\nu_{\mathrm{C}}}{dt} = -\frac{d\nu_{\mathrm{D}}}{dt} \,. \tag{24.2}$$

We consider the rate of change of molecules of species A. Molecules of type A are produced in backward reactions at rate r_b and are consumed in forward reactions at rate r_f. The rate of change is just the difference between backward and forward reactions,

$$\frac{d\nu_{\mathrm{A}}}{dt} = r_b - r_f \,. \tag{24.3}$$

Each rate, r_b or r_f, is proportional to the probability to find the reaction partners at the same location, and to the probability that a reaction will actually take place when the partners meet. An intuitive choice for the rates is

$$r_b = k_b \nu_{\mathrm{C}} \nu_{\mathrm{D}} \quad , \quad r_f = k_f \nu_{\mathrm{A}} \nu_{\mathrm{B}} \,, \tag{24.4}$$

where $\nu_C\nu_D$ and $\nu_A\nu_B$ are measures for the probability to find the reaction partners C – D or A – B, and k_b, k_f measure the reaction probability (and include scaling factors for the probability). The above are the simplest meaningful choices for the reaction rates; realistic reactions will have more complicated relations between reaction rates and mole densities.

According to the model, the rate of change of the mole density of species A is

$$\frac{d\nu_A}{dt} = k_b\nu_C\nu_D - k_f\nu_A\nu_B \; . \tag{24.5}$$

This equation describes the approach to thermodynamic equilibrium. In the final equilibrium state $\frac{d\nu_A}{dt} = 0$ and the rate equation gives the law of mass action of this reaction as

$$\frac{\nu_C\nu_D}{\nu_A\nu_B} = \frac{k_f}{k_b} \; . \tag{24.6}$$

$\frac{k_f}{k_b}$ is the chemical constant. Indeed, since for this reaction the total number of moles stays constant, and since all stoichiometric coefficients are ± 1, we can write

$$\frac{\nu_C\nu_D}{\nu_A\nu_B} = \frac{X_C X_D}{X_A X_B} = \prod_\sigma X_\alpha^{\gamma_\alpha} = \frac{k_f}{k_b} = K_X = e^{-\frac{\Delta \bar{g}_R}{RT}} \; ; \tag{24.7}$$

the last identity follows from the law of mass action for ideal mixtures (23.30).

24.3 Gibbs Free Energy of Activation

The law of mass action (24.7) does not allow to identify the reaction rate coefficients k_b and k_f individually, it only relates their ratio to the Gibbs free energy of reaction. We split the latter into two contributions, the Gibbs free energies of activation for backward and forward reactions, $\bar{g}_b$ and $\bar{g}_f$, as

$$-\Delta \bar{g}_R = \bar{g}_b - \bar{g}_f \; . \tag{24.8}$$

The law of mass action (24.7) is fulfilled for the coefficients

$$k_f = k_0 e^{-\frac{\bar{g}_f}{RT}} \quad , \quad k_b = k_0 e^{-\frac{\bar{g}_b}{RT}} \; , \tag{24.9}$$

where k_0 is a dimensional factor.

Both coefficients are of the form $\exp\left[-\frac{\bar{e}}{RT}\right]$, where $\bar{e}$ is an energy (here, the Gibbs free energy of activation). The barometric formula is of the same form, only that there $\bar{e}$ is the potential energy Mgz. The factor $\exp\left[-\frac{\bar{e}}{RT}\right]$ is know as the *Boltzmann factor*, it is the typical form of a probability in a thermally activated system.

Figure 24.2 shows the Gibbs free energies of activation in an energy landscape that refers back to Fig. 24.1. The Gibbs free energy of reaction, $-\Delta \bar{g}_R$, is a measure for the energy difference between the reactants (A, B) and the

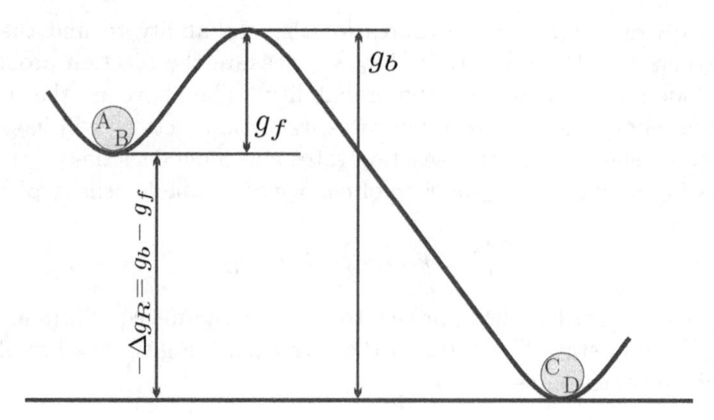

Fig. 24.2 Gibbs free energies of activation

products (C, D). The equilibrium of the reaction lies towards the products. For the forward reaction (A, B) $\rightarrow$ (C, D) to occur the activation energy $\bar{g}_f$ must be overcome, for the backward reaction (C, D) $\rightarrow$ (A, B) the larger activation energy $\bar{g}_b$ must be overcome.

In equilibrium, the same number of reactions takes place in both directions, $r_f = r_b$, or $k_f \nu_A \nu_B = k_b \nu_C \nu_D$, so that the mole densities of all constituents do not change in time. Due to the larger activation energy $\bar{g}_b$, the likelihood for a backward reaction is less ($k_b = k_0 e^{-\frac{\bar{g}_b}{RT}}$) than the likelihood for the forward reaction ($k_f = k_0 e^{-\frac{\bar{g}_f}{RT}}$). With $k_b < k_f$ and $k_f \nu_A \nu_B = k_b \nu_C \nu_D$ follows $\nu_A \nu_B < \nu_C \nu_D$: there must be more particles on the product side.

The activation energies $\bar{g}_f, \bar{g}_b$ do not affect the equilibrium, which is described by the Gibbs free energy of reaction $\Delta \bar{g}_R$ in the law of mass action. However, the activation energies determine how fast the equilibrium will be reached. When the activation energies are large, reactions are unlikely and the reactants will persist in a metastable state, and make no progress towards the final equilibrium state.

The smaller the activation energies divided by temperature, $\frac{\bar{g}_{f,b}}{RT}$, are, the faster the equilibrium will be reached. Thus, equilibrium will be reached faster at higher temperatures. Fuel in contact with oxygen must be ignited—that is heated to higher T—to overcome the energy barrier at the point of ignition. The released heat then provides the energy to overcome the barrier in the neighborhood, a fast chain reaction occurs.

Catalysts do not take part in the reaction, but their presence lowers the activation energies. Ammonia synthesis provides an example for a reaction that is stuck in a metastable state, and will come to reaction only with the help of catalysts. Even then, to overcome the energy barrier, the temperature must be increased.

24.4 Entropy Generation

We show that with the reaction laws (24.5,24.9) the entropy generation is always positive as the reaction approaches towards equilibrium, and that it is zero in equilibrium. For simplicity of the argument, we consider only the reaction (24.1) as before.

When the reaction takes place in a closed system the first and second laws read

$$\frac{dU}{dt} = \dot{Q} - \dot{W} \quad , \quad \frac{dS}{dt} - \frac{\dot{Q}}{T} = \dot{S}_{gen} \geq 0\,. \tag{24.10}$$

We restrict the discussion to the case where temperature and pressure are kept constant throughout the reaction. Then $\dot{W} = p\frac{dV}{dt}$ and the two laws can be combined by elimination of the heat $\dot{Q}$ to

$$\frac{dG}{dt} = -T\dot{S}_{gen} \leq 0\,. \tag{24.11}$$

The total Gibbs free energy of the mixture is

$$G = \sum n_\alpha \bar{\mu}_\alpha \tag{24.12}$$

where $n_\alpha = \nu_\alpha V$ is the mole number and ν_α is mole density.

For the time derivative of the Gibbs free energy we have to remember the Gibbs-Duhem relation (20.22). With (24.2) we obtain

$$\frac{dG}{dt} = \sum_\alpha \bar{\mu}_\alpha \frac{dn_\alpha}{dt} = \left(\bar{\mu}_{\mathrm{C}} + \bar{\mu}_{\mathrm{D}} - \bar{\mu}_{\mathrm{A}} - \bar{\mu}_{\mathrm{B}}\right) \Lambda \; < \tag{24.13}$$

where $\Lambda = (r_f - r_b)\, V$ is the net reaction rate in the system.

For an ideal mixture, the chemical potential is

$$\bar{\mu}_\alpha = \bar{g}_\alpha + \bar{R}T \ln X_\alpha \tag{24.14}$$

and thus

$$\frac{dG}{dt} = \Lambda \left(\Delta\bar{g}_R + \bar{R}T \ln \frac{X_{\mathrm{C}} X_{\mathrm{D}}}{X_{\mathrm{A}} X_{\mathrm{B}}} \right) = -T\dot{S}_{gen} \leq 0\,. \tag{24.15}$$

Here, $\Delta\bar{g}_R = \sum_\alpha \gamma_\alpha \bar{g}_\alpha = \bar{g}_{\mathrm{C}} + \bar{g}_{\mathrm{D}} - \bar{g}_{\mathrm{A}} - \bar{g}_{\mathrm{B}}$ is the Gibbs free energy of reaction. Since the total particle number is conserved, we can write the entropy generation rate as

$$\dot{S}_{gen} = -\frac{\Lambda}{T}\left(\Delta\bar{g}_R + \bar{R}T \ln \frac{X_{\mathrm{C}} X_{\mathrm{D}}}{X_{\mathrm{A}} X_{\mathrm{B}}} \right) = -\frac{\Lambda}{T}\left(\Delta\bar{g}_R + \bar{R}T \ln \frac{\nu_{\mathrm{C}} \nu_{\mathrm{D}}}{\nu_{\mathrm{A}} \nu_{\mathrm{B}}} \right) \geq 0\,. \tag{24.16}$$

In thermodynamic equilibrium the term in the bracket, and thus the entropy generation rate, vanishes: this gives the law of mass action (24.7).

With $\Delta \bar{g}_R = \bar{g}_f - \bar{g}_b$ we can split this term into contributions referring to the forward and backward reactions,

$$\dot{S}_{gen} = \frac{\Lambda}{T}\left[\left(-\bar{g}_f + \bar{R}T \ln \nu_{\mathrm{A}}\nu_{\mathrm{B}}\right) - \left(-\bar{g}_b + \bar{R}T \ln \nu_{\mathrm{C}}\nu_{\mathrm{D}}\right)\right] \geq 0\,. \tag{24.17}$$

We find the same expressions in the overall reaction rate, for which we find from (24.9)

$$\frac{\Lambda}{V} = k_f \nu_{\mathrm{A}}\nu_{\mathrm{B}} - k_b \nu_{\mathrm{C}}\nu_{\mathrm{D}} = k_0 \left(e^{\frac{-\bar{g}_f + \bar{R}T \ln \nu_{\mathrm{A}}\nu_{\mathrm{B}}}{\bar{R}T}} - e^{\frac{-\bar{g}_b + \bar{R}T \ln \nu_{\mathrm{C}}\nu_{\mathrm{D}}}{\bar{R}T}} \right) . \tag{24.18}$$

With (24.17,24.18) the entropy generation is of the form $\frac{V}{T}(x-y)(e^x - e^y) \geq 0$. Therefore the entropy generation rate is always non-negative, while it vanishes in thermodynamic equilibrium ($x = y$, i.e., the law of mass action). This confirms that the reaction rate model is thermodynamically sound. Note that the model relies on simplifying assumptions that correspond to ideal mixtures.

Problems

24.1. Velocity of Reactions

Consider a reaction of the type $\mathrm{A} \mapsto \mathrm{B} + \mathrm{C}$ which occurs according to $\frac{dn_{\mathrm{A}}}{dt} = -k\, n_{\mathrm{A}}$ (decay of A). Compute the mole density of A as a function of time, when the initial mole density of A is n_{A}^0, and neither B nor C are present initially. Assume constant volume.

24.2. Velocity of Reactions

Consider a reaction of the type $\mathrm{B} + \mathrm{C} \leftrightharpoons 2\mathrm{A}$ which occurs according to

$$\frac{d\nu_{\mathrm{A}}}{dt} = k_b \nu_{\mathrm{B}}\nu_{\mathrm{C}} - k_f \nu_{\mathrm{A}}\nu_{\mathrm{A}}\,.$$

The initial mole densities of A, B, C are $\nu_{\mathrm{A}}^0 = 0$, and $\nu_{\mathrm{B}}^0 = \nu_{\mathrm{C}}^0 = \nu_0$, that is B and C are mixed in equal parts when the reaction commences. The rate constants have the values $k_b = 1 \frac{\mathrm{m}^3}{\mathrm{mol\,s}}$, $k_f = 0.1 \frac{\mathrm{m}^3}{\mathrm{mol\,s}}$. The reaction takes place at constant volume.

1. Derive the law of mass action for this reaction, and compute the equilibrium mole densities.
2. Compute the mole density of A as a function of time, and also compute the mole densities of B and C as functions of time.
3. Plot $\frac{\nu_{\mathrm{A}}}{\nu^0}$ and $\frac{\nu_{\mathrm{B}}}{\nu^0}$ as functions of time.

Chapter 25
Combustion

25.1 Fuels

Combustion describes the exothermic reaction between a fuel and an oxidizer, most often oxygen from air. Normally, the reactive equilibrium lies almost completely on the product side (~99.99%,), so that for computations it can be assumed that all fuel is consumed, as long as enough oxygen is present. However, by the Le Chatelier principle, the reaction will be less complete when the product temperature is high: in very hot combustion one might have to account for the law of mass action.

Many fuels are hydrocarbons, C_xH_y, of different compositions. Of these, the best known are methane (natural gas) CH_4, propane C_3H_8, n-octane C_8H_{18}, n-dodecane $C_{12}H_{26}$, and hydrogen H_2. Octane is the main ingredient in gasoline which is a mixture of various hydrocarbons. Similarly, Diesel fuel is a mixture of heavier hydrocarbons such as dodecane. Other fuels contain additional elements as well, e.g. methyl alcohol CH_3OH, ethyl alcohol C_2H_5OH, or coal which is mainly carbon (C) and other elements (S, O, H, N, ash) in varying amounts.

The basic reactions occurring are the formation of water and carbon dioxide

$$\mathrm{C}_x\mathrm{H}_y + \left(x + \frac{y}{4}\right)\mathrm{O}_2 \rightarrow x\mathrm{CO}_2 + \frac{y}{2}\mathrm{H}_2\mathrm{O}\,. \tag{25.1}$$

In the above, $\left(x + \frac{y}{4}\right)$ is the stoichiometric amount of oxygen required to fully oxidize a hydrocarbon fuel. With the equation written for one mole of fuel, the stoichiometric coefficients are $\gamma_{\mathrm{C}_x\mathrm{H}_y} = -1$, $\gamma_{\mathrm{O}_2} = -\left(x + \frac{y}{4}\right)$, $\gamma_{\mathrm{CO}_2} = x$, $\gamma_{\mathrm{H}_2\mathrm{O}} = \frac{y}{2}$.

In case that not enough oxygen is present, the combustion is incomplete, and some carbon monoxide is formed,

$$\mathrm{C}_x\mathrm{H}_y + \left(x + \frac{y}{4} - \frac{z}{2}\right)\mathrm{O}_2 \rightarrow (x - z)\,\mathrm{CO}_2 + \frac{y}{2}\mathrm{H}_2\mathrm{O} + z\mathrm{CO}\,. \tag{25.2}$$

H. Struchtrup, *Thermodynamics and Energy Conversion*,
DOI: 10.1007/978-3-662-43715-5_25, © Springer-Verlag Berlin Heidelberg 2014

Carbon monoxide is an odorless and highly toxic gas that is a fuel itself ($CO + \frac{1}{2}O_2 \rightleftharpoons CO_2$). Combustion processes take place in many reaction steps, which involve the splitting of larger molecules into smaller units, and the formation of new species. The formation of H_2O and CO is relatively fast, while the formation of CO_2 is slow, hence CO is formed first, and then reacts to CO_2 later.

25.2 Combustion Air

Most combustion processes use air as an oxidizer, which is freely available for most applications, e.g., stationary power plants or air breathing engines for cars and airplanes. Rockets fly at high altitudes or in space where the oxygen density is low or zero and thus they must bring their oxidizer along. To reduce payload, rocket engines are fed with fuel and pure oxygen, or with solid oxidizer-fuel compounds, e.g., ammonium perchlorate and aluminium powder.

For combustion with air, the oxygen is accompanied by the other components of air. Since oxygen and nitrogen are the two main ingredients, in combustion analysis one normally ignores the other components, and considers air as a mixture of oxygen and nitrogen with $X_{O_2} = 0.21$ and $X_{N_2} = 0.79$. Thus, in dry air for one mole of oxygen $\frac{1}{X_{O_2}} = 4.76$ moles of air are required, that is each mole of oxygen is accompanied by 3.76 moles of nitrogen.

Due to the presence of nitrogen, nitrogen oxides (NO_x) may form in the combustion processes which are toxic. Since the amounts are small, the formation of NO_x will be ignored below, but due to the toxicity the formation of NO_x must be monitored, and suppressed, in practice.

The amount of oxygen that is required for complete combustion of a fuel is known as the stoichiometric amount, the corresponding amount of air is denoted as *stoichiometric air* or as *theoretical air*. For the general hydrocarbon reaction (25.1) the stoichiometric air is $\left[4.76\left(x + \frac{y}{4}\right)\right]$ moles of air per mole of fuel. Often the air amount is given as *percent of theoretical air*, for instance 150% of theoretical air corresponds to $\left[7.14\left(x + \frac{y}{4}\right)\right]$ moles of air per mole of fuel, or to 50% of *excess air*.

25.3 Example: Mole and Mass Flow Balances

As an example we consider the combustion of a mass flow $\dot{m}_F = 5\frac{\text{kg}}{\text{h}}$ of octane (C_8H_{18}) with 150% theoretical air. We ask for the mass flow of air required, and the mass flows of the products. Since chemical balances concern mole numbers rather than masses, we first determine the mole flow of fuel as ($M_{C_8H_{18}} = 114\frac{\text{kg}}{\text{kmol}}$)

$$\dot{n}_F = \frac{\dot{m}_F}{M_{C_8H_{18}}} = 0.0122\frac{\text{mol}}{\text{s}} \, .$$

The required amount of air, and the resulting amount of product, is determined by writing a chemical equation for all involved components as

$$C_8H_{18} + 1.5a\,(O_2 + 3.76N_2) \rightarrow 0.5aO_2 + bCO_2 + cH_2O + dN_2\,.$$

Here, a is the amount of oxygen required for the stoichiometric combustion per mole of fuel, the factor 1.5 on the left accounts for 150% of theoretical air. The first term on the right is the unused portion of oxygen, which is total oxygen coming in ($1.5a$) minus the stoichiometric amount (a). The numbers a, b, c, d must be determined from balancing the elements C, H, O and N, on both sides of the equation. This gives

$$8 = b \;\;,\;\; 18 = 2c \;\;,\;\; 1.5a \times 2 = 0.5a \times 2 + 2b + c \;\;,\;\; 1.5a \times 3.76 \times 2 = 2d\,,$$

so that

$$a = 12.5 \;\;,\;\; b = 8 \;\;,\;\; c = 9 \;\;,\;\; d = 70.5\,.$$

The resulting mole flows entering the combustion process are

$$\begin{aligned}
\dot{n}_{O_2} &= 1.5a\,\dot{n}_F = 0.229\frac{\text{mol}}{\text{s}}\,,\\
\dot{n}_{N_2} &= d\,\dot{n}_F = 0.86\frac{\text{mol}}{\text{s}}\,,\\
\dot{n}_{\text{air}} &= \dot{n}_{O_2} + \dot{n}_{N_2} = 1.089\frac{\text{mol}}{\text{s}}\,,
\end{aligned}$$

and the outgoing flows are

$$\begin{aligned}
\dot{n}_{CO_2} &= b\,\dot{n}_F = 0.098\frac{\text{mol}}{\text{s}}\,,\\
\dot{n}_{H_2O} &= c\,\dot{n}_F = 0.110\frac{\text{mol}}{\text{s}}\,,\\
\dot{n}_{O_2} &= 0.5a\,\dot{n}_F = 0.076\frac{\text{mol}}{\text{s}}\,,\\
\dot{n}_{N_2} &= d\,\dot{n}_F = 0.86\frac{\text{mol}}{\text{s}}\,.
\end{aligned}$$

The mass flows of incoming air and outgoing carbon dioxide and water are ($M_{\text{air}} = 29\frac{\text{kg}}{\text{kmol}}$, $M_{CO_2} = 44\frac{\text{kg}}{\text{kmol}}$, $M_{H_2O} = 18\frac{\text{kg}}{\text{kmol}}$)

$$\begin{aligned}
\dot{m}_{\text{air}} &= M_{\text{air}}\dot{n}_{\text{air}} = 31.58\frac{\text{g}}{\text{s}} = 113.7\frac{\text{kg}}{\text{h}}\\
\dot{m}_{CO_2} &= M_{CO_2}\dot{n}_{CO_2} = 4.31\frac{\text{g}}{\text{s}} = 15.5\frac{\text{kg}}{\text{h}}\\
\dot{m}_{H_2O} &= M_{H_2O}\dot{n}_{H_2O} = 1.98\frac{\text{g}}{\text{s}} = 7.13\frac{\text{kg}}{\text{h}}
\end{aligned}$$

The mass-based air-fuel ratio is

$$AF = \frac{\dot{m}_{\text{air}}}{\dot{m}_{\text{F}}} = \frac{M_{\text{air}}}{M_{\text{F}}}\frac{\dot{n}_{\text{air}}}{\dot{n}_{\text{F}}} = \frac{M_{\text{air}}}{M_{\text{F}}} 1.5a\,(1+3.76) = 22.7\,.$$

One litre of octane ($\rho_{C_8H_{18}} = 0.703\frac{\text{kg}}{\text{litre}}$) has the mass $m_{\text{F}} = 703\,\text{g}$ which corresponds to the mole number $n_{\text{F}} = 6.17\,\text{mol}$. Thus, the amount of CO_2 produced in combustion of 1 litre of octane is $n_{CO_2} = bn_{\text{F}} = 49.3\,\text{mol}$ that is $m_{CO_2} = M_{CO_2} n_{CO_2} = 2.17\,\text{kg}$ of CO_2 per litre of fuel. A car with a gas mileage of $14\frac{\text{litres}}{100\,\text{km}}$ travelling at $50\frac{\text{km}}{\text{h}}$ requires approximately $5\frac{\text{kg}}{\text{h}}$ of octane and expels 15.5 kg of carbon dioxide per hour.

25.4 Example: Exhaust Water

For the analysis of combustion processes it is important to know whether some of the product water is liquid. Liquid water can lead to corrosion in the system, and should be avoided; it will form when the temperature of the combustion product sinks below the dewpoint of the product. For the analysis, the moisture content of the incoming combustion air must be considered as well to obtain accurate results.

We study this by means of an example. Fuel gas with a volumetric analysis of 60% CH_4 and 40% H_2 is burnt with moist air at temperature 20 °C and relative humidity $\phi = 0.7$ with 60% excess air; the pressure is 1 bar. We ask for the dew point temperature of the combustion product.

The mole fraction of water in the incoming air is

$$X_{H_2O} = \frac{p_{\text{v}}}{p} = \phi\frac{p_{\text{sat}}\,(20\,°\text{C})}{p} = 0.0164\,,$$

so that each mole of oxygen is accompanied by $X_{H_2O}/X_{O_2} = 0.079$ moles of water.[1] Then, the overall mole balance for the combustion per mole of fuel reads

$$0.6CH_4 + 0.4H_2 + 1.6a\,(O_2 + 3.76N_2 + 0.079H_2O) \rightarrow 0.6aO_2 + bCO_2 + cH_2O + dN_2\,,$$

which yields, from balancing elements (C, H, O, N),

$$0.6 = b \quad , \quad 0.6 \times 4 + 0.4 \times 2 + 1.6a \times 0.079 \times 2 = 2c\,,$$
$$1.6a\,(2 + 0.079) = 0.6a \times 2 + 2b + c \quad , \quad 12.032a = 2d\,,$$

with the solution

$$a = 1.4 \quad , \quad b = 0.6 \quad , \quad c = 1.78 \quad , \quad d = 8.42\,.$$

[1] The calculation is as follows: Mole fraction of water is $X_{H_2O} = \frac{n_{H_2O}}{n_{O_2}+n_{N_2}+n_{H_2O}}$; with $n_{N_2} = 3.76n_{O_2}$ follows $\frac{n_{H_2O}}{n_{O_2}} = \frac{X_{H_2O}}{X_{O_2}} = \frac{4.76}{\frac{1}{X_{H_2O}}-1}$.

The mole fraction of water in the product is

$$X_{H_2O}^{prod} = \frac{c}{0.6a + b + c + d} = 0.153\,.$$

Thus, the partial pressure of the vapor in the product is $p_v = 0.153\,\text{bar}$ which corresponds to a dewpoint temperature $T_d = T_{sat}(p_v) = 54.4\,°\text{C}$.

When the reaction product is cooled below the dewpoint, c_{liq} moles of water per mole of fuel will condense, while the remaining $c_{vap} = c - c_{liq}$ moles per mole of fuel will be in the vapor phase. When the product temperature is 25 °C, the partial pressure of the vapor in the gas phase is $p_v = p_{sat}(25\,°\text{C}) = 3.169\,\text{kPa}$, and the mole fraction of the vapor is

$$X_{H_2O} = \frac{p_v}{p} = \frac{c_{vap}}{0.6a + b + c_{vap} + d}\,,$$

so that

$$c_{vap} = \frac{0.6a + b + d}{\frac{p}{p_v} - 1} = 0.323 \quad \text{and} \quad c_{liq} = 1.452\,.$$

Thus, for each mole of fuel there will be 1.452 mol of liquid water in the system, which have to be removed.

25.5 First and Second Law for Combustion Systems

The goal of combustion is to produce heat, either for heating purposes or for conversion into work in heat engines. Since the changes in chemical energy are reflected in the proper values for enthalpies (23.20) and entropies (23.22), the first and second law for a combustion system have the usual form (9.9, 9.10),

$$\frac{dE}{dt} + \sum_{\alpha,out} \dot{m}_\alpha \left(h_\alpha + \frac{1}{2}\mathcal{V}_\alpha^2 + g z_\alpha \right) - \sum_{\alpha,in} \dot{m}_\alpha \left(h_\alpha + \frac{1}{2}\mathcal{V}_\alpha^2 + g z_\alpha \right) = \dot{Q} - \dot{W}\,, \tag{25.3}$$

$$\frac{dS}{dt} - \sum_k \frac{\dot{Q}_k}{T_k} + \sum_{\alpha,out} \dot{m}_\alpha s_\alpha - \sum_{\alpha,in} \dot{m}_\alpha s_\alpha = \dot{S}_{gen} \geq 0\,. \tag{25.4}$$

We shall consider mainly open systems, e.g., combustion chambers, at steady state and ignore kinetic and potential energies, so that, now written with mole flows instead of mass flows, and with molar enthalpy and entropy, the first and second law read

$$\sum_{\alpha,out} \dot{n}_\alpha \bar{h}_\alpha - \sum_{\alpha,in} \dot{n}_\alpha \bar{h}_\alpha = \dot{Q} - \dot{W} \, , \tag{25.5}$$

$$\sum_{\alpha,out} \dot{n}_\alpha \bar{s}_\alpha - \sum_{\alpha,in} \dot{n}_\alpha \bar{s}_\alpha - \sum_k \frac{\dot{Q}_k}{T_k} = \dot{S}_{gen} \geq 0 \, . \tag{25.6}$$

The mole flows $\dot{n}_\alpha$ must be determined by analysis of the combustion process as shown in the previous sections.

25.6 Adiabatic Flame Temperature

If the combustion system is adiabatic, and no power is exchanged, the first law for isobaric combustion reduces to

$$\sum_{\alpha,out} \dot{n}_\alpha \bar{h}_\alpha (T_f) = \sum_{\alpha,in} \dot{n}_\alpha \bar{h}_\alpha (T_{in}) \, . \tag{25.7}$$

The temperature T_f of the combustion product is the adiabatic flame temperature.

25.7 Example: Adiabatic Flame Temperature

As an example, we compute the adiabatic flame temperature for the combustion process studied in Section 25.3, when the incoming fuel and airstreams are at reference temperature and pressure, $T_{in} = T_0$, $p = p_0$. Then, the enthalpies for the incoming streams are

$$\bar{h}_\text{F} (T_{in}) = \bar{h}^0_{f,\text{C}_8\text{H}_{18}} = -249.95 \frac{\text{kJ}}{\text{mol}} \, ,$$
$$\bar{h}_{\text{O}_2} (T_{in}) = \bar{h}^0_{f,\text{O}_2} = 0 \, , \; \bar{h}_{\text{N}_2} (T_{in}) = \bar{h}^0_{f,\text{N}_2} = 0 \, .$$

After division by $\dot{n}_F$, and with the results of Sec. 25.3, the energy balance (25.7) becomes

$$b\bar{h}_{\text{CO}_2} (T_f) + 0.5a\bar{h}_{\text{O}_2} (T_f) + c\bar{h}_{\text{H}_2\text{O}} (T_f) + d\bar{h}_{\text{N}_2} (T_f) = \bar{h}_\text{F} (T_{in}, p) \, .$$

When written with tabulated enthalpy values $\tilde{h}_\alpha (T)$ instead, this gives, by means of (23.20),

$$\begin{aligned} b\tilde{h}_{\text{CO}_2} (T_f) + 0.5a\tilde{h}_{\text{O}_2} (T_f) + c\tilde{h}_{\text{H}_2\text{O}} (T_f) + d\tilde{h}_{\text{N}_2} (T_f) \\ = \bar{h}^0_{f,\text{C}_8\text{H}_{18}} + b \left[\tilde{h}_{\text{CO}_2} (T_0) - \bar{h}^0_{f,\text{CO}_2} \right] \\ + 0.5a \left[\tilde{h}_{\text{O}_2} (T_0) \right] + c \left[\tilde{h}_{\text{H}_2\text{O}} (T_0) - \bar{h}^0_{f,\text{H}_2\text{O}} \right] + d \left[\tilde{h}_{\text{N}_2} (T_0) \right] \, . \end{aligned}$$

Here we used that the products are ideal gases, so that the enthalpies depend only on temperature. From the tables we find the values

$$\bar{h}^0_{f,CO_2} = -393.52\frac{\text{kJ}}{\text{mol}} \quad , \quad \bar{h}^0_{f,H_2O} = -241.82\frac{\text{kJ}}{\text{mol}}$$

and

$$\begin{aligned}\tilde{h}_{CO_2}(T_0,p_0) &= 9.624\frac{\text{kJ}}{\text{mol}} \quad , \quad \tilde{h}_{O_2}(T_0,p_0) = 8.903\frac{\text{kJ}}{\text{mol}} \,, \\ \tilde{h}_{H_2O}(T_0,p_0) &= 9.783\frac{\text{kJ}}{\text{mol}} \quad , \quad \tilde{h}_{N_2}(T_0,p_0) = 8.672\frac{\text{kJ}}{\text{mol}} \,,\end{aligned}$$

so that,

$$\begin{aligned}H(T_f) := 8\tilde{h}_{CO_2}(T_f) + 6.25\tilde{h}_{O_2}(T_f) + 9\tilde{h}_{H_2O}(T_f) + 70.5\tilde{h}_{N_2}(T_f) \\ = 5906.65\frac{\text{kJ}}{\text{mol}} \,.\end{aligned}$$

The flame temperature must be determined by trial and error from tabulated data. We find $H(1800\,\text{K}) = 5810.9\frac{\text{kJ}}{\text{mol}}$ and $H(1850\,\text{K}) = 5994.6\frac{\text{kJ}}{\text{mol}}$. Linear interpolation between these values gives the flame temperature as $T_f = 1826\,\text{K}$.

25.8 Closed System Combustion

For a combustion process in a closed system, the integrated first law gives (kinetic and potential energies ignored)

$$U_2 - U_1 = Q_{12} - W_{12} \,. \tag{25.8}$$

As for enthalpies, the internal energies must be taken with respect to proper reference data, and the best way to ensure this is to determine them from enthalpies.

For ideal mixtures, which exhibit neither enthalpy nor volume of mixing, we have

$$U_1 = \sum_{\alpha,\,\text{react}} n_\alpha \bar{u}_\alpha = \sum_{\alpha,\,\text{react}} n_\alpha \left[\bar{h}_\alpha(T_1,p_1) - p_1\bar{v}_\alpha(T_1,p_1)\right] \,, \tag{25.9}$$

$$U_2 = \sum_{\alpha,\,\text{prod}} n_\alpha \bar{u}_\alpha = \sum_{\alpha,\,\text{prod}} n_\alpha \left[\bar{h}_\alpha(T_2,p_2) - p_2\bar{v}_\alpha(T_2,p_2)\right] \,. \tag{25.10}$$

Here, $\bar{v}_\alpha(T,p)$ is the mole volume of α alone at (T,p) (Amagat model). Application of the ideal gas laws $p\bar{v} = \bar{R}T$, $\bar{h} = \bar{h}(T)$, gives

$$U_1 = \sum_{\text{react}} n_\alpha\left(\bar{h}_\alpha(T_1) - \bar{R}T_1\right) \,, \quad U_2 = \sum_{\text{prod}} n_\alpha\left(\bar{h}_\alpha(T_2) - \bar{R}T_2\right) \,. \tag{25.11}$$

25.9 Example: Closed System Combustion

As an example we consider the isochoric and adiabatic combustion of $n_{CH_4} = 1\,\text{mol}$ of methane with $n_{O_2} = 9\,\text{mol}$ of oxygen ($X^1_{CH_4} = 0.1, X^1_{O_2} = 0.9$) in a closed container, the initial temperature is $T_1 = 25°C$ and the initial pressure is $p_1 = 10\,\text{atm}$. Reactants and products are considered as ideal gases.

The chemical equation for this reaction is

$$CH_4 + 9O_2 = CO_2 + 2H_2O + 7O_2 \,,$$

which implies $n_{CO_2} = 1\,\text{mol}$, $n_{H_2O} = 2\,\text{mol}$, $n_{O_2} = 7\,\text{mol}$. The final temperature must be determined from the first law, which reduces to $U_2 = U_1$. With $\bar{h}^0_{f,CH_4} = -74.85\frac{\text{kJ}}{\text{mol}}$ and $\bar{h}^0_{f,O_2} = 0$, we find

$$U_1 = n_{CH_4}\left(\bar{h}_{CH_4}(T_1) - \bar{R}T_1\right) + n_{O_2}\left(\bar{h}_{O_2}(T_1) - \bar{R}T_1\right) = -99.64\,\text{kJ}\,.$$

The final energy is given by

$$U_2 = n_{CO_2}\left(\bar{h}_{CO_2}(T_2) - \bar{R}T_2\right) + n_{H_2O}\left(\bar{h}_{H_2O}(T_2) - \bar{R}T_1\right) + n_{O_2}\left(\bar{h}_{O_2}(T_2) - \bar{R}T_2\right)\,.$$

Again we must use trial and error to determine the final temperature. We find $U_2\,(T = 2800\,\text{K}) = -107.6\,\text{kJ}$ and $U_2\,(T = 2900\,\text{K}) = -70.98\,\text{kJ}$ so that, from $U_2 = U_1$ we find by linear interpolation, $T_2 = 2822\,\text{K}$.

The total number of moles stays constant over the process, and the ideal gas equation gives the final pressure

$$p_2 = p_1\frac{T_2}{T_1} = 9.464\,p_1 = 94.64\,\text{atm}\,.$$

The partial pressures are obtained from the mole ratios, $X_\alpha = \frac{p_\alpha}{p} = \frac{n_\alpha}{n}$ as

$$p^1_{CH_4} = 1\,\text{atm} \quad , \quad p^1_{O_2} = 9\,\text{atm} \quad , \quad p^2_{CO_2} = 9.464\,\text{atm}\,,$$
$$p^2_{H_2O} = 18.93\,\text{atm} \quad , \quad p^2_{O_2} = 66.25\,\text{atm}\,.$$

25.10 Entropy Generation in Closed System Combustion

In a combustion process, the entropy changes due to temperature and compositions change. and are affected by the different entropies of formation $\bar{s}^0_{f,\alpha}$ of the various constituents. One will expect entropy generation in combustion, since combustion processes cannot be controlled once started. As always, the generation of entropy is related to a work loss.

The computation of the entropy changes in steady state combustion in open systems will be seen in Sec. 25.12. Here we have a look at the entropy generation in closed system combustion, for the process described in the preceding section.

Since the process is adiabatic, the second law can be integrated to

$$S_2 - S_1 = S_{gen} \geq 0 \tag{25.12}$$

where $S_{gen} = \int_{t=0}^{t_f} \dot{S}_{gen} dt$ is the overall entropy generated over the duration of the process. We obtain

$$S_1 = \sum_{\text{reactants}} n_\alpha \bar{s}_\alpha \left(T_1, p_\alpha^1\right) \quad , \quad S_2 = \sum_{\text{products}} n_\alpha \bar{s}_\alpha \left(T_2, p_\alpha^2\right) \, , \tag{25.13}$$

where the entropies must be evaluated as

$$\bar{s}_\alpha \left(T, p_\alpha\right) = \bar{s}_\alpha^0 \left(T, p_0\right) - \bar{R} \ln \frac{p_\alpha}{p_0} \tag{25.14}$$

with the partial pressures $p_\alpha = X_\alpha p$.

The $\bar{s}_\alpha^0 \left(T, p_0\right)$ are tabulated, and we find from the tables (with some interpolation)

$$\bar{s}_{CH_4} \left(T_1, p_0\right) = 186.16 \frac{\text{kJ}}{\text{kmol K}} \quad , \quad \bar{s}_{O_2} \left(T_1, p_0\right) = 205.04 \frac{\text{kJ}}{\text{kmol K}} \, ,$$
$$\bar{s}_{CO_2} \left(T_2, p_0\right) = 331.09 \frac{\text{kJ}}{\text{kmol K}} \quad , \quad \bar{s}_{H_2O} \left(T_2, p_0\right) = 283.19 \frac{\text{kJ}}{\text{kmol K}} \, ,$$
$$\bar{s}_{O_2} \left(T_2, p_0\right) = 282.08 \frac{\text{kJ}}{\text{kmol K}} \, .$$

Then we have

$$S_1 = 2.032 \frac{\text{kJ}}{\text{K}} \quad , \quad S_2 = 2.872 \frac{\text{kJ}}{\text{K}} \quad \text{and} \quad S_{gen} = 0.841 \frac{\text{kJ}}{\text{K}} \, .$$

25.11 Work Potential of a Fuel

As we have seen in the last section, combustion processes are accompanied by entropy generation. Combustion is irreversible, and work potential is lost to the irreversibility. In order to quantify the work loss, we ask how much work one could obtain from a fuel in a reversible process.

Figure 25.1 shows the mass and energy flows of a fuel consuming power plant: fuel and oxidizer enter the plant at reference conditions (T_0, p_0) and are processed inside the plant, which produces the net power $\dot{W}_{\text{net}}$ and rejects the heat $\left|\dot{Q}_{\text{out}}\right|$. The reaction products leave the plant through the exhaust, and we assume that they are in thermal and mechanical equilibrium with the outer environment, i.e., they are at (T_0, p_0) as well. This guarantees that most work available from the exhaust with respect to the environment is harvested. Note that, in principle, some work could be obtained by reversible

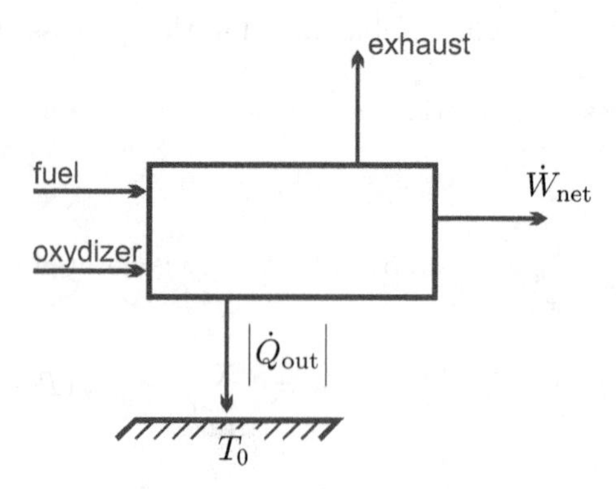

Fig. 25.1 Mass and energy flows for a fuel consuming power plant

mixing of the exhaust and the atmosphere. If the exhaust is just ejected into the environment, the mixing process is irreversible with an associated work loss; this loss will not be quantified.

Since nothing is said about the processes inside the plant, this set-up is applicable to a wide array of systems, in particular to all combustion based heat engines, that is Otto, Diesel, Brayton, Rankine cycles, and their variants, and to electrochemical power devices, i.e., fuel cells.

Elimination of the heat rejected into the environment between (25.5) and (25.6) gives the net work of the plant as

$$\dot{W}_{\rm net} = -T_0 \dot{S}_{gen} + \sum_{\substack{\text{reactants}\\ \text{at } T_0, p_0}} \dot{n}_\alpha \left(\bar{h}_\alpha - T_0 \bar{s}_\alpha\right) - \sum_{\substack{\text{products}\\ \text{at } T_0, p_0}} \dot{n}_\alpha \left(\bar{h}_\alpha - T_0 \bar{s}_\alpha\right) . \tag{25.15}$$

All irreversible processes occurring inside the plant, e.g., combustion, heat transfer over finite temperature differences, or friction losses in turbines, contribute to the entropy generation $\dot{S}_{gen}$, and thus diminishes the net work output of the plant.

We evaluate (25.15) for the stoichiometric combustion of a fuel with pure oxygen, with the chemical equation

$$|\gamma_{\rm fuel}| \, \text{fuel} + |\gamma_{O_2}| \, O_2 = \gamma_{CO_2} CO_2 + \gamma_{H_2O} H_2O . \tag{25.16}$$

We choose $\gamma_{\rm fuel} = -1$, and thus the mole flow of oxygen entering is $\dot{n}_{O_2} = |\gamma_{O_2}| \, \dot{n}_{\rm fuel}$ and the mole flows of the products, carbon dioxide and water, are $\dot{n}_{CO_2} = \gamma_{CO_2} \dot{n}_{\rm fuel}$ and $\dot{n}_{H_2O} = \gamma_{H_2O} \dot{n}_{\rm fuel}$.

Since all flows are entering and leaving at atmospheric conditions (T_0, p_0), we have $\bar{h}_\alpha (T_0, p_0) - T_0 \bar{s}_\alpha (T_0, p_0, X_\alpha) = \bar{g}_\alpha (T_0, p_0) + \bar{R} T_0 \ln X_\alpha$, where we have assumed ideal mixtures; as always $\bar{g}_\alpha (T_0, p_0)$ is the Gibbs free energy of α alone at standard conditions.

For this case, (25.15) can be written as

$$\dot{W}_{\text{net}} = -T_0\dot{S}_{gen} - \dot{n}_{\text{fuel}}\Delta\bar{g}_R\left(T_0,p_0\right) - \dot{n}_{\text{fuel}}\bar{R}T_0 \ln\left(X_{CO_2}^{\gamma_{CO_2}} X_{H_2O}^{\gamma_{H_2O}}\right) . \quad (25.17)$$

The last term is the entropy of mixing of the products. Since fuel and oxidizer enter the system separately, there is no mixing term for the reactants.

The first term—which is always negative—is the work lost to the irreversible processes inside the system, and the last term is the work that could be obtained by reversible mixing of the combustion products CO_2 and H_2O. The second term—which is positive, since for a fuel $\Delta\bar{g}_R < 0$—is just the Gibbs free energy of reaction, which is the dominant contribution to the work potential of a fuel. The maximum work obtainable from the fuel in a reversible process results from setting $\dot{S}_{gen} = 0$.

In case that the combustion products (CO_2, H_2O) leave the plant in separate streams at (T_0, p_0) the mixing term vanishes (last term in (25.17)), and the maximum work that can be extracted per mole of fuel in a reversible process is just the Gibbs free energy of the reaction,

$$w_{\text{fuel}}^{\text{rev}} = \frac{\dot{W}_{\text{net}}}{\dot{n}_{\text{fuel}}} = -\Delta\bar{g}_R\left(T_0,p_0\right) . \quad (25.18)$$

Additional work could be obtained by mixing the exhaust streams reversibly (last term in (25.17)), and by reversible mixing of the exhaust with the environmental air. The exhaust is in thermal (T_0) and mechanical (p_0) equilibrium with the environment, but not in chemical equilibrium ($\bar{\mu}_{\alpha,0}$). For combustion with (excess) air, additional mixing terms arise due to the presence of N_2 and excess O_2; these will be studied in the next section.

The question arises whether the maximum work obtainable from a fuel can be obtained from actual engineering devices. The answer is no, obviously, since all realistic processes are somewhat irreversible. It is, however, important to understand and quantify the different causes for irreversible losses, since only this understanding can lead to the design of better devices, with smaller losses.

In a classical combustion power plant losses are due to combustion, heat transfer, irreversible mixing, and friction. We shall study the relative importance of these in an extended example in the next section.

In fuel cells the flow of electrons is controlled, and thus combustion losses do not occur, but there are other causes for irreversible losses. Fuel cells and the losses within will be discussed in Chapter 26.

25.12 Example: Work Losses in a CH_4 Fired Steam Power Plant

The work produced in a combustion power plant differs significantly from the work available from the fuel, due to entropy generation within the system. In the present section we study and compare the contributions to work loss from all elements of a combustion driven Rankine cycle, where we consider different amounts of excess air. Figure 25.2 shows a simplified picture of the energy and mass flows in the considered system; neither reheat nor regeneration are considered. Fuel and air enter the plant at (T_0, p_0), the fuel is burned at constant pressure in the combustion chamber so that the flame temperature is T_2, and then the hot flue gas passes the steam generator that delivers heat to the Rankine cycle. The exhaust leaves the plant at (T_3, p_0).

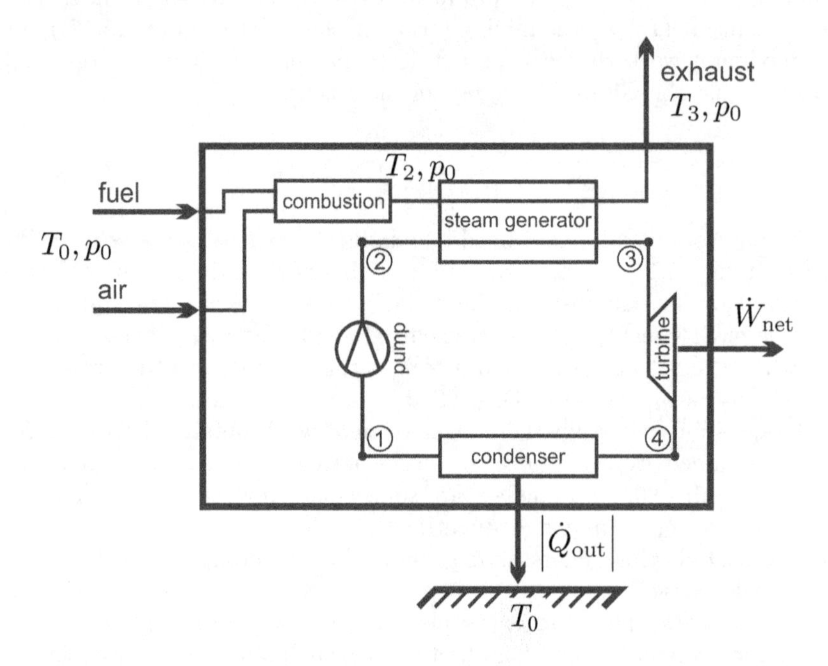

Fig. 25.2 A combustion steam power plant

For simplicity we consider the fuel to be methane (CH_4). The heat of reaction and the Gibbs free energy of reaction for the combustion of CH_4 are (at T_0, p_0, product water is vapor)

$$\Delta \bar{h}_R = -802.31 \frac{\text{kJ}}{\text{mol}} \quad , \quad \Delta \bar{g}_R = -800.89 \frac{\text{kJ}}{\text{mol}} \, .$$

The combustion equation reads

$$\mathrm{CH_4} + 2\,(1+x)\,(\mathrm{O_2} + 3.76\mathrm{N_2}) \longrightarrow \mathrm{CO_2} + 2\mathrm{H_2O} + 2x\mathrm{O_2} + 2\,(1+x)\,3.76\mathrm{N_2}\ ,$$

where x is the excess air in percent.

The corresponding incoming and outgoing mole flows are, expressed as multiples of the fuel flow $\dot{n}_\mathrm{F}$,

inflows:

$$\begin{aligned} &\text{fuel} : \dot{n}_\mathrm{F}\ ,\\ &\mathrm{O_2}\ : \dot{n}_{\mathrm{O_2}} = 2\,(1+x)\,\dot{n}_\mathrm{F}\ ,\\ &\mathrm{N_2}\ : \dot{n}_{\mathrm{N_2}} = 7.52\,(1+x)\,\dot{n}_\mathrm{F}\ , \end{aligned}$$

outflows:

$$\begin{aligned} &\mathrm{CO_2} : \dot{n}_{\mathrm{CO_2}} = \dot{n}_\mathrm{F}\ ,\\ &\mathrm{H_2O} : \dot{n}_{\mathrm{H_2O}} = 2\dot{n}_\mathrm{F}\ ,\\ &\mathrm{O_2}\ \ : \dot{n}_{\mathrm{O_2}} = 2x\dot{n}_\mathrm{F}\ ,\\ &\mathrm{N_2}\ \ : \dot{n}_{\mathrm{N_2}} = 7.52\,(1+x)\,\dot{n}_\mathrm{F}\ . \end{aligned}$$

The mole fractions in the outflow are

$$X_{\mathrm{CO_2}} = \frac{1}{10.52 + 9.52x}\ ,\quad X_{\mathrm{H_2O}} = \frac{2}{10.52 + 9.52x}\ ,$$
$$X_{\mathrm{O_2}} = \frac{2x}{10.52 + 9.52x}\ ,\quad X_{\mathrm{N_2}} = \frac{7.52\,(1+x)}{10.52 + 9.52x}\ .$$

The amount of excess air, x, follows from the first law for a given value of the flame temperature T_2, for which we consider three different values,

$$T_2 = \{1000\,\mathrm{K},\ 1600\,\mathrm{K},\ 2200\,\mathrm{K}\}\ .$$

The first law for adiabatic combustion reads

$$\sum_{in} \dot{n}_\alpha \bar{h}_\alpha\,(T_0) = \sum_{out} \dot{n}_\alpha \bar{h}_\alpha\,(T_2)\ ,$$

which yields (with $\bar{h}^0_{f,\mathrm{O_2}} = \bar{h}^0_{f,\mathrm{N_2}} = 0$)

$$x = \frac{\bar{h}^0_\mathrm{F} - \bar{h}_{\mathrm{CO_2}}\,(T_2) - 2\bar{h}_{\mathrm{H_2O}}\,(T_2) - 7.52\bar{h}_{\mathrm{N_2}}\,(T_2)}{2\bar{h}_{\mathrm{O_2}}\,(T_2) + 7.52\bar{h}_{\mathrm{N_2}}\,(T_2)}\ ,$$

and thus, for the three chosen values of the flame temperature,

$$x = \{2.684,\ 0.776,\ 0.091\}\ .$$

Next we ask for the maximum amount of work that could be obtained from the fuel with the given amounts of excess air, i.e., in a reversible process with exhaust at T_0, p_0. For fully reversible operation (25.15) reduces to

$$\dot{W}_{\max} = \sum_{in} \dot{n}_\alpha \left(\bar{h}_\alpha - T_0 \bar{s}_\alpha\right) - \sum_{out} \dot{n}_\alpha \left(\bar{h}_\alpha - T_0 \bar{s}_\alpha\right) ,$$

where the appropriate values of enthalpies and entropies must be inserted.

Since the gas mixture is always at p_0, the entropies are $\bar{s}_\alpha (T, p_0, X_\alpha) = \bar{s}^0_\alpha (T) - \bar{R} \ln X_\alpha$. One has to be careful with the choice of the mole fractions for the inflow: since the fuel enters the plant unmixed with the air, which is a mixture of oxygen and nitrogen, the entropies for the incoming flows are

$$\bar{s}_{\mathrm{F}} = \bar{s}^0_{\mathrm{F}} (T_0) \quad , \quad \bar{s}_{\mathrm{O}_2} = \bar{s}^0_{\mathrm{O}_2} (T_0) - \bar{R} \ln X^{\mathrm{air}}_{\mathrm{O}_2} \quad , \quad \bar{s}_{N_2} = \bar{s}^0_{\mathrm{N}_2} (T_0) - \bar{R} \ln X^{\mathrm{air}}_{\mathrm{N}_2}$$

with $X^{\mathrm{air}}_{\mathrm{O}_2} = 0.21$, $X^{\mathrm{air}}_{\mathrm{N}_2} = 0.79$. For the computation of reversible work, the exhaust temperature is T_0, while the mole fractions of the products are given above in terms of the amount of excess air.

Using tabulated data for $\bar{s}^0_\alpha (T)$ and $\bar{h}_{\mathrm{CO}_2} (T)$, the following values for the maximum work per mole of fuel are found:

$$\frac{\dot{W}_{\max}}{\dot{n}_{\mathrm{F}}} = \left\{ 822.9 \frac{\mathrm{kJ}}{\mathrm{mol}}, \ 816.5 \frac{\mathrm{kJ}}{\mathrm{mol}}, \ 810.8 \frac{\mathrm{kJ}}{\mathrm{mol}} \right\} .$$

More work could be obtained for larger amounts of excess air. Part of the reversible work is related to reversible mixing of the products, and this contribution is larger for larger amounts of excess air.

Now we proceed with the evaluation of the actual process, which is irreversible.

The combustion chamber is adiabatic, and exchanges no work with the surroundings. The entropy generation in the combustion process follows from applying the second law between inlet and combustion chamber exit as

$$\dot{S}^{\mathrm{comb}}_{gen} = \sum_{out,2} \dot{n}_\alpha \bar{s}_\alpha - \sum_{in} \dot{n}_\alpha \bar{s}_\alpha .$$

The corresponding power loss is $T_0 \dot{S}^{\mathrm{comb}}_{gen}$ and we compute the ratio between the loss and the maximum work as

$$\frac{T_0 \dot{S}^{\mathrm{comb}}_{gen}}{\dot{W}_{\max}} = \{51.9\%, \ 38.3\%, \ 30.5\%\} .$$

Thus, depending on the flame temperature (or the amount of excess air), a large amount of the work that is available from the fuel is lost to the irreversibility of the combustion process. The loss is larger for lower flame temperature, since the hotter flow has increased work potential (or exergy): heat at higher temperature is more valuable. A closer look shows that only a small part of this loss (2.67%, 1.91%, 1.23%) can be attributed to mixing loss.

After the combustion chamber, the hot flue gas enters the heat exchanger where the heat $\dot{Q}_{\mathrm{SE}}$ is withdrawn from the gas and transferred to the steam engine. The heat exchange follows from the first law for the gas as

$$\dot{Q}_{\mathrm{SE}} = \sum_2 \dot{n}_\alpha \bar{h}_\alpha - \sum_3 \dot{n}_\alpha \bar{h}_\alpha \, .$$

To determine the values, we need to specify the exhaust temperature, and we assume an exhaust temperature of 400 K (127 °C). Depending on the flame temperature, we find the following amounts of heat transmitted to the steam cycle (per mole of fuel):

$$\frac{\dot{Q}_{\mathrm{SE}}}{\dot{n}_{\mathrm{F}}} = \left\{ 692.7 \frac{\mathrm{kJ}}{\mathrm{mol}}, \; 746.9 \frac{\mathrm{kJ}}{\mathrm{mol}}, \; 766.3 \frac{\mathrm{kJ}}{\mathrm{mol}} \right\} .$$

Since some thermal energy leaves with the exhaust, the heat transmitted lies below the available heat of reaction, $\Delta \bar{h}_R$. Since the total mass flow is higher at high excess air (low flame temperature), more heat (13.6% $\Delta \bar{h}_R$) is lost than in the case for lower amounts of excess air (6.9%, 4.5%).

For the further calculation we have to specify the details of the steam engine. We choose a standard Rankine cycle operating at condenser temperature of $T_c = 40\,°\mathrm{C}$, turbine inlet conditions are $T_T = 550\,°\mathrm{C}$, $p_T = 80\,\mathrm{bar}$; pump and turbine have isentropic efficiencies of 80% and 85%, respectively. The corresponding specific enthalpies and entropies of the vapor stream are (the subscripts refer to the the encircled numbers in Fig. 25.2)

$$\begin{aligned} h_1 &= 167.6 \tfrac{\mathrm{kJ}}{\mathrm{kg}} \, , & s_1 &= 0.5725 \tfrac{\mathrm{kJ}}{\mathrm{kg\,K}} \, , \\ h_{2s} &= 175.6 \tfrac{\mathrm{kJ}}{\mathrm{kg}} \, , & s_{2s} &= 0.5724 \tfrac{\mathrm{kJ}}{\mathrm{kg\,K}} \, , \\ h_2 &= 177.7 \tfrac{\mathrm{kJ}}{\mathrm{kg}} \, , & s_2 &= 0.5792 \tfrac{\mathrm{kJ}}{\mathrm{kg\,K}} \, , \\ h_3 &= 3521 \tfrac{\mathrm{kJ}}{\mathrm{kg}} \, , & s_3 &= 6.878 \tfrac{\mathrm{kJ}}{\mathrm{kg\,K}} \, , \end{aligned}$$

$$\begin{aligned} h_{4s} &= 2142 \tfrac{\mathrm{kJ}}{\mathrm{kg}} \, , & s_4 &= 6.878 \tfrac{\mathrm{kJ}}{\mathrm{kg\,K}} \, , \\ h_4 &= 2348 \tfrac{\mathrm{kJ}}{\mathrm{kg}} \, , & s_4 &= 7.488 \tfrac{\mathrm{kJ}}{\mathrm{kg\,K}} \, . \end{aligned}$$

The thermal efficiency of the steam cycle is

$$\eta = \frac{h_1 - h_2 + h_3 - h_4}{h_3 - h_2} = 34.8\%$$

for irreversible operation, and $\eta_{rev} = 41.0\%$ when pump and turbines are reversible (states $2s$, $4s$ instead of states 2, 4).

The energy balance for the heat exchanger gives the mass flow of steam per mole of fuel as

$$\frac{\dot{m}_{\mathrm{steam}}}{\dot{n}_{\mathrm{F}}} = \frac{\dot{Q}_{\mathrm{SE}}}{\dot{n}_{\mathrm{F}} \left(h_3 - h_2 \right)} \, .$$

The entropy generated in the heat exchanger, which is adiabatic to the outside, is computed from the entropy flows as

$$\dot{S}_{gen}^{\text{HE}} = \sum_3 \dot{n}_\alpha \bar{s}_\alpha - \sum_2 \dot{n}_\alpha \bar{s}_\alpha + \dot{m}_{\text{steam}} \left(s_3 - s_2 \right) .$$

The relative amounts of work lost are

$$\frac{T_0 \dot{S}_{gen}^{\text{HE}}}{\dot{W}_{\text{max}}} = \{9.33\%,\ 20.7\%,\ 27.5\%\} .$$

Entropy generation in heat transfer is larger when the temperature difference between the flows is large. Since the temperatures for the steam cycle are fixed, there are higher heat transfer losses for larger flame temperature. The heat transfer loss could be reduced by using a heat engine that operates at higher maximum temperature. This would reduce heat transfer losses, and increase the thermal efficiency.

Another source of irreversibility is the thermal equilibration of the hot exhaust (state 3) with the atmosphere,

$$\dot{S}_{gen}^{\text{exhaust}} = \sum_{\text{exhaust at } T_0} \frac{\dot{n}_\alpha}{T_0} \left(\bar{h}_\alpha - T_0 \bar{s}_\alpha \right) - \sum_3 \frac{\dot{n}_\alpha}{T_0} \left(\bar{h}_\alpha - T_0 \bar{s}_\alpha \right) .$$

The relative contributions of the exhaust loss are

$$\frac{T_0 \dot{S}_{gen}^{\text{exhaust}}}{\dot{W}_{\text{max}}} = \{1.85\%,\ 0.95\%,\ 0.62\%\} .$$

The condenser contributes to work loss, due to the temperature difference between the condensing steam and the environment. The entropy generation for this process is

$$\dot{S}_{gen}^{\text{cond}} = \left| \dot{Q}_C \right| \left(\frac{1}{T_0} - \frac{1}{T_C} \right) = (1 - \eta) \dot{Q}_{\text{SE}} \left(\frac{1}{T_0} - \frac{1}{T_C} \right) ,$$

and this corresponds to the relative work loss

$$\frac{T_0 \dot{S}_{gen}^{\text{cond}}}{\dot{W}_{\text{max}}} = \{7.62\%,\ 8.28\%,\ 8.56\%\} .$$

Finally, we determine the actual power delivered by the power plant relative to the maximum work

$$\frac{\dot{W}_{\text{SE}}}{\dot{W}_{\text{max}}} = \frac{\eta \dot{Q}_{\text{SE}}}{\dot{W}_{\text{max}}} = \{29.3\%,\ 31.8\%,\ 32.9\%\} .$$

The efficiency of the power plant with respect to the heat available from combustion is

$$\frac{\dot{W}_{\text{SE}}}{\dot{n}_{\text{F}} \Delta \bar{h}_R} = \{30.0\%,\ 32.4\%,\ 33.2\%\} .$$

The actual work delivered and the various losses add up to the maximum work,

$$\dot{W}_{\max} = \dot{W}_{\rm SE} + T_0 \left(\dot{S}_{gen}^{\rm comb} + \dot{S}_{gen}^{\rm HE} + \dot{S}_{gen}^{\rm exhaust} + \dot{S}_{gen}^{\rm cond} \right) .$$

The above analysis gives insight into the relative importance of the different entropy generating processes.

For this standard steam cycle it becomes evident that more than 60% of the available work are lost in combustion and heat transfer ($T_0 \dot{S}_{gen}^{\rm comb} + T_0 \dot{S}_{gen}^{\rm HE}$). The heat transfer loss is reduced when the flame temperature is smaller, and more excess air is used, but this also increases the combustion loss.

For the low exhaust temperature chosen, the exhaust loss ($T_0 \dot{S}_{gen}^{\rm exhaust}$) is relatively small, but still amounts to up to 6% of the work produced ($\dot{W}_{\rm SE}$). Obviously this contribution will increase when the exhaust is hotter. While we did not explicitly account for regeneration of the exhaust, which might be used for preheating of the combustion air, see Sec. 12.1, it is clear that a regenerator will reduce the final exhaust temperature, T_3. When the exhaust temperature is lowered to 350 K, the exhaust loss is $\{0.51\%,\ 0.27\%,\ 0.18\%\}$ of $\dot{W}_{\max}$, and the overall efficiency is increased, so that $\dot{W}_{\rm SE}/\dot{W}_{\max} = \{31.6\%,\ 33.0\%,\ 33.7\%\}$.

Due to the large amount of heat transferred, the condenser loss ($T_0 \dot{S}_{gen}^{\rm cond}$) is relatively large. While the temperature difference between environment and condenser ($T_c - T_0 = 15\,^{\circ}{\rm C}$) cannot be reduced further, this contribution to the irreversible losses will become smaller when a more efficient heat cycle (reheat and regeneration) is used, where more of the incoming heat is converted to work, and less heat is rejected in the condenser.

The entropy generation in turbine and pump is

$$\dot{S}_{gen}^{\rm irr} = \dot{m}_{\rm steam} \left(s_2 - s_1 + s_4 - s_3 \right) ,$$

which relates to the relative work loss

$$\frac{T_0 \dot{S}_{gen}^{\rm irr}}{\dot{W}_{\max}} = \{5.36\%,\ 5.40\%,\ 5.44\%\} .$$

This loss is already included in the above calculation. Improved turbine and pump increase the thermal efficiency and thus increase the net work, and reduce the heat rejection—and thus the entropy generation—in the condenser. Indeed, for reversible operation of the steam cycle (use of h_{2s}, h_{4s} instead of h_2, h_4, $\eta = \eta_{\rm rev} = 41\%$), the relative condenser loss reduces to $\{2.40\%,\ 2.61\%,\ 2.70\%\}$,while the relative amount of power produced, $\dot{W}_{\rm SE}/\dot{W}_{\max}$, increases to $\{34.5\%,\ 37.5\%,\ 38.7\%\}$.

In summary, most of the heat available from the fuel, $\Delta \bar{h}_R$, is supplied to the steam cycle, and converted to work with the thermal efficiency of that cycle. Improving the thermal efficiency of the cycle leads to better conversion of the thermal energy of the hot combustion gas. Modern combined cycle power plants, in which high temperatures are reached, have a thermal efficiency of close to 60%.

It must be noted, though, that the combustion process and the subsequent heat transfer process are accompanied by a substantial work loss, due to irreversible combustion and heat transfer. The combustion irreversibility is due to the uncontrolled reorganization of the molecules in the reaction, which involves movement of electrons. In fuel cells the paths of the electrons are controlled, and combustion entropy is avoided. Thus they promise to harvest larger portions of the theoretically available work. However, also in fuel cells irreversible processes occur, which reduce their efficiencies, see Chapter 26.

Problems

25.1. Combustion Analysis
One kmol of ethane (C_2H_6) is burned with an unknown amount of air. An analysis of the combustion products reveals that the combustion is complete, and there are 3 kmol of free O_2 in the products. Determine the air-fuel ratio and the percentage of theoretical air.

25.2. Combustion Analysis
Octane (C_8H_{18}) is burned with dry air. The mole fractions of the products on a dry basis are 9.21 percent CO_2, 0.61 percent CO, 7.06 percent O_2, and 83.12 percent N_2.

Determine the air–fuel ratio and the percentage of theoretical air used.

25.3. Dewpoint Temperature
In a closed system, 3 kg of C_4H_{10} (butane) is burned with 75 kg of saturated moist air at 30 °C, 90 kPa. Determine the air-fuel ratio and the dewpoint temperature of the combustion product when the products are at 0.5 bar.

25.4. Dewpoint Temperature
A mass flow of $5\frac{\text{kg}}{\text{s}}$ of $C_{12}H_{26}$ (dodecane) is burned with a mass flow of $150\frac{\text{kg}}{\text{s}}$ of moist air at 30 °C, 90 kPa with relative humidity of 80%. Determine the relative amount of excess air, and the dewpoint temperature of the combustion product when the products are at 0.7 bar.

25.5. Combustion of Liquid Fuel Mixture
A liquid mixture of 90 mol octane (C_8H_{18}) and 10 mol ethyl alcohol (C_2H_5OH) at 25 °C, 1 atm is burned isobarically at 150% theoretical air with dry air at 25 °C. Heat is transferred to the surroundings and the final product temperature is 25 °C.

1. Determine the mole numbers of the combustion products.
2. Determine the amount of liquid water in the product.
3. Determine the heat transferred to the surroundings.

25.6. Incomplete Combustion

When hydrocarbon fuels are burnt with less than theoretical air, the products may contain carbon monoxide, carbon (as soot) and hydrogen. When there is only little deficiency of oxygen, all hydrogen in the fuel will form water, but some carbon monoxide will exist. This is due to reaction kinetics: water and carbon monoxide are formed earlier in the combustion process, while carbon dioxide is formed later from the reaction between CO and O_2.

Benzene gas, C_6H_6 at 25 °C is burned in a steady flow process with 95% of theoretical air that enters at 25 °C as well. The products leave at 1000 K. Determine the mole fraction of CO in the product and the heat transfer from the combustion chamber.

25.7. Soot Formation

2 mole of propylene gas (C_3H_6) react with 6 moles of oxygen gas (O_2) to form a mixture of water (H_2O), carbon dioxide (CO_2), carbon monoxide (CO) and soot (i.e., pure carbon, C). Determine the mole numbers of the products.

25.8. Non-adiabatic Flame Temperature of Acetylene

Acetylene gas (C_2H_2) at 25 °C is burned with 30% excess air at 27 °C, combustion is complete. The combustion chamber loses 75 kJ of heat per mole of fuel. Determine the temperature of the combustion products.

25.9. Adiabatic Combustion of Methanol

Liquid methanol (CH_3OH) at 25 °C is burned adiabatically with excess air that enters the combustion chamber at a temperature of 47 °C, combustion is complete. The temperature of the combustion products is 1500 K. Determine the relative amount of excess air.

25.10. Combustion

Liquid Ethanol (C_2H_5OH) at 25 °C, 1 atm is burned isobarically with 50% excess air (dry air) at 25 °C. Heat is transferred to the surroundings and the final product temperature is 600 K.

1. Determine the mole fractions of the combustion products.
2. Determine the heat transferred to the surroundings for a mass flow of ethanol of 15 kg/ h.

25.11. Combustion of Dodecane

Dodecane ($C_{12}H_{26}$) is burned adiabatically with 150% excess air.

Determine the balanced reaction equation and compute the upper heating value.

25.12. Combustion of Ethane

Consider the combustion of ethane (C_2H_6) with 100% of excess air at a pressure of 1.74 bar. Consider the air as dry air.

1. Set up the chemical equation.
2. Compute the higher and the lower heating value.

3. Compute the dewpoint of the combustion products.
4. Compute the heat of reaction at the dewpoint.

25.13. Combustion of Diesel Fuel
Diesel fuel (modelled as dodecane, $C_{12}H_{26}$, enthalpy of formation at 25 °C: $\bar{h}_f^0 = -291.01\frac{\text{kJ}}{\text{mol}}$) is burned in an adiabatic steady-flow combustion chamber with 50% excess air. Fuel and air enter at 25 °C. The hot combustion gas flows through an heat exchanger where heat is transferred to an environment at 750 K, the combustion gas leaves the heat exchanger at 800 K. Assume complete combustion and determine the required mass flow rate of diesel fuel to supply heat at a rate of 3000 kW. Compute the production of entropy in the combustion chamber and in the heat exchanger.

25.14. Combustion: Heat and Entropy Generation
In a technical process, a mass flow of $2\frac{\text{kg}}{\text{s}}$ of liquid ethyl alcohol C_2H_5OH is burned with 50% excess air; both incoming flows are at standard conditions (1 bar, 25 °C). The exhaust leaves at 500 K. Determine the heat provided by this process, and the entropy generation rate, assuming that the heat is received at 500 K.

25.15. Isochoric Combustion
Consider the adiabatic combustion of methyl alcohol vapor CH_3OH with the stoichiometric amount of air in an 0.8 litre combustion chamber. Initially, the mixture is at 25 °C and 98 kPa.

Determine the maximum pressure that can occur in the combustion chamber if the combustion takes place at constant volume.

25.16. Combustion Analysis
An equimolar mixture of carbon monoxide (CO) and methane (CH_4) is burned with 200% theoretical air (dry air). The mole flow of fuel is $2\frac{\text{kmol}}{\text{s}}$. Fuel stream and air enter the combustor at 25 °C, 1 atm, and the reactants leave at 127 °C. Determine:

1. The air fuel ratio on per mass basis.
2. The dew point of the products.
3. The heat transfer out of the combustion chamber.
4. Entropy generation and work loss.
5. The work to isothermally separate the CO_2 from the exhaust gas.

25.17. Combustion: Ammonia as a Fuel
By means of catalysts, gaseous ammonia (NH_3) reacts with oxygen (O_2) to water (H_2O) and molecular nitrogen (N_2). Consider the oxidation of ammonia with with 200% excess air (dry) at a pressure of 3 bar.

1. Set up the combustion equation.
2. Determine the dewpoint of the combustion products.
3. Determine the heat of reaction per mole of NH_3 at 25 °C for the cases that all product water is liquid, or all water is vapor, respectively.

25.18. Combustion Plant: Heat and Entropy Generation

In a technical process, a mass flow of $20\frac{\text{g}}{\text{s}}$ of liquid ethyl alcohol C_2H_5OH is burned with 100% excess air; both incoming flows are at standard conditions (1 bar, 25 °C). The process produces a power of 300 kW, and the exhaust leaves the plant at standard conditions.

1. Determine the heat rejection rate of the plant.
2. Determine the entropy generation rate, assuming that the plant rejects waste heat into the environment at 25 °C.

Hint: start with a good sketch.

Chapter 26
Thermodynamics of Fuel Cells

26.1 Fuel Cells

We have seen before that entropy generation in combustion, and subsequent heat transfer, leads to considerable irreversible work losses. The combustion loss can be attributed to the uncontrolled movement of electrons as new molecules are formed in the reaction. Fuel cells offer a process in which the electron movement is controlled, and thus no combustion losses occur. The performance of real life fuel cells is diminished by irreversible losses due to resistance, reaction activation and mass flow restrictions, and these will be discussed after the basic discussion of fuel cells and their efficiencies.

In fuel cells an electrochemical process allows to directly convert the energy stored in a fuel into electrical energy. There are many types of fuel cells that consume different fuels, e.g. direct methanol fuel cells, direct carbon fuel cells, and hydrogen fuel cells. Large parts of the discussion below are valid for all types, however, when it comes to evaluation of the equations, we shall consider only hydrogen fuel cells with the overall reaction

$$H_2 + \frac{1}{2}O_2 \rightleftharpoons H_2O \; . \tag{26.1}$$

Figure 26.1 shows a schematic of a hydrogen fuel cell. Hydrogen and oxygen (pure or with air) are supplied to the two sides of the fuel cell in transport channels, and enter gas diffusion layers (GDL) through which they travel towards the catalyst layer. The GDL shields the gas from the electrolyte, and contributes to the management of water flows within the cell. The electrochemical reactions take place in the catalyst layers at anode and cathode. These are separated by an electrolyte through which electrons cannot pass. The electrons move from anode to cathode through the electrical device (with resistance R_d) and provide the power $\dot{W}$ to run the device. At the same time ions move through the electrolyte.

The reactions at anode and cathode, and the transport processes through the electrolyte depend on the type of electrolyte.

H. Struchtrup, *Thermodynamics and Energy Conversion*,
DOI: 10.1007/978-3-662-43715-5_26, © Springer-Verlag Berlin Heidelberg 2014

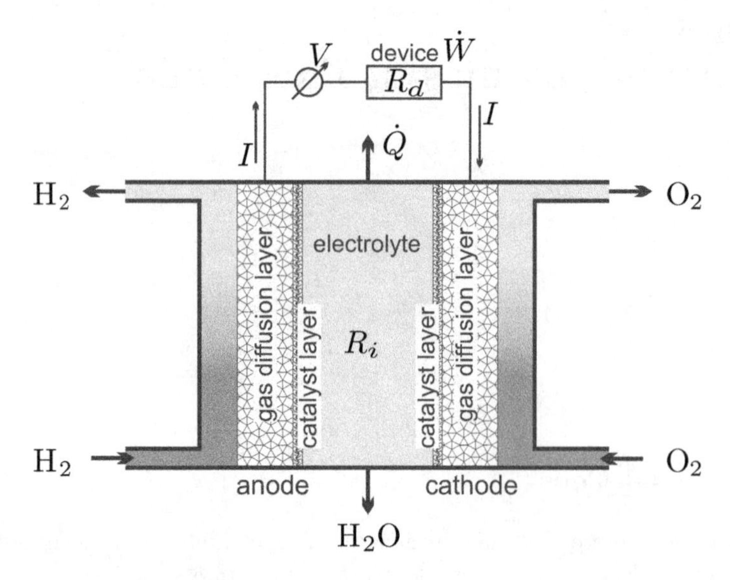

Fig. 26.1 Schematic of a fuel cell

Figure 26.2 sketches the reactions for an acidic electrolyte, e.g., phosphoric acid ($PO_4^{(3-)} + 3H^+ +$ water), or polyelectrolyte membranes (PEM), which are polymers with sulfuric acid groups (polymer $+SO_3^- + H^+ +$ water). At the anode, the incoming hydrogen is split into electrons, e^-, and protons, H^+. The protons travel through the electrolyte to the cathode, while the electrons pass through the device. At the cathode, protons, electrons and the incoming oxygen react to water, which must be removed from the electrolyte.

In alkaline electrolyte fuel cells, see Fig. 26.3, the electrolyte is a base, e.g., potassium hydroxide ($K^+ + OH^- +$ water). At the anode, the incoming hydrogen reacts with hydroxide, OH^-, to form water and electrons, which travel through the electrical device to the cathode. At the cathode, the electrons and the incoming oxygen react with water to form hydroxide, that then travels through the electrolyte to the anode, while half of the water produced at the anode travels through the electrolyte to replenish the water consumed in the cathode reaction.

In both types of fuel cells, the ion and electron movement is forced by the electric potential V between cathode and anode.

26.2 Fuel Cell Potential

For the thermodynamic analysis of fuel cells, it is not necessary to distinguish between acidic and alkaline fuel cells. We consider the mass and energy flows as in Fig. 26.1, and apply the first and second law, which read for steady state operation

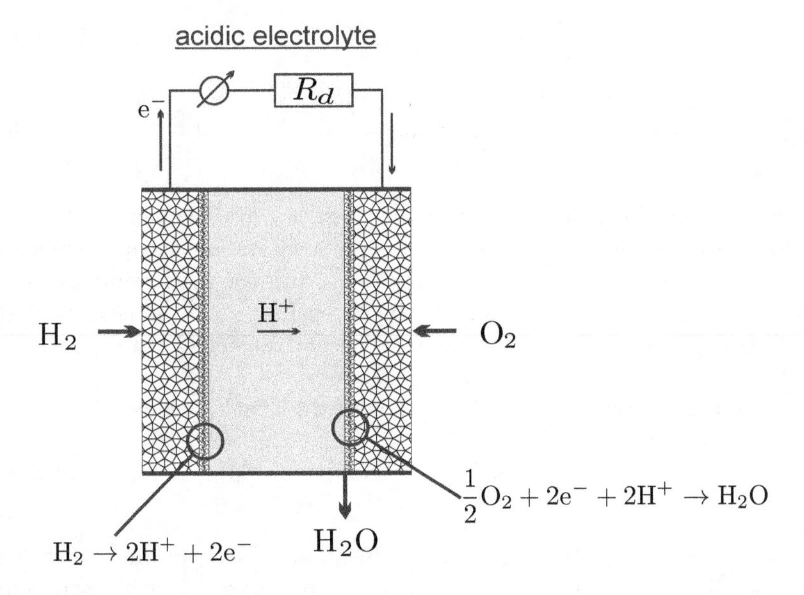

Fig. 26.2 Acidic electrolyte fuel cell

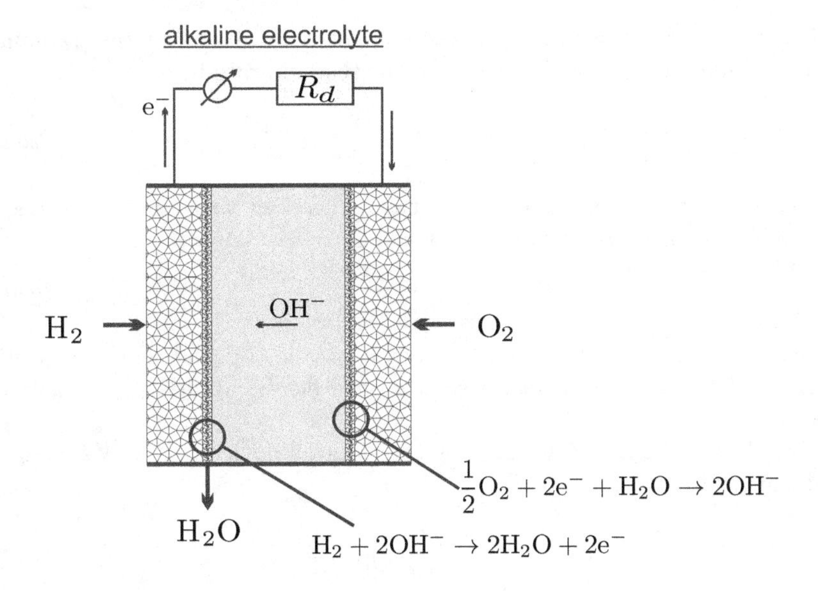

Fig. 26.3 Alkaline electrolyte fuel cell

$$\sum_{\alpha,out} \dot{n}_\alpha \bar{h}_\alpha - \sum_{\alpha,in} \dot{n}_\alpha \bar{h}_\alpha = \dot{Q} - \dot{W} , \tag{26.2}$$

$$\sum_{\alpha,out} \dot{n}_\alpha \bar{s}_\alpha - \sum_{\alpha,in} \dot{n}_\alpha \bar{s}_\alpha - \frac{\dot{Q}}{T} = \dot{S}_{gen} \geq 0 . \tag{26.3}$$

For most of our discussion, the system to be considered is just the fuel cell. For the evaluation we shall assume that all inflows and outflows take place at the homogeneous fuel cell temperature T. Additional irreversible processes (external losses) might occur outside the fuel cell, e.g., in the heating of the incoming oxygen and hydrogen, or in the heat transfer $\dot{Q}$ between the fuel cell and its exterior environment.

Combining the two laws by elimination of the heat $\dot{Q}$ yields

$$\dot{W} = \sum_{\alpha,in} \dot{n}_\alpha \bar{g}_\alpha - \sum_{\alpha,out} \dot{n}_\alpha \bar{g}_\alpha - T\dot{S}_{gen} . \tag{26.4}$$

In order to guarantee sufficient supply of fuel and oxidizer to the gas diffusion layers at all locations along the gas channels, one normally will have excess hydrogen and oxygen which are circulated back to the inlet (at least the hydrogen).

The rate of reactions taking place is denoted by Λ so that the incoming and outgoing mole flows are related to the reaction rate by

$$\dot{n}^{in}_{\mathrm{H}_2} - \dot{n}^{out}_{\mathrm{H}_2} = \Lambda \quad , \quad \dot{n}^{in}_{\mathrm{O}_2} - \dot{n}^{out}_{\mathrm{O}_2} = \frac{1}{2}\Lambda \quad , \quad \dot{n}^{out}_{\mathrm{H}_2\mathrm{O}} = \Lambda . \tag{26.5}$$

For each reaction there are two electrons traveling through the electrical device which corresponds to the electrical current

$$I = 2F\Lambda . \tag{26.6}$$

Here $F = 96485 \frac{\mathrm{Cb}}{\mathrm{mol}}$ denotes Faraday's constant (Michael Faraday, 1791-1867), which gives the absolute value of the electrical charge per mole of electrons.

The electrical power consumed by the external device is $\dot{W} = VI = R_d I^2$ and thus, with $\Lambda = \frac{I}{2F}$,

$$\dot{W} = VI = -\frac{I}{2F}\Delta\bar{g}_R - T\dot{S}_{gen} , \tag{26.7}$$

where $\Delta\bar{g}_R$ is the Gibbs free energy of reaction at T. The fuel cell potential follows as

$$V = -\frac{\Delta\bar{g}_R}{2F} - \frac{T\dot{S}_{gen}}{I} . \tag{26.8}$$

The last term, $\frac{T\dot{S}_{gen}}{I} = V_{\mathrm{over}}$ is the overpotential, i.e., the potential loss due to irreversible processes within the fuel cell. In Section 25.11 it was shown that the Gibbs free energy of reaction is the maximum amount of work that can be obtained per mole of fuel. This work could be obtained from a fully reversible fuel cell, where $V_{\text{over}} = 0$.

When the electron circuit is interrupted, reactions cannot take place, and accordingly no irreversible processes occur. In this case one measures the open circuit voltage between anode and cathode,[1]

$$V_0 = -\frac{\Delta \bar{g}_R}{2F} \,. \tag{26.9}$$

At standard reference conditions, the open circuit potential of a hydrogen fuel cell is $V_0 = 1.23V$. Higher voltages, and thus higher powers, are obtained by connecting several cells in series, to form fuel cell stacks.

26.3 Fuel Cell Efficiency

Efficiencies of engines are normally defined as

$$\eta = \frac{\text{gain}}{\text{expense}} \,, \tag{26.10}$$

where for power producing heat engines the gain is the power produced, and the expense is the heat put into the engine through the combustion of fuel, nuclear reactions, etc. Fuel cells are not heat engines, and thus the question arises how to best define their efficiency. The gain is the power produced, hence this is a question of defining the expense.

For comparison of heat engines and fuel cells the following definition is often used: In a combustion process one could obtain the heat $\dot{Q}_C = -\Lambda \Delta \bar{h}_R = -\frac{I}{2F} \Delta \bar{h}_R$. Considering this heat as the expense, one defines the "thermal efficiency" of a fuel cell as

$$\eta_{\text{th}}^{\text{FC}} = \frac{\dot{W}}{\dot{Q}_C} = \frac{-\Delta \bar{g}_R - \frac{2F}{I} T \dot{S}_{gen}}{-\Delta \bar{h}_R} \,. \tag{26.11}$$

With this definition, the perfect—i.e., reversible—fuel cell has the efficiency

$$\eta_{\text{th}}^{\text{FC}} = \frac{-\Delta \bar{g}_R}{-\Delta \bar{h}_R} = \frac{-\left(\Delta \bar{h}_R - T \Delta \bar{s}_R\right)}{-\Delta \bar{h}_R} = 1 - \frac{T \Delta \bar{s}_R}{\Delta \bar{h}_R} \,. \tag{26.12}$$

For all fuels $\Delta \bar{h}_R < 0$, but there is no definite sign for the entropy of reaction $\Delta \bar{s}_R$. For the hydrogen reaction, $\Delta \bar{s}_R < 0$ and thus $\eta_{\text{th}}^{\text{FC}} < 1$ for reversible fuel cells (irreversible fuel cells have even smaller efficiencies, of course). However, there are reactions, in particular the reaction between carbon and oxygen, $C + O_2 \rightleftharpoons CO_2$, in direct carbon fuel cells, for which the entropy of reaction

[1] There is no current, so that $I = 0$ and $\dot{W} = 0$.

is positive, $\Delta\bar{s}_R > 0$, so that $\eta_{\rm th}^{\rm FC}$ as given in (26.11) becomes larger than unity. A proper efficiency measure should always assume values between zero and unity. It follows that the efficiency definition (26.11) is not suitable for the evaluation of fuel cells.

In order to understand why this efficiency measure can be above unity, one needs to consider that, according to the second law, the heat exchanged with the surroundings for the reversible fuel cell is $T\Delta\bar{s}_R$. If $T\Delta\bar{s}_R < 0$, heat is rejected into the surroundings, but if $T\Delta\bar{s}_R > 0$, heat is imported from the surroundings.

One should not be surprised that this efficiency definition leads to problems, since the heat of reaction, $-\Delta\bar{h}_R$, is not relevant in the thermodynamics of fuel cells, as is apparent in that it does not appear in the discussion of fuel cell power and voltage. Heat of reaction is a quantity relevant only for combustion systems.

A more meaningful efficiency is the ratio of the actual power produced by the fuel cell and the maximum power that could be obtained from the fuel, in a fully reversible process. This leads to the second law efficiency,

$$\eta_{FC}^{\rm II} = \frac{\dot{W}}{\dot{W}_{\rm rev}} = \frac{V}{V_0} = 1 - \frac{T\dot{S}_{gen}}{(-\Delta\bar{g}_R)\frac{I}{2F}}\,, \tag{26.13}$$

which is always positive, and becomes unity only when all irreversibilities vanish.

When only the fuel cell is considered, $\Delta\bar{g}_R$ is to be evaluated at the fuel cell temperature T. The following table shows the Gibbs free energy, the open circuit voltage V_0, and the ratio $\frac{-\Delta\bar{g}_R}{-\Delta\bar{h}_R}$ of the hydrogen fuel cell as a function of temperature (product water is liquid for $T = 25\,°\mathrm{C}, 80\,°\mathrm{C}$ and vapor for higher temperatures[2]) at standard pressure p_0:

T	$-\Delta\bar{g}_R$	$V_0 = \frac{-\Delta\bar{g}_R}{2F}$	$\frac{-\Delta\bar{g}_R}{-\Delta\bar{h}_R}$
25 °C	$237.2\frac{\rm kJ}{\rm mol}$	1.23 V	0.83
80 °C	$228.2\frac{\rm kJ}{\rm mol}$	1.18 V	0.80
100 °C	$225.2\frac{\rm kJ}{\rm mol}$	1.17 V	0.79
200 °C	$220.4\frac{\rm kJ}{\rm mol}$	1.14 V	0.77
400 °C	$210.3\frac{\rm kJ}{\rm mol}$	1.09 V	0.74
600 °C	$199.6\frac{\rm kJ}{\rm mol}$	1.04 V	0.70
800 °C	$188.6\frac{\rm kJ}{\rm mol}$	0.98 V	0.66
1000 °C	$177.6\frac{\rm kJ}{\rm mol}$	0.92 V	0.62

All three quantities decrease with increasing temperature. Thus, if one considers a fuel cell alone, one will gain more work at lower temperatures. High temperature fuel cells, e.g. solid oxide fuel cells, operate at temperatures between 700 and 1000 °C, and thus have lower open circuit voltages than low

[2] For the Gibbs free enthalpy, this makes no difference!

temperature fuel cells. However, due to the high temperatures, the activation losses (see below) are lower, and expensive catalysts are not required. Moreover, the heat rejected from high temperature fuel cells can be used to drive heat engines in a combined cycle to produce additional power. If the heat is barely rejected into the environment, there is an external loss.

In other words, fuel cells must be imbedded into a system, where the fuel cell is at the center, but additional systems for extracting work, heating and cooling must be considered as well. In order to clarify this, Fig. 26.4 shows a (reversible) fuel cell operating at temperature T as part of a fully reversible external system, where the incoming fuel and oxidizer are heated to T by a series of infinitesimal Carnot heat pumps, the heat $\dot{Q}_{FC}$ rejected from the fuel cell drives a Carnot heat engine, and the exhaust is cooled through infinitesimal Carnot heat engines so that it leaves at T_0. Of course, other fully reversible set-ups are possible, e.g., some heat could be exchanged between the incoming reactant streams and the product stream, using reversible counter-flow heat exchangers. Heat is only exchanged at environmental temperature, and since all incoming and outgoing mass flows are at T_0 and since the system exchanges heat only at T_0, and when all flows are at p_0, the work produced per mole of fuel is given by (25.18),

$$\dot{W}_{\text{rev}} = \dot{W}_{\text{FC}} + \dot{W}_{\text{C}} = -\dot{n}_{\text{fuel}} \Delta \bar{g}_R \left(T_0, p_0 \right) = -\frac{I}{2F} \Delta \bar{g}_R \left(T_0, p_0 \right) . \tag{26.14}$$

Thus, the obvious definition of a second law efficiency for a fuel consuming system is the relative amount of the work that is actually produced from the available work, i.e.,

$$\eta_{\text{system}}^{\text{II}} = \frac{\dot{W}}{\dot{W}_{\text{rev}}} = 1 - \frac{T_0 \dot{S}_{gen}}{\dot{W}_{\text{rev}}} = 1 - \frac{T_0 \dot{S}_{gen}}{-\dot{n}_{\text{fuel}} \Delta \bar{g}_R \left(T_0, p_0 \right)} . \tag{26.15}$$

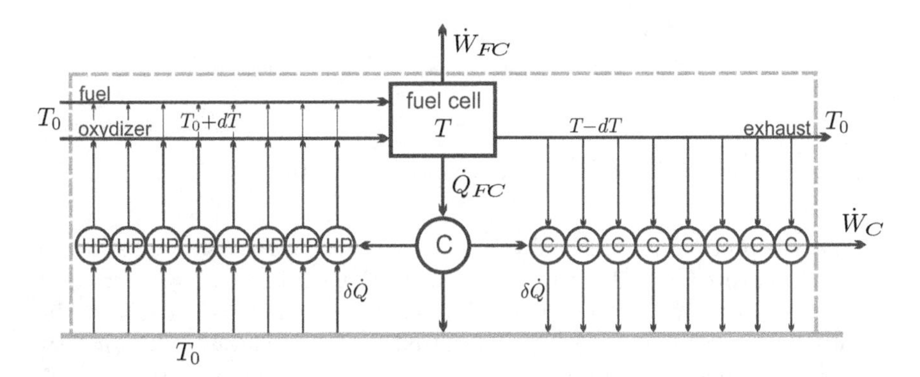

Fig. 26.4 Fuel cell embedded into a fully reversible system of Carnot heat engines and Carnot heat pumps. Heat is only exchanged at environmental temperature T_0.

This definition can be used for the evaluation of *any* fuel consuming power generation device that is embedded in the environmental at T_0. The relation between the second law efficiency and the thermal efficiency is

$$\eta_{\text{system}}^{\text{II}} = \frac{\dot{W}}{\dot{W}_{\text{rev}}} = \frac{\dot{W}}{\dot{Q}_C}\frac{\dot{Q}_C}{\dot{W}_{\text{rev}}} = \eta_{\text{th}}\eta_{\text{heat}}\frac{\Delta\bar{h}_R\left(T_0, p_0\right)}{\Delta\bar{g}_R\left(T_0, p_0\right)} . \tag{26.16}$$

Here we introduced the heat utilization factor

$$\eta_{\text{heat}} = \frac{\dot{Q}_C}{-\dot{n}_{\text{fuel}}\Delta\bar{h}_R\left(T_0, p_0\right)} \tag{26.17}$$

to account for losses in heat exchanger and exhaust; $\dot{Q}_C$ is the heat actually transmitted into the heat engine, while $-\Delta\bar{h}_R$ is the available heat.

To finish this section we present and criticize a misleading figure that is sometimes found in the fuel cell literature. Figure 26.5 shows the thermal efficiency (26.11) of a reversible hydrogen fuel cell, $\eta_{\text{th}}^{\text{FC}} = \frac{\Delta\bar{g}_R(T)}{\Delta\bar{h}_R(T)}$, and the efficiency of a Carnot heat engine, $\eta_C = 1 - \frac{T_0}{T}$, both plotted over temperature T. The Carnot efficiency η_C grows with temperature, while $\frac{\Delta\bar{g}_R(T)}{\Delta\bar{h}_R(T)}$ decreases. The figure seems to imply that at higher temperatures fuel cell efficiency might not be as good as the efficiency of a heat engine.

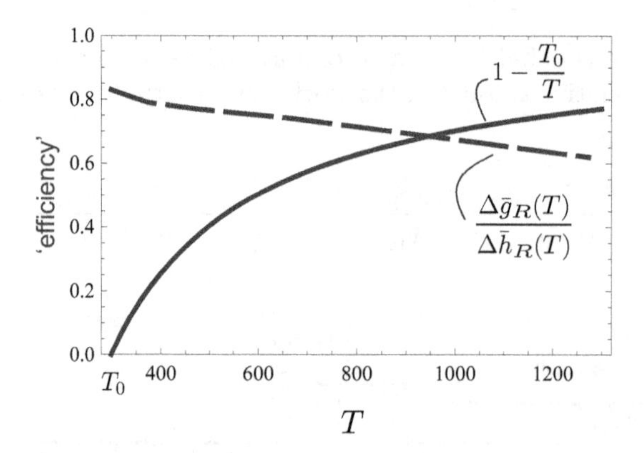

Fig. 26.5 A misleading figure ... see the discussion in the text

This interpretation is misleading for two reasons: (a) As discussed above, a high temperature fuel cell must be seen as one element in a larger system that includes heat engines to convert the high temperature heat rejected from the fuel cell and its exhaust. (b) The Carnot efficiency is relevant only for heat engines operating between two reservoirs at constant temperatures T, T_0. A heat engine that is driven by the reaction of a fuel does not belong into that category, and thus the Carnot efficiency is not necessarily relevant for a fuel driven engine process.

The proper efficiency measure to compare fuel cell systems and heat engines is the second law efficiency $\eta^{II}_{\text{system}}$, Eq. (26.15), which relates the actual work produced to the maximum amount that could be produced in a fully reversible process. For real engines, this efficiency is below unity, due to entropy generation in irreversible processes. As discussed in Sec. 25.12, combustion engines suffer from losses in combustion, heat transfer, friction and mixing. Fuel cells suffer from a different array of irreversible losses that will be discussed below. Whether or not a fuel cell system will have a higher efficiency than a combustion system depends on the details of the design, materials used, and the processes within the *system.*

26.4 Nernst Equation

The Nernst equation describes how the Gibbs free energy of reaction, and thus open circuit voltage and power generation of a fuel cell, depends on the pressures of the reactant and product streams. Indeed, so far we have only considered the temperature dependence of $\Delta\bar{g}_R$ and implicitly assumed that the flows are at reference pressure. Now we consider the more general case, where the in- and outflows have different pressures p_α.

To simplify the argument, we assume that all streams are ideal gases, so that $\bar{s}_\alpha(T, p_\alpha) = \bar{s}^0_\alpha(T) - \bar{R}\ln\frac{p_\alpha}{p_0}$. As always, $\bar{s}^0_\alpha(T)$ is the tabulated entropy at reference pressure p_0. We recall the definitions $\Delta\bar{g}_R = \Delta\bar{h}_R - T\Delta\bar{s}_R$ with $\Delta\bar{h}_R = \sum_\alpha \gamma_\alpha \bar{h}_\alpha(T)$ and $\Delta\bar{s}_R = \sum_\alpha \gamma_\alpha \bar{s}_\alpha(T, p_\alpha)$. This gives the Nernst equation

$$\Delta\bar{g}_R(T, p_\alpha) = \Delta\bar{g}_R(T, p_0) + \bar{R}T\ln\prod_\alpha\left(\frac{p_\alpha}{p_0}\right)^{\gamma_\alpha}, \qquad (26.18)$$

where the argument (T, p_α) indicates that the in- and outflows are all at the same temperature T, but at different pressures p_α.

For a hydrogen fuel cell in which the product is steam, the Nernst equation gives the open circuit potential

$$V_0 = \frac{1}{2F}\left[-\Delta\bar{g}_R(T, p_0) + \bar{R}T\ln\frac{p_{H_2}\sqrt{p_{O_2}}}{p_{H_2O}\sqrt{p_0}}\right]. \qquad (26.19)$$

If the product water is liquid, the entropy of the water is independent of pressure due to incompressibility, $\bar{s}_{H_20}(T, p_{H_2O}) = \bar{s}_{H_20}(T)$, and the open circuit potential is

$$V_0 = \frac{1}{2F}\left[-\Delta\bar{g}_R(T, p_0) + \bar{R}T\ln\frac{p_{H_2}\sqrt{p_{O_2}}}{\sqrt{p_0}^3}\right]. \qquad (26.20)$$

These two equations show that the open circuit voltage can be increased by supplying fuel (H_2) and oxidizer (O_2) at elevated pressures. When the product is steam, lowering the steam pressure increases the open circuit voltage.

26.5 Mass Transfer Losses

The pressure considered in the Nernst equation above are the pressures at which the inflows are supplied, and the outflows are removed, that is the pressures in the transport channels. The transport of reactants and products through the porous gas diffusion layers leads to friction losses, and thus generation of entropy as discussed in Sec. 9.7. Entropy generation and overpotential associated with this loss will be determined next.

The flow through the porous medium can be described by Darcy's law (9.30) which we can write for mole flows as

$$\dot{n}_\alpha = K\left(p_{\alpha,1} - p_{\alpha,2}\right) , \tag{26.21}$$

where flow goes from $p_{\alpha,1}$ to $p_{\alpha,2}$; K is an overall transport parameter.

For ideal gases, the corresponding entropy generation rate (9.29) can be written in the equivalent forms[3]

$$\dot{S}_{gen} = -\dot{n}_\alpha \bar{R} \ln \frac{p_{\alpha,2}}{p_{\alpha,1}} = -\dot{n}_\alpha \bar{R} \ln \left[1 - \frac{\dot{n}_\alpha}{K p_{\alpha,1}}\right] = \dot{n}_\alpha \bar{R} \ln \left[1 + \frac{\dot{n}_\alpha}{K p_{\alpha,2}}\right] . \tag{26.22}$$

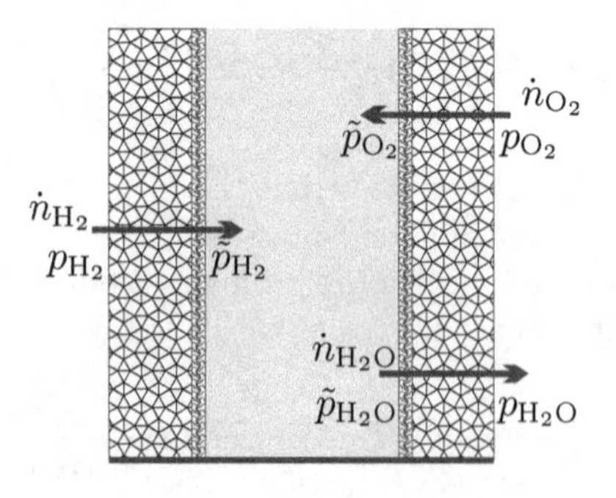

Fig. 26.6 Pressures in a fuel cell

The relevant mole flows and pressures for a hydrogen fuel cell are depicted in Fig. 26.6, where p_α denotes the pressures in the gas channels, and $\tilde{p}_\alpha$ denotes the pressures at the catalyst layers. Note that hydrogen and oxygen flow from the channels to the catalyst layer, while water flows in the opposite direction. The mole flows are related to current as

[3] Check the differences between these forms carefully!

$$\begin{aligned}
\dot{n}_{H_2} &= \frac{I}{2F} = K_{H_2}\left(p_{H_2} - \tilde{p}_{H_2}\right) , \\
\dot{n}_{O_2} &= \frac{I}{4F} = K_{O_2}\left(p_{O_2} - \tilde{p}_{O_2}\right) , \\
\dot{n}_{H_2O} &= \frac{I}{2F} = K_{H_2O}\left(\tilde{p}_{H_2O} - p_{H_2O}\right) .
\end{aligned} \tag{26.23}$$

Thus, the respective entropy generation rates for the flows are

$$\begin{aligned}
\dot{S}_{gen}^{H_2} &= -\frac{I}{2F}\bar{R}\ln\left[1 - \frac{I}{I_{H_2}}\right] , \\
\dot{S}_{gen}^{O_2} &= -\frac{I}{4F}\bar{R}\ln\left[1 - \frac{I}{I_{O_2}}\right] , \\
\dot{S}_{gen}^{H_2O} &= \frac{I}{2F}\bar{R}\ln\left[1 + \frac{I}{I_{H_2O}}\right] .
\end{aligned} \tag{26.24}$$

For compact notation we used the abbreviation $I_\alpha = 2FK_\alpha p_\alpha / |\gamma_\alpha|$ for the so-called limiting currents. Note that all pressures in these are taken in the supply channels. The different forms of the expressions are due the fact that H_2 and O_2 are entering the device, while the water leaves; all three expressions are positive.

In the fuel cell literature, one sometimes finds these contributions to loss, or overpotential, subsumed into just one expression of the form

$$\dot{S}_{gen}^{mass} = -IB\ln\left[1 - \frac{I}{I_{\text{limit}}}\right] , \tag{26.25}$$

where B and I_{limit} are suitable parameters that describe the overall mass transfer losses. With this, the fuel cell potential (26.30) assumes the form

$$V = -\frac{\Delta\bar{g}_R(T, p_\alpha)}{2F} - BT\ln\left[1 - \frac{I}{I_{\text{limit}}}\right]^{-1} - \left[\frac{T\dot{S}_{gen}}{I}\right]_{\text{other}} . \tag{26.26}$$

Here, the last term refers to other contributions to entropy generation which will be discussed below.

In acidic fuel cells, water is produced at the cathode, and must be removed. In low temperature fuel cells the produced water is liquid, and might clog the pores of the gas diffusion layer, and even the gas channels. This reduces the transport parameter K and the limiting current I_{limit}. The air flow that provides the oxygen must be dry enough, so that the product water can evaporate into the exhaust. The use of excess air increases the water intake, and also guarantees sufficient oxygen pressures everywhere (see Nernst equation).

It is worth noting that the Nernst equation and the entropy generation terms for transport can be combined to give the fuel cell potential as[4]

[4] This simple exercise is left to the reader.

$$V = -\frac{\Delta \bar{g}_R(T, \tilde{p}_\alpha)}{2F} - \left[\frac{T\dot{S}_{gen}}{I}\right]_{\text{other}} . \tag{26.27}$$

This form of the equation shows that it is really the pressures $\tilde{p}_\alpha$ at the catalyst layers that are important. These however cannot be controlled, rather they depend on the pressures p_α in the gas channels and the current I as expressed in (26.23). This dependence is explicit in the form (26.26), which therefore must be used.

26.6 Resistance Losses

The ions travelling through the electrolyte between anode and cathode, and the electrons forming the electrical current that provides electrical power, experience resistance in the media they move in. The overall internal resistance of the fuel cell is denoted by R_i.

Electrical resistance is an irreversible process, and we proceed by determining the corresponding entropy generation, for a resistor at steady state. The resistor consumes work in form of electrical power $\dot{W} = -VI = -R_i I^2$, where I is the current and V is the voltage. The temperature of the resistor is T, and first and second law reduce to

$$0 = \dot{Q} - \dot{W} \ , \quad -\frac{\dot{Q}}{T} = \dot{S}_{gen} \geq 0 \ , \tag{26.28}$$

respectively. As done often before, first and second law are combined by eliminating the heat, which here gives[5]

$$\dot{S}_{gen} = -\frac{\dot{Q}}{T} = -\frac{\dot{W}}{T} = \frac{VI}{T} = \frac{R_i I^2}{T} \geq 0 \ . \tag{26.29}$$

Electrical resistance produces entropy by downgrading electrical work to heat.

Thus, the voltage (26.8) of a fuel cell with resistance R_i is

$$V = -\frac{\Delta \bar{g}_R(T, p_\alpha)}{2F} - R_i I - BT \ln\left[1 - \frac{I}{I_{\text{limit}}}\right]^{-1} - \left[\frac{T\dot{S}_{gen}}{I}\right]_{\text{other}} , \tag{26.30}$$

where the last term refers to other contributions to entropy generation.

In PEM fuel cells, a particular contribution to resistance loss is the drying-out of the membrane. The protons travelling from anode to cathode drag some water along, and this water is removed together with the product water. Thus, the membrane becomes somewhat dryer at the anode, and this reduces its conductivity, i.e., increases the membrane resistance. A common method to

[5] Also here, the entropy generation can be considered as the product of a force (the potential V) and a flux (the current I). The linear relation between flux and force, $I = V/R$, guarantees positive entropy generation.

deal with this problem is to moisturize the incoming hydrogen fuel, so that new water is available at the anode.

26.7 Activation Overpotential

The third main cause for overpotential in fuel cells is activation loss. This irreversible effect is related to finite reaction rates and activation barriers at the reaction sites. The following discussion is based in part on the ideas discussed in Sec. 24.

At interfaces between different substances one observes electric potentials, due to different charge distribution at the interface. Figure 26.7 shows schematically the electric double layers that result at the interfaces between catalyst layers and electrolyte at anode and cathode in a fuel cell. For the sake of simplicity, the electrolyte at the interface is assumed to be electrically neutral.

The upper part of the figure shows anode and cathode potentials at open circuit, denoted as $V_{a,0}$ and $V_{c,0}$, with respect to an arbitrary reference, chosen such that the anode potential is negative and the cathode potential is positive. The overall cell potential at open circuit is $V_0 = V_{c,0} - V_{a,0}$.

The lower part of the figure sketches the conditions at closed circuit: a current flows and negative charges are removed from the anode, which becomes less negative, $V_a - V_{a,0} > 0$. On the other side, the additional electrons weaken the cathode potential, $V_c - V_{c,0} < 0$.

Due to reactions, new electrons are produced constantly at the anode, and consumed at the cathode. When the current is low, the electrons withdrawn are replaced through reactions, and the resulting change in the potentials is small. When the current is large, at the anode electrons are not replaced fast enough by reactions, and at the cathodes electrons are not consumed fast enough, so that the absolute values of the potentials drop. Thus, the overall potential of the cell, $V = V_c - V_a$, depends on the rate of reactions relative to the current drawn.

The activation overpotentials for anode and cathode are defined as

$$\eta_a = V_a - V_{a,0} \quad , \quad \eta_c = V_{c,0} - V_c \; , \tag{26.31}$$

so that they both are positive, and the total overpotential is

$$\eta = \eta_a + \eta_c = V_0 - V \; . \tag{26.32}$$

From the above discussion follows that the activation overpotential η should be small for small current, and large for large current. We proceed to find the relation between overpotential η and current I.

The anode and cathode potentials are related to the Gibbs free energies of reaction at anode and cathode as

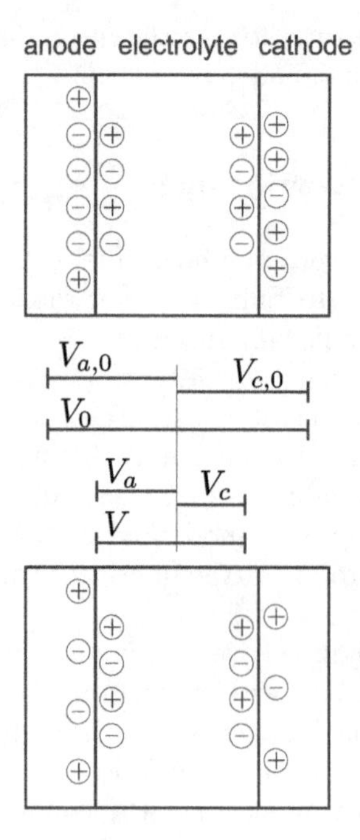

Fig. 26.7 Interface potentials and cell potential in a fuel cell. Top: open circuit. Bottom: closed circuit.

$$V_a = \frac{\Delta \bar{g}_a}{2F} = \frac{1}{2F}\left(-\bar{g}_{H_2} + 2\bar{g}_{H^+} + 2\bar{g}_{e^-}\right) , \tag{26.33}$$

$$V_c = -\frac{\Delta \bar{g}_c}{2F} = -\frac{1}{2F}\left(-\frac{1}{2}\bar{g}_{O_2} - 2\bar{g}_{H^+} - 2\bar{g}_{e^-} + \bar{g}_{H_2O}\right) . \tag{26.34}$$

The cell potential is, with $\Delta \bar{g}_R = \Delta \bar{g}_a + \Delta \bar{g}_c$,

$$V = V_c - V_a = -\frac{\Delta \bar{g}_R}{2F} = -\frac{1}{2F}\left(-\bar{g}_{H_2} - \frac{1}{2}\bar{g}_{O_2} + \bar{g}_{H_2O}\right) . \tag{26.35}$$

Adapting the results from Sec. 24.4, we write the entropy generation rate for the anode reaction as

$$\dot{S}_{gen,a} = -\frac{\Lambda}{T}\left[\bar{g}_f\left(V_a, T\right) - \bar{g}_b\left(V_a, T\right) + \bar{R}T \ln \prod X_\alpha^{\gamma_\alpha}\right] , \tag{26.36}$$

where $\bar{g}_f\left(V_a, T\right)$ and $\bar{g}_b\left(V_a, T\right)$ are activation barriers for the reactions, with $\Delta \bar{g}_a = \bar{g}_b - \bar{g}_f$. At open circuit the entropy generation vanishes, which implies the law of mass action in the form

$$\bar{g}_f\left(V_{a,0}, T\right) - \bar{g}_b\left(V_{a,0}, T\right) + \bar{R}T \ln \prod X_{\alpha,0}^{\gamma_\alpha} = 0 . \tag{26.37}$$

By subtracting this from the entropy generation, and multiplying with T/I we find the anode overpotential as

$$\begin{aligned}\eta_a &= \frac{T\dot{S}_{gen,a}}{I} \\ &= \frac{1}{2F}\left[\left(\bar{g}_f\left(V_{a,0},T\right)-\bar{g}_f\left(V_a,T\right)+\bar{R}T\ln\prod_{\text{reactants}}\frac{X_\alpha^{|\gamma_\alpha|}}{X_{\alpha,0}^{|\gamma_\alpha|}}\right)\right. \\ &\quad \left.-\left(\bar{g}_b\left(V_{a,0},T\right)-\bar{g}_b\left(V_a,T\right)+\bar{R}T\ln\prod_{\text{products}}\frac{X_\alpha^{\gamma_\alpha}}{X_{\alpha,0}^{\gamma_\alpha}}\right)\right] .\end{aligned} \tag{26.38}$$

The overpotential is larger the further the system is away from equilibrium. The activation barriers $\bar{g}_f\left(V_a,T\right)$ and $\bar{g}_b\left(V_a,T\right)$ depend on the current, and we proceed with their determination.

As a first step we note that the above equation (26.38) is satisfied by

$$\begin{aligned}\bar{g}_f\left(V_{a,0},T\right)-\bar{g}_f\left(V_a,T\right)+\bar{R}T\ln\prod_{\text{reactants}}\frac{X_\alpha^{|\gamma_\alpha|}}{X_{\alpha,0}^{|\gamma_\alpha|}} &= 2F\left(1-\beta_a\right)\eta_a , \\ \bar{g}_b\left(V_{a,0},T\right)-\bar{g}_b\left(V_a,T\right)+\bar{R}T\ln\prod_{\text{products}}\frac{X_\alpha^{\gamma_\alpha}}{X_{\alpha,0}^{\gamma_\alpha}} &= -2F\beta_a\eta_a ,\end{aligned} \tag{26.39}$$

for arbitrary coefficients β_a. For interpretation we can say that the parameter β_a distributes the overpotential η_a between the forward and the backward reactions. In principle, β_a could be a complicated function of current, temperature and other parameters, but experimental measurements show that it is a constant.

Reaction rates and current are related as

$$I = 2F\Lambda = 2F\left(r_f - r_b\right) , \tag{26.40}$$

where r_f and r_b are the forward and backward reaction rates, respectively. From the results of Chapter 24 we can write the reaction rates as

$$\begin{aligned}r_f &= k_0\exp\left[-\frac{\bar{g}_f\left(V_a,T\right)}{\bar{R}T}\right]\prod_{\text{reactants}}X_\alpha^{|\gamma_\alpha|} , \\ r_b &= k_0\exp\left[-\frac{\bar{g}_b\left(V_a,T\right)}{\bar{R}T}\right]\prod_{\text{products}}X_\alpha^{\gamma_\alpha} ,\end{aligned} \tag{26.41}$$

where k_0 is a rate constant. We introduce the exchange current I_0 as the current associated with the number of reactions taking place at open circuit in either direction; recall that forward and backward reaction rates are equal at open circuit. For the anode

$$I_{0,a} = 2Fr_f\left(V_{a,0}\right) = 2Fr_b\left(V_{a,0}\right) \; , \tag{26.42}$$

or, in more detail,

$$\begin{aligned} I_{0,a} &= 2Fk_0 \exp\left[-\frac{\bar{g}_f\left(V_{a,0},T\right)}{\bar{R}T}\right] \prod_{\text{reactants}} X_{\alpha,0}^{|\gamma_\alpha|} \\ &= 2Fk_0 \exp\left[-\frac{\bar{g}_b\left(V_{a,0},T\right)}{\bar{R}T}\right] \prod_{\text{products}} X_{\alpha,0}^{\gamma_\alpha} \, . \end{aligned} \tag{26.43}$$

With this, we can write the total current as

$$\begin{aligned} I &= I_{0,a}\frac{2F\left[r_f - r_b\right]}{I_{0,a}} \\ &= I_{0,a}\left(\exp\left[\frac{\bar{g}_f\left(V_{a,0},T\right) - \bar{g}_f\left(V_a,T\right)}{\bar{R}T} + \ln \prod_{\text{reactants}} \frac{X_\alpha^{|\gamma_\alpha|}}{X_{\alpha,0}^{|\gamma_\alpha|}}\right]\right. \\ &\qquad \left. - \exp\left[\frac{\bar{g}_b\left(V_{a,0},T\right) - \bar{g}_b\left(V_a,T\right)}{\bar{R}T} + \ln \prod_{\text{products}} \frac{X_\alpha^{\gamma_\alpha}}{X_{\alpha,0}^{\gamma_\alpha}}\right]\right) . \end{aligned} \tag{26.44}$$

The expressions in the exponentials are just those that occur in the 2nd law expression for the overpotential (26.38). Replacing them with (26.39) finally leads to the desired relation between current and overpotential, the Butler-Volmer equation (Max Volmer, 1885-1965; John Butler, 1899-1977):

$$I = I_{0,a}\left(\exp\left[\frac{2F\left(1-\beta_a\right)\eta_a}{\bar{R}T}\right] - \exp\left[-\frac{2F\beta_a\eta_a}{\bar{R}T}\right]\right) . \tag{26.45}$$

Due to the sign conventions used here, the Butler-Volmer equation for the cathode is obtained simply by switching signs, as

$$I = I_{0,c}\left(\exp\left[\frac{2F\beta_c\eta_c}{\bar{R}T}\right] - \exp\left[-\frac{2F\left(1-\beta_c\right)\eta_c}{\bar{R}T}\right]\right) . \tag{26.46}$$

For large overpotentials the exponential with negative argument can be ignored against the exponential with positive argument; in this limit the Butler-Volmer equation reduces to the so-called Tafel equation (Julius Tafel, 1862-1918), e.g., for the anode,

$$\eta_a = \frac{\bar{R}T}{2F\left(1-\beta_a\right)} \ln \frac{I}{I_{0,a}} \qquad \text{for} \qquad \frac{2F\left(1-\beta_a\right)\eta_a}{\bar{R}T} \gg 1 \, . \tag{26.47}$$

Figure 26.8 shows the Butler-Volmer and the Tafel equation in a logarithmic plot (*Tafel plot*), that is overvoltage η as function of $\ln I$. The curves are plotted for a constant value for the parameter β. Both curves coincide at larger η, where the approximation (26.47) is valid. The overpotential can be measured (not shown), and the resulting curve agrees with the prediction of the Butler-Volmer equation when β and I_0 are adjusted properly. Indeed, β and I_0 can be read of the experimental Tafel plot as slope and intercept as indicated in the figure.

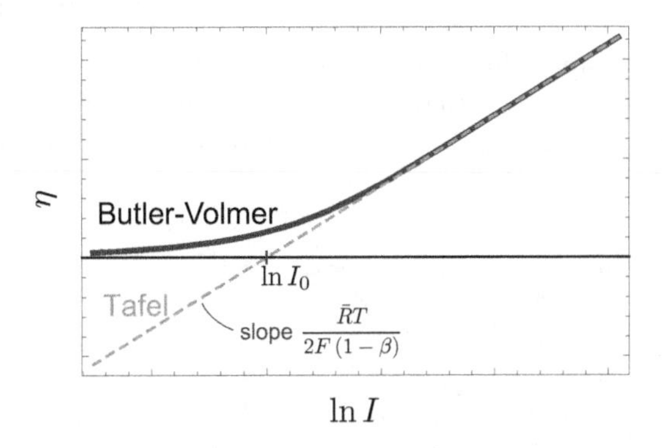

Fig. 26.8 Tafel plot: $\eta(\ln I)$ according to the Butler-Volmer equation (continuous line) and the Tafel equation (dashed). Both equations agree for large η.

According to the Butler-Volmer equation the activation overpotential will be smaller for large exchange currents I_0. For an efficient fuel cell one must aim to make the exchange current large. According to (26.43), the exchange current depends on an activation energy and on temperature. It grows with temperature, so that activation losses are smaller for high temperature fuel cells. The activation barriers $\bar{g}_f$, $\bar{g}_b$ depend strongly on the electrode material: good catalysts have low activation energies and thus low overpotential. For a hydrogen electrode at T_0, the following data for the exchange current per unit area can be found in the literature:

$$\begin{aligned}
&\text{lead:} && \hat{I}_0 = 2.5 \times 10^{-13} \frac{\mathrm{A}}{\mathrm{cm}^2}\,, \\
&\text{nickel:} && \hat{I}_0 = 6 \times 10^{-6} \frac{\mathrm{A}}{\mathrm{cm}^2}\,, \\
&\text{platinum:} && \hat{I}_0 = 5 \times 10^{-4} \frac{\mathrm{A}}{\mathrm{cm}^2}\,, \\
&\text{palladium:} && \hat{I}_0 = 4 \times 10^{-3} \frac{\mathrm{A}}{\mathrm{cm}^2}\,.
\end{aligned}$$

The data indicates that expensive catalysts must be used at low temperatures. The most common catalyst for low temperature fuel cells is platinum. No catalysts are required for high temperature fuel cells. The total exchange current of a fuel cell is proportional to the surface of the catalyst layer, $I_0 = \hat{I}_0 A_{\text{catalyst}}$. To reach sufficient catalyst area, the catalyst must be distributed well within the catalyst layer, e.g., as extremely small spheres.

In hydrogen fuel cells, typically the reaction at the anode is considerably slower than the reaction at the cathode. Then, the activation overpotential at the cathode is small and can be ignored against the anode overpotential.

We end the discussion with a short look on the activation energies for forward and backward reactions, for which the reaction rates can be written as

$$r_f = I_{0,a} \exp\left[\frac{2F(1-\beta)\eta_a}{\bar{R}T}\right] \quad , \quad r_b = I_{0,a} \exp\left[-\frac{2F\beta\eta_a}{\bar{R}T}\right] . \tag{26.48}$$

These equations imply that the overpotential leads to a change in energy barriers and reaction rates. Figure 26.9 illustrates the influence of the overpotential on the energy landscape, similar to Fig. 24.2 and the discussion around it.

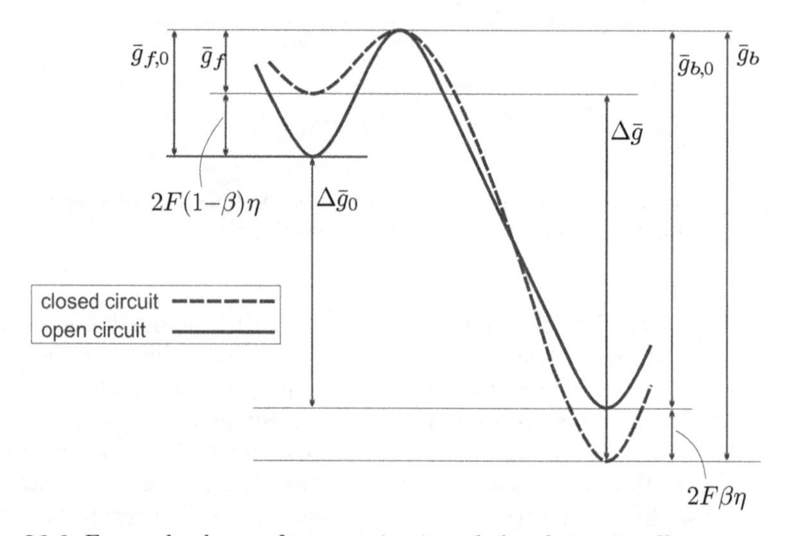

Fig. 26.9 Energy landscape for open circuit and closed circuit cell

26.8 Voltage/Current and Power/Current Diagrams

Summarizing the results of the last sections, the fuel cell potential is

$$V = -\frac{\Delta\bar{g}_R}{2F} - R_i I - BT \ln\left[1 - \frac{I}{I_{\text{limit}}}\right]^{-1} - \eta(I) , \tag{26.49}$$

where the first term is the open circuit potential, and the following three terms describe the irreversible losses due to resistance, mass transfer and activation. Figure 26.10 compares the actual potential (continuous) with the open circuit potential (grey), and also shows the individual losses, all in dimensionless quantities where the open circuit potential is unity.

The activation loss (short dashes) causes the sharp drop of the potential for small currents, it grows only slightly for larger currents. The resistance

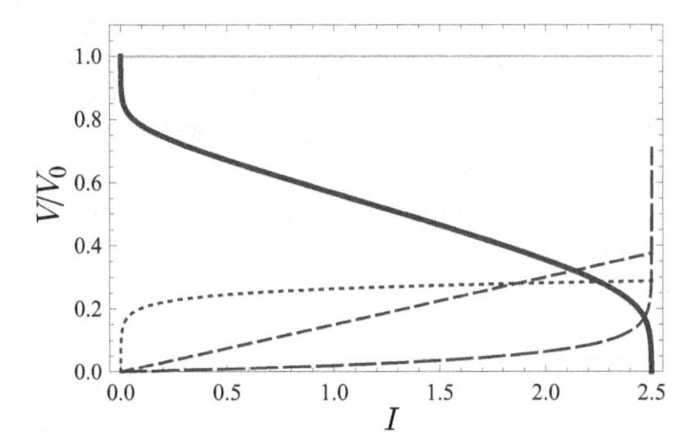

Fig. 26.10 Fuel cell potential (continuous), open circuit potential (grey), resistance loss (dashes), mass transfer loss (long dashes), activation loss (short dashes) as functions of current; $T/T_0 = 1$

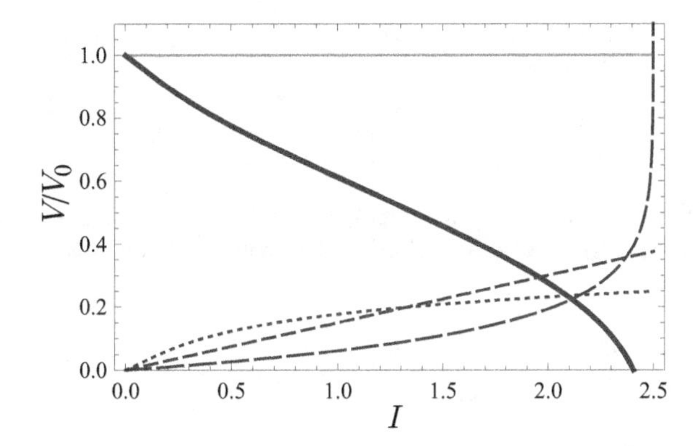

Fig. 26.11 Same as Fig. 26.10, but higher temperature; $T/T_0 = 3$

loss (dashes) grows linearly with current and causes the linear drop of the potential in the middle. The mass transfer loss (long dashes) is relatively small until the current approaches the limiting current which causes a sharp increase, and the sudden drop of the potential.

These curves agree qualitatively with curves found in the specialist literature on fuel cells. The relative contribution of the different losses depends on design and materials, and on temperature.

Figure 26.11 shows the same curves at a higher temperature. Now the activation losses are reduced, but the mass transfer losses are increased. It should be noted that low and high temperature fuel cells are fundamentally different in materials, physical processes and design, and thus this is only a qualitative comparison.

The power-current characteristic is shown in Fig. 26.12, based on the same data as the first voltage curve, Fig. 26.10. A reversible fuel cell would deliver the power $V_0 I$ (continuous), but due to the various losses, the power curve exhibits a maximum, and drops to zero at the limiting current. In order to minimize the losses, a fuel cell should operate at currents well left of the maximum, where the losses are relatively small. Note that the power loss is the difference between the actual power curve and the reversible curve, and grows non-linearly with current!

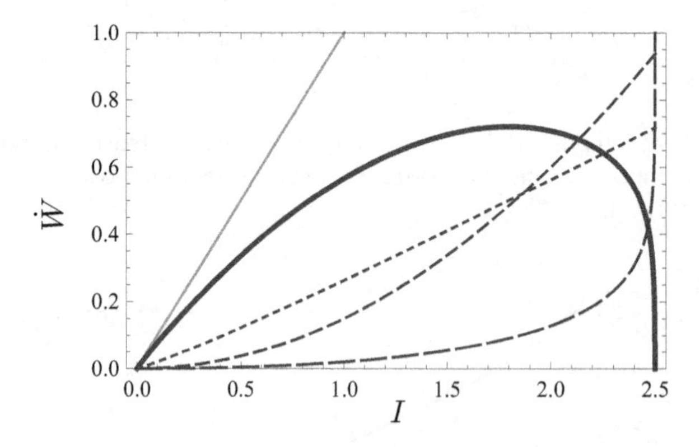

Fig. 26.12 Fuel cell power (continuous), reversible power (grey), resistance loss (dashes), mass transfer loss (long dashes), activation loss (short dashes)–all as function of current drawn

26.9 Crossover Losses

Even when the circuit is open, the observed potential lies below the open circuit potential (which describes reversible operation). The common explanation for this drop is the occurrence of electron crossover losses, due to electrons that find a path through the electrolyte and travel from anode to electrode without delivering electrical work. Thus, there is a certain number of net reactions taking place which do not contribute to the useful current I_{used}. The overall current $I = 2F\Lambda$, where Λ is the fuel consumption rate, can be split into the useful current and the crossover current I_{lost}, so that the voltage-current relation reads

$$V = -\frac{\Delta \bar{g}_R}{2F} - R_i \left(I_{\text{used}} + I_{\text{lost}}\right) - BT \ln\left[1 - \frac{I_{\text{used}} + I_{\text{lost}}}{I_{\text{limit}}}\right]^{-1} - \eta\left(I_{\text{used}} + I_{\text{lost}}\right) . \tag{26.50}$$

When plotted over the useful current for constant I_{lost}, the current scale is merely shifted, Fig. 26.13 shows an example.

Additional losses can occur due to fuel and oxidizer entering the electrolyte, and reacting directly, without flow of electrons involved. This does not affect

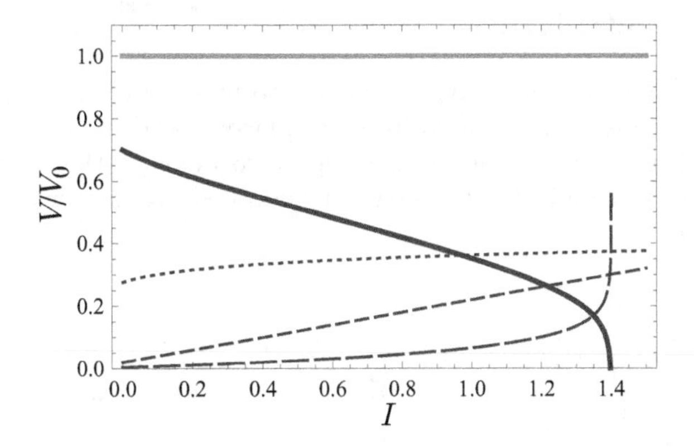

Fig. 26.13 Same as Fig. 26.10, but with crossover loss I_{loss}

the potential V of the fuel cell, nor the power drawn which still is $\dot{W} = VI$, but leads to additional heat developed by the fuel cell, and additional fuel consumption.

If the reaction rate of these reactions is Λ_c, the corresponding heat to be removed from the fuel cell is $\dot{Q}_c = -\Lambda_c \Delta\bar{h}_R$. This heat, if unused, leads to external entropy generation, and waste of fuel.

The work potential of the fuel consumed in crossover is $-\Lambda_c \Delta\bar{g}_R$, and, since this potential is not used, this is just the entropy generated in fuel crossover, $T\dot{S}^c_{gen} = -\Lambda_c \Delta\bar{g}_R$. The relation between power and entropy generation is $\dot{W} = -\Lambda\Delta\bar{g}_R - T\dot{S}_{gen}$ where Λ counts all reactions, i.e., the fuel consumption. Accounting explicitly for the loss due to crossover, we have

$$\dot{W}_{FC} = -\Lambda\Delta\bar{g}_R + \Lambda_c \Delta\bar{g}_R - T\dot{S}_{gen} = -\Lambda_I \Delta\bar{g}_R - T\dot{S}^I_{gen} \,, \tag{26.51}$$

where $\Lambda_I = \Lambda - \Lambda_c = \frac{I}{2F}$ is the reaction rate for electrochemical reactions. Moreover, $T\dot{S}^I_{gen}$ denotes the power loss due to all mechanisms discussed above, excluding crossover loss. Thus, the expressions for fuel cell work of the previous sections remain valid. Fuel crossover does not affect the voltage-current curve, but leads to increased fuel consumption, and increased heat transfer from the cell.

The influence of crossover is best seen in the second law efficiency for the fuel cell. The reversible work available from the fuel is $\dot{W}_{rev} = -\Lambda\Delta\bar{g}_R$ where the reaction rate Λ measures the amount of fuel used. With $\Lambda_I = \Lambda - \Lambda_c$ the second law efficiency becomes

$$\eta_{II} = \frac{\dot{W}_{FC}}{\dot{W}_{rev}} = 1 - \frac{\Lambda_c}{\Lambda_I + \Lambda_c} - \frac{T\dot{S}_{gen}}{(\Lambda_I + \Lambda_c)\,(-\Delta\bar{g}_R)} \,. \tag{26.52}$$

26.10 Electrolyzers

In fuel cells, hydrogen and oxygen combine to produce electrical energy and water. In electrolyzers, the opposite takes place: electrical energy is used to split water into hydrogen and oxygen. Figure 26.14 shows the basic reactions taking place, and indicates the flows of hydrogen, oxygen, water, protons, and electrons.

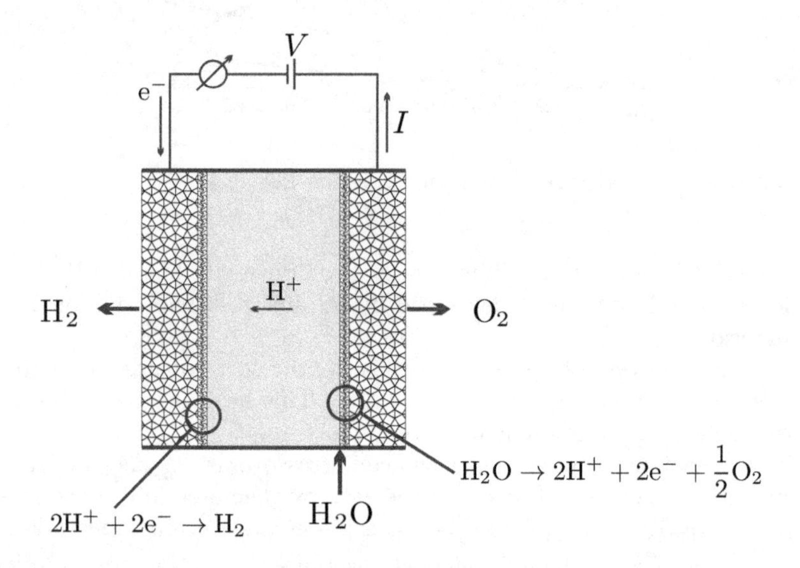

Fig. 26.14 Reactions and flows in an electrolyzer

The power consumption of the electrolyzer is $\dot{W}_E = -V_E I$ and the amount of hydrogen produced is equal to the reaction rate $\Lambda = I/2F$. Thus, the work required for the production per moler of hydrogen, w_{H_2}, is directly proportional to the electrolyzer potential,

$$w_{H_2} = \frac{\dot{W}_E}{\Lambda} = -2FV_E \; . \tag{26.53}$$

Since all flows are just in opposite direction as in a fuel cell, the potential and power consumption for an electrolyzer follow from the equations for fuel cells simply by inverting the sign of the electrical current. Then, from (26.49), the electrolyzer potential becomes

$$V_E = -\frac{\Delta \bar{g}_R}{2F} - R_i\left(-I\right) - BT\ln\left[1 - \frac{-I}{I_{\text{limit}}}\right]^{-1} - \eta\left(-I\right) \; , \tag{26.54}$$

where $\eta\left(-I\right)$ solves the Butler-Volmer equation for negative current,

$$-I = I_0 \left[\exp\left[\frac{2F(1-\beta)\eta}{\bar{R}T}\right] - \exp\left[-\frac{2F\beta\eta}{\bar{R}T}\right]\right] ; \tag{26.55}$$

obviously, $\eta(-I) = -\eta_E(I)$ must be negative.

With the positive activation potential η_E, the electrolyzer equations become

$$V_E = -\frac{\Delta\bar{g}_R}{2F} + R_i I + BT\ln\left[1 + \frac{I}{I_{\text{limit}}}\right] + \eta_E , \tag{26.56}$$

$$I = I_0 \left[\exp\left[\frac{2F\beta\eta_E}{\bar{R}T}\right] - \exp\left[-\frac{2F(1-\beta)\eta_E}{\bar{R}T}\right]\right] . \tag{26.57}$$

Irreversible processes due to internal resistance, mass transfer, and activation lead to an increase of the potential above the open circuit potential $V_0 = -\frac{\Delta\bar{g}_R}{2F}$. The proper second-law efficiency measure for an electrolyzer is

$$\eta_E^{II} = \frac{V_0}{V_E} \leq 1 . \tag{26.58}$$

Figure 26.15 compares the voltage-current curves for fuel cell and electrolyzer to the open circuit potential. The gap between the curves is the loss to irreversible processes.

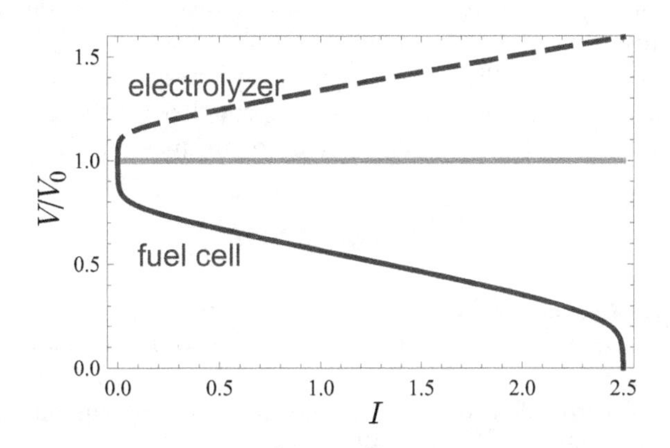

Fig. 26.15 Voltage-current curves for fuel cell and electrolyzer

26.11 Hydrogen

A common method to produce hydrogen is steam methane reforming, in which hydrogen (H_2) is split from natural gas (that is mainly methane, CH_4) according to the overall reaction $CH_4+2H_2O \leftrightarrow 4H_2+CO_2$. Larger quantities of H_2 are produced this way for NH_3 production (see Sec. 23.10). This process generates CO_2 and thus is not carbon neutral. If the natural gas is used in a combined cycle power plant with thermal efficiency of 60% (Sec. 13.6), there will be less CO_2 produced per kWh of electricity than if H_2 from steam methane reforming is used in a fuel cell.

By means of electrolyzers, hydrogen can be produced from any primary source of electrical power, including carbon neutral power sources like nuclear, solar, wind, or tidal. As opposed to electricity, hydrogen can be stored and distributed in pipelines and tanks, and therefore offers a means to store power produced from intermittent sources (solar, wind, tidal).

The stored hydrogen can be reconverted into electricity either in traditional combustion systems (e.g. combined cycle power plants, Atkinson cycle), or in fuel cells. The latter offer an elegant and efficient means to use power produced by stationary non-carbon power plants (solar, wind, ...) for transportation (cars, trucks, busses, trains, ...). The use of hydrogen produced by carbon neutral energy sources could play a role in the future energy system, which must aim to reduce the emission of greenhouse gases, including carbon dioxide.

Due to irreversibilities, only a portion of the energy fed into the electrolyzer is retrieved from the fuel cell. Moreover, one needs to account for the efficiency of processes required to store and distribute the hydrogen. Indeed, at normal conditions hydrogen is a gas which, due to its low molar mass, assumes a relatively large specific volume. In particular for use as transportation fuel, the hydrogen needs to be compacted, either by compression of the gas, or by liquefaction. The (irreversible) processes involved can be described by a storage and distribution efficiency measure η_{SD}^{II}.

With the second law efficiencies for electrolyzer, storage and distribution, and fuel cell, the ratio between the power provided by the fuel cell, $\dot{W}_{FC}$, and the power $\dot{W}_E$ which was consumed to produce the hydrogen in the electrolyzer, is

$$\frac{\dot{W}_{FC}}{\dot{W}_E} = \eta_{FC}^{II}\eta_{SD}^{II}\eta_E^{II} \,. \tag{26.59}$$

Typical values for these efficiencies are $\eta_{FC}^{II} = 0.6$, $\eta_{SD}^{II} = 0.75$, $\eta_E^{II} = 0.6$, so that only 27% of the energy provided at the source is finally recovered. Doubtless, new materials and better designs will lead to efficiency improvements in the future. Meanwhile, other storage concepts might have higher efficiencies, e.g., batteries or pumped hydro.

Problems

26.1. Fuel Cell Potential

Compute the maximum voltage (reversible operation) for the fuel cell depicted below under the assumption that the oxygen supplied is five times the stoichiometric amount. Assume isothermal operation and assume that the water leaves as vapor.

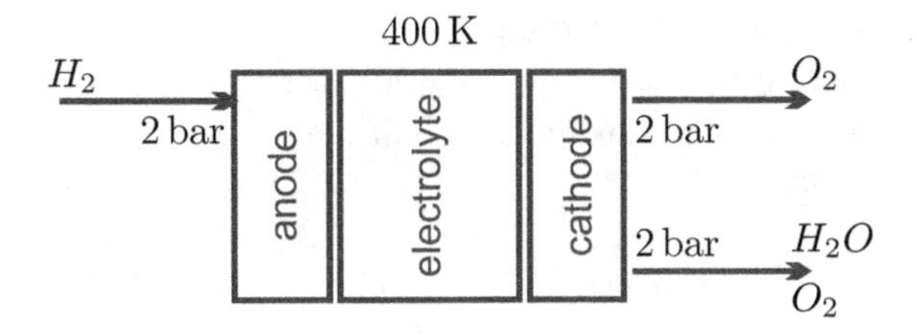

26.2. Fuel Cell Potential and Power
We can write the fuel cell voltage as $V = V_0 - V_R - V_{tr} - \eta$, where V_0 is the open circuit potential, $V_R = \mathcal{R}I$ are the ohmic losses due to the internal resistance $\mathcal{R}$ of the fuel cell, and $V_{tr} = -BT\ln(1 - I/I_{\text{limit}})$ is the potential drop due to mass transfer loss. Moreover the activation overpotential η is related to the current I through the Butler-Volmer equation

$$I = I_0 \left\{ \exp\left[\frac{2F(1-\beta)\eta}{R_u T}\right] - \exp\left[-\frac{2F\beta\eta}{R_u T}\right] \right\} ,$$

where the exchange current is modelled as (constant factor k_I, activation energy E_a)

$$I_0 = k_I T \exp\left[-\frac{E_a}{R_u T}\right] .$$

1. Introduce dimensionless quantities, and show that the dimensionless voltage can be written as

$$v = 1 - \rho i - \gamma \hat{T} \ln[1 - \varepsilon i] - \hat{\eta}$$

with

$$i = \hat{T} \exp\left[-\frac{e_a}{\hat{T}}\right] \left\{ \exp\left[\frac{\alpha(1-\beta)\hat{\eta}}{\hat{T}}\right] - \exp\left[-\frac{\alpha\beta\hat{\eta}}{\hat{T}}\right] \right\} ,$$

Identify the dimensionless quantities that appear in the above.

2. Chose $\alpha = 95$, $\beta = 0.6$, $\gamma = 0.04$, $\rho = 0.15$, $\varepsilon = 0.4$, $e_a = 10$, and plot dimensionless voltage v and power $w = vi$ for some values of the dimensionless temperature $\hat{T}$ between 1 and 5. Also plot the individual voltage losses into the same diagram.

26.3. Electrolyzer
An electrolyzer can be considered as an "inverted fuel cell": it consumes electric power and produces hydrogen and oxygen gases. For the following, ignore mass transfer contributions. Obviously, the current in the electrolyzer flows in the opposite direction: replace I by $-I$ in the fuel cell equations of the previous problem. Show that now resistance and activation increase the fuel cell potential. Use the same constants as for the fuel cell, and plot the electrolyzer potential over the current, and also the power consumed as a function of current.

26.4. Direct Methanol Fuel Cell
In a direct methanol fuel cell, liquid methanol CH_3OH reacts with oxygen to form water and carbon dioxide. The half-reactions at anode and cathode follow the equations

$$CH_3OH + H_2O \mapsto CO_2 + 6H^+ + 6e^- \quad \text{and} \quad \frac{3}{2}O_2 + 6H^+ + 6e^- \mapsto 3H_2O$$

Determine the open circuit voltage at 330K. Assume methanol is an incompressible liquid with constant specific heat $\bar{c}_{CH_3OH} = 0.082\frac{kJ}{mol\,K}$.

26.5. Borohydride Fuel Cell
The reactions in a direct borohydride fuel follow the equations
Anode: $NaBH_4 + 8OH^- \mapsto NaBO_2 + 6H_2O + 8e^-$,
Cathode: $O_2 + 4H_2O + 8e^- \mapsto 8OH^-$.
Determine the open circuit voltage at standard conditions (Remark: Actual devices operate at elevated temperatures). Assume that all participating components enter or leave in separate streams. Use the following data:
$NaBH_4$: $\bar{h}_f^0 = -192\frac{kJ}{mol}$, $\bar{s}_f^0 = 101\frac{J}{mol\,K}$; $NaBO_2$: $\bar{h}_f^0 = -960\frac{kJ}{mol}$, $\bar{s}_f^0 = 83\frac{J}{mol\,K}$.

26.6. Electrolyzer
Compute the voltage of a reversible electrolyzer that splits liquid water into hydrogen and oxygen. Assume that all incoming and outgoing streams are at reference pressure $p_0 = 1\,\text{atm}$ and have a temperature of 360 K.

26.7. Direct Carbon Fuel Cell
The anode and cathode reactions in a direct carbon fuel cell follow the equations

$$C + 2O^{2-} \mapsto CO_2 + 4e^- \quad \text{and} \quad O_2 + 4e^- \mapsto 2O^{2-}$$

Determine the open circuit voltage at 900 K, which is the typical operating temperature of actual DCFC's. Assume carbon (graphite) is a incompressible substance with specific heat $c_p = 0.71\frac{J}{kg\,K}$.

26.8. Magnesium Fuel Cell
The reactions in a Magnesium fuel cell with salt water electrolyte follow the equations
Anode: $2Mg \mapsto 2Mg^{2+} + 4e^-$
Cathode: $O_2 + 2H_2O + 4e^- \mapsto 4OH^-$
Electrolyte: $2Mg^{2+} + 4OH^- \mapsto 2Mg\,(OH)_2$
Determine the open circuit voltage at standard conditions (Remark: Actual devices operate at elevated temperatures).

26.9. Molten Carbonate Fuel Cell
Molton carbonate fuel cells (MCFC) employ a mixture of salt and molten carbonate as electrolyte, and operate at temperatures around 900 K.

MCFC can work with carbon monoxide (CO) as fuel. Then the anode reaction occurs in two steps, first the water gas shift reaction, followed by reaction of the generated hydrogen with the carbonate ion, CO_3^{2-}. The reactions occurring are:

Anode: $CO + H_2O \mapsto H_2 + CO_2 \quad , \quad H_2 + CO_3^{2-} \mapsto H_2O + CO_2$,

Cathode: $\frac{1}{2}O_2 + CO_2 + 2e^- \mapsto CO_3^{2-}$.

1. Make a sketch of a fuel cell, where you indicate the relevant flows.
2. Determine the open circuit voltage at standard conditions and at 900 K.

Index

V

W

Z

CPSIA information can be obtained
at www.ICGtesting.com
Printed in the USA
LVHW06*0507040918
589024LV00001B/55/P

9 783662 437148